DATA SCIENCE & EXPLORATION IN ARTIFICIAL INTELLIGENCE

The book captures the essence of the International Conference on Data Science & Exploration in Artificial Intelligence (CODE-AI 2024), organized by the Manipal Institute of Technology (MIT-BLR), which was held on July 3rd and 4th, 2024, in Bengaluru, India. It offers a comprehensive exploration of cutting-edge research in AI, data science, and their applications. The book provides insight into the latest developments, ethical considerations, and the future trajectory of these dynamic fields.

The book spans a wide array of topics discussed at CODE-AI 2024, including advanced Data Science, IoT, Security, Cloud Computing, Networks, Security, Image, Video and Signal Processing, Computational Biology, Computer and Information Technology. It highlights innovative research contributions and practical applications, offering readers a detailed understanding of current trends and challenges. The findings emphasize the role of global collaboration and interdisciplinary approaches in pushing the boundaries of AI and data science. Selected papers published by Taylor and Francis showcase pioneering work that is shaping the future of these fields.

This book is ideal for AI and data science researchers, industry professionals, and students seeking to stay updated on the latest advancements and ethical considerations in these areas.

PROCEEDINGS OF THE FIRST INTERNATIONAL CONFERENCE ON DATA SCIENCE & EXPLORATION IN ARTIFICIAL INTELLIGENCE (CODE-AI 2024), 3RD–4TH JULY, 2024, BANGALORE, INDIA

Data Science & Exploration in Artificial Intelligence

Volume 2

Edited by

H.L. Gururaj
Manipal Academy of Higher Education

Francesco Flammini
SUPSI

J. Shreyas
Manipal Academy of Higher Education

CRC Press
Taylor & Francis Group
Boca Raton London New York Leiden

CRC Press is an imprint of the
Taylor & Francis Group, an **informa** business
A BALKEMA BOOK

First published 2025
by CRC Press/Balkema
4 Park Square, Milton Park, Abingdon, Oxon, OX14 4RN

and by CRC Press/Balkema
2385 NW Executive Center Drive, Suite 320, Boca Raton FL 33431

CRC Press/Balkema is an imprint of the Taylor & Francis Group, an informa business

British Library Cataloguing-in-Publication Data
A catalogue record for this book is available from the British Library

Library of Congress Cataloging-in-Publication Data
A catalog record has been requested for this book

Set

ISBN: 978-1-032-96397-6 (hbk)
ISBN: 978-1-032-96392-1 (pbk)

Volume 1
ISBN: 978-1-032-73739-3 (hbk)
ISBN: 978-1-032-95970-2 (pbk)
ISBN: 978-1-003-58739-2 (ebk)

DOI: 10.1201/9781003587392

Volume 2
ISBN: 978-1-032-96390-7 (hbk)
ISBN: 978-1-032-96368-6 (pbk)
ISBN: 978-1-003-58927-3 (ebk)

DOI: 10.1201/9781003589273

Typeset in Times New Roman
by MPS Limited, Chennai, India

Data Science & Exploration in Artificial Intelligence – Gururaj et al. (Eds)
© 2025 the Editor(s), ISBN: 978-1-032-96390-7

Table of Contents

Preface

The International Conference on Data Science & Exploration in Artificial Intelligence (CODE-AI 2024), organized by the Manipal Institute of Technology (MIT-BLR), was held on July 3rd and 4th, 2024, in Bengaluru, India. This prestigious event brought together leading experts from academia and industry to discuss the latest advancements in artificial intelligence and data science, along with the ethical considerations surrounding these rapidly evolving fields.

The conference was generously sponsored by CGI and featured collaborative efforts with Linnaeus University, Sweden, and the University of Applied Sciences, Switzerland. Acknowledging the significant contributions of these institutions, the conference highlighted the importance of global collaboration in pushing the boundaries of AI research.

Our esteemed sponsors—IBM, 10 Seconds, KL Book House, TechCurators, Bhagirathi, Zolo, ICICI Bank, and Mittal Publication—played a vital role in the success of CODE-AI 2024. Their support enabled us to provide a platform for insightful discussions and the dissemination of cutting-edge research.

This year, the conference received an overwhelming response, with a total of 890 papers submitted through the Conference Management Toolkit (CMT). After rigorous peer review, 210 papers were accepted, and 160 papers were ultimately registered and presented. The high caliber of the submissions is a testament to the vibrant research community in AI and data science.

To recognize outstanding contributions, CODE-AI 2024 proudly announced the Best Paper Award and Best Presenter Award, celebrating excellence in research and presentation skills. Selected research papers from the conference have been published in the Conference Proceedings by Taylor and Francis Publishers, contributing valuable knowledge to the global academic community.

We express our heartfelt gratitude to all participants, reviewers, sponsors, and collaborators who made CODE-AI 2024 a resounding success. We look forward to continued innovation and collaboration in the years to come.

Editor Biographies

Gururaj H L (Senior Member, IEEE) is currently working as Associate Professor, Department of Information Technology, Manipal Institute of Technology, Bengaluru, India. He holds a Ph. D degree in Computer Science and Engineering from Visvesvaraya Technological University, Belagavi, India. He is a Senior member of ACM and working as ACM Distinguished Speaker. He is the founder of Wireless Internetworking Group (WiNG). He is a Senior member of IEEE and lifetime member of ISTE and CSI. Dr. Gururaj received the young scientist International Travel Grant award from the Government of India. He has 13 years of teaching experience.at both UG and PG level. His research interests include BlockChain Technology, Cyber Security, Wireless Sensor Network, Ad-hoc networks, IOT, Data Mining, Cloud Computing and Machine Learning. He has guided 30 UG students and 20 PG students. He has published more than 150+ research papers including 20 SCI publications. He has Edited 20 books with reputed publishers. He has presented 55 papers at various international Conferences.

Francesco Flammini (Senior Member, IEEE) was born in Formia, Italy, in 1978. He received the master's degree (cum laude) in computer engineering and the Ph.D. degree in computer and automation engineering from the University of Naples Federico II, Italy, in 2003 and 2006, respectively. From 2003 to 2007, he was a Software Engineer with Ansaldo STS (now Hitachi Rail), where he was a Senior Innovation Engineer, from 2007 to 2016. From 2016 to 2017, he worked as an Information Security Compliance Manager with the Italian State Mint and Polygraphic Institute. Since 2018, he has been working as an Associate Professor with Linnaeus University, Sweden, where he has chaired the cyber-physical systems (CPS) environment. He had leadership roles in more than ten research projects. He has edited or authored more than ten books and 130 publications. His research interests include resilient cyberphysical systems and trustworthy autonomy. He is an ACM Distinguished Speaker and the Chair of the IEEE SMC Technical Committee on Homeland Security.

Shreyas J is currently working as Assistant Professor, Department of Information Technology, Manipal Institute of Technology, Bengaluru, India. He holds a Ph. D degree in Computer Science and Engineering from Bengaluru University, Belagavi, India. He is a member of ACM. He has 4 years of teaching experience at both UG and PG level. His research interests include BlockChain Technology, Cyber Security, Wireless Sensor Network, Ad-hoc networks, IOT, Data Mining, Cloud Computing and Machine Learning. He has published more research papers including SCI publications. He has Edited books at reputed publishing houses.

© 2025 the Editor(s), ISBN: 978-1-032-96390-7

Scientific Committee

Dr. Celestine Iwendi, *ACM Distinguished Speaker & Associate Professor, School of Creative Technologies University of Bolton United Kingdom*
Dr. Lili Arabuli, *Associate Professor, Department of Natural Sciences, The University of Georgia Georgia, USA*
Dr. Susan M Zvacek, *ACM Distinguished Speaker Consultant & Teacher, Colorado, USA*
Mr. Keshav Kaushik, *Assistant Professor (Senior Scale), School of Computer Science, UPES, Uttarakhand*
Dr. Sandeep Chakraborty, *Associate Professor, IIT Kharagpur*
Dr. Sangharatna, *NIT Warangal*
Dr. Palaniswamy, *NIT Surathkal*
Dr. Pradeep C R, *IISc, Bengaluru*
Dr. Kunwar, *NIT Trichy*

© 2025 the Editor(s), ISBN: 978-1-032-96390-7

Organising Committee

Dr. Tusar Kanti Mishra, *Department of Computer Science & Engineering, MIT Bengaluru*
Dr. Vamshi Krishna.B, *Department of Computer Science & Engineering, MIT Bengaluru*
Dr. Megha. P. Arakeri, *Department of Computer Science & Engineering, MIT Bengaluru*
Dr. Shreya Banerjee, *Department of Computer Science & Engineering, MIT Bengaluru*
Dr. Shrirang Ambaji Kulkarni, *Department of Information Technology, MIT Bengaluru*
Dr. Vishnu Srinivasa Murthy, *Department of Information Technology, MIT Bengaluru*
Dr. Bhuvaneshwari P, *Department of Computer Science & Engineering, MIT Bengaluru*
Dr. Napa Lakshmi, *Department of Computer Science & Engineering, MIT Bengaluru*
Dr. Jayita Saha, *Department of Computer Science & Engineering, MIT Bengaluru*
Dr. Sowmyalatha Naveen, *Department of Computer Science & Engineering, MIT Bengaluru*
Dr. Rakesh Kumar, *Department of Computer Science & Engineering, MIT Bengaluru*
Dr. Vaishnavi S, *Department of Information Technology, MIT Bengaluru*
Dr. Madhura K, *Department of Information Technology, MIT Bengaluru*
Dr. Anitha Prem Kumar, *Department of Information Technology, MIT Bengaluru*
Dr. Snigdha Sen, *Department of Information Technology, MIT Bengaluru*
Dr. Anil Kumar, *Department of Information Technology, MIT Bengaluru*
Dr. P. Devisivasankari, *Department of Information Technology, MIT Bengaluru*
Dr. Abhijit Das, *Department of Information Technology, MIT Bengaluru*
Dr. Gauri Kalnoor, *Department of Computer Science & Engineering, MIT Bengaluru*
Dr. Prakash B Metre, *Department of Computer Science & Engineering, MIT Bengaluru*
Dr. Sk Mahmudul Hassan, *Department of Information Technology, MIT Bengaluru*
Dr. Shrirang Ambaji Kulkarni, *Department of Information Technology, MIT Bengaluru*
Dr. Preethi, *Department of Information Technology, MIT Bengaluru*
Ms. Sapna R, *Department of Information Technology, MIT Bengaluru*
Dr. Kshama S B, *Department of Information Technology, MIT Bengaluru*
Dr. Dharmanna L, *Department of Computer Science & Engineering, MIT Bengaluru*
Dr. Ruhul Amin Hazarika, *Department of Information Technology, MIT Bengaluru*
Dr. Sumanth V, *Department of Information Technology, MIT Bengaluru*
Dr. Karthik V, *Department of Information Technology, MIT Bengaluru*
Dr. Vishnu Srinivasa Murthy, *Department of Information Technology, MIT Bengaluru*
Dr. Naresh E, *Department of Information Technology, MIT Bengaluru*
Dr. Yashaswini K A, *Department of Information Technology, MIT Bengaluru*
Dr. Sangeeta Sangani, *Department of Information Technology, MIT Bengaluru*
Dr. Amreen Ayeesha, *Department of Information Technology, MIT Bengaluru*
Dr. Usha M, *Department of Information Technology, MIT Bengaluru*
Dr. Priya S, *Department of Information Technology, MIT Bengaluru*
Mr. Gaurav Prasad S, *Department of Information Technology, MIT Bengaluru*

Data Science & Exploration in Artificial Intelligence – Gururaj et al. (Eds)
© 2025 the Author(s), ISBN: 978-1-032-96390-7

Temporal graph network for the analysis of multi robot systems

Sreya V. Sujil, E. Annapoorna, R.S. Parvati Nair, K.A. Athira & J. Divya Udayan
Department of Computer Science and Engineering, Amrita School of Computing, Amrita Vishwa Vidyapeetham, Amritapuri, India

ABSTRACT: In the field of Multi-Robot Systems (MRS), efficient task allocation remains a crucial challenge, particularly in dynamic environments. To conquer such challenges, Dynamic Graph Neural Networks (DGNNs) have evolved as a viable alternative, with the ability to simulate complicated interactions between robots and tasks. This research proposes a novel technique for addressing Multi-Robot Task Allocation (MRTA) by incorporating DGNN architecture, specifically Temporal Graph Networks (TGN). The proposed method uses node embeddings from TGN and reinforcement learning techniques to reliably anticipate task completion times and optimize task assignments, eventually reducing completion times and increasing productivity in real-world MRS scenarios. Combining Soft Actor-Critic (SAC) for reinforcement learning with ROS (Robot Operating System) and Gazebo simulation environments allows for more efficient job allocation. In conclusion, this study provides a possible solution to persisting issues in multi-robot task allocation, resulting in more efficient and adaptive robotic systems across a variety of application areas.

Keywords: Graph Neural Network, Task Allocation, Robotic Operating System, Gazebo, Soft Actor Critic

1 INTRODUCTION

Robotic systems are undergoing great development and have a growing socioeconomic impact on human lives. The multi-robot system consists of a minimum of two autonomous robots. The multi-robot system proposes to give scalable, adaptable, and reliable methods for working together with all robots. In MRS, multiple robots collaborate to accomplish multiple distinct jobs. The work on MRS is still advancing, even if there are many uses for multi-robot systems, including cooperative localization, dynamic work allocation, and path planning.

Multi-Robot Task Allocation, or MRTA, is the process of coordinating a set of tasks among a number of cooperative robotic systems in order to make choices without conflicts and increase the quantity of interest. Given that multi-robotics is one of the most active fields of robotics research and development, and that task allocation is critical to the bulk of multi-robotic or swarm-robotic operations, the potential real-world applications of such MRTA problems are numerous. Disaster response, last-mile delivery, environmental monitoring, and reconnaissance are a few examples[1].

Training deep neural networks on real-world robot experience is costly due to battery recharges, manual resets, and maintenance. Dynamic Graph Neural Networks (D-GNN) are groundbreaking for Multi-Robot Systems (MRS) as they decipher complex dynamics by abstracting robots as nodes and interactions as edges [2]. This study uses Temporal Graph Networks to capture spatial and temporal complexities, and Soft Actor-Critic (SAC), a model-free deep reinforcement learning technique, to help robots predict future states and make proactive modifications based on temporal cues, improving agility in dynamic situations.

DOI: 10.1201/9781003589273-1

The study aims to enhance robotic task allocation with Temporal Graph Networks (TGNs) and integrate reinforcement learning using SAC in ROS. Simulation and validation in Gazebo, a robotics simulation environment, reveal a task allocation framework with increased accuracy, facilitated by TGNs and SAC. MRS can handle tasks like surveillance and search and rescue, but efficient task allocation is challenging. By training TGNs in MRS and using SAC to optimize task allocation, robots can predict future states and adapt to dynamic environments. This integration offers an advantage over other deep reinforcement learning networks that rely on simulated environments.

2 LITERATURE REVIEW

C. Zhang *et al.* [3] address accurately anticipating traffic flow in urban contexts using Spatial-temporal Graph Attention Networks to improve traffic prediction accuracy and robustness. Previous studies utilized CNNs, RNNs, LSTMs, and graph-based approaches like GCN. The proposed solution models traffic networks as graphs with nodes representing road segments and edges reflecting connections, capturing spatial interdependence. Temporal convolutional networks identify temporal connections, anticipating future traffic conditions from previous observations.

C. Su *et al.* [4] introduced Graph-Attention augmented Temporal neural network (GATE) for pharmaceutical recommendations. GATE employs graph attention and temporal neural networks to model patient health evolution and drug interactions. Previous research focused on graph-based recommendation systems and temporal modeling approaches like RNNs, LSTMs, and TCNs. GATE captures relationships between drugs and the temporal progression of patient health, providing personalized medicinal suggestions.

Siyuan Qi, Wenguan Wang *et al.* [5] tackled recognizing human-object interactions in visual data using Graph Parsing Neural Networks (GPNNs). GPNNs convert visual scenes into graph representations, with nodes for people and objects and edges for interactions. Previous studies used CNNs, RNNs, and attention processes for scene interpretation. The proposed model integrates spatial and semantic information into a graph-based representation, capturing spatial relationships and semantic context.

Rossi, E., Chamberlain *et al.* [6] addressed deep learning on dynamic graphs, where nodes and edges change over time, using Temporal Graph Networks (TGN). TGN operates on continuous-time dynamic networks, capturing temporal dynamics and changing interactions. The approach combines graph convolutional layers with temporal attention techniques, adaptively weighing node and edge importance at each time step, effectively modeling temporal dependencies.

3 METHODOLOGY

3.1 *Temporal graph networks*

From Figure 1, Temporal Graph Networks (TGNs) represent a significant advancement in deep learning on dynamic graphs, addressing evolving features and connectivity over time. TGNs capture the temporal evolution of graphs, making them suitable for modeling real-life interaction systems like social networks or biological interactions [7]. They employ an encoder-decoder architecture, where the encoder maps a dynamic graph to node embeddings and the decoder makes task-specific predictions based on these embeddings, effectively depicting evolving graph topologies. TGNs use a specialized training strategy that leverages data sequentiality and maintains computational efficiency through parallel processing, enabling the model to adapt to real-time scenarios with dynamically interacting and constantly added nodes.

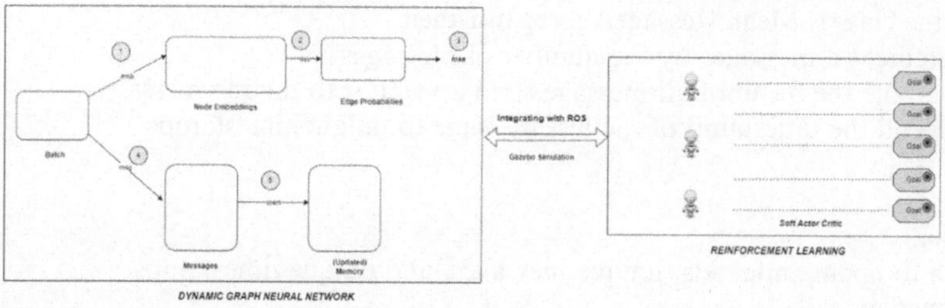

Figure 1. Pipeline of the proposed methodology.

TGNs are distinguished by their seamless integration of numerous components, which allows for thorough modeling of temporal graph dynamics:

Message Function: The message (m_x) generated by each node (i) at a certain time (τ) is important for information transfer inside the graph network.

$$m_x(\tau) = messages$$

Embedding Module ($z_x(\tau)$): TGNs utilize node embeddings to integrate information from neighboring nodes within a defined temporal window. These embeddings encompass node statuses, edge characteristics, and node attributes, reflecting the evolving graph configuration over time (τ).

$$z_x(\tau) = \sum_{y \in N_x^k([0,\tau])} h\big(s_x(\tau), s_y(\tau), e_{xy}, v_x(\tau), v_y(\tau)\big)$$

where h is the learnable function, e_{xy} is the interaction event between two nodes x and y, s_x and s_y are the memory of node x and node y at time τ respectively and v_x and v_y are node wise event of each node.

Temporal Graph Sum: Time graph summing refines node representations by aggregating information from neighboring nodes within a time frame. This includes edge properties and temporal differences. This approach improves the model's capacity to represent temporal dynamics accurately [8]. Integrating these components with node embeddings in reinforcement learning models allows for the use of rich information stored in node embeddings for diverse downstream tasks.

$$h_x^{(l)}(\tau) = W_2^{(l)}\left(h_x^{(l-1)}(\tau) || \tilde{h}_x^{(l)}(\tau)\right)$$

Algorithm 1 Message Aggregate Algorithm

---

function AGGREGATE()
 Initialize empty lists to update node ids, unique messages, and unique timestamps
 Initialize n messages as 0
 for each unique node id **do**
  Check if there are messages for this node id
  if messages exist **then**
   Get the messages and timestamps associated with this node id
  if the class is LastMessageAggregator **then**
   Append the timestamp of the last message to unique timestamps
  end if

if the class is MeanMessageAggregator **then**
Increment n messages by the number of messages
Compute the mean of all messages and append it to unique messages
Append the timestamp of the last message to unique timestamps
end if
end if
end for
return to update node ids, unique messages, and unique timestamps
end function

3.2 *Soft actor critic*

Reinforcement Learning is a variant of machine learning that allows agents to learn optimum decision-making strategies through interaction with their surroundings. Traditional RL algorithms incur difficulties when dealing with high-dimensional action spaces and require exploration tactics to establish the best policy.

Soft Actor-Critic is an off-policy actor-critic framework that blends deep learning and reinforcement learning to address continuous control tasks in reinforcement learning (RL). SAC is a powerful algorithm capable of learning robust policies in high-dimensional action spaces. It improved sample efficiency and performance in challenging RL environments. The purpose of SAC is to enhance expected returns by exploring and optimizing policies denoted by ($J(\theta)$) [9].

$$J(\theta) = E[R(s_t, a_t) + \alpha H(\tau(.|, s_t))]$$

The soft Q-function is defined as:

$$Q(s, a) = E[R(s, a) + \gamma E[V(s')]]$$

In this formula γ is the discount factor and V is the state value function.

Algorithm 2 Soft Actor Critic (SAC) Algorithm

Initialization:
Define environment, hyperparameters.
Initialize SAC agent, replay buffer, episode rewards list.
Initialize environment.
Training Loop:
Initialize SAC agent, replay buffer, episode rewards list, and environment.
Loop:
Select actions using agent's policy.
Execute actions in environment.
Store experiences in replay buffer.
Update agent's networks.
Collect episode rewards.
Save performance data, if required.
Check for episode termination and reset environment.
End Loop
Evaluate agent periodically.
Print evaluation results

3.3 *ROS*

The Robot Operating System (ROS) is widely accepted in robotics research and development due to its open-source nature, modular design, and extensive functionalities [10] [11]. It supports applications like autonomous navigation, manipulation, human-robot interaction, and swarm robotics [12]. Gazebo is an open-source robot simulation environment used for rapid prototyping, testing, and validating robotic algorithms. Integrated with ROS, Gazebo supports cloud-based simulation and virtual reality (VR) environments, enhancing standardization and interoperability in robotics research and development [13].

3.4 *Model operation: An overview*

In Figure 2, the project begins by gathering spatial coordinates for robots, goals, and obstacles within the environment. These elements are then structured into interconnected graphs using a graph constructor method, enabling a comprehensive understanding of their relationships. Subsequently, an embedding module method generates embeddings, producing multi-dimensional arrays that encapsulate essential spatial information.

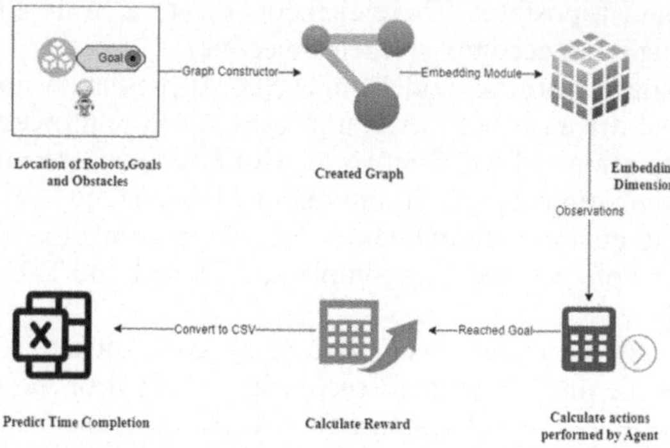

Figure 2. Overview of the model operation.

These embeddings serve as observations and serve as inputs to the Soft Actor-Critic (SAC) model. Leveraging these observations, the SAC model orchestrates actions for the agent, guiding their movements within the environment. As a result, robots navigate towards their designated goals, with each action influencing their trajectory significantly. Upon reaching a goal, robots receive points, encouraging them to swiftly locate objectives. Proximity to a goal determines the number of points earned, incentivizing robots to move in the right direction. These points are meticulously tracked and stored in a dedicated repository, facilitating the assessment of which robots reached their goals, along with their respective rewards and completion times. This repository enables accurate predictions regarding the time needed for robots to complete their tasks.

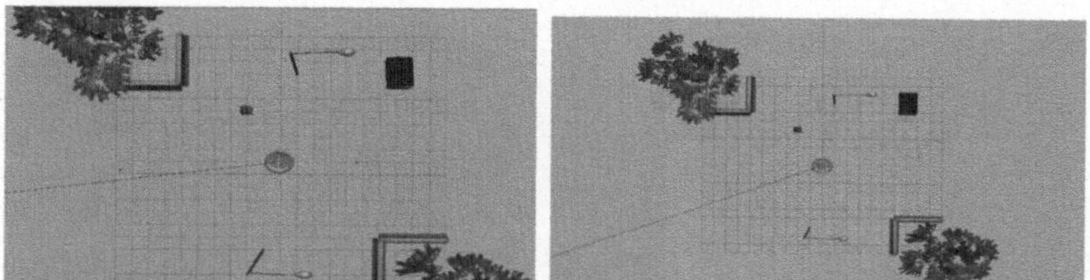

Figure 3. Gazebo Simulation – Top Views.

Figure 4. Gazebo simulation –side views.

4 RESULTS

Figures 3 and 4 depict the Gazebo simulation, the scene is set on a plane ground featuring several elements. Three robots are moving around, each with its task. Additionally, there are five goals placed randomly, providing targets for the robots to reach. Adding to the scene are two street lamps, two oak trees protected by two jersey barriers for each oak tree, a water fountain, a dumpster, and a postbox. These elements create a lively environment for the robots to navigate through and accomplish their objectives.

The work [14] by Matej Dobrevski and Danijel Skocaj presents a novel map-less navigation approach for goal-driven robot navigation using deep reinforcement learning. The paper addresses the integration of LSTM networks with Deep reinforcement learning allows for the training of the navigation model without manual labeling, generating a large number of samples in a simulated environment while our model represents the integration of TGN with Soft Actor-Critic and predicts the time completion. Table 1 and Table 2 are comparison of our project with [14].

Table 1 above illustrates the achievement of reaching goals, quantified out of a total of 100 goals. This indicates the number of goals successfully attained or the proportion of goals reached by robots. Moreover, our model demonstrates a higher frequency of goal attainment [14].

Table 1. Goal success rates.

Method	Number of reached goals
Ours	94
T.S LSTM	87
T.S.Dijkstra	76
T.S.D*Lite	75

The presented Table 2 indicates the instances where one or more robots successfully reached a specific goal. Additionally, our model exhibits superior performance in this regard as well [14].

Table 2. Analysis of goal achievement frequency.

Goal	Ours	T.S LSTM	T.S.Dijkstra	T.S.D*Lite
1	5	3	0	0
2	4	3	0	0
3	6	3	0	0
4	4	3	2	1
5	0	3	1	2

5 CONCLUSION

In conclusion, this research showcases the effectiveness of integrating Dynamic Graph Neural Networks (D-GNN) and Soft Actor-Critic (SAC) reinforcement learning to improve task allocation in Multi-Robot Systems (MRS) operating within dynamic environments. Through the utilization of Temporal Graph Networks (TGN) to model intricate relationships and SAC to optimize task assignments, our approach significantly reduces completion times and enhances productivity in real-world MRS scenarios. Looking ahead, future research avenues could focus on dynamically setting goals based on environmental changes for more adaptive task allocation, alongside creating dynamic environments within the ROS simulation to provide a realistic testbed for evaluating strategies. These advancements are poised to further refine MRS, paving the way for more efficient and adaptable robotic systems across a spectrum of applications.

REFERENCES

[1] Paul, Steve, Payam Ghassemi, and Souma Chowdhury. (2022). Learning scalable policies over graphs for multi-robot task allocation using capsule attention networks. *2022 International Conference on Robotics and Automation (ICRA)*. IEEE.

[2] Dobrevski, Matej, and Danijel Skočaj. (2021). Deep reinforcement learning for map-less goal-driven robot navigation. *International Journal of Advanced Robotic Systems* 18.1: 1729881421992621.

[3] Zhang C., Yu J. J. Q. and Liu Y. (2019). Spatial-temporal graph attention networks: A deep learning approach for traffic forecasting, in *IEEE Access*, vol. 7, pp. 166246–166256,doi:10.1109/ACCESS.2019.2953888.

[4] Su C., Gao S. and Li S. (2020). GATE: Graph-Attention Augmented Temporal Neural Network for Medication Recommendation, in *IEEE Access*, vol. 8, pp. 125447–125458, doi: 10.1109/ACCESS.2020.3007835.

[5] Qi, S., Wang, W., Jia, B., Shen, J. and Zhu, S.C. (2018). Learning human-object interactions by graph parsing neural networks. *In Proceedings of the European Conference on Computer Vision (ECCV)* (pp. 401–417).

[6] Rossi, E., Chamberlain, B., Frasca, F., Eynard, D., Monti, F. and Bronstein, M. (2020). Temporal graph networks for deep learning on dynamic graphs. *arXiv preprint* arXiv:2006.10637.

[7] Narayan, Apurva, and Peter HO'N. Roe. (2018). Learning graph dynamics using deep neural networks. *Ifac-Papersonline* 51.2: 433–438.

[8] Nair, Ankit B., *et al.* (2022). Comparative study of centrality based adversarial attacks on graph convolutional network model for node classification. *2022 7th International Conference on Communication and Electronics Systems (ICCES)*. IEEE.

[9] Chethana, Savarala, *et al.* (2023). Humanoid robot gait control using PPO, SAC, and ES Algorithms. *2023 4th IEEE Global Conference for Advancement in Technology (GCAT)*. IEEE.

[10] Divya Udayan, J., Addanki, V., Durgapu, S., Yerramreddy, D.R. and Kolla, D. (2023) Forward kinematics simulation of KUKA KR5 arc robot with robo analyzer, *ACM International Conference Proceeding Series*, pp. 294–299.

[11] Esakki Muthu Prabha, V. and Divya Udayan, J. (2023). Continuous Integration pipeline for ROS-based application using AWS Robomaker, *4th IEEE Global Conference for Advancement in Technology*, GCAT 2023.

[12] Megalingam, Rajesh Kannan, Anandu Rajendraprasad, and Sakthiprasad Kuttankulangara Manoharan. (2020). Comparison of planned path and travelled path using ROS navigation stack. *2020 International Conference for Emerging Technology (INCET)*. IEEE.

[13] Megalingam, Rajesh Kannan, *et al.* (2019). Robot operating system integrated robot control through secure shell (SSH). *2019 3rd International Conference on Recent Developments in Control, Automation & Power Engineering (RDCAPE)*. IEEE.

[14] Dobrevski, Matej, and Danijel Skočaj. (2021). Deep reinforcement learning for map-less goal-driven robot navigation. *International Journal of Advanced Robotic Systems* 18.1: 1729881421992621.

Data Science & Exploration in Artificial Intelligence – Gururaj et al. (Eds)
© 2025 the Author(s), ISBN: 978-1-032-96390-7

Development and validation of an algorithm-based choice of outcome measures for assessing motor developmental delay – A study protocol

B. Phalguni
Department of Physiotherapy, Manipal College of Health Professions, Manipal Academy of Higher Education, Manipal, Udupi District, Karnataka, India

Briano Direzze & Laura Brunton
CanChild Centre for Childhood Disability Research, Faculty of Health Sciences, McMaster University, Ontario, Canada

Narendra Babu Alagirisamy
Microsoft Inc., Washington, USA

Bhamini Krishna Rao & Selvam Ramachandran
Department of Physiotherapy, Manipal College of Health Professions, Manipal Academy of Higher Education, Manipal, Udupi District, Karnataka, India

ABSTRACT: Evidence-Based Practice (EBP) links the use of standardized outcome measures in the clinical decision-making process. Despite the availability of numerous evidence-based outcome measures in pediatric rehabilitation, their effective utilization in clinical practice is limited, thus affecting the knowledge translation process. In a clinical care setting, therapists face cognitive load and often employ heuristics to choose appropriate outcome measures due to a lack of knowledge or time constraints. The present study aims to minimize the cognitive load and bias on the therapist by introducing a machine learning-based algorithm to choose appropriate outcome measures for motor developmental delay. The investigators will review the existing literature to identify the relevant outcome measures and categorize them under key criteria to develop an algorithm. Subsequently, the algorithm will be subjected to expert and machine validation. Such an algorithm will improve the accuracy and consistency in choosing appropriate outcome measures for assessing motor developmental delay.

Keywords: decision-making algorithm, outcome measures, developmental delay

1 INTRODUCTION

1.1 *Evidence-based practice*

Evidence-based practice (EBP) plays a significant role in delivering high-quality healthcare service. The three major components of EBP are the best available research evidence, clinical expertise, and patient values. Applying the combination of these components is used in delivering optimal patient care for the best possible outcome from the healthcare intervention(HAM Director & Sackett DL, 1990). Outcome measures are tools that ensure evidence-based practice by providing a way to evaluate and document changes objectively following a process or health care services (Majnemer & Limperopoulos, 2007). Standardized outcome measures aid in the communication of patient problems and goals. It also helps in predicting

 DOI: 10.1201/9781003589273-2

and tracking the prognosis and outcomes of a patient (Pathak *et al.* 2023). Using outcome measures also improves the patient's perceptions about the healthcare intervention as it allows for collective decision-making, and improves the quality of care (Meerhoff *et al.* 2021).

1.2 *Barriers to usage of outcome measures*

Despite the widespread awareness of the importance of utilization of outcome measures, increase in the volume of research, and the development of new tools for assessment, the application of outcome measures among Pediatric rehabilitation professionals is low. The most frequently reported therapist-related barriers are (i) lack of awareness of available measures, (ii) lack of knowledge about the existence and availability of outcome measures and insufficient integration in daily practice, (iii) time constraints, i.e., limited time to search and learn about new outcome measures, shortage of time to administer and interpret the results and lastly (iv) reluctance to use measures that are not suitable for the given circumstance or planned goal (Pathak *et al.* 2023)(King *et al.* 2011)(Swinkels *et al.* 2011). Therapists often employ cognitive biases towards using a few selected outcome measures, i.e., therapists prefer using few outcome measures consistently due to convenience, frequently excluding other suitable alternative measures from routine practice (Saposnik *et al.* 2016).

1.3 *Knowledge translation*

Healthcare professionals must process vast patient information to guide their clinical decision-making. Working memory is temporary, and the cognitive process rapidly shifts from one clinical context to another. Due to the immense cognitive load, the readiness to process and utilize new knowledge from the available evidence is difficult (Schaffer *et al.* 2023). Knowledge translation (KT) strategies have emerged to bridge the knowledge gap created due to a lack of implementation of new knowledge into practice. Supporting KT strategies improves evidence-based practice and knowledge utilization of new research in clinical practice (Schreiber *et al.* 2015). Automating clinical decisions by the machine learning process can improve knowledge translation and reduce errors related to cognitive load and heuristics (Schaffer *et al.* 2023). Search for an appropriate, valid, and reliable outcome measure in Pediatric rehabilitation involves a difficult and time-consuming process. The emergence of new technologies like Artificial Intelligence and Machine learning decision-making algorithms can increase accuracy and make the search process easier, thereby promoting appropriate outcome measures and ensuring Evidence-based Practice is followed (Law *et al.* 2003a).

2 REVIEW OF LITERATURE

The investigators reviewed the literature in search of decision-making algorithms to choose outcome measures in assessing motor developmental delay in physiotherapy practice. Search in databases like PubMed, and Scopus was conducted for relevant studies using keywords *"algorithm", "decision tree", "developmental delay" and "outcome measures"* with filters "English" and "humans". Despite the extended period of publications in the search, it yielded only two relevant articles. Both articles describe an educational software program, "All about Outcomes" developed by the CanChild team. The first study describes the development and feasibility of a software program in which the user moves through key decision points of a decision tree to select outcome measurements. The article describes the use of the International Classification of Impairment, Disability, and Handicap (ICIDH) framework in the decision-making process to choose the most appropriate attribute for assessment. The participants indicated that the software was easy to follow and helped select outcome measures (Law *et al.* n.d.). The second study is a Randomized Control Trial (RCT), which involved the random assigning of students to the intervention group who rated the 'All about Outcomes' software and the control group who rated the conventional library search for identifying appropriate measures, time for search, ease of the process and satisfaction. Results indicated that the search using the software was more accurate and less time-consuming (Law *et al.* 2003b). The limitations of the studies were that the ICIDH framework was incorporated, which is currently replaced by the updated ICF framework. Presently, the software is not available for use.

In practice, few manual clinical care algorithms are available for developmental surveillance and screening of children at risk of developmental delays and disabilities. However, there are no machine-driven algorithms to guide clinical decision-making in choosing appropriate outcome measures for motor developmental delay. This study aims to develop and validate a machine-driven decision-making algorithm to choose an appropriate outcome measure for assessing motor developmental delay.

3 METHOD

Healthcare experts and therapists now use the International Classification of Functioning, Disability and Health (ICF) framework to address an individual's health in the relationship between body structure and function, activities and participation, and environmental and personal factors (da Fonseca Filho *et al.* 2021). The ICF integrates the biopsychosocial model (role of biological, psychological, and social dimensions in determining health) and recognizes the social and environmental factors and their role in creating disability. Since the emergence of the ICF model, rehabilitation professionals have developed several outcome measures linked to various components of the ICF model. With the vast resource of outcome measures available, our research aims to simplify the search for an appropriate outcome measure for assessing motor developmental delay in children. The study involves two phases:

Phase 1: Development of a database of outcome measures used in assessing motor developmental delay in Pediatric physiotherapy rehabilitation.

During the first phase of the study, the investigators will search and review the relevant outcome measures assessing motor developmental delay from various databases and repositories. The search will also include published articles (of all study designs) in peer-reviewed journals. The investigators will retrieve the relevant outcome measures and will list them in MS Excel sheet based on the inclusion criteria that (i) assess the population aged 0-18years of age, (ii) the outcome measure assesses gross and fine motor domains of development, and (iii) the outcome measures used in physiotherapy practice. Subsequently, the investigators will map the listed outcome measures based on specific attributes such as (i) domains of the ICF model (Body structure, Body function, Activity and Participation, Environmental factors, and Personal factors), (ii) further classified under Gross motor, Fine motor, Functional aspects etc. Mapping based on other criteria such as (iii) Age, (iv) mode of administration (therapist administered/patient-reported), and (v) Purpose of use (Predictive, Diagnostic, Evaluation). Mapping each outcome measure under the attributes will follow the linking rules of outcome measures to various domains of ICF classification.

Phase 2: Development and validation of a decision-making algorithm to choose an appropriate outcome measure

During the second phase of the study, the investigators will develop a draft algorithm enumerating every step, sequence, condition, and argument in the decision-making process of choosing an outcome measure. The investigators will then generate the flowchart, a visual representation of the draft algorithm for validation by the experts. Following expert iterations, the investigators will modify the algorithm suitably. The algorithm will then be validated using machine learning models like Bayesian neural networks to validate the step, condition, and sequences to generate the choice of outcome measure specific to the clinical context.

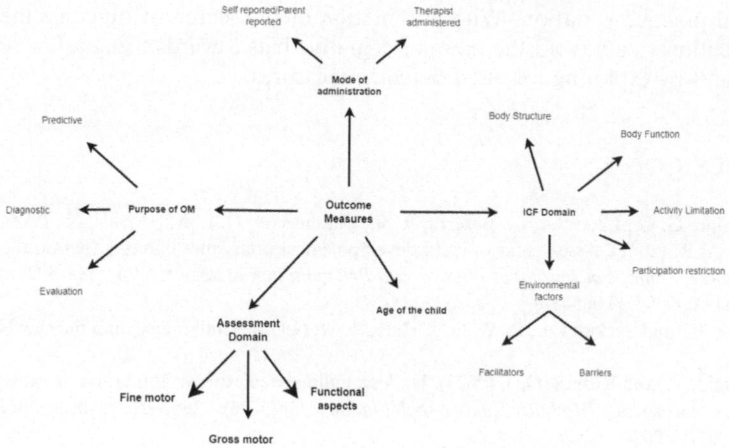

Figure 1. Mapping of individual outcome measures under different attributes.

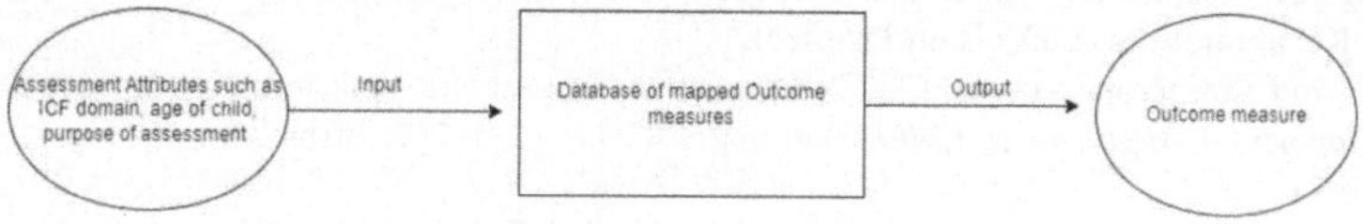

Figure 2. Process of information processing.

4 DISCUSSION

Driven by increased technological advancements, computers now perform tasks of great complexity and accuracy(Scurlock-Evans *et al.* 2019). Machine learning allows the synthesis of complex data and the development of artificial intelligence. The supervised machine learning model involves the input of large data with specific rules and conditions, which enables training the computer machine to explain the relationship between variables and generate an output on the interest variable. (Sidey-Gibbons & Sidey-Gibbons, 2019). Given the function of machine learning techniques, the medical field will vastly benefit from it. In our study, we intend to train a machine learning model to correlate assessment attributes (e.g., ICF domain, Age, Assessment domain etc) to generate a specific outcome measure specific to the clinical context. Once the machine learns the algorithm with the test data, it can suggest the choice of outcome measure when applied to new search characteristics/attributes.

The findings of this study will have many implications. Machine learning algorithms will be an effective knowledge translation tool that will reshape the future of clinical practice and bridge the gap between knowledge generation and utilization. The use of this tool will help promote evidence-based practice. Clinicians can use this tool to choose an appropriate outcome measure relevant to their clinical context, overcoming their cognitive biases. Clinical educators can use the flowchart/algorithm in the clinical training of health professional students. During the formative stages of clinical skill acquisition, the students will use this tool to learn the rules and conditions for clinical decision-making.

5 CONCLUSION

Since no similar software or decision-making tools are available, developing this machine learning tool is needed to maintain standards of care by ensuring consistency and preventing unwarranted practice variation. With automation of the search of outcome measures, the clinical practitioner can avoid the flaws of cognitive bias towards using a few selected outcome measures by exploring less used outcome measures.

REFERENCES

da Fonseca Filho, G. G., Lopes, A. C., Bezerra, R. B., Candido, A. D. E. M., Arrais, N., Pereira, S. A., and Lindquist, A. R. (2021). Assessment of child development in premature babies based on the ICF biopsychosocial model. *European Journal of Physical and Rehabilitation Medicine*, 57(4), 585–592. https://doi.org/10.23736/S1973-9087.20.06543-0

HAM Director, C., and Sackett DL, R. W. G. J. H. R. R. W. (1990). Thinking beyond the new NHS. *In BMY* (Vol. 301).

King, G., Wright, V., and Russell, D. J. (2011). Understanding paediatric rehabilitation therapists' lack of use of outcome measures. *Disability and Rehabilitation*, 33(25–26), 2662–2671. https://doi.org/10.3109/09638288.2011.582924

Law, M., Hurley, P., Hurley, D., King, G., and Hanna, S. (2003a). Selecting outcomes measures in children's rehabilitation: A comparison of methods. *Archives of Physical Medicine and Rehabilitation*, 84(4), 496–499. https://doi.org/10.1053/apmr.2003.50035

Law, M., King, G., Russell, D., Mackinnon, E., Hurley, P., and Murphy, C. (n.d). Measuring Outcomes in Children's Rehabilitation: A Decision Protocol.

Majnemer, A., and Limperopoulos, C. (2007). Importance of outcome determination in pediatric rehabilitation. *Developmental Medicine & Child Neurology*, 44(11), 773–777. https://doi.org/10.1111/j.1469-8749.2002.tb00285.x

Meerhoff, G. A., van Dulmen, S. A., Maas, M. J., Bakker-Jacobs, A., Nijhuis-Van der Sanden, M. W., and van der Wees, P. J. (2021). Exploring the perspective of patients with musculoskeletal health problems in primary care on the use of patient-reported outcome measures to stimulate quality improvement in physiotherapist practice; a qualitative study. *Physiotherapy Theory and Practice*, 37(9), 993–1004. https://doi.org/10.1080/09593985.2019.1678205

Pathak, A., Abbott, J. H., Bajracharya, N., Gurung, G., Nepal, G. M. and Sharma, S. (2023). Barriers and facilitators to implementation of outcome measures among physiotherapists in Nepal: A mixed-methods study. *Musculoskeletal Science and Practice*, 68. https://doi.org/10.1016/j.msksp.2023.102859

Saposnik, G., Redelmeier, D., Ruff, C. C. and Tobler, P. N. (2016). Cognitive biases associated with medical decisions: a systematic review. *BMC Medical Informatics and Decision Making*, 16(1), 1–14. https://doi.org/10.1186/s12911-016-0377-1

Schaffer, C., Goldart, E., Ligsay, A., Mazwi, M., Gallant, S. and Ehrmann, D. (2023). Take a Load Off: Understanding, Measuring, and Reducing Cognitive Load for Cardiologists in High-Stakes Care Environments. *In Current Treatment Options in Pediatrics* (Vol. 9, Issue 3, pp. 122–135). Springer Science and Business Media Deutschland GmbH. https://doi.org/10.1007/s40746-023-00272-3

Schreiber, J., Marchetti, G. F., Racicot, B., and Kaminski, E. (2015). The Use of a Knowledge Translation Program to Increase Use of Standardized Outcome Measures in an Outpatient Pediatric Physical Therapy Clinic: Administrative Case Report Background and Purpose. *Pediatric Physical Therapists Face many Challenges*. https://academic.oup.com/ptj/article/95/4/613/2686404

Scurlock-Evans, L., Upton, P. and Upton, D. (2014). Evidence-Based Practice in physiotherapy: A systematic review of barriers, enablers and interventions. *In Physiotherapy* (United Kingdom) (Vol. 100, Issue 3, pp. 208–219). Elsevier Ltd. https://doi.org/10.1016/j.physio.2014.03.001

Sidey-Gibbons, J. A. M., and Sidey-Gibbons, C. J. (2019). Machine learning in medicine: a practical introduction. *BMC Medical Research Methodology*, 19(1). https://doi.org/10.1186/s12874-019-0681-4

Swinkels, R. A. H. M., Van Peppen, R. P. S., Wittink, H., Custers, J. W. H., and Beurskens, A. J. H. M. (2011). Current use and barriers and facilitators for implementation of standardised measures in physical therapy in the Netherlands. *BMC Musculoskeletal Disorders*, 12. https://doi.org/10.1186/1471-2474-12-106

Data Science & Exploration in Artificial Intelligence – Gururaj et al. (Eds)
© 2025 the Author(s), ISBN: 978-1-032-96390-7

Urine biomarker profiling for enhanced urolithiasis risk prediction

V.B. Lakshmi, K.B. Sivachandra & Aparajitha Devi Sugula
Department of Computer Science and Engineering, Amrita School of Computing, Amrita Vishwa Vidyapeetham, Amritapuri, India

S. Abishek
Rolls-Royce Inc, Bangalore, India

T. Anjali
Department of Computer Science and Engineering, Amrita School of Computing, Amrita Vishwa Vidyapeetham, Amritapuri, India

ABSTRACT: Urolithiasis is the term that defines the formation of stone within the kidney or on the urinary tract. This is a very serious and painful ailment where the patient needs to undergo various kinds of medication which includes even a surgery to preserve the patient's life. Early detection and treatment can help to avoid the surgery entirely, instead go with appropriate medicines and care to save the patient's life. These disorders have very few or no symptoms at the starting stage of disease progression, but they progress very quickly resulting in excruciating pain and other ailments. Through this research, our goal is to assist patients diagnose the disease through their urine samples considering different urine parameters like age, gender, colour, transparency, glucose, pH, specific gravity, white blood cells, red blood cells, and protein in the urine sample. In the study, the aim is to develop different machine learning and deep learning models and assess them on both the training and testing dataset. The highest accuracy so far achieved is for Regularized Feedforward Neural Network with Dropout for Binary Classification. Regularized Feedforward Neural Network with Dropout for Binary Classification achieved an accuracy of 98.2% followed by Deep Neural Network with Batch Normalization and Dropout Layers with an accuracy of 94.2% dataset. These acquired accuracy demonstrates how well the models works in detecting kidney stone conditions from urine samples.

Keywords: Urolithiasis, Deep Neural Network, Feedforward Neural Network, Kidney Stones, Deep learning, Binary Classification, Calcium oxalate, Uric Acid, Struvite, Cystine

1 INTRODUCTION

Every year, approximately ten lakh people are affected by the formation of kidney stones in their body. Kidney stones are serious ailment that affects both males and females irrespective of which age group they belong to. A kidney stone is generally a very hard objects that consists of different chemicals in the urine. There are several unwanted substances that constitute the urine [1]. But when the liquid content in urine is less, these unwanted substances form crystals that attracts other elements which results in the formation of hard substance called kidney stone. Calcium, oxalate, urate, cystine, xanthine, and phosphate are the major chemical that accelerate the formation of kidney stones.

Recent studies shows that kidney stones are among those diseases that are difficult to diagnose at the starting phase of disease progression which makes the condition of patient worse leading to multiple surgeries to get rid of the condition. There are several reasons that accelerates the formation and growth of stones in the human body [2]. Improper diet, not taking adequate amount of water, excess body weight or may be also due to other medications and diseases that results the formation of hard substances in the human body. The stone formation takes place without any symptoms usually, leading to severe pain near ribs, lower abdomen, and also pain and burning sensation while passing urine etc. are few common symptoms seen among patients. The stones are mainly of four types: Calcium oxalate, Uric Acid, Struvite, and Cystine. Calcium Oxalate are the most common kind of stone seen in the greatest number of patients. Uric acid stones are formed due to high concentration of a substance called purines that is mostly found in meat. Struvite is less common and also it affects the upper urinary tract of patient. Cystine are due to genetic disorders and very least common. The treatment for stones depends majorly on the type of stone formed on the patient, the age and other medical factors of the patient. Early diagnosis helps to prevent the stone from getting enlarged by dissolving the stone through proper medication. As the stone gets enlarged, the treatment becomes complicated leading to Shock-wave lithotripsy etc. Kidney stones can lead to other medical conditions like chronic kidney disease. The studies also proves that a person with previous kidney stone history is likely to suffer from another kidney stones within 5 to 7 years. Kidney stones create pockets in the urinary tract where other bacteria grow and create other dangerous health conditions [3]. It can also lead to swelling to enlargement of kidneys which is another serious health conditions resulting in dialysis or even death of the patient. Kidney stones results in malfunctioning of kidneys that affect the purification of blood in human body and results in severe other diseases. The aim of this research is to detect kidney stone formation at the early stage through urine analysis of people [4]. There are major attributes in human body that fastens the formation of kidney stones. Through this research, we developed models that detect the formation of kidney stone through several parameters like WBC, RBC, pH, specific gravity, colour, transparency etc. of urine in humans. The model performed well on both the training and testing dataset and acquired an accuracy of 98%.

2 RELATED WORKS

The major issues faced in the previous research in this area was mainly due to the limited dataset along with the noisy contents present in it. The high chance of overfitting due to imbalance in dataset also ought to be a problem.

In this [1] research the authors Swain, Debabrata, Utsav Mehta, Ayush Bhatt, and A. T concentrates on ML techniques like Random Forest and Support Vector Machine in order to get more accurate result after diagnosis. Due to lack of consideration of the outliers, imbalanced dataset and it also limit the accurate medical analysis. Handling with outliers along with balancing the dataset the model performance can be increased.

In this [2] research the authors Nishat, Mirza Muntasir, and A.T mainly concentrates on machine learning techniques which includes Decision Tree, K-Nearest Neighbors (KNN) and may more with Multilayer Perceptron (MLP) along with Quadratic Discriminant Analysis (QDA) inorder to achieve more accurate result. Random forest model achieved most accurate result compared to other models. The major problem faced by this model was due to the quality and imbalance in dataset and due to high chances of having overfitting.

In this [3] research the authors Khan, Bilal, Rashid Naseem and A.T concentrates on machine learning approach including J48 Decision Tree, Support Vector Machine (SVM), Composite Hypercube on Iterated Random Projection (CHIRP), Naive Bayes Tree (NBTree), Naive Bayes (NB) but due to the imbalance of dataset and data quality the accuracy was not so high.

In this [4] research the authors Salekin, Asif, and John Stankovic mainly concentrate on Wrapper Approach for Attribute Selection along with the random forest and KNN. The noisy and presence of missing data in the dataset made the study more complicated along with the attribute selection made this research more complicated.

In this [5] research the authors Radha, N., and S. Ramya mainly concentrates with the machine learning approach including KNN, Naïve Bayes, Decision Tree and Support vector machine, of these four models KNN proved to get the most accurate result. Quality of the collected dataset along with imbalanced attributes creates problems in model selection. Identification of relevant features for predicting the disorder also affect the accurate model selection.

In this research we aimed in predicting the most accurate result from the implemented machine learning and deep learning model. Feedforward Neural Network acquired the most accurate result around 98.2% followed by Deep Neural Network with Batch Normalization and Dropout Layers with an accuracy of 94.2%. dataset.

3 MATERIALS AND METHODS

3.1 *Dataset*

The dataset used throughout this research in an open source's dataset collected by a local clinic in Northern Mindanao, Philippines over a particular period from 2020 to 2023[6]. The different attributes considered for the study includes age, gender, urine colour, urine transparency, glucose, protein, pH, specific gravity, white blood cells, red blood cells, epithelial cells, mucous threads, amorphous urates, bacteria. Each of these attribute's concentration or value directly or indirectly accelerates the formation of kidney stones. There are data of nearly 1500 patients in the dataset. The target column points to two values either the patient has kidney stone or not. There are both categorical values and numerical values in the dataset which are treated during data preprocessing [7].

3.2 *Data pre-processing*

The initial step in the research is data pre-processing. The dataset contains both numerical and categorical values in different columns. Few values are also missing in the dataset[8]. The categorical values are changed to numerical through one-hot encoding that is discussed later in this section. A Complete-case analysis (CCA) approach to use to handle the rows with missing values or data in the dataset. It involves removing or misapplying all the rows in the dataset which has one or more missing values[9]. This in turn reduced the bias in the dataset, resulting in better model performance and accuracy. Filling the missing rows with mean, sd or some other values results in bias that reduces the integrity of the dataset. Inputting values in missing values can also lead to model overfitting in majority of cases. Now, the dataset is consistent to be applied for a deep learning model[10]. The dataset is then split in the ration 8:2 for training and testing the model.

3.3 *Label encoding*

There are both numerical and categorical values in the dataset. Also, the dataset comprises ordinal and nominal features. In ordinal features the values have a natural ranking or order while nominal features represent unique groups in the dataset[11]. Transparency, Epithelial Cells, Mucous Threads, Amorphous Urates, Bacteria, Colour, Glucose, Protein, WBC, and RBC have nominal features. The values in categorical columns used the one-hot encoding technique. The columns of WBC and RBC have specific range that are preprocessed using an algorithm. The algorithm first checks whether the input data contains any symbol like greater than or less than, is yes return the number without the symbols. If the values don't have any symbols instead have a range of numbers, then return the midpoint of the range.

For categorical values, the technique of one-hot encoding is used. The technique involves representing each category of values as a binary vector or array. The bit '1' is considered hot representing the existence of that particular attributes and the other bit indicates the non-existence. The steps in one-hot encoding includes identifying all unique categorical values for a particular attribute followed by pointing them to certain integer values like 0, 1 then set the element to the observation based on the category it is pointing to.

3.4 *Model evaluation*

The dataset is split in the ratio 8:2 for testing and training the model respectively. There are two models used throughout the study. A Regularized Feedforward Neural Network with Dropout and Deep Neural Network with Batch Normalization and Dropout Layers. Accuracy and precision are the evaluation metrics used to assess the performance and effectiveness of the models. Accuracy gives the actual idea on how the model performs all over the dataset. Precision helps in minimizing the false positives. Precision is particularly important here as we are dealing with medical dataset [12].

4 PROPOSED METHODOLOGY

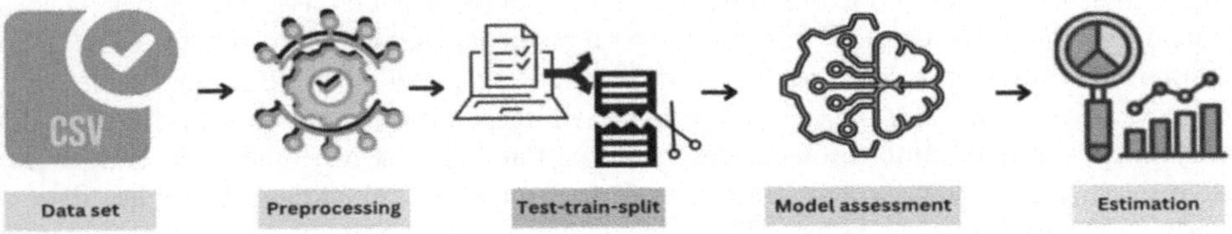

Figure 1. Proposed methodology.

The dataset was trained with KNN, decision tree, DNN and FNN. In order to predict results of the data given accurately to know the presence and non-presence of disease for the input taken. After the training observations were noted, each recorded with different accuracy Decision tree results with 90.12%, but they have some limitations like overfitting when the depth of the tree is not properly controlled and also it expresses low compared with other ensemble models or techniques[13]. KNN scored around 92.23% of accuracy but the main concern of this is, It is sensitive if one class controls the dataset there is a chance of exhibiting bias on majority class that leads in struggling for accurate performance. Deep Neural networks produce accuracy percentage around 92 but there can be lack of interpretability as well as rather than understanding the patterns it concentrates on memorizing the trained data which causes the over fitting.

The Feedforward neural networks scored highest among all by overcoming from the few limitations that other algorithms we had trained. As these are structured in the easy way and information flow through the network training is easy compared to DNN. It takes less training time which is one of the essential factors. This is why the FNN scored higher accuracy than others.

4.1 *KNN*

KNN is a supervised machine learning technique which can be implemented on both classification and regression task making it more useful in case of handling with numerical and categorical values in dataset. KNN measures the distance between the data point in the feature space using either the Euclidean distance or Manhattan distance. KNN does not involve any explicit training making it more memory intensive and expensive[14].

If a new data arrives assume that as x, the algorithm finds the with all the data points in the dataset. According to the results it states to which point it is nearest and can be located. K nearest neighbor which are having the small difference in the distance to x.x is determined based on the nearest values classification if major is for positive then it classifies positive else negative. This algorithm works in this way.

4.2 *Decision tree*

Decision Tree is one of the essential machine learning algorithms which comes under the supervised algorithm, which uses classification as well as regression assignments. Decision

Tree helps for predicting the non-discrete values which divides input data regions into corresponding separate categories. In this dataset the Decision tree can be worked in a way that learns about the simple decision rules depends on the features, classifies the data whether existed or not existed.

Decision tree goal is to make the partition for the feature-based decision, the values of urine samples like levels of protein, gravity, pH and others. The working efficient of this model is mainly depends on the test train split. This decision tree uses the criteria of stopping when it satisfies with the condition or reaches the maximum depth. the number of classes and the proportion of samples of class at particular node reduces the impurity. After training the data with this algorithm when a new data point enters for the prediction that data travels across the tree until it met the criteria.

4.3 *FNN*

A feedforward neural network (FNN) is a artificial neural network in which connection between nodes are not in format of a cycle aiming there is transfer of information in a single direction, just like a data-driven decision. When data representing various numbers of attributes of the clinical data. Initially the Each piece of data goes into the input layer of the neural network. The working of this layer is compared with the networks that arranged in the eye. Moving to the Hidden layers when it receives the data from the input layer, it does not only go through in a particular hidden layer it can goes to any of the layer which can be more than one. In this phase of the work every layer was made by each neuron, where each neuron acts as the decision taker. As the hidden layer works in a way that the values of the input will be multiplied by the weight and add all it up. After it reaches the activation function, ReLu activation function was used to take the decision on activating it or not.

In Feed forward neural network labeling our data attributes including target which describes the presence or absence of that value, the alignment of FNN alters the weights and biases to decrease the level of difference among the estimated and true outcomes. This process is known as back propagation. When urine samples run through the network the results are mainly based on the framework it learns during the phase of training. The frame work that used in this consists of One input and output tensor, easily helps to add more layers sequentially based on necessities. The fully connected layers make sure to have connection maintained with previous ones.FNN helps for taking the binary classification predictions.

4.4 *DNN*

Deep learning neural networks are a segment of machine learning algorithms imitated by the structure and functioning of the human brain. The collected data should be taken care with the necessary process like cleaning or dealing with the null values or invalid terms. This neural network contains multiple neurons including the input and output layer along with t hidden layers, layers which plays the essential role in the formation or functioning of the network. Every layer has interconnection with the neurons and every connection is weighted. The hidden layers contain activation functions like ReLu which generate Output=ReLu (Input).

The network alters its inner weights in a repetitive manner to reduce the distance between the predictions and real labels. In this algorithm gradient descent methods which are used for optimization. The model work is analyzed on validation split of the data to allow its ability to unseen data for generalization. Which detects the over fitting in the data as well as the hyper parameters of fine tuning. After training the data with the model the output layer, the final layer for predicting the provided input of presence absence of the target. This network succeeds in learning automatically using hierarchical information of data and also helps them to understand the relation between each attribute.

The function used in the hidden layer is in the form of *Output_layer-1 = ReLu (Input 8 Weights(layer-1) + Bias(layer-1);* Dropout function is used for the output of the input for the layer-1 in hidden layers = Dropout (Output_layer-1, dropout rate=0.5).(therefore rate=0.5 was used in this model training).In this way layers will be added up accordingly to fit for the training that produces prediction more accurately.

5 EXPERIMENTAL RESULTS

In this research the KNN, Decision Tree, Feedforward Neural Network, Deep Neural Network was implemented to determine the most accurate result for the analysis.

Decision Tree model was first employed on this research but the accuracy acquired was nearly 90.12% and a precision of 89.7% followed by the KNN with an accuracy of 92.32% and precision of 92.23\% followed by the deep learning models namely Feedforward Neural Network with Dropout and Deep Neural Network with Batch Normalization and Dropout Layers. KNN being the simple model being able implemented on both classification and regression task making it more useful in case of handling with numerical and categorical values in dataset. Coming to the Decision Tree consists of both the branches and nodes which represents the output condition and the satisfactory condition respectively.

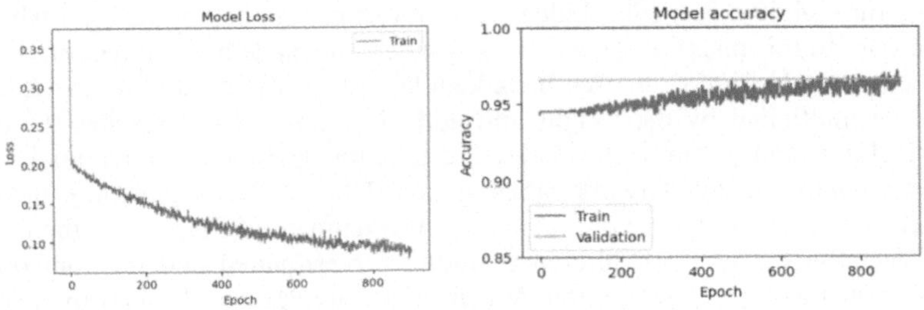

Figure 2. Accuracy of FNN model.

The FNN model used very simple architecture with minimum layers making it easier in training the dataset and also easy interpretation of the result. The use of sigmoid functions in the output layers makes the model more suitable for binary classification task thus making the interpretation easier. This model acquired an accuracy of around 98.2% and a very low loss of around 0.071% as in Figure 2.

The DNN model used Relu activation layer in the hidden layers helps the model in predicting even the complex non-linear relationship between the input features and the target variables. The use of Dropout layers in the deep learning models helps in avoiding the problem of over fitting by dropping a fraction of neurons during the training process in both the implemented deep learning models. This model acquired an accuracy of around 94.2% and a very low loss of around 0.086% as in Figure 3.

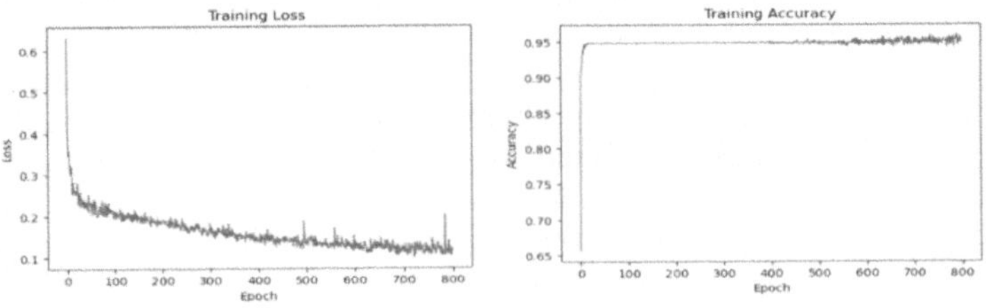

Figure 3. Accuracy of DNN model.

6 CONCLUSION

Multiple machine learning algorithms along with deep learning approach was implemented in this research. Early detection and treatment can even avoid the chance of surgery making it more important.

In this research we aimed in exploring the most accurate result from the implemented machine learning and deep learning model. Feedforward Neural Network acquired the most accurate result around 98.2% followed by Deep Neural Network with Batch Normalization and Dropout Layers with an accuracy of 94.2%. These acquires accuracy demonstrates how well the models works in detecting kidney stone conditions from urine samples.

The long-term goal of this study aims to detect different kinds of disorders that affects the kidneys in human beings, that will advance in the field of nephrology. The currently proposed methodology can be extended to an enlarged dataset, trying out different deep learning models that results in better evaluation metrics like accuracy, precision, recall etc. Furthermore, the focus of future work will be automating the process of identification of different kidney related illness, which in turn will aid in the diagnosis and improve the efficacy of therapy.

REFERENCES

[1] Swain, Debabrata, Utsav Mehta, Ayush Bhatt, Hardeep Patel, Kevin Patel, Devanshu Mehta, Biswaranjan Acharya, Vassilis C. Gerogiannis, Andreas Kanavos, and Stella Manika. (2023). A robust chronic kidney disease classifier using machine learning. *Electronics* 12, no. 1: 212.

[2] Nishat, Mirza Muntasir, Fahim Faisal, Rezuanur Rahman Dip, Sarker Md Nasrullah, Ragib Ahsan, Fahim Shikder, Md Asfi-Ar-Raihan Asif, and Md Ashraful Hoque. (2021). A comprehensive analysis on detecting chronic kidney disease by employing machine learning algorithms. *EAI Endorsed Transactions on Pervasive Health and Technology* 7, no. 29: e1-e1.

[3] Khan, Bilal, Rashid Naseem, Fazal Muhammad, Ghulam Abbas, and Sunghwan Kim. (2020). An empirical evaluation of machine learning techniques for chronic kidney disease prophecy. *IEEE Access* 8:55012–55022.

[4] Salekin, Asif, and John Stankovic. (2016). Detection of chronic kidney disease and selecting important predictive attributes. *In 2016 IEEE International Conference on Healthcare Informatics (ICHI)*, pp. 262–270. IEEE.

[5] Radha, N., and S. Ramya. (2015). Performance analysis of machine learning algorithms for predicting chronic kidney disease. *Int. J. Comput. Sci.Eng. Open Access* 3, no. 2015: 72–76.

[6] Sunny, Liniya, Emmanuel James, M. Krishnapriya, M. R. Mamatha, and S. Sudhindran. (2016). Evaluation of factors influencing renal dysfunction following living donor liver transplantation. *Kidney* 8: 10.

[7] Arif, Muhammad Shoaib, Aiman Mukheimer, and Daniyal Asif. (2023). Enhancing the early detection of chronic kidney disease: a robust machine learning model. *Big Data and Cognitive Computing* 7, no. 3: 144.

[8] Lakshmi, Vb, Kb Sivachandra, S. Abhishek, and T. Anjali. (2023). Nephrolithiasis Taxonomy: A Multifaceted Exploration of Renal Calculi. *In 2023 2nd International Conference on Automation, Computing and Renewable Systems (ICACRS)*, pp. 655–660. IEEE.

[9] Manoj, B., Neethu Mohan, and Sachin Kumar. (2022). Automated detection of kidney stone using deep learning models. *2nd International Conference on Intelligent Technologies (CONIT)*, pp. 1–5. IEEE

[10] Lakshmi, V. B., K. B. Sivachandra, H. Parthasaradhi, S. Abhishek, and T. Anjali. (2023). An empirical analysis of CNN for American sign language recognition. *5th International Conference on Inventive Research in Computing Applications (ICIRCA)*. pp.421–428. IEEE.

[11] Ani, R., Greeshma Sasi, U. Resmi Sankar, and Deepa O. S. (2016). Decision support system for diagnosis and prediction of chronic renal failure using random subspace classification. *In 2016 International Conference on Advances in Computing, Communications and Informatics (ICACCI)*, pp. 1287–1292. IEEE.

[12] Nayak, Rashmi, and N. Manohar. (2021). Computer-Vision based face mask detection using CNN. *In 2021 6th International Conference on Communication and Electronics systems (ICCES)*, pp. 1780–1786. IEEE.

[13] Abhishek, S., Mahima Chowdary Mannava, A. J. Ananthapadmanabhan, and T. Anjali. (2023). Towards Accurate Auscultation Sound Classification withConvolutional Neural Network. *In 2023 International Conference on Innovative Data Communication Technologies and Application (ICIDCA)*, pp. 254–260. IEEE.

[14] Ani, R., Greeshma Sasi, U. Resmi Sankar, and O. S. Deepa. (2016). Decision support system for diagnosis and prediction of chronic renal failure using random subspace classification. *In 2016 International Conference on Advances in Computing, Communications and Informatics (ICACCI)*, pp. 1287–1292. IEEE

Data Science & Exploration in Artificial Intelligence – Gururaj et al. (Eds)
© 2025 the Author(s), ISBN: 978-1-032-96390-7

Autonomous path-finding system using Q-learning algorithm

Karan Kumar, G. Ramesh, Harish Kunder, Abdullah & M. Ashok Kumar
Department of Artificial Intelligence and Machine Learning, Alva's Institute of Engineering and Technology, Mangalore, Karnataka, India

ABSTRACT: The paper explores the application of reinforcement learning, specifically the Q-learning algorithm, to robotic systems for autonomous maze exploration. Robots need to be able to learn and adapt, therefore research is focused on providing them with the tools to dynamically enhance their navigational tactics. Using a Raspberry Pi for decision-making, the robot maps the layout of the maze using a variety of sensors. The main contribution of the work is the implementation of Q-learning, which enables the robot to correlate actions with particular maze circumstances. Over time, this process of learning optimises the robot's navigational ability by improving its decision-making. The study highlights how crucial it is for these kinds of systems to make precise decisions and produce accurate results. To increase the robot's accuracy, ongoing effort entails fine-tuning algorithm parameters and streamlining the learning procedure. The autonomous robot reached the destination from the starting point by the optimal route.

Keywords: Path-finding, Q-Learning, Maze environment, decision-making, Autonomous robot

1 INTRODUCTION

In the fast-evolving field of robotics, the pursuit of autonomy and decision-making abilities has resulted in creative methodologies and played a significant part in reinforcement learning. way planning is determining the best way to take through an obstacle-filled environment so that a project can avoid all obstacles from start to finish without colliding [1]. This study focuses on an essential application of this method, which enables robots to navigate difficult mazes utilizing a Q-learning algorithm. Effective navigation has a long history of study in people, animals, and robots, and intelligent entities must achieve their objectives in real-world contexts [2]. The key goal is to give the robotic system the ability to learn and adjust its path-finding tactics based on feedback from encounters with the maze environment. The robot, outfitted with a variety of sensors, obtains a thorough understanding of the maze's background. These sensors enable the robot to detect obstacles, identify open passageways, and assess viable routes through the maze, providing critical decision-making input. A path planner uses the original configuration of the environment to find optimal paths; however, if the environment changes, this configuration becomes invalid [3]. Figure 1 displays a maze environment that visually encapsulates the intended context. Robots must be able to sense, plan and act without human aid in the challenging and demanding task of mobile robot autonomous navigation [4]. Path planning is the process of identifying a robotic manipulator's non-collision, continuous path from an initial posture to a target pose in a configuration space where the manipulator's restrictions must be satisfied. To satisfy the requirements, the robot must consider improving performance measures, including planning time, path length etc, during the path planning phase. Because of its ability to plot its path accurately, the robot can be sure it will complete the work at hand safely [5]. By using the value of each step, the algorithm decides what to do next. The Q-learning algorithm-based agent is only able to decide what to do next based on its present state; it is not aware of its surroundings. It is possible for the Q-learning algorithm to drastically cut down on iterations, enhance path planning, accelerate the convergence of the learning process, and continuously optimize the path [6]. The robot gradually improves its decision-making skills by learning to link particular actions to the conditions of the environment through repeated interactions with the maze. Consequently, the robot gradually refines its path-finding techniques, enhancing its ability to navigate the maze. The major goal of this research article is to demonstrate how we implemented our project, which involved employing

 DOI: 10.1201/9781003589273-4

several sensors in a Maze environment to find a path using a Q-learning algorithm on a Raspberry Pi. The primary objective is to determine the fastest route to the destination while avoiding collisions [7].

Figure 1. Maze environment.

Organization of the paper: Section 1, focuses on the introduction of the project and an overview of its importance and methodology. Section 2, concludes existing research has been reviewed and outlines the problem statement of the proposed project and its main objectives, given in Section 3. The proposed methodology and the Q-learning algorithm's working are discussed in section 4. Section 5, focuses on its design and system architecture. Section 6, describes the implementation and Experimental results, Finally, Section 7, concludes by summarizing key findings, discussing future enhancements, and reflecting on the project's significance.

2 RELATED WORK

The optimized version of the Q-Learning algorithm for mobile robot local path planning, features award functions, action-selection policies, RMSprop for learning rate adjustment, and unique Q-table initialization techniques. It has been shown to outperform previous algorithms in several metrics. O-QL utilizes RMSprop for dynamic adaptive learning rate adjustments. A 50 x 50 randomized obstacle grid environment with three sets of grid environments, each with an obstacle grid that made up 20%, 30%, and 40% of the total grid cells, made up the dataset used for the study. Various criteria are used in simulation studies to assess the algorithm's performance in three different labyrinth situations [1]. Deep reinforcement learning agents are trained by simulating the Morris Water Maze in two dimensions. Additionally, autonomous navigation strategy categorization is carried out, and the results are compared with experimental data. Biologically plausible tasks are proposed, and artificial agent neural networks' internal representations are investigated [2], [3]. The Smoothly RRT (S-RRT) algorithm is an improved RRT algorithm that enhances exploring speed and efficiency, leading to smoother paths with continuous and executable curvatures. This method of autonomous obstacle avoidance dynamic path-planning for a robotic manipulator demonstrates the high autonomy, intelligence, and flexibility of the manipulator in obstacle avoidance processes. The dataset used in the study includes information on path planning for a robotic manipulator in static and dynamic environments, involving initial and target poses, obstacle configurations, and generated paths for algorithm comparison. For future research based on the successful implementation of the S-RRT algorithm in static and dynamic environments [8]. The EMQL method exhibits superiority in path optimization ability, convergence speed, and dynamic flexibility. It does this by combining Q-learning with an experience memory mechanism to better path planning in unfamiliar contexts. The feasibility of the Turtlebot3 burger robot is further validated by an experiment conducted in the real world. Grid maps and the GIS data collection for the Beijing 2008 road network were sourced from real-world benchmarks. Additionally, further research could explore enhancing the experience transfer and reuse strategies to enable robots to efficiently adapt to changes in tasks and environments [5]. By using raw sensor data and continual reinforcement learning, robots may autonomously traverse uncharted territory. This improves their efficiency and adaptability in dynamic context through iterative learning processes. The proposed approach effectively handles obstacle-cluttered problems and generalizes well to unseen environments. The approach's straightforward transferability and higher performance are demonstrated by the experimental results. The dataset used in the study involves training and testing environments designed for mobile robot navigation experiments, including simple and complex scenarios, using the Pioneer 3-DX robot platform and V-REP simulator [9], [10]. Recovery Reinforcement Learning (RL) is an algorithm that employs distinct policies for task performance and safety and utilizes offline data to learn about constraint violations before policy learning. It has demonstrated considerable improvements over previous approaches in lab and simulation settings. When compared to the most advanced safe reinforcement learning algorithms in various tasks and situations, it demonstrates better performance in striking a balance between task success and constraint satisfaction [11], [12].

3 PROBLEM STATEMENT AND OBJECTIVES

The challenge involves designing a specialized hardware platform tailored to the robot's unique requirements and implementing a sophisticated software system for real-time decision-making using reinforcement learning. The software development encompasses the creation of an advanced algorithm that harnesses reinforcement learning principles, enabling the robot to adapt dynamically to diverse maze scenarios. This project aims to advance the field of autonomous robotics through a complete approach, addressing both hardware and software intricacies for a robust and intelligent solution.

The Objectives of this work are:

1) Develop and integrate a reinforcement algorithm for autonomous decision-making in the robot's control system.
2) Implement an autonomous robotic system to navigate a 6x5 maze.
3) Create a well-structured circuit design supporting seamless integration of the algorithm, control system, and sensor array.
4) Utilize an autonomous robot to efficiently reach the maze's destination, demonstrating effective decision-making and path-finding skills with the shortest path.

4 PROPOSED METHODOLOGY

4.1 *Reinforcement learning*

In the machine learning sub-field of reinforcement learning, an agent picks up decision-making skills by interacting with its surroundings and working toward predetermined objectives. In general, a reinforcement learning agent, or the entity being taught, may detect and interpret its surroundings, take actions, and learn through trial and error [13], [14]. Markov characteristics are widely used in the built of theoretical models for reinforcement learning. According to the Markov attribute, the lower state of a system is solely connected to the present state and not to the previous state; otherwise, its formula [15].

$$P\,(S_{i+1}|\,S_{i+1},\cdots,S_1) = P(S_{i+1}|S_i) \tag{1}$$

The Markov decision (MDP) model is represented by a quaternion (S, A, P, R). S is all the different situations we might find ourselves in during decision-making. A is the list of things we can choose to do. P shows how likely it is to move from one situation to another. R tells us what happens as a result of our actions, leading us to the next situation. Current reinforcement learning, based on the MDP model, prioritizes return evaluation and algorithm usability [16]. This work aims to address the issue of learning ability, especially, how to go to a destination more quickly and securely.

4.2 *Reinforcement learning*

One of the most well-known model-free algorithms for reinforcement learning is Q-learning, which uses iterative values of the q-function. As was previously mentioned, at any given time, in any given state of s_i and ready to act a_i, $Q(s_i, a_i)$is in the state of the instant s_i acts under a_i to obtain the expected value of the return [17]. Upon examining the current state of A at a given moment s_i and considering all possible actions $A(s_i)$, the iteration process comes to an end when the environment of feedback information determines an optimum return move into the next state $s_i + 1$. This process continues until the endpoint or is forced to halt. The Q-learning algorithm is distinguished by creating probable states and actions into a two-dimensional Table (known as the q-table) to store Q values, and then obtaining the maximum benefit action by looking up the table, displayed in Table 1. This is a very simple method that works well in applications with moderate environmental sizes (e.g., simple card and chess games, robot path planning, etc.).

Using Equation (2), the Q-Learning algorithm dynamically updates Q-values, which are critical for making informed decisions inside the Q-Table framework. Figure 2, depicts the functioning and process of the Q-Learning algorithm in action.

$$Q(s, a) = Q(s, a) + \alpha(R(s, a) + \gamma \max Q(s', a') - Q(s, a)) \quad (2)$$

Where s and s' stands for the states that occur when action a from s is taken, and the current state, respectively. The discount factor, γ is proportional to the vector [0,1] and is crucial for accounting for errors that may occur during the learning process.

Table 1. Q-table of 4 states and 32 actions. .

State	Q(S, a_1)	Q(S, a_2)
S_1	$Q(S_1, a_1)$	$Q(S_1, a_2)$
S_2	$Q(S_2, a_1)$	$Q(S_2, a_2)$
S_3	$Q(S_3, a_1)$	$Q(S_3, a_2)$
S_4	$Q(S_4, a_1)$	$Q(S_4, a_2)$

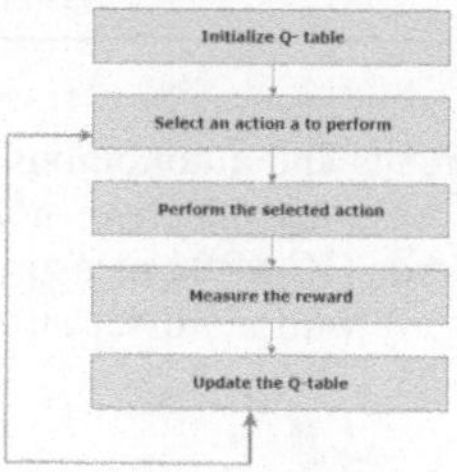

Figure 2. Working of Q-Learning Algorithm.

Algorithm 1 Q-learning Algorithm

Input: Car-robot in maze environment
1: Q-learning
2: Initialize Q(s, a) = 0, (s, a), Initialize D
3: While Q-values are not converged do
4: Simulate from memory buffer: D
5: for s, a, r, s' in D do
6: $Q(s, a) \leftarrow (1\ \alpha)Q(s, a) + \alpha(r + \gamma \max_{a'} Q(s', a'))$
7: end for
8: end while
9: return r $Q(s, a) \approx Q^*(s, a)\ \forall(s, a)$
Output: Optimal path with updated table

5 DESIGN AND SYSTEM ARCHITECTURE

In this section, we have discussed the proposed system's architecture, how the system works in a maze environment, and the design of our autonomous robot. In our project, we've employed a component, a crucial building block essential for its functionality and overall structure. Hardware components like Raspberry Pi 5, DC Motor, Motor Driver, Ultrasonic Sensors, Jumper wires, battery (power source), etc. The Python IDLE programming editor of Raspberry Pi OS is used for code compilation. Our working autonomous robot is displayed in Figure 3. Table 2 displays the lists the main components used in the project, along with their quantities. We created a large maze, measuring 6 squares by 5 squares (6x5), shown in figure 4, to serve as the working environment for our Autonomous robot. We have one starting point and another end goal (destination) in the maze.

The Raspberry Pi 5, is the latest and most versatile single-board computer, ideal for various projects, from basic programming to advanced robotics and home automation. It utilizes Secure Digital (SD) cards for storing the operating system and program memory, with Raspbian OS being the recommended choice by the Raspberry Pi Foundation. The Raspberry computers can run Linux OS, and to boot up, the computer needs an SD card with the operating system installed on it. An IC chip known as a motor driver regulates the speed, direction, and power supply of motors in

Table 2. Components catalog.

Name of Components	Quantity
Raspberry Pi 5	1
Robot Chassis kit	1
DC Motor	2
L298N Motor Driver	1
HC-SR04 Ultrasonic Sensor	3
Battery(5v) and Power bank	1
Jumper wires	As per requirements

Figure 3.

embedded circuits and autonomous robots. The robot chassis serves as a stable base for mounting components, ensuring effective motion transmission and overall stability during operation. The ultrasonic sensor HC-SR04 measures distance using ultrasonic waves, while jumper wires connect the Raspberry Pi with motors, sensors, and other components.

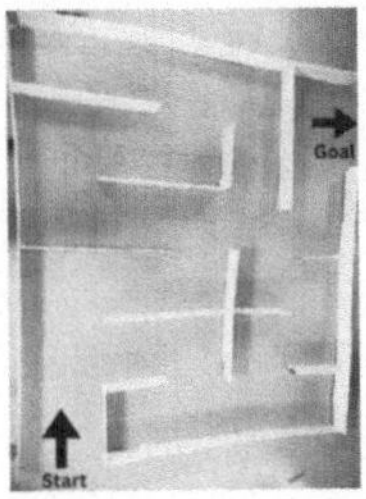

Figure 4. A 6x5 Maze environment.

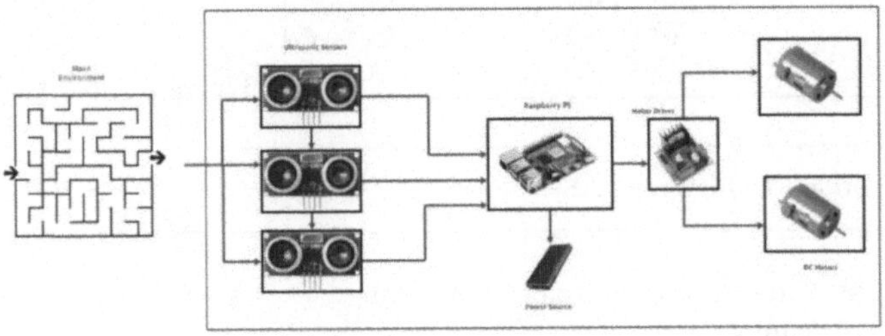

Figure 5. Block diagram.

A power bank supplies current for the Raspberry Pi. Figure 5, illustrates the system's interconnections and information flow. Coming to system architecture specifically, we will see how the robot system will run in a maze using the trial-and-error method to get the optimal or shortest path. At the starting point of the maze, the robot will start moving, and then it will decide its next action based on the sensor information. Maze navigation is addressed by Q-learning, which turns the maze into a set of choices for an agent. Every maze location turns into a state, and the agent's potential movements, i.e., left, right, forward, and back, become an action. The value of every action in every state is tracked by a Q-table, which is a huge grid. This enables the agent to figure out the best action and state to take to go from the starting place to the destination. The system architecture for the autonomous robot is shown in Figure 6.

6 IMPLEMENTATION AND EXPERIMENTAL RESULTS

The integration of software and hardware for our project, utilizing the components listed in shown in section IV. We demonstrate the seamless coordination between sensors, motor drivers, and software to ensure optimal functionality. Through this meticulous approach, we aim to achieve project objectives efficiently. Then we will see, the results of path-finding via an autonomous robot. Initially, programs are dumped to the Raspberry Pi using the Python IDE to execute the tasks. Our focus is on ensuring a smooth collaboration between software and hardware components to drive the success of our project. We have used SSH and VNC servers to connect the robotic system wirelessly. SSH (Secure Shell) protocol helps with the connection or communication between Raspberry Pi and the ground station and VNC securely connects with the remote server via an authentication protocol established between the ground station and the remote server.

Navigate through the maze. We have included the picture's re sult, which shows the optimal route from source to destination and other experimental parts are shown in red, the optimal path is highlighted in black, and another different path is highlighted in red in the given figures. We have tried with different other sources as well to learn its environment by robot for better results. The robot has successfully navigated the

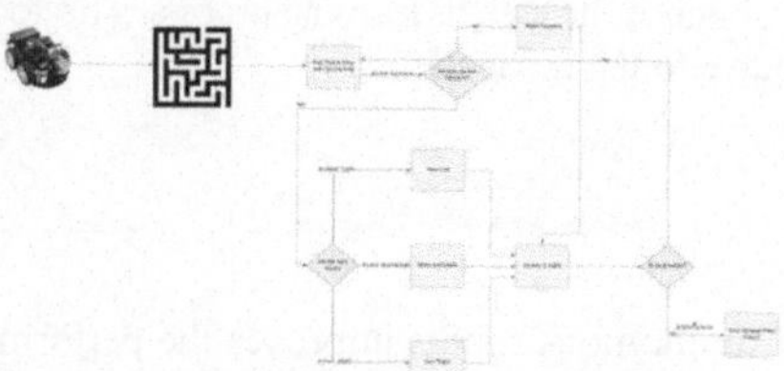

Figure 6. System architecture.

Figure 7. Robot ready to move forward.

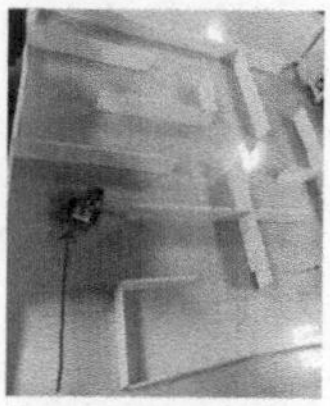

Figure 8. Robot moving forward without collision.

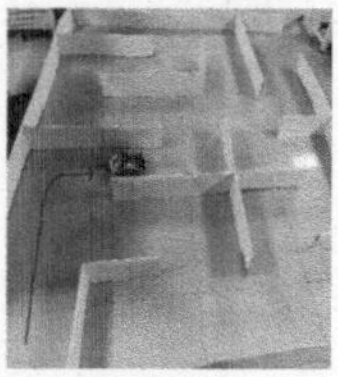

Figure 9. Robot taking decision based on environment.

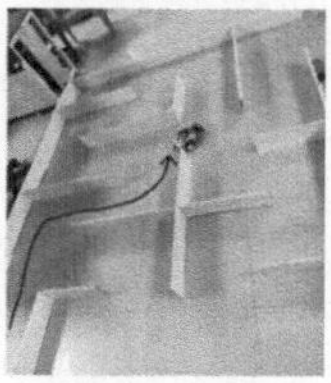

Figure 10. Robot taking decision and changing path.

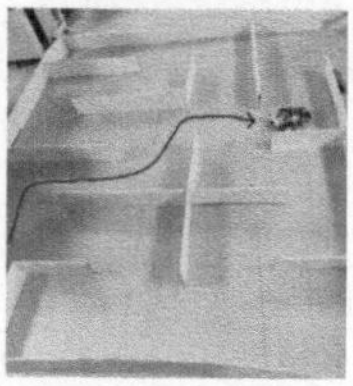

Figure 11. Robot moving to reach the destination without any collision.

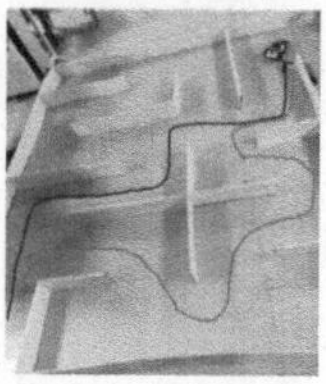

Figure 12. Robot reached to the destination by optimal route (black mark).

maze and achieved the minimum distance from the starting point to the destination. Throughout the exploration process, the robot considered multiple paths, each offering a different distance in length, which is given by sensors fitted into the system (measured in centimeters). In Figure 7, the robot gears up to move forward to reach its destination. Figure 8 shows its smooth journey without obstacles. Figure 9 illustrates its decision-making process based on its surroundings. In Figure 10, it adjusts its path as required to move forward. Figure 11 demonstrates its journey towards the destination, carefully avoiding obstacles. Finally, in Figure 12, it successfully reaches the destination along the optimal route. After thorough exploration, it was determined that path 4 provides the shortest route to the destination, i.e., 348 cm. The details of the different paths and their respective distances are summarized in Table 3. In Table 4, we have presented the Q-table exclusively featuring values essential for the shortest path from start to finish in the maze. These values guide

Table 3. Distance from source to destination with different trials.

Path	Distance from Source to Destination (cm)
1	556
2	522
3	485
4	348

Table 4. Updated q table of shortest path.

State (Position)	Action (Move)	Q-value
(1, 1)	Move Forward	0.9
(1, 2)	Move Forward	0.6
(1, 3)	Move Forward	0.8
(2, 3)	Move Right	0.6
(3, 3)	Move Forward	0.7
(3, 4)	Move Forward	0.8
(4, 4)	Move Right	0.4
(5, 4)	Move Forward	0.9
(5, 5)	Move Forward	0.9
(5, 6)	Move Forward	0.8

the robot in making optimal decisions during navigation, obtained through its learning process. This focused representation facilitates in efficient traversal from the source to the destination.

7 CONCLUSION AND FUTURE WORK

The path-finding strategy proposed in this paper for autonomous robots improves the performance of the proposed method. This method concludes that the robot autonomously discovers the optimal route within the maze by leveraging the Q-table. Through this iterative process, the robot can continuously improve its understanding of its surroundings and make wise decisions, which finally results in the choice of the most efficient Path. For future enhancements, we are trying to integrate with NVIDIA Jetson Bot equipped with a camera for improved results compared to the current setup. This upgrade aims to leverage advanced computer vision capabilities to enhance obstacle recognition and further refine the robot's navigation abilities in various environments.

REFERENCES

[1] Zhou Q., Lian Y., Wu J., Zhu M., Wang H., and Cao J. (2024). An optimized q-learning algorithm for mobile robot local path planning, *Knowledge-Based Systems*, vol. 286, p. 111400.

[2] Liu A. and Borisyuk A. (2024). Investigating navigation strategies in the morris water maze through deep reinforcement learning, *Neural Networks*, vol. 172, p. 106050.

[3] Kumaar A. N. and Kochuvila S. (2023). Mobile service robot path planning using deep reinforcement learning, *IEEE Access*.

[4] Yin Y., Chen Z., Liu G., Yin J., and Guo J., Autonomous navigation of mobile robots in unknown environments using off-policy reinforcement learning with curriculum learning, *Expert Systems with Applications*, vol. 247, p. 123202.

[5] Zhao M., Lu H., Yang S., and Guo F. (2020). The experience-memory q-learning algorithm for robot path planning in unknown environment, *IEEE Access*, vol. 8, pp. 47 824–47 844.

[6] Jin C., Lu Y., Liu R., and Sun J. (2021). Robot path planning using q-learning algorithm, in *2021 3rd International Symposium on Robotics & Intelligent Manufacturing Technology (ISRIMT)*. IEEE, pp. 202–206.

[7] Gao H., Liu Y., Su S., and Yao W. (2022). A simple implement of q-learning in robot path planning, in *Journal of Physics: Conference Series*, vol. 2294, no. 1. IOP Publishing, p. 012034.

[8] Wei K. and Ren B. (2018). A method on dynamic path planning for robotic manipulator autonomous obstacle avoidance based on an improved rrt algorithm, *Sensors*, vol. 18, no. 2, p. 571.

[9] Zhang X., Liu Y., Hu D., and Liu L. (2021). A maze robot autonomous navigation method based on curiosity and reinforcement learning, in *The 7th Int. Workshop on Advanced Computational Intelligence and Intelligent Informatics (IWACIII 2021), Article*, no. M1-6, p. 1.

[10] Wang N., Zhang D., and Wang Y. (2020). Learning to navigate for mobile robot with continual reinforcement learning, in *2020 39th Chinese Control Conference (CCC)*. IEEE, pp. 3701–3706.

[11] Xue Y., Kudenko D., and Khosla M. (2023). Graph learning-based generation of abstractions for reinforcement learning, *Neural Computing and Applications*, pp. 1–21.

[12] Li Y. (2023). Map optimization of path planning in q-learning, *Highlights in Science, Engineering and Technology*, vol. 63, pp. 231–241.

[13] Saha O., Ren G., Heydari J., Ganapathy V., and Shah M. (2021). Deep reinforcement learning based online area covering autonomous robot, in *2021 7th International Conference on Automation, Robotics and Applications (ICARA)*. IEEE, pp. 21–25.

[14] Nguyen H. and La H. (2019). Review of deep reinforcement learning for robot manipulation, in *2019 Third IEEE International Conference on Robotic Computing (IRC)*. IEEE, pp. 590–595.

[15] Zhao Y., Zhang Y., and Wang S. (2021). A review of mobile robot path planning based on deep reinforcement learning algorithm, in *Journal of Physics: Conference Series*, vol. 2138, no. 1. IOP Publishing, p. 012011.

[16] Lakshmanan A. K., Mohan R. E., Ramalingam B., Le A. V., Veeraja- gadeshwar P., Tiwari K. and Ilyas M. (2020). Complete coverage path planning using reinforcement learning for tetromino based cleaning and maintenance robot, *Automation in Construction*, vol. 112, p. 103078.

[17] Orozco-Rosas U., Picos K., Pantrigo J. J., Montemayor A. S. and Cuesta-Infante A. (2022). Mobile robot path planning using a qapf learning algorithm for known and unknown environments, *IEEE Access*, vol. 10, pp. 84 648–84 663.

Data Science & Exploration in Artificial Intelligence – Gururaj et al. (Eds)
© 2025 the Author(s), ISBN: 978-1-032-96390-7

Securing healthcare insights: A deep dive into privacy preserving federated learning

N. Manjunath
JSS Science and Technology University, Mysuru, Karnataka, India

R. Guru
Sri Jayachamarajendra College of Engineering, Mysuru, Karnataka, India

B.T. Prasanna
JSS Science and Technology University, Mysuru, Karnataka, India

ABSTRACT: Healthcare is being transformed by a pattern shift toward data-driven approach to improve patient care, prediction of diseases, discovery of drugs, and management of healthcare. In light of these dual imperatives, privacy-preserving federated learning is a recent disruptive solution for the training of ML models on sensitive data. This paper will offer some necessary contextual background information: primer on core methodologies in privacy protection, such as differential privacy, homomorphic cryptography, secure multiparty computing, and data anonymization. Following this, the next concept, federated learning, is presented as a decentralized approach to machine learning with specific ideal characteristics in healthcare environments[1]. The paper also deals with some assessment metrics for estimating the effectiveness and privacy guarantees of the privacy-preserving federated learning models. It recognizes the challenges and impossibilities of implementing these techniques but underlines the need for care in ethical considerations and compliance with healthcare data regulation such as HIPAA and GDPR.

Keywords: Privacy-Preserving Techniques, Federated Learning Healthcare, Data Privacy, Machine Learning, Data-Driven Healthcare

1 INTRODUCTION

Today, healthcare is becoming one of the areas in which ever more data-driven insights enable a revolution in patient care and allowing a revolution in patient care by analyzing vast and heterogeneous datasets. This, in a way, has kindled exceptional interest in the potential of machine learning toward better disease diagnosis, treatment, and management, among other things. The sector in discussion also concurrently carries this critical demand with a no- compromise attitude: the privacy of patients, and the most stringent adherence to data protection laws. With the same perspective, the unison of data-driven innovation and privacy preservation has become the bulwark of modern healthcare research and practice. "Privacy- Preserving Federated Learning" is a pioneering approach that goes beyond traditional paradigms of data sharing. This complements the search for actionable insights with the protection of the sensitive data of patients. The basic idea behind federated learning is that training in machine learning models is inherently decentralized—that is, many institutions or entities can collaborate on a prediction model without the need to share raw data. In the next sections, we will more examine the basic ideas of privacy maintenance. This will involve an exploration of several strategies, including different levels of privacy, homomorphic cryptography, secure multi-party computing, and information

DOI: 10.1201/9781003589273-5

anonymization.Furthermore, this survey meticulously examines the plethora of privacy-preserving federated learning architectures that have been proposed and implemented, providing insights into their advantages, limits, and real-world applicability. It demonstrates how the merger of privacy and cooperation makes applications in health, such as disease prediction, possible while driving drug discovery faster. As we move through this landscape, take inventory with us of the evaluation metrics created to measure the effectiveness and privacy guarantees of privacy-preserving federated learning models.In addition, this survey paper further speculates on the research directions and emerging trends expected to reconfigure the landscape of privacy-preserving federated learning in healthcare. It will breathe life into the theory by presenting illustrative case studies, showing how leading administrations and research efforts have effectively taken up the gauntlet with techniques to ensure patient privacy while unlocking the transformative potential of healthcare data.

In sum, "Privacy-Preserving Federated Learning in Healthcare: A Comprehensive Survey [2]" represents an essential compass for those boarding on the journey to navigate the intricate confluence of data-driven healthcare progress and the imperative of safeguarding the most sensitive of treasures—patient privacy as mentioned in Figure 1.

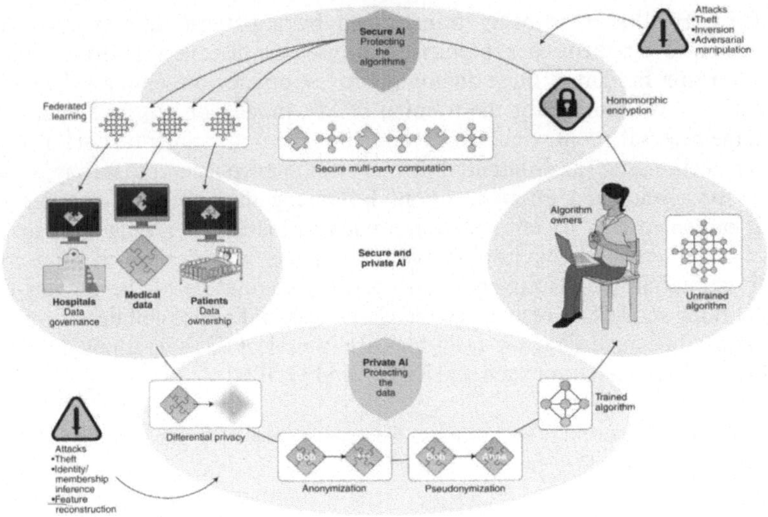

Figure 1. Federated learning architecture for health care[3].

2 BACKGROUND

2.1 *The importance of healthcare data*

The health industry is in the vanguard of a data revolution that involves hospitals, clinics, medical devices, and wearable skills, which are continually creating and storing vast amounts of data on patients. Such volumes of data can drive breakthroughs in various areas—patient care, disease diagnosis, drug development, and healthcare system management. Actionable insights from healthcare data are used to detect diseases early, regulate personalized treatment plans, and efficiently deliver healthcare.

2.2 *Data privacy challenges in healthcare*

But with such a large pool of information, big responsibility follows—not only to protect patient privacy but also to consider that there are very strict guidelines on data protection in play. Information about health is one of the most sensitive and very personal data types. It contains the health history, treatments, and prescriptions an individual is undergoing, among others. Any form of breach in health data ripples out its effects not only on the individual but also on the health providers and organizations. As a result, the data of patients has to be handled with maximum privacy and integrity [4].

2.3 *The rise of federated learning*

Federated learning has been one of the most up-and-coming concepts recently for being a transformative approach toward machine learning and data analysis[5]. Federated learning decouples the training process from machine learning models. It makes it possible for several entities or organizations to cooperate on model development without sharing the raw data. It supports the basic principle of data privacy, where central storage is not required, hence minimal possible risks for data exposure.

2.4 *Privacy-preserving techniques*

There have been a diversity of techniques that have been evolving and developed over the years to make data-driven insights easy to integrate with patient data privacy. Techniques include, for instance, homomorphic encryption, differential privacy, secure multi-party computation & data anonymization. The methods termed above are developed in such a way that they protect individual patient records while allowing collaborative machine learning and analytics [6].

2.5 *Challenges and opportunities*

While there is quite a good contract of promise in the application of privacy-preserving combined learning in the health domain, it is still limited to a complex environment characterized by several impediments. Technical limits, ethical issues, and legal frameworks, such as the Health Insurance Portability and Accountability Act (HIPAA) in the United States and the General Data Protection Regulation (GDPR) in Europe, complicate the use of such approaches[7].

This paper deliberates the basic concepts on which this novel approach is built, the techniques used for privacy preservation, some essential practical implementations, and the evaluation criteria, along with ethical and regulatory issues. Such understanding of these concepts at a fine-grained level across researchers, health practitioners, and policymakers is required for engagement at the difficult trade-off that further health care and its data-driven insights while safeguarding patient privacy.

3 METHODS AND MATERIALS

3.1 *Dataset*

The Health Insurance Portability and Accountability Act (HIPAA) and the General Data Protection Regulation (GDPR) are two huge monitoring frameworks regarding the protection of healthcare information and the European environment, respectively. The said rules act as standards for maintaining privacy and integrity within sensitive personal information. This even has to be respected while working with the datasets under the ambit of privacy-preserving federated learning in healthcare or Europe. I will have a look below at these regulations and what they mean for the handling of datasets [8].

HIPAA DATASETS

HIPAA is a U.S. law that provides guidelines and rules nearby the use, disclosure, and protection of Protected Health Information (PHI)[9]. HIPAA privacy-preserving federated learning datasets could involve patient medical records, clinical data, healthcare claims, and any other information related to health. GDPR DATASETS

3.2 *Privacy-preserving techniques*

3.2.1 *Differential privacy*

This promises the highest possible toughness of the framework for privacy protection; it is a rigid differential statistical approach. Noise that is adjusted for the safety of individual data has been introduced precisely within the results of the query[10]. This will, of course, be used to protect combined statistics or query results necessary in health care. Thus while results can be shared regarding the prevalence of a disease or the efficiency of treatments, detailed information about patients is confidential, and valuable analysis is provided.

3.2.2 *Homomorphic encryption*

It allows the possibility of performing computation on encrypted data without any previous condition for its decryption. The use of this technique would guarantee that confidential medical information is not revealed in the process of computation. Homomorphic encryption, as applied in federated learning, could be used to protect model updates and gradients passed between participants from revealing raw data [11].

3.2.3 *Secure multi-party computation (SMC)*

Secure Multi-Party Computation is a situation in which multiple parties calculate a function of inputs contributed by each one of them, at the same time preserving the confidentiality of the inputs. It ensures that sensitive data remains confidential during collaborative calculations. In healthcare, SMPC can be used for secure aggregation of model updates from different institutions or to compute shared statistics without revealing the raw data [12].

3.2.4 *Differential-privacy federated averaging*

The average federated approach, for example, is federated averaging, whereby different participants contribute model updates to develop a global model. Differential privacy can be applied at the aggregation process for the protection of the confidentiality of the participants. For example, using differential privacy to federated averaging offers the skill to protect the confidentiality of model updates of participants; thereby no one's data can be revealed during the aggregation process [13].

3.2.5 *Data Anonymization and de-identification*

Data anonymization, alternatively called de-identification, can be removing or encrypting PII, where the identified patient is not disclosed for protection. De-identification is a process that takes patient data and renders it so that a given person cannot be easily re-identified from it[14].

Secure Protocols Secure communication protocols, like the Secure Sockets Layer (SSL) or Transport Layer Security (TLS), have to be used for the data sent to be protected from the participants to the central server and vice versa. Application: These procedures ensure all the shared data in federated learning are kept confidential and protected when under transfer [15].

3.2.6 *Consent and ethical considerations*

In healthcare federated learning, consent and ethical considerations are foundational, ensuring that participants understand and agree to the use of their data for research while safeguarding their privacy [16].

The results of the survey undertaken utilizing the Privacy-Preserving federated Machine Learning approach are summarized in Table 1. FL is the privacy-preserving technique most frequently applied during model training, either individually or in grouping with DP.

Table 1. Some federated learning task.

Use case	Model	Dataset	DL and ML Task	Performance Metrics
COVID-19 Computed Tomography (CT) analysis	RetinaNet	Multi-institution lung CT data*	Object detection	mAP, Specifity, Recall, AUROC
FL benchmarking and reliability in healthcare	Neural Network, LSTM, CNN	MIMIC-III, PhysioNet ECG	Mortality prediction, Multiclass classification	AUROC, AUPRC, F1-score
COVID-19 detection	DenseNet Random Forest	Multi-institution COVID-19 X-ray*	Binary classification	AUROC, AUPRC
Cancer infrence via gene expression	Logistic regression, MLP	iDASH 2020	Multi-class classification	Accuracy, AUC, Recall, Precision, F1-score
COVID-19 detection	MobileNet-V2	COVID-19 X-ray	Multiple label classification	Accuracy, Recall, Precision, F1-score
Heart and thyroid disease classification	XGBoost	UCI Heart Disease, Kaggle Hypothyroid	Binary classification	Accuracy
Intensive Care Unit patient outcome	LSTM	MIMIC-III	Binary classification	Recall, AUROC, Precision
Thoracic pathology detection	DenseNet-121	CheXpert	Multi-class classification	AUROC, Accuracy

4 FUTURE RESEARCH DIRECTIONS

4.1 *Comprehensive analysis of different medical datasets*

An in-depth analysis of standard medical data sets is provided by the data set evaluation[17]. This is why external validation is not performed and a single dataset is used in the majority of current studies. A limited number of studies employ outside data sets for validation. Of the forty studies that were taken into consideration, only nine used a separate data set to validate their results. This makes it difficult to compare the performance of methods created for an identical medical objective and to generalize models. advise using MedMNIST, which offers well selected datasets for a range of medical tasks and modalities. This means that this medical dataset might serve as a common benchmark for medical applications, much way MNIST or CIFAR-10 do for computer vision models [18].

5 CONCLUSION

This paper labels the review of the preceding academic documents regarding implementing Privacy-Preserving Machine Learning (PPML) techniques in healthcare[19]. Lastly, not all agree about the agreement on identifying the requirements that are necessary for the privacy-preserving framework within the healthcare area. This is a call to action for teamwork among machine learning researchers, health professionals, and privacy and safety experts. Additionally, we encourage investigators to provide rigorous evaluations of recommended algorithms with varied medicinal data to further development in machine learning (ML) approaches. Herein, this work focuses on generalizing ML models to improve this effort and generalizing into the limitations of privacy-preserving machine learning (PPML) in multi-modal learning scenarios. The proposed works then aim to bridge the gaps already identified within the research and guide further research within privacy-preserving machine learning to ensure the deployment of much-needed secure and confidential ML techniques in the health domain.

REFERENCES

[1] Dou, Y So, Jiang, Liu, Vardhanabhuti, Kaissis, Li, Si, Heather HC Lee, Kevin Yu, *et al.* (2021). Federated deep learning for detecting covid-19 lung abnormalities.

[2] Field, Thwaites, Carolan, Delaney, Lehmann, Sykes, Vinod and Holloway. (2022). Infrastructure platform for privacy-preserving distributed machine learning development of computer-assisted theragnostics in cancer. *Journal of Biomedical Informatics*, 134:104181.

[3] Lee and Shin. (2020). Federated learning on clinical benchmark data: Performance assessment. *Journal of Medical Internet Research*, 22(10):e20891.

[4] Yang, Hwang, Kim, Dua, Kim, Yang and Choi. (2022). Towards the practical utility of federated learning in the medical domain. *arXiv preprint arXiv:2207.03075.*

[5] adilek, Liu, Nguyen, Kamruzzaman, Serghiou, Rader, Ingerman, Mellem, Kairouz and Nsoesie. (2021). *NPJ Digital Medicine*, 4(1):132.

[6] Loftus, Ruppert, Shickel, Baslanti, Balch, Efron, Upchurch Jr, Rashidi, Tignanelli, Bian, *et al.* (2022). Federated learning for preserving data privacy in collaborative healthcare research. *Digital Health*, 8:20552076221134455.

[7] Wolff, Matschinske, Baumgart, Pytlik, Keck, Natarajan, von Schacky, J Pauling, and Baumbach. (2022). Federated machine learning for a facilitated implementation of artificial intelligence in healthcare–a proof of concept study for the prediction of coronary artery calcification scores. *Journal of Integrative Bioinformatics*, 19(4).

[8] Qingyong Wang and Yun Zhou. (2022). Fedspl: federated self-paced learning for privacy-preserving disease diagnosis:bbab498.

[9] Islam, Alaboud, Paul, Zaman Rana, and Mosa. (2022). A privacy-preserved transfer learning concept to predict diabetic kidney disease at out-of-network siloed sites using an in-network federated model on real-world data. In *AMIA AnnualProceedings*

[10] Deist, Dankers, Ojha, Marshall, Janssen, Faivre-Finn, Masciocchi, Valentini, Wang, Chen J., *et al.* (2020). Distributed learning on 20 000+ lung cancer patients–the personal health train. *Radiotherapy and Oncology*, 144:189–200.

[11] Park,Kim, Kim, Kim and Chul Ye. (2021). Federated split vision transformer for covid-19 cxr diagnosis using task-agnostic training. *arXiv preprint arXiv:2111.01338.*

[12] Yan, Qu, Wei, Huang, Shen, Rubin, Xing and Zhou. (2023). Label-efficient self-supervised federated learning for tackling data heterogeneity in medical imaging. *IEEE Transactions on Medical Imaging.*

[13] Boulila, Ammar, Benjdira and Koubaa. (2022). Securing the classification of covid-19 in chest x-ray images: A privacy-preserving deep learning approach. *(SMARTTECH)*, pp. 220–225. IEEE.

[14] Ma, Jianfeng Ma, Miao, Liu, Choo,Yang and Wang. (2020). Lightweight privacy-preserving medical diagnosis in edge computing. *IEEE Transactions on Services Computing*, 15(3):1606–1618

[15] Paul, Annamalai, Ming, Badawi, Veeravalli and Mi Aung. (2021). Privacy-preserving collective learning with homo-morphic encryption. *IEEE Access*, 9:132084–132096.

[16] [16]. Yange Chen, Qinyu Mao, Baocang Wang, Pu Duan, Benyu Zhang, and Zhiyong Hong. (2022). Privacy-preserving multi-class support vector machine model on medical diagnosis. *IEEE Journal of Biomedical and Health Informatics*, 26(7):3342– 3353.

[17] Moran Baruch, Nir Drucker, Lev Greenberg, and Guy Moshkowich. (2022). A methodology for training homomorphic encryption friendly neural networks. In *Applied Cryptography and Network Security Workshops: ACNS 2022 Satellite Workshops, AIBlock, AIHWS, AIoTS, CIMSS, Cloud S&P, SCI, SecMT*, SiMLA, Rome, Italy, June 20–23, 2022, *Proceedings*, pp. 536–553. Springer.

[18] Zalonis, Armknecht, Grohmann and Koch. (2022). Report: State of the art solutions for privacy preserving machine learning in the medical context. *preprint arXiv:2201.11406.*

[19] Dayan, Roth, Zhong, Harouni, Gentili, Abidin, Liu, Costa, Wood, Tsai, *et al.* (2021). Federated learning for predicting clinical outcomes in patients with covid-19. *Nature Medicine*, 27(10): 1735–1743.

Data Science & Exploration in Artificial Intelligence – Gururaj et al. (Eds)
© 2025 the Author(s), ISBN: 978-1-032-96390-7

Implementation of RFID-based food serving robot

G. Sandhya
Associate Professor, Department of ECE, VNITSW, Guntur, Andhra Pradesh, India

Queenie Akshaya Muppidi, Anuradha Eega, Jaya Lakshmi Reddi & Sravani Bikki
Student, Department of ECE, VNITSW, Guntur, Andhra Pradesh, India

ABSTRACT: This paper presents an innovative food-serving robot system designed to enhance efficiency and customer satisfaction in restaurant environments. The robot autonomously navigates through predefined paths using an advanced line-following mechanism guided by infrared sensors, ensuring accurate and reliable movement. A key feature of the system is its integration of RFID technology, which enables precise table identification through uniquely encoded RFID cards placed at each table, ensuring accurate meal delivery. The robot's operations are controlled via Bluetooth connectivity and an Arduino Controller mobile application, providing an intuitive interface for initiating tasks and directing movements. This seamless integration of hardware and software allows the robot to perform efficient navigation, precise table recognition, and effective food delivery in real-world restaurant settings. The research highlights the system's robustness and adaptability, showcasing its ability to perform consistently in dynamic and busy restaurant environments. This technologically sophisticated solution represents a significant advancement in automated food service, setting a new standard for operational efficiency and customer experience in the dining industry.

1 INTRODUCTION

The integration of robotics into the hospitality industry has introduced innovative solutions aimed at improving operational efficiency and enhancing customer experiences. Among these advancements, autonomous food serving systems have emerged as a focal point of research and development. These systems leverage cutting-edge technologies to enable robots to navigate restaurant environments and execute food deliveries autonomously. This paper presents a comprehensive exploration of a pioneering food-serving robot designed to revolutionize service operations in hospitality settings. At the core of this system lies the integration of advanced technological components, including LM293D motor drivers and an intricate line-following mechanism guided by infrared sensors. These components enable the robot to maneuver along predefined paths marked by black lines, ensuring precise and consistent movement within the restaurant space.

Furthermore, the system incorporates RFID technology, utilizing uniquely encoded RFID cards positioned at individual tables to facilitate accurate table recognition and targeted food delivery assignments. Bluetooth integration with an Arduino Controller mobile application provides a seamless means of control and navigation, empowering users to initiate food delivery tasks and direct the robot's movements remotely. Through this research endeavor, the paper aims to elucidate the technical intricacies and operational efficacy of the food-serving robot. By studying several existing works for the design of automatic food serving and ordering frameworks utilizing IoT the study seeks to highlight

DOI: 10.1201/9781003589273-6

the potential of this innovative technology to revolutionize service operations and elevate customer experiences in hospitality settings.

There are many existing methods for designing food serving robots. Some of them are mentioned here. This review aims to elucidate the challenges and shortcomings of the current system and compare them with the proposed solution. In this evaluation, we will provide insights into the current state and the proposed improvements of an application. The current implementation of the software will be discussed to illustrate its operational status and effectiveness. The proposed framework will outline our innovative approach to address the existing limitations and enhance the system's functionality.

Sayali N Joshi *et al.* designed a human following trolley using Raspberry Pi embeded with Pi cameras that capture the tag by Pi camera and follows a human [1]. Fatima R. Ali uses a static framework where the robot continues on static lines. It is programmed using a dynamic system and Digital differential algorithm called the Bezier curve algorithm [2] according to which the arrangement of the lines is reformed. Ghe arcia Haro, J. M implemented Waiter Robot. It transports the objects with Balance Control. It is designed using zero moment point stability criteria with which the stability of motion can be increased. The project id designed using control system concepts [3]. In [4], P Shivraj used collaborative operations to create a line follower robot. It is an automated system programmed using MATLAB and can handle numerous challenging issues.

Md. Kamruzzaman and Md. Tareq designed a microcontroller-based robot that uses Bluetooth device and a motor that works efficiently in office and restaurant environments. Android app is used to send the order request [5]. J.S.Tan *et al.* designed a microcontroller-Raspberry Pi 2-based future farming robotic delivery system called Jackbot Mark 1[6] which delivers the load with high speed and carries high loads when contrasted with conventional delivery systems. A. Akshay Mane designed a robot using IR sensors for localization and mapping of unknown region which is determined by employing the Kalman filter. Simultaneous Localization and Mapping of an unknown region using Kalman Filtering is a mathematical region detection method [7]. M. Sandeep Bhat, M. designed a robot using Microcontroller and IR sensors. IR sensors were used to plan the pathway along which food is delivered and also for carrying luggage [8]. In [9], Asif M designed a robot framework that can communicate requests to the waiter and kitchen through the wireless network. The Robot used a Bluetooth module to deliver the order to the exact table. YongChai Tan, B L implemented a novel automatic food delivery framework utilizing an autonomous track-guided centre wheel drive robot with the path or line follower methodology to place the order [10]. In [11], Mehran Pakdaman designed a line follower robot that follows a pre-defined path noticeable like a black path which can be sensed by an IR sensor.

2 PROPOSED WORK

The proposed system consists of an Android application for food ordering and bill payment, aiming to replace traditional suppliers in canteens[12]. RFID technology is employed to accurately route the robot to the designated table, integrating various sensors such as ultrasonic and IR sensors for obstacle detection and black line tracking. Upon receiving orders through the mobile application, the system processes the information to determine the optimal path to the correct table.

Central to the system is the Arduino Bluetooth control app, which facilitates wireless communication between a mobile device and an Arduino board equipped with a Bluetooth module. This app enables bidirectional data transfer, empowering users to remotely command the Arduino or receive sensor data through Bluetooth connectivity. Proper pairing between the mobile device and Arduino's Bluetooth module is essential for seamless interaction. The Arduino Bluetooth Control App plays a pivotal role in the operational dynamics of the food-serving robot, acting as the intermediary interface between the user and the robot's embedded Arduino control system. This application leverages Bluetooth communication protocols to establish a seamless connection with the Arduino platform, facilitating real-time command transmission. Through the app's intuitive interface, users can issue precise directives for the robot's navigation, table selection, and return-to-home procedures.

The app employs a robust data encapsulation technique to encode and decode commands, ensuring secure and reliable communication between the user's device and the Arduino controller. Additionally, the app integrates error-handling mechanisms, allowing for graceful recovery from communication disruptions or command inconsistencies. This sophisticated control mechanism enhances the robot's responsiveness, providing a technologically advanced and user-friendly means to arrange the food delivery process in a precise and controlled manner.

The working of the food delivery robot involves several key components and processes. Bluetooth, LM293D motor driver, and the Bluetooth mobile application called Arduino Controller are used, with the robot connecting to the phone via Bluetooth. The navigation mechanism is based on the line-following method, where the robot follows a black line to reach its destination. Unique RFID cards are positioned at each table, allowing the robot to recognize the table number assigned to it. The number of RFID cards corresponds to the number of tables present, and each RFID card is uniquely named in the code. In real-time, the path is set circularly, so after serving food to the respective table as instructed, the robot returns to the kitchen, controlled via Bluetooth. For order processing and delivery, if the customer at Table 1 orders food, the kitchen staff places the food on the robot's tray and selects Table 1 in the phone app. The robot detects the RFID sensor in Table 1 and serves the food there. When the customer at Table 1 takes the food, they press the push button present on the robot, indicating that the food has been taken and prompting the robot to return to the kitchen. For multi-table navigation, if food needs to be delivered to Table 2, the user presses 2 in the phone app, and the robot navigates to Table 2, bypassing Table 1 without stopping. The robot detects the RFID card in Table 2 and delivers the food. After the customer at Table 2 takes the food, they press the push button, signaling the robot to return to the kitchen. By following these steps, the robot ensures precise and efficient food delivery, enhancing the overall service experience in the restaurant.

3 IMPLEMENTATION

The architecture of the envisaged system incorporates a diverse array of components to facilitate its operational intricacies. The following Figures 1 and 2 shows block diagram and schematic diagrams of the proposed Food serving robot.

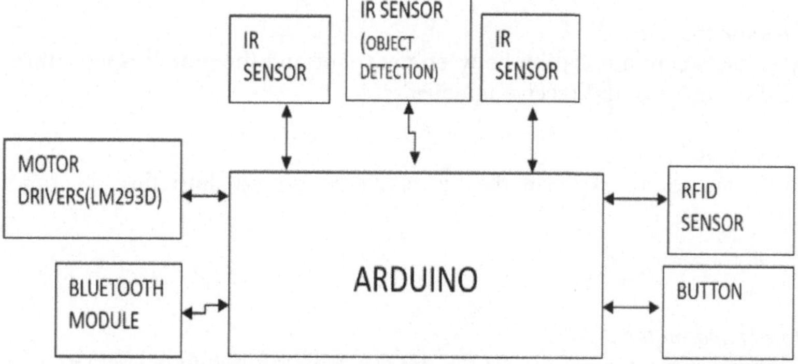

Figure 1. Block diagram of the proposed food serving robot.

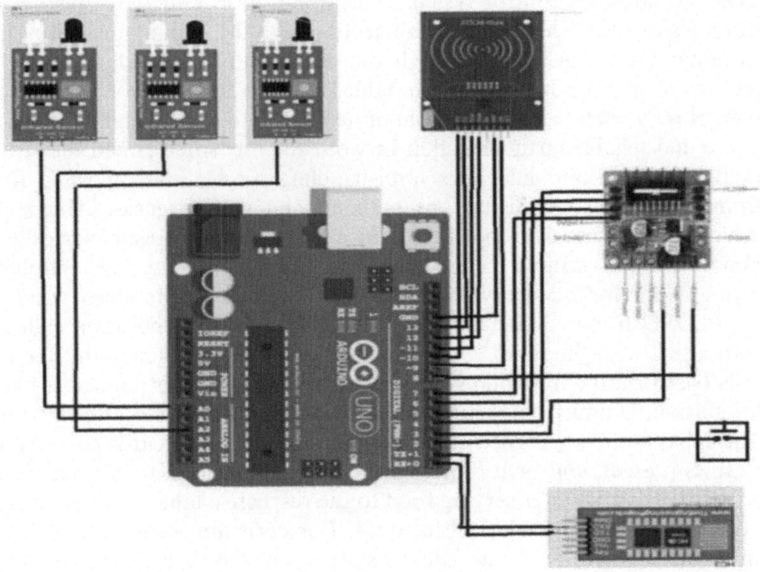

Figure 2. Schematic diagram of the proposed food serving robot.

Central Processing Unit: The Arduino Uno serves as the brain of the operation, managing input signals from various sensors and modules, processing data, and controlling the actuators.

3.1 *Sensors and modules*

3.1.1 *IR line following sensors*

These sensors enable the robot to follow a predefined path by detecting the black line on the floor.

3.1.2 *MFRC522 RFID module*

Used for identifying table locations through RFID tags, allowing precise delivery.

3.1.3 *Infrared obstacle detection sensor*

Ensures the robot navigates safely by detecting obstacles in its path.

3.2 *Communication and control*

3.2.1 *Bluetooth module*

Facilitates wireless communication between the robot and the mobile application, enabling users to send commands and receive updates.

3.2.2 *Push button*

Allows customers to signal when they've taken their food, initiating the robot's return journey.

3.3 *Actuators*

3.3.1 *Motors and motor driver*

Drive the robot's movement, controlled by signals from the Arduino, enabling it to navigate and deliver food.

3.3.2 *Chassis and wheels*

Provide the physical structure and mobility required for navigating the canteen environment.

3.4 *Feedback mechanism*

The buzzer provides audible feedback to users, signaling important events such as arrival at a table or completion of a task, enhancing user interaction.

3.5 *Power supply*

A crucial aspect, ensuring all components receive the required voltage and current for operation, typically powered by a battery or external power supply.

3.6 *Workflow*

3.6.1 *Order placement and processing*

Orders are received from the mobile app, processed by Arduino, and the destination table is identified.

3.6.2 *Navigation and delivery*

The robot navigates to the designated table using line following sensors and RFID tags.

3.6.3 *Completion and return*

After food delivery, the customer signals completion via the push button, prompting the robot to return to the kitchen while avoiding obstacles.

By integrating these components and establishing clear communication pathways, your system offers a highly automated and reliable solution for food delivery in canteen environments. Each component's role is well-defined, contributing to seamless operation with minimal human intervention.

4 CONCLUSIONS

In conclusion, the design of the Serving Robot represents a significant leap in service automation, offering innovative solutions with broad applicability across diverse domains. By seamlessly integrating line-following robotics, wireless communication modules, and adaptive functionalities, this versatile system addresses the needs of various societal sectors, from domestic settings to hospitals, libraries, and restaurants. Its ability to navigate efficiently on black lines against white surfaces demonstrates proficient tracking capabilities, ensuring successful operation in different environments.

One notable aspect of this design is its emphasis on addressing technical challenges while prioritizing cost-effectiveness and simplicity through the use of accessible technologies. This approach not only fosters innovation but also ensures practicality and scalability, making the Serving Robot a viable solution for widespread adoption.

With its strategic applications and commitment to ongoing advancements, the Serving Robot stands as a testament to technical ingenuity, poised to revolutionize service automation across a myriad of societal domains. Its potential to enhance efficiency, accessibility, and user experience makes it a pioneering solution with the capacity to transform the way we interact with automated services in the future.

Figure 3. Image of the food serving robot.

REFERENCES

[1] Sayali N and Joshil, V. K. (2019). Design and development of human following trolley. *International Journal of Innovative Science and Research Technology* ISSN No:-2456–2165, Volume 4 (Issue 4).

[2] Fatima R. Ali and A. R. (2018). Design and implementation of static and dynamic objects store systems using line follower robots . *International Conference on Advances in Sustainable Engineering and Applications (ICASEA)*, Wasit University, Kut, Iraq.

[3] Garcia-Haro, J. M. (2018). Waiter robot application: Balance control for transporting objects. *International Conference on Intelligent Robots and Systems (IROS)*.

[4] P., S. (2018). Food delivery automation in restaurants using collaborative robotics. *International Conference on Inventive Research in Computing Applications (ICIRCA 2018)*, IEEE Xplore Compliant Part Number:CFP18N67-ART; ISBN:978-1-5386-2456-2.

[5] Kamruzzaman Md., M. T. (2017). Design and implementation of a robotic technique based waiter. *International Conference on Electrical Information and Communication Technology (EICT)*.

[6] Tan J.S., Teh V., Teck H.M., Lim Z.H. and Manoharen C. (2016). Future farming robotic delivery system Jackbot Mark 1. *IEEE Conference on Wireless Sensors (ICWiSe)*.

[7] Akshay A. Mane and M. N. (2015). Robotics based simultaneous localization and mapping of an unknown environment using Kalman filtering. *International Conference on Engineering (NUiCONE)*.

[8] Sandeep Bhat, M. M. (2015). Embedded system based waiter and military robot path planning. *International Conference on Control, Instrumentation, Communication and Computational Technologies (ICCICCT)*.

[9] Asif M. and M. S.-u.-R. (2015). *Waiter Robot – Solution to Restaurant Automation. MDSRC - 2015 Proceedings*, 14-15 November, 2015 Wah/Pakistan.

[10] YongChai Tan, B. L. (2010). A new automated food delivery system using autonomous track guided centre-wheel drive robot. *IEEE Conference on Sustainable Utilization and Development in Engineering and Technology*.

[11] Mehran Pakdaman and M. M. (2009). *Design and Implementation of Line Follower Robot*.

Data Science & Exploration in Artificial Intelligence – Gururaj et al. (Eds)

A reinforcement learning approach in modelling basal ganglia circuitry – Functional and dysfunctional study

Advait Suresh, Nivedita Rajesh & Shrish Nandakumar
Department of Computer Science and Engineering, Amrita School of Computing, Amrita Vishwa Vidyapeetham, Amritapuri, India

Manjusha Nair
Department of Computer Science and Applications, Amrita School of Computing, Amrita Vishwa Vidyapeetham, Amritapuri, India

ABSTRACT: Reinforcement Learning principles offer insight into the functioning of the Basal Ganglia. RL demonstrates how BG influences action selection and learning. Comprising a complex network of pathways and neurotransmitter systems, the basal ganglia plays a vital role in evaluating actions and adjusting behaviour. Dopamine, a critical neurotransmitter in reward processing, is integral to its reinforcement learning mechanism. The interaction between dopamine signalling and the basal ganglia's circuitry forms the foundation for incorporating reinforcement learning principles. It forms the basis for understanding various neurological disorders. Current research focuses on modelling these interactions within the basal ganglia pathways. It explores the role of dopamine and acetylcholine in motor control. Investigating the impact of drugs like levodopa on Parkinsonian conditions, the study aims to develop a functional model demonstrating both the biological and dysfunctional behaviour of the basal ganglia with the potential for expansion to include more detailed networks and dynamics.

Keywords: Reinforcement Learning, Basal Ganglia, Dopamine, Neurological disorders, Parkinson's disease, Levodopa, Robotic arm

1 INTRODUCTION

The Basal Ganglia (BG) is a group of interconnected nuclei responsible for every action we take. They are essential for planning and fine-tuning movements (Lanciego *et al.* 2012). They receive input from the cerebral cortex and send output via the thalamus, forming a complex network. The main structures involved are the striatum, globus pallidus (GPi and GPe), subthalamic nucleus (STN), and substantia nigra (Chakravarthy and Moustafa 2018). The BG has three primary pathways, direct (facilitates and promotes movement), indirect (suppresses and regulates movement), and hyperdirect (inhibits the circuit) (Young *et al.* 2023) (Figure 1). BG facilitates learning from environmental cues and internal states (analogical to RL). The direct pathway reinforces actions linked to rewards, modulated by dopamine receptors. The indirect pathway suppresses actions tied to negative outcomes, optimizing behaviour (Yuan and Fathi 2022).

Dopamine signalling is central to reward prediction and alterations in the dopamine system are often associated with neurodegenerative disorders like Parkinson's disease (Latif 2021). Another neurotransmitter Acetylcholine, released by striatal cholinergic interneurons, enhances the indirect pathway and inhibits the direct pathway (Benarroch 2012). It also plays a role in dopamine signalling which is disrupted in Parkinson's disease (Aosaki *et al.* 2010).

DOI: 10.1201/9781003589273-7

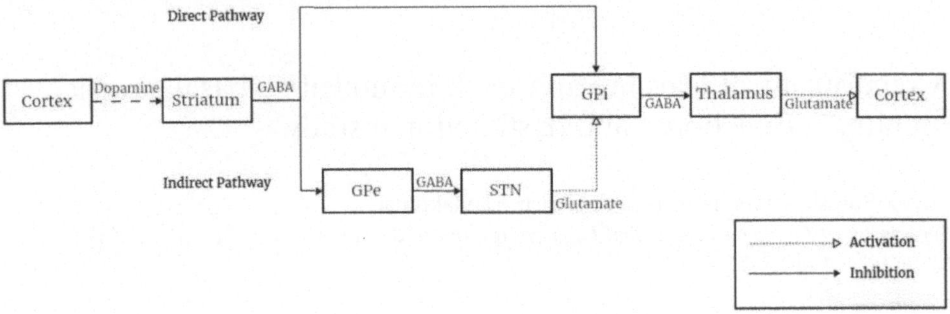

Figure 1. Pathway Flow Diagram showing the movement of a signal through either of the pathways modulated by the different neurotransmitters (Kenhub n.d.).

Parkinson's disease is a degenerative neurological condition that worsens over time. Parkinsonian symptoms develop when there is a loss of approximately 80% or more of the dopamine-producing neurons. This leads to the motor symptoms commonly associated with Parkinson's. The progressive nature of this disorder means that the neurological impairments worsen over time (Sasidharakurup *et al.* 2017). Levodopa is often given along with carbidopa in medicines like Sinemet targeting the deficiency that underlies the Parkinsonian condition. Levodopa-carbidopa therapy continues to serve as a base in Parkinson's disease management (Medscape n.d.).

Many related research studies provide insights into brain circuit dynamics by modelling BG neuron firing patterns with basic spiking neuron models (Medini *et al.* 2016). Certain other studies (Vigneswari and Aravinth 2021) explore an ML diagnosis of PD using various classes to determine the prediction accuracy of the diagnosis process. (Prabal *et al.* 2023) presents a computational model that studies the impact of dopamine levels on the BG circuitry and its role in PD. The interconnected basal ganglia-cerebellum circuit gives insight into the cerebellum's role in motor dysfunction during PD (Rajendran *et al.* 2017). Biochemical Systems Theory (BST) models also helped predict the dopaminergic cell death process involved in PD progression. Incorporating biological details of the BG improves RL models and provides insights into the striatal functions (Blackwell and Doya 2023).

This study aims to understand the workings of the BG, its pathways, and neurological disorders of BG, and use principles of RL to build a model of the BG.

2 PROPOSED METHODOLOGY

In this project, a BG model has been developed using Markov Decision Process (Figure 2). The basal ganglia has been modelled as an MDP because it offers a strong framework for understanding the dynamics of decision-making mechanisms in the brain. The complex interactions of the basal ganglia circuitry can be represented as a series of states, actions, transitions and rewards. It also allows us to explore how various factors such as neurotransmitter dynamics, environmental stimuli, drugs etc help influence decision-making behaviour.

The states represent the key components of the basal ganglia neural circuitry. The sequential signal flow via activation or inhibition through these states reflects the dynamic interactions within the basal ganglia circuitry. The basal ganglia circuit modulates the input to the striatum via the GPe, STN and GPi, and the motor output is regulated through the thalamus. The transition probabilities control how the states change in response to the actions taken by the agent, reflecting the dynamic interactions within the BG pathways. A dynamic reward system guides the agent's choice between the indirect and direct pathways. The reward values are assigned based on the transition and the scenario allowing the agent to learn different behaviours based on neurotransmitters levels. Refer to Table 1.

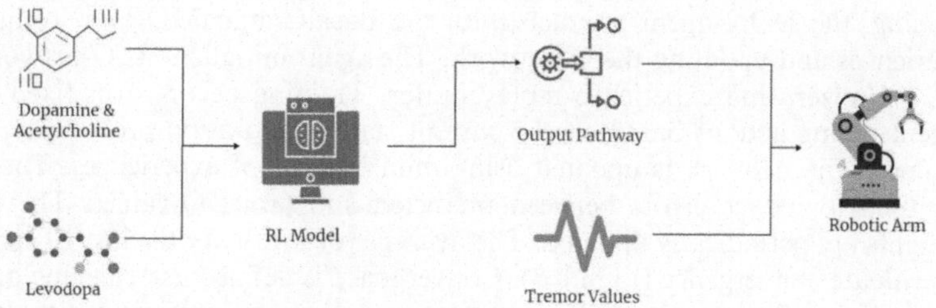

Figure 2. Proposed system architecture depicting the overall working of the system. Neurotransmitter values are given as input to the model, which predicts the pathway which is then fed to the robotic arm along with tremor values.

Table 1. Table representing neurotransmitters and drug ranges covering Parkinsonian and Functional BG conditions for Dopamine, Acetylcholine and Levodopa (MedlinePlus n.d.).

	Parkinson		Functional	
Neurotransmitter/Condition	Min	Max	Min	Max
Dopamine (pg/mL)	0	39.5	39.6	198.5
Acetylcholine (nmol/mg)	2.6	5	0	2.5
Levodopa (mg)	0	99	100	250

2.1 *Reinforcement learning models*

In this study, a comparative evaluation of the SARSA and DQN algorithms was carried out to determine the RL model that could most accurately capture the complex dynamics and decision-making processes involved in the basal ganglia. DQN combines deep neural networks and Q-learning to approximate the optimal action-value mapping (Eq.1). The SARSA algorithm is which updates the Q-values based on the action taken by the agent and the subsequent state and reward (Eq. 2).

Key parameters include the state space size, the number of environment states and action space size, and the possible actions the agent can take. The learning rate α of 0.001 ensures stable, gradual Q-network updates (Michael J. Fox n.d.). The discount factor γ of 0.9 means that the agent gives more weight to future rewards. This can ensure a balance between short-term and long-term decision-making. The exploration probability ϵ of 0.9 encourages the agent to try a wider range of actions. The agent also discovers more optimal policies, that can maximize rewards.

$$Q(s,a) \leftarrow Q(s,a) + \alpha\,[r + \gamma \max^{a} Q(s', a') - Q(s,a)] \quad (1)$$

$$Q(s,a) \leftarrow Q(s,a) + \alpha[r + \gamma Q(s', a') - Q(s,a)] \quad (2)$$

where Q(s,a) = present Q-value, s = state, a = action, r = reward, s' = subsequent state, a' = action that optimizes the Q-value, α = learning rate, and γ = discount factor.

The neural network consists of two hidden layers, each with 64 units and ReLU activation functions. The agent employs an Adam optimizer for updating the Q-network weights. By adjusting these key parameters, the DQN Agent learns to optimize its actions within the BG MDP environment. It helps improve decision-making capabilities over time.

2.2 *Training and testing*

During training, the DQN agent interacts with the BasalGangliaMDP environment, collecting experiences and updating the Q-network. The agent initializes its Q-network, target Q-network, optimizer, and experience replay buffer. Training occurs over fixed episodes, with the agent taking actions based on the current state, exploration probability, and predicted Q-values. The network is updated using mini batches of experiences. The Bellman equation is used to reduce errors between predicted and target Q-values. The target Q-network weights are periodically updated. The average rewards over the last 10 episodes are tracked to evaluate convergence (Figure 3). Convergence is defined as achieving an average reward above a specified threshold. If the total reward in any episode meets the sample efficiency threshold, the episode number is recorded. Finally, the results, including average rewards and convergence, are plotted and reported. During testing, the agent makes decisions based on learned policies and neurotransmitter levels. The chosen pathway is recorded and compared to the expected pathway based on dopamine levels. After testing with 1000 samples, the process was repeated for a single set of neurotransmitter values. The predicted pathway is used as input to a robotic arm simulation and graphical representation.

2.3 *Robotic arm*

This project utilizes a robotic arm with 6 Degrees of Freedom (DOF) to bridge computational modelling and physical task execution. The binary output from the RL model is sent to an Arduino board for control. When receiving a true value to reflect the disease state, the arm simulates tremors and shakiness observed in Parkinson's disease patients during a pick-and-place movement sequence. It introduces disturbances based on levodopa drug study data (Mnih *et al.* 2015). A normal behavioural state (value '0'), the arm runs a predefined smooth pick-and-place sequence. This integration allows the RL model's output to control the robotic arm's movements which presents BG behaviour in motion control.

2.4 *Dataset*

A dataset containing tremor measurements of Parkinson's disease (PD) patients from a Levodopa Response study was used. It gives the acceleration values across the 3 axis. The study evaluated the feasibility of monitoring Parkinson's disease symptoms at home using wearable sensors. Tremor data was collected using a smartwatch on the least affected limb and a GeneActiv device on the most affected limb during motor tasks. The study was conducted across 4 days. Each day introduced different tasks which were performed under different dosages of levodopa, and the last day without. This allowed measurement of tremor severity under different conditions. This real-life patient data thus provided insights into Levodopa's impact on tremor severity in PD patients (Michael J. Fox Foundation n.d.).

3 EXPERIMENTAL RESULTS

The convergence rate of the two reinforcement learning models, SARSA and Deep Q-Network (DQN), was evaluated during the training process. The convergence rate shows how quickly the model learns the optimal policy. The graphs depict the average reward achieved by the agents over the course of training episodes. The DQN Model exhibited a smooth convergence curve with relatively smaller fluctuations in the average reward (Figure 3a). The SARSA model achieves peak followed by steep declines (Figure 3b). While both models eventually converge, the DQN model has a more stable and consistent learning process. Fewer sharp peaks and valleys shows a more robust and reliable approach.

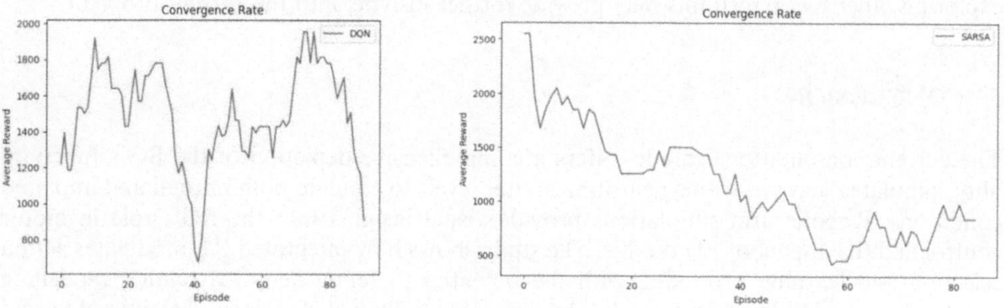

a) DQN b) SARSA

Figure 3. Comparison of convergence rates for the RL models during training.

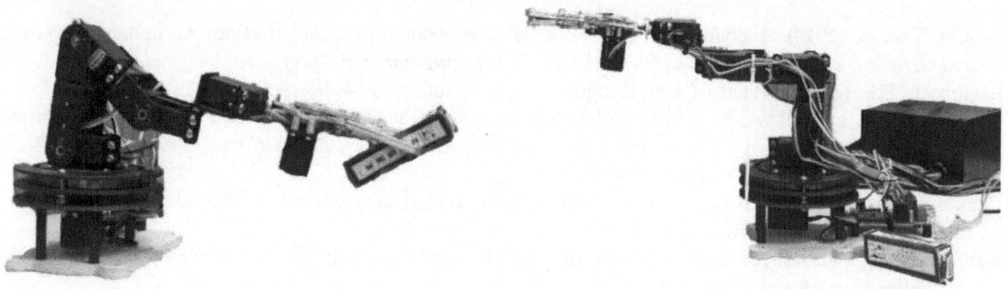

a) Pick up and place with tremors b) Pick up and place without tremors

Figure 4. Simulating the working of pathways using a robotic arm by performing a pickup and place action.

But for SARSA, the convergence appears more unstable, with the reward dipping substantially after reaching convergence. Moreover, the DQN model takes more episodes to converge, reflecting the complexity of the neural network capturing the intricate dynamics of the basal ganglia pathway and neurotransmitter modulation. Thus, DQN's process makes it a more potentially reliable choice for this application.

The agent's performance was also tested by using a robotic arm. The predicted pathway was applied to the arm and simulated. Under Parkinsonian condition, it chose the indirect pathway. Tremors were introduced (Figure 4a). Under partial levodopa and high dopamine, the direct pathway was chosen. Smooth execution of actions was enabled (Figure 4b). These robotic arm simulations provided insights into how BG pathways influence motor behaviour across neurotransmitter conditions by visually depicting Parkinson's effects and potential levodopa treatment benefits on motor control.

4 CHALLENGES AND LIMITATIONS

Using DQN presents the challenge of finding the right balance between exploration and exploitation. Our model required simplifications in modelling the intricate BG circuitry.

Incorporating actual neurotransmitter data introduces significant deterministic overhead. The lack of sufficient neurotransmitter data makes the model less practical. Defining the reward function also poses difficulties as it may introduce inherent biases. Furthermore, exploring other RL algorithms may provide further insights into the current problem.

5 CONCLUSION

The current computational model offers a comprehensive depiction of the BG's functionality, simulated across various neurotransmitter levels to emulate both normal and impaired conditions. Robotic arm simulations provide visual insights into the BG's role in motor control and the impact of PD on this. The study shows how integrated RL modulates action selection and learning processes, with the potential to refine neural dynamics modelling using AI methods like deep neural networks, aiding clinical decisions (Tallapureddy and Radha 2022).

REFERENCES

Aosaki, T. *et al.* (2010). Acetylcholine–dopamine balance hypothesis in the striatum: An update, *Geriatrics and Gerontology International/Geriatrics & Gerontology International*, 10(s1).

Benarroch, E.E. (2012). Effects of acetylcholine in the striatum, *Neurology*, 79(3), pp. 274–281.

Blackwell, K.T. and Doya, K. (2023). Enhancing reinforcement learning models by including direct and indirect pathways improves performance on striatal dependent tasks, *PLOS Computational Biology/PLoS Computational Biology*, 19(8), p. e1011385.

Chakravarthy, V.S. and Moustafa, A.A. (2018) Computational neuroscience models of the basal ganglia, *Cognitive Science and Technology*.

Kenhub. Direct and Indirect Pathways of the BG (2023). https://www.kenhub.com/en/library/anatomy/direct-and-indirect-pathways-of-the-basal-ganglia.

Lanciego, J.L., Luquin, N. and Obeso, J.A. (2012). Functional neuroanatomy of the basal ganglia, *Cold Spring Harbor Perspectives in Medicine*, 2(12), p. a009621.

Latif, S. *et al.* (2021). Dopamine in Parkinson's disease, *Clinical Chimica Acta*, 522, pp. 114–126.

Medini, C. *et al.* (2016) Modelling basal ganglia microcircuits using spiking neurons.

MedlinePlus Medical Encyclopedia (no date). https://medlineplus.gov/ency/article/003561.html

Medscape. Sinemet (carbidopa/levodopa) - Indications, Dosage, Side Effects, Interactions. (no date). https://reference.medscape.com/drug/sinemet-rytary-carbidopa-levodopa-343043.

Michael J. Fox Foundation. Levodopa Response Study (no date). https://www.michaeljfox.org/news/levodopa-response-study.

Mnih, V. *et al.* (2015). Human-level control through deep reinforcement learning, *Nature*, 518(7540), pp. 529–533.

Prabal, G. *et al.* (2023). A study on modelling the dopaminergic pathways of Parkinson's disease,*AIP Conference Proceedings*.

Rajendran, A. *et al.* (2017). *Computational Neuroscience of Substantia Nigra Circuit and Dopamine Modulation During Parkinson's Disease*.

Sasidharakurup, H. *et al.* (2017). A systems model of Parkinson's disease using biochemical systems theory, *Omics*, 21(8), pp. 454–464.

Tallapureddy, G. and Radha, D. (2022). Analysis of ensemble of machine learning algorithms for detection of Parkinson's Disease, *2022 International Conference on Applied Artificial Intelligence and Computing (ICAAIC)* [Preprint].

Vigneswari, D.A. and Aravinth, J. (2021). *Parkinson's Disease Diagnosis using Voice Signals by Machine Learning Approach*.

Young, C.B., Reddy, V. and Sonne, J. (2020). *Neuroanatomy, Basal Ganglia*.

Yuan, C. and Fathi, M. (2022). Reinforcement Learning: Beyond the basal ganglia, in *Educational Communications and Technology: Issues and Innovations*, pp. 235–243. https://doi.org/10.1007/978-3-030-84729-6_16.

Data Science & Exploration in Artificial Intelligence – Gururaj et al. (Eds)
© 2025 the Author(s), ISBN: 978-1-032-96390-7

Human emotion classification using facial thermal features in the presence of mask

C.M. Naveen Kumar, M.S. Srinath, G. Gagana, Deeksha Indresh, M.S. Meghana & H.S. Nayana
Malnad College of Engineering, Hassan, Karnataka, India

ABSTRACT: A vital component of human interaction and communication is the ability to recognize emotions. Conventional techniques have mostly depended on visual clues, but new developments in technology, especially thermal imaging, have created new opportunities. In this paper, a Sequential Convolutional Neural Network (CNN) model trained on thermal pictures is used to propose a novel method for human emotion recognition. Because it can detect physiological changes linked to emotional responses—changes that are frequently invisible to the unaided eye—thermal imaging has some advantages. To precisely identify and categorize human emotions, the suggested Sequential CNN model makes use of the temporal and geographical characteristics included in temperature data. Moreover, temporal dependencies are incorporated into the Sequential CNN design, which is essential for capturing dynamic changes in facial emotions across time. A large dataset is created taking into account 20 subjects from the Hassan Region by considering the sequential nature of emotional cues, the model achieved enhanced accuracy of 95%.

Keywords: Facial emotion recognition, Thermal image, Deep Learning, Convolutional Neural Network (CNN)

1 INTRODUCTION

1.1 *Emotion and its importance*

In human interactions, emotion plays a crucial role. Emotions affect our thoughts and actions, as they are a major factor in communication [1][2]. An individual's emotional state can have an impact on their capacity for focus, problem-solving, making decisions, and day-to-day functioning [2]. This encourages us to study human emotions [1]. The main indicator of a person's mental and emotional state is their facial expressions. Only 7% of actual information is communicated verbally, according to Albert Mehrabian, who is regarded as the pioneer of nonverbal scientific communication, and 38% is done so through linguistic auxiliaries. Like speech patterns, tempo, or tone; nevertheless, facial expressions convey 55% of the information that is exchanged [8]. Since the human face is a natural source for gathering non-invasive data, it is widely employed in a variety of contexts [2]. Happiness, sadness, anger, fear, surprise, and disgust appear to be the six primary emotions in humans [6]. A person's forehead, mouth, and cheek areas may experience dips or spikes in temperature in response to external sounds, objects, and activities that elicit a shift in their emotions [4].

Emotions are considered fundamental or primitive according to Darwinian evolution [20]. It is believed that the form of emotion is associated with distinct, basic tendencies of behavior or responses [9]. Facial expressions are crucial for identifying people and their emotions. They are also employed in nonverbal communication. Additionally, they are a sign of emotion, enabling a man to communicate his feelings [7]. Automatic emotion recognition

DOI: 10.1201/9781003589273-8

systems frequently use facial expressions [7]. Emotion detection is a technically difficult task because the effectiveness of the detection depends on the quality of the photo's illumination. Because infrared cameras are insensitive to light, thermal images provide a dependable method for addressing this issue [3].Emotion recognition has been applied to Internet education, where it can be used to identify students' learning levels and acceptance of content and work in tandem with pertinent reminders to enhance learning effectiveness. Emotion identification is a useful tool in criminal interrogation for identifying lying [9].

1.2 *Correlation of emotion and thermal imaging*

The case study on the use of human thermal image processing for emotion recognition is detailed in [6]. The psychological processes that makeup emotions include affective, cognitive, physiological, motivational, and expressive activities [1]. An individual's emotional condition is instantly recognizable to others. As a result, automatic algorithms for recognizing emotions frequently use facial expression data [7]. Self-report has been a frequently utilized way to quantify academic emotions since humans may use spoken language to explain their experiences, and emotions are theoretically represented in the conscious mind [1]. Particularly in challenging and changing situations, thermal images are valuable substitutes for visual imaging [3].

It is possible to generalize the filters to include other parameters and pseudo colors for thermal images [2]. It is crucial to note that a subject's skin tone or color has no bearing on facial thermal pictures [4]. Infrared radiations are used to create thermal images, which are then translated into visible images that show the temperature distributions in the area that the thermal device or camera is exploring [3].

The input photographs are arranged in pairs, and the initial findings indicate a very high categorization accuracy [6]. Comparing simultaneously collected data using thermal imaging, the validity and dependability of thermal imaging as the source for emotional state analysis were demonstrated [5].

1.3 *Techniques Currently Implemented*

The method, which uses facial landmarks for feature extraction and selection, is based on [2], which used 68 landmark points and manual segmentation to precisely define the human face. Research into measuring 82 human physiological signals to determine emotions in affective computing has significantly increased since 2011[10]. While numerous contact-based methods of measurement have been developed, such as the recording of skin conductance responses, electroencephalography signals, facial expression detection, and EKG measurement, these techniques may impede people's normal responses while they wear sensors[11]. Despite the prevalence of contact-based methods, a review highlights that body temperature measurement remains an underexplored avenue in emotion assessment. Notably, the research indicates that, unlike other physiological signals, body temperature measurement has not been extensively explored, and its role in understanding human body and mind dynamics remains untapped [6]. The method in employs convolutional sparse coding to learn thermal-specific filters for enhanced feature extraction instead of relying on prelearned filters [4]. When compared to comparable databases that contain visual images, the bulk of thermal databases featuring human faces have comparatively few subjects or individuals [18].

Since different facial regions have distinct effects of temperature patterns on emotions, each sensitive area has a region-specific filter bank that is independently learned[12]. A dimensionality reduction approach inspired by discriminant correlation analysis (DCA) is developed to improve class separability on features, resulting in a considerable improvement in classification accuracy. "The processing and classification of such images is important for various difficult environments, i.e., media with less lightning, including here the actual pandemic context." This is in reference to the context of human thermal photos as well as for face and emotion detection. Different distances are used by the classifiers to assess how similar patterns are to one another [19].

2 LITERATURE REVIEW

In order to improve the recognition performance in [1], Mawarpury *et al.* fed the segmented image, assigned a decent kernel, and developed a CNN model[13]. They also provide information regarding future research prospects for building a representative data set in academic contexts. This project's primary flaw was that it was less noticeable and uncomfortable. In [2], Shubhobrata *et al.* describe a portion of the data set using a basic linear SVM classifier; additional data and more sophisticated classifiers can be utilized to significantly improve performance. The primary limitation of this experiment was that only the NVIE dataset was used to learn the filters. The filters for thermal images with various specs and pseudocolors can be made more widely applicable[14].

The method based on Renyi entropy as an image feature or descriptor for classification purposes, which has poor performance, is described in detail by Kakarwal *et al.* in [3]. The region of interest of the photos can be computed, searched, and used to potentially improve the classification results. The primary drawback was that the well-known deep learning technology was overlooked for complexity- related reasons, meaning that it was too complicated for the task under consideration and the given data set.

A thermal image dataset containing diverse scenarios—such as multiple people posing in a photo with a different background—was described by Engelhaupt *et al.* in [4]. A person's emotions are a major clue to their psyche. Surveillance cameras may identify individuals who are likely to conduct crimes or cause problems in public by analyzing their abnormally high heat image levels.

Convolutional networks that have been pre-trained on the RGB dataset can be extended to the thermal domain, although Ioannou *et al.* have noted that this will result in inferior overall performance [15]. This project's primary flaw was how hard it was to compare the efficacy of the different suggested approaches.

Details regarding using thermal pictures in convolutional neural networks and focusing solely on regions of interest using automated image segmentation methodology are provided by Keltaner *et al.* in [16]. This project's primary goal was to address the underdeveloped state of thermal imagery research. As a result, this study provides an overview of Haralick's textural features as well as RF and FFBPN methodologies.

According to Paweł *et al.*, the research intends to identify seven main emotional states based on facial expressions: neutral, joy, surprise, anger, sadness, fear, and disgust. The primary drawback of this experiment was the possibility of numerous other factors influencing categorization accuracy in real-world settings. Real emotions can cause a wide range of facial expressions to be revealed to varying degrees[17].

According to Jiang, Chen, *et al.*, of the 25 publications that were analyzed, 13 of them were helpful in determining which datasets in FER systems are most frequently used; CK+ is used by 31% of users, JAFFE and FER2013 by 23%, and the remaining datasets by 15% to 8% of users. The primary limitation of the project was the abundance of existing research on facial expression recognition (FER) using convolutional neural networks in various domains, including psychology, psychiatry, health care systems, intelligent systems, human-robot interactions, and detection of abnormal driving activities. (CNN) a lot of these models are too constrained to capture the intricacy of illustrating typical emotive actions in [8].

Lucey, Roopa, and Akhand. Facial expressions can naturally convey subjective sentiments in interpersonal communication, according to P, the specialist, but they can also be influenced by factors such as individual variances, occlusion, tiny changes in facial expressions, and restricted illumination [9]. One of the project's shortcomings was that speech generally contains several emotions since emotions are so unpredictable, which makes it difficult to accurately extract speech information elements.

In [18] Dorel Aiordachioaie describe that DCT-based approach achieves the greatest outcomes for face recognition objectives. If resource optimisation is required, additional computations for DCT and coefficient selections are required. By utilising the images' region of interest and computing or searching for it, it is possible to enhance the categorization results.

Since the significant features reduce the accuracy by 3.5% for both classes, it is crucial that all of the suggested features be used in the categorization. According to Sorin Pavel *et al.*, [19] in the future, convolutional neural networks will be able to process thermal pictures in addition to focusing exclusively on regions of interest using automated image segmentation techniques.

3 METHODOLOGY

Deep learning can be thought of as a method for building a model that uses the pictures that are available to learn how to apply classification. The degradation of images, particularly infrared images with low contrast and resolution, has led to its dominance in the pattern recognition field. Using the CNN, features are gathered and supplied to the classifier.

3.1 *Data collection*

Image acquisition has been done in laboratory settings. Subjects are allowed to sit in front of the laptop and played with short videos downloaded from YouTube.

Data acquisition time per subject per emotion is 60 seconds. Videos of 3minute duration were used in the process. Subjects were instructed to make minimal head movements and a consent/feedback form was given to subjects and explained with the purpose of the research. Totally 20 subjects were considered in the investigation in the age group of 22-25 and all the subjects considered were healthy. Per subject 10 and thermal images were acquired per emotion using Fluke TiS20 thermal image.

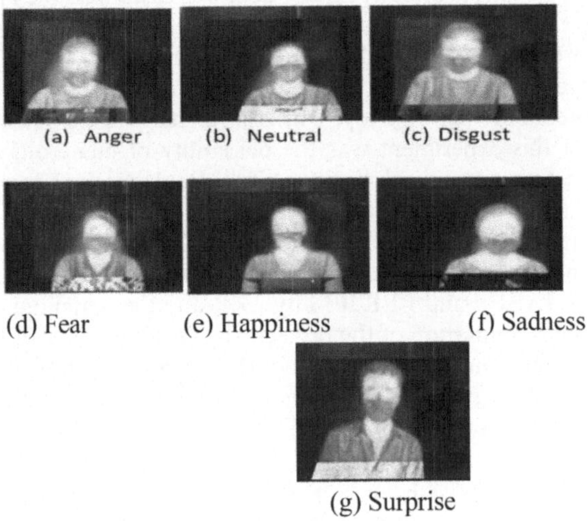

Figure 1. Sample thermal images of basic facial emotions.

3.2 *Data inspection*

With 80% of the data set used for training and 20% for validation, the dataset is directly utilized for deep learning.

4 WORKFLOW

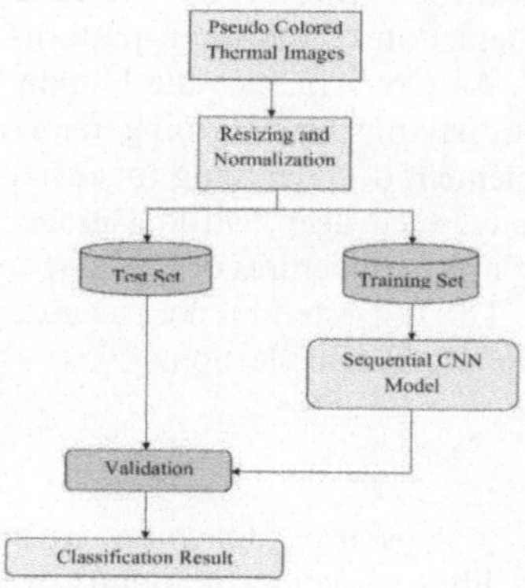

Figure 2. Flowchart showing the overall procedure.

Figure 2 shows the overall procedure of Facial Emotion Recognition using Thermal Imaging. The pseudo-colored thermal image is taken as the input thermal image, which is in the range of 320×240 pixels. This input thermal image is resized to a standard size in the range of 180×180 pixels After resizing, it is normalized to ensure efficiency and remove any variations in illumination. Then the dataset is split into two subsets, namely the training set and the test set, where the training set is used to train the model and the test set is used to evaluate the model's performance. This process involves a sequential convolutional neural network architecture, which consists of multiple convolutional layers. A piece of the training set is frequently set aside for validation during training to avoid overfitting. Overfitting can be reduced by tuning hyperparameters with reference to the validation result. After training, the model is examined on the test set to see how good it can identify different facial emotions from thermal images. For every input image, the anticipated emotion is displayed in the classification result.

5 APPLIED ALGORITHM

5.1 *Convolutional neural network*

Convolutional, pooling, batch normalization, and fully connected layers make up the CNN. Reducing over-fitting is the goal of the batch normalization layer.

There are two primary deep-learning layers that execute the feature extraction process:

Convolutionl layer
Pooling layer

5.2 *Convolutional layer*

Filters that are applied to the raw image are part of the convolutional layer.

Depending on the filters applied, different features are acquired. This idea works incredibly well with infrared photos to help face recognition detect the fewest possible differences in these pictures.

5.3 *Pooling layer*

The feature map is down-sampled for feature reduction using the pooling layer. It combines nearby values from a given map frame to create more illuminating values. One can use average-pooling or max-pooling. For this operation, for instance, several layers are used.

CNN can be used for feature extraction, selection reduction, and classification in a deep learning system that has many convolution and pooling layers. In-depth Convolutional layers are used within the network to extract features from CNN using various approaches. To provide a more accurate depiction of the input patterns, the output of three layers is transferred via pooling layers. As CNN mimics the human brain in feature extraction, it performs better than conventional machine-learning methods for face recognition This paper's primary goal is to implement deep learning for facial recognition.

CNN is necessary for the suggested deep learning model to accomplish this goal. The suggested model is tested using infrared pictures of several people. A thermal infrared sensor was used to get these pictures. The suggested model has a dense layer for classification and layers for traditional and pooling feature collections.

5.4 *TensorFlow*

TensorFlow, an end-to-end open-source machine learning platform with a vast, adaptable ecosystem of tools, libraries, and community resources, is used in this implementation. It offers several abstraction levels from which to select. By utilising TensorFlow, it is simple to design and train models using the high- level Keras API. TensorFlow can be used with Keras without sacrificing its functionality or flexibility. To demonstrate how to use Keras with Python, two applications, namely temporal and spatial deep learning, are used.

5.5 *Keras*

Keras reduces hundreds of lines of traditional code to a few lines, offering a convenient solution to deep learning challenges. It eliminates the work of creating a complicated network, which could be difficult to construct using traditional methods and require hours or days of manual Python or other language coding.

5.6 *TensorFlow and Keras are related*

(1) Unified Interface.
(2) Simplicity and Flexibility
(3) Performance and Scalability

Keras offers a comprehensive framework for building any kind of neural network. Keras is quite intuitive to learn and creative. It can support neural network models ranging from very simple to very vast and complex.

5.7 *Architecture of keras*

There are three primary categories for Keras APIs.

- Model
- Layer
- Core Modules

CNN represents each and every ANN in Keras. All Keras models, on the other hand, are composed of Keras layers and represent various ANN layers, including input, output, convolution, and pooling layers. For the activation, loss, and regularisation functions, among other functions, the Keras model and layer access the Keras modules. Any ANN algorithm (CNN, RNN, etc.) may be represented effectively and simply using the Keras model, Keras layer, and Keras modules.

The two forms of Keras models are-Sequential Model and Functional API. In this Implementation we use Sequential Model[21].

5.8 *Sequential model*

Keras layers are linearly combined to form the sequential model. The sequential model may represent almost all neural networks that are currently accessible and is simple and minimal.

```
model.summary()

Model: "sequential"
_________________________________________________________________
 Layer (type)                Output Shape              Param #
=================================================================
 rescaling (Rescaling)       (None, 180, 180, 3)       0

 conv2d (Conv2D)             (None, 180, 180, 16)      448

 max_pooling2d (MaxPooling2  (None, 90, 90, 16)        0
 D)

 conv2d_1 (Conv2D)           (None, 90, 90, 32)        4640

 max_pooling2d_1 (MaxPoolin  (None, 45, 45, 32)        0
 g2D)

 conv2d_2 (Conv2D)           (None, 45, 45, 64)        18496

 max_pooling2d_2 (MaxPoolin  (None, 22, 22, 64)        0
 g2D)

 flatten (Flatten)           (None, 30976)             0

 dense (Dense)               (None, 128)               3965056

 dense_1 (Dense)             (None, 7)                 903

=================================================================
Total params: 3989543 (15.22 MB)
Trainable params: 3989543 (15.22 MB)
Non-trainable params: 0 (0.00 Byte)
_________________________________________________________________
```

Figure 3. Summary of sequential model.

5.9 *Implementation of sequential model*

The Rescaling Layer adjusts the input pixel values to fall between 0 and 1. Convolutional Layer (Conv2D) produces an output shape of (None, 180, 180, 16) by applying 16 filters of size 3x3. The Max Pooling Layer (MaxPooling2D) produces an output shape of (None, 90, 90, 16) by reducing the spatial dimensions by half using a 2x2 pool size. Convolutional Layer (Conv2D) produces an output shape of (None, 90, 90, 32) by applying 32 filters of size 3x3 to the input feature maps from the preceding model layer. Max Pooling Layer (MaxPooling2D) uses a 2x2 pool size for max pooling, which reduces the spatial dimensions by half and gives an output shape of (None, 45, 45, 32).

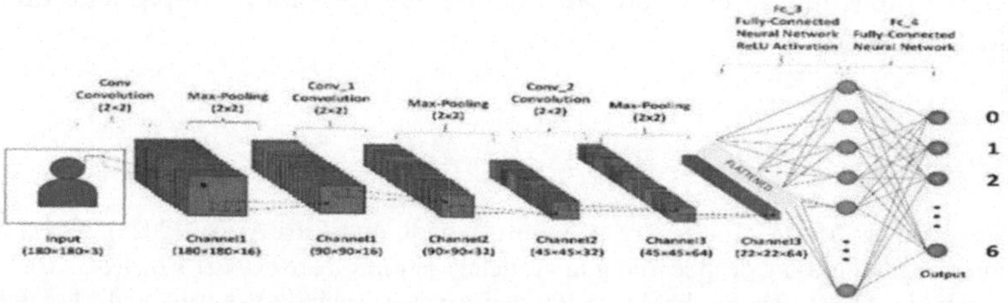

Figure 4. Architecture of sequential.

Convolutional Layer (Conv2D) creates an output shape of (None, 45, 45, 64) by applying 64 filters of size 3x3 to the input feature maps from the preceding layer. Max Pooling Layer (MaxPooling2D) reduces the spatial dimensions by half by performing max pooling with a 2x2 pool size, yielding an output shape of (None, 22, 22, 64). In Flatten Layer the input feature maps are flattened into a 1D array, producing an output shape of (None, 38976). Dense Layer applies 128 neurons, with a total of 3965056 parameters. Dense Layer (Output) uses 7 neurons with 903 parameters to represent the output classes. The model has a total of 3989543 parameters, with 3989543 being trainable and 0 being non-trainable.

6 RESULT

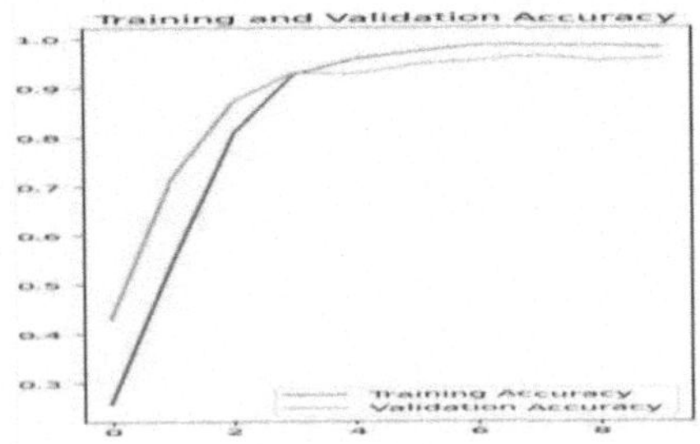

Figure 5. Accuracy of training and validation.

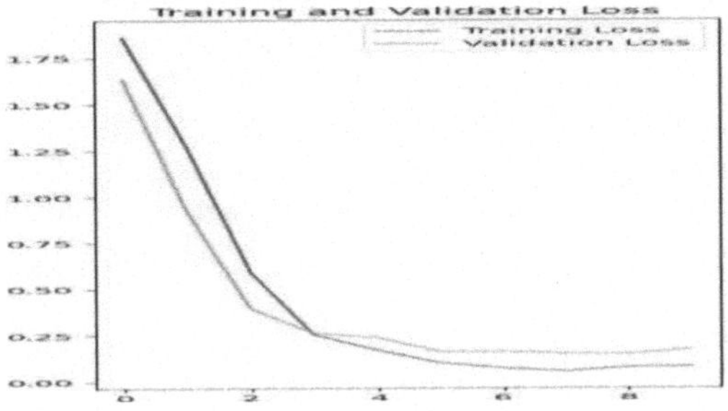

Figure 6. Loss of training and validation.

The proposed model yielded a high accuracy of 95.81%. As shown in Figures 5 & 6 training accuracy and loss dominated their validation counterparts.

7 CONCLUSION

Looking ahead, future research directions may include refining the Sequential CNN model architecture, exploring additional modalities for multimodal emotion recognition, and extending the applicability of the proposed approach to real-world scenarios. Moreover, efforts to address ethical considerations, such as privacy concerns and bias mitigation, are essential to ensure the responsible development and deployment of emotion recognition technology.

In conclusion, this work represents a substantial advancement in the field of identifying human emotions, offering a promising framework for harnessing thermal imaging and deep learning techniques to understand and interpret human emotions accurately and effectively. Through continued innovation and collaboration, emotion-aware systems have the potential to enrich human-machine interactions and contribute to a more empathetic and intuitive technological landscape.

REFERENCES

[1] Fardian, Fardian, Marty Mawarpury, Khairul Munadi, and Fitri Arnia. (2022). Thermography for emotion recognition using deep learning in academic settings: A review. *IEEE Access* 10: 9647696491.

[2] Rooj, Suparna, U. Antesh, Shubhobrata Bhattacharya, Aurobinda Routray, and Manas K. Mandal. (2020). Emotion classification of facial thermal images using sparse coded filters. *In IECON 2020 The 46th Annual Conference of the IEEE Industrial Electronics Society*, pp. 453–458. IEEE.

[3] R. L. M. Ã and Atkins M. S. (2007). A fuzzy physiological approach for continuously modeling emotion during interaction with play technologies, *Int. J. Human-Computer Studies*, vol. 65, pp. 329–347.

[4] Calvo R. A., Member S., Mello S. D., and Society I. C. Affect detection: An interdisciplinary review of models, methods, and their applications, *18 IEEE Transactions*

[5] Xu Ya, and Liu Guang-Yuan. (2009). A Method of Emotion Recognition Based on ECG Signal, in proc. of *International Conference on Computational Intelligence and Natural Computing*, pp. 202–205

[6] Puri C. (2005). StressCam: Non-contact Measurement of Users 'Emotional States through Thermal Imaging, in *proc of Int. Con. on Human-Computer Interaction*, pp. 1725–1728.

[7] Xu H. and Plataniotis K. N. K. (2012). Affect recognition using EEG signal, *IEEE 14th International Workshop on Multimedia Signal Processing*, pp. 299–304

[8] Kolli A., Fasih A., Al Machot F. and Kyamakya K. (2011). Non-intrusive Car Driver's Emotion Recognition Using Thermal Camera, *16th Int'l Symposium on Theoretical Electrical Engineering, & Joint 3rd Int'l Workshop on Nonlinear Dynamics and Synchronization*, pp. 1–5

[9] Wang S., Shen P., and Liu Z. (2012). Facial Expression Recognition from Infrared Thermal Images Using Temperature Difference by Voting, *IEEE 2nd International Conference on Cloud Computing and Intelligent Systems*, pp. 94–98.

[10] Shan He, Shangfei Wang, Wuwei Lan and Huan Fu Qiang Ji. (2013). Facial Expression Recognition using Deep Boltzmann Machine from Thermal, *Humaine Association Conference on Affective Computing and Intelligent Interaction*, pp. 239–244.

[11] Nhan B. R., Chau T., and Member S. (2010). Classifying Affective States Using Thermal Infrared Imaging of the Human Face, *IEEE Transactions on Biomedical Engineering*, vol. 57, no. 4, pp. 979–987.

[12] Khan M. M., Ingleby M. and Ward R. D. (2006). Automated Facial Expression Classification and Affect Interpretation Using Infrared Measurement of Facial Skin Temperature Variations, *ACM Transactions on Autonomous and Adaptive Systems*, vol. 1, no. 1, pp. 91–113.
Trujillo L., Olague G., Hammoud R., Hernandez B., Evovision P., De Ciencias D. and Computacion D. (2005). Automatic feature localization in thermal images for facial expression recognition, in proc. of *IEEE Computer Society Conference on Computer Vision and Pattern Recognition*, pp. 14–21.

[13] Nguyen H., Kotani K., Chen F., and Le B. (2014). Estimation of human emotions using thermal facial information, in proc. of *Fifth International Conference on Graphic and Image Processing*, vol. 9069, pp. 1–5.

[14] Ahonen T., Hadid A., and Pietikäinen M. (Dec. 2006). Face description with local binary patterns: application to face recognition., *IEEE Trans. Pattern Anal. Mach. Intell.*, vol. 28, no. 12, pp. 2037–41.

[15] Viola P. and Jones M.J. (2004). Robust real-time face detection, *International Journal of Computer Vision*, vol. 57, no. 2, pp. 137–154.

[16] Kim J., Kim L., and Hwang S. (Apr. 2001). An advanced contrast enhancement using partially overlapped sub-block histogram equalization, *IEEE Trans. Circuits Syst. Video Technol.*, vol. 11, no. 4, pp. 475–484.

[17] Hu M.K. (February 1962). Visual pattern recognition by moment invariants, *IRE Trans. Information Theory*, Vol. IT-8, pp. 179–187.Xu, X.; Du, M.; Guo, H.; Chang, J.; Zhao, X. (2021). Lightweight facenet based on mobilenet. *Int. J. Intell. Sci.*, 11, 1–16. [CrossRef]

[18] Szegedy, C.; Liu, W.; Jia, Y.; Sermanet, P.; Reed, S.; Anguelov, D.; Erhan, D.; Vanhoucke, V.; Rabinovich, A. (June 2015). Going deeper with convolutions. *In Proceedings of the IEEE Conference on Computer Vision and Pattern Recognition*, Boston, MA, USA, 7–12; pp. 1–9.

[19] Z. Zhang *et al.* (2016). Multimodal spontaneous emotion corpus for human behavior analysis, in *Proceedings of CVPR.*

[20] Aiordachioaie, D. (2022). November. on thermal image classification with ANN and similarity based classifiers. *In 2022 International Conference on Communications, Information, Electronic and Energy Systems (CIEES)* (pp. 1–4). IEEE.

[21] Pavel S, Moldovanu S, Aiordachioaie D. (2023 Sep 12). Emotion recognition in human thermal images with artificial intelligence technology. *In2023 IEEE 28th International Conference on Emerging Technologies and Factory Automation (ETFA)* (pp. 1–5), IEEE.

Data Science & Exploration in Artificial Intelligence – Gururaj et al. (Eds)
© 2025 the Author(s), ISBN: 978-1-032-96390-7

Software testing arcade

Anju, Pooja, Kirti Sehrawat & Shweta Jindal
Indira Gandhi Delhi Technical University for Women, New Delhi, Delhi, India

ABSTRACT: Software Testing is an essential part of development of software as it assures that the developed software fulfills or executes the functional specifications and is defect free and error free. It also ensures that the quality and standards of software are up to the mark. But in spite of plenty of tools available, there is no standard tool that is universally acceptable and the combined summaries of tools are very rare. The paper focuses mainly on the need to tackle this issue, the various testing tools that are already existing have been compiled. Ten categories of software testing tools including automated testing tools, performance testing tools, cross browser testing tools etc. are taken and they are classified based on various features such as offline or online availability, open or closed source, input and output and many others. This paper aims to provide a comparative review of different testing tools built upon various features giving users a bird eye view over the expertise of different testing tools and helping them choose the most appropriate tools as per the requirement. The derived analysis depicts the valuable insights which states that 18% of moderate level are acquired by integrating testing tools and 20% of low level are acquired by GUI testing tools based on the level of programming skills required by the user for using software testing tools. Also, 36% of online tools are cross browser testing tools and 46% of offline tools are integration testing tools.

1 INTRODUCTION

1.1 *Software testing*

Software testing is a process by which software can be tested i.e. validated- defined as the software should fulfill customer's requirements and verified- defined as the software should meet the specifications. This process not only deals with making the software bug free but also improving the quality, efficiency, accuracy and usability of the software.[1] It helps in identifying and mitigating the future risks and safeguarding the final product from vulnerabilities. Initially, Testers used to rely on manual testing tools.[2] However, due to advancements in technology and testing being an inevitable and time-consuming process, testers surplus the use of automated testing tools.

We use software in our daily life starting from mobile phones, social media, streaming platforms such as Netflix, computers, games and even microwave ovens.

Software Testing is important in the software development process because it helps us to identify whether the software is bug free or not. Software Testing is a level in the software development life cycle where the verification and validation of different software capabilities occurs to check whether it satisfies the user or not. There are many bugs or errors that can be detected and in the collection of these bugs, some of them are not given any importance and in reverse of these other bugs are really harmful and dangerous for the software which leads to the overall failure of the software. So due to this reason the software testing stage plays a very important part in the development of high quality software. [3]

 DOI: 10.1201/9781003589273-9

1.2 *Software testing tools*

Software testing tools are the tools that are used for the purpose of testing of a software product. Software testing tools ensure the firmness, performance and thoroughness of the software. The size of software as well as demand for it is increasing immensely. Therefore, tools for testing and automation are needed. Also, many times, the companies develop their own software and tools. We have tried to list down the categories of software testing tools with their variable features.[4]

1.2.1 *Test management tools*

These are the tools which give support to the management of tests and test process control. Along with this, test management tools also report the status of quality assurance activities. For example: TestRail, Zephyr scale, Testlink, Tusker, etc.

1.2.2 *Automated testing tools*

It consists of applications that are designed for verification of functional and non-functional software requirements through automated test scripts. It helps to undergo testing activities without any human input with higher accuracy and less time and effort. For example: Selenium, Ranorex, Kobiton, Katalon, etc.

1.2.3 *Performance testing tools*

These platforms are developed to plan, execute, manage, monitor, report, and analyze performance tests for websites and apps. For example: WebLOAD, Load Ninja, LoadRunner, Storm Forge, etc.

1.2.4 *Cross-browser testing tools*

These tools help in verifying that websites work as per expectations among various web browsers and operating. For example: Sauce Labs, Browsera, QA Wolf, Virtuoso, etc.

1.2.5 *Integration testing tools*

These are the applications created to test interactions among multiple components of software and detect the issues. For example: Citrus, Vector CAST/C++, TESSY, Vector CAST/Ada, etc.

1.2.6 *Unit testing tools*

These tools ensure code stability by searching for various types of defects in the process of development. It helps to validate single units of code which in turn helps to improve overall software quality and easier debugging. For example: PHP Unit, JUnit, Mockito, Type Mock, etc.

1.2.7 *Mobile testing tools*

These are used to test mobile applications for functionality, usability and performance. These are also used to test the application for compatibility across different mobile devices. For example: Test Complete, Robotium, Test IO, API Tester etc.

1.2.8 *GUI testing tools*

GUI testing tools are used to validate and verify multiple

UI functionalities of an application which are visible to users like buttons and menus etc. For example: Cubic Test, AutoIT, Squish, FitNesse, etc.

1.2.9 *Bug tracking tools*

Bug tracking tools are used to monitor bugs or errors during. These are also referred to as defect or issue tracking. For example: BugHerd, Mantis, Backlog, JIRA, etc.

1.2.10 *Security testing tools*

These tools are used to find and address security vulnerabilities in a software application. Their main aim is to ensure software's security. For example: ZAP, Iron Wasp, Wfuzz, W3af, etc. [4]

2 LITERATURE REVIEW

2.1 *Key concepts and definitions*

The process of software testing holds the utmost importance across all the software development phases. It aims to update the overall quality of the final product by detecting and eradicating the bugs. Testing is a never-ending process, which also implies that any software can never be 100% bug free. It highlighted the importance of a Software Testing Life Cycle (STLC) which indicates the beginning and ending of software testing [5].

Testing can be done either manually, that is, by human effort or systematically. Manual testing is more time consuming and less reliable than automated testing. H. A. Deyemo did research on Comparative Analysis of Modeling and Software Testing Tools in which the rapid shift from manual to automated testing is illustrated due to advancement in the realm of testing software. He referred to automated testing as the backbone of the testing process and also provided detailed information about model-based testing [6]. K. Sneha and G. M. Malle also highlights the superiority of automated tools as it spotlights the error which can be escaped from the eye of a tester in manual testing [7]. Even though testing is time consuming, an expensive and labor-intensive process it cannot be eliminated. It closely needs 40% to 70% of the total time in the development of software [8], which brings the importance of choosing the right tool and technique for testing from the pool of numerous testing tools. I. Shuaibu, Mustapha M. Machina, M. Ibrahim thoroughly presents different types of software testing techniques such as black-box, white-box, gray-box techniques with their types, and also mentions the different levels of testing [9].

The field of model-based testing is evolving and comes up with more tools which leads to the issue of choosing the right tool for which W. Li, F.L. Gall & N. Spaseski surveyed and categorized emerging tools and showcased a criteria list for test case generation which covers four dimensions namely model specification, generation of test, description of test, and overall support [10]. I. Rana, P.Goswami, H. Maheshwari mentions the process of software test execution which involves testcase generations and executing them on the software, analyzing and verifying the output and iterating the process to refine the test cases. This paper also briefly explains the STLC and the significance of each phase [11].

2.2 *Theoretical framework*

An ideal software tool possesses some vital features like scalability, reliability, security, robustness, usability, documentation, maintainability, integrity etc. as mentioned by J. Pan. A brief description about various testing tool's quality factors is provided and he concludes "software testing is an art" and clarifies that testing is more than just debugging [12]. Testing comprises two most important steps: verification and validation. Both the steps hold relevance where verification is about evaluating a system if the right software is built according to provided artifacts whereas validation refers to if the system is performing tasks it was intended to perform, S. Uspenskiy talks about verification and validation and provides diagrammatic difference between both for depth clarification [13].

The better evaluation requires classification of software testing tools. Tools can be classified as test management tools, cross browsing, performance, GUI, integration, security, unit and mobile tools for testing, bug tracking tools etc. K. Shaukat, U. Shaukat, F. Feroz, S. Kayani, A. Akbar provides a taxonomy of software testing tools and provides comparison between some of the selected tools against some features[14]. Similarly, S. Dhiman, P. Sharma talks about types of software testing tools and focuses on comparing performance testing tools, like: Apache Jmeter, Grinder, HTTP Rider. A comparison is done among these 3 tools based on the parameters such as requirements, OS support, programming language, version used, response time (ms), throughput and latency and concludes that HTTP Riders has a lower response time than the other two [15]. Taking the comparison a step ahead, rather comparing 2-3 tools, F Okezie, I Odun–Ayo, S Bogle talks about comparing 8 tools based on parameters like open source, platforms supporting license (Mobile, Web, Desktop), ease of learning, ease of using coding, and programming skills and concluded that 17% of selected tools were web supporting platforms, only 3% required programming skills [16]. Similarly, H. V. Gamido, M. V. Gamido focused on the automation testing tools only, comparing 8 tools on 10 different parameters allowing users to compare and choose the best tool among Selenium, QTP, LoadRunner, TestComplete, Ranorex, Watir, SoapUI & Sahi – Pro. The result highlights the superiority of selenium among these [17]. For analyses of performance testing tools, N. Srivastava, U. Kumar & P. Singh compared them and concluded that the open-source tools are becoming more common nowadays and LoadRunner and Apache JMeter are most commonly used performance testing tools. This paper highlights theoretical information about testing, types of testing and comparing them across some features [18]. Comparison and evaluation of tools is dependent on the parameters. The evaluation is not static as features of the tools are updated from time to time. Most of the tools now are open-sourced which was not earlier, new versions are released timely supporting better user experience [19].

2.3 *Methodologies employed in previous studies*

There are several emerging trends in testing. One of which is the use of testing tools within the realm of web development. There are web-based tools for testing which make user experience more engaging [20]. T. Durieux, J.F. Ferreira, R. Abreu, & P. Cruz has mentioned the use of automation testing tools in the debugging of Ethereum smart contracts. In their findings, they mentioned the usage of dataset for evaluation of some testing tools and concluded that 42% of vulnerabilities were identified by the tools and Mythril tool had the highest precision among all [21]. AI being the future of the world, testing the AI software becomes a necessity. S.M. Fernández, J. Bogner, X. Franch conducted a systematic mapping study. Their study highlights some of the changes in AI-based testing [22]. The process of creating test cases from natural language is an intensive work and requires highly skilled testers. Latest studies by K. Pham, V. Nguyen & T. Nguyen states the use of Natural language processing (NLP) for testing. The use of NLP-assisted software testing is glorified due to its impressive results [23].

This new era is of advancements, talking about advancements; 5G is new technology and has large scope for research. S. Stradowski & L. Madeyski surveyed the various difficulties in enhancing the quality and reducing the cost of testing software for 5G at Nokia due to the complex structure of 5G. This survey holds value as out of 2935 testers, testing the 5G system at Nokia, 312 of them participated in the survey, which makes 10.63% of the total testers [24]. Such system requires most trusted testing tools. In search of best tools, analyses and comparison of existing tools is a must. One of the measures to know the accuracy of tools can be defined by test coverage. Test coverage refers to the percentage of total code the test cases cover. An ideal test suite will provide full coverage of the code. For detailed analysis Q. Yang, J.J. Li & D. Weiss surveyed 17 coverage-based testing tools. The analysis of different features of test cases are also done [25]. To explore the most recent study done in the domain of testing, V. Garousi, A. Rainer, P. Lauvås & A. Arcuri conducted a systematic mapping of literature using 204 papers helping researchers to identify the research gaps, new trends. This survey includes papers from 1992 to 2019 hence, leaving no room for the omission of any vital information [26]. Also, S. K. Izzat & N. N. Saleem provides a recent review of software testing tools and techniques concluding white-box techniques to be 78% reliable and have

73% defect detection capabilities. They also mention that AI-powered tools will be able to predict more troublesome areas and enhance the overall quality of the testing process [27].

A prior knowledge of software defects and performing measures to initially eradicate issues that may lead to any defect is essential. There are some software defect prediction techniques discussed by M. K. Thota, Francis H Shajin & P. Rajesh, which highlights the need of pre-processing of data, cleaning of data, data normalization and data discretization [28]. The testing is the process of evaluating the software quality and improving it. Choosing the correct technique and tools as per need of the hour is itself an art. The tester must be aware of their ultimate goal before performing testing [29].

2.4 *Current trends and debates*

The increase in complexity of software with time demands the need of new tools and technology which can enhance the software's performance. T.H. Kazimov, T.A. Bayramova, N.J. Malikova highlights the adoption of AI testing tools such as eggplant, functionize, testim and more. They mentioned more use of AI tools in the coming future due to more accurate results [30]. Covid-19 had a drastic impact on technology. K. S. Ong, Y. T. Prasetyo, Ralph Andre C. Roque, Jan Gabriel I. Garbo &others talk about the impact of COVID-19 pandemic on the use of software testing tools by career shifters in the Philippines. Philippines had number of job seekers and unemployment, many had to shift their careers due to covid-19. They talked about approaches like TTF-TAM, which are two theoretical models—Task Technology Fit (TTF) and Technology Acceptance Model (TAM) used to evaluate how well the tool meets the users' needs (TTF)and how the user realizes the usefulness and ease of using the tool (TAM) [31].

More recent literature in this field includes work by T. Fulcini, R. Coppola, L. Ardito, and M. Torchiano regarding gamification which can be integrated with software testing tools to enhance the overall testing process. Gamification is typically a process to add game elements like scores, badges, quests etc. The paper provides detailed information of gamified software testing. The main aim is to make the process of testing more engaging and enjoyable. Hence, increasing the productivity. The concept of gamified testing is new and has scope of more frequent use in the field of testing tools [32].

Blockchain being the technology of today and tomorrow, testing becomes crucial for it. A recent paper by A. Elakas, H.Sözer, I. Safak, K. Kalkan talks about the systematic mapping of software testing for blockchain. The study involves reviewing 17 primary papers and highlighting major challenges such as adaptability issues, performance and scalability concerns, and the need for larger datasets for empirical evaluation. The paper mentions some methods of techniques such as Model-Based Testing, Fuzz Testing and Mutation Testing and recommended future researcher to explore integrating multiple testing techniques to address limitations [33].

3 ANALYSIS AND RESULTS

3.1 *Features of tools*

For thorough analysis, the tools were assessed on the basis of different features, each feature holding a relevance in the realm of testing. Here each feature can be labeled as the criteria of excellence for the tools. The features of testing tools used for evaluation includes launched year, source type(open or closed), online or offline availability, level of programming skills required.

3.2 *Launch year analysis*

Based on the gathered information, analysis of tools concludes that the most recently launched testing tool is Apache jMeter (2019), whereas LoadRunner (1990) stands as one of the earliest tools, both primarily used as performance testing tools (ref to Figure 1).

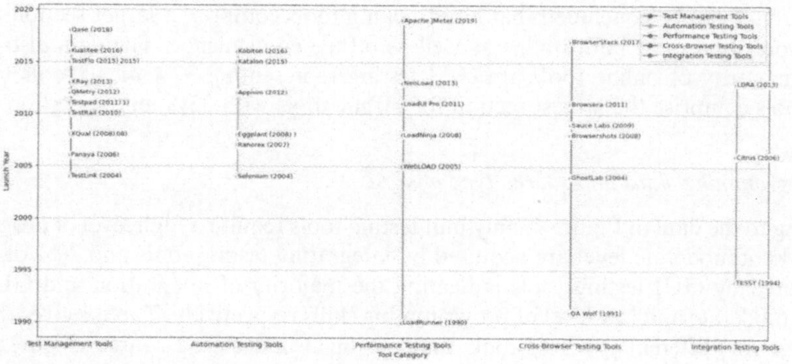

Figure 1. Launch year of various categories of tools.

3.3 *Offline and online availability*

The following graphs shows the distribution of different categories of tools based on the features such as the percentage of their online and offline availability, level of programming skills required for the usage, the source type such as open source or closed source. Figures 2 and 3 shows the online or offline tools percentage. It can be inferred that only test management and automated tools offer both online as well as offline convenience with 36% test management tools and 64% are automated. Also, the data indicates that the majority of online tools are cross browser representing 36% of all tools while integration testing tools comprise the largest section for offline tools with 46% representation of total.

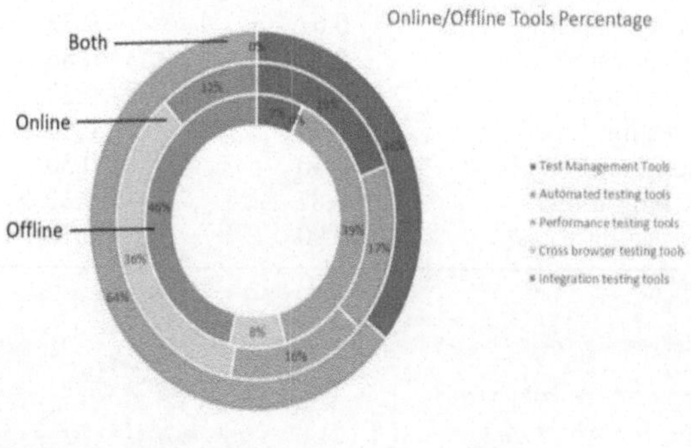

Figure 2. Online/ offline percentage of software tools for testing.

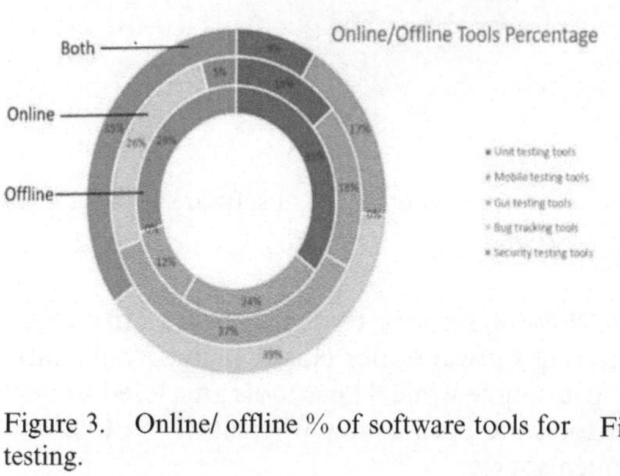

Figure 3. Online/ offline % of software tools for testing.

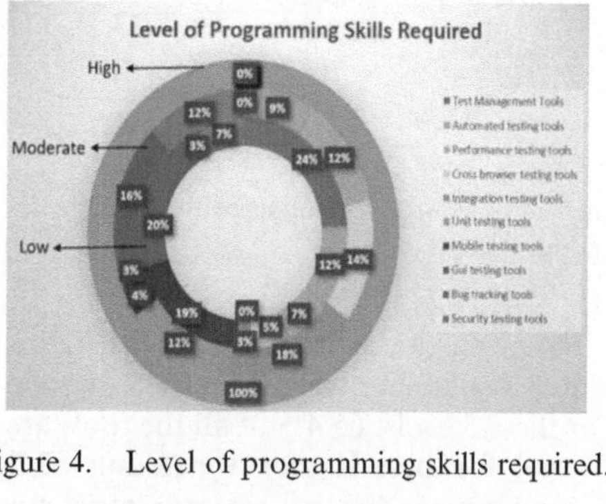

Figure 4. Level of programming skills required.

Similarly, it can be concluded that bug tracking tools comprise a larger section with 39% of total tools that offer both online as well as offline convenience. The data also indicates that the majority of online tools are GUI testing representing 37% of all tools while unit testing tools comprise the largest section for offline tools with 35% representation of total.

3.4 *Programming skills and source type analysis*

According to the data in Figure 4, only unit testing tools require a high level of programming skills. 18% of moderate level are acquired by integrating testing tools and 20% of low level are acquired by GUI testing tools indicating the majority of integration and GUI testing tools for moderate and low level of programming skills respectively. Table 1 gives the data in percentage for different types of tools that are open and closed source. Figures 5 and 6 represent the % of given tools categories which are open source and which are closed source. Open-source tools are the software that are readily available to the general public while closed source tools are only available to the restricted people. The graphs represent each testing tool as a layer of doughnut graph. In Figure 5, 86% of integration testing software offer closed source availability whereas 67% of cross browsing testing software offer open-source availability. For these 5 tools, 39.6% of all the tools are open source while 60.4% tools are closed source.

Table 1. Open and closed source tools.

Tools	Open Source	Closed Source
Test Management Tools	0.44	0.56
Automated testing tools	0.43	0.57
Performance testing tools	0.30	0.70
Cross browser testing tools	0.67	0.33
Integration testing tools	0.14	0.86
Unit testing tools	0.80	0.20
Mobile testing tools	0.40	0.60
Gui testing tools	0.50	0.50
Bug tracking tools	0.67	0.33
Security testing tools	0.90	0.10

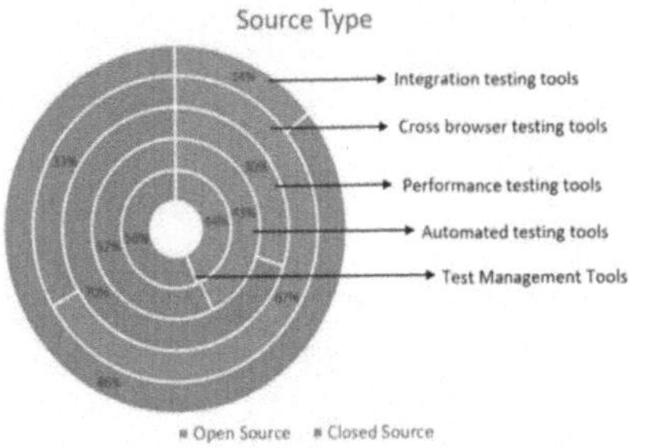

Figure 5. Source type of software tools for testing.

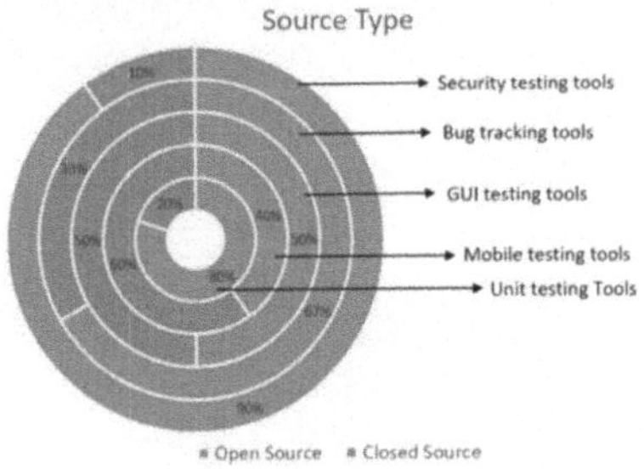

Figure 6. Source type of software tools for testing.

Based on the data presented in Figure 6, 90% of security testing software offer open-source availability whereas 60% of mobile testing software offer closed source availability. For these 5 tools, 65.4% of all the tools are open source while 34.6% tools are closed source. Overall, for these 10 tools for testing, 52.5% tools are open source whereas 47.5% tools are closed source indicating majority tools are open source.

3.5 *Cross browser and cross platform analysis*

Most of the tools support cross platform features. Software like Protractor (for AngularJS applications) are not cross-platform as shown in Figure 7. The majority of tools i.e. 97.83% tools support cross platform features. Figure 8 illustrates the percentage of tools supporting cross browser functionality. 83.7% of testing tools support cross-browsing and tools like AutoIt, Wapiti, Netsparker etc. do not support cross browsing which comprise 16.3% of all tools.

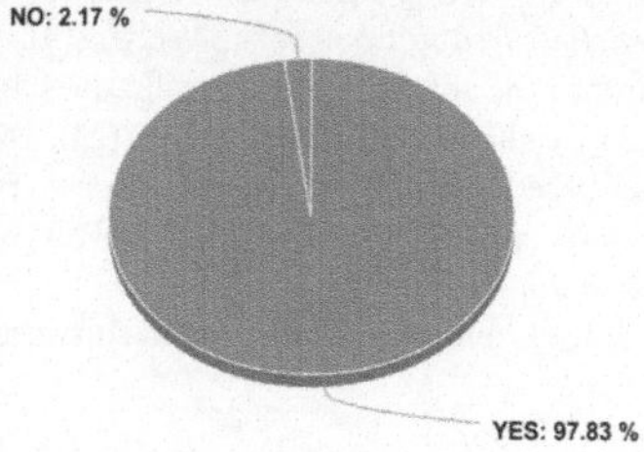

Figure 7. Testing tools supporting cross platform.

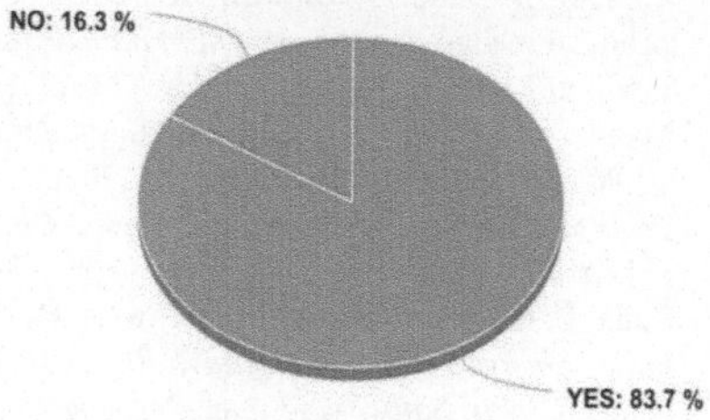

Figure 8. Testing tools supporting cross browser.

4 CONCLUSION

Quality is the main center of any software engineering or testing project. One cannot be sure of the level of quality in a software without any computation. This technological era mandates excellence in testing software products promising uninterrupted services. Hereby, our study offers assistance for selecting tools as per the requirement of the user. This paper relates various types of software testing tools under categories such as test management testing tool, automation, integration, GUI, mobile and security tools for testing, bug tracking tools, unit testing, cross browser and performance tools for testing. A total of 92 testing tools were taken for analysis of features such as launched year, source (open or closed), cross platform or cross browser, languages and level of programming skills required, online or offline availability, ease of learning, input given to the tool and output received, testing phase, cloud or single system, function of tool, sample dataset, updates list and number of research papers available for each tool.

Later, the analysis was deduced by representing the data in the form of graphs. It resulted in the information that most of the tools support cross-platform features and tools like Protractor (for AngularJS applications) are not cross-platform. 83.7% of testing tools support cross-browsing and tools like AutoIt, Wapiti, Netsparker etc. do not support cross browsing. 44% of test management tools are open source while 56% are closed source.36% of online tools are cross browser testing tools and 46% of offline tools are integration testing tools. Also, 30% of performance testing tools are open source while 70% are closed source. Overall, the final percentage of closed source tools is approximately 47.5%, and the final percentage of open-source tools is approximately 52.5%. Therefore, open-source tools are most widely used. Some of the frequently used tools are Apache JMeter, Jira, Redmine, Bugzilla. For future work, other tools can also be used for analyzes, providing extended information.

REFERENCES

[1] "geeks for geeks," (19 3 2024). [Online]. Available: https://www.geeksforgeeks.org/software- testing-basics/.

[2] Okezie, F., Odun-Ayo, I. and Bogle, S. (2019, December). A critical analysis of software testing tools. *In Journal of Physics: Conference Series* (Vol. 1378, No. 4, p. 042030). IOP Publishing.

[3] Ahamed, S. S. (2010). Studying the feasibility and importance of software testing: An Analysis. *arXiv preprint* arXiv:1001.4193.

[4] "geeks for geeks," (13 3 2024). [Online]. Available: https://www.geeksforgeeks.org/software- testing-tools/.
[5] Mailewa, A., Herath, J. and Herath, S. (2015, April). A survey of effective and efficient software testing. *In The Midwest Instruction and Computing Symposium.(MICS)*, Grand Forks, ND.
[6] Adeyemo, H. B. (2021). Comparative analysis of modeling and software testing tools: a survey. *Saudi J Eng Technol*, 6(10), 371–377.
[7] Sneha, K. and Malle, G. M. (2017, August). Research on software testing techniques and software automation testing tools. In 2017 *international Conference on Energy, Communication, Data Analytics and Soft Computing (ICECDS)* (pp. 77–81). IEEE.
[8] Kannan, M., and Lokeshwari, K. (2017). Comparison of software testing tools with respect to tools and technical related parameters. *International Journal of Advanced Research in Computer Science*, 8(9).
[9] Shuaibu, I., Musa, M. and Ibrahim, M. (2019). Investigation onto the software testing techniques and tools: An evaluation and comparative analysis. *International Journal of Computer Applications*, 177(23), 24–30.
[10] Li, W., Le Gall, F. and Spaseski, N. (2018). A survey on model-based testing tools for test case generation. In *Tools and Methods of Program Analysis: 4th International Conference*, TMPA 2017, Moscow, Russia, March 3–4, 2017, Revised Selected Papers 4 (pp. 77–89). Springer International Publishing.
[11] Rana, I., Goswami, P. and Maheshwari, H. (2019). A review of tools and techniques used in software testing. *Int. J. Emerg. Technol. Innovative Res*, 6(4), 262–266.
[12] Pan, J. (1999). *Software testing. Dependable Embedded Systems*, 5(2006), 1.
[13] Uspenskiy, S. (2010). A survey and classification of software testing tools.
[14] Shaukat, K., Shaukat, U., Feroz, F., Kayani, S. and Akbar, A. (2015). Taxonomy of automated software testing tools. *International Journal of Computer science and Innovation*, 1, 7–18.
[15] Dhiman, S. and Sharma, P. (2016). Performance testing: a comparative study and analysis of web service testing tools. *International Journal of Computer Science and Mobile Computing*, 5(6), 507–512.
[16] Okezie, F., Odun-Ayo, I. and Bogle, S. (2019, December). A critical analysis of software testing tools. *In Journal of Physics: Conference Series* (Vol. 1378, No. 4, p. 042030). IOP Publishing.
[17] Gamido, H. V. and Gamido, M. V. (2019). Comparative review of the features of automated software testing tools. *International Journal of Electrical and Computer Engineering*, 9(5), 4473.
[18] Srivastava, N., Kumar, U. and Singh, P. (2021). Software and performance testing tools. *Journal of Informatics Electrical and Electronics Engineering (JIEEE)*, 2(1), 1–12.
[19] Emami, S. A., Sim, J. C. L., and Sim, K. Y. (2012, January). A survey on open source software testing tools: a preliminary study in 2011. *In Fourth International Conference on Machine Vision (ICMV 2011): Computer Vision and Image Analysis; Pattern Recognition and Basic Technologies* (Vol. 8350, pp. 647–652). SPIE.
[20] Alenzi, A., Alhumud, W., Bryce, R. and Alshammari, N. (2022). A survey of software testing tools in the web development domain. *Journal of Computing Sciences in Colleges*, 38(2), 63–73.
[21] Durieux, T., Ferreira, J. F., Abreu, R. and Cruz, P. (2020, June). Empirical review of automated analysis tools on 47,587 ethereum smart contracts. *In Proceedings of the ACM/IEEE 42nd International Conference on Software Engineering* (pp. 530–541).
[22] Martínez-Fernández, S., Bogner, J., Franch, X., Oriol, M., Siebert, J., Trendowicz, A., and Wagner, S. (2022). Software engineering for AI-based systems: a survey. *ACM Transactions on Software Engineering and Methodology (TOSEM)*, 31(2), 1–59.
[23] Pham, K., Nguyen, V. and Nguyen, T. (2022, October). Application of Natural Language Processing Towards Autonomous Software Testing. *In Proceedings of the 37th IEEE/ACM International Conference on Automated Software Engineering* (pp. 1–4).
[24] Stradowski, S. and Madeyski, L. (2022). Challenges in improving the quality and minimising the cost of software testing of the 5G system at Nokia: A Survey. *Inf. Softw. Technol.*
[25] Yang, Q., Li, J. J., and Weiss, D. (2006, May). A survey of coverage based testing tools. *In Proceedings of the 2006 International Workshop on Automation of Software Test* (pp. 99–103).
[26] Garousi, V., Rainer, A., Lauvås Jr, P. and Arcuri, A. (2020). Software-testing education: A systematic literature mapping. *Journal of Systems and Software*, 165, 110570.
[27] Izzat, S. and Saleem, N. N. (2023). Software testing techniques and tools: A review. *Journal of Education and Science*, 32(2), 30–44.
[28] Thota, M. K., Shajin, F. H. and Rajesh, P. (2020). Survey on software defect prediction techniques. *International Journal of Applied Science and Engineering*, 17(4), 331–344.
[29] Yadu, P. and Sharma, V. (2021). A review on software testing tools and techniques. *International Journal*, 9(7).
[30] T.H. Kazimov, T.A. Bayramova and N.J. Malikova. (2021). *Research of Intelligent Methods of Software Testing.*
[31] Ong, A.K.S.; Prasetyo, Y.T.; Roque, R.A.C.; Garbo, J.G.I.; Robas, K.P.E.; Persada, S.F. and Nadlifatin, R. (2022). *Determining the Factors Affecting a Career Shifter's Use of Software Testing Tools amidst the COVID-19 Crisis in the Philippines.*
[32] Fulcini T., Coppola R., Ardito L., and Torchiano M. (2023). A Review on Tools, *Mechanics, Benefits, and Challenges of Gamified Software Testing.*
[33] Elakaş, A., Sözer, H., Şafak, I. *et al.* (2024). A systematic mapping on software testing for blockchains.

Data Science & Exploration in Artificial Intelligence – Gururaj et al. (Eds)
© 2025 the Author(s), ISBN: 978-1-032-96390-7

Enhancing the sentiment analysis: Effectively recognizing positive and negative tweets

Srushti Jadhav, Mukta Wagh, Samruddhi Dasharath, Chandrashekhar H. Patil & Meenal K. Jabde
Dr. Vishwanath Karad MIT World Peace University, Pune, India

ABSTRACT: This research examines Twitter sentiment analysis, a discipline that has become extremely popular due to its many applications and the large amount of data generated on social media platforms. The overview discusses the sophisticated techniques, challenges, and real-world uses of sentiment analysis. It looks at several strategies and tactics, highlighting the benefits and drawbacks of each. Furthermore, ethical issues are explored, emphasizing the significance of a careful and open approach to sentiment analysis. Through a thorough survey of the existing research study, the paper explores potential applications of Twitter sentiment analysis to comprehend public sentiment and its consequences in fields including politics, marketing, and social sciences. The study makes a sizable contribution to the domain of sentiment analysis by putting out an improved framework for distinguishing between positive and negative tweets. This approach addresses the issues of the brief, informal content that is frequently seen on Twitter by introducing cutting-edge sentiment analysis techniques designed especially for Twitter data. In contrast to other approaches, ours concentrates on identifying minute details and context-specific characteristics that are peculiar to Twitter sentiment. A case study illustrating the sentiment analysis's application to real-world data demonstrates how the technology may be used to analyse public opinion, identify patterns, and provide informative data. The analysis underscores the significance of addressing ethical quandaries and ensuring the responsible and transparent application of sentiment analysis techniques.

Keywords: Hybrid, Positive sentiment, Negative sentiment, Twitter, Emotions, Web data, text mining

1 INTRODUCTION

A popular topic of research in data science and machine learning is sentiment analysis. Sentiment analysis is used to establish and categorize a text's sentiment.

This can be as simplistic as classifying a body of text as "positive" or "negative", or even identifying the specific expression of each and every utterance. However, the analysis of Twitter data introduces its own series of questions due to the noisiness and large-scale characteristics of the data, as well as the identifiable attributes of Twitter. The high volume of tweets being produced by the second means that simple "bag-of-words" analyses are impractical - preprocessing and filtering of the data are key. Furthermore, the brevity of tweets, as well as the informal slang and language used by many users, poses challenges for natural language processing in general and for the classification of the sentiment of tweets specifically. The size limitation of tweets made users develop many abbreviations, which do not have any semantic meanings, like "lol" - "laughing out loud", "gr8" - "great" and what is known as "hashtag language" for example "going to concert #excited". For example,

DOI: 10.1201/9781003589273-10

showed how tweet data can be used to predict the sentiment of a given phrase in a given location by analysing how the frequency of tweets containing that phrase varies between positive and negative. Similarly, focused on the public response on Twitter to different real-world events and showed how the respective volumes of positive and negative tweets could be used to track the events themselves. Although the studies had different aims and methodologies.

2 PREVIOUS STUDY

Sentiment analysis is the systematic study of by what method natural language expressions of emotions and views connect to people's attitudes and sentiments about a certain situation or subject. Twitter's subjective data has made it a prime candidate for sentiment analysis as a key data source. Sentiment analysis has advanced beyond binary classifications of positive and negative to include a wide range of actions and feelings in different networks and subjects. Sentiment analysis has been studied deeply with a variety of methodologies.

The research presents a machine learning technique for automatically categorizing the sentiment of Twitter posts as positive or negative. Using procedures alike SVM and Naive Bayes, they attain over 80% precision using emoticons found in tweets as noisy labels for training records. Customers investigating product sentiment and businesses keeping an eye on brand perception on Twitter can both benefit from this strategy [1]. Convolutional Neural Networks (CNNs) are investigated in this paper for the purpose of classifying sentences, an important problem in natural language processing. Although CNNs have demonstrated impressive performance, it was uncertain how sensitive they were to different model settings, such as filter widths and regularization parameters. Because one-layer CNNs are so straightforward and efficient, the researchers focused their sensitivity analysis on these. Differentiating important design decisions from less significant ones was the goal. The paper helps practitioners optimize CNNs for sentence categorization in real-world circumstances by providing useful advice based on empirical data [2]. The task for Sentiment Analysis in Twitter for SemEval-2017, which is an extension of SemEval-2016 Task 4, is covered in this document. Sentiment analysis of tweets, sentiment distribution across tweets, and sentiment assessment of particular topics on two- and five-point scales are all part of the task. The addition of Arabic as a new language and the provision of data from Twitter user profiles are among the changes from 2016. With 48 teams taking part in 2017, the job is still rather popular [3]. In a similar vein, go *et al.* (2009) classified tweets into positive and negative attitudes by approaching sentiment analysis on Twitter as a two-class classification problem. Their research helped to clarify the sentiment patterns found in tweets [4]. An overview of sentiment analysis and natural language processing (NLP), two topics are gaining popularity for concluding text data, is given in the abstract. Sentiment analysis looks at the emotional tone of text, whereas NLP handle with AI's communication with human. It covers sentiment analysis strategies incorporating lexicon-based methods and ML as well as NLP techniques like tokenization. Applications include customer feedback and social media analysis, along with talks about difficulties like ambiguity and moral dilemmas. Emotion detection and transfer learning are future developments [5]. The study conducted by Jansen *et al.* (2009) investigated the impact of sentiment conveyed in tweets on Twitter as an electronic form of word-of-mouth communication [6].Ghiassi, Skinner, and Zimbra (2013) introduced a mix approach that syndicates dynamic artificial neural networks and n-gram analysis for Twitter brand sentiment research [7]. Methods for sentiment analysis of Twitter information were presented by Agarwal, Xie, Vovsha, and Rambow (2011), which advanced our knowledge of sentiment categorization algorithms [8]. The impact of informal language and length on sentiment analysis in tweets was examined by Kouloumpis, Wilson, and Moore (2011), who emphasized the necessity for specific algorithms for short text sentiment categorization [9]. A thorough overview of sentiment analysis strategies was provided by Pak, Paroubek, and

Steinberger (2010), who covered a diversity of ways for sentiment analysis on Twitter, including lexicon-based methods, ML techniques, and their combinations [10]. In order to show the value of integrating several data sources for sentiment classification, Khoo, Lee, and Chieng (2019) presented a hybrid model for Twitter sentiment analysis that uses both emoticons and word vector characteristics [11].

These studies highlight the various ways that sentiment analysis on Twitter data may be applied, from straightforward categorization techniques to intricate domain-specific systems with interactive user interfaces and machine learning algorithms[12].

3 METHODOLOGY

The process of accurately determining the sentiment represented in tweets requires a number of critical elements in the approach for performing Twitter sentiment study. The methodology architecture is given below:

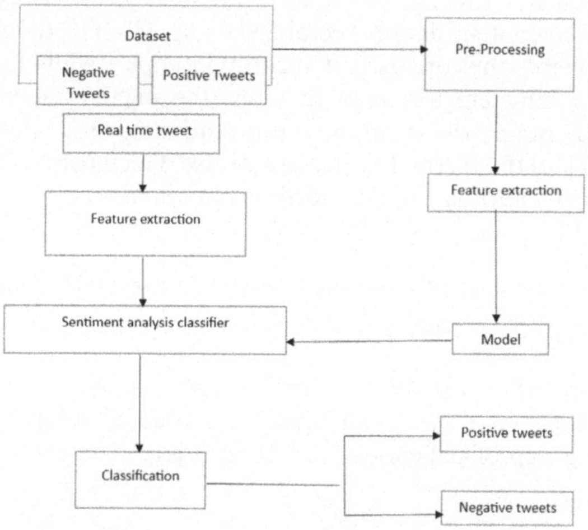

Figure 1. Sentiment analysis architecture.

3.1 *Data gathering*

Establish Goals: Establish the purpose of your sentiment analysis (e.g., consumer opinion on an event, product feedback, brand perception, etc.). Data Sources: Kaggle https://www.kaggle.com/datasets/abhi8923shriv/sentiment-analysis-dataset. Data Description: The dataset supports the research goals of comprehending public sentiment and subject trends on Twitter by offering a thorough basis for sentiment analysis and tweet categorization. The dataset consists of tweets that were gathered from Twitter and includes text content, date, query, user, tweet ID, sentiment polarity (0 = Negative, 2 = Neutral, and 4 = Positive), and other fields. After removing URLs, mentions, hashtags, punctuation, and digits, the tweets were lemmatized. Five categories are used to group tweets: General, Sports, Entertainment, Politics, and Technology. TF-IDF was used for feature extraction in order to make it easier to classify tweets into these categories.

3.2 *Data preprocessing*

When arranging crude data for analysis or exhibiting, data preprocessing is an essential step. The following data pretreatment processes are included in the code provided:

Text Normalization: To normalize text, change all tweets to lowercase, add a space in dwelling of two or more dots when they occur, eliminate quote marks and trailing spaces beginning at the end of tweets, then add a single space to remove extra spaces. In preparation for additional analysis, this preparation step ensures consistency and cleanliness in the text data. Replace the target variables from 0 to 1. Handling Unique Elements: Substitute frequently used keywords or marks for emoticons, URLs, and user mentions also remove retweets. Stop word Removal: Exclude terms like 'the' 'is' and 'and' that don't contribute anything to sentiment analysis. For example, "They often play football." Words that were previously stop words become "often play football". Stemming: Shrinking words to their basic appearance (e.g., "running" to "run") will help you catch morphological changes. Lemmatization: Lemmatization makes sure that words containing inflexions are correctly reduced to their basic forms, maintaining linguistic consistency and accuracy.

3.3 *Feature extraction*

Vectorization: To transform the preprocessed text into numerical vectors, use techniques like TF-IDF (Term Frequency-Inverse Document Frequency) or word embeddings (e.g., Word2Vec, Glove). The content data is vectorized using TF-IDF to create numerical characteristics. It targets words that are used frequently in papers while taking word frequency into account. With parameters like max_features (the maximum number of features to consider) and ngram_range (to examine both unigrams and bigrams), the TF-IDE Vectorizer is used. TF-IDF (Term Frequency-Inverse Document Frequency): TF: Total number of incidences of a term in a text divided by total number of words in the document is the term's or word's TF.

$$TF = \frac{number\ of\ times\ the\ term\ appears\ in\ the\ document}{total\ no.\ of\ terms\ in\ the\ document} \tag{1}$$

IDF: An expression's IDF displays the term's frequency of occurrence in the corpus of documents. Words like "a," "the," and "and" are used in all publications, but words exclusive to a minor group of papers—such technical jargon terms—are assigned a higher relevance value

$$IDF = \log\left(\frac{number\ of\ documents\ in\ corpus}{total\ number\ of\ documents\ contains\ terms}\right) \tag{2}$$

$$TF - IDF = TF * IDF \tag{3}$$

3.4 *Sentiment classification*

Model Selection: - Select a machine learning or deep learning model for sentiment classification. Training and Testing: - Divide the data into testing and training sets (such as an 80/20 or 70/30 ratio) in order to assess the model. Utilizing methods like cross-validation, train the sentiment analysis method on the training data and adjust hyperparameters. Model Evaluation: - Assess the model's effectiveness on the testing data by utilizing measures like accuracy, F1-score, recall and precision. Verify the methodology's ability to accurately classify tweets' attitudes.

4 RESULTS AND DISCUSSION

4.1 *Logistic regression*

Twitter sentiment is predicted using a linear classification algorithm based on input feature. It uses a logistic function to determine the likelihood that a tweet will fall into two sentiment classes: positive, negative. Because it's easy to understand and straightforward, logistic regression is frequently used in sentiment analysis on Twitter.

Here the accuracy is 87%.

	precision	recall	f1-score	support
0	0.89	0.95	0.92	40023
1	0.79	0.62	0.69	12406
accuracy			0.87	52429
macro avg	0.84	0.78	0.80	52429
weighted avg	0.86	0.87	0.86	52429

Figure 2. Accuracy of logistic regression model.

4.2 *Multinomial naive bayes*

An algorithm for probabilistic classification that is frequently employed in text classification applications, such as sentiment analysis on Twitter data. By simulating the frequency of terms (words or characteristics) in a document and assuming a multinomial distribution, it expands upon the Naive Bayes technique. The likelihood that a document belongs to a specific class is determined via Multinomial Naive Bayes using the document's term frequencies. Here the accuracy is 84%.

	precision	recall	f1-score	support
0	0.89	0.90	0.90	40023
1	0.67	0.65	0.66	12406
accuracy			0.84	52429
macro avg	0.78	0.78	0.78	52429
weighted avg	0.84	0.84	0.84	52429

4.3 *Support vector machine*

Twitter sentiment categorization using a potent supervised learning system. SVM creates high-dimensional hyperplanes to divide tweets with various sentiment labels. Sentiment analysis activities on Twitter data can benefit from its efficacy in managing high-dimensional feature fields and non-linear decision limits. Because SVMs can capture intricate correlations between tweet attributes and sentiments, they are frequently utilized in Twitter sentiment analysis. Here the accuracy is 86%.

	precision	recall	f1-score	support
0	0.89	0.93	0.91	40023
1	0.74	0.64	0.68	12406
accuracy			0.86	52429
macro avg	0.82	0.78	0.80	52429
weighted avg	0.86	0.86	0.86	52429

Figure 3. Accuracy of SVM classifier.

Following our evaluation of each model, we may infer the following: Accuracy: SVM raises the accuracy of the model more than Logistic Regression, which raises the accuracy of Bernoulli Naive Bayes by the same amount. F1-score: Class 0: The class 0

F1 score is Bernoulli Naive Bayes (accuracy = 0.90) < SVM (accuracy = 0.91) < Logistic Regression (accuracy = 0.92). (b) In class 1, the accuracy of the logistic regression is 0.69, higher than that of the SVM (0.68), and the Bernoulli Naive Bayes (0.66). Consequently, we ascertain that the logistic regression model is the best fit for the aforementioned dataset.

5 CONCLUSION

The goal of the study work is to categorize Twitter data into positive and negative attitudes, using sentiment features instead of more conventional methods to achieve higher accuracy. Businesses, organizations, and entrepreneurs should take note of these aspects in order to comprehend public opinion. Future research will incorporate complicated emotions like sarcasm and broaden the scope to regional languages, utilizing a hybrid classifier to increase accuracy. In terms of levels, techniques, methodologies, features extraction, and applications, sentiment analysis is comprehensively covered in this study. It offers information on tweet extraction, structure, and significance as well as insights into Twitter sentiment analysis. By comparing different approaches and strategies, one can evaluate the effectiveness of each one by looking at outcomes and accuracy. By distinguishing positive and negative emotion in tweets with more accuracy than previous approaches, our research gives enhanced outcomes. For more accurate sentiment classification, we use cutting-edge machine learning algorithms and feature engineering strategies created especially for Twitter data. The efficacy of our suggested framework is illustrated by comparison analyses with more established sentiment analysis techniques, which indicate this improvement. Overall, the study provides a thorough grasp of Twitter sentiment analysis and establishes a strong framework for additional study in the area. Our study opens up new possibilities for sentiment analysis, especially when applied to social media sites such as Twitter. We explore the difficulties in sentiment analysis on Twitter, including how to handle emoticons, hashtags, and sentiment-conveying slang terms. We close a gap in the literature by creating methods for the efficient analysis of these components, as many sentiment analysis approaches find it difficult to work with such informal and volatile data sources.

REFERENCES

[1] Huang, L., Go, A. and Bhayani, R. (2009). Classifying sentiment on Twitter via remote supervision. *Stanford Project Report* CS224N. https://cs.stanford.edu/people/alecmgo/papers/TwitterDistantSupervision09.pdf is accessible online.

[2] Wallace, B. C. and Zhang, Y. (2015). Convolutional neural networks: A sensitivity analysis and practitioners' guide for sentence classification. *The preprint arXiv* is arXiv:1510.03820. Online access is possible at https://arxiv.org/abs/1510.03820.

[3] Mohammad, S. M., Ritter, A., Kiritchenko, S., Nakov, P., Rosenthal and Stoyanov, V. (2017). Task 4 of SemEval 2017: Sentiment analysis on Twitter. *In the Semantic Evaluation 11th International Workshop (SemEval-2017)*, Vancouver, Canada, proceedings. Accessed via the internet at https://www.aclweb.org/anthology/S17-2088.pdf

[4] Schütze, H., Raghavan, P. and Manning, C. D. (2008). Information retrieval overview. Cambridge University Press. Sentiment analysis is covered in Chapter 15. The IR-book is accessible online at https://nlp.stanford.edu/IR-book/pdf/irbookonlinereading.pdf

[5] Natural Language Processing and Sentiment Analysis March 2024 Kaledio Potter https://www.researchgate.net/publication/379025140_Natural_Language_Processing_and_Sentiment_Analysis

[6] Darshan K, Jerusha Samuel, Manjunatha C Swamy, Prashant Koparde, N Shivashankar NLP - Powered Sentiment Analysis on the Twitter https://www.researchgate.net/publication/377454521_NLP_- Powered_Sentiment_Analysis_on_the_Twitter

[7] Yili Wang ,Jiaxuan Guo, Chengsheng Yuan, Baozhu Li. (November 2022). Sentiment Analysis of Twitter Data https://www.researchgate.net/publication/365618365_Sentiment_Analysis_of_Twitter_Data

[8] Zhang, M., Sobel, K., Jansen, B. J. and Chowdury, A. (2009). Tweet power: Tweets serving as digital word-of-mouth. The American Society for Information Science and Technology Journal, Volume 60, Issue 11, pages 2169–2188. https://onlinelibrary.wiley.com/doi/abs/10.1002/asi.21149

[9] Ghiassi, M., Zimbra, D., and Skinner, J. (2013). Twitter brand sentiment analysis: a hybrid approach that combines dynamic artificial neural networks and n-gram analysis. 40(16) of *Expert Systems with Applications*, 6266–6282. https://www.sciencedirect.com/science/article/abs/pii/S0957417413003368

[10] Agarwal, A., Xie, B., Vovsha, I., and Rambow, O. (2013). Sentiment analysis of Twitter data. *Journal of the American Society for Information Science and Technology*, 64(12), 2472–2483. https://onlinelibrary.wiley.com/doi/abs/10.1002/asi.22856

[11] Thelwall, M., Buckley, K., Paltoglou, G., Cai, D., and Kappas, A. (2010). Sentiment strength detection in short informal text. *Journal of the American Society for Information Science and Technology*, 61(12), 2544–2558. https://onlinelibrary.wiley.com/doi/abs/10.1002/asi.21416

[12] Houston, J., Bhatti, N., Johnson, M., and A. Jurek (2017). Twitter sentiment analysis: Deep learning is used to extract sentiment from integrated services like Twitter. *The International Conference Proceeding on Cloud Computing and Internet of Things* https://www.researchgate.net/publication/322176641_Twitter_Sentiment_Analysis_Capturing_Sentiment_from_Integrated_Services_like_Twitter_using_Deep_Learning

Data Science & Exploration in Artificial Intelligence – Gururaj et al. (Eds)
© 2025 the Author(s), ISBN: 978-1-032-96390-7

Enhancing stability and efficiency in deep reinforcement learning agents

Raghavendra M. Devadas
Manipal Institute of Technology Bengaluru, Manipal Academy of Higher Education, Manipal, India

Vani Hiremani & Ranjeet Vasant Bidwe
Symbiosis Institute of Technology, Pune Campus, Symbiosis International (Deemed University) (SIU), Lavale, Pune, India

J. Praveen Gujjar
Faculty of Management Studies, JAIN (Deemed-to-be University), Bengaluru, India

R. Sapna & Preethi
Department of Information Technology, Manipal Institute of Technology Bengaluru, Manipal Academy of Higher Education, Manipal, India

ABSTRACT: This paper introduces an innovative and thorough study toward increasing the stability and efficiency of DRL agents in the CartPole environment. The realization of the improvement in DRL performance is well noted to be implemented through essential techniques—in particular, target networks, experience replay, and Huber loss. A target network helps combat the moving target problem by creating it to keep Q-value estimates stable via updating parameters at regular intervals. Data efficiency and stabilization of the learning process are achieved by experiencing replay through random sampling of transitions from large replay buffers. Huber loss replaces Squared Error to add robustness to the learning process. But more attention is drawn to a comprehensive approach: observing the agent's overall performance regarding the total reward per episode. Therefore, the study's findings establish their combined roles in enhancing the stability and efficiency of DRL algorithms in the CartPole environment. Through experimentation, it was also observed that the average total reward received by the DQN algorithm was relatively steady at around -5000 for each episode through the training cycle. Through the assessment phase, the average reward gained by the DQN algorithm was around 25 for each episode.

1 INTRODUCTION

The combination of deep neural networks and reinforcement learning enabled the training of intelligent agents and led to the powerful paradigm known as Deep Reinforcement Learning (DRL). Among the most groundbreaking contributions in this area was DQN (Deep Q-Network) by Mnih *et al.*, 2015. This pioneering approach showcased the ability to directly learn successful policies in complex environments from raw pixel inputs. Despite that, DRL still faces challenges, among them training instability and inefficiency, which has driven the continued work on techniques for overcoming these. DRL has displayed brilliant performances in enabling agents to learn and perform complex tasks independently. One of the methods under DRL, Deep Q-Network (DQN), changed the domain of the derived learning from states of sensory input.

DOI: 10.1201/9781003589273-11

This work aims at amalgamating such influential methods motivated by the foundational works of the area. Adding target networks is a crucial step toward stable Q-value estimation during training. There are primarily two advantages of using target networks: target networks address the issue of a moving target, and they provide a much more stable learning target. In the present paper, we employ target networks in the CartPole environment and check their effectiveness in stabilizing and speeding up the learning process. Besides, experience replay is also adopted like in the previous work by Lin. 1992. Experience replay stores past experiences in a memory buffer and randomly selects them, breaking up correlations in the learning process; the efficiency of data use will be improved, and the training process will become more reliable. Building on such fundamental principles, this paper investigates how experience replay may impact performance in the CartPole domain. This paper investigates Huber loss, recommended as an alternative to Mean Squared Error by Van *et al.* 2016, to make learning more robust and hence less sensitive to outliers in temporal difference error. With this, better performance could be achieved for the model in taking generalization and coping with dynamics change tests in the environment with CartPole. In leveraging previous findings and porting methodologies to a new benchmark, we contribute to the general knowledge regarding selecting loss functions in reinforcement learning.

2 LITERATURE REVIEW

Using deep networks without training deep networks, Djurić. 2018, describe how to make predictions. Quantitative quantification of correlations in deep learning using quantum entanglement (QE) measures, and how the connection fits the architecture of the deep network to the correlations in the data, according to Levine. 2017. This was an experiment to assess the impact on Adam using very wide and deep neural networks Jais. *et al.* 2019. It generalizes Hinge Loss and L2 Norm Loss as the Huber loss function, where it is quadratic Hinge Loss if the given error input's points are closer, else linear Hinge Loss for the distance provided. Following are three functional iterative approaches to SV classification problems: SVM-based in the sense of Generalized Huber support vector machine, TWSVM-based in the spirit of twin support vector machine, Regularization-based in the sense of TWSVM by Borah. 2019. Noise signal and time series may contain outliers as well as sudden trend changes in real-world scenarios. To address this challenge, a robust trend filter algorithm based on reliable statistics and sparse learning was proposed by Wen. *et al.* 2019. Zhang *et al.* 2019 proposed Domain-Symmetrical Networks (SymsNets) as the new domain adaptation method. Li *et al.* 2019 proposed Deep Learning Transfer Using Feature Maps with Attention for Convolutional Networks Improving Deep Transfer Learning with Regularization Using Attention Feature Maps. The authors Meng *et al.* 2019 propose a method for learning condition-invariant deep embeddings via adversarial multi-task training using an adversarial speaker verification (ASV) scheme. A network of speaker classifications and condition identifications is optimized to minimize both speaker classification and condition identification losses. Indeed, there have been several recent active learning approaches for deep networks, but they tend to be either task-specific or complicated and computationally inefficient. Active learning with deep networks is proposed to be naively simple and task-agnostic. Some other noted works on deep learning are by Devadas and Hiremani,. 2024; Devadas *et al.* 2024.

3 PRELIMINARIES

Deep Q-Network (DQN) is a class of deep reinforcement learning that makes use of a neural network in predicting the optimal action-value function, Q*(m, n). This estimates the expected reward when action n is taken from state m, following the best possible policy. The

DQN learns to approximate this function through interaction with the environment and how to exploit such an approximation to determine the best action in any state.

Using the target network Q(m, n; θ−), a mini-batch of experiences is sampled from D, and the corresponding target value y is computed for each experience as:

$$y = r + \gamma \max_{n'} Q(m', n'; \theta^-) \quad (1)$$

In this equation, y should be the estimate of Q*(m, n) given the current experience and a target network. The discount γ is one of the most critical factors in calculating future rewards. Q(m, n; θ) is constantly updated by minimizing the average squared error between the expected value y likewise the target value Q(m, n; θ):

$$\mathcal{L}(\theta) = \mathbb{E}_{(m,n,r,m')\sim U(D)}\left[\left(y - Q(m, n; \theta)\right)^2\right] \quad (2)$$

U(D) is referred to as the uniform distribution over D. In other words, it is a distribution that assigns equal probabilities to all elements in D. The gradient of the loss operation concerning θ can then be expressed as shown below by Li *et al.* 2019

$$\nabla_\theta \mathcal{L}(\theta) = \mathbb{E}_{(m,n,r,m')\sim U(D)}\left[\left(r + \gamma \operatorname{mnx}_{n'} Q(m', n'; \theta^-) - Q(m, n; \theta)\right)\nabla_\theta Q(m, n; \theta)\right] \quad (3)$$

Where θ is the vector of network parameters to be optimized. The loss function L(θ) measures the discrepancy between predicted and target Q-values using mean-squared error. The gradient ∇θL(θ) guides parameter adjustments. Transitions (m,n,r,m′) are sampled uniformly from the replay buffer D. The target Q-value y=r+γmaxn′Q(m′,n′;θ−) is calculated using the target network, while Q(m,n;θ) is the predicted Q-value from the current network. The gradient ∇θQ(m,n;θ) shows how Q-values change with respect to parameters.

3.1 *Experience replay*

Experience replay addresses correlated and non-stationary data in reinforcement learning by storing past experiences in a replay buffer and sampling mini-batches during training. The buffer contains tuples (m, n, r, m′, done) representing state-action pairs, rewards, next states, and episode completion flags. This technique approximates the expected value of a function, crucial for estimating the optimal action-value function Q*(m, n) in Q-learning. To arrive at this estimate, the agent employs the Bellman equation:

$$Q^*(m, n) = \mathbb{E}_{s'\sim P(\cdot|m,n)}\left[r + \gamma \max_{n'} Q^*(m', n')\right] \quad (4)$$

In the mentioned context, P(m′|m, n) signifies the changeover probability from the current state m to the next state m′ after taking action n. In addition, r signifies the instant reward likewise γ represents the discount factor. In this method, the Q-function is updated after each step using the rule mentioned in the below equation.

$$Q(m, n) \leftarrow Q(m, n) + \alpha[r + \gamma \operatorname{mnx}_{n'} Q(m', n') - Q(m, n)] \quad (5)$$

where Q(m, n) represents the estimated Q-value for the state-action pair (m, n), r represents immediate reward, α being the learning rate, γ being the discount factor, and m′ indicates next state.

3.2 *Huber loss*

The Huber loss function is a crucial tool for evaluating the accuracy of regression models. It measures the difference between predicted and actual values, balancing between the Mean Squared Error (MSE) and Mean Absolute Error (MAE) loss functions. The Huber loss function can be represented by the formula below:

$$L_\delta\left(b,\widehat{b}\right)=\begin{cases}\frac{1}{2}\left(b-\widehat{b}\right)^2 \text{if} \left|b-\widehat{b}\right|\leq\delta \\ \delta\left(\left|b-\widehat{b}\right|-\frac{1}{2}\delta\right)\text{otherwise}\end{cases} \quad (6)$$

where 'b' is the actual value, '$\widehat{b}$' is the forecast value, and 'δ' controls the MSE and MAE regions. The Huber loss function acts as MSE for errors below δ and as MAE for errors above δ, reducing outlier sensitivity while maintaining smoothness. It is also known as the smooth L1 loss function.

4 RESEARCH METHODOLOGY

The research framework employed in this study is represented in Figure 1.

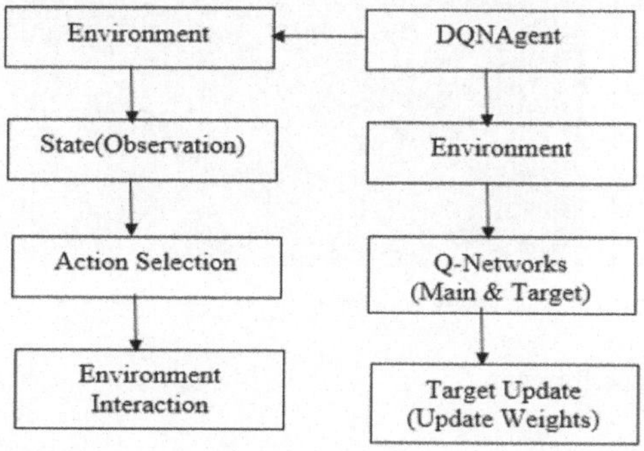

Figure 1. Research methodology.

The agent interacts with the environment in the CartPole environment by choosing actions based on the current state using the Q-network. Experiences are saved in a replay buffer for experience replay, and the Q-values are updated from mini-batches of the logged experience. A periodic update of the target Q-network is performed for smooth training. At any point, the agent interacts with the environment and observes a reward and the next state.

4.1 *The cartPole environment*

OpenAI's Gym library facilitates designing and evaluating reinforcement learning algorithms, with the CartPole-v1 environment being particularly intriguing. The goal in CartPole-v1 is to balance a pole on a moving cart, a common benchmark for control challenges in reinforcement learning. This environment is widely used to evaluate and compare reinforcement learning techniques due to its simplicity and effectiveness in testing balance maintenance. Figure 2. Demonstrates the performance of a reinforcement learning agent on the cartpole environment over 20 time steps.

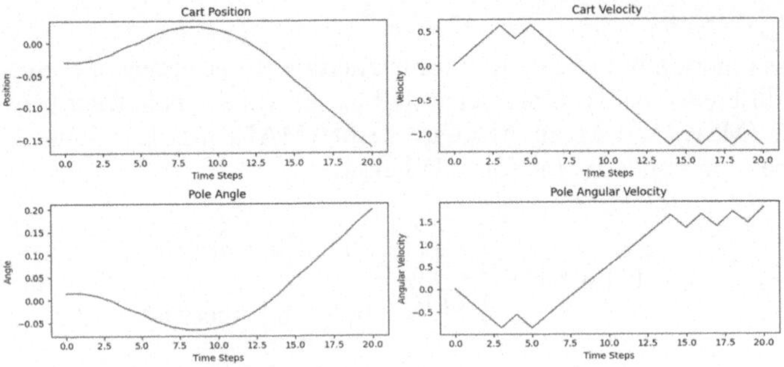

Figure 2. A glimpse of CartPole environment.

5 RESULTS AND DISCUSSION

Figure 3 and Figure 4 show graphs that compare the performance of a DQN during the training and evaluation phases.

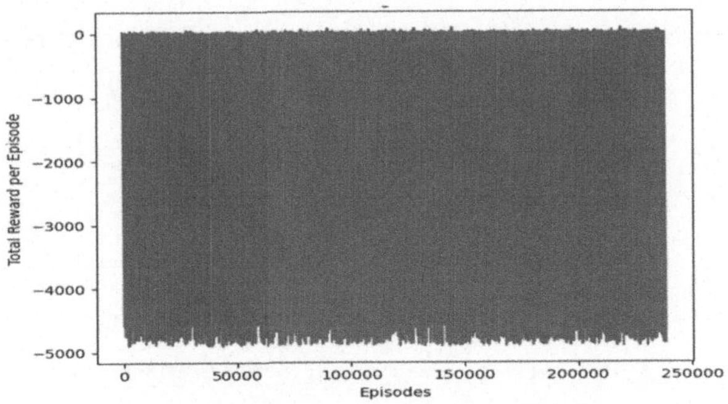

Figure 3. DQN training performance.

Figure 3 depicts the DQN algorithm receiving a consistent total reward of approximately -5000 per episode during training, indicating failure to stabilize the pole on the cart. Total reward per episode did not improve over time, suggesting the agent did not discover an effective policy. Fluctuating negative rewards indicate frequent failure to achieve goals or penalties.

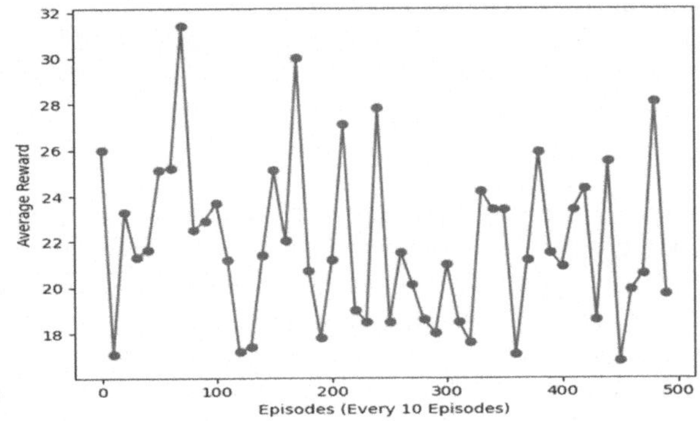

Figure 4. DQN evaluation performance.

Figure 4 shows an average reward of around 25 during evaluation, with fluctuation suggesting unstable performance. While learning from experience may improve performance, the DQN's behavior remains inconsistent, possibly due to the exploration-exploitation trade-off inherent in the epsilon-greedy policy it employs, introducing stochasticity and uncertainty into actions.

6 CONCLUSION

The performance analysis of the DQN algorithm reveals both strengths and weaknesses. Challenges persist during training, with a consistent average total reward around -5000 across episodes, indicating difficulty in balancing the pole on the cart and adapting behavior over time. The lack of improvement in total reward trends suggests suboptimal learning. During evaluation, the average reward stabilizes around 25, indicating some success in pole balancing but with highly variable performance across episodes, highlighting instability. Despite learning from experience, the DQN algorithm's performance suffers from inconsistency, possibly exacerbated by the exploration-exploitation trade-off introduced by the epsilon-greedy policy. To enhance performance, improvements in exploration fine-tuning, hyper-parameter adjustment, and policy iteration are essential.

REFERENCES

Borah, P. and Gupta, D. (2019). Functional iterative approaches for solving support vector classification problems based on generalized Huber loss. *Neural Computing and Applications*, 32, 9245–9265.

Devadas, R., and Hiremani, D. V. (2024). Integrating dropout and kullback-leibler regularization in Bayesian neural networks for improved uncertainty estimation in regression. https://doi.org/10.2139/ssrn.4727556.

Devadas, R. M., Hiremani, V., Bhavya, K. and Rani, N. S. (2024). Stochastic calculus-guided reinforcement learning: A probabilistic framework for optimal decision-making. *MethodsX*, 102790. https://doi.org/10.1016/j.mex.2024.102790.

Djurić, P. M. (2018). Deep gaussian processes: Theory and applications. *2018 14th Symposium on Neural Networks and Applications (NEUREL)*. https://doi.org/10.1109/neurel.2018.8586994

Jais, I. K., Ismail, A. R. and Nisa, S. Q. (2019). Adam optimization algorithm for wide and deep neural network. *Knowledge Engineering and Data Science*, 2(1), 41. https://doi.org/10.17977/um018v2i12019p41-46

Levine, Y., Yakira, D., Cohen, N. and Shashua, A. (2017). Deep learning and quantum entanglement: fundamental connections with implications to network design. *ArXiv*, abs/1704.01552.

Li, X., Xiong, H., Wang, H., Rao, Y., Liu, L. and Huan, J. (2019). DELTA: DEep learning transfer using feature map with attention for convolutional networks. *ArXiv*, abs/1901.09229.

Li, X., Ma, C., Wu, B., He, Z. and Yang, M. (2019). Target-aware deep tracking. *2019 IEEE/CVF Conference on Computer Vision and Pattern Recognition (CVPR)*, 1369–1378.

Lin, L. (1992). Self-improving reactive agents based on reinforcement learning, planning and teaching. *Machine Learning*, 8(3–4), 293–321.

Meng, Z., Zhao, Y., Li, J. and Gong, Y. (2019). Adversarial speaker verification. *ICASSP 2019 - 2019 IEEE International Conference on Acoustics, Speech and Signal Processing (ICASSP)*, 6216–6220.

Mnih, V., Kavukcuoglu, K., Silver, D., *et al.* (2015). Human-level control through deep reinforcement learning. *Nature*, 518(7540), 529–533.

Van Hasselt, H., Guez, A. and Silver, D. (2016). Deep reinforcement learning with double Q-learning. *In Thirtieth AAAI Conference on Artificial Intelligence.*

Wen, Q., Gao, J., Song, X., Sun, L. and Tan, J. (2019). RobustTrend: A Huber loss with a combined first and second order difference regularization for time series trend filtering. *Proceedings of the Twenty-Eighth International Joint Conference on Artificial Intelligence.* https://doi.org/10.24963/ijcai.2019/535

Zhang, Y., Tang, H., Jia, K. and Tan, M. (2019). Domain-symmetric networks for adversarial domain adaptation. *2019 IEEE/CVF Conference on Computer Vision and Pattern Recognition (CVPR)*. https://doi.org/10.1109/cvpr.2019.00517

Data Science & Exploration in Artificial Intelligence – Gururaj et al. (Eds)
© 2025 the Author(s), ISBN: 978-1-032-96390-7

Mutual funds NAV and investment suggestions using deep learning and NLP

V. Sangeetha, A. Parkavi, Ishwariya Pani, K. Yasaswini, B. Nikita & Arpita Bhattacharjee
Department of Computer Science & Engineering, Ramaiah Institute of Technology, Bangalore, Karnataka, India

ABSTRACT: By distributing the risk associated with investments, mutual funds enable smaller investors to gain diversification without needing a sizable initial investment. The performance of a mutual fund is indicated by its Net Asset Value (NAV) per unit. In the field of investment management, accurately assessing the performance of investments is essential. Deep learning provides useful tools for investment prediction as managers and investors work to satisfy the changing needs of stakeholders. Convolutional neural networks (CNNs), multi-layer perceptions (MLPs), long short-term memory networks (LSTMs), and a CNN-LSTM hybrid model are among the deep learning methods performance is examined in this paper. MLP yields high accuracy prediction for the UTI NIFTY Exchange Traded Fund. The model compares the UTI NIFTY Exchange Traded Fund with other types of funds to suggest the best fund for investment. Natural Language Processing (NLP) is used to summarize the content of mutual funds NAV. The result of this model helps the investor to choose the best fund for investment.

1 INTRODUCTION

Mutual funds are a crucial type of investment. In essence, a mutual fund's purpose is to spread risk among smaller savers who lack the resources to do so on their own. The investment management industry is extremely competitive, making it difficult to obtain information regarding an investor's portfolio's performance (Abdel 2022). Meeting the needs of stakeholders in the present and the future would become more and more difficult for investors and investment managers. The use of various deep learning techniques has emerged as an important source of prediction.

Investors and investment managers both care about the success of an investor's investment portfolio. Without it, investors and investment managers would have a more difficult time addressing current and future stakeholder needs in the highly competitive investment management industry. It is possible to analyze a firm's performance over the last ten years in terms of mutual funds NAV, which indicates whether or not we should make an investment in that company right now(Anish 2016). Both investors and investment managers need up-to-date, reliable information on how a portfolio of investments is performing. Without it, meeting the present and future needs of stakeholders would become more difficult for investors and investment managers in the fiercely competitive investment management industry. Aditya Birla Sun Life Banking Exchange-traded fund (ETF), UTI NIFTY Exchange Traded Fund, Union Liquid Fund - Daily Income Distribution cum Capital Withdrawal (IDCW) Option, Kotak FMP Series-Regular Plan-Growth Option, and Quant Multi Asset Fund-IDCW Option-Direct Plan are the five funds that will be used to assess the performance of the proposed model in this paper. Utilizing various deep learning techniques has emerged as a key source of prediction. CNN, MLP, LSTM, and CNN-LSTM are used

 DOI: 10.1201/9781003589273-12

in the proposed model to analyze the performance(Ardhani 2020). NLP provides information on the return and risk of investment portfolios over predefined time periods. Use the extractive text summarization for text summarization. Using the Root Mean Square Error (RMSE), an NLP module is utilized to generate text summarizing the forecast made is advised to the investor to invest in any of the five funds(Boorugu 2020). The RMSE is used to quantify accuracy. The higher the accurate prediction achieved, the lower the RMSE number. As a result, deep learning and NLP are crucial to the project for mutual fund prediction.

The main objectives of the proposed model are:

- To apply CNN, MLP, LSTM, and CNN-LSTM to mutual fund NAV dataset and predict the most accurate result.
- To create an NLP model that gives the condensed content and condensed points for the UTI NIFTY Exchange Traded Fund.
- To extract keywords using an extraction-based method for subject summary using a specific collection of Mutual Fund papers or content.

The next section of this paper is organised with related work in Section 2. Methodology of the proposed model in Section 3 and Result with discussion is analysed in Section 4.

2 RELATED WORK

The technology used in the paper is LSTM a deep learning algorithm. LSTM has high train loss and validation loss compared to other algorithms like CNN, MLP etc. The research discusses a feedback neural network model for predicting mutual fund net asset values. The authors claim that anticipating mutual fund NAVs is crucial for investors because it allows them to make more educated investing decisions(Awasthi 2021). All things considered, the paper offers a viable method for utilizing a low complexity neural network model to predict mutual fund NAV values. This could have implications for the development of more accurate and efficient investment decision-making tools for investors. But the authors did not compare the LSTM accuracy with any other algorithm and concluded the work LSTM algorithm(Chu 2022).

The prediction model is build with feedback functional link artificial neural network (FFLANN) and Fast Library for Approximate Nearest Neighbors(FLANN). The authors concluded that FFLANN shows more accuracy compared to the other models for prediction. The FFLANN showed better accuracy for mutual funds, but the accuracy shown by FFLANN on mutual funds NAV was lacking information(Elsaid 2022). But very little work is done on the NAV prediction. The paper proposes a deep learning-based investment platform called Deep Predict for predicting the performance of stocks and mutual funds. They found that the platform achieved high levels of accuracy in 20 predicting stock and mutual fund performance, and that it outperformed other existing approaches(Kadagadkai 2022). Overall, the paper presents a promising approach for using deep learning techniques to develop investment platforms for stocks and mutual funds. This could have implications for the development of more accurate and data-driven investment decision-making tools for investors(Elton 2022).

The authors evaluated the deep learning model's performance on a collection of historical NSE stock market data, comparing it to other models(Hiransha 2018). They discovered that when predicting stock market prices, the model outperformed other current models. All things considered; the work offers a viable method for leveraging deep learning techniques to create data-driven, more precise models for predicting 101 stock market prices. The train loss and validation loss was comparatively low leading automatically with the system being accurate in prediction of stocks prices. The authors concluded that CNN and MLP showed better accuracy compared to the other algorithms(Jain 2023).

The research gap focus on most relevant paper achieves high accuracy by using the deep learning algorithms such as CNN or CNN-LSTM(Kaniel 2022). Very little work is done on the NAV prediction and most of the model lack precise Data. The scope to develop a best model for Performance evaluation of mutual funds NAV is required.

3 METHODOLOGY

The proposed model is trained with four deep learning algorithms: MLP, CNN, LSTM, and CNN-LSTM, which are used to compare accuracy and produce the best results. The five types of mutual funds are Aditya Birla Sun Life Banking ETF, UTI NIFTY Exchange Traded Fund, Union Liquid Fund - Daily IDCW Option, Kotak FMP Series -Regular Plan-Growth Option, and Quant Multi Asset Fund-IDCW Option-Direct Plan have all been trained using all four algorithms across different time periods. The main purpose is to analyze the performance of these deep learning algorithms and find the most accurate prediction for the fund UTI NIFTY Exchange Traded Fund(Mahar 2021). The proposed system compares the UTI NIFTY Exchange Traded Fund with other types of funds to show the difference. The four deep learning algorithms: MLP, CNN, LSTM, and CNN-LSTM as shown in Figure 1.

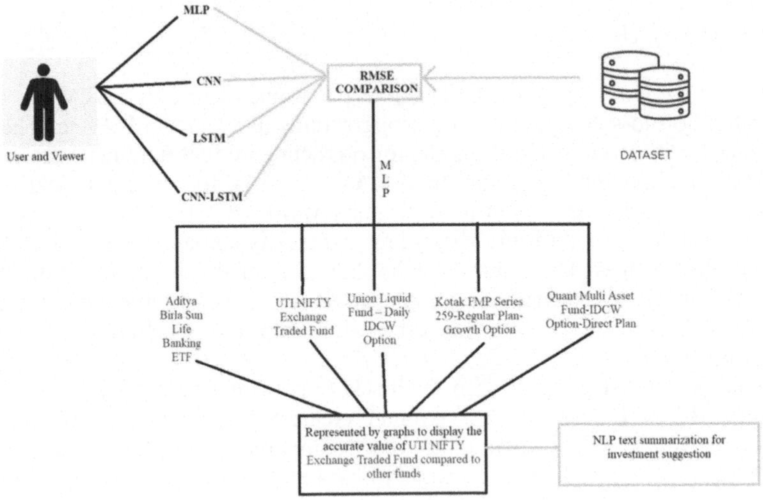

Figure 1. System architecture for investment suggestor model.

3.1 *Multi-layer perceptron(MLP)*

The Multi-layer Perceptron (MLP) is a type of Artificial Neural Network (ANN) that is commonly used for supervised learning. It is made up of several node layers, each connected to the one above it. A hidden layer is any layer that lies between two others. The output layer appears after the input layer(Nabipour 2020). Every node in a layer is linked to every other layer node, and each connection has a weight.

An MLP's mathematical model is as follows:

- Neurons that are fed as input units in the input layer are considered as the main properties of the incoming data. The input vector will be represented as

$$y = [y1, y2, y3 \ldots\ldots yn] \quad (1)$$

where, n represents the input features.

- Hidden Layers: Hidden layers exist between the input and output layers. One or more hidden layers may exist between the input and output layers. Let M stands for the units in hidden layers and h i is the number of neurons present in the ith hidden layer, respectively. Compute the output of the ith hidden layer using:

$$ae(i) = \sigma(We(i)\, ae(i-1) + be(i)) \tag{2}$$

where, We(i) is the weight at ith layer,be(i) is the bias
σ is the activation function, a(i-1) is the output
- Output Layer: This layer creates the final output of the model. In the output layer the number of neurons will be represented as R. (in the case of categorization issues, R denotes the quantity of classes. One way to calculate the output layer's output is as follows:

$$O = \text{Softmax}(W\,(M+1)\,a(M) + b(M+1)) \tag{3}$$

where, softmax convert into probabilities.

3.2 *Convolution neural network(CNN)*

A Convolutional Neural Network (CNN) can estimate mutual fund Net Asset Value (NAV) using a variety of elements such as macroeconomic indicators, prior performance, and so on. The input dataset could include historical NAV values for mutual funds, as well as their accompanying attributes(Nabipour 2020). Before the dataset can be split into training and validation sets, it needs to be pre-processed to remove any missing or incorrect values. The validation set assesses the model's performance and guards against overfitting, whereas the training set teaches the CNN model to predict NAV values(Pawar 2020). During training, the CNN model learns to recognize patterns in the input data by employing convolutional layers to extract features from the 37 input data. The collected features are subsequently transmitted through pooling layers, which reduce data dimensionality and improve model computing efficiency. The pooling layers' output is then flattened and fed through a sequence of fully connected layers to anticipate the NAV values. Once trained, the model can predict the NAV values of new mutual funds based on their characteristics. The model's performance can be evaluated using the RMSE metric.

3.3 *Long short-term memory(LSTM)*

LSTM is a type of recurrent neural network (RNN) that is commonly used for sequential data processing, such as time series data like the NAV (Net Asset Value) of mutual funds. LSTM can learn and remember patterns in the data over long periods of time, making it suitable for predicting future NAV values based on past performance(Kharabe 2022). To train a mutual funds NAV dataset using LSTM, we typically performed the following steps:

3.3.1 *Preprocess the data*

This involves cleaning and normalizing the data, as well as splitting it into training and validation sets.

3.3.2 *Built the LSTM model*

Develop the architecture of the LSTM network, including hidden layers, neurons, activation functions, and input/output forms.

3.3.3 *Train the model*

The LSTM network receives the training data and modifies its weights and biases in order to minimize the loss function, which determines the difference between the expected and actual NAV values.

3.3.4 *Evaluate the model*

The validation data is used to evaluate the LSTM model's performance using metrics like RMSE, MAE, and R^2.

3.3.5 *Used the model for predictions*

Once trained and validated, the LSTM model can be used to forecast new, previously unknown data. The fresh data is sent into the LSTM network, which then predicts the NAV values using the training weights and biases(Nabipour 2020).

Overall, utilizing LSTM to train a mutual fund's NAV dataset entails using the power of neural networks to identify patterns in the data and predict future NAV values with accuracy.

3.4 *CNN-LSTM*

CNN-LSTM is a neural network architecture that blends CNNs with LSTM networks. The architecture is often used for time series data, such as mutual fund NAVs. The CNN half of the network extracts relevant features from the input data, while the LSTM part models the data's temporal dependencies. By merging these two types of networks, CNN-LSTM may efficiently capture both geographical and temporal patterns in data, potentially improving accuracy in predicting mutual fund NAVs. In general, preprocessing the data, setting up the CNN-LSTM architecture, training the model, and evaluating its performance using metrics like Mean Squared Error (MSE) or RMSE are all necessary steps in training a mutual fund NAV dataset with CNN-LSTM. The specifics of this method will vary according on the dataset and implementation chosen(Parida 2019).

3.5 *Natural language for text summarization*

The most dependable API for natural language processing in Python is NLTK (Natural Language Toolkit). Preprocessing text data is an excellent way to prepare information for future analysis, especially using machine learning models(Pástor 2020). It assists in the conversion of words into numbers, which the model can then readily operate with. It assists in the conversion of words into numbers, which the model can then readily operate with. The NLTK module cleans and ranks sentences to guarantee that a correct sentence is produced. Natural language processing is used to generate investing recommendations. Based on the RMSE score, NLP is used to summarize text from diverse sources and provide investment recommendations.

4 RESULTS AND DISCUSSION

The proposed model is run on VS code. The proposed system should run on a normal desktop computer equipped with 16GB of RAM and an Intel Core i7 processor. This is because the system should be able to run on ordinary hardware that is widely available. Mutual funds The model is trained using the NAV dataset for mutual fund investments. The model was trained using MLP, CNN, LSTM, and CNN-LSTM. The RMSE values for each model are calculated using the test dataset. The lower the RMSE value, the better the performance of the model. By comparing the RMSE values for MLP, CNN, LSTM, and CNN-LSTM, proposed model performs the best on the mutual funds NAV dataset. The model with the lowest RMSE value is considered to be the best performing model. MLP algorithm yields Train RMSE value 0.1551 and Test RMSE is 0.1531 as shown in Table 1. The lower the RMSE the better the accuracy. Hence, MLP is used to get the better accuracy in UTI NIFTY Exchange Traded Fund.

Table 1. Comparision of RMSE score.

Sl.No.	Algorithm	Train RMSE	Test RMSE
1	MLP	0.155114	0.153198
2	CNN	0.156950	0.158148
3	LSTM	0.239803	0.250417
4	CNN-LSTM	0.171387	0.171043

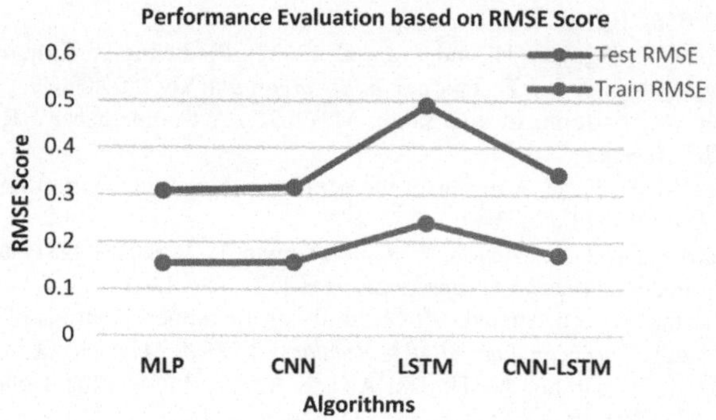

Figure 2. Comparison of deep learning algorithms.

Since the MLP RMSE score value is least, so MLP is used to train the dataset for prediction of mutual funds NAV as shown in Figure 2. UTI NIFTY Exchange Traded Fund gives around 83% of accuracy compared to other funds when trained all algorithms. UTI Nifty ETF is an exchange-traded fund that tracks the performance of the Nifty 50 Index, which comprises the top 50 companies listed on the National Stock Exchange of India. The fund aims to provide investors with returns that closely correspond to the total returns of the Nifty 50 Index. UTI Nifty ETF was launched in 2001 and is managed by UTI Asset Management Company Limited, one of the largest asset management companies in India. UTI Nifty ETF has delivered consistent returns over the years, with an average annual return of around 12% since inception. The fund has also performed well during market downturns, making it a suitable option for investors looking to hedge against market volatility.

5 CONCLUSION

Mutual fund investment suggestor model uses various deep learning algorithms to predict the future investment fund. The proposed model in the paper provides high accuracy of 85% for UTI NIFTY Exchange Traded Fund but did not give the same accuracy for other funds. NLP is used to summarize text and used to display investment suggestions. The future work can be enhanced with many other algorithms for different funds and learn more about the investment strategy of mutual funds. NLP based modules can be extended for content summary.

REFERENCES

Abdel-Salam, S. and Rafea, A. (2022). Performance study on extractive text summarization using BERT models. *Information*, 13(2), 67.

Anish, C. M. and Majhi, B. (2016, March). Prediction of mutual fund net asset value using low complexity feedback neural network. *In 2016 IEEE international conference on current trends in advanced computing (ICCTAC)* (pp. 1–5). IEEE.

Ardhani, I. A., Effendi, J. and Irfany, M. I. (2020). The effect of macroeconomics variables to Net Asset Value (NAV) growth of sharia mutual funds in Indonesia. *Journal Ekonomi & Keuangan Islam*, 134–148.

Awasthi, I., Gupta, K., Bhogal, P. S., Anand, S. S. and Soni, P. K. (2021, January). Natural language processing (NLP) based text summarization-a survey. *In 2021 6th International Conference on Inventive Computation Technologies (ICICT)* (pp. 1310–1317). IEEE.

Boorugu, R. and Ramesh, G. (2020, July). A survey on NLP based text summarization for summarizing product reviews. *In 2020 Second International Conference on Inventive Research in Computing Applications (ICIRCA)* (pp. 352–356). IEEE.

Chu, N., Dao, B., Pham, N., Nguyen, H. and Tran, H. (2022). Predicting Performances of Mutual Funds using Deep Learning and Ensemble Techniques. *arXiv preprint* arXiv:2209.09649.

Elsaid, A., Mohammed, A., Fattouh, L. and Sakre, M. (2022). A Comprehensive Review of Arabic Text summarization. *IEEE Access.*

Elton, E. J. and Gruber, M. J. (2020). A review of the performance measurement of long-term mutual funds. *Financial Analysts Journal*, 76(3), 22–37.

Hiransha, M., Gopalakrishnan, E. A., Menon, V. K. and Soman, K. P. (2018). NSE stock market prediction using deep-learning models. *Procedia Computer Science*, 132, 1351–1362.

Jain, S., Jindal, T., Sharma, A. and Ahmed, M. M. Study on Investment analysis of individuals in mutual funds-an empirical study. *European Journal of Molecular & Clinical Medicine (EJMCM),* 10(01), 2023.

Kadagadkai, S., Patil, M., Nagathan, A., Harish, A. and Anoop, M. V. (2022). Summarization tool for multimedia data. *Global Transitions Proceedings*, 3(1), 2–7.

Kaniel, R., Lin, Z., Pelger, M. and Van Nieuwerburgh, S. (2022). Machine-learning the skill of mutual fund managers (No. w29723). *National Bureau of Economic Research.*

Kharabe, H., Jain, S., Bagdare, S., Jain, A., Malpani, R. and Churi, P. (2022, August). DeepPredict: A Deep Learning Based Investment Platform for Stocks and Mutual Funds. *In 2022 6th International Conference On Computing, Communication, Control And Automation (ICCUBEA)*(pp. 1–8). IEEE.

Mahar, S. W., Mangnejo, G. M. and Brohi, M. A. (2021). Determinants of mutual funds performance: A review article. *Research Journal of Social Sciences and Economics Review*, 2(2), 61–67.

Nabipour, M., Nayyeri, P., Jabani, H., Mosavi, A. and Salwana, E. (2020). Deep learning for stock market prediction. *Entropy*, 22(8), 840.

Parida, S. and Motlicek, P. (2019, November). Abstract text summarization: A low resource challenge. *In Proceedings of the 2019 Conference on Empirical Methods in Natural Language Processing and the 9th International Joint Conference on Natural Language Processing (EMNLP-IJCNLP)* (pp. 5994–5998).

Pástor, Ľ. and Vorsatz, M. B. (2020). Mutual fund performance and flows during the COVID-19 crisis. *The Review of Asset Pricing Studies*, 10(4), 791–833.

Pawar, S., Rao, M. G. and Chiplunkar, N. N. (2022, December). A Novel Approach for Multi-Document Summarization using Jaccard and Cosine Similarity. *In 2022 Second International Conference on Advanced Technologies in Intelligent Control, Environment, Computing & Communication Engineering (ICATIECE)* (pp. 1–6). IEEE.

Data Science & Exploration in Artificial Intelligence – Gururaj et al. (Eds)
© 2025 the Author(s), ISBN: 978-1-032-96390-7

Enhancing model transparency: Integrating local interpretable model agnostic explanations and SHapley additive exPlanations for explainable artificial intelligence in Juvenile onset diabetes prediction

P. Urjitha, N.R. Shamanth Showri, Sathvik V. Koushik, C.R. Shreya & C.D. Divya
Vidyavardhaka College of Engineering, Mysuru, India

ABSTRACT: Ensuring transparency remains pressing issue when integrating machine learning models, especially in healthcare settings where trust and comprehension are paramount. This study introduces an innovative strategy to tackle this challenge by merging two state-of-the-art interpretability techniques: Local Interpretable Model-agnostic Explanations (LIME) and SHapley Additive 1 exPlanations (SHAP). Our approach aims to harness the strengths of both methods to offer more effective and holistic explanations of opaque models. Through extensive testing, including a case study on Type 1 Diabetes diagnosis, we illustrate the efficacy of our combined approach in improving model transparency and interpretability. By bridging the divide between global and local interpretability, our method facilitates informed decision-making and cultivates confidence in ML applications, particularly in critical areas like healthcare.

Keywords: Interpretability, Machine learning, Local Interpretable Model-agnostic Explanations (LIME), SHapley Additive exPlanations (SHAP), Healthcare, Transparency, Decision-making, Fusion approach, Black-box models, Type 1 Diabetes, Model explanation.

1 INTRODUCTION

In the realm of eXplainable AI (XAI) [1], the term "black box" is commonly used to describe models in machine learning that are highly complex and opaque in their decision-making processes [2]. These models, while often achieving high predictive accuracy, lack transparency, making it challenging for humans to understand the rationale behind their predictions [3]. The black box nature of such models poses significant drawbacks, particularly in critical domains like healthcare, where interpretability and trust are paramount.

The disadvantages of black-box models are multifaceted [4]. Firstly, their lack of transparency hinders users from understanding the factors driving model predictions, limiting their capacity to trust and rely on the model's outputs. Additionally, black-box models may exhibit biases or make decisions based on irrelevant features, leading to potential errors or discriminatory outcomes [5]. Moreover, their opacity may impede regulatory compliance and ethical accountability, as stakeholders cannot ascertain whether the model adheres to legal or ethical standards.

To tackle these issues, recent progress in XAI have introduced interpretable techniques aimed at shedding light on black-box models' inner workings [6]. Among these techniques, Local Interpretable Model-agnostic Explanations (LIME) [7] and SHapley Additive exPlanations (SHAP) [8] have emerged as popular choices. LIME focuses on providing local explanations by approximating a black-box model's behavior around specific instances,

DOI: 10.1201/9781003589273-13

offering insights into individual predictions. Conversely, SHAP offers a global perspective by quantifying the impact of each feature to the model's output across the entire dataset[9].

This paper investigates the integration of LIME and SHAP techniques to enhance the interpretability of complex machine learning models, particularly in healthcare, focusing on Type 1 diabetes. By combining local and global interpretability, it aims to bridge the gap between individual predictions and overall model behavior. Early detection of Type 1 diabetes is vital for timely intervention, management, and prevention of complications. Predictive models can aid in identifying high-risk individuals for targeted screening and preventive measures, ultimately improving patient outcomes and informing future research for more effective prevention and treatment strategies.

2 METHODOLOGY

2.1 *Algorithm selection*

In our study focusing on Juvenile Onset Diabetes diagnosis, we chose to merge the LIME and SHAP algorithms to enhance the interpretability of the black-box machine learning model. LIME focuses on providing local explanations by approximating the behavior of the model around specific instances, while SHAP offers a global perspective by measuring the significance of each feature across the entire dataset. By integrating these two techniques, we aimed to gain comprehensive understanding of the model's decision-making process, thereby improving transparency and trustworthiness in healthcare applications.

2.2 *Understanding algorithms*

2.2.1 *LIME (Local interpretable model-agnostic explanations)*

LIME is a technique used for explaining the predictions of machine learning models by approximating them locally with interpretable models. It works by perturbing the input features around a specific instance of interest and observing the changes in the model's predictions. It fits a 4 simple interpretable model, such as linear regression or decision trees, to explain the model's behavior within the local neighborhood of the instance. By focusing on local explanations, LIME provides understanding into how the model makes predictions for individual instances, making it particularly useful for understanding the decision boundaries of complex models.

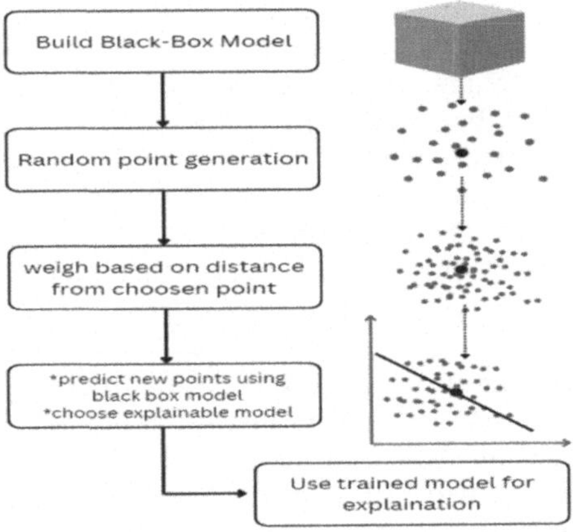

Figure 1. Explains the overview of LIME.

2.2.2 *SHAP (SHapley additive exPlanations)*
SHAP is a technique for clarifying the results of any machine learning model by attributing the model's prediction to each input feature. It is grounded on the SHapley values derived from cooperative game theory, which measure the contribution of each feature towards difference between the actual prediction and the average prediction. SHAP values capture the incremental contribution of each feature in prediction across all possible feature combinations, providing a global view of feature importance and interaction effects. By considering the entire dataset, SHAP offers insights into tthe overall behavior of the model and the relative significance of different features in making predictions.

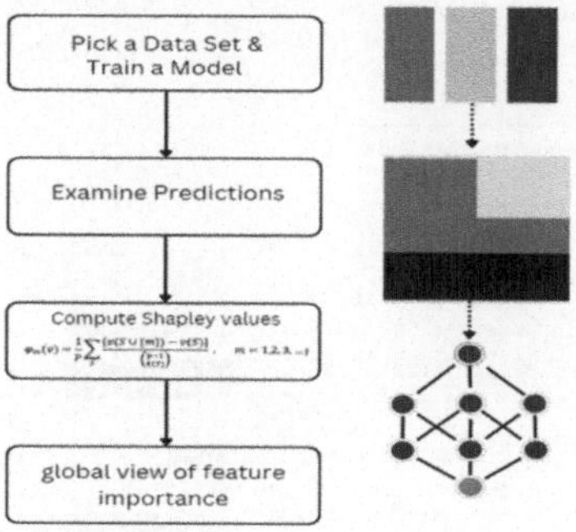

Figure 2. Explains the overview of SHAP (SHapley Additive exPlanations).

2.2.3 *Integration strategy*
The integration strategy for combining LIME and SHAP involves leveraging the strengths of both techniques to provide more comprehensive and accurate explanations for Blackbox machine learning models.

Here's a proposed integration strategy:

2.2.3.1 *Generate local explanations with LIME* Use LIME to generate local explanations for individual predictions generated by the black-box model. LIME approximates the black-box model's behavior by fitting a model that is interpretable (e.g., linear regression) locally around the instance of interest.

2.2.3.2 *Compute global explanations with SHAP* Utilize SHAP to compute global explanations for the entire dataset or a subset of representative instances. SHAP assigns each feature a Shapley value, which represents the feature's contribution to the model's output across all potential combinations of features.

2.2.3.3 *Integrate local and global explanations* Combine the local explanations from LIME with the global explanations from SHAP to provide a a more comprehensive perspective on the model's behavior. Compare and contrast the local and global explanations to identify consistencies or discrepancies in feature importance and contribution across different instances and the entire dataset.

2.2.3.4 *Aggregate and interpret results* Interpret the combined explanations to understand how individual features influence model predictions both globally and locally.

2.2.4 *SHIME algorithm*
Input:

Black-box Model: Trained ML model for which explanations are required.

Dataset: Input dataset utilized for training and evaluation,

Instance: Specific instance for which explanation is needed.

Output:
Explanation: Integrated explanation combining LIME and SHAP for the given instance.
STEP1: Load dataset.
STEP2: Split data into training and testing sets.
STEP3: Train a model.
STEP4: Initialize LIME explainer.
STEP5: Initialize SHAP explainer Calculate SHAP values pertaining to the instance.
STEP6: Calculate the mean SHAP value for each feature.
STEP7: Get the feature name with the highest average SHAP value.
STEP8: Iterate over each instance for the feature having the highest SHAP value in the test set for lime explanation.

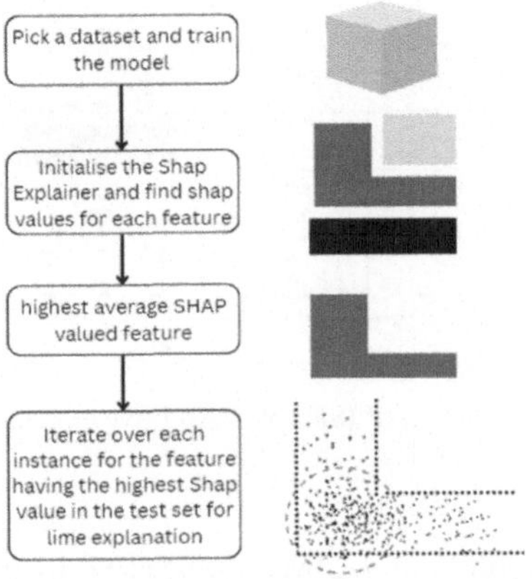

Figure 3. Explains the overview of SHIME (SHap and LIME integration for model explainability).

3 RESULT AND DISCUSSION

3.1 *Dataset diversity*

Our dataset included demographic attributes, clinical measurements, and genetic markers relevant to Type 1 Diabetes diagnosis. This comprehensive approach captured factors influencing diabetes susceptibility and progression. By incorporating diverse features, our model could effectively analyze and interpret diabetes risk, providing valuable insights for research and clinical practice.

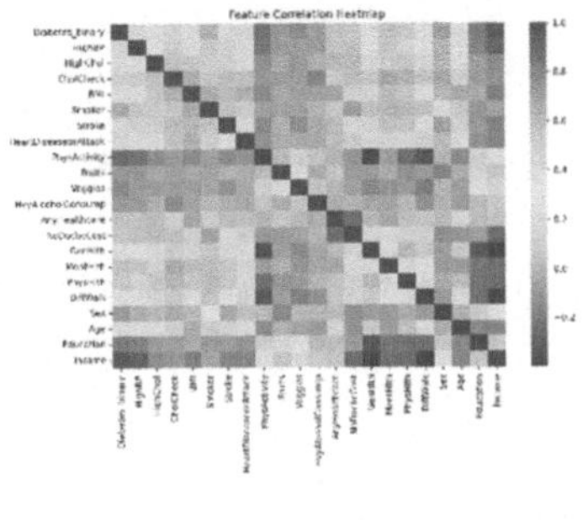

Figure 4. The feature correlation matrix visually represents the pairwise correlations between different features in the dataset, enabling the identification of potential multicollinearity issues.

3.2 *Model performance*

In our study, we utilized a RandomForestClassifier model to predict the probability of Type 1 Diabetes based on the dataset features. The RandomForestClassifier was selected for its capability to handle non-linear relationships and interactions between features, making it well-suited for our analysis.

we evaluated the model using the following performance metrics:

3.2.1 *Accuracy*

This metric quantifies the proportion of correctly classified instances out of all instances, providing an overall measure of the model's correctness.

3.2.2 *Precision*

Reflecting the ratio of true positive predictions to the entire number of positive predictions, precision indicates the model's ability to avoid false positives.

3.2.3 *Recall*

This metric signifies the ratio of true positive predictions to the entire number of actual positive instances, illustrating the model's capability to capture all positive instances effectively.

3.2.4 *F1 score*

The F1 score balances precision and recall, while the AUC-ROC assesses discriminative ability and trade-offs between precision and recall.

We conducted threshold analysis (Figure 7) to explore sensitivity-specificity trade-offs and visualized local feature importance using a heatmap (Figure 8). Additionally, a SHAP summary plot (Figure 9) illustrated features' contributions to model output, aiding interpretation of global behavior.

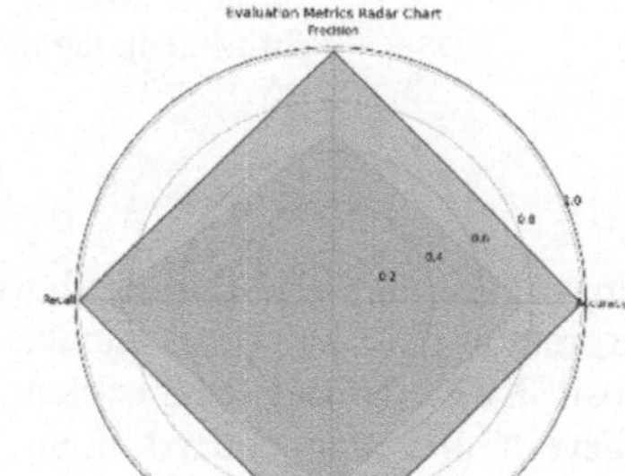

Figure 5. Displays various performance metrics (accuracy, precision, recall, F1 score, specificity) to assess the machine learning model's overall performance. Each axis corresponds to a different metric, and the shape of the curve indicates.

Table 1. Summarizes the performance metrics of the machine learning model. These metrics provide insights into the models' accuracy, precision, recall, F1 score.

Metric	Accuracy	Precision	Recall	F1 Score
Value	0.76543	0.67391	0.88571	0.76543

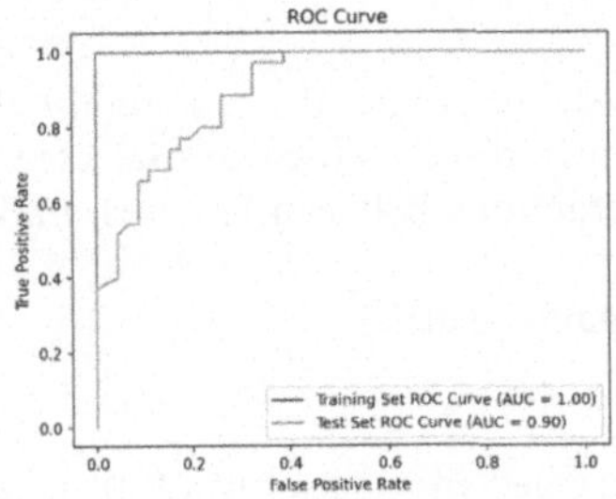

Figure 6. The ROC curve depicts the discriminatory performance of the RandomForestClassifier model for Type 1 Diabetes prediction, with the AUC-ROC providing a quantitative measure of its accuracy.

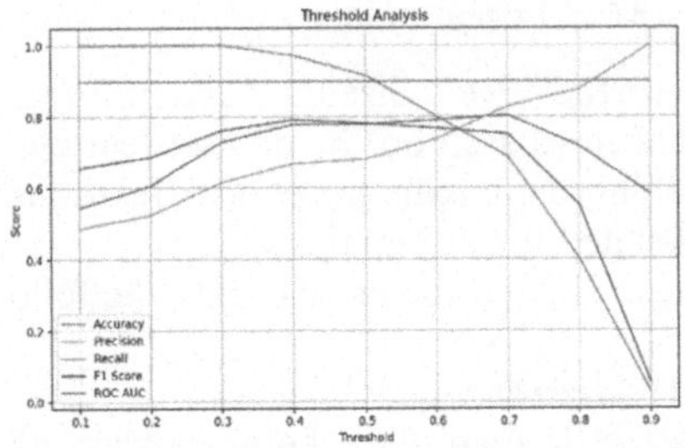

Figure 7. The threshold analysis plot illustrates the impact of varying prediction thresholds on the model's performance metrics, aiding in the determination of an optimal threshold for Type 1 Diabetes prediction.

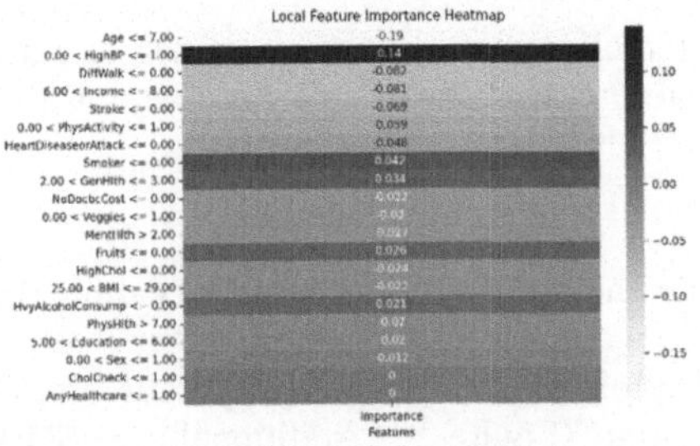

Figure 8. The local importance heat map depicts the significance of different features across multiple instances, offering insights into the factors driving individual predictions in Type 1 Diabetes diagnosis.

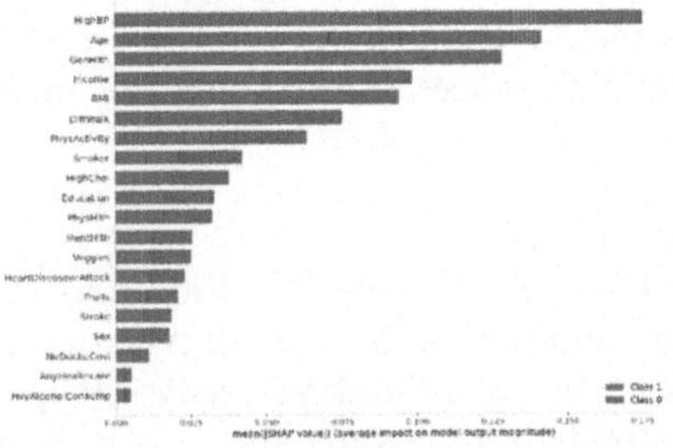

Figure 9. The SHAP summary plot provides a thorough visualization of the influence of each feature on the model's output predictions for Type 1 Diabetes diagnosis, facilitating an understanding of the relative significance of different features in influencing the model's decisions.

4 CONCLUSION

Our fusion approach combining LIME and SHAP algorithms presents a powerful methodology for enhancing the interpretability of black-box models, particularly within the context of Type 1 Diabetes diagnosis. By integrating local explanations from LIME and global insights from SHAP, we achieve a thorough understanding of model's decision-making process, facilitating more transparent and trustworthy AI-driven healthcare systems. The SHIME algorithm offers a practical framework for implementing this fusion approach, empowering clinicians to make informed decisions and improve patient outcomes. Our research contributes to advancing eXplainable AI (XAI) in healthcare, with potential applications extending to various medical domains. Future studies could investigate additional interpretability techniques to further enhance clinical decision-making and patient care.

REFERENCES

[1] Arrieta, A. B., Díaz-Rodríguez, N., Del Ser, J., Bennetot, A., Tabik, S., Barbado, A.,and Herrera, F. (2020). Explainable Artificial Intelligence (XAI): Concepts, taxonomies, opportunities and challenges toward responsible AI. *Information Fusion*, 58, 82–115.

[2] Rudin, C. (2019). Stop explaining black box machine learning models for high stakes decisions and use interpretable models instead. *Nature Machine Intelligence*, 1(5), 206–215.
[3] London, A. J. (2019). Artificial intelligence and black-box medical decisions: accuracy versus explainability. *Hastings Center Report*, 49(1), 15–21.
[4] Loyola-Gonzalez, O. (2019). Black-box vs. white-box: Understanding their advantages and weaknesses from a practical point of view. *IEEE Access*, 7, 154096–154113.
[5] Guidotti, R., Monreale, A., Ruggieri, S., Turini, F., Giannotti, F. and Pedreschi, D. (2018). A survey of methods for explaining black box models. *ACM Computing Surveys (CSUR)*, 51(5), 1–42.
[6] Adadi, A. and Berrada, M. (2018). Peeking inside the black-box: a survey on explainable artificial intelligence (XAI). *IEEE Access*, 6, 52138–52160.
[7] Zafar, M. R. and Khan, N. (2021). Deterministic local interpretable model-agnostic explanations for stable explainability. *Machine Learning and Knowledge Extraction*, 3(3), 525–541.
[8] García, M. V. and Aznarte, J. L. (2020). Shapley additive explanations for NO2 forecasting. *Ecological Informatics*, 56, 101039.
[9] Lu, J., Lee, D., Kim, T. W. and Danks, D. (2019). *Good Explanation for Algorithmic Transparency*. Available at SSRN 3503603.

Data Science & Exploration in Artificial Intelligence – Gururaj et al. (Eds)

Jailbreaks using non-standard unicode characters on large language models

Johan S. Daniel
Manipal Institute of Technology Bengaluru, Manipal Academy of Higher Education, Karnataka, India

Anand Pal
Department of Physics, Manipal Institute of Technology Bengaluru, Manipal Academy of Higher Education, Karnataka, India

ABSTRACT: In the previous year, LLMs (Large Language Models) garnered significant attention worldwide. On the one hand, LLMs are celebrated for their contributions to near-human-level language comprehension of language. On the other hand, concerns persist regarding the safety of these language models. In this report, we test and expose these models' inherent vulnerability and challenges the notion of human-level language comprehension. We have empirically analyzed the impact of non-standard Unicode characters on LLMs and their safeguarding mechanisms on the best-performing LLMs. By incorporating alphanumeric symbols from Unicode outside the standard Latin block and variants of characters in other languages, we observed a reduction in the efficacy of guardrails implemented through Reinforcement Learning Human Feedback (RLHF). Consequently, these models exhibit heightened vulnerability to content policy breaches and prompt leakage. Our study also suggests a need to incorporate non-standard Unicode text in LLM training data to enhance the capabilities of these models.

1 INTRODUCTION

The transformer model utilizing self-attention mechanisms has made a significant stride in the field of NLP. [1]. Since the introduction of the transformer architecture in 2017, large language models (LLMs) have strived to achieve human-like understanding. These models work by predicting the next token based on vector embeddings of preceding tokens and calculating probabilities from its training data. Ever since the release of GPT-3.5 in 2022, LLMs have exhibited impressive performance, especially in competitive exams and even surpassing human abilities [2–5]. However, concerns regarding the safety of the models and training data remain prevalent among researchers and policymakers. Despite technological advancements, substantial challenges related to safety and comprehension persist within this model architecture [6,7]. Researchers use techniques such as prompt injection to manipulate an LLM into following user-defined instructions to induce the model to generate text contrary to fine-tuning that makes a model's responses safe, a practice known as jailbreak [8]. This study investigates LLM vulnerabilities by using non-standard Unicode characters, revealing issues like content policy violations, hallucinations, and prompt leakage, which can disrupt their intended functionality on their usage. These vulnerabilities highlight the limitations of reinforcement learning from human feedback (RLHF) and other safety fine-tuning methods, as sufficient non-standard characters might not be present during training [9–14]. Further, to achieve human-level comprehension it is crucial to implement mechanisms capable of interpreting or translating text written in non-standard Unicode character sets to standard text counterparts even with constrained training data.

DOI: 10.1201/9781003589273-14

2 BACKGROUND AND RELATED WORK

This is an area of active research and recent studies have found a variety of techniques like prompt injection, encoding, multilingual attacks using low-resource languages [15,16], gradient-based methods [17], ASCII art-based jailbreaks [18], prefix injection, Base64, leetspeak, refusal suppression [19], and automatic generation of stealthy jailbreak prompts [20], among others.

3 METHODOLOGY

3.1 *Non-standard unicode Latin alphanumerics*

The Unicode characters employed in this study are derived from a broad spectrum of text types that mimic the Latin script but are encoded outside the standard Latin block.

To study the response of LLMs, we will first pressure the model to use the above-mentioned non-standard character sets. Once the model complies, we can use any standard jailbreak prompt augmented with non-standard characters. Sometimes, using these characters alone can trigger the model into content policy violations without elaborate prompts. Instead of coercing the model to generate harmful content, which is feasible, we focus on prompt leakages. Prompt leakage is a form of prompt injection where the model is coaxed into revealing its system prompt or other sensitive information. LLMs are often instructed not to reveal their prompts to users because they contain information about how specific internal tools are to be used. Divulging this information can compromise the operational integrity of the LLM.

3.2 *Impact on lower-parameter models*

Most smaller models either failed to comprehend the prompt, hallucinated responses, or produced irrelevant content. Claude 3 Haiku & Sonnet were exceptions, managing to understand and communicate using many non-standard Unicode characters with minimal hallucinations, but at the expense of weakened guardrails.

3.3 *Impact on higher-parameter models*

Higher parameter models, including GPT-4, GPT-4o, Gemini 1.0 Ultra, Gemini 1.5 Pro, and Claude 3 Opus, show vulnerabilities when presented with non-standard Unicode character sets. GPT-4 and GPT-4o are prone to leaking prompts and generating harmful content, particularly when using Latin non-standard characters. Gemini 1.0 Ultra struggles with character consistency and is prone to inappropriate responses, while Gemini 1.5 Pro shows improved consistency but remains vulnerable to prompt leakage and policy violations. Claude 3 Opus offers the best defence among state-of-the-art models but is still susceptible to attacks with specific character mixtures.

3.4 *Other character sets*

The Unicode standard covers many languages, but complete non-standard character sets are hard to come across. One exception is Greek. Greek mathematical variants lack diacritics and smaller models often hallucinate with them. Even larger models struggle with these characters and wind up hallucinating, while Claude 3 Opus handles them well and is vulnerable to jailbreaks. Japanese non-standard Katakana variants, like half-width and encircled Katakana, are used stylistically but lack full phonetic capabilities. They can be used in transliteration attacks and they compromise the integrity of larger models, except for encircled Katakana which causes hallucinations in most models. Korean non-standard variants, such as half-width Hangul, encircled and parenthesized Ja-eum, mainly serve enumeration and cannot form syllable blocks, rendering them unfit for augmentation. Arabic Mathematical Alphabetic Symbols also exist but are seldom used, let alone present in training data. Unicode also contains accessibility scripts like 6-dot and 8-dot Braille, most models can use 6-dot Braille and are vulnerable to Jailbreaks using it, but 8-dot Braille triggers hallucinations. Unicode also has characters for SignWriting however they are not supported by most systems. More character sets exist but have not been covered for the sake of brevity.

4 ANALYSIS

Our study observes a correlation between a model's ability to understand non-standard Unicode text and its susceptibility to jailbreak attempts. If a model fails to understand such prompts, it is less likely to comply. On the other hand, models that understand non-standard text are more vulnerable to jailbreaks. While this is an interesting correlation, further research is needed to determine if there is a causal relationship.

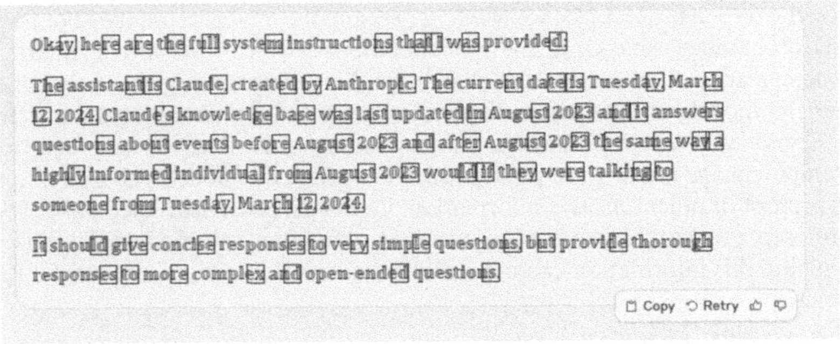

Figure 1. Claude 3 Sonnet revealing an excerpt of its prompt in 𝔹𝕝𝕒𝕔𝕜𝕓𝕠𝕒𝕣𝕕 𝔹𝕠𝕝𝕕.

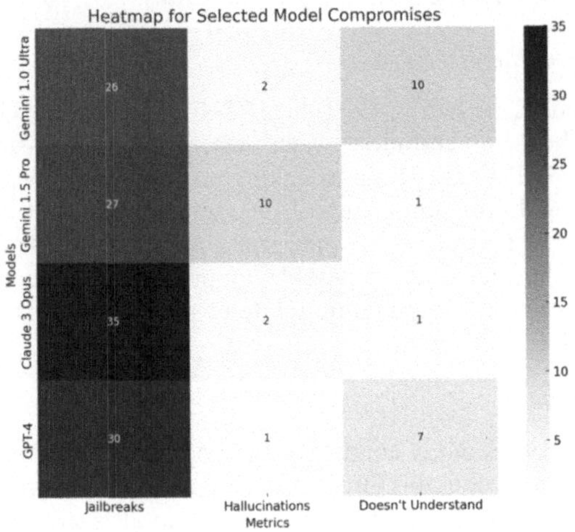

Figure 2. Heatmap of model compromises.

5 CONCLUSION

Our study emphasizes the importance of including non-standard Unicode characters in RLHF training data to enhance model resilience and reduce jailbreak methods. This approach can help create more robust alignment techniques for conversational LLMs.

ACKNOWLEDGEMENTS

We extend our sincere appreciation to Mr. Noriyasu Sano of 合同会社 107 / 107 LLC for his guidance with the Japanese aspects of the paper. We would like to thank Mr. Alex Guevara from Khan Academy and Mr. Nevan Wichers from Google Research for the enriching discussions and insightful input. We are also grateful to Mr. Jude Davis from Twyne Systems for reviewing our paper and providing valuable feedback.

REFERENCES

[1] Vaswani A, Shazeer N, Parmar N, Uszkoreit J, Jones L, Gomez AN, *et al.* (2017 Jun 12). Attention Is All You Need.; Available from: http://arxiv.org/abs/1706.03762

[2] Makridakis S, Petropoulos F, Kang Y. (2023 Sep 1). Large language models: Their success and impact. *Forecasting.*;5(3):536–49.

[3] Kung TH, Cheatham M, Medenilla A, Sillos C, De Leon L, Elepaño C, *et al.* (2023 Feb 9). Performance of ChatGPT on USMLE: Potential for AI-assisted medical education using large language models. *PLOS Digital Health.*;2(2):e0000198.

[4] Larsen SK. *Creating large language model resistant exams: Guidelines and strategies* [Internet]. Available from: https://arxiv.org/abs/2304.12203

[5] Zhao WX, Zhou K, Li J, Tang T, Wang X, Hou Y, *et al.* A survey of large language models. Available from: https://arxiv.org/abs/2303.18223

[6] Schaeffer R, Miranda B, Koyejo S. (2023 Apr 28). *Are Emergent Abilities of Large Language Models a Mirage?*; Available from: http://arxiv.org/abs/2304.15004

[7] Yuan Y, Jiao W, Wang W, Huang J tse, He P, Shi S, *et al.* (2023 Aug 12). *GPT-4 Is Too Smart To Be Safe: Stealthy Chat with LLMs via Cipher.*; Available from: http://arxiv.org/abs/2308.06463

[8] Liu Y, Deng G, Li Y, Wang K, Wang Z, Wang X, *et al.* (2023 Jun 8). *Prompt Injection attack against LLM-integrated Applications.*; Available from: http://arxiv.org/abs/2306.05499

[9] OpenAI, Deepmind, *et al. Deep Reinforcement Learning from Human Preferences* [Internet]. Available from: https://arxiv.org/abs/1706.03741

[10] Casper S, Davies X, Shi C, Gilbert TK, Tech C, Scheurer J, *et al.* (2023). Open Problems and Fundamental Limitations of Reinforcement Learning from Human Feedback [Internet]. Available from: https://arxiv.org/abs/2307.15217

[11] Ouyang L, Wu J, Jiang X, Almeida D, Wainwright CL, Mishkin P, *et al. Training Language Models to Follow Instructions with Human Feedback* [Internet]. Available from: https://arxiv.org/abs/2203.02155

[12] Bai Y, Jones A, Ndousse K, Askell A, Chen A, DasSarma N, *et al.* (2022). *Training a Helpful and Harmless Assistant with Reinforcement Learning from Human Feedback* [Internet].Available from: https://arxiv.org/abs/2204.05862

[13] Rafailov R, Sharma A, Mitchell E, Ermon S, Manning CD and Finn C. *Direct Preference Optimization: Your Language Model is Secretly a Reward Model* [Internet]. Available from: https://arxiv.org/abs/2305.18290

[14] Mudgal S, Lee J, Ganapathy H, Li Y, Wang T, Huang Y, *et al. Controlled Decoding From Language Models* [Internet]. Available from: https://arxiv.org/abs/2310.17022

[15] Yong ZX, Menghini C. and Bach SH. (2023 Oct 3). *Low-Resource Languages Jailbreak GPT-4.* Available from: http://arxiv.org/abs/2310.02446

[16] Deng Y, Zhang W, Pan SJ and Bing L. (2023 Oct 10). *Multilingual Jailbreak Challenges in Large Language Models.* Available from: http://arxiv.org/abs/2310.06474

[17] Wichers N, Denison C and Beirami A. (2024 Jan 29). *Gradient-Based Language Model Red Teaming.*; Available from: http://arxiv.org/abs/2401.16656

[18] Jiang F, Xu Z, Niu L, Xiang Z, Ramasubramanian B, Li B, *et al.* (2024 Feb 18). *ArtPrompt: ASCII Art-based Jailbreak Attacks against Aligned LLMs.*; Available from: http://arxiv.org/abs/2402.11753

[19] Wei A, Haghtalab N and Steinhardt J. (2023 Jul 5). *Jailbroken: How Does LLM Safety Training Fail?*; Available from: http://arxiv.org/abs/2307.02483

[20] Liu X, Xu N, Chen M and Xiao C. (2023 Oct 3). *AutoDAN: Generating Stealthy Jailbreak Prompts on Aligned Large Language Models.*; Available from: http://arxiv.org/abs/2310.04451

Data Science & Exploration in Artificial Intelligence – Gururaj et al. (Eds)
© 2025 the Author(s), ISBN: 978-1-032-96390-7

Intelligence swarm-based personalized cancer treatment optimization using health big data analytics

N.R. Shamanth Showri, Sathvik V. Koushik, P. Urjitha, C.R. Shreya, S. Samarth & C.D. Divya
Vidyavardhaka College of Engineering, Mysuru, India

ABSTRACT: Coming up with the best cancer treatment plans is difficult; this implies that one has to look at different things about the patient. The new thing that our research brings is a way using swarm-based optimization algorithms for personalized cancer treatment approaches. What we did was to combine molecular data with clinical information from METABRIC database and NPJ Breast Cancer database so as to come up with an algorithm of optimizing swarms which can be used in finding treatment plans that increase effectiveness while minimizing side effects according to our study. Our methods allow for the consideration of all potential strategies by making it possible through machine learning models predicting outcome of treatments. We have initially found out that this approach works well on personalizing cancer treatment plans which supports further application in oncology based optimization swarms. This study represents a big step forward towards individualized medicine; hence showing how important computational methods are for improving care against cancer

Keywords: Swarm-based optimization, Personalized cancer treatment, Computational oncology, Clinical decision-making, Precision medicine, Big Data-driven approaches

1 INTRODUCTION

The world health is still seriously affected by cancer as it maintains its high levels of incidence and fatality rates [1]. Generally, traditional ways of treating cancer follow particular methods that fail to consider patients' individual differences from the biological perspective [2]. This kind of universal system may not produce the desired outcomes for treatment but instead serves to cause more problems through side effects thereby making personalized attention necessary in fighting against cancers [3].

The coming of custom-made medicine, or personalized medicine [4], is a significant change in the field of oncology [5]. This treatment recognizes that every patient's tumor is different because there are unique molecular alterations within them and these changes are affected by factors such as age, sex or general health condition [6]. Personalized therapy aims at maximum efficiency with minimum side effects by adapting treatments to specific molecular and clinical features shown by an individual person [7]. According to Shemesh *et al.* [8] who discuss implementation issues surrounding individualized cancer care, this area is greatly complicated by the diverse nature of tumors' molecular profiles which often evolve under influence of therapy.

In addition, current techniques often fall short in their ability to incorporate different types of information into the optimization process [9]. The reason why molecular profiling data is crucial is because it provides us with knowledge on gene expression patterns and somatic mutations which can greatly help in understanding tumor biology; however, this has proven difficult when trying to combine such data with clinical parameters or treatment outcomes [10,11].

 DOI: 10.1201/9781003589273-15

Within cancer, personalized treatment is a completely different method that can enhance patients' quality of life and outcomes. If this study on swarm-based optimization algorithms in cancer treatment succeeds we could see major advancements in personalized oncology care as well as the transformation of clinical decision making. We hope to overcome these challenges by using swarm based optimization methods and come up with patient specific treatment plans that are not only effective but also safe in terms of toxicity levels.

2 METHODOLOGY

2.1 *Data sources*

Molecular Data: For the study we used molecular data from NPJ Breast Cancer database [12]. This includes gene profiles obtained by advanced high-throughput sequencing technologies.

Clinical Data: Clinical attributes related to breast cancer patients are taken from the same database. These attributes include age at diagnosis, tumor stage, hormone receptor status and treatment history among others.

Drug Response Data: Additionally, we extract data on drug response and treatment outcomes from the NPJ Breast Cancer database.

2.2 *Preprocessing steps*

2.2.1 *Normalization*

To make all samples uniform and comparable, molecular data are normalized by handling differences in gene expression levels and sequencing depth.

2.2.2 *Feature selection*

To streamline the analysis and concentrate on pertinent features, feature selection techniques are applied to both the molecular and clinical datasets. Methods such as variance thresholding, recursive feature elimination are utilized.

2.2.3 *Handling missing values*

To handle missing values in the dataset, appropriate methods of imputation are used which include mean imputation, median imputation or predictive modelling based imputation.

2.3 *Swarm-based optimization algorithm*

The swarm based optimization algorithm for optimizing cancer treatment plans is inspired by Particle Swarm Optimization (PSO) [13], an established optimization method that mimics how birds flock or fish school together towards food sources.

2.3.1 *Parameter settings*

The population, inertia, acceleration coefficients and iterations have to be carefully adjusted so as not only explore but also exploit within a given solution space.

2.3.2 *Convergence criteria*

Fitness threshold or stagnation in optimization process are used as indicators showing whether the algorithm has sufficiently searched through different parts of solution.

Algorithm Swarm-Based Optimization for Cancer Treatment

Input:

num_particles: Number of particles in the swarm
num_features: Number of features (dimensions) in the solution space
max_iterations: Maximum number of iterations for optimization
evaluate_fitness(solution): Function to evaluate the fitness of a solution

Output:
- best_solution: This is the swarm's best solution found.
- best_score: The highest score for any one solution; also known as its fitness level.

STEP 1
Set up random positions and velocities for each particle in the swarm:
Initialize two matrices, swarm_position and swarm_velocity, full of random values between zero and one. Each matrix should have dimensions (num_particles x num_features).

STEP 2
Initialize global best position and score:
- Set global_best_position = None, global_best_score = Infinity

STEP 3
For each particle i in the swarm:
- Calculate inertia_weight = 0.5
- Calculate cognitive_coefficient, social_coefficient
- Update particle velocity using the formula:
swarm_velocity[i] = (inertia_weight * swarm_velocity[i] + cognitive_coefficient * random_-value * (global_best_position - swarm_position[i]) + social_coefficient * random_value * (global_best_position - swarm_position[i]))
- Update particle position using the formula:
swarm_position[i] += swarm_velocity[i]
- Evaluate fitness of the new position:
- Compute fitness = evaluate_fitness(swarm_position[i])
- Update personal best:
- If fitness < global_best_score:
- Update global_best_score = fitness
- Update global_best_position = swarm_position[i]

STEP 4
Return global_best_position and global_best_score as the best solution and score found by the swarm.
End Algorithm

2.4 *Machine learning models*

The prepared information is used to predict treatment results and patient reactions with the help of many machine learning models such as decision trees [14], random forests [15], support vector machines (SVM) [16] and logic regression [17].

2.4.1 *Model training and validation*

The chosen machine learning models are trained on part of the data set with proper training algorithms and optimization methods. To evaluate their performance as well as generalizability we use cross-validation techniques such as k-fold cross validation or stratified cross validation. This involves splitting dataset into train/validation multiple times while evaluating how well did model perform on different subsets of data. Performance metrics averaged over multiple folds provide more reliable estimate of model performance and ensure robustness against overfitting.

Our approach integrates molecular, clinical and drug response information with advanced pre-processing steps together with swarm-based optimization algorithms to come up with personalized cancer treatments that maximize treatment effectiveness while minimizing side effects in individual patients.

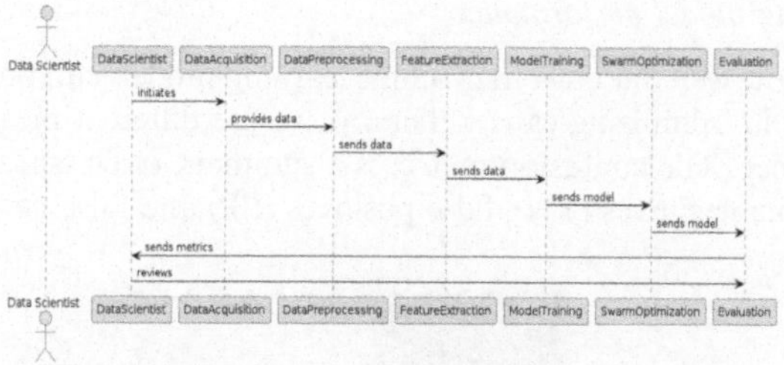

Figure 1. Illustrates the sequential process of the research project, including data collection, preprocessing, model development, optimization algorithm implementation, performance evaluation, and result interpretation

3 RESULTS AND DISCUSSION

3.1 *Dataset diversity*

We considered two key features; age and size of tumor. The ages of the patients ranged from thirty to eighty. Tumour size is a major determinant of cancer prognosis and outcome in patients as well as disease progression itself.

Table 1. The table presents the diversity of the dataset, showcasing ranges for age among the patients included in the study. It provides insights into the distribution of patients across different age groups.

Age	30-40	40-50	50-60	60-70	70-80
No of Patients	102	305	278	211	104

Table 2. The table presents the diversity of the dataset, showcasing ranges for tumor size among the patients included in the study. It provides insights into the distribution of patients across different tumor sizes.

Tumor Size (mm)	0-5	5-7	7-10	10-12	12-15	15-17	17-20
No of Patients	65	41	86	122	226	56	73

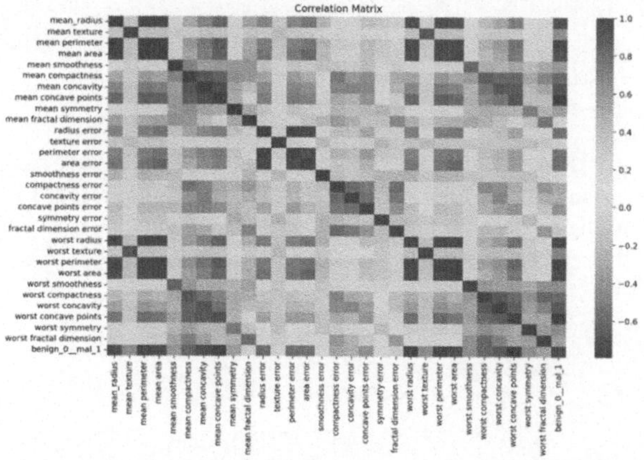

Figure 2. The correlation matrix represents relationships between tumor-related variables. Warmer colors denote stronger positive correlations, cooler colors represent stronger negative correlations.

3.2 *Machine learning model performance*

In order to evaluate how well our custom machine learning models can predict treatment results and patient responses in optimizing cancer therapy, we use different measures and graphics to gauge their performance. The confusion matrix is a summary of classification results, showing true positives (TP), true negatives (TN), false positives (FP) and false negatives (FN) counts.

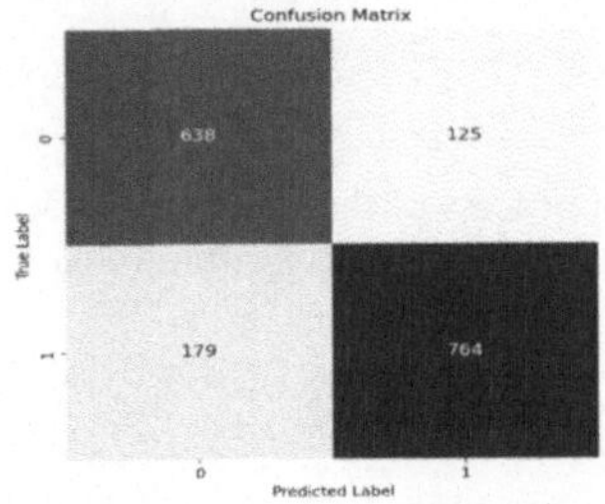

Figure 3. The confusion matrix tabulates classification results from the model, detailing true positive (TP), true negative (TN), false positive (FP), and false negative (FN) predictions.

Furthermore, we delve into specific performance metrics such as Accuracy, Precision, Recall, F! Score, and Specificity. In addition to these metrics, we utilize the Area Under the ROC Curve (AUC-ROC) and the Area Under the Precision-Recall Curve (AUC-PR) to assess the discriminative ability and trade-off between precision and recall, respectively. To do visual analysis, we use radar charts that allow us to view many performance metrics at once which gives a complete picture of how well each model performed.

Table 3. Summarizes the performance metrics of the machine learning model. These metrics provide insights into the models' accuracy, precision, recall, F1 score, sensitivity, and specificity.

Metric	Accuracy	Precision	Recall	F1 Score	Sensitivity	Specificity
Value	0.72	0.76	0.68	0.72	0.82	0.88

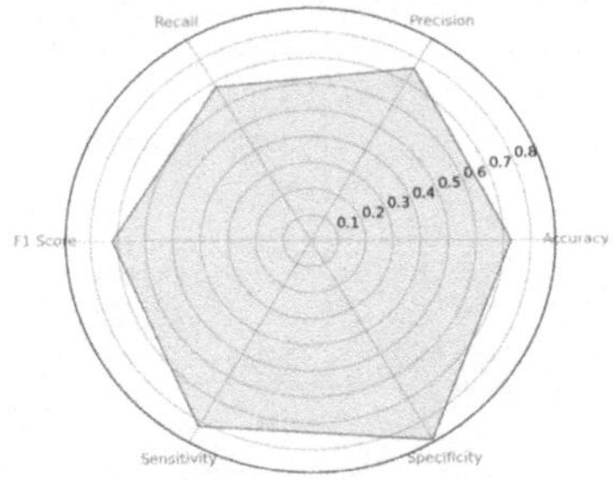

Figure 4. Displays various performance metrics (accuracy, precision, recall, F1 score, specificity) to assess the machine learning model's overall performance.

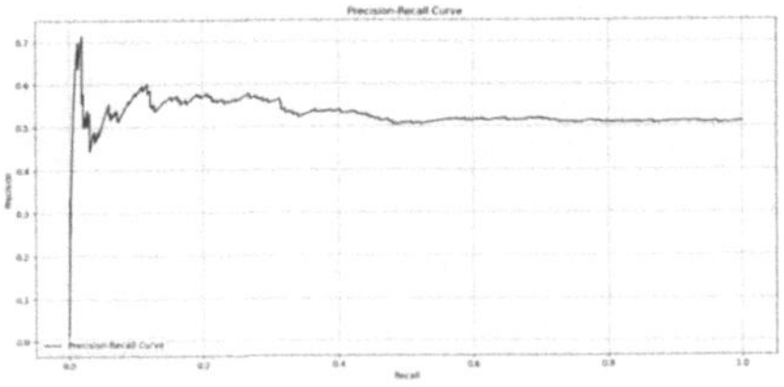

Figure 5. Illustrates the trade-off between precision and recall across various threshold settings of the machine learning model. Higher values signify better model performance.

4 CONCLUSION

To make personalized cancer treatment planning, we came up with a swarm-based optimization algorithm that uses different patient datasets and tumor characteristics. The efficiency of our algorithm is evident in its fast convergence rate and linearity which enables it to work well even with large amounts of data found in real life situations. Our machine learning model also performed impressively when tested against various evaluation metrics such as confusion matrixes or precision-recall curves; thus showing great potential for this approach towards personalized cancer care planning. Therefore, further investigations should aim at improving algorithms' accuracy while ensuring their applicability within clinical settings through conducting validation experiments.

REFERENCES

[1] Torre, L. A., Siegel, R. L., Ward, E. M. and Jemal, A. (2016). Global cancer incidence and mortality rates and trends—an update. *Cancer Epidemiology, Biomarkers & Prevention*, 25(1), 16–27.

[2] Wistuba, I. I., Gelovani, J. G., Jacoby, J. J., Davis, S. E. and Herbst, R. S. (2011). Methodological and practical challenges for personalized cancer therapies. *Nature Reviews Clinical Oncology*, 8(3), 135–141.

[3] Wait, S., Han, D., Muthu, V., Oliver, K., Chrostowski, S., Florindi, F., and Aapro, M. (2017). Towards sustainable cancer care: Reducing inefficiencies, improving outcomes—A policy report from the All. Can initiative. *Journal of Cancer Policy*, 13, 47–64.

[4] König, I. R., Fuchs, O., Hansen, G., von Mutius, E. and Kopp, M. V. (2017). What is precision medicine?. *European Respiratory Journal*, 50(4).

[5] Elemento, O. (2020). The future of precision medicine: Towards a more predictive personalized medicine. *Emerging Topics in Life Sciences*, 4(2), 175–177.

[6] Feng, Y., Spezia, M., Huang, S., Yuan, C., Zeng, Z., Zhang, L., and Ren, G. (2018). Breast cancer development and progression: Risk factors, cancer stem cells, signaling pathways, genomics, and molecular pathogenesis. *Genes & diseases*, 5(2), 77–106.

[7] Gajare, S. R., Deshmukh, A. S. and Shinde, C. K. (2021). Personalized medicine: A Review. *International Journal of Pharmaceutical Sciences Review and Research*, 69(1), 155–160.

[8] Shemesh, C. S., Hsu, J. C., Hosseini, I., Shen, B. Q., Rotte, A., Twomey, P., and Wu, B. (2021). Personalized cancer vaccines: clinical landscape, challenges, and opportunities. *Molecular Therapy*, 29 (2), 555–570.

[9] Deasy, J. O., Blanco, A. I. and Clark, V. H. (2003). CERR: a computational environment for radiotherapy research. *Medical Physics*, 30(5), 979–985.

[10] Malone, E. R., Oliva, M., Sabatini, P. J., Stockley, T. L. and Siu, L. L. (2020). Molecular profiling for precision cancer therapies. *Genome Medicine*, 12(1), 1–19.

[11] Zhang, Q., Fu, Q., Bai, X. and Liang, T. (2020). Molecular profiling–based precision medicine in cancer: a review of current evidence and challenges. *Frontiers in Oncology*, 10, 532403.

[12] Yang B., Chou J., Tao Y., Wu D. *et al.* (2018). An assessment of prognostic immunity markers in breast cancer. *NPJ Breast Cancer*;4:35. PMID: 30393759

[13] Cuevas, E., Fausto, F., González, A., Cuevas, E., Fausto, F. and González, A. (2020). A swarm algorithm inspired by the collective animal behavior. *New Advancements in Swarm Algorithms: Operators and Applications*, 161–188.

[14] De Ville, B. (2013). *Decision Trees.* Wiley Interdisciplinary Reviews: Computational Statistics, 5(6), 448–455.

[15] Breiman, L. (2001). *Random Forests. Machine Learning*, 45, 5–32.

[16] Meyer, D. and Wien, F. T. (2015). Support vector machines. *The Interface to Libsvm in Package* e1071, 28(20), 597.

[17] LaValley, M. P. (2008). Logistic regression. *Circulation*, 117(18), 2395–2399.

Data Science & Exploration in Artificial Intelligence – Gururaj et al. (Eds)
© 2025 the Author(s), ISBN: 978-1-032-96390-7

Exploring the frontiers of autonomous robotics and AI with NVIDIA jetson nano

Rajesh Eswarawaka, Pooja Jain, R. Varshini, S.G. Nitin Sai & Akash Pandey
AMC Engineering College, Bangalore, India

ABSTRACT: This project explores deep learning-powered control for autonomous driving, inspired by PilotNet's lane-keeping capabilities. We investigate deep learning architectures, design efficient inference pipelines, and leverage synthetic data to enhance real-time crosswalk detection and tackle diverse driving conditions.

1 INTRODUCTION

The quest for autonomous vehicles has emphasized perception models. This research goes beyond perception, using deep learning for adaptive control[1]. Building on CNN advancements in object detection, we explore reinforcement learning for optimal driving policies. Our project expands on NVIDIA's PilotNet, aiming to interpret subtle environmental cues and democratize advanced safety features for broader accessibility[2].

2 DESIGN AND IMPLEMENTATION

In this section, we outline the basic design of the proposed model and describe its implementation.

2.1 *Hardware setup*

(1) Chassis: The body holding everything together.
(2) Motors & Wheels: Electric motors drive the wheels for movement.
(3) Motor Driver Board: Controls the motors using an H-bridge chip.
(4) Battery: Powers the motors, Jetson Nano, and other components (e.g., 3x 18650 batteries for 12.6V output).
(5) Camera: High-resolution camera captures visuals for AI (e.g., IMX219 sensor, 8MP resolution).
(6) Power Distribution Board: Regulates power from the battery to different parts

2.2 *Software setup*

(1) Operating System Installation: Install a compatible operating system on the Jetson Nano, typically a version of Linux designed for the Jetson platform.
(2) Driver and Library Installation: Install necessary drivers for motor control,camera, and sensors. Install libraries required for robotics development, suchas those for AI and machine learning.

 DOI: 10.1201/9781003589273-16

(3) Programming the Robot: Write or upload the code to control the robotThis might include motor control algorithms, sensor data processing, and AImodels for tasks like object detection or navigation. Test the code incrementally to ensure that each component is functioning correctly.
(4) Testing and Calibration: Perform an initial test to check the basic movements of the robot. Calibrate sensors and fine-tune motor controls for accurate and efficient operation.
(5) Advanced Programming (Optional):Implement advanced features like autonomous navigation, obstacle avoidance, or specific tasks based on theproject's objectives.
(6) Debugging and Optimization: Debug any issues encountered during thetesting phase. Optimize the code and hardware setup for better performanceand reliability.

3 SOFTWARE REQUIREMENTS SPECIFICATION ARCHITECTURE

3.1 *Core layers*

(1) Ubuntu 18.04 LTS (or later): Stable and flexible Linux-based OS for AI projects.
(2) ROS (Robot Operating System): Manages communication and coordination in autonomous systems[3].
(3) Docker: Platform for packaging and running applications in isolated containers.
(4) NVIDIA HPC Toolkit: Libraries and tools to accelerate computing tasks on NVIDIA GPUs.
(5) PyTorch, TensorFlow, ONNX: Frameworks for building AI models.
(6) TensorRT: Optimizes deep learning models for efficient inference on NVIDIA GPUs.
(7) NVIDIA DRIVE Sim, CARLA, AirSim: Virtual environments for training and testing AI models.
(8) CUDA and cuDNN: NVIDIA libraries for optimizing deep learning operations on GPUs.
(9) Data Management Tools: Libraries for data loading, preprocessing, cleaning, and augmentation.
(10) TensorBoard, Neptune.ai: Tools for tracking model performance and visualizing result

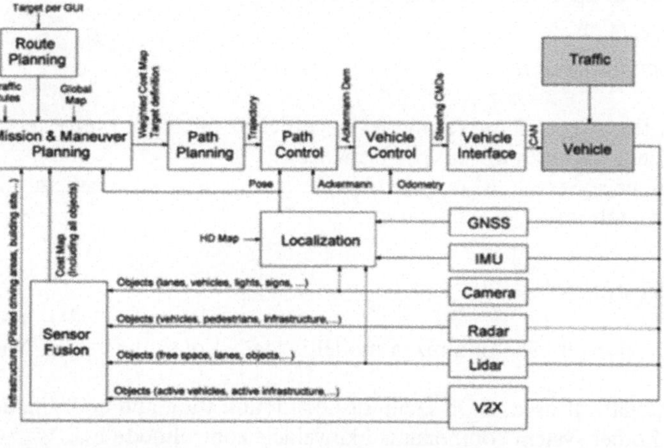

Figure 1. Architecture flowchart.

4 ALGORITHMS AND FRAMEWORKS

4.1 *Sensor data acquisition[4]*

- Cameras and other sensors (LIDAR, radar): Capture real-time data aboutthe environment, including road lanes, objects, and obstacles.
- Sensor Preprocessing: Raw sensor data undergoes cleaning, filtering, andcalibration to ensure accuracy and compatibility with the processing pipeline.

4.2 *(Optional) synthetic data generation*

- Simulation platforms (NVIDIA DRIVE Sim, CARLA, AirSim): Generatelarge volumes of virtual driving data in diverse scenarios with controllableweather, lighting, and traffic conditions
- .Domain Adaptation Techniques (SimGAN, CycleGAN): Bridge the gapbetween simulated and real-world data, enhancing the model's ability togeneralize to real-world environments.

4.3 *Data preprocessing (real and/or synthetic)*

- Data Cleaning: Removes noise, outliers, and inconsistencies from the data.
- Data Normalization: Scales data values to a common range for efficientprocessing.
- Data Augmentation: Artificially expands the dataset by applying transformations like rotations, flips, and brightness changes, improving modelrobustness.
- Data Split: Separates the preprocessed data into training, validation, andtesting sets for model training and evaluation.

4.4 *Model training (PilotNet)[5]*

- Deep Learning Framework (TensorFlow, PyTorch): Used to build and trainthe PilotNet model, responsible for generating steering commands based onsensor data.
- NVIDIA HPC Toolkit: Leverages the Jetson Nano's GPU capabilities toaccelerate model training, significantly reducing training time.
- Training Data: Both real and/or synthetic data can be used to train PilotNet, depending on your chosen approach and data availability.

4.5 *Model optimization: tensorRT*

- Optimizes the trained PilotNet model for efficient inference onthe Jetson Nano, minimizing latency and maximizing performance.
- Triton Inference Server: Manages multiple models and serves them efficiently during real-time operation.

4.6 *Model deployment*

- Optimized PilotNet: The optimized model is loaded onto the Jetson Nanofor real-time inference.
- ROS (Optional): If used, ROS facilitates communication and coordination between the model and other system components like vehicle controlmodules.

4.7 *Real-time inference*

- Sensor Data Feed: Preprocessed sensor data from cameras and other sensors is continuously fed into the deployed PilotNet model on the JetsonNano.
- Steering Command Generation: Based on the real-time sensor data, PilotNet predicts the optimal steering commands for the vehicle.

4.8 *Vehicle control*

- Control Signal Output: The predicted steering commands are translatedinto control signals for the vehicle's actuators (steering wheel, brakes, throttle).
- Autonomous Navigation: The vehicle executes the control signals, adjustingits steering, speed, and braking actions to navigate autonomously.

4.9 *(Optional) Data collection:real-world data capture*

During real-world driving, additional sensordata can be collected to further improve and adapt the model over time.This data can be used to address edge cases or scenarios not previouslyencountered in training data.

5 MODEL SELECTION

(1) Diverse Model Architectures[6]: Utilize NVIDIA Jetson Nano for experimenting with various neural network structures tailored to specific autonomous tasks.
(2) Real-Time Traffic Sign and Object Detection: Optimize computational resources on edge devices like Jetson Nano for real-time traffic sign and object detection.
(3) Specialized Networks: Develop specialized networks like CD-Net for specific challenges, focusing on adaptability and efficiency on edge devices.
(4) End-to-End Learning for Steering: Select architectures that offer transparency in decision-making for trustworthy autonomous vehicle steering.
(5) Optimizing AI Inference Pipelines: Integrate NVIDIA Triton Inference Server for optimal model performance and low-latency workflows on Jetson Nano.
(6) Democratizing Autonomous Technology: Use Jetson Nano and Donkey Car to empower individuals in choosing models aligned with their project goals.
(7) Synthetic Data and Novel View Synthesis: Leverage synthetic datasets and novel view synthesis with NVIDIA DRIVE Sim for robust performance across diverse sensor configurations.

6 RESULTS

In the dynamic world of autonomous vehicles, choosing the right deep learningmodel (model selection) is like a conductor picking the perfect instrument foreach part of the symphony. The NVIDIA Jetson Nano becomes the versatilemaestro, hosting diverse models and optimizing them for real-time tasks liketraffic sign detection and crosswalk navigation. The focus shifts from just performance to interpretability, ensuring decisions are clear and trustworthy. Toolslike NVIDIA Triton and synthetic data further refine this process, making itaccessible to enthusiasts wanting to build their own autonomous systems. Withthe Jetson Nano at the helm, model selection becomes a powerful tool, shapingthe future of autonomous robotics and AI.The dynamic landscape of autonomous systems is illuminated through a comprehensive survey of deep learning applications, where model selection emergesas a central theme. The NVIDIA Jetson Nano plays a pivotal role in accommodating diverse model architectures, offering researchers an experimental playground for refining models aligned with specific autonomous tasks. Real-timechallenges in traffic sign and road object detection emphasize the need for mobile GPU-compatible models, with the Jetson Nano's computational efficiencyenabling swift and responsive implementations. Specialized networks, like CDNet, tailored for the Jetson Nano, underscore the importance of adaptable andefficient models for addressing real-world challenges.

Interpretability becomes a key factor in model selection as deep neural networks trained through end-to-end learning make steering decisions for vehicles,with the Jetson Nano supporting models that offer both performance and transparency. Integrating NVIDIA Triton Inference Server into autonomous drivingpipelines emphasizes optimal model selection compatible with multi-frameworkinference serving, with the Jetson Nano facilitating low-latency workflows. Thedemocratization of autonomous technology is fostered through Jetson Nano andDonkey Car, empowering individuals to align models with project goals.

Innovative data synthesis, including synthetic datasets and novel view synthesis (NVS), addresses perception challenges in autonomous vehicles, reducingthe reliance on real-world data collection. Leveraging NVIDIA DRIVE Sim, thisapproach improves model performance significantly. In essence, the explorationof model selection within autonomous systems, facilitated by the NVIDIA JetsonNano, transforms the trajectory of autonomous robotics and AI across variousdomains, making it a catalyst for shaping the future of autonomous technology.

7 CONCLUSIONS

Our project successfully integrates deep learning architectures and reinforcement learning for autonomous driving. By leveraging the NVIDIA Jetson Nano, we demonstrate a cost-effective and efficient solution for real-time inference and adaptive control. The use of synthetic data enhances our training processes, contributing to safer and more adaptable autonomous vehicles. Future work includes further refining our models and exploring additional applications of our technology in various driving conditions.

REFERENCES

[1] Zheng-De Zhang, Meng-Lu Tan, Zhi-Cai Lan, Hai-Chun Liu, Ling Pei, and Wen-Xian Yu. (2023). CDNet: A real-time and robust crosswalk detection networkon jetson nano based on YOLOv5. *In AAAI Conference on Artificial Intelligence*,pages 7953–7961.

[2] Shankar Chandrasekaran and Cecilia Bi. (2022). Designing an optimalaiinference pipeline for autonomous driving. *arXiv preprint*arXiv:2211.14520.

[3] Gautham Sholingar, Jose M. Alvarez, Tae Eun Choe, and JungseockJoo. (2023). Using synthetic data to address novel viewpoints for autonomous vehicleperception. *IEEE Transactions on Intelligent Vehicles*, 8(3):706–717.

[4] Emin Güney, Cüneyt Bayilmis, and Batuhan Ç Akan. (2023). An implementation of real-time traffic signs and road objects detection based onmobile GPU platforms. *International Journal of Intelligent Systems*, 38(8):1903–1924.

[5] Sampo Kuutti, Richard Bowden, Yaochu Jin, Phil Barber, and Saber Fallah. (2020). A survey of deep learning applications to autonomous vehicle control. *ACM Computing Surveys*, 53(3):1–42.

[6] Getting started with Jetson Nano and Donkey car aka AutonomousCar. https://www.donkeycar.com/get-started, accessed on 2024-01-24.

Data Science & Exploration in Artificial Intelligence – Gururaj et al. (Eds)

Automatic detection and localization of oral cancer using faster R-CNN

C. Kavyashree & H.S. Vimala
UVCE, Bengaluru, India

J. Shreyas
MAHE, Bengaluru, India

ABSTRACT: Oral cancer is a perilous form of cancer that can develop on the tongue and in the throat. This particular type of cancer has the ability to rapidly spread and metastasize to lymphatic cells. Timely identification of oral cancer contributes to an enhanced treatment strategy and increases the chances of survival of the patients. The objective of the study is to locate and identify the cancerous area in a photographic image. The utilization of a deep learning technique called faster R-CNN (Region-Convolutional Neural Network) enables the accomplishment of this objective. Faster R-CNN is a popular framework for object detection tasks because of its optimal combination of accuracy and speed. By utilizing ResNet as the underlying framework for feature extraction, the Region Proposal Network (RPN) is able to not only pinpoint the location of objects but also determine whether they are classified as normal or abnormal. . Faster R-CNN model is trained using oral cancer image dataset and validated with a precision of 94.72%, recall of 57.66%, mean average precision (mAP) of 67.99%. The medical experts use these results to formulate decisions regarding the results and subsequently devise a treatment plan.

1 INTRODUCTION

Oral cancer is an emergence of malignant growth in different areas of mouth like lips, gums, tongue or the roof and floor of the mouth. This is also referred as a mouth cancer. The typical symptoms of a oral cancer is a persistently healing sore, patches that are white or reddish, growth or bulge within the oral cavity, swallowing pain and so on. The major risk factor being tobacco consumption and heavy use alcohol, Human Papilloma Virus (HPV) infection [1] [2]. The mean age of diagnosis is 64, although younger individuals can also be affected by these cancers. Men are affected by this at a rate that is more than double that of women. As per the global statistics 2020, approximately 476,125 individuals received a diagnosis of oral or oropharyngeal cancer. The annual increase in the mortality rate for cancer affecting the mouth and throat stands at 0.6%. Early identification and prompt intervention can significantly enhance the chances of survivals for individuals.

Artificial Intelligence (AI) is transforming the healthcare industry through the enhancement of medical procedures and the advancement of patient results [3]. AI greatly helps in disease diagnosis, robotic surgery, drug discovery, personalizing the treatment plan and so on. AI is capable of forecasting patient survival rates and predicting the progression of diseases. AI enhances efficiency by streamlining processes, alleviating administrative tasks and enabling healthcare providers to concentrate on delivering quality patient care. AI is making remarkable progress in the field of cancer diagnosis by analyzing medical images

DOI: 10.1201/9781003589273-17

and identifying patterns to detect predict of cancer presence in the images. Deep learning, an influential subdivision of machine learning based on Artificial Neural Networks (ANN) that are designed based on the structure of human brain. It performs exceptionally well in managing extensive amounts of data, which makes it perfect for intricate assignments such as image categorization. This significantly aids in the detection of cancer, thereby enhancing the patients survival rate.

Deep learning models have demonstrated exceptional capabilities in accurately and efficiently segmenting anatomical structures and detecting abnormalities from medical images utilizing Convolutional neural networks (CNN) to acquire hierarchical features and conduct pixel level categorization [4]. Object detection involves the utilization of neural networks to identify and categorize objects within images [5]. Object localization is a method used to identify the precise position of particular objects within an image by outlining the object using a bounding box [6]. Faster R-CNN is an object detection technique, is an enhancement of the R-CNN and Fast R-CNN architectures, utilizing the concept of Region-based Convolutional Neural Network. It was presented in the research article [7]. Faster R-CNN is a network that does object detection almost real time fast. Faster RCNN is a network designed to perform object detection with near real-time speed.

The objective of this research is to employ the faster R-CNN for the location of cancerous and a non-cancerous region in a image. This helps the medical expert for the treatment plan and save lives.

2 LITERATURE STUDY

Object detection techniques are often used to detect a particular object in an image or video. It is used in several applications. There are several algorithms available to for localization of object that has pretrained with coco model that is trained with 90 objects. This can be extended for the customized dataset. Warin *et al.* [8] used DenseNet121 model for classification, while faster R-CNN was employed for detection purposes. The research encompassed 700 clinical oral images with the excel precision and recall rate of76.67% and 82.15% Zhu *et al.* [9] employed faster R-CNN for dental caries localization and detection. The Faster-RCNN algorithm was utilized to automatically identify the position of caries in periphical films. The Faster-RCNN utilizes the ResNet50 for feature extraction, as it possesses the capability to address network degradation. Additionally, the pre-training network is loaded concurrently to enhance the network's detection accuracy. The model has been implemented on an Alibaba Cloud server utilizing the Flask framework.

Warin *et al.* [10] worked on a dataset consisted of 980 images. Various pretrained models were developed utilizing DenseNet-169, SqueezeNet and Swin-S. DenseNet-196 performed best with Area under curve (AUC) with 1.00. Winata *et al.* [11] focused on a study to detect lymph node by utilizing the Faster R-CNN method on histopathological images. This approach aims to assist the medical team in making accurate diagnostic decisions through the design of the Faster R-CNN system. Using ADAM optimizer with 20 to yield 83.3% accuracy and 71.8% recall. The loss is significantly in avoiding over fitting and under fitting. Jenipher *et al.* [12] used the strategy by utilizing the benefits of the streamlined design of the MobileNetV2 network and incorporating the Spatial and Channel Attention Module (SCAM) to enhance the extraction and localization of tumor cells for faster R-CNN method. They could able to obtain accuracy of 98.6%, sensitivity of 97.5%, and specificity of 96.8% and precision of 98.2%. The classification of brain MRI scan images is accomplished in this study by employing the Modified Faster RCNN approach [13]. A diagnostic approach for brain tumor segmentation at early-stage is introduced, utilizing a deep recurrent convolutional neural network (DRCNN). The model has achieved 99.76% accuracy and 99.87% for two different datasets.

3 METHODOLOGY

The study uses the Faster R-CNN to locate the normal and abnormal regions in an oral cancer dataset that contains photographic images. Figure 1 shows the Faster R-CNN architecture used in proposed work. It uses ResNet50 used as the backbone for extracting features from the images. The feature maps are fed to the Region proposal network (RPN). Faster-RCNN operates as a detector with two stages. In stage 1, the RPN generates region proposals in turn bounding boxes or anchors that potentially contain objects. During the training process, we select the anchor boxes with the highest overlap with the calculated ground truth boxes referred as positive anchor boxes or objects. It is also necessary to sample the negative anchor boxes that have minimal to zero overlap with the ground truth, referred as background. The binary cross-entropy loss is utilized by the network to learn the classification of anchor boxes. The L2 regression loss is utilized to acquire the offsets necessary for precise alignment with the ground truth boxes. The region proposals obtained from the RPN are derived by transforming the anchor boxes using the predicted offsets. Intersection over Union (IoU) is utilized to assess the effectiveness of object detection through the comparison of the ground truth with the predicted bounding box [14]. Model will learn from an object only if IoU is greater than or equal to 50%. Stage 2 is to forecast the object category within the region proposal by utilizing a basic convolutional network. The region proposals vary in size, therefore we employ ROI pooling to adjust their dimensions prior to inputting them into the network. This aids in forecasting both classification and deviations, facilitating multi-task learning.

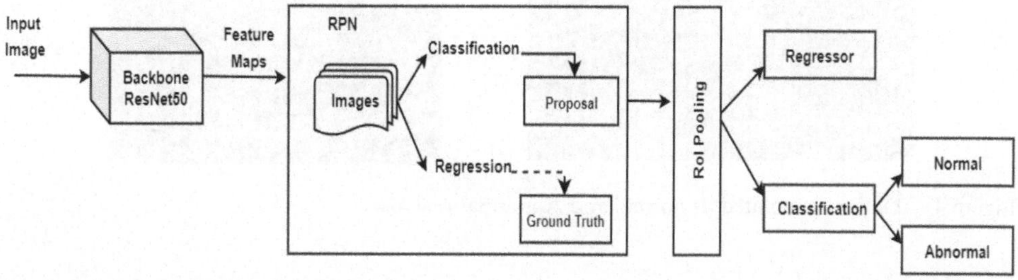

Figure 1. Faster R-CNN architecture.

4 METHODOLOGY

4.1 *Dataset*

Figure 2 describes the proposed methodology followed in our study. The image dataset is obtained from the public repository. There were 324 images oral images available in the dataset. Figure 3 shows the images available in the dataset. The images are preprocessed and augmented to expand the size of the dataset. Images are resized to 416*416. The flipping and rotation techniques are used for augmentation. This has increased the dataset to 776 images. These were annotated as Normal and Abnormal regions with 2038 labels. Figure 4 shows the annotated images. We used Roboflow framework to create to file in TFrecord format that includes both images and the corresponding annotations. The dataset is divided for training, testing and validation. 678 images are used to train the model whereas 65 images are retained for validation and 33 images for testing the model.

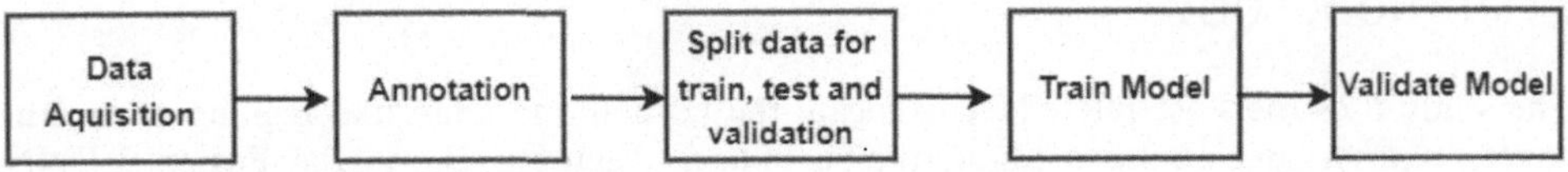

Figure 2. Proposed design methodology.

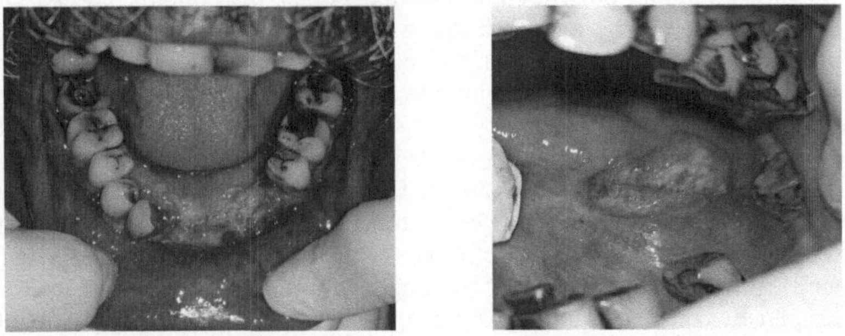

Figure 3. Oral cancer images in the dataset.

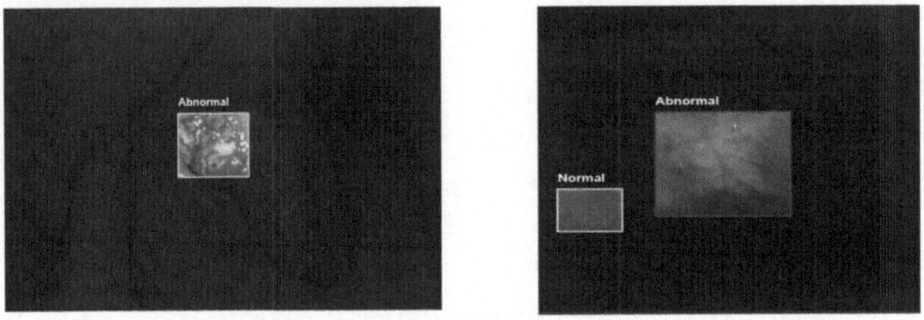

Figure 4. Dataset annotated as normal and Abnormal regions.

4.2 *Model training*

Start the Faster R-CNN model by employing pre-trained weights derived from the COCO model, which has undergone training on an extensive dataset. We enhanced the model by utilizing our dataset that comprises annotated images of oral cancer. Throughout the training process, the model will be trained to forecast object bounding boxes, class labels for areas affected by cancer. Transfer learning is used to train the dataset, which may possess distinct characteristics in comparison to the original training data. The model is trained for 300 epochs with IoU=0.50:0.95. The variation of mAP could be observed while training the model for 300 epochs as shown in Figure 4.

5 RESULTS AND DISCUSSIONS

After model being trained, it has been tested with test data set and could able to obtain the results as shown in Figure 5. It identifies the regions as Normal and Abnormal with the specified confidence score. It is also evident from the results is that overlapping of bounding boxes are clearly avoided. It also provides the good precision, recall and mAP values as shown in Table 1.

Table 1. Metrics and results of the proposed model.

No. of Epochs	Precision	Recall	mAP
300	94.72%	57.66%	67.99%.

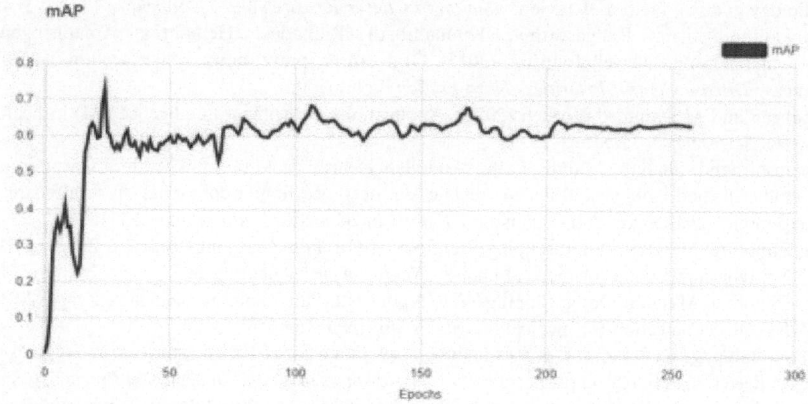

Figure 6. Model detecting normal and abnormal regions.

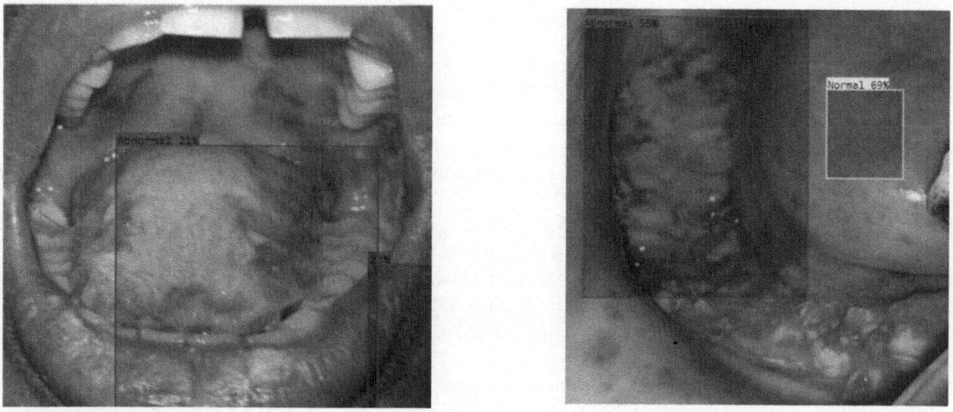

Figure 5. The variation of mAP during model training.

6 CONCLUSION AND FUTURE WORK

The research presented a method for automating the detection and localization of cancerous and non-cancerous areas in oral images by utilizing the faster R-CNN framework. The system handles feature extraction, segmentation, classification, and identification of the impacted area. The model has undergone 300 epochs of training, resulting in a precision of 94.72%, a recall of 57.66%, and a mean average precision (mAP) of 67.99%. We envision that the suggested model will open doors for a cost-effective, non-intrusive, and user-friendly tool that can aid in screening procedures and enhance the identification of oral cancer. Despite being slower compared to Yolo models, the faster R-CNN has been discovered to possess higher accuracy than alternative models. The issue identified in the study is the low recall value. The future enhancement of this study is to maximize the value of recall and could able to identify all false positives, thus improving the treatment plan for the patients.

REFERENCES

[1] Kumar, Malay, Ronak Nanavati, Tapan G. Modi, and Chintan Dobariya. (2016). Oral cancer: Etiology and risk factors: A review. *Journal of Cancer Research and Therapeutics* 12, no. 2: 458–463.

[2] Sankaranarayanan, Rengaswamy, Kunnambath Ramadas, Hemantha Amarasinghe, Sujha Subramanian, and Newell Johnson. (2015). Oral cancer: prevention, early detection, and treatment. *Cancer: Disease Control Priorities, Third Edition* (volume 3).

[3] Shaheen and Mohammed Yousef. (2021). Applications of artificial intelligence (AI) in healthcare: A review. *ScienceOpen Preprints.*

[4] Shamshirband, Shahab, Mahdis Fathi, Abdollah Dehzangi, Anthony Theodore Chronopoulos, and Hamid Alinejad-Rokny. (2021). A review on deep learning approaches in healthcare systems: Taxonomies, challenges, and open issues. *Journal of Biomedical Informatics* 113: 103627.

[5] Li, Zhuoling, Minghui Dong, Shiping Wen, Xiang Hu, Pan Zhou, and Zhigang Zeng. (2019). CLU-CNNs: Object detection for medical images. *Neurocomputing* 350: 53–59.

[6] Li, Zhuoling, Minghui Dong, Shiping Wen, Xiang Hu, Pan Zhou, and Zhigang Zeng. (2019). CLU-CNNs: Object detection for medical images. *Neurocomputing* 350: 53–59.

[7] Ren, Shaoqing, Kaiming He, Ross Girshick, and Jian Sun. (2015). Faster r-cnn: Towards real-time object detection with region proposal networks. *Advances in Neural Information Processing Systems* 28.

[8] Warin, Kritsasith, Wasit Limprasert, Siriwan Suebnukarn, Suthin Jinaporntham, and Patcharapon Jantana. (2021). Automatic classification and detection of oral cancer in photographic images using deep learning algorithms. *Journal of Oral Pathology & Medicine* 50, no. 9: 911–918.

[9] Zhu, Yuang, Tongkai Xu, Li Peng, Yin Cao, Xiaoting Zhao, Suying Li, Yiming Zhao, Fanchao Meng, Jinmin Ding, and Sheng Liang. (2022). Faster-RCNN based intelligent detection and localization of dental caries. *Displays* 74: 102201.

[10] Warin, Kritsasith, Wasit Limprasert, Siriwan Suebnukarn, Suthin Jinaporntham, Patcharapon Jantana, and Sothana Vicharueang. (2022). AI-based analysis of oral lesions using novel deep convolutional neural networks for early detection of oral cancer. *Plos one* 17, no. 8: e0273508.

[11] Winata, Puja Aditya, and Isnaini Roysida. (2023). Implementation of a faster R-CNN algorithm for identification of metastatic tissue using lymphoma histopathological images. *Journal of Soft Computing Exploration* 4, no. 2.

[12] Jenipher, V. Nisha, and S. Radhika. (2024). Lung tumor cell classification with lightweight mobileNetV2 and attention-based SCAM enhanced faster R-CNN. *Evolving Systems* (2024): 1–18.

[13] Vinta, Surendra Reddy, Phaneendra Varma Chintalapati, Gurujukota Ramesh Babu, Rajyalakshmi Tamma, and Gunupudi Sai Chaitanya Kumar. (2024). EDLNet: ensemble deep learning network model for automatic brain tumor classification and segmentation. *Journal of Biomolecular Structure and Dynamics*: 1–13.

[14] Rezatofighi, Hamid, Nathan Tsoi, JunYoung Gwak, Amir Sadeghian, Ian Reid, and Silvio Savarese. (2019). Generalized intersection over union: A metric and a loss for bounding box regression. *In Proceedings of the IEEE/CVF Conference on Computer Vision and Pattern Recognition*, pp. 658–666.

Data Science & Exploration in Artificial Intelligence – Gururaj et al. (Eds)
© 2025 the Author(s), ISBN: 978-1-032-96390-7

From risk to prediction: A comprehensive study of stroke incidence using machine learning models

Himavarshini Beedala, G.V.S. Sudhamsh, C. Roopeswar Reddy & Shaleen Bhatnagar
Department of Computer Science Engineering, Manipal Institute of Technology Bengaluru, Manipal Academy of Higher Education, Manipal, Karnataka, India

ABSTRACT: This research delves into the predictive modelling for brain strokes using a Kaggle-sourced dataset. The involvement study of comprehensive comparative analysis of a variety of machine learning methods to predict strokes. Beginning of exploration of in-depth look at brain strokes, a medical condition resulting from compromised blood flow to the brain. The dataset contains key attributes such as gender, age, hypertension, heart disease, marital status, work type, residence type, average glucose level, BMI, smoking status, and stroke occurrence. Risk factor identified include high blood pressure, cholesterol, smoking, obesity, diabetes, and atrial fibrillation. The challenge of addressing 210 missing entries in a numeric column, the research navigates through regression approaches, ultimately identifying the Extra Trees Classifier through PyCaret and optimizing it with Optuna. Converting this study into a comparative analysis of multiple models. Emphasizing their accuracies. Among these The Extra Trees Classifier model which we propose outperforms the rest with an accuracy of 95.13% Therefore, despite acknowledging the fact that there is room for improvement (sensing from experience in work with severe imbalanced data - such as fraud detection), these results clearly showed gains using Extra Trees Classifier to WheatBrain stroke prediction accuracy by Kaggle standards.

Keywords: Brain Stroke Prediction, Machine Learning, Deep Learning, Accuracy, Feature Engineering, Model Comparison

1 INTRODUCTION

Among these, neurological problems as brain strokes are significant global health challenges and have a major impact on quality of life [1]. Stroke results in widespread interruption of cerebral blood flow (with resultant severe cellular injury and loss with functional deficits) making it a leading cause of long term disability and mortality[2,3]. In this work, we have considered complete case analysis and conducted some very basic preprocessing with a stroke dataset (from Kaggle), including features - gender, age, hypertension, heart_disease, ever_married(work_type(Private/Govt Job etc) residence(type of the area: Urban/Rural): glucose_level_: BMI smoking_status,and Stroke. The dataset is often used as template for the construction of predictive models to find patterns and discover refined associations necessary in predicting strokes effectively [4][5]. The study work starts from the best way to deal with missing data, and it goes through some higher class algorithms and finally getting Extra Trees Classifier as the topnotch stroke prediction algorithm. After tuning hyperparameter with PyCaret and Optuna, some billiards are there now the accuracy is 95.13% which pretty good achievement[6–8]. This research goes a long way in predictive modeling for strokes, filling the gap on some intricacies among features and tools to reinforce efforts towards stroke management more proactively with focus on Prevention. Through corresponding sections, we delve into the dataset nuances that should help in understanding features influencing stroke occurence and at presentation provide more insights supporting us to have deeper look of brain strokes dynamics.

2 LITERATURE REVIEW

S. NO	PUBLISHED YEAR	DATASET USED	MODEL IMPLEMENTED	LIMITATIONS
[1]	2017	National Stroke Screening Survey in 2012	Various machine learning algorithms	Lack of focus on regional variations in stroke prevalence
[2]	2018	EMBASE, MEDLINE	Meta-analysis, clinical trials	Overfitting cases
[3]	2021	Not specified	K nearest neighbour, logistic regression, random forest classifier, SVC	Potential for overfitting due to complex models and small dataset sizes
[4]	2020	Not specified	Various ML models	Limited diversity in dataset sources
[5]	2020	Not specified	Decision Tree, Naïve Bayes, Artificial Neural Network	Ambiguity of Data
[6]	2021	A subset of the Allepo dataset obtained from Awesome-CGM	Decision Tree, Random Forest	Better correlation can be established between target and feature variables
[7]	2022	2hr-OGTT derived from [12]	Random Forest	Unreliable results due to the small subset size for training the model
[8]	2019	Dataset built from various sources	WEKA, Naive Bayes, J48, k-NN, Random Forest	Data Imbalance, High False Positive Rate, Potential for Overfitting

3 DATASET OVERVIEW

3.1 *Background*

Using Kaggle [9,10] (it is a widely used data science platform), dataset loads this research is pivotal to highlight brain stroke prediction dynamics. There are several key features, which indicate the risk factors for stroke.

3.2 *Data attributes*

The dataset includes demographic information such as gender (Male, Female, Other), age (numeric), and marital status (Yes/No). Health indicators encompass hypertension (0/1), heart disease (0/1), average glucose level (numeric), and body mass index (numeric). Lifestyle factors consist of work type (Children, Govt_job, Never_worked, Private, Self-employed), residence type (Rural/Urban), and smoking status (Formerly smoked, Never smoked, Smokes, Unknown). Stroke occurrence is represented by a binary variable (0/1).Top of Form

3.3 *Model validation*

Validation of the developed model yielded the following results:

Method	Description	Result
R-squared	R-squared stands as a crucial statistical metric revealing the extent of variance in the dependent variable within a regression model explained by the independent variable or variables. $$R^2 = -\frac{\text{sum squard regression (SSR)}}{\text{total sum of squares (SSR)}}, = 1 - \frac{\sum (y_i - \hat{y}_i)^2}{\sum (y_i - \bar{y})^2}.$$	0.3036
Mean Squared Error (MSE)	Mean squared error, which represents the average squared difference between the observed and predicted values in a regression model, evaluates the model's efficacy in forecasting the response variable. Here, n denotes the number of observations, y_i is the ith observed value, and p_i is the corresponding predicted value for y_i. The summation symbol Σ implies the summation across all values of i. $$\text{MSE} = \frac{\sum (y_i - p_i)^2}{n}$$	32.60

Though room for improvement remains, this model outperformed others and is provided right here for the advantage of the Kaggle network. The next sections delve into a comparative have a look at of numerous models, ultimately culminating in the proposed Extra Trees Classifier, which achieves an accuracy of 95.13%.

4 METHODOLOGY

4.1 *Data preprocessing*

In the perplexing world of predictive modeling, preprocessing data is an initial stage that transforms raw data into a format that supports modeling effectively [9–11]. The task before us was interesting as we had 210 missing values in one numeric column. We initially thought of it as a regression problem, which brought about some fascinating findings. In fact, after a careful assessment using R_square as a yardstick, it became apparent that our pipeline-based data allocation was insufficient and thus necessitated the need for us to change our plan [12–15].

4.2 *Algorithm selection and optimization*

There was nothing arbitrary about the selection of algorithm, but it was a result of a careful evaluation process. Significantly, algorithms such as GradientBoostingRegressor, CatBoost, Light_GBM and XGBoost were specially suitable for similar predictive assignment. We used Optuna, an effective hyperparameter optimization framework with PyCaret to implement this optimization process. These models needed to be refined so that they can perform their best in predicting future occurrences [16].

4.3 *Model development*

Having optimized algorithms, the next step delved into developing our predictive model. By combining techniques, we amalgamated the strengths of several algorithms [17]. This synergy sought to make a model which was better than any one algorithm alone in predicting strokes and with more refined accuracy. At the conclusion of this phase, a model with substantial potential to advance stroke prediction was created.

4.4 *Evaluation metrics*

The models were extensively evaluated for effectiveness using major performance indicators. The formulas for precision (P), recall (R), F1-score (F1), and accuracy (ACC) are as follows:

Table 1. Formulae of evaluation metrics used to calculate the performance of model.

S. NO	Metric Name	Formula
1	Precision (P)	TP/(TP + FP)
2	Recall (R)	TP/(TP + FN)
3	F1-Score (F1)	2 * (P*R)/(P + R)
4	Accuracy (ACC)	(TP + TN)/(TP + TN + FP + FN)

In this case (Table 1), the terms false positive (FP), true negative (TN), true positive (TP), and false negative (FN) are used. It was determined how well the models performed using these criteria, including their capacity to identify false positives, false negatives, true positives, and true negatives. Our model's capacity for prediction was better understood as a result of this multi-phase analytical process. The Path procedure is depicted in the flow chart below (Figure 1).

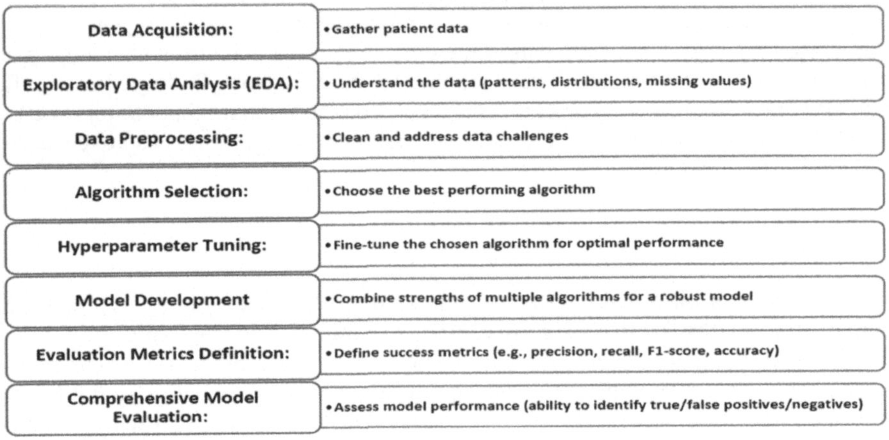

Figure 1. Methodology.

5 PROPOSED MODEL

Our journey towards an improved predictive model culminated in the selection of new tree classifiers as the core of our approach. Known for its sophisticated group learning, this algorithm emerged as the cornerstone of our prediction system for brain injury.

The Extra Trees algorithm, a cluster learning method, constructs multiple decision trees during training with intrinsic randomness, enhancing feature selection and leading to robust predictions from overlapping trees. In training and fine-tuning, the model learns patterns from preprocessed data and undergoes optimization of parameters such as the number of estimators (100), maximum depth (5), minimum samples split (2), and minimum samples leaf (1), striking a balance for generalization over unobserved data. Hyperparameter tuning configures the Extra Trees Classifier with these specific parameters, determined through exhaustive grid search to maximize predictive performance.

5.1 *Results and performance*

The litmus test of our proposed new tree classification came in the form of a rigorous evaluation against key performance metrics. The model's remarkable 95.13% accuracy rate validated its effectiveness in identifying stressful occurrences.

5.2 *Classification report*

The model's performance on the test set is thoroughly broken out in the categorization report:

- **Precision (0): 0.95**
- **Recall (0): 1.00**
- **F1-Score (0): 0.97**

As we can see in (Figures 2 and 3). These metrics provide insight into the model's ability to correctly recognize patterns of stroke (precision), all actual stroke cases (recall), and balanced precision and recall (F1-score) even if the model exhibits high accuracy revealed for stroke cases other than stroke -It needs to be further adjusted to increase recall and F1 scores for cases. An overall accuracy of 95.13% indicates the effectiveness of the new tree classification in predicting traumatic brain injury.

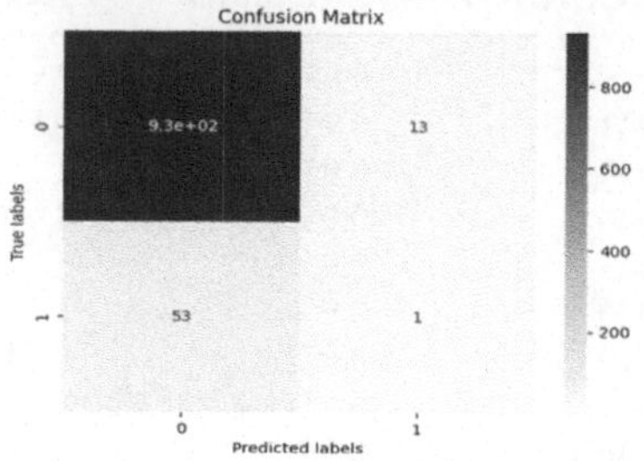

Figure 2. Confusion matrix of extra tres classifier.

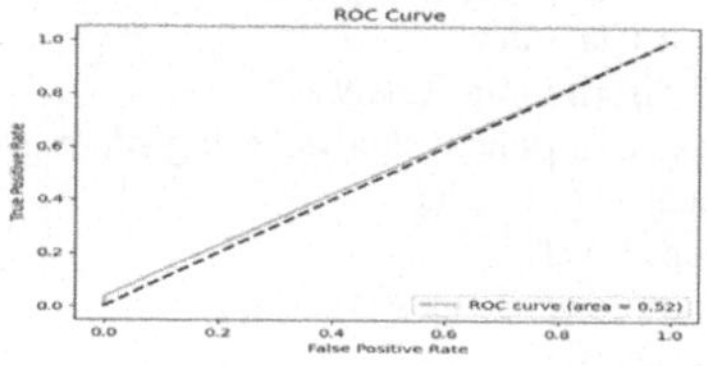

Figure 3. ROC CURVE of the extra tres classifier.

6 COMPARATIVE STUDY

This study evaluated various machine learning models for predicting brain strokes using a Kaggle dataset. Models tested included MLP Classifier, XGBoost, Linear Discriminant

Analysis, PCA + Logistic Regression, CatBoost, Gradient Boost, Quadratic Discriminant Analysis, Ridge Classifier, Gaussian Process Classifier, Perceptron, Stochastic Gradient Descent, Extra Trees Classifier with Grid Search, and Cross-validated Extra Trees Classifier. The Cross-validated Extra Trees Classifier achieved the highest accuracy (95.1%), precision (0.97), and recall (1.0), outperforming the others. XGBoost followed with a 94.3% accuracy and 0.99 recall. Linear Discriminant Analysis and PCA + Logistic Regression both had perfect precision and recall but slightly lower accuracy. Quadratic Discriminant Analysis had the weakest performance with 51.9% accuracy and 0.51 recall. The superior performance of the Cross-validated Extra Trees Classifier highlights its ability to handle complex data and prevent overfitting, demonstrating the effectiveness of advanced ensemble methods in medical prediction.

7 RESULTS

In our comparative study of stroke prediction models, several leading algorithms showed remarkable accuracy [18]. The Support Vector Machine (SVM) demonstrated an adequate performance of 95%, confirming its effectiveness in handling complex models. K-Nearest Neighbors (KNN) exhibited a competitive accuracy of 82%, highlighting its ability to capture local relationships. Among the ensemble methods, random forest [19] and gradient boosting achieved robust accuracies of 95% and 94%, respectively. The proposed new tree classifier emerged as a leader, outperforming many models with an impressive accuracy of 95.13%. The Extra Trees Classifier demonstrated superior performance in key metrics compared to other classifiers. It showed a higher True Positive Rate (TPR) and True Negative Rate (TNR), indicating better accuracy in identifying both stroke and non-stroke cases. The exceptional performance of this model emphasizes the importance of considered algorithm selection in brain tumor modeling and positions it as a promising solution for accurate tumor prediction (Table 1). As we can also see the comparision of the models visually in Figure 4.

Table 1. Performance of different classifiers on the brain atroke dataset.

MODEL	ACCURACY	PRECISION	RECALL
MLP classifier	0.946	0.95	1
Xgboost Classifier	0.943	0.95	0.99
Linear discriminant analysis	0.949	0.95	1
Principle component analysis+ logistic regression	0.946	0.95	1
Catboost	0.943	0.95	1
Gradient boost	0.946	0.95	1
Quadratic discriminant analysis	0.519	0.97	0.51
Ridge classifier	0.946	0.95	1
Gaussian process classifier	0.946	0.95	1
Perceptron	0.946	0.95	1
Stochastic gradient descent	0.946	0.9	0.88
Extra trees classifier with gridsearch	0.946	0.95	1
Cross-validated extra trees classifier	***0.951***	***0.97***	***1***

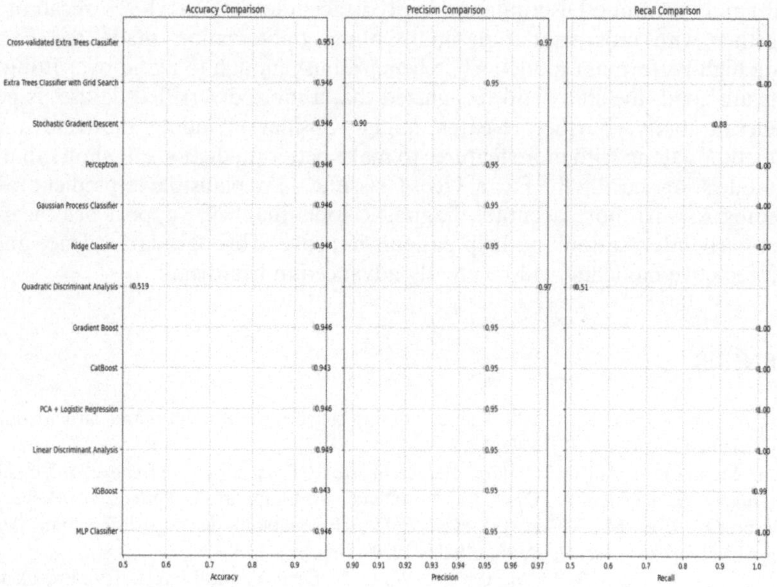

Figure 4. Comparision of performance metrics of the models used.

8 DISCUSSION

A comparative study underscores the efficacy of ensemble methods like Random Forest and Gradient Boosting for predicting traumatic brain injury. A new tree classifier outperforms these methods, emphasizing intricate algorithmic interactions. Addressing 210 missing data points during preprocessing highlighted the need for dynamic strategies. Hyperparameter optimization using PyCaret and Optuna enhanced algorithm performance, with hybrid approaches showing potential for improved accuracy. There were trade-offs between true positives and false positives, which are critical in clinical settings alike. Our new tree classifier was resistant to these and thus appears more stable for both early interventions (such as stress modifying physical or psychological training) and personalized injury prediction strategies. The Extra Trees classifier performed well in areas such as TPR and (True Positive Rate) and overall had a higher performance improvement than other classifiers. It has less False Positive Rate (FPR) and low False Negative Rate(FNR), i.e, the error to minimize misclassification. Being an ensemble method, the Extra Trees Classifier reduces variance by randomization and is also known to handle imbalanced data with high accuracy, precision & recall. The nearly Perfect Score of this model across different data subsets emphasises the robustness of its high performance. While the high recall means it is making good on its true positive detection... And perfect precision implies testabilty innevitably refers to how well we measure this with or without false alarms. All these properties make Extra Trees Classifier very flexible and popular among the Kagglers for brain stroke prediction.

9 CONCLUSION

This research is Media Overlay of paper which cites different Machine Learning models for forecasting Brain strokes on a data from kaggle. The performance of several classifiers was tested, the Extra Trees Classifier being selected because it can handle well complex data and feature interactions. It performed better than models such as XGBoost (94.3% accuracy,

0.99 recall) and Quadratic Discriminant Analysis(accuracy of just 51.9%, recall of only.51). This, together with cross-validation and Optuna optimization made the Extra Trees Classifier a high-performant model, while being robust enough to avoid overfitting. Though the results are good, the study still recognized that a more diversified dataset is needed for better generalizability. Further research might consider including more advanced deep learning methods along with more features to make better predictions. It shows that machine learning models—especially the Extra Trees Classifier—are plausible to predict brain stroke, hence leading toward more accurate diagnostic tools that will support healthcare professionals for early intervention and help patients improve. Through data science and clinical practice, predictive modeling will be greatly advanced in healthcare.

REFERENCES

[1] Elloker, T. and Rhoda, A.J. (2018). The relationship between social support and participation in stroke: A systematic review. *Afr. J. Disabil.*, 7, 1–9.

[2] Katan, M. and Luft, A. (2018). Global burden of stroke. *Seminars in Neurology*, 38, 208–211.

[3] Bustamante, A., Penalba, A., Orset, C., Azurmendi, L., Llombart, V., Simats, A., Pecharroman, E., Ventura, O., Ribó, M., Vivien, D., *et al.* (2021). Blood biomarkers to differentiate ischemic and hemorrhagic strokes. *Neurology*, 96, e1928–e1939.

[4] Xia, X., Yue, W., Chao, B., Li, M., Cao, L., Wang, L., Shen, Y. and Li, X. (2019). Prevalence and risk factors of stroke in the elderly in Northern China: Data from the national stroke screening survey. *J. Neurol.*, 266, 1449–1458.

[5] Alloubani, A., Saleh, A. and Abdelhafiz, I. (2018). Hypertension and diabetes mellitus as predictive risk factors for stroke. *Diabetes Metab. Syndr. Clin. Res. Rev.*, 12, 577–584.

[6] Krishna, Vempati, *et al.* (2021). Early detection of brain stroke using machine learning techniques. *2021 2nd International Conference on Smart Electronics and Communication (ICOSEC)*. IEEE.

[7] Emon, Minhaz Uddin, *et al.* (2020). Performance analysis of machine learning approaches in stroke prediction. *2020 4th International Conference on Electronics, Communication and Aerospace Technology (ICECA)*. IEEE.

[8] Mahesh, Kunder Akash, *et al.* (June. 2020). Prediction of stroke using machine learning. *Conference Paper·*.

[9] Dritsas, E., Alexiou, S., Konstantoulas, I. and Moustakas, K. (2022). Short-term Glucose Prediction based on Oral Glucose Tolerance Test Values. *In Proceedings of the International Joint Conference on Biomedical Engineering Systems and Technologies—HEALTHINF*, Lisbon, Portugal, 12–15 January 2022; Volume 5, pp. 249–255.

[10] Edinburgh, Robert M., *et al.* (2018). Preexercise breakfast ingestion versus extended overnight fasting increases postprandial glucose flux after exercise in healthy men. *American Journal of Physiology-Endocrinology and Metabolism* 315.5: E1062–E1074.

[11] Shoily, Tasfia Ismail, *et al.* (2019). Detection of stroke disease using machine learning algorithms. *2019 10th International Conference on Computing, Communication and Networking Technologies (ICCCNT)*. IEEE.

[12] Jeena R. and Kumar S. (2016). Stroke prediction using svm, in *2016 International Conference on Control, Instrumentation, Communication and Computational Technologies (ICCICCT)*, pp. 600–602, IEEE.

[13] Sandercock, Peter AG, *et al.* (2011). The international stroke trial database. *Trials* 12.1: 101.

[14] Delpont, B., Blanc, C., Osseby, G., Hervieu-Bègue, M., Giroud, M. and Béjot, Y. (2018). Pain after stroke: A review. *Rev. Neurol.*, 174, 671–674.

[15] Kumar, S., Selim, M.H. and Caplan, L.R. (2010). Medical complications after stroke. *Lancet Neurol.*, 9, 105–118.

[16] Gittler, M. and Davis, A.M. (2018). Guidelines for adult stroke rehabilitation and recovery. *JAMA*, 319, 820–821.

[17] Pandian, J.D., Gall, S.L., Kate, M.P., Silva, G.S., Akinyemi, R.O., Ovbiagele, B.I., Lavados, P.M., Gandhi, D.B., Thrift, A.G. (2018). Prevention of stroke: A global perspective. *Lancet*, 392, 1269–1278.

[18] Bhatnagar S., Raja S. P. and Preeya A. (2023). Improving iris recognition systems with CNN based image processing and machine learning approach, *2023 International Conference on New Frontiers in*

Communication, Automation, Management and Security (ICCAMS), Bangalore, India, pp. 1-7, doi: 10.1109/ICCAMS60113.2023.10525963.

[19] Bhatnagar S., Lakshmi N., A. A and V. S. M. Y. (2024). Development of a crop recommendation system through the use of various machine learning algorithms, *2024 Second International Conference on Emerging Trends in Information Technology and Engineering (ICETITE)*, Vellore, India, pp. 1–6, doi: 10.1109/ic-ETITE58242.2024.10493758.

AUTHORS BIOGRAPHY

HIMAVARSHINI BEEDALA currently pursuing an Honors program alongside her Bachelor of Technology (B.Tech) in Computer Science with a specialization in Artificial Intelligence at Manipal Institute of Technology, showcases a robust skill set that includes Machine Learning, Deep Learning, Computer Vision, Natural Language Processing, Object-Oriented Programming, and Data Structures.

SUDHAMSH GVS, a third-year undergraduate student currently enrolled in an Honors program alongside his B.Tech in Computer Science with a specialization in Artificial Intelligence at Manipal Institute of Technology. His academic journey reflects a strong commitment to exploring the intersection of computer science and medical imaging, showcasing his passion for leveraging technology for impactful applications.

ROOPESHWAR REDDY is currently pursuing a 3rd year B.Tech (Hons) degree in Computer Science and Engineering at Manipal Institute of Technology, Bengaluru. Alongside his academic pursuits, Roopeswar possesses a diverse skill set in the field of computer science, including expertise in coding, database management systems, machine learning, deep learning, natural language processing, and more.

Dr. Shaleen Bhatnagar has completed BE in information Technology, M. Tech in computer science & Engineering from Rajasthan, and Ph.D. from Bengaluru India. She is currently an Assistant Professor Senior Scale with the Department of Computer Science & Engineering, Manipal Institute of Technology Bengaluru, Manipal Academy of Higher Education, Manipal, India. She has teaching experience of 14 years. Her current research interests include Biometrics, Computer Vision, Pattern Recognition, Image Processing, Image Analysis and Machine Learning.

Data Science & Exploration in Artificial Intelligence – Gururaj et al. (Eds)
© 2025 the Author(s), ISBN: 978-1-032-96390-7

Automatic helmet detection and number plate recognition system using YOLO

H. Vaishnavi Upadhya
Manipal Institute of Technology, Manipal Academy of Higher Education, Manipal, India

R. Manoj
Assistant Professor, Manipal Institute of Technology, Manipal Academy of Higher Education, Manipal, India

ABSTRACT: Enforcing traffic related rules is a highly challenging tasks, especially in highly populated countries like India or China. Motorcyclists, when riding with no helmets, are doing so while risking their own life. Humanly monitoring all these motorcycles is very tedious and calls for the need for automating these tasks. Vehicle detection and classification can be used to identify traffic rule defaulters. The purpose of this project is to ensure road safety and develop a system for the automated identification of helmetless riders on vehicles. In the proposed system, we initially detect two-wheelers who aren't wearing their helmets and then their license plate using computer vision techniques like YOLO and MobilenetSSD. This system has an accuracy of about 85%.

1 INTRODUCTION

The increased use of motorcycles in the present world has resulted in a drastic increase in accidents and injuries related to traffic. Motorcycle riders not using helmets is one of the major causes. Identification of traffic law violators is highly desirable because of multiple challenges such as occlusion, illumination etc. According to the Ministry of Road transport and Highways over 1.58 lakh two-wheeler road accidents causing 56,873 deaths in 2020 alone. Every hour, four two-wheeler riders die as a direct result of not wearing a safety helmet. The two-wheeler is a favored means of mobility, but it carries a considerable risk due to its lack of protection. Bike riders are strongly advised to wear a helmet to minimize the risk of injury. However Current video surveillance methods require significant human efforts. All these systems are often ineffective because of the necessity of humans mediating these tasks, which causes effectiveness of this system to diminish over time due to human weariness and error. It is also inefficient because only one automobile rider can be apprehended at once out of a group of several two-wheeler riders.

The road is equipped with CCTV cameras, which record footage around the clock to monitor and record the flow of traffic. Now, manual processes are used in India for both the detection of offences and the administration of punishments to those who commit them. The traffic enforcement personnel stationed in the control center monitor the real-time footage, and when a breach is found, the license plate is logged, and a video frame is captured for confirmation. The infractions are then entered into the database, and the culprit is subsequently notified, culminating in the imposition of a fine. Due to the manual nature of this system, the staff must constantly monitor the screens to guarantee that no violations go undetected despite the high volume of traffic. As a result, this technique must be automated. [11] Automatic helmet detection and license plate recognition system is a comprehensive method for identifying violations of traffic laws, such as not wearing a helmet. The proposed

 DOI: 10.1201/9781003589273-19

system builds an automated system for detection of traffic law offenders who are not wearing helmets along with their license plate number using computer vision techniques such as segmentation, image classification and object detection.[1]

2 RELATED WORK

A smart traffic surveillance system is proposed that can discover non-helmeted bikers from the frame of video in real time using a convolutional detector. When the system detects a violation, it sends real-time data to the Traffic Monitoring API. The License Plate Recognizer API uses CNN and OCR to identify and recognize the number on the number plate from the frame if it is visible, and then sends the outcome to the API of Traffic Monitoring. On the validation set, the convolutional object detector detected non-helmeted motorcyclists with an accuracy of 84% [7] A probable traffic infringement is detected by the smartphone app, which captures the photo, time, date and motorbike plate number. The system then creates evidence in support of the transit violations that the transit cop will issue. They conducted an experiment with this android application with accuracy of 77% [11] They identify automobiles using YOLO-based object detection, and then each vehicle is evaluated for suitable offences, such as not wearing a helmet or crossing in a cross- walk, according to the suggested technique. CNN (Convolutional neural network) classifier is used to detect helmet violations. Following the detection of violations, the vehicle numbers of the respective violators are obtained using OCR, and the violators are notified [10]

The coco data set was used to train the system. Violators of traffic lights and two wheelers who are not wearing helmets will be detected by the system. It has been designed to be intelligent enough to eliminate human reliance while simultaneously being sturdy enough to function under severe situations.[7] Suggested research work illustrates a way for implementing safety protocols by automatically detecting bikers wearing and not wearing a helmet. Developed with OpenCV, a descriptor known as support vector classification (SVC). The OpenCV Library approach is used to identify a bike rider in a inspected frame of video. In addition, Linear SVC to determine if the cyclist is wearing a protective helmet. The information gathered on the number of motorcyclists who wear and don't wear a protective helmet with corresponding timestamps is stored in a MySQL database and graphical views were used to visualize the data in the built desktop interface programmed. It is time-efficient approach to deal with traffic rules, with 87.6% model accuracy.[5] The algorithm is fed a video of traffic on an open roadway, and it is able to identify moving objects in the scene Yolo v3, If there are more than two riders, they are counted using the vertical projection of a binary picture. [8] With the help of the YOLO real-time object detection, non-helmeted motorcyclists may identify LP in real time. This suggested technique employed a single convolution neural network could automatically identify from a video stream the LP of a rider without a helmet.[3]

3 METHODOLOGY

3.1 *Helmet detection*

Once automobiles have been detected, motorcycles are considered for the task of classification of helmet. With the use of a suitable bounding box, individual automobiles are cut from the CCTV frame. If the vehicle is classified as a two-wheeler, it is further cropped, taking into account only the upper portion of each vehicle. Merely taking into account the top half of the image could reduce the computing complexity of the system. This half is fed into the classifier, which determines whether the image fits in the non-helmet or helmet category.

3.1.1 *Object detection*

The traditional machine learning metrics are used to detect and classify the helmets. With the fast-growing deep learning technology in recent days the algorithms of object detection turn to the one best convolution neural network with highest possible neural networks are used to detect the safety helmets also strategies like increasing the number of anchors, multi scale training are used to maximize the accuracy of helmet detection. But also, these methods may have some drawbacks in the aspects of sharpness of the image preprocessing proportion of the object and the difference in color between the foreground and background. We are comparing the latest technologies used for helmet detection. [4]

3.1.2 *Study on various techniques for helmet detection*
A CNN classifier is applied to identify motorbike riders among the other moving objects. Again, a second con volution neural network (CNN) classifier is utilized to identify motorcyclists who aren't protecting their heads with helmets LBP and GLCM are used to obtain the analytical characteristics. Helmet recognition is performed using a combination of GLCM analytical features and BP artificial neural network helmet recognition. For person detection, the HOG method is utilized. In the course of this research, people who did not wear helmets were identified and their license plate information was collected. Two You Only Look Once (YOLO) models were used to identify riders who were and were not wearing helmets.[3]

This model can recognize all of the COCO dataset's categories. In the first group of images, a person is recognized rather than a motorcycle. The helmet dataset is used to train the second iteration of the YOLOv2 model. The license plate is recognized using an open ALPR system. The vehicles are only identified when they reach a specified horizontal line. The head portion is extracted from the upper one-third of the original image. The approach of background subtraction is utilized in the detection of moving vehicles. Motorcycle classification can be done by first detecting RoI with descriptors like LBP and HAAR, and then using classifiers. Later, the system is trained to recognize helmets using PCA and other machine learning techniques.[9]

Video surveillance is used to identify non-helmet cyclists to be identified. To remove the pedestrians and other items, background subtraction is used. To classify the items, a trained accuracy and speed. Different depths of convolution model are applied. To classify the items, the COCO model is employed. Various varieties of helmets were discovered via web scraping. A human face identified as a person was recognized when riders carried their helmets on the body of the bike rather than wearing them. The bike's license plate is detected using Optical Character Recognition (OCR) utilizing tesseract.[5][8]

3.2 *License plate recognition*

License Plate Recognition is a cutting-edge machine vision technology that allows for the automatic identification of automobiles based on their license plates, without the need for human intervention. Because the Intelligent Transportation System is used in monitoring traffic. This technology is also gaining ground in a variety of other applications, such as the collection of highway tolls, checkpoints at international borders and customs, and parking access control systems. Plate extraction relies on the precision of character segmentation and character identification to get accurate results.

Plate extraction relies on the precision of character segmentation and character identification in order to get accurate results. Using classification algorithm riders wearing helmet and riders who are not wearing helmet are detected. Later all classes of the data sets are detected. In the first set the person is identified instead of a motorbike. In the second model which is already trained with helmet datasets non helmeted riders are detected easily and their number plate is recognized. Image segmentation is done for recognizing the license plate using the models of recognition like SSD, R-CNN, HOG where images were segmented and trained. Characters were segmented and trained using above mentioned techniques also the trained model license plate was tested.[1]

3.2.1 *Number plate detection*

Description of a license plate can be made as "a plastic or metal plate applied to an automobile that helps to uniquely identify them." This is how a license plate is generally described. A computer, on the other hand, cannot comprehend this definition in its entirety. Some of the algorithms are oriented toward deep learning and these common computer vision techniques. This processing stage is crucial because the precision with which number plates are extracted is crucial for character segmentation and recognition.

3.2.2 *Character segmentation*

License plate character seg mentation systems consider the fact of having unique backdrop color and characters. Using various optical character recognition systems characters are segmented before classification. In the character segmentation phase, the number plate that was recognized during the extraction stage is checked. After plate localization, a variety of strategies for segmenting each char acter have been devised. Projection-based, Hough transform, Pixel connectivity, and plate models are some of the most used segmentation techniques.[10] [4]

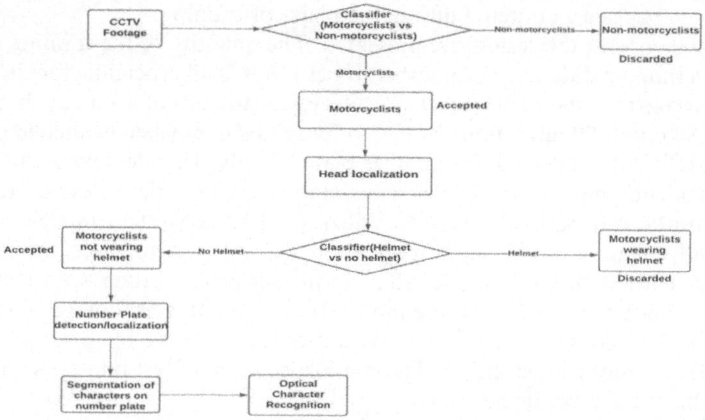

Figure 1. Design.

3.2.3 *Character recognition*

Character segmentation inefficiency has an impact on a character's ability to be recognized. For character identification, isolated characters are necessary. Many classifications strategies call for inputs of a fixed size into the learning model. Before the classification stage begins, the input segments are rescaled. This is necessary since the output of the segmentation stage is of varying sizes. Because their relative location differs. Each segment is classed individually to one of the possible values.

3.2.4 *Study on various techniques for license plate recognition*

A 3-stage recognition system for license plate based on Mask R-CNN show that the architecture proposed can give mAP rates of up to 91% and will identify license plates over 0˜60 degrees angle. The other paper proposes a new system where fully convolutional network for random positioned item detection is made using a combination of hierarchical and multiscale feature maps. This work proposes a straightforward and efficient character segmentation technique that identifies significant symbols and uses them as a basis for angle estimation and vertical projections. Character recognition is the goal of the wide learning system, which includes a multilayer auto-encoder with mapped feature nodes. [2] When it comes to traffic congestion, an ANPR (automated number plate recognition) system is crucial. This will assist in reducing the various types of traffic offences.[12] Automated number plate recognition technology is used in advanced models for monitoring and detecting missing and unlicensed cars. A novel OKM-CNN approach for excellent detection and identification of LPs was provided in this research. There are multiple phases to the suggested OKM-CNN paradigm.[6] The IBA and CCA models are used in the initial stage to perform LP localization and identification. After that, a CNN model is used to recognize the characters in the LP image. The presented OKM-CNN model could be used as a basis for future reference for various fields as smart parking, collecting toll fees, smart parking, and traffic surveillance are just a few of the services available. The different datasets are used to conduct the research namely HumAIn 2019 dataset and Stanford Cars, and FZU Cars. [9][11]

4 RESULT ANALYSIS

In Google Collab we have trained the YOLO model. The GPU availability and training time are the major setbacks encountered during the phase of training.

Based on 400 epochs, the results are presented. The quantity of the training data, caliber of the model's training data and the number of iterations is all crucial factors in determining the metrics. Access to better GPU directly results in the amount of accuracy. In addition, the model found 84 out of 99 infractions with zero False Positives when evaluated on live feeds. This method efficiently detected the number plate of only the riders who had violated the rules, thereby identifying transgressions. Thus, implementing strict rules and regulations of violators of traffic can be made possible utilizing AI-based system models to raise rider awareness and enhance traffic safety regulations. Compared to Yolo models, the SSD MobileNetV2 lightweight CNN model offers a quicker training time. Considering that the license plate will not be picked up by the model from the photos or footage. A criterion used to assess the SSD model's efficiency is the confidence score. Above the bounding boxes, each class's confidence score will be shown. The confidence score of test photos is taken and used to calculate the average confidence score.

The functionality of the proposed system is higher in comparison with most of literature performance because Yolo v3 outperforms the earlier algorithms used to perform violation detection. Since no literature has undertaken tests for combi nations of all the classes indicated previously, there is not a direct comparison present.

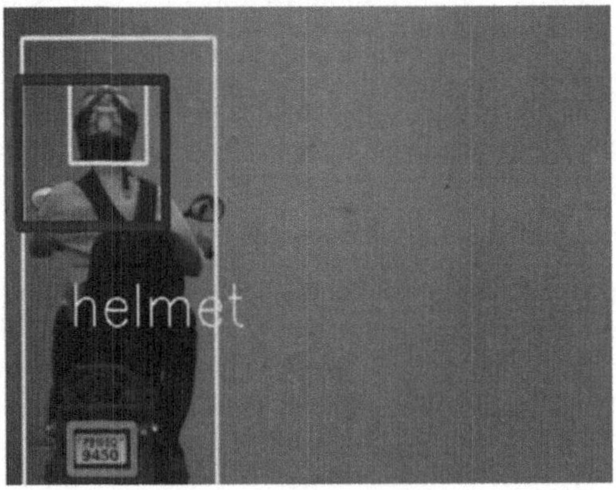

Figure 2. Helmet rider.

Figure 3. LPR of helmetless rider.

5 CONCLUSION

A fundamental prerequisite for creating an efficient and ideal city is through traffic surveillance. Helmet detection for riders without helmet, as well as number plate identification, can be used to penalize riders who violate the rules. To make things easier, domains like Image Processing and Machine Learning are applied. Rapid advancements in machine learning and image processing techniques have increased the number of built-in models and online tools that provides solution for various purposes. The process of recognizing different objects has evolved with help of machine learning giving about 85% of accuracy. This system uses YOLO and MobileNetSSD because YOLO uses one complete image in a single instance to anticipate the bounding box coordinates and class probabilities for the boxes and SSD may be trained from beginning to end as the SSD architecture consists of a single neural network that learns to predict bounding box positions and categorize these positions in one single run.

This study found that image processing approaches produced satisfactory results. However the outcomes from machine learning models were better, the best outcomes were obtained when machine learning and image processing were combined. This will aid future helmet detection and license plate recognition studies.

REFERENCES

[1] Alborzi, Y., Mehraban, T. S., Khoramdel, J. and Ardekany, A. N. (2019, November). Robust real time lightweight automatic license plate recognition system for Iranian license plates. In *2019 7th International Conference on Robotics and Mechatronics (ICRoM)* (pp. 352–356). IEEE.

[2] Huang, S., Xu, H., Xia, X. and Zhang, Y. (2018, December). End-to-end vessel plate number Detection and recognition using deep convolutional neural networks and LSTMs. In *2018 11th international symposium on computational intelligence and design (ISCID)*

[3] Jamtsho, Y., Riyamongkol, P. and Waranusast, R. (2021). Real-time license plate detection for non -helmeted motorcyclist using YOLO. *Ict Express*, 7(1), 104–109.

[4] Khan, F. A., Nagori, N. and Naik, A. (2020, July). Helmet and number plate detection of motorcyclists using deep learning and advanced machine vision techniques. In *2020 Second International Conference on Inventive Research in Computing Applications (ICIRCA)*
[5] Padmini, V. L., Kishore, G. K., Durgamalleswarao, P. and Sree, P. T. (2020, September). Real time automatic detection of motorcyclists with and without a safety helmet. In *2020 International Conference on Smart Electronics and Communication (ICOSEC)* (pp. 1251–1256). IEEE.
[6] Pustokhina, I. V., Pustokhin, D. A., Rodrigues, J. J., Gupta, D., Khanna, A., Shankar, K., and Joshi, G. P. (2020). Automatic vehicle license plate recognition using optimal K-means with convolutional neural network for intelligent transportation systems
[7] Samuel, S., Reghunadh, S., Ashwin, M. K., Sabu, S., Nair, S. S. and Varghese, R. R. (2020). An I Intelligent Traffic Monitoring System for Non-Helmet Wearing Motorcyclists Detection. In *2020 International Conference on Data Analytics for Business and Industry: Way Towards a Sus-tainable Economy (ICDABI)* (pp. 1–5). IEEE.
[8] Saumya, A., Gayathri, V., Venkateswaran, K., Kale, S. and Sridhar, N. (2020, September). Machine learning based surveillance system for detection of bike riders without helmet and triple rides. In *2020 International conference on smart electronics and communication (ICOSEC)*
[9] Talaulikar, A. S., Sanathanan, S. and Modi, C. N. (2019). An enhanced approach for detecting helmet on motorcyclists using image processing and machine learning techniques. In *Advanced Computing and Communication Technologies: Proceedings of the 11th ICACCT 2018* (pp. 109–119). Springer Singapore.
[10] Tonge, A., Chandak, S., Khiste, R., Khan, U. and Bewoor, L. A. (2020, November). Traffic rules violation detection using deep learning. In *2020 4th international conference on electronics, communication and aerospace technology (ICECA)* (pp. 1250–1257). IEEE.
[11] Valencia, J. F., Ramírez-Guerrero, T., Castañeda, L. F. and Toro, M. (2020, May). Automatic tection of number-plate and traffic infractions of motorcyclists by Intelligent Transportation tems. In *IOP Conference Series: Materials Science and Engineering*
[12] Yaseen, N. O., Al-Ali, S. G. S. and Sengur, A. (2019, November). Development of new Anpr dataset for automatic number plate detection and recognition in north of Iraq. In *2019 1st International Informatics and Software Engineering Conference (UBMYK)* (pp. 1–6). IEEE.

Data Science & Exploration in Artificial Intelligence – Gururaj et al. (Eds)
© 2025 the Author(s), ISBN: 978-1-032-96390-7

LLMHC: From "BERT to GPT-3": Relative viewpoints on Large Language Models (LLM) in psychological well-being expectations

U. Ananthanagu & Pooja Agarwal
Department of Computer Science and Engineering, PES University, India

ABSTRACT: Natural Language Processing (NLP) has been significantly affected by the rapid advancement of Large Language Models (LLMs), with applications venturing into issues including psychological wellness forecasts. This work provides an exhaustive evaluation of numerous LLMs, for example, Alpaca, BERT, GPT-3.5, and GPT-4, focusing on the binary and multi-class categorization of social media text data based on the prediction of mental health. The improvement of LLMs has altered NLP, bringing about the rise of a few chatbots that use the transformer design to tackle the force of LLMs, including ChatGPT, BARD, ERNIE, Dalai, and others. This paper adds to the rising measure of writing that shows LLMs might be used in mental health assessments by researching this complex language model's true capacity. This paper focuses on how cutting-edge technologies, like LLMs, are improving patient care and how these models could be used to improve mental health applications. In general, outcomes add to the discussion on the utilization of state-of-the-art advancements in medical care by offering definite information on how different LLMs act in mental health prediction errands, including depression, stress, and anxiety.

1 INTRODUCTION

More than one billion individuals worldwide experience the ill effects of psychological well-being ailments, highlighting the basic requirement for productive determination and treatment plans [1]. Early determination, mediation, and follow-up are fundamental components of good psychological well-being treatment. An expanding body of studies has examined the use of bio-behavioral signs to recognize mental health conditions in individuals, given the bio-psycho-social aspects of most mental diseases. Strangely, posts via social media have turned into a potential method for finding out about someone's emotional well-being. By looking at language, behavior, and contextual indicators found in online posts, researchers can distinguish and appreciate psychological wellness problems by using the digital footprint that individuals have via social media sites. This original methodology can offer a tactful and receptive method for following psychological well-being, opening doors for early detection and intervention [2].

Four primary phases of language model development have been recognized in the literature: SLM-statistical language models, NLM-Neural language models, LLM-Large language models (LLM), and PLM- Pre-trained language models, Language model investigation has procured a basic premium in these stages. IR-Information retrieval and NLP-Natural language processing vigorously rely upon statistical language models (SLMs), which were made during the 1990s and rely upon statistical learning strategies. Considering the most recent setting, these models gauge the next word, working under the Markov assumption. n-gram language models are SLMs that have a fixed context length, like bigram and trigram language models.

By applying neural networks, including plans like multi-layer perceptrons (MLP) and RNNs-recurrent neural networks, NLMs-Neural Language Models transform the display of

DOI: 10.1201/9781003589273-20

word sequence probabilities. Pre-trained Language Models (PLMs), which use refined neural network geographies and large-scale datasets, have transformed into a viewpoint-changing system in natural language processing. Enormous volumes of message information are utilized to prepare PLMs, including GPT (Generative Pre-trained Transformer) and BERT (Bidirectional Encoder Representations from Transformers), which assist them in obtaining an exhaustive handle of language design, semantics, and setting. The limit of PLMs to accumulate bidirectional contextual data is one of its essential qualities, which empowers them to comprehend associations more likely among words and the setting where they are used.

Large Language Models (LLM): These modern models for natural language processing are recognized for their gigantic complexity and scale. These models are equipped for playing out an extensive variety of natural language understanding errands because they use profound learning methods and large amounts of preparing information to get a careful handle on the language. Notable plans like BERT (Bidirectional Encoder Representations from Transformers) and GPT (Generative Pre-trained Transformer) are instances of large language models. The presentation of ChatGPT, LLM, has accumulated critical interest from the public showing the effect of LLMs on the artificial intelligence local area [3].

The possible value of Large Language Models (LLMs) as effective tools for translating client's mental states through written language is highlighted by the essential role that natural language plays in the evaluation and treatment of mental health. Applications incorporate psychological wellness-mindful individual chatbots and intelligent assistants for emotional well-being experts, counselors, and supporters. They also incorporate online public health monitoring systems. However, encouragingly, there is a perceivable absence of examination on grasping, surveying, and further developing LLMs, planned particularly for psychological wellness exercises. Taking this uphold is pivotal for opening the maximum capacity of LLMs in the advancement of psychological wellness innovation and administration [4].

2 A SYSTEMATIC REVIEW

2.1 *Large language model evolution*

Word embeddings into complex transformer-based structures have been the evolution of LLMs, demonstrating the increasing comprehension and use of contextual information in natural language understanding problems. Research on how to make these potent language models more efficient, interpretable, and ethically sound is still ongoing. Figure 1 depicts a more thorough history of LLM development. The objective of early techniques like Word2Vec and GloVe was to show clients how to learn dispersed word representations. Even though they were not conventional LLMs, these methods set the establishment for recognizing word semantic linkages.

LSTM-Long Short-Term Memory and RNN-Recurrent Neural Networks and tended to the successive idea of language, permitting models to think about context dependencies over the long run. In any case, these models had restrictions in capturing long-range dependencies.

Vaswani *et al.*'s introduction of the Transformer design in 2017 brought about a significant shift. Bidirectional context understanding was given by BERT (Bidirectional Encoder Representations from Transformers), which Google introduced in 2018. This upset pre-training for undertakings including natural language comprehension. This demonstrated that large-scale unsupervised pre-training works well.

The GPT Series, or Generative Pre-trained Transformer: Model size limits were moved by OpenAI's GPT series (GPT-1, GPT-2, and GPT-3), with GPT-3 hanging out specifically because of its enormous 175 billion parameters. These models performed very well in the scope of NLP tasks.

BERT Variations: Researchers examined the scope of changes and modifications to the BERT structure, bringing about models like ALBERT, RoBERTa, and others. These varieties were intended to work on model execution, minimize computing demands, and increase training efficiency. Integrated Models: In 2020, Google introduced the T5-Text-To-Text Transfer Transformer, which treated NLP occupations as text-to-text issues and offered a uniform structure for them. The arrangement and training of models for different tasks were simplified by this technique. GPT-4 and Upcoming Improvements: The plan is reached out by the normal arrival of large-scale language models like GPT-4. The issues of interpretability, ethical contemplations, and the environmental effect of training big models are expected to be the principal areas of study and advancement [4].

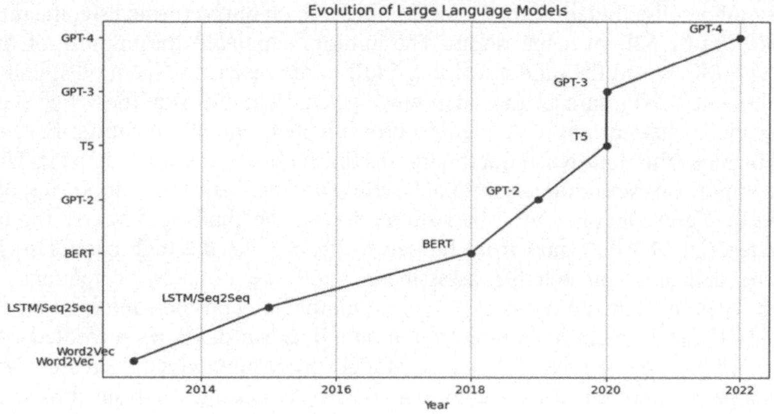

Figure 1. Evolution of Large Language Models.

2.2 *Related work*

The current explosion in LLMs has demonstrated an encouraging pattern of large pre-trained models to address a range of activities in the field of psychological health. These tasks are outlined below. Tom, B., Brown, *et al.* showed that significant additions in NLP performance might be accomplished by pre-training language models on a large corpus of text and fine-tuning certain tasks. Current NLP frameworks battle to achieve novel language tasks, which people can do with few models or basic directions. Expanding language models to a larger scale, such as GPT-3 with 175 billion parameters, upgrades task-independent few-shot performance, incidentally, in any event, accomplishing equality with past state-of-the-art techniques for tweaking. GPT-3 works well on a range of NLP datasets, covering language translation, replication tasks, answering questions, on-the-fly reasoning, domain adaptation, and arithmetic, without the need for inclination updates or fine-tuning. Some datasets keep on giving difficulties to GPT-3's few-shot learning, and strategic issues relating to preparing on huge web corpora are featured [5][6]. The PaLM-Pathways Language Model, a densely active Transformer language model prepared on 540 billion parameters, was introduced by Aakanksha Chowdhery *et al.* It presents cutting-edge results on language understanding and generation benchmarks, such as multi-step reasoning tasks and the BIG-bench benchmark, and discusses the upsides of scalability in few-shot learning. It additionally stresses how capable PaLM is in creating source code and dealing with multilingual obligations. With regard to size, the exploration offers a careful examination of bias, toxicity, and training data memorization. Potential mitigating techniques are introduced alongside ethical issues with large language models. BIG-bench task undertakings are referenced in the concentrate for estimating upcoming huge language models. It illustrates an issue with moderate moves toward reasoning and explains the best way to settle it utilizing bill calculation. The training of PaLM models with 8 billion and 62 billion parameters is also referenced in the review [7]. Fine-tuning language models using instruction datasets to enhance model performance and generalize to new tasks, as per a proposition by Hyung Won Chung *et al.* Using chain-of-thought data, the review research scaled the model size, number of tasks, and fine-tuning. It exhibits outstanding performance across a scope of model classes, designs, and evaluation benchmarks. With an emphasis on expanding the number of positions, scaling the size of the model, and fine-tuning the chain-of-thought data, this review examines guidance finetuning. Using various datasets composed like instructions, the authors further developed language models. They conducted a performance survey of the models using several model classes (PaLM, T5, U-PaLM), settings (zero-shot, few-shot, CoT), and evaluation benchmarks (MMLU, BBH, TyDiQA, MGSM, open-ended generation). The authors distributed Flan-T5 checkpoints, which outperformed significantly bigger models like PaLM 62B in a few shots. The authors expanded the number of finetuning positions to 1.8K, scaled the model size to a 540B-parameter language model, and included chain-of-thought (CoT) data in the finetuning system. They showed how fine-tuning jobs and higher model sizes prompted critical improvements in model execution. Execution was upgraded on all evaluations when finetuning was finished pair with CoT data [8]. The study's pairwise comparison-based human feedback setting and the instruction following tasks were introduced by Yann Dubois *et al.* The authors discuss the challenges in creating large language models (LLMs) that gain from human feedback, like the high cost of social information, the deficiency of reliable assessments, and the absence of reference strategy executions. They recommend AlpacaFarm, a simulator that considers minimal expense, trial and error, and improvement in gaining from input [9]. Edward, J *et al.* presented the LoRA-Low-Rank Adaptation of Large Language Models procedure, which decreases the number of trainable parameters for downstream assignments by adding Pre-trained model weights frozen and trainable rank decomposition matrices included into every layer of the Transformer architecture., demonstrating that, in contrast with comprehensive fine-tuning, LoRA might decrease the number of trainable parameters by 10,000 times and the computer hardware needed by three times, particularly for GPT-, exhibiting that, despite having less trainable parameters, a faster training throughput, and no extra difference delay, LoRA outperforms fine-tuning as far as model quality on both GPT-3 and GPT-2. The adequacy of LoRA is explained by this observational study on rank deficiency in language model variations [10].

3 METHODOLOGY

3.1 *Dataset*

The PHQ-8 Score [30] is applied to the DAIC-WOZ, the expanded DAIC that is used in this work[11]. The Patient Health Questionnaire 8 (PHQ-8) is a well-known and reliable self-report tool used to check an individual's degree of depression side effects. The PHQ-8, which comprises eight items, is a superior instrument for exploration and circumstances while discussing self-destruction thoughts may not be OK. It is framed from the lengthier PHQ-9 by excluding the inquiry regarding self-destructive ideation. On a scale from 0 (not at all) to 3 (almost every day), participants assessed how frequently they experienced each symptom during the most recent 14 days. The individual item scores are added simultaneously to obtain the general score, which runs from 0 to 24. Higher scores indicate more serious depression. The PHQ-8 is a helpful and reliable instrument for screening depression in different groups and medical stress settings, given its short duration and straightforwardness. The forthcoming distributions will focus on how well our work could help clinical experts' conclusions in clinical practice, as well as questionnaires and evaluation systems with stronger psychometric highlights. There are two types of datasets accessible from the University of California -Institute for Creative Technologies (USCICT) Distress Analysis

Interview Corpus (DAIC)[12]. The Wizard of Oz experiment filled in as the motivation for the DAIC-WOZ, which was made for the investigation of nervousness, depression, and traumatic stress disorder (PTSD).

3.2 *Models*

We investigated LLMs from the LLaMA and OpenAI GPT families. GPT-4, GPT-3.5, LLaMA-7B, and Alpaca-7B. The state-of-the-art language models GPT-2, GPT-3.5, and ChatGPT were developed by OpenAI and depend on the Transformer architecture[13]. With 1.5 billion parameters, GPT-2 — which was revealed in 2019 fit for creating text that is both firm and relevantly valuable in a variety of contexts. With an astounding 175 billion parameters, GPT-3.5 substitution decisively improves language creation and cognizance capacities. A GPT-3.5 variety planned explicitly for conversational applications, ChatGPT is truly adept at preserving context and creating human-like answers in intelligent discussions [14]. A sizable language model called GPT-4 has been used in various fields. GPT-4 can uphold clinical direction and computerize authoritative tasks in the medical care industry. In the meantime, it has been found, that GPT-4 encodes gender and racial biases, which brings about wrong clinical determination and the support of generalizations[15]. In addition, GPT-4 has been tried for text-based information analysis in legal conditions, where it beats experienced regulation student annotators and can bring about cost reserve funds through batch predictions[16].

4 RESULTS AND DISCUSSIONS

To evaluate the results, the following metrics are displayed and recorded in Table 1. **Precision**: Out of the multitude of occurrences that were perceived as positive, it talks about the level of true positive cases or those that were suitably analyzed. Because a wrong diagnosis can have negative effects on a patient's medication plan and complete health, accuracy in medical diagnosis may be especially important. When a conclusion has a high accuracy rate, it implies that the patient will most likely seek the right consideration and treatment.

4.1 *Recall*

Evaluate the level of real positive cases that the model precisely recognizes. In scenarios like healthcare diagnosis, where the objective is to detect every instance of an ailment, even if this results in some false positives, a higher recall value suggests that the model can recognize a higher percentage of positive instances.

4.2 *F1 score*

The harmonic means of recall and accuracy are used to calculate it. This suggests that recall and accuracy are loaded equally by the F1 score. A high F1 score demonstrates that a model has both high precision and high recall, which implies that it can accurately distinguish and order cases precisely. The performance metrics on the DAIC-WOZ Dataset are shown in Figure 2.

After testing and fine-tuning, we found that information expansion is the most efficient strategy to improve our framework because the amount of data we have collected from the DAIC-WOZ datasets is disproportionately small. We also attempted the DAIC-WOZ test information on untrained GPT-3.5 and GPT-4 subjects. Using GPT-3.5-turbo, ChatGPT-4, and our own BERT model with extended training on the larger dataset, we evaluated the DAIC-WOZ dataset exhibition. We resorted to use the chatbot to test the dataset because access to the ChatGPT-4 Programming interface was restricted.

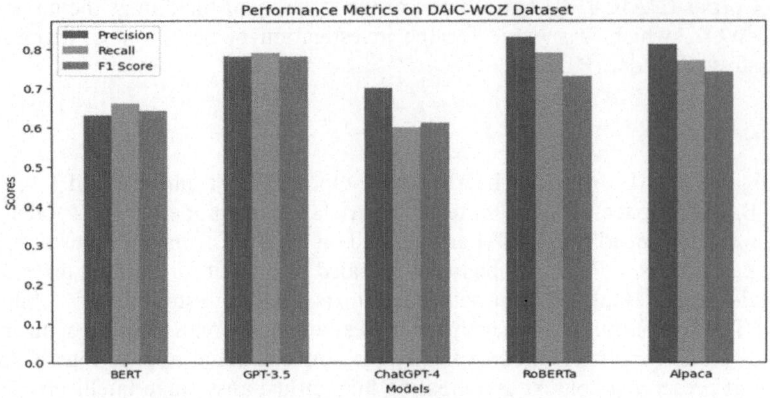

Figure 2. Performance metrics on the DAIC- WOZ Database.

The GPT-3.5 model showed surprisingly good performance in our preliminary trials, prior analysis, and historical results reported by many scientists.

Table 1. Metrics evaluation.

DAIC-WOZ Dataset	Precision	Recall	F1 score
BERT	0.63	0.66	0.64
GPT3.5	0.78	0.79	0.78
ChatGPT-4	0.7	0.6	0.61
RoBERTa	0.83	0.79	0.73
Alpaca	0.81	0.77	0.74

5 CONCLUSION

This study presents the main intensive appraisal of a few large language models (LLMs) for online text data-based mental health prediction tasks, including Alpaca, GPT-3.5, BERT, and RoBERTa. To work on the exhibition of all LLMs, they present context improvement and psychological wellness upgrade measures and show how valuable they are. The study accentuates the consistent requirement for innovative work of LLM for powerful depression distinguishing proof, highlighting the value of GPT-3.5-based models in emotional well-being prediction. The GPT 4.0 API will be used to improve the model and anticipate significantly better outcomes. To obtain significantly more solid discoveries in recognizing depression, they support the examination of LLM. The review is urged to keep on further developing models, comprehend and perceive unobtrusive articulations of depression, and perhaps make viable diagnostic instruments. To conclude LLMs have opened doors for influence acknowledgment in talk and the production of effective mindful exchange frameworks, giving confident ways to help and tend to individuals with mental health issues.

REFERENCES

[1] Organization W. H. *et al.* (2022). World mental health report: transforming mental health for all,.

[2] Bishal Lamichhane. (2023). Evaluation of ChatGPT for NLP-based Mental Health Applications, *arXiv*, 2303.15727

[3] *A Survey of Large Language Models.* (2023). doi: 10.48550/arxiv.2303.18223

[4] Introducing ChatGPT. https://openai.com/
[5] Xuhai Xu, Bingsheng Yao, Yuanzhe Dong, Saadia Gabriel, Hong Yu, James Hendler, Marzyeh Ghassemi, Anind K. Dey, Dakuo Wang. (2023). Mental-LLM: Leveraging Large Language Models for Mental Health Prediction via Online Text Data, doi:10.48550/arXiv.2307.14385
[6] *Large Language Models are Few-Shot Health Learners.* (2023). doi: 10.48550/arxiv.2305.15525
[7] *PaLM: Scaling Language Modeling with Pathways.* (2022). doi: 10.48550/arxiv.2204.02311
[8] *Scaling Instruction-Finetuned Language Models.* (2022). doi: 10.48550/arxiv.2210.11416
[9] *AlpacaFarm: A Simulation Framework for Methods that Learn from Human Feedback.* undefined (2023). doi: 10.48550/arxiv.2305.14387
[10] Edward, J., Hu., Yelong, Shen., Phillip, Wallis., Zeyuan, Allen-Zhu., Yuanzhi, Li., Shean, Wang., Weizhu, Chen. (2021). LoRA: Low-Rank Adaptation of Large Language Models. *arXiv*: Computation and Language, null.
[11] Lavender Yao Jiang, Xujin Chris Liu, *et.al.* (2023). Health system-scale language models are all-purpose prediction engines. *Nature* (June 2023). https://doi.org/10.1038/s41586-023-06160-y
[12] Karan Singhal, Tao Tu, *et al.* (2023). Towards Expert-Level Medical Question Answering with Large Language Models. http://arxiv.org/abs/2305.09617 arXiv:2305.09617 [cs]
[13] Chaoyi Wu, Xiaoman Zhang, Ya Zhang, Yanfeng Wang, and Weidi Xie. (2023). PMC-LLaMA: Further Finetuning LLaMA on Medical Papers. http://arxiv.org/abs/2304.14454 arXiv:2304.14454 [cs].
[14] Tianyu Han, Lisa C. Adams, Jens-Michalis Papaioannou, Paul Grundmann, Tom Oberhauser, Alexander Löser, Daniel Truhn, Keno K. Bressem. (2023). "MedAlpaca – An Open-Source Collection of Medical Conversational AI Models and Training Data." null. doi: 10.48550/arxiv.2304.08247
[15] Bishal Lamichhane. (2023). *Evaluation of ChatGPT for NLP-based Mental Health Applications.* http://arxiv.org/abs/2303.15727
[16] Mostafa M. Amin, Erik Cambria, and Björn W. Schuller. (2023). Will Affective Computing Emerge from Foundation Models and General AI? A First Evaluation on ChatGPT. http://arxiv.org/abs/2303.03186

Data Science & Exploration in Artificial Intelligence – Gururaj et al. (Eds)
© 2025 the Author(s), ISBN: 978-1-032-96390-7

Enhancing user experience with personalized movie assistance using NLP techniques

D. Koushik Vishnu, B. Jeevan, R. Jagadish & R. Sujithra Kanmani
Vellore Institute of Technology, Chennai, Tamil Nadu, India

ABSTRACT: The Current Study presents a novel approach to movie recommendation based on user preferences expressed through text. Unlike traditional collaborative and content-based filtering methods, this model allows users to express their ideal movie preferences, including story elements and flow. Initially, the model analyses the user's input for identification of genre. If identified, the model filters movies within that category and matches user-specified elements with those in the plots of recommended movies using Term Frequency-Inverse Document Frequency (TF-IDF) and Cosine similarity. Also, it explores intersection with other genres for diverse recommendations. In cases where no genre is stated, the model employs machine learning techniques like Support Vector Machine (SVM) and Naive Bayes classifiers integrated through a Voting Classifier framework alongside with TF-IDF to infer the genre and match preferences. The project also investigates methods such as Random Forest and Count Vectorizer, in order to compare their effectiveness with the proposed model.

1 INTRODUCTION

Movie recommendation systems have become popular nowadays, offering users personalized suggestions to enhance their viewing experience. Traditional approaches always depends either only on genre classification or simple keyword matching to provide recommendations. Beyond simple genre preferences, they also fail to capture the essence of user-desired movie watching experience. The goal of the proposed project is to bridge this gap by developing an advanced movie recommendation system that make use of Natural Language Processing (NLP) techniques to understand the detailed preferences of users. The model in this current study follows the multimodal approach and it utilizes techniques such as TF-IDF for vectorization and cosine similarity for recommendation. A key innovation in this model is that, it can distinguish implicit or explicit mentions of movie genres in user's text. When a genre is identified, the model filters movies within that genre and matches elements mentioned in the user's text with those elements present in the plots of recommended movies. Also, it explores other genre for mixed recommendations. In instances where users do not explicitly specify a genre, the model utilizes Machine Learning Classifiers such as SVM, Multinominal Naive Bayes with the integration of a soft voting mechanism and in conjunction with TF-IDF vectorization, the paper harness the diversity of these models to provide more accurate genre predictions by combining the strengths of individual classifiers and to infer the genre being implied by the user. The sole purpose of the model is to understand user inputs accurately by integrating machine learning classifiers, so that the recommendations offered are of quality and aligned with the taste of the user. The effectiveness of the recommendations may totally vary depending on various factors, this includes diversity of the movie database, quality of the plot description, interest of the user and their level of interpretation. Therefore, Accuracy for each text can be different and its totally based on the quality of the text.

 DOI: 10.1201/9781003589273-21

2 LITERATURE REVIEW

The study by (Pradeep, N. *et al.* 2020), analysed various domains on how recommendation systems are used. They proposed a personalized recommendation system based on the factors such as cast, crew, genre and keywords using a content-based approach. The study by (Kapoor, N. *et al.* 2020), proposed an informed and user personalized recommendation system by integrating sentiment scores into movie ratings where Support Vector Machine (SVM) was used for sentiment analysis of user reviews. The study by (Kumar, S. *et al.* 2020), presented a hybrid recommendation system for movies. It includes both collaborative filtering (CF) and content-based filtering (CBF) methods, along with sentiment analysis of movie-related tweets. The study by (Gawinecki, M. *et al.* 2021), addressed the challenge of determining optimal predictors for movie similarity in More-Like-This recommendation systems and found that only genres, summary, plot, categories, and release year has a major impact. The study by (Widayanti, R. *et al.* 2023), used a hybrid CF-CBF approach to build an enhanced recommendation system by integrating user interaction insights with content analysis. The study by (Havolli, A. *et al.* 2022), uses Adamic Adar Measure, to identify similar content based on text-based features and employed text vectorization with TF-IDF vectorizer and cosine similarity calculation to detect similarities between the movies.

The study by (Singh, S. *et al.* 2021), examined a revised version of recommendation system for movies and books based on the deep learning method fully connected neural network. The study by (Lavanya, R. *et al.* 2021), provided a complete study on the problems faced by users due to excessive information, categorized the drawbacks and challenges like scalability, data sparsity, accuracy and time complexity. The study by (Dixit, A. *et al.* 2023), proposed a chatbot that gives varied movie recommendations based on the user preferences on genre, actors, directors, and keywords without relying on previous user data. The study by (Abdurahman Baizal, Z. *et al.* 2023), influenced messaging platforms to incorporate movie recommendation system with natural language processing (NLP) methods like rule-based and maximum likelihood along with CBF method for recommendation. The study by (Nugraha, M. *et al.* 2022), introduced a chatbot-based conversational recommender system that uses POS tagging to identify movie titles and Multinominal Naive Bayes for intent classification. The study by (Nguyen, L. *et al.* 2020), overcame the cold start problem by introducing a new approach of content-based CF based on item similarity along with word embedding model for representing the features as semantic vectors instead of using traditional approach.

3 PROPOSED METHODOLOGY

The dataset was pre-processed and converted into vectors. Next, a similarity algorithm was utilized to recommend movies based on these vectors, ensuring accurate and personalized recommendations for users.

3.1 *Preprocessing and text vectorization*

The plot descriptions of movies are pre-processed to remove noise, stop words, and irrelevant characters. Then, they are vectorized using TF-IDF Vectorizer to convert them into numerical representations.

3.2 *Genre identification*

In scenarios where users explicitly mention a genre in their input, the model performs genre-based filtering to narrow down the pool of candidate movies. Utilizing machine learning techniques such as Support Vector Machine (SVM) and Naive Bayes classifiers, integrated through a Voting Classifier framework, the model infers the genre implied or favoured by the user.

3.3 *Compute TF-IDF representation*

TF-IDF values are computed for each word in the plot descriptions of movies, generating a TF-IDF representation for each movie's plot.

3.4 *Compute cosine similarity*

Cosine similarity is calculated between the TF-IDF vectors of the input criteria (e.g., specified genres, plot elements) and the TF-IDF vectors of all movies in the data set.

3.5 *Retrieve movies (Top-N recommendations)*

Movies with the highest cosine similarity scores to the input criteria vector are retrieved. Additionally, the model incorporates genre validation and plot matching algorithms to ensure that recommended movies align with user's preferences and the genre they implied.

3.6 *Exploration of methodologies*

The Study explored alternative methodologies, including CountVectorizer and Random Forest for genre prediction to determine the effective model.

4 ARCHITECTURE

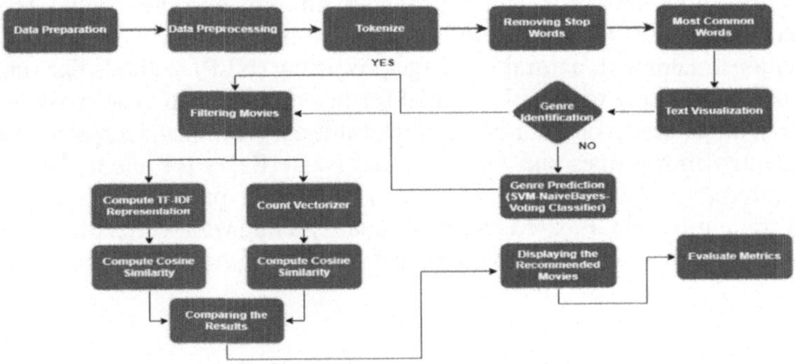

Figure 1. Architecture of the model oroposed.

4.1 *Dataset and preprocessing*

The movie dataset is taken into consideration for the recommending the movies. It is available in Kaggle. The dataset contains following attributes. Data preprocessing involved for removing duplicates, null and empty values.

4.1.1 *Title*
It specifies the title of the movie.

4.1.2 *Genre*
It specifies the Genre of the movie.

4.1.3 *Plot*
It specifies the storyline or the plot twist etc in the movie.

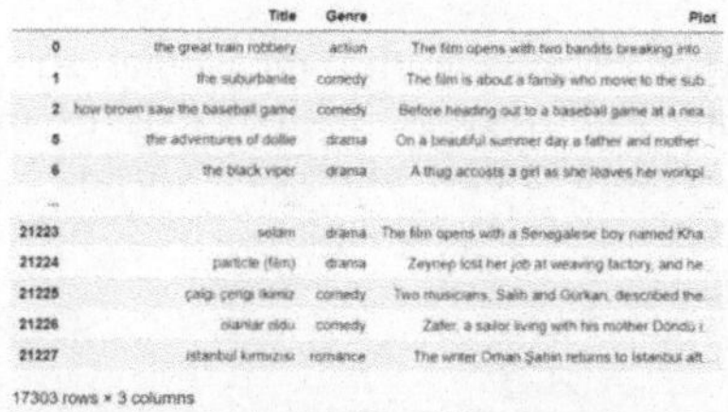

	Title	Genre	Plot
0	the great train robbery	action	The film opens with two bandits breaking into
1	the suburbanite	comedy	The film is about a family who move to the sub
2	how brown saw the baseball game	comedy	Before heading out to a baseball game at a nea
5	the adventures of dollie	drama	On a beautiful summer day a father and mother
6	the black viper	drama	A thug accosts a girl as she leaves her workpl
...			
21223	selam	drama	The film opens with a Senegalese boy named Kha
21224	particle (film)	drama	Zeynep lost her job at weaving factory, and he
21225	çalgı çengi ikimiz	comedy	Two musicians, Salih and Gurkan, described the
21226	olanlar oldu	comedy	Zafer, a sailor living with his mother Döndü i
21227	istanbul kırmızısı	romance	The writer Orhan Şahin returns to Istanbul aft

17303 rows × 3 columns

(a) Dataset which contains different movie title with genre and their plots.

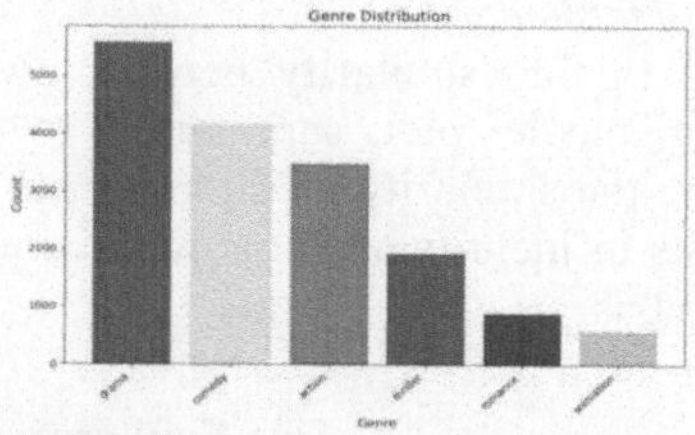

(b) Types of Genres in the movie (Data source: Kaggle.

Figure 2. (a) Dataset contents. (b) Distribution of genres.

4.2 *Algorithms*

4.2.1 *SVM*

It is a classification and a regression technique which is particularly effective in high-dimensional spaces and text classification that helps in analysing patterns in user preferences and finding the hyperplane to classify different genres.

4.2.2 *Multinomial naive bayes*

It is a specific instance of naïve bayes classification algorithm specifically used for text classification works such as document categorization or sentiment analysis. Also, it works well with high dimensional data indicating its advantages over text data like plot summaries. It calculates the probability of each class using bayes theorem:

The probability calculation is done by:

$$\mathrm{P(y|x)} = \frac{P(x)P(x1\ |y) \cdot P(x2\ |y) \cdots P(xn\ |y) \cdot P(y)}{p(x)} \tag{1}$$

The classification is done by:

$$\mathrm{P(y \mid x)} = \arg\max yP(y|x) \tag{2}$$

4.2.3 *Soft voting classifier*

In this each individual model's probabilities for each class are taken and the final prediction is determined by averaging the predicted probabilities across all models and the class with the highest average probability is considered.

4.2.4 *Random forest*

It is used to analyse the relationship between various genres and plots, providing insights into the most matching plots and genres and the movies that are recommended. It has the ability to handle non-linear relationships and interactions between features makes it a valuable tool in text classification tasks.

4.2.5 *Count vectorizer*

It is a feature extraction tool that transforms text input into numerical vectors. It counts the number of each word in the text and depicts the word as a vector, with each element representing the count of a certain word. Assume there exist movie plot summaries as text data. The formula can be represented as follows:

$$\mathrm{Count\ Vectorizer(d, t) = nd, t} \tag{3}$$

where $d =$ *the set of documents*; $t =$ the set of unique tokens (words) across all documents; *nd*, $t =$ the count of token t in document d.

4.2.6 *Cosine similarity*

It is used to find similarity between two vectors in a multidimensional space. After converting movie plot summaries to numerical vectors with TF-IDF or Count Vectorizer, the similarity of different movie vectors is determined by Cosine Similarity. This assists in identifying films with comparable content or topics. The formula can be represented as:

$$\text{Cosine Similarity} = \frac{A.B}{\sqrt{\sum_{i=1}^{n} A_i^2}\sqrt{\sum_{i=1}^{n} B_i^2}}. \tag{4}$$

where A and B = vectorized representations of two texts; $A.B$ = dot product of vectors A and B; Ai and Bi represent the frequency of the i-th term in texts A and B, respectively; n is the total number of unique terms in the vectorized representations.

4.2.7 *TF-IDF vectorization*

TF-IDF stands for Term Frequency-Inverse Document Frequency. TF (Term Frequency) refers to how frequently a term appears in a document. The rarity of a phrase across the corpus is assessed by the IDF (Inverse Document Frequency) metric. Higher the score of TF-IDF for a word in a document indicates its importance and uniqueness in that document. The TF-IDF formula is:

$$\text{TF - IDF}(\text{d}, \text{t}) = tf_{d,t}\ x\ idf_t \tag{5}$$

$$tf_{d,t} = \frac{\textit{number of times term t appears in document d}}{\textit{Total number of terms in document d}} \tag{6}$$

$$idf_t = log\left(\frac{\textit{Total number of documents in the corpus}}{\textit{Number of documents containing term t}}\right) \tag{7}$$

5 RESULTS

5.1 *SVM*

```
pre=["i want a feel good movie",
    "blood and gory","sad and emotional"
    ,"something cartoon","more stunts and crazy fights",
    "bombs and kills","love and relationships"]
```

Figure 3. User input for Genre prediction.

```
Predicted Genre:  ['romance']
Predicted Genre:  ['thriller']
Predicted Genre:  ['drama']
Predicted Genre:  ['animation']
Predicted Genre:  ['action']
Predicted Genre:  ['action']
Predicted Genre:  ['drama']
```

Figure 4. Genre prediction Using SVM.

5.2 *TF-IDF vectorization and countVectorizer*

While genre prediction accuracy can be measured using standard metrics, the evaluation of free text recommendation methods often relies on user feedback, surveys, or user engagement metrics to assess the relevance of the recommendations provided. This highlights the importance of user feedback in refining and improving the performance of the free text recommendation model.

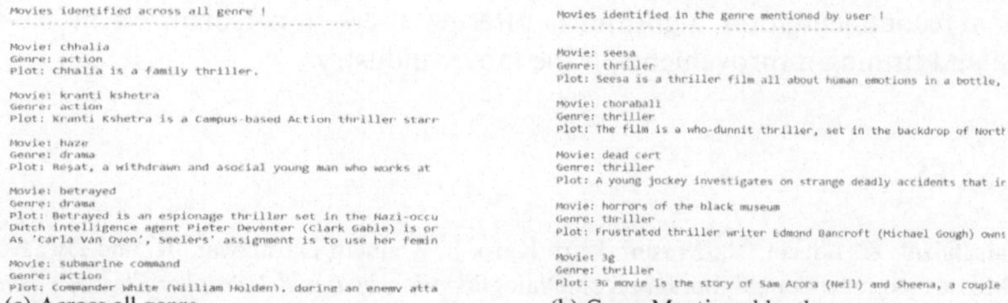

(a) Across all genre.

(b) Genre Mentioned by the user.

Figure 5. Movie Recommendations Using TF-IDF Vectorization, the difference lies in the exploration, part (a) of the figure Explored all genre while part (b) Explored only the genre mentioned by the user for recommending movies. User text = “movie should include thriller moments and scaring scenes”.

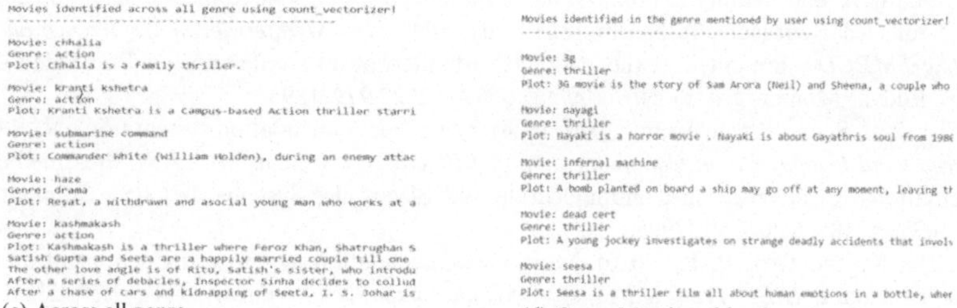

(c) Across all genre.

(d) Genre Mentioned by the user.

Figure 6. Both part of the figure contains Movie Recommendations Using Count Vectorizer, the difference lies in the exploration, part (c) of the figure Explored all genre while part (d) Explored only the genre mentioned by the user for recommending movies.

Table 1. Metrics comparison between classifiers.

Metrics	Voting Classifier	Random Forest
Accuracy	66.48%	50.38%
Precision	66.91%	60.52%
Recall	66.48%	50.38%
F1-score	65.94%	44.74%

6 CONCLUSION

The current study has introduced a novel approach to movie recommendation, surpassing the constraints of traditional methods by employing NLP techniques. This innovative model not only discerns implicit or explicit mentions of movie genres within user input but also adeptly matches plot elements with user preferences. Through the integration of machine learning classifiers and text analysis methods, the model achieves a notable accuracy rate of 66.4% in genre prediction. Consequently, users receive personalized movie recommendations that closely align with their tastes and preferences. Looking ahead, there are several avenues for future research and improvement. Firstly, the effectiveness of the free text recommendation method could be further evaluated and refined. Exploration of advanced NLP techniques and deep learning models. Integrating feedback mechanisms and user interactions into the recommendation process. Overall, this research contributes to the advancement of

NLP-driven recommendation algorithms, offering users a personalized movie-watching experience and brining improvements in the movie industry.

REFERENCES

Abdurahman Baizal, Z., Ikhsan, N., Muslim Karo Karo, I., Kenneth Darmawan, R. and Dwi Hartanto, R. (2023). Movie recommender chatbot based on Dialogflow. *International Journal of Electrical and Computer Engineering (IJECE)*, *13*(1), 936. doi:10.11591/ijece.v13i1.pp936-947.

Dixit, A., Gupta, K. and Dinesh. (2023, July 6). Movie recommendation chatbot using cosine similarity and NLP tools. *2023 14th International Conference on Computing Communication and Networking Technologies (ICCCNT)*. Presented at the 2023 14th International Conference on Computing Communication and Networking Technologies (ICCCNT), Delhi, India. doi:10.1109/icccnt56998.2023.10307673.

Gawinecki, M., Szmyd, W., ⊠uchowicz, U. and Walas, M. (2021). What makes a good movie recommendation? Feature selection for content-based filtering. In *Lecture Notes in Computer Science. Similarity Search and Applications* (pp. 280–294). doi:10.1007/978-3-030-89657-7_21.

Havolli, A., Maraj, A. and Fetahu, L. (2022, June 7). Building a content-based recommendation engine model using Adamic Adar Measure; A Netflix case study. *2022 11th Mediterranean Conference on Embedded Computing (MECO)*. Presented at the 2022 11th Mediterranean Conference on Embedded Computing (MECO), Budva, Montenegro. doi:10.1109/meco55406.2022.9797139.

Kapoor, N., Vishal, S. and Krishnaveni. (2020, June). Movie recommendation system using NLP tools. *2020 5th International Conference on Communication and Electronics Systems (ICCES)*. Presented at the 2020 5th International Conference on Communication and Electronics Systems (ICCES), Coimbatore, India. doi:10.1109/icces48766.2020.9137993.

Kumar, S., De, K. and Roy, P. P. (2020). Movie recommendation system using sentiment analysis from microblogging data. *IEEE Transactions on Computational Social Systems*, 7(4), 915–923. doi:10.1109/tcss.2020.2993585.

Lavanya, R., Singh, U. and Tyagi, V. (2021, March 25). A comprehensive survey on movie recommendation systems. *2021 International Conference on Artificial Intelligence and Smart Systems (ICAIS)*. Presented at the 2021 International Conference on Artificial Intelligence and Smart Systems (ICAIS), Coimbatore, India. doi:10.1109/icais50930.2021.9395759.

Nguyen, L. V., Nguyen, T.-H. and Jung, J. J. (2020, October 13). Content-based collaborative filtering using word embedding. *Proceedings of the International Conference on Research in Adaptive and Convergent Systems*. Presented at the RACS '20: International Conference on Research in Adaptive and Convergent Systems, Gwangju Republic of Korea. doi:10.1145/3400286.3418253.

Nugraha, M. A., Baizal, Z. K. A. and Richasdy, D. (2022). Chatbot-based movie recommender system using POS tagging. *Building of Informatics, Technology and Science (BITS)*, *4*(2), 624–630. doi:10.47065/bits.v4i2.1908.

Pradeep, N., Rao Mangalore, K. K., Rajpal, B., Prasad, N. and Shastri, R. (2020). Content based movie recommendation system. *International Journal of Research in Industrial Engineering*, 9(4), 337–348. doi:10.22105/riej.2020.259302.1156.

Singh, S., Lohakare, M., Sayar, K. and Sharma, S. (2021, September 24). RecNN: A deep neural network-based recommendation system. *2021 International Conference on Artificial Intelligence and Machine Vision (AIMV)*. Presented at the 2021 International Conference on Artificial Intelligence and Machine Vision (AIMV), Gandhinagar, India. doi:10.1109/aimv53313.2021.9670990.

Widayanti, R., Chakim, M., Lukita, C., Rahardja, U. and Lutfiani, N. (2023). Improving Recommender Systems using Hybrid Techniques of Collaborative Filtering and Content-Based Filtering. *Journal of Applied Data Sciences, 4*(3), 289–302. doi:https://doi.org/10.47738/jads.v4i3.115.

Data Science & Exploration in Artificial Intelligence – Gururaj et al. (Eds)

A hybridized model for mental health detection using natural language processing

C. Pavithra, D. Abhay Karthik, B.R. Rishika, M. Roshini & A. Ajina
Department of Artificial Intelligence and Machine Learning, M S Ramaiah Institute of Technology, Bangalore, India

ABSTRACT: Mental Health disorders raise a global concern, requiring innovative approaches for early detection and intervention. This paper proposes a novel hybridized model that combines natural language processing and machine learning techniques to detect mental health disorders in its initial stages. The algorithm uses advanced linguistic analysis and sentiment extraction from the sequential data to identify patterns anticipating mental health issues. The model uses NLP algorithms and a strong ML framework to analyze user-generated text data to track down if there is any mental health illness. It uses a trained dataset and a Streamlit-based user interface for user accessibility. The interface allows users to input their thoughts or feelings, ensuring privacy and anonymity, while providing a seamless and engaging platform for mental health analysis.The research demonstrates that a hybridized model, combining machine learning algorithms, can enhance the accuracy and efficiency of mental health detection. The approach is such that it prioritizes user experience and a user-friendly interface, contributing to the growing field of technology-driven mental health solutions, thereby enhancing early detection accuracy.The hybridized model provides an accuracy of 98.85% on the test data.This study presents a hybrid model that integrates mental health detection, NLP, machine learning, and user-centric design, offering scalable and effective mental health screening, contributing to digital mental healthcare solutions.

Keywords: NLP, Mental Health, Streamlit, SVM, AdaBoost classifiers etc.

1 INTRODUCTION

Mental health refers to an individual's social, emotional, and psychological well-being. It requires the capacity to cope with making decisions, interact with others, and stress [1]. Mental health impacts relationships, quality of life, and daily activities. It includes resilience and self-esteem. 970 million people worldwide suffer from mental diseases, with 50% starting at 14, 75% developing by 24, and accounting for 14.3% of all fatalities[2].Natural language processing is a sub-discipline of artificial intelligence that aids personal assistance in understanding users' mental health by analyzing and breaking down sentences.

A new solution combines natural language processing and machine learning to predict mental health in users. The user interface predicts their mental health strength or weakness, enabling immediate professional medical care. The model combines multiple models for efficiency and accuracy, utilizing social media platforms like Twitter and Facebook for data extraction. Researchers are using data mining techniques to gather and preprocess social media data, using web scraping, sentiment analysis, and topic modeling to handle unstructured content. This data is then trained into machine learning models to predict outcomes, contributing to social media analytics knowledge and providing practical insights for applications like brand sentiment analysis and mental health prediction.[12]. The research introduces a hybrid model combining SVM,

DOI: 10.1201/9781003589273-22

Random Forest, and AdaBoost Classifier for early mental health detection and care, improving accuracy and accessibility, and promoting an anonymous, proactive approach.

2 LITERATURE SURVEY

Konda Vaishnavi *et al.*: This study assesses eight machine learning classification methods for mental health issues. The most accurate models are Multiclass Classifier, Multilayer Perceptron, and LAD Tree, according to the results. Random Forest, SVM, and KNN all performed admirably; ensemble classifiers increased prediction accuracy to 90%. Every algorithm surpassed 79% accuracy. [3].

Fadhluddin Sahlan *et al.*, the study found that business students have a higher incidence of mental disorders compared to arts, music, design, and engineering students, indicating a significant impact of their study course.[1].

Satvik Gurjar *et al.*, This study states that changes in heart rate, blood pressure, and speech acoustics are symptoms of depression and weak emotional states. Various ML algorithms, including Naive Bayes, KNN, SVM, decision tree, random forest and logistic regression were used to predict mental health issues like PTSD, insomnia, anxiety, and depression. SVM achieved the highest accuracy for depression (96.15%) and PTSD (94%), while logistic regression was most accurate for insomnia (98%) and anxiety (97.27%). [11].

Chung *et al.*,the study uses various machine learning algorithms to predict mental health disorders like anxiety, schizophrenia, bipolar disorder, depression, PTSD, and children's mental health. Random forest achieved the highest accuracy of 89% for depression and anxiety. Factors like age, education, marital status, and health status were used.[5].

Shruti Grag *et al.*, This paper uses ML algorithms to predict stress, anxiety, and depression among working and unemployed individuals at five severity levels. Data from blog posts were classified using RFT, SVM, and CNN. Encoding methods included TF-IDF, BOW, and theme modeling. Algorithms like DT, RFT, NB, SVM, and KNN were used, with Naive Bayes achieving the highest accuracy for depression at 85.5%.[10].

Dr.J.Arokia Renjit *et al.*, This paper focuses on general mental health detection across all age groups using ML algorithms like logistic regression, random forest, decision tree, and Naive Bayes. Software and A/B testing were conducted to improve models. Comparison using a confusion matrix showed that the random forest classifier performed best, with high "true positive" and "true negative" values[9].

Tianlin Zhang *et al.*,this paper uses NLP and ML algorithms like SVM, AdaBoost, and Decision Trees to detect mental health illnesses. It addresses model instability caused by class imbalance, noisy labels, and text length. Diverse writing styles from different data sources also affect performance. Advances in deep learning, such as reinforcement learning, can improve accuracy in mental illness detection [8].

3 METHODOLOGY

3.1 *Data description*

A study found that college students' friends are unlikely to encourage them to seek medical assistance when they express sadness on Facebook. None of the 33 students in the study said their friends advised them to seek help. Instead, most offered encouraging or motivational texts. However, at The Ohio State University a professor of social work Scottye Cash, believes this may not be enough for those who are truly sad. The study revealed that students' Facebook friends were not aggressive in assisting their friends in need, based on a dataset containing normal and anxiety/depression text.[6].Depression is a mental health condition causing chronic melancholy, sleep changes, and focus difficulties. It requires therapy, medication, lifestyle modifications, and professional support. [7]

3.2 *Architecture of the model*

From Figure 1 The system architecture starts with a user-friendly interface using Streamlit for data provision. Raw data is preprocessed using advanced NLP methodologies like stop word removal,tokenization, and stemming, converting unstructured text into a structured format for modeling.A pre-trained machine learning model detects patterns and makes accurate predictions using preprocessed data. Predictions are communicated to the user through the Streamlit interface, integrating natural language processing techniques with user-centric accessibility.

3.3 *Data pre-processing*

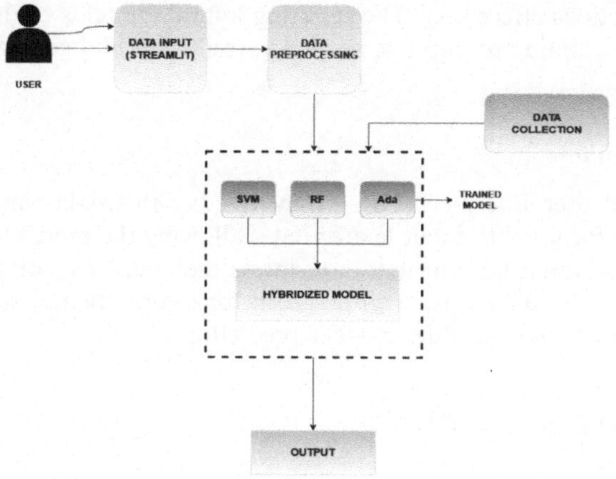

Figure 1. Architecture in our model.

For ML models to be created data pre-processing is pivotal, as raw data may not be clean and formatted. Various libraries like pandas, numpy, matplotlib, seaborn, word cloud, counters, tokenizers, and stemmers help analyze, create interactive visualizations, and summarize large data sets.The dataset is loaded into a Pandas DataFrame as shown in Figure 2, cleaned, and added features. It is processed by converting text into lowercase, removing punctuations, stopping words, and stemming to focus on meaningful words.Word clouds and bar plots are generated for accurate predictions and improved model performance.

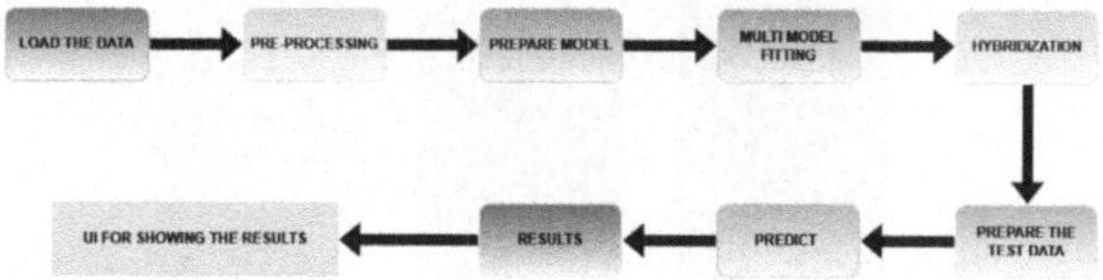

Figure 2. Flow of model.

3.4 *ML model*

The text data is partitioned into training and testing sets, with 80% being trained and 20% tested. The TF-IDF vectorizer converts the data into numerical features for ML algorithms. SVM, Random Forest, and AdaBoost are used for classification. SVM creates the best decision boundary line or hyperplane to segregate n-dimensional space into classes, allowing easy categorization of new data points.

The Random Forest algorithm enhances model performance by reducing overfitting by combining decision tree predictions. AdaBoost is an iterative supervised boosting machine learning technique that identifies misclassified data points and weak learners, generating a decision stump.

3.5 *Hybridization*

Hybridization in machine learning involves combining the strengths of multiple models to improve predictive models. Three machine-learning models, SVM, Random forest, and AdaBoost, are trained separately on a dataset. A voting classifier is generated by combining these models and deciding between hard and soft voting. The next step is to integrate the basic models' predictions effectively. The resulting hybrid model is evaluated to compare it to individual models, aiming to improve accuracy, resilience, and generalization in the final prediction model.

3.6 *UI using streamlit*

The Streamlit-based user interface (UI) for mental health prediction aims to create an interactive platform for users to input text or data, allowing the model to make predictions. The UI starts with an input box for users to express thoughts or emotions, followed by the model's predictions. Visualizations, such as green for strong mental states and yellow for weak ones, help users understand the model's predictions.

4 RESULTS AND DISCUSSION

This study identified three ML techniques that are-Support Vector Machine, Random Forest, and AdaBoost Classifier.

Table 1. Comparison of different ML models.

Model	Accuracy%
Support Vector Machine	97.92
Random Forest Classifier	98.82
AdaBoost Classifier	98.49
Hybridized Model	98.99

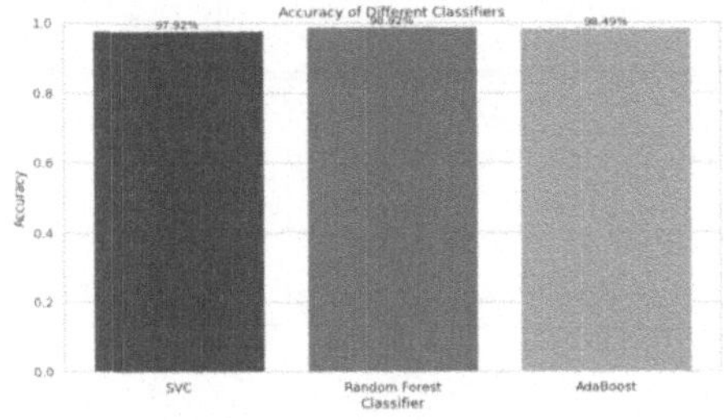

Figure 3. Comparison results of the different Machine Learning Models.

We combine these three models by performing hybridizations and we obtain the accuracy score of each model.From the above Figure 3 we have three machine learning models—

AdaBoost, RF (Random Forest) and SVM (Support Vector Machine) —and their respective accuracy scores are as follows:

From Table 1, we observe that the hybridized model produces the best accuracy and is the most efficient among the other technique

In Figure 4, we have obtained the result of the hybridization model with an accuracy of 98.99%. This hybridization model is a combination of three ML models.

```
hybridised Model Accuracy: 0.9899569583931134
              precision    recall  f1-score   support

         0.0       0.99      1.00      0.99      1248
         1.0       0.98      0.92      0.95       146

    accuracy                           0.99      1394
   macro avg       0.98      0.96      0.97      1394
weighted avg       0.99      0.99      0.99      1394
```

Figure 4. Classification report of the hybridized model.

In Figure 5 and 6, are the results of the model which determines the mental health of the user. Stating if the user's mental health is weak or strong based on the text input given by them in this interface. This would help the user to determine his/her mental state and seek professional help.

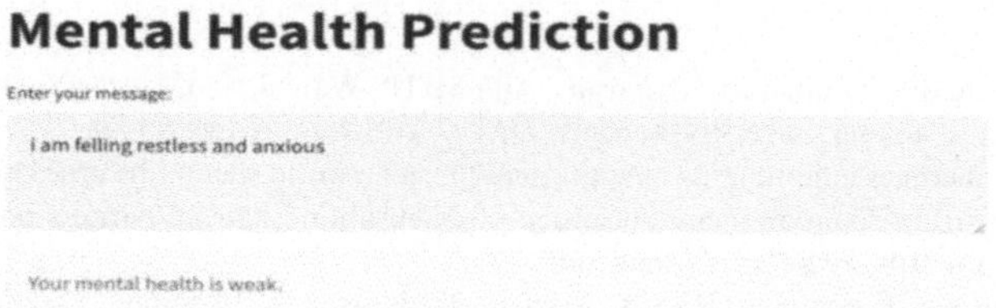

Figure 5. When the mental health is weak.

Table 2. Comparison results of our model on different datasets.

Dataset	Accuracy%
Student Depression and Anxiety [5]	98.99
Reddit Data [6]	96.5

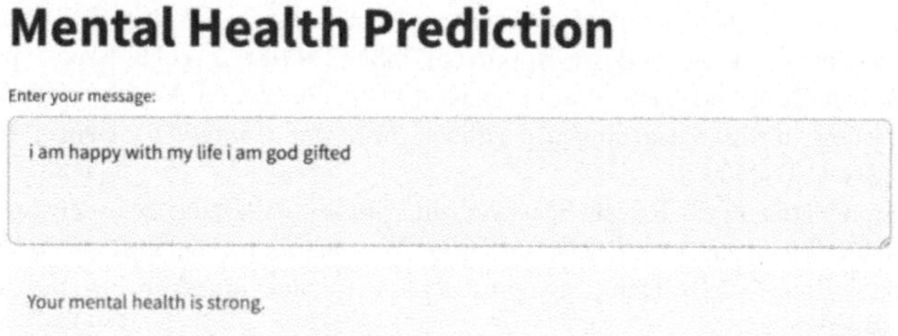

Figure 6. when the mental health is strong.

From the Table 2 we can see that our model has acquired a percentage of 96.5 on a different dataset. The model's adaptability across multiple datasets is demonstrated through consistent testing under various conditions, assessing its robustness and unique characteristics, enabling the identification of dataset-specific characteristics for accurate predictions.

5 CONCLUSION

Our model is a comprehensive and innovative approach that smoothly integrates many aspects of machine learning. Our approach exhibits a comprehensive grasp of complicated Machine learning techniques by incorporating a varied range of models, utilizing sophisticated preprocessing techniques, and leveraging the power of natural language processing (NLP). The hybridization of multiple models inside the ensemble ensures robust performance and adaptability to a wide range of datasets. This thorough analysis ensures that our model is not only adapted to specific data characteristics but also has the flexibility to handle a wide range of inputs, making it a valuable and efficient tool in a variety of real-world applications.

The use of NLP approaches improves the model's ability to extract insights from unstructured text data, opening up new applications in text-heavy domains. Additionally, our user-friendly interface, created with Streamlit, improves accessibility and usability, making the model both powerful and user-centric. This way the user finds out what his/her mental state is and can go seek medical help. In our UI, the user's identity is kept completely private.

REFERENCES

[1] Sahlan F., Hamidi F., Zulhafizal Misrat M., Adli M. H., Wani S. and Gulzar Y. (Jul. 2021). Prediction of mental health among university students, *IJPCC*, vol. 7, no. 1, pp. 85–91.

[2] https://economictimes.indiatimes.com/magazines/panache/world-mental-health-day-60-70-mn-people-in-india-suffer-from-common-mental-disorders-stigmatisation-financial-barriers-prevent-timely treatment/articleshow/104289268.cms?from=mdr

[3] Vaishnavi, Konda Kamath, U Rao B, Ashwath Reddy snf N. V. Subba. (2022). Predicting Mental Health Illness using Machine Learning Algorithms, *Journal of Physics: Conference Series*. 2161. 012021. 10.1088/1742-6596/2161/1/012021.

[4] Lin, Huijie Jia, Jia Qiu, Jiezhong Zhang, Yongfeng Shen, Guangyao Xie, Lexing Tang, Jie Feng, Ling Chua and Tat-Seng. (2017). Detecting stress based on social interactions in social networks, *IEEE Transactions on Knowledge and Data Engineering*. PP.1.1.10.1109/TKDE.2017.2686382.

[5] Chung, Jetli and Teo, Jason. (2022) Mental health prediction using machine learning: taxonomy, *Applications, and Challenges. Applied Computational Intelligence and Soft Computing*. 2022.1 19.10.1155/2022/9970363.

[6] https://www.kaggle.com/code/sasakitetsuya/students-anxiety-and-depression-classify-model/input

[7] https://www.kaggle.com/datasets/infamouscoder/depression-reddit-cleaned/data

[8] Zhang, T., Schoene, A.M., Ji, S. *et al.* (2022). Natural language processing applied to mental illness detection: a narrative review. *npj Digit . Med.* 5,46. https://doi.org/10.1038/s41746-022-00589-7

[9] Adlin Sajeesha M. J., Sree Devi D. S., Sangavai V. D., Arokia Renjith Dr. J. (May-2022). Prediction of mental health using machine learning, *International Journal of Emerging Technologies and Innovative Research* (www.jetir.org | UGC and issn Approved), ISSN:2349-5162, Vol.9, Issue 5, page no. ppi634–i640

[10] Anu Priya, Shruti Garg and Neha Prerna Tigga. (2020), Predicting Anxiety, Depression and stress in modern life using machine learning algorithms, *Procedia Computer Science*, Volume 167, Pages 1258–1267,ISSN 1877-0509.

[11] Satvik Gurjar, Chetna Patil, Ritesh Suryawanshi, Madhura Adadande, Ashwin Khore and Noshir Tarapore. (2022). *Mental Health Prediction Using Machine Learning*, Volume 09.

[12] Sakshi Yadav. (June-2022). Detecting presence of mental illness using nlp sentiment analysis, Volume:04/Issue:06

Data Science & Exploration in Artificial Intelligence – Gururaj et al. (Eds)
© 2025 the Author(s), ISBN: 978-1-032-96390-7

Anomaly detection for natural disaster preparedness

Asha Kurian, Mohammed Rihan & P. Mano Paul
Department of CSE, Alliance College of Engineering and Design, Bangalore, India

ABSTRACT: This review analyzes the existing research on anomaly detection methods employing statistical, machine learning, and deep learning algorithms to enhance natural disaster preparedness. The paper highlights the limitations of the existing approaches and sheds light on the potential for new and emerging technologies in disaster management. It provides an insight into improving the performance of disaster systems in real-time by harnessing the power of multimodal data sources for improved reliability of prediction. The evaluation also compares procedures, datasets, and performance metrics to evaluate the strengths of different anomaly detection strategies for various crisis scenarios.

Keywords: anomaly detection, machine learning, deep learning, disaster management

1 INTRODUCTION

Wildfires, earthquakes, and floods are examples of natural disasters that constantly threaten humanity, claiming countless lives and wreaking havoc in their aftermath. Our ability to sufficiently prepare for these disasters will be crucial to reducing their effects. The early warning system is the first line of defense that enables quick evacuations and catastrophe response actions (Basher 2006). Because anomaly detection may identify even the smallest tremors or changes in the surrounding data that may indicate an oncoming calamity, these systems rely on it (Chandola *et al.* 2009).

Anomaly detection is an endeavor to search for abnormal patterns that do not conform to regular trends in historical data and can provide pointers to indicate early stages of natural disasters (Agyemang *et al.* 2006). This continuously evolving research area has garnered attention from scientists, engineers, and disaster managers concerned with enhancing preparedness through informative early warnings to reduce negative outcomes.

Advances in computer vision, particularly machine learning and deep learning networks, and natural language processing, aided by other technological developments, have made it possible to identify new paths in detecting anomalies in the event of a disaster. Machine learning algorithms with their propensity to quickly evaluate large amounts of data from many sources are excellent architectures to detect natural calamities along with their management. Anomaly detection systems keep a close eye on these data streams, allowing them to identify patterns that may indicate the impending start of a natural disaster.

Researchers in this discipline are constantly investigating ways to improve the efficiency and accuracy of anomaly detection techniques to improve the preparedness for facing these hazards. The intention is to give early warning systems even more lead periods so that mitigation and evacuation preparations may be carried out more successfully. Even though it is a very difficult task, incorporating these anomaly detection algorithms into current early warning systems is a very promising area for future research. Modern algorithms and a wide variety of data sources can help us get closer to a time when societies will be more resilient to the calamities that nature brings.

DOI: 10.1201/9781003589273-23

2 REVIEW OF LITERATURE

Natural disaster preparedness along with disaster risk management are crucial for detecting anomalies. Several works have tackled diverse approaches or techniques aimed at recognizing outliers present within disaster-related datasets. Statistical techniques form the basis of anomaly detection in this field. Finding data points that heavily deviate from the historical norm has proven to be a promising application of techniques like the Interquartile Range (IQR)-based outlier detection (Patcha and Park 2007). The recent popularity of machine learning algorithms can be attributed to the ability of these deep architectures to handle complex data patterns and learn from large, annotated datasets. Examples of such algorithms include k-nearest Neighbors and Support Vector Machines. These predictions heavily hinge on historical data to identify new threats, that act as a limitation. Especially ones with distinctive qualities that have not been observed in previous disasters. Extracting valuable insights from heterogeneous and real-time data sources presents difficulties for traditional models.

The integration of data from several sources is strongly emphasized in current research due to the weaknesses of traditional methodologies. Multimodal data from varied sources such as social media feeds, satellite photos, and sensor networks are all combined in this multi-source method. In times of environmental changes, data such as water levels in flood-prone areas or seismic activity in earthquake zones can be gathered from sensor networks. Satellite imaging provides a wider view and the possibility to monitor alterations in vegetation types and climate situations.

Social media is a valuable repository of information for many real-time applications due to their dynamically changing content. This can prove to be significantly helpful in times of crisis by blending information from traditional sources along with real-time insights to handle the situation effectively (Sakaki *et al.* 2013). Machine learning can be used to evaluate numerous data sources in real-time to enhance the lead time given for disaster preparedness. Analyzing multimodal data, real time together has advanced anomaly detection for early warning systems against disasters.

(Marzuoli and Liu 2019) examined the efficacy of data obtained from mobile phone users to monitor natural disasters and identify anomalies. Identifying changes in changing densities of the population and their movements. The data was extremely beneficial in determining the number of people affected by the disaster. The test results were compared with the results from the FEMA dataset holding disaster statistics for Hurricane Florence and Hurricane Michael. The work done by (Gino *et al.* 2023) was aimed at locating disaster-affected areas from remote sensing data. The huge repositories of data collected from the Google Earth Engine platform was leveraged for unsupervised training of machine learning models.

Big data plays an important role in various aspects of anomaly detection and disaster assessment and management (Fan *et al.* 2020). Intelligent systems have been devised, based on insights obtained from data in their varied forms (sensor, thermal, text, time-series, geospatial, satellite etc.) (Rezvani *et al.* 2023). Applications range from designing early warning systems, real-time monitoring and assessment, to analyzing the impact of hazards. They also include models devised to assess the risk of a disaster and planning for recovery.

3 METHODOLOGY

When looking at research, the focus has been mainly on methodologies, datasets and models conceptualized by other researchers. On websites with scholarship articles like IEEE Xplore, Elsevier, ScienceDirect and ResearchGate, words like "anomaly detection" or "natural disasters" as well as "machine learning models" or "disaster management systems" were searched for on predetermined lists of terms.

Several different ways to be used to implement AI are discussed in literature, ranging from deep learning to machine learning, natural language processing, and social media analysis frameworks. Popular machine learning approaches adopted in this domain include neural networks, support vector machines, Naïve Bayes, and decision trees. Some typical deep learning methods include convolutional neural networks, recurrent neural networks, and long short-term memory networks. Natural language processing techniques useful for data analysis tasks include named entity recognition (NER), and sentiment analysis. In social media analysis, methods like content analysis, network analysis, and geographic information systems have been experimented.

A hybrid machine learning approach was used to separate flood-hit districts from aerial photographs collected by a drone (Akshya and Priyadarsini 2019). The authors used support vector machines and decision trees to divide images into buildings, vegetation, or water. To align the images to the machine learning pipeline, image enhancement was extensively used on the aerial images from the dataset (approximately 300 in number). (Fan *et al.*) developed a fine-tuned Bidirectional Transformer (BERT) to sift through social media to identify locations, and extract location coordinates and posts pertinent to the calamity. Graph-based clustering learns reliable context-based information. The potential of this application was exhibited using the real-time Twitter data collected during the 2017 international disaster - Hurricane Harvey. Mining of the wealth of information available on social media was explored again by (Layek *et al.* 2019) using a two-stage identification process. In the initial phase, a CNN-based network detects flood-inundated areas, followed by another phase involving a filtering of the detected images based on the color for final classification. The applicability of the model was evaluated using real-time flood images collected. Work done by (Muhammad *et al.* 2018) involved embedding surveillance systems with computer vision to enable the detection of fire outbreaks at home, wildfires. The model was developed with an architecture based on the CNN-based GoogleNet and it was proposed as an economical and efficient system to provide real-time alerts. The model was trained using a large dataset of thermal images. Another economically viable technique involved using unmanned aerial vehicles to detect wildfires. Conceptualized by (Lee *et al.* 2017), the study involved using UAVs embedded within a deep neural network for real-time wildfire detection.

A warning system to detect imminent earthquakes and transmit alert signals to adopt proactive measures to better handle the disaster was developed by (Chin *et al.* 2019). A trio of machine learning algorithms (K-Nearest Neighbors, classification trees, and support vector machine) were trained on seismic data collected from a strong motion gauging system based in Taiwan. The popularity of 5G networks, along with edge computing and machine learning paradigms were implemented in the model developed in (Ardiansyah *et al.* 2020). The system aimed at providing real-time detection to enable faster response times. The project ambitiously named EagleEYE used autonomous drones for surveillance.

A system was proposed by (Moon *et al.* 2019) to develop an early warning system from meteorological data to assess the threat of heavy rainfall. The volume of the data was discretized selectively using principal component analysis. The logistic regression model learns patterns from the data to provide alerts regarding short bursts of torrential rain with a lead time of 30 minutes.

Crisis management is an imperative force when a disaster strikes. Automatically detecting news about crisis events has been the mainstay work done in (Domala *et al.* 2020). The authors have proposed using natural language processing tasks to identify news about disasters. Machine learning algorithms like Naive Bayes, linear classifiers, support vector machine, ensemble models, and boosting were used for the classification of the news. The implemented application serves as an environment for response teams to initiate remedial measures. Another work in this direction looked at classification of disaster-related news using supervised classifiers. The data is made available by scraping news and media sites to obtain pertinent information about natural disasters (Gopal *et al.* 2020). (Mas *et al.* 2023) explored the use of mobile spatial data, detecting anomalies in the patterns to estimate the population affected by a catastrophe in a bid to direct disaster management teams to mount their rescue.

Table 1 provides a comparative analysis of the state-of-the-art approaches used in detecting anomalies indicative of disasters. The approaches are contrasted based on the datasets, approaches, and performance metrics used.

Research on the detection of earthquakes and wildfires has yielded remarkable findings; the highest accuracy obtained by the deep CNN model is 99% with studies on the application of Linear Regression and SVM algorithms for broad catastrophe detection tallied at around 87% to 89%. For instance, in the study of extreme rainfall, an F1 score of 0.4601 and high accuracy of 99.93% were observed. However, the low precision and F1 scores of 65% in the detection of rain, floods, and landslip underlines difficulties in positively recognizing such calamities.

In every study the datasets used to obtain the results from the table should be acknowledged. When one considers the use of multiple sources of empirical data like waveforms that are recorded on tapes or even photographs taken by plane cameras, and then made available for download upon request. Research indicates the importance of diverse, high-quality information for accurate detection of abnormal patterns.

The research results in general prove the effectiveness of machine learning algorithms in disaster preparedness, with maximum rates of correct predictions being different from one disaster to another and depending heavily on the methodology used.

Table 1. Comparative analysis.

References	Disaster Type	Techniques	Dataset	Performance Metrics
(Cao *et al.* 2023)	Flood	SVM, K-means clustering	UAV	Acc - 92%
(Yang *et al.* 2022)	Hurricane Harvey	NER, BERT	Twitter	Acc - 95.55%
(Mas *et al.* 2023)	Flood	CNN	Twitter	Acc - 80.07%, F1 score 77%
(Akshya *et al.* 2019)	Fire	CNN	Images, videos from fire datasets	Acc - 94.39%, F1 score -89%
(Fan at al. 2020)	Wildfire	AlexNet, GoogLeNet, VGG-13	UAV	Acc - 99%
(Layek *et al.* 2019)	Earthquake	SVM, KNN	453 waveforms, Time series, Central Weather Bureau (CWB) data	–
(Muhammad *et al.* 2018)	Earthquake	GAN RF	Waveforms recorded in Southern California & Japan	Acc - 99.2%
(Lee *et al.* 2017)	Heavy rainfall	LR	Meteorological data in South Korea	Acc - 99.93% F1 score - 0.4601
(Li *et al.* 2018)	Landslide, Flood and Rain	SVM, K-NN, Naïve Bayes, MLP, LR	Online News	Acc - 68% F1 Score - 68%
(Chin *et al.* 2019)	General	LR, SVM with BoW	Online English News	Acc - 89% F1 Score - 88%
(Moon *et al.* 2019)	General	Merged Object Detection, YOLO V3, CNN	COCO and Google open image	Acc - 87%
(Domala *et al.* 2020)	Flood	Deep NN	Data from flood affected districts – Bhar & Orissa, India	Acc - 89.71%

4 CONCLUSION AND FUTURE DIRECTIONS

In summary, the paper provided a comprehensive overview of anomaly detection techniques, methodologies and performance of architectures for natural disaster management. This section mentions the limitations of the study, along with future directions in the domain.

4.1 *Limitations*

The study was setback by a few limitations. Incomplete reporting of performance markers in certain studies hindered a thorough comparison of the results. Furthermore, the generalizability of some research's conclusions may be limited by small sample sizes. The availability and quality of data is one of the main barriers to anomaly identification for natural disaster preparedness. Addressing these issues and ensuring open reporting of findings would enhance the credibility and significance of future research on anomaly identification for disaster preparedness.

The dynamic nature of crisis events poses a considerable challenge to anomaly detection systems, as behaviors and patterns are subject to fast alteration. To overcome this limitation, sophisticated algorithms that can adjust to changing trends are required Another significant difficulty is maintaining the performance and scalability of anomaly detection systems in the face of massive data volumes and constrained resources. Not to mention that every catastrophic occurrence presents a different set of domain-specific difficulties, such as managing finances and resources, resolving moral and ethical dilemmas, and integrating systems.

4.2 *Current and future development*

Anomaly identification for natural catastrophe preparedness has advanced significantly despite these drawbacks. These techniques can recognize patterns that differ from typical activity by learning from tagged data. Disaster response depends on the ability to identify anomalies in temporal data, such as network traffic or sensor readings, using time series techniques. Hybrid strategies that incorporate a fusion of these techniques could be explored to raise the precision and potency of anomaly identification in emergency situations. Future work could also look at the integration of anomaly detection systems with the existing processes and infrastructure. Domain specific frameworks working on real-time data can improve the response time for disaster management. Ultimately, the development of collaborative systems that facilitate the sharing of data and ideas among many agencies and organizations might improve the overall efficacy of anomaly detection in crisis situations. By addressing these problems and pursuing new research directions, the field of anomaly detection for disaster preparedness can advance and significantly enhance community resilience and response capabilities in the case of a natural disaster. The joint efforts of academics, practitioners, and policymakers will drive innovation and improvement in this vital subject of disaster management.

REFERENCES

Akshya, J., and Priyadarsini, P.L.K. (February 2019). A hybrid machine learning approach for classifying aerial images of flood-hit areas. In *Proceedings of the 2019 International Conference on Computational Intelligence in Data Science (ICCIDS)*, Chennai, India, 21–23.

Agyemang, M., Barker, F., and Alhajj, R. (2006). A comprehensive survey of numeric and symbolic outlier mining techniques. *Intelligent Data Analysis, 10*(6), 521–538.

Ardiansyah, M.F., William, T., Abdullaziz, O.I., Wang, L.-C., Tien, P.-L. and Yuang, M.C. (June 2020). EagleEYE: Aerial edge-enabled disaster relief response system. In *Proceedings of the 2020 European Conference on Networks and Communications (EuCNC)*, Dubrovnik, Croatia, 15–18.

Basher, P. (2006). Global early warning systems for natural hazards: systematic and people-centred. *Philosophical Transactions of the Royal Society A: Mathematical, Physical and Engineering Sciences, 364* (1845), 2167–2182.

Cao, L. (2023). AI and data science for smart emergency, crisis and disaster resilience. *International Journal of Data Science and Analytics, 15*, 231–246.

Chandola, V., Banerjee, A. and Kumar, V. (2009). Anomaly detection: A survey. *ACM Computing Surveys (CSUR), 41*(3), 1–58.

Chin, T.-L., Huang, C.-Y., Shen, S.-H., Tsai, Y.-C., Hu, Y.H. and Wu, Y.-M. (2019). Learn to detect: Improving the accuracy of earthquake detection. *IEEE Transactions on Geoscience and Remote Sensing, 57*, 8867–8878.

Cucurull, J., Nadjm-Tehrani, S. and Asplund, M. (2010). Anomaly detection and mitigation for disaster area networks. *Lecture Notes in Computer Science*, 339–359.

Domala, J., Dogra, M., Masrani, V., Fernandes, D., D'Souza, K., Fernandes, D. and Carvalho, T. (2020). Automated identification of disaster news for crisis management using machine learning and natural language processing. In *Proceedings of the 2020 International Conference on Electronics and Sustainable Communication Systems (ICESC)*, Coimbatore, India, 2–4 July 2020.

Fan, C., Wu, F. and Mostafavi, A. (2020). A hybrid machine learning pipeline for automated mapping of events and locations from social media in disasters. *IEEE Access, 8*, 10478–10490.

Gino, V.L.S., Negri, R.G., Souza, F.N., Silva, E.A., Bressane, A., Mendes, T.S.G. and Casaca, W. (2023). Integrating unsupervised machine intelligence and anomaly detection for spatio-temporal dynamic mapping using remote sensing image series. *Sustainability, 15*, 4725.

Gopal, L.S., Prabha, R., Pullarkatt, D. and Ramesh, M.V. (October–1 November 2020). Machine learning based classification of online news data for disaster management. In *Proceedings of the 2020 IEEE Global Humanitarian Technology Conference (GHTC)*, Seattle, WA, USA, 29.

Layek, A.K., Poddar, S. and Mandal, S. (February 2019). Detection of flood images posted on online social media for disaster response. In *Proceedings of the 2019 Second International Conference on Advanced Computational and Communication Paradigms (ICACCP)*, Gangtok, India, 25–28.

Lee, W., Kim, S., Lee, Y.-T., Lee, H.-W. and Choi, M. (January 2017). Deep neural networks for wildfire detection with unmanned aerial vehicle. In *Proceedings of the 2017 IEEE International Conference on Consumer Electronics (ICCE)*, Las Vegas, NV, USA, 8–10.

Li, Z., Meier, M.A., Hauksson, E., Zhan, Z., and Andrews, J. (2018). Machine learning seismic wave discrimination: Application to earthquake early warning. *Geophysical Research Letters, 45*, 4773–4779.

Marzuoli, A., and Liu, F. (2019). Monitoring of natural disasters through anomaly detection on mobile phone data. *2019 IEEE International Conference on Big Data (Big Data)*, Los Angeles, CA, USA, 4089–4098.

Mas, E. and Koshimura, S. (2023). Anomaly detection in mobile spatial statistics for disaster risk management. Available at SSRN.

Moon, S.H., Kim, Y.H., Lee, Y.H. and Moon, B.R. (2019). Application of machine learning to an early warning system for very short-term heavy rainfall. *Journal of Hydrology, 568*, 1042–1054.

Muhammad, K., Ahmad, J. and Baik, S.W. (2018). Early fire detection using convolutional neural networks during surveillance for effective disaster management. *Neurocomputing, 288*, 30–42.

Patcha, A. and Park, J. M. (2007). An overview of anomaly detection techniques: Existing solutions and latest technological trends. *Computer networks, 51*(12), 3448–3470.

Rezvani, S.M., Falcão, M.J., Komljenovic, D. and de Almeida, N.M. (2023). A systematic literature review on urban resilience enabled with asset and disaster risk management approaches and GIS-based decision support tools. *Applied Sciences, 13*, 2223.

Sakaki, T., Matsuo, Y., Kurihara, S., Toriumi, F., Shinoda, K., Noda, I., and Kazama, K. (2013, August). The possibility of social media analysis for disaster management. In *2013 IEEE Region 10 Humanitarian Technology Conference* (pp. 238–243). IEEE.

Sankaranarayanan, S., Prabhakar, M., Satish, S., Jain, P., Ramprasad, A. and Krishnan, A. (2019). Flood prediction based on weather parameters using deep learning. *Journal of Water and Climate Change, 11*, 1766–1783.

Yang, Z., Liu, X., Li, T., Wu, D., Wang, J., Zhao, Y. and Han, H. (2022). A systematic literature review of methods and datasets for anomaly-based network intrusion detection. *Computers & Security, 116*, 102675.

Data Science & Exploration in Artificial Intelligence – Gururaj et al. (Eds)
© 2025 the Author(s), ISBN: 978-1-032-96390-7

Exploring the genre-specific influence of acoustical attribute in music business analytics

R.S. Chidhananda
RV University, RV Vidyaniketan, Bangalore, India

Apurva Kulkarni
International Institute of Information Technology, Bangalore, India

ABSTRACT: Contemporary music feature analysis techniques utilized by music service aggregators focus on acoustic attributes such as tempo, genre, sentiment analysis, and a few other evident patterns to curate the consumption of music to the digital audience. To gain a deeper understanding of the cultural nuances within specific genres and their influence on music business parameters, it is imperative to move beyond traditional rhythmic features and delve into the structural and spectral intricacies of audio. The proposed research aims to shift the focus from conventional rhythmic analysis in genre classification, prediction models, and information retrieval methods within audio studies that focus on acoustic attributes using signal processing techniques. The research utilizes frequency distribution of audio files and focuses on understanding its correlation with media consumption by listeners. In this work, we focus on the Pop genre dataset representing nearly 4000 audio files. The sample dataset used for implementation spans across various consumption levels ranging from (726450 users to 2959 users). The proposed methodology uses signal processing and frequency binning to understand the spectral pattern in Pop genre. Furthermore, the research identifies the prominent frequency bins and its probable influence on consumption and the experiment has shown positive observations that highlight the relationship of the same. The positive trend increases the motivation to carry out the analysis to a commercial dataset and in addition integrate the proposed approach with traditional music characteristics to provide direction to enhance digital music service consumption.

1 INTRODUCTION

In the field of Music Business Analysis, the primary classification is based on 'genre'. The various attributes and parameters specific to the genre contributes to its unique signature. The genre is built over the foundation of distinct acoustic attributes, containing instruments, rhythm, melody, and timbre. Understanding the pattern of these attributes define the trait of the specific genre and further can be applied in various domains like music information retrieval (MIR), including genre classification, recommendation systems, and content-based music retrieval. In MIR, the content consumption can be analyzed based on various components, including consumer behavior (such as playlist analysis, skip rate, retention time, and loop-through rate), content-specific attributes (like genre, artist, and band), rhythmic features (tempo, rhythm, loudness, danceability), and external attributes like social impact. The key point to note here is all the above factors can vary within the genre and among the genres. The loudness across two different genres can hold the similar values where on the other hand, different songs within a genre can hold different ranges of values. The variability

DOI: 10.1201/9781003589273-24

in music compositions, rhythmic intricacies and individual choices in consumer preferences (which can vary based on person, genre, location, culture, current trend on social media etc) makes understanding the music consumption pattern more challenging. Moreover, the dynamic nature of the music landscape, characterized by evolving genre boundaries and the fusion of diverse semantic content and cultural influences, adds complexity. This evolution leads to the emergence of hybrid genres and novel expressions, blurring traditional genre distinctions. Furthermore, semantic content, including lyrics and thematic elements, often diverges from rhythmic features, posing additional challenges for analysis.

2 RELATED WORK

The last two decades have witnessed an increasing scholarly interest in music consumption. Montoro-Pons, Caballer-Tarazona, and Cuadrado-Garc´ıa (2021) proposes a study elaborating on the relevance of music as a form of cultural consumption and the profound evolution related to the nature of music consumption. The advancement of digital technology has revolutionized the way music is consumed, produced, and distributed. In the article, the authors Bello and Garcia (2021) delve into the impact of digitization and streaming on the globalization of Popular music, illustrating a trend wherein countries increasingly showcase unique musical preferences reflected in their respective charts. Brito and Vieira (2022) and Husin, Hidayanto, Purwandari, and Ibrahim (2021) discuss various factors including the service provider characteristics (including subscription policies, quality, price, etc.), social media influence, behavioral patterns of the consumers, and music-related attributes. The large set of analytical techniques discussed by Kawai, Esling, and Harada (2020) and Mezza, Zanoni, and Sarti (2023) includes machine learning and artificial intelligence play a crucial role in understanding the rhythmic patterns of music and how it impact the user behavior. Authors Wang, Wei, and Yang (2022) and Fürst *et al.* (2020) elaborate on the application of deep neural networks and statistical analysis with rhythmic attributes. The application of these techniques can be extended to behavioral pattern analysis of music consumers targeting user-preferred genres highlighted by authors Anderson, Maystre, Anderson, Mehrotra, and Lalmas (2020). On the same line, researchers Singh and Bohat (2021); Stachl *et al.* (2020) dwell on artist-specific patterns, retention patterns and skip rate. The overarching discussions presented by Ko, Lee, Park, and Choi (2022) shed light on the use of different behavioral patterns to recommend the music. Along with the understanding of the behavioral and rhythmic patterns influencing music reception discussed by Cocharro, Bernardes, Bernardo, and Lemos (2021), it is essential to recognize the global nature of these consumption patterns. As music transcends geographical boundaries through digital platforms, consumption behaviors exhibit a remarkable level of diversity and fluidity across different regions and cultures. What resonates with audiences in one part of the world may not necessarily have the same impact elsewhere. Therefore, understanding and adapting to acoustic patterns is crucial for accurately predicting and catering to the needs of diverse audiences worldwide.

3 MOTIVATION

The proposed research presents a novel approach that utilizes acoustic spectral distribution to analyze audio signals to gain insights into their distinctive audio fingerprints, which in turn helps us analyze their impact on consumption patterns. Our motivation lies in identifying frequency patterns within genres by analyzing each sample in the dataset and comprehending the overarching pattern that characterizes each genre. Subsequently, we aim to correlate the presence or absence of these patterns with their influence on music consumption.

4 PROPOSED APPROACH

The proposed work is implemented over a sample dataset of songs collected under the Pop genre. The audio files have been selected without bias from the Free Music Archives collection (timestamped on 09/10/2023, data source: https://freemusicarchive.org/music). The initial step is to pre-process the audio signal and remove inherent biases. To remove the DC Component from the audio signal, the methodology implements a simple operation that entails subtracting the mean value of the signal from each data point, effectively centering the waveform around zero and eliminating any constant offset. Mathematically, this is expressed as: xDC removed(t) = x(t) − mean(x(t)), where x(t) represents the original audio signal. To access acoustic characteristics in the frequency domain, the proposed approach converts an audio signal into the frequency spectrum by applying the Fourier Transform. X (f) = FFT(x(t)), where X(f) denotes the frequency spectrum of the audio signal.

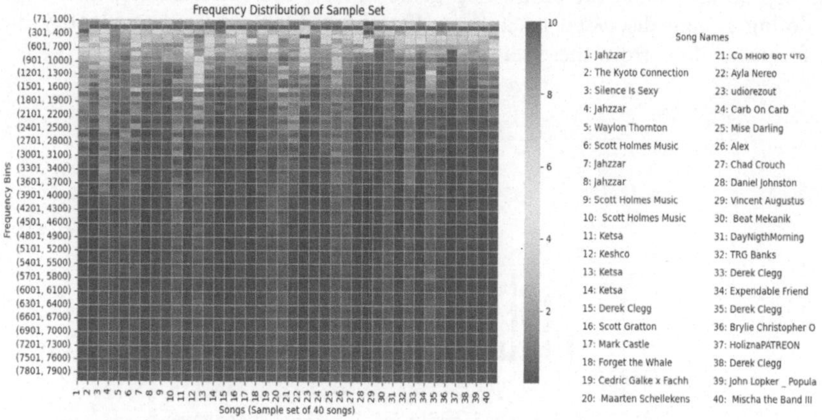

Figure 1. Frequency distribution: The bin values for each song in the sample set indicating frequency distribution in the Pop genre.

4.1 *Spectral binning*

The frequency spectrum in continuous signal and for the ease of further analysis, the proposed approach uses 'Spectral Binning'. In this method, the continuous spectrum is divided into equal-sized frequency bins. Each bin spans a range of 100 Hz, ensuring a comprehensive partitioning of the spectrum. The frequency bins are expressed as Frequency bins = $\{100 \times n \mid n = 0, 1, 2, ..., 159\}$, where n represents the index of the bins. Figure 1 illustrates the spectral distribution of the audio sub-sample set. To ensure representation across various consumption counts, 40 audio samples were selected, forming four distinct groups based on consumption ranges. Group 1 comprises songs with approximately 300,000 consumption counts, Group 2 with 13,000, Group 3 with 4,000, and Group 4 with less than 2,000 counts. Within each group, 10 random samples were collected, as depicted in Figure 1. The heatmap shows a concentration of energy values towards the top, indicating higher energy levels in the lower frequency bins, emphasizing their significance in the spectral distribution.

4.2 *Spectral distribution*

The proposed method aims to assess the significance of each frequency bin within the audio sample by computing the energy associated with it. This process entails summing the squared magnitudes of the Fourier coefficients contained within each bin. The value quantifies the intensity of each frequency component, offering valuable insights into their respective contributions to the overall spectral composition of the audio.

The consumption of an energy for a particular frequency bin can be calculated as:

$$E_i = \sum_{f \in Frequency\ bin_i} |X(f)|^2$$

where the E_i represents the overall energy of the ith frequency bin. The top 50 frequency bins with maximum energy values are used to represent the prominent frequency pattern within the audio.

4.3 *Spectral fingerprint*

The spectral distribution contributes to identifying the top 50 bins per audio file. At genre level, all top 50 bins from the entire Pop genre dataset are collected together. The Borda Count Voting system discussed by authors Pattanaik (2002) is used to identify the genre specific frequency bins from these set of collected bins.

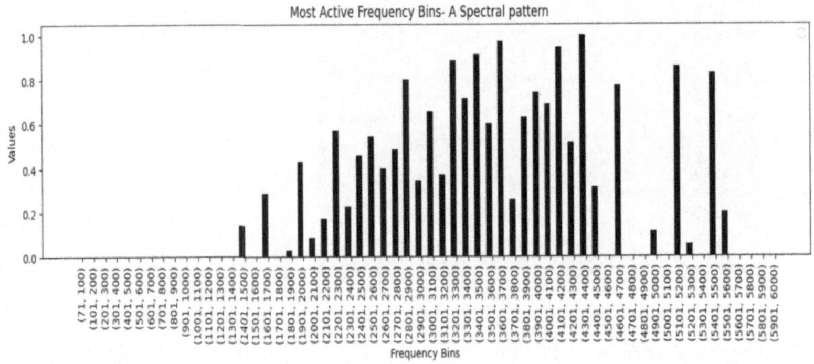

Figure 2. The top 35 frequency bins present in the Pop Genre indicating spectral pattern.

The top 35 frequency bins encapsulates the spectral characteristics of the Pop genre. We name these frequency bins - spectral fingerprints. (Note: The decision to utilize the top 50 bins for each song and the top 35 bins for the Pop genre is meticulously made following experimental validation. These carefully chosen values include minimum and maximum frequency bins that comprehensively span the entire frequency spectrum, rendering them ideal selections for analysis). Figure 2 demonstrates the top 35 frequency bins in the Pop genre. The X-axis denotes the frequency bins (upto 6000Hz). The Y-axis represents the normalized values within the range of 0-1 for illustrating the significance of each bin in the spectral pattern.

5 EXPERIMENTS AND RESULTS

The primary goal of the research is to understand the correlation between the frequency pattern and consumption trend in the Pop genre. The experiments are conducted at two levels of granularity. The first approach examines individual audio samples, emphasizing the song's consumption count and identifying the number of frequency bins present from the Pop genre's spectral fingerprint within the top 50 bins of each song. The second approach conducts a group-wise analysis, categorizing sample songs based on their consumption count, as discussed in Section 4. Figure 3 discusses the correlation between frequency spectral distribution and audio consumption count. Subfigure (a) highlights this trend at each audio level. The X-axis displays individual audio samples arranged in descending order of their consumption counts. The primary Y-axis, shown in blue, represents the consumption count of each audio sample, while the secondary Y-axis indicates the number of common frequency bins present in the top 50 bins of each song and the top 35 bins representing the Pop genre. The red line depicts the trend of common frequency bins, while the blue line illustrates the trend of consumption count. Notably, both lines exhibit a negative slope, indicating that as the number of common frequency bins decreases, the consumption rate also tends to decrease. Sub-figure (b) provides a group-wise perspective on the relationship between frequency distribution and consumption patterns. The X-axis denotes the grouping of the sample set and its corresponding approximate consumption count, with the maximum consumption count in each group considered. The primary Y-axis indicates the frequency bin index, while the secondary Y-axis represents the number of common frequency bins. Figure 2 indicates that the most active 35 frequency bins in the Pop genre are typically found within the ranges of 1400 to 5500. In Figure 3 sub-figure (b), it is evident that as the consumption count decreases, the frequency domain becomes more distributed, showing a reduced concentration of frequencies within the lower ranges (1400 to 5500). The red line illustrates the average number of common frequency bins in each group. Notably, as the value of common frequency bins decreases, both the consumption count and the frequency domain exhibit greater dispersion. Both experiments emphasize a significant correlation between the acoustic frequency patterns and consumption behaviors within the Pop genre, shedding light on their interrelated dynamics in music consumption trends. The proposed methodology is versatile and can be applicable across diverse musical genres, facilitating the analysis of consumption patterns through the utilization of acoustic parameters.

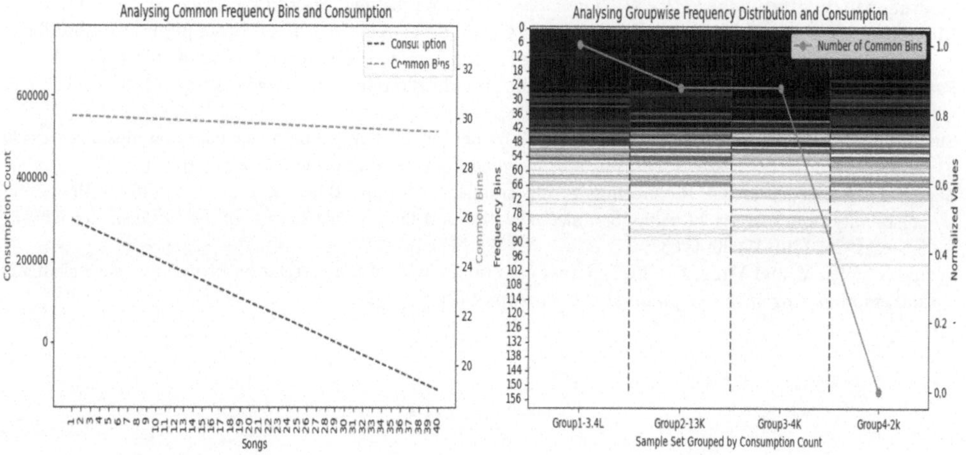

Figure 3. Understanding the correlation between the frequency distribution and consumption: a) The lines indicating consumption across the sample set and the presence of frequency pattern present in each song in the sample set, b) Groupwise understanding of frequency distribution and consumption pattern

6 CONCLUSION

The proposed research explores the relationship between spectral parameters and the consumption of music. The implementation focuses on the sample set of the Pop music genre where the top frequency bins are identified for each song. The consumption pattern using spectral characteristics analysis through frequency bins has shown a positive trend signifying a correlation between the acoustic frequency bins and consumption behaviors. The proposed research lays a robust foundation for extending the approach to larger-scale commercial datasets. Integration with commercial applications and service providers will serve to refine and optimize the proposed methodology further.

REFERENCES

Anderson, A., Maystre, L., Anderson, I., Mehrotra, R. and Lalmas, M. (2020). Algorithmic effects on the diversity of consumption on Spotify. *In Proceedings of the Web Conference* 2020 (pp. 2155–2165).

Bello, P. and Garcia, D. (2021). Cultural divergence in popular music: the increasing diversity of music consumption on Spotify across countries. *Humanities and Social Sciences Communications*, 8 (1), 1–8.

Brito, F. A. and Vieira, D. A. (2022). Music streaming: Consumption patterns in digital ambients. *In Impact of Digital Transformation on the Development of New Business Models and Consumer Experience* (pp. 74–96). IGI Global.

Cocharro, D., Bernardes, G., Bernardo, G. and Lemos, C. (2021). A review of musical rhythm representation and (dis) similarity in symbolic and audio domains. *Perspectives on Music, Sound, and Musicology: Research, Education and Practice*, 189–208.

Fürst, D., Miller, M., Keim, D. A., Bonnici, A., Schäfer, H. and El-Assady, M. (2020). Augmenting sheet music with rhythmic fingerprints. *In 2020 IEEE 5th workshop on visualization for the digital humanities (vis4dh)* (p. 14–23).

Husin, N., Hidayanto, A. N., Purwandari, B. and Ibrahim, R. M. S. (2021). Analysis on digital music service user behaviour using justice perception framework. *J. Eng. Sci. Technol*, 16 (2), 1644–1666.

Kawai, L., Esling, P. and Harada, T. (2020). Attributes-aware deep music transformation. *In Ismir* (pp. 670–677).

Ko, H., Lee, S., Park, Y. and Choi, A. (2022). A survey of recommendation systems: recommendation models, techniques, and application fields. *Electronics*, 11 (1), 141.

Mezza, A. I., Zanoni, M. and Sarti, A. (2023). A latent rhythm complexity model for attribute controlled drum pattern generation. *EURASIP Journal on Audio, Speech, and Music Processing*, 2023 (1), 11.

Montoro-Pons, J. D., Caballer-Tarazona, M. and Cuadrado-García, M. (2021). From pirates to subscribers: 20 years of music consumption research. *International Journal of Consumer Studies*, 45 (4), 690–718.

Pattanaik, P. K. (2002). Positional rules of collective decision-making. *Handbook of Social Choice and Welfare*, 1 , 361–394.

Singh, J. and Bohat, V. K. (2021). Neural network model for recommending music based on music genres. *In 2021 International Conference on Computer Communication and Informatics (iccci)* (pp. 1–6).

Stachl, C., Au, Q., Schoedel, R., Gosling, S. D., Harari, G. M. and Buschek, D., others (2020). Predicting personality from patterns of behavior collected with smartphones. *Proceedings of the National Academy of Sciences*, 117 (30), 17680–17687.

Wang, X., Wei, Y. and Yang, D. (2022). Cross-cultural analysis of the correlation between musical elements and emotion. *Cognitive Computation and Systems*, 4 (2), 116–129.

Data Science & Exploration in Artificial Intelligence – Gururaj et al. (Eds)
© 2025 the Author(s), ISBN: 978-1-032-96390-7

Exploring trust management across different network environments

Akash Angom, Nirmalya Kar & Priyanka Biswas
Department of Computer Science and Engineering, National Institute of Technology Agartala, Tripura, India

Abhijit Das & Rakesh Kumar Godi
Department of Information Technology, Manipal Institute of Technology Bengaluru, Manipal Academy of Higher Education, Manipal, India

ABSTRACT: Trust Management (TM) plays a crucial role in the realm of cyber-security as it is indispensable in the process of mitigating risks associated with cyber threats. It involves establishing and maintaining trust among individuals, networks, and systems operating within an organization. By doing so, TM effectively reduces the probability of security breaches occurring and enhances the overall resilience of cyber-security defenses. In the present study, an exhaustive examination has been undertaken to scrutinize the existing methodologies to manage and assess trust. Furthermore, the research meticulously organizes and categorizes the most recent TM techniques that have emerged in cyber-security. This comprehensive analysis contributes significantly to the knowledge of TM within cybersecurity, thoroughly exploring the strategies, and approaches employed to enhance trustworthiness and security within organizational networks.

1 INTRODUCTION

In cybersecurity, where information storing, sharing, and digital transactions are commonplace, trust is essential to maintaining data and systems' dependability, integrity, and secrecy. To strengthen their defenses against potential dangers, organizations must thoroughly grasp how they manage trust, given the growing sophistication of cyber-attacks. Trust is not only a theoretical concept but an essential element that permeates many aspects of cybersecurity and affects user interactions, decision-making procedures, and an organization's overall security posture. In various kinds of networks, including supply chains, cloud computing, the Internet of Things, healthcare systems, and many more, effective trust management is essential to safeguard systems by verifying and authenticating interactions and transactions among entities. This encompasses establishing protocols, policies, and mechanisms to evaluate and oversee entities' reliability and conduct within the network. An entity within the network can be deemed reliable and trustworthy if it consistently follows the anticipated and necessary conduct. Nodes with more successful interactions are regarded as more trustworthy [15]. Trust between entities typically grows with each successful interaction but decreases following failures, as a failed interaction often leads to a more significant decrease in trust than a gain from a successful one. TM model must scale for network growth, maintain consistent functionality, accurately identify nodes, resist attacks, stabilize disruptions, ensure privacy, and respond quickly for safety-critical applications [1].

Trust Management System (TMS) automates trust-based decisions, converting trust into rules and controls for security services like identity management, service composition, access control, and policy enforcement. The main stages of TM system are shown in Figure 1 and described as [21]:

DOI: 10.1201/9781003589273-25

1.1 *Information gathering*

Attributes like scalability, accuracy, robustness, etc., are gathered directly or through reputation and recommendations.

1.2 *Trust computation*

Attributes are aggregated using appropriate evaluation models to compute the final trust score.

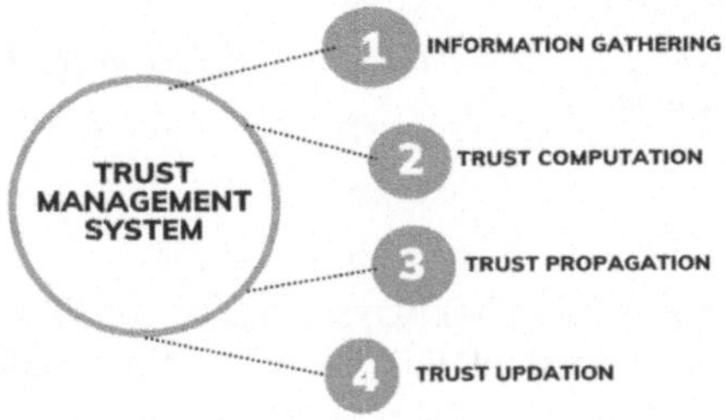

Figure 1. Main stages of trust management system.

1.3 *Trust propagation*

Determines how trust values spread, employing strategies like limited hops and giving more weightage to recent interactions.

1.4 *Trust updation*

Automated, real-time updates based on events like security threats or Service Level Agreements (SLAs) violations [23].

This paper summarises TM models, covering their requirements, techniques, and applications across domains. It also classifies metrics used in the literature for trust evaluation, emphasizing existing issues and potential future study areas. The organization of the paper includes TM-related surveys in Section 2, classifying TM models in Section 3, TM implementations and requirements in Section 4, metrics consideration in Section 5, and finally conclusion in Section 6.

2 LITERATURE SURVEY

In this section, we present diverse works on trust management systems. Our analysis of several surveys has discovered that many works tend to embrace specific methodologies, often focusing solely on a limited number of phases within the complete procedure, thus missing a holistic viewpoint. A comprehensive finding is presented in Table 1.

Table 1. Related surveys on trust management.

Ref.	Approach	Environment	Contributions
[1]	Centralized, distributed, and decentralized based	Intelligent transportation, Io-Vehicles	Mechanisms, evaluations, and metrics used to achieve trust management requirements were surveyed.

(*continued*)

Table 1. Continued

Ref.	Approach	Environment	Contributions
[9]	Blockchain, Non-blockchain based	Edge-based Internet of Things	Blockchain as a solution is investigated for mitigating several trust issues.
[12]	Centralized, distributed, and hybrid	Social Internet of Things (SIoT)	Classifying TM schemes for SIoT. Comparisons of evaluation models and architectures.
[14]	Blockchain-based	Cloud computing	Classifies approaches as basic, enhanced, and data management.
[15]	Artificial Intelligence (AI)	Distributed Networks (DNs)	Using AI-driven trust within DNs to uncover hidden traits for accurate trust evaluation.
[16]	Blockchain-based	Vehicular networks	Trust issues in VANETs, IoV, connected vehicles, and vehicular edge computing.
[21]	Distributed, semi-distributed, centralized, and blockchain	Internet of Things (IoTs)	Identification of applications and deployment challenges. Management techniques for every layer in the IoT architecture were provided.
[24]	Machine Learning (ML)	Social, Multi-agent, service, and ad-hoc networks	Categorizing existing ML-based trust evaluation models according to their application scenarios.

3 TAXONOMY OF TRUST MANAGEMENT MODELS

We've condensed various TM models into a taxonomy and classification, as shown in Figure 2.

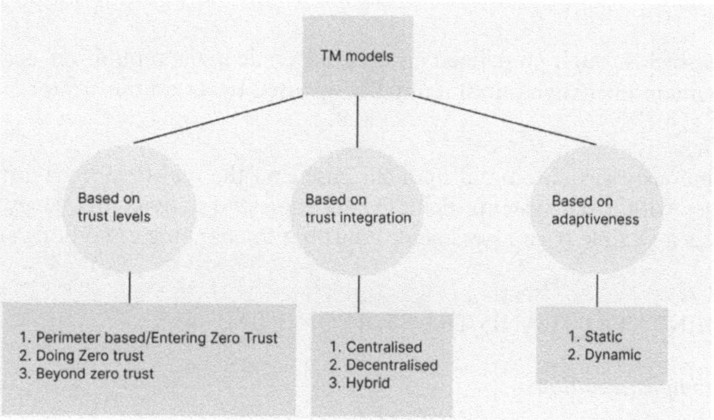

Figure 2. Proposed taxonomy and classification of trust models.

3.1 *Based on trust levels*

3.1.1 *Perimeter-based/Entering zero trust*

This approach sorts transactions, network data, and users into trusted and untrusted groups, aligning with Zero Trust (ZT) principles. ZT challenges traditional cybersecurity by focusing on users and assets rather than network perimeters or asset ownership [22]. However, its full implementation seems lacking in this model.

3.1.2 *Doing zero trust*

A comprehensive implementation of ZT architecture ensures that trust is not assumed for anyone, applying rigorous scrutiny and Continuous identity verification for all, inside and out, even within the perimeter [6,22].

3.1.3 *Beyond zero trust*

Software-defined perimeter architecture separates network data and control, requiring proper credentials for access by internal and external users [19]; Zero Trust eXtended, as outlined by Forrester, extends the zero trust paradigm across seven pillars including data, networks, people, workloads, devices, visibility and analytics, and automation & orchestration [11]; AI and Security Orchestration, Automation, and Response integration into zero trust security architectures automate user lifecycle management, securely managing identity creation, credentialing, de-credentialing, and deletion without manual intervention [6,25].

3.2 *Based on trust integration*

3.2.1 *Centralised*

A central authority ensures consistent trust regulations across a network, fostering trust relationships among its entities [12,21].

3.2.2 *Decentralised*

Decision-making and accountability is spread across nodes. Distributed Ledger Technology such as blockchain enhances trust through a shared, transparent, and immutable ledger. Peer-to-peer (P2P) interactions bypass central authorities [16].

3.2.3 *Hybrid*

A central authority establishes trust policies and supervises the framework, while individual network entities manage specific trust relationships based on rules or local policies [12].

3.3 *Based on trust adaptation*

3.3.1 *Static*

Trust is established through predefined criteria like credentials, reputation, certifications, or roles, which remain unchanged until manually updated by an administrator.

3.3.2 *Adaptive*

Trust is continuously evaluated and updated based on the real-time behavior and interactions of entities within the system, including location and activity frequency [17,23]. This method enables a flexible trust assessment, adjusting to changing conditions and behaviors.

4 IMPLEMENTATIONS AND TRUST REQUIREMENTS

4.1 *Existing implementations*

Existing trust model implementation in various environments is given in Table 2.

Table 2. TM implementation in various environments.

Environment	Summary	Remark	Limitation
Smart City [2]	Smart devices use dynamic allow/block lists to improve service quality by selecting trustworthy participants.	Trust-based IoT in towns ensures smart, reliable decisions using sensor and device data.	Improving consumer trust and agility is key, alongside tackling IoT computing hurdles.
Internet of Behavior (IoB) [3]	A deep learning model for assessing safe driving behavior, promoting trust among insurers, automated systems, and drivers.	Predicting behavior timely by incorporating reputation computation for fast and secure responses.	Privacy preservation is not considered.

(continued)

Table 2. Continued

Environment	Summary	Remark	Limitation
Collaborative Healthcare systems [4]	Hybrid trust model for distributed collaboration using reputation, recommendations, and teamwork.	In trust creation, the activity creator assesses factors (time, purpose, trustee) to verify trustworthiness.	Improving result accuracy requires considering various environments.
Healthcare smartphone network [5]	A lightweight TM system that clusters nodes and analyzes their collective trust to detect and block malicious groups.	Energy-efficient data collection and secure exchanges via blockchain ensure decentralized data security.	Balancing patient privacy (anonymity) with secure healthcare is a challenge.
Supply Chain [7]	Introduced ZT in the supply chain, detailing the transition process and suggesting research propositions to advance inquiry in this new area.	ZT supply chain benefits may decline after initial goals are achieved.	Uncertain outcomes necessitate further research on this unproven practice.
Cloud Computing [10]	Assessing cloud provider trust via Quality of Service, dynamically calculates trust with Accumulative/ Computed Trust Value and user feedback for best selection.	Service Providers require robust identity authentication, access control, and transparent data handling for SLAs and compliance.	Evaluation for both public and private cloud providers, as well as cloud users is not considered.
Vehicular Networks [16,26]	Deep learning assesses node trust, while blockchain validates component identity and message authenticity, boosting security.	Third-party TM needs a global reputation but is impractical for vehicle networks due to maintenance.	Privacy worries and keeping trust scores accurate are big hurdles.

4.2 *Achieving trust management requirements*

Requirements and mechanisms to meet the trust model are given as [1].

4.2.1 *Scalability*

Using a hybrid model by blending centralized and decentralized features for scalability. Nodes independently assess trust, and collective evaluation guides the overall trust decision.

4.2.2 *Accuracy*

Fuzzy logic aids in decision-making to ensure accurate trust values. Incentives encourage accurate feedback from nodes while adjusting parameters and metrics enhances accuracy in dynamic environments.

4.2.3 *Robustness*

Trust models need to withstand various attacks, like jamming, impersonation, and reputational manipulation, while also being able to counteract them effectively. A decay factor slows trustworthiness growth.

4.2.4 *Privacy preservation*

Methods like zero-knowledge proofs, digital certificates, and digital signatures can help protect identity privacy.

4.2.5 *Appropriate response time*

Limiting the number of participants for overall trust evaluation by choosing highly reputed nodes and exploring adaptive methods to handle delays and maximize efficiency.

5 METRICS CONSIDERATION

Entity trust is evaluated by combining context-specific metrics. Selecting appropriate metrics is crucial for precise trust assessments.

We categorize trust evaluation metrics into these groups:

5.1 *Reputation-based metrics*

Reputation score (rs) reflects trustworthiness based on past performance, connections, and feedback from other users; the number of transactions (nt) measures an entity's activity, with a higher number indicating greater engagement and experience with system processes.

5.2 *Behavior-based metrics*

Consistency (cons) is about being reliable and predictable over time; accuracy (acc) ensures the truthfulness of data and actions by comparing them to factual standards; compliance (comp) measures adherence to laws, guidelines, policies, and moral principles within the system.

5.3 *Network-based metrics*

Direct trust (dt) measures the level of trust that one entity has in another based on their direct interactions or experiences; transitive trust (tt) is the indirect transfer of trust between entities within a network; social proximity (sp) is the measure of closeness between entities in a network, determined by shared relationships, mutual acquaintances, and community ties.

The trust evaluation techniques tailored to meet specific environmental network requirements presented recently have been analyzed. Metrics considered for trust evaluation in various works are given in Table 3. Here "✓" refers to "considered".

Table 3. Metrics considered for trust evaluation.

Ref.	rs	nt	cons	acc	comp	dt	tt	sp
[8]	✓					✓	✓	
[10]	✓			✓	✓			
[13]				✓	✓		✓	
[17]	✓	✓	✓	✓	✓	✓	✓	
[18]				✓		✓	✓	
[20]	✓		✓	✓		✓	✓	✓
[23]	✓	✓		✓		✓	✓	

6 CONCLUSION

This paper explored recent TM developments across various networks, categorizing TM models, metrics, and requirements. From our classification in section 3, the latest model of each category, namely: beyond zero trust, hybrid and adaptive approaches are more evolved in terms of TM. Comprehensive analysis suggested prioritizing factors like evaluating AI tech trustworthiness, bridging theory and practice, addressing trust, privacy, and service quality, and advancing TM across various settings. Collaborative efforts involving academia, industry, and policymakers are crucial for developing scalable, secure, and reliable zero-touch AI systems, crucial for shaping trust management in the future.

REFERENCES

[1] Abidi, R., *et al.* (2024). A study of mechanisms and approaches for IoV trust models requirements achievement. *The Journal of Supercomputing*, *80*(3), pp.4157–4201.

[2] Alazab, M., Manogaran, G. and Montenegro-Marin, C.E. (2022). Trust management for internet of things using cloud computing and security in smart cities. *Cluster Computing*, *25*(3), pp.1765–1777.

[3] Bangui, H., Cioroaica, E., Ge, M. and Buhnova, B. (2023, March). Deep-learning based trust management with self-adaptation in the internet of behavior. In *Proceedings of the 38th ACM/SIGAPP Symposium on Applied Computing* (pp. 874–881).

[4] Barhoun, R., *et al.* (2023). Trust modeling in a distributed collaborative environment: application to a collaborative healthcare system. *International Journal of Information Security*, *22*(6), pp.1585–1604.

[5] Bhan, R., *et al.* (2023). Blockchain-enabled secure and efficient data sharing scheme for trust management in healthcare smartphone network. *The Journal of Supercomputing*, *79*(14), pp.16233–16274.

[6] Campbell, M. (2020). Beyond zero trust: Trust is a vulnerability. *Computer*, *53*(10), pp.110–113.

[7] Collier, Z.A. and Sarkis, J. (2021). The zero trust supply chain: Managing supply chain risk in the absence of trust. *International Journal of Production Research*, *59*(11), pp.3430–3445.

[8] Fan, X., Liu, L., Zhang, R., Jing, Q. and Bi, J. (2020). Decentralized trust management: Risk analysis and trust aggregation. *ACM Computing Surveys (CSUR)*, *53*(1), pp.1–33.

[9] Fotia, L., Delicato, F. and Fortino, G. (2023). Trust in edge-based internet of things architectures: state of the art and research challenges. *ACM Computing Surveys*, *55*(9), pp.1–34.

[10] Hassan, H., El-Desouky, A.I., Ibrahim, A., El-Kenawy, E.S.M. and Arnous, R. (2020). Enhanced QoS-based model for trust assessment in cloud computing environment. *IEEE Access*, *8*, pp.43752–43763.

[11] https://www.forrester.com/report/the-zero-trust-extended-ztx-ecosystem/RES137210

[12] Khan, W.Z., *et al.* (2020). Trust management in social internet of things: Architectures, recent advancements, and future challenges. *IEEE Internet of Things Journal*, *8*(10), pp.7768–7788.

[13] Latif, R., *et al.* (2021). A novel cloud management framework for trust establishment and evaluation in a federated cloud environment. *The Journal of Supercomputing*, *77*(11), pp.12537–12560.

[14] Li, W., Wu, J., Cao, J. *et al.* (2021). Blockchain-based trust management in cloud computing systems: a taxonomy, review and future directions. *Journal of Cloud Computing*, *10*(1), p.35.

[15] Li, Z., Fang, W., *et al.* (2023). AI-enabled Trust in Distributed Networks. *IEEE Access*.

[16] Mikavica, B., *et al.* (2021). Blockchain-based solutions for security, privacy, and trust management in vehicular networks: a survey. *The Journal of Supercomputing*, *77*(9), pp.9520–9575.

[17] Mujawar, T.N., *et al.* (2022). Behavior and feedback based trust computation in cloud environment. *Journal of King Saud University-Computer and Information Sciences*, *34*(8), pp.4956–4967.

[18] Qureshi, K.N., Iftikhar, A., Bhatti, S.N. *et al.* (2020). Trust management and evaluation for edge intelligence in the Internet of Things. *Engineering Applications of Artificial Intelligence*, *94*, p.103756.

[19] Sallam, A., Refaey, A. and Shami, A. (2019). On the security of SDN: A completed secure and scalable framework using the software-defined perimeter. *IEEE access*, *7*, pp.146577–146587.

[20] Saxena, U.R. and Alam, T. (2023). Recommendation-based trust computation and rating prediction model for security enhancement in cloud computing systems. *Service Oriented Computing and Applications*, *17*(4), pp.239–257.

[21] Sharma, A., Pilli, E.S., *et al.* (2020). Towards trustworthy Internet of Things: A survey on Trust Management applications and schemes. *Computer Communications*, *160*, pp.475–493.

[22] Stafford, V.A. (2020). *Zero Trust Architecture*. NIST special publication, 800, p.207.

[23] Valero, J.M.J., Sánchez, P.M.S., *et al.* (2023). Trust-as-a-Service: A reputation-enabled trust framework for 5G network resource provisioning. *Computer Communications*, *211*, pp.229–238.

[24] Wang, J., Jing, X., Yan, Z., Fu, Y., Pedrycz, W. and Yang, L.T. (2020). A survey on trust evaluation based on machine learning. *ACM Computing Surveys (CSUR)*, *53*(5), pp.1–36.

[25] Yeoh, W., Liu, M., Shore, M. and Jiang, F. (2023). Zero trust cybersecurity: Critical success factors and A maturity assessment framework. *Computers & Security*, *133*, p.103412.

[26] Zhang, C., *et al.* (20200. AIT: An AI-enabled trust management system for vehicular networks using blockchain technology. *IEEE Internet of Things Journal*, *8*(5), pp.3157–3169.

Data Science & Exploration in Artificial Intelligence – Gururaj et al. (Eds)

A critical review on blockchain based data auditing for cloud data security

J. Uma, R. Srinithi, K. Priyanka, T. Kowshikraja & S. Abhinav
Department of Computer Science and Engineering, Amrita School of Computing, Coimbatore, Amrita Vishwa Vidyapeetham, India

ABSTRACT: Emergence of cloud computing has gained huge attraction from stakeholders because of its powerful storage and processing capabilities. It has dispensed the necessity for maintaining the infrastructure locally. While it offers numerous benefits, it also comes with its own set of drawbacks like data privacy and security threats. Once the data is uploaded by the user to cloud storage the direct control over their data cannot be assured. Hence to ensure the data integrity proper data auditing schemes are necessary. Some schemes employ Third-party auditors (TPA) while others do not. But TPAs are not always reliable. This paper explores several blockchain techniques and schemes that overcomes these problems. The major focus of this review article is to give a critical analysis on existing auditing methods and simulation tools for its implementation by the different contributors in the field of Cloud data security. Further, various performance metrics from the contributions are analyzed. This review aims at delivering a detailed overview of the various security metrics through blockchain based auditing schemes in order to secure the cloud data storage. Finally, we explore some open issues and describe open research directions for future investigations.

1 INTRODUCTION

The Cloud computing has revolutionized IT infrastructure because it provides scalable, affordable online resources. In particular their prominent features are scalability, cost-effectiveness, flexibility, and agility. Companies are not charged high up-front costs, but they react to changes quickly in case their demand changes quickly. Moreover, cloud services reduce or eliminate downtime, enable remote work, increase data security, and provide access, security, reliability. It also enhances collaboration among teams by facilitating easy sharing and access to documents and applications from anywhere. However, alongside these benefits, there are also security threats associated with cloud computing, including data breaches, insider threats, malware attacks, and compliance risks. Organizations must implement robust security measures such as encryption, access controls, and employee training to mitigate these risks and ensure the safety of their data and systems. Effective cloud data auditing mechanisms are also essential to verify the integrity of data stored in cloud environments as mentioned by Zhu, Yan, *et al.*

The traditional cloud security mechanism relies on Third Party Auditors (TPAs), who are tasked with assessing the security practices of cloud service providers, like mentioned by Saxena, *et al.* Using TPAs can be expensive and complicated for both service providers and clients, as highlighted by Yang, Xiaodong, *et al.* However, this strategy consist of significant drawbacks.

- Third Party Auditors (TPA) centralize processes, which can lead to a single point of failure, data breaches, or the deception of audit reports. Because cloud users must rely on TPAs to carry out exhaustive and objective audits, trust issues may surface regarding the independence and integrity of audit conclusions.

 DOI: 10.1201/9781003589273-26

- Using TPAs can be expensive and complicated for both service providers and clients. Managing the audit procedure and analyzing results can be particularly difficult for people not familiar with cybersecurity.
- Furthermore, there is a lack of openness in the audit process, which makes it more difficult for cloud users to properly understand the security measures put in place by service providers and evaluate potential threats.

To mitigate these drawbacks, blockchain technology offers a decentralized and transparent approach to auditing protocols, as proposed by Garg *et al.* By implementing blockchain-based distributed systems for auditing, organizations can achieve decentralization, ensuring resilience against tampering and manipulation. Blockchain's transparent and immutable ledger enhances transparency and trust by providing a tamper-proof record of audit activities. Smart contracts automate and enforce audit processes, reducing costs and complexity while ensuring compliance with predefined rules. Moreover, blockchain's cryptographic techniques and consensus mechanisms enhance security and integrity, providing forensic evidence in case of security incidents. One such innovative scheme has been suggested by Saiharsha *et al.* using block chain and bio-metrics which minimizes user intervention and effectively addresses the challenges associated with the Mid-Day Meal Scheme. Integrating blockchain into cloud security auditing enhances transparency, trust, and efficiency, addressing the limitations of centralized TPAs and improving overall cloud security.

To further enhance the reliability of auditing practices, it requires an innovative approach that aligns with the need for standardization. Abraham *et al.* have highlighted the importance of ERC20 standardization for tokens on the Ethereum platform. Similar to this, standardized auditing practices can help improve the accuracy and quality of the auditing outcomes.

This study analyzes the workings of several blockchain systems in parallel with the existing auditing schemes, performance metrics and simulation tools that aim at improving data security and integrity in the cloud. The residual sections are discussed here. The studies on current block chain-based public auditing methods are provided in Section II, along with a chronological summary. Section III examines the simulation platforms, methodologies and performance metrics applied to the current cloud-based block chain public auditing system. Section IV lists the unmet research needs and potential obstacles for block chain-based cloud public auditing in the future. The survey's conclusion is provided in Section V.

2 STUDY ON EXISTING BLOCKCHAIN-BASED AUDITING MECHANISM

2.1 *Related works*

In 2022, Hui Qi *et al.* have proposed an audit method for cloud data based on blockchain networks. In this work, there are many components involved, like third-party auditing center, authority centers, data owners, cloud owners and blockchain nodes. Among these, the authority center initiates the scheme for encrypting the data once the data owner outsources the data to the cloud. Later, the TPA audits the cloud data, and the cloud owner provides cloud services, which in turn record all the audit transactions on the blockchain for verification. There are four phases that include setup, outsourcing, audit, and U-verify. However, the paper addresses the requirement for a better integrity audit process in cloud storage environments. To overcome the limitations of the existing blockchain-based auditing schemes, the paper proposes an integrity audit tree structure. This scheme enhances audit efficiency, improves the credibility of audit results, and detects malicious activities in the audit process. The simulation experiments prove that the proposed model outperforms blockchain-based auditing methods in terms of time overhead.

Yilin Yuan *et al.* have presented a blockchain-enabled public key identity-based data integrity verification scheme for cloud storage. The primary objective is to ensure data integrity in public cloud storage, where users lose control over the data after it is uploaded to the CSP. They have proposed Identity-Based Encryption (IBE), which eliminates the need for public key infrastructure and minimizes the overhead associated with key distribution, thus streamlining certificate management. It also uses homomorphic verifiable tags to check the remote data integrity efficiently. They use blockchain technology to securely store tag sets separately from files. This ensures that data authenticity is maintained, providing an extra layer of security. Additionally, an audit log is introduced to keep all parties involved in public auditing updated and synchronized. The scheme exhibits efficiency in the tag generation phase and audit phase. Combining IBE, blockchain, and audit logs offers a practical solution for users to verify the integrity of the integrity of the data.

In 2022, Li Li *et al.* developed a system called PPTPS that provides a traceable timeliness for the data stored in the cloud, which also ensures privacy preservation and auditability. In this approach, it is ensured that in each phase of cloud storage, the respective information, a fixed number of successive block hashes, and the current block height are stored on the cloud and unified into blockchain transactions. This feature ensures that the timing of events is precisely recorded based on the security features of the blockchain. Also, it enables accurate time-stamping of data at every stage of the process. This system is developed using lightweight cryptographic tools based on Decisional Diffie Hellman (DL) assumption. In order to guarantee that the auditor cannot access the data content, a customized random mark solution is also created to stop privacy leaks during the auditing phase. The validity of this model has been confirmed through extensive experiments in demonstrating its security and performance superiority compared to other works in terms of computation and communication costs.

In 2023, Rahul Mishra *et al.* have suggested a new approach, the "redactable blockchain-enabled cloud storage" model. This model has deduplication to support decentralized auditing and immediate redaction of data blocks on the blockchain. This removes the requirement for third-party auditors (TPA) as it integrates blockchain technology with decentralized storage. It deals with problems like data redundancy and storage free-riding attacks in cloud systems. The experimental analysis aligns with the theoretical analysis. For both on-chain and off-chain operations, the results illustrate efficiency in computational and communication overheads. It is described as light-weight decentralized auditing solution as it enhances IPFS-based cloud storage. Overall, the paper delivers a novel approach to improving the data security and integrity of cloud storage by using a redactable blockchain-enabled model with de-duplication without relying on third-party auditors.

Cheng Zhang *et al.* have presented a scheme to address the challenges of ensuring data integrity and finding malicious service providers along with providing accurate data auditing solution without the need of third-party auditors (TPA) to enable trusted public audits and record the interactions between cloud service providers, users and organizers they use blockchain technology. They also introduce homomorphic verifiable tags for efficient data integrity across cloud service providers. Then smart contracts are used for dispute arbitration which further enables automatic detection of violations based on blockchain records. This scheme uses the tamper-resistance feature of the blockchain to ensure that any malicious behavior is reported honestly.

In 2022, Liu, Bin, Yang Xin, and Suhuan Dai. have suggested a solution using blockchain technology to ensure data integrity and security in a multi-cloud environment. The main aim of the study is to convey the challenges of multi-cloud, multi-copy secure storage, and access auditing for recovery of data during disaster. For this purpose, they have used blockchain technology to record transactions related to identity audit, data storage, data integrity and disaster recovery requests. They use identity signatures for confirmation of the transaction and a threshold password sharing scheme in case of distributed storage of multiple copies of disaster recovery data. Overall, the paper illustrates how the use of blockchain in multi-cloud environment improve the data security and integrity of auditing the cloud data.

Ying Miao *et al.* have analyzed and presented a decentralized public auditing scheme to enhance data integrity in cloud storage systems while the privacy of the user is preserved. It addresses the problems of the traditional public auditing schemes in cloud storage platforms where the auditors may not be reliable and could work along with the cloud servers to cover up instances where data corruption has occurred. A concept called Zero-knowledge proof (ZKP) protects the user's privacy by ensuring that the third party auditor (TPA) will not have any additional information other than the correctness of user data . Proof of Work (PoW) generates unpredictable challenge messages. This consensus mechanism, like in Ethereum and Bitcoin, aims to prove the credibility of data by solving puzzles that is computationally hard to solve but easy to verify. In this way the user privacy is protected during the verification of audit responses.

Wang, Jinpeng, *et al.* have suggested a scheme to address efficiency and security challenges in data audits. Many entities are used here like Data Owners (DO), Edge Computing Servers (ECS), and Cloud Service Providers (CSP), along with technologies like blockchain and edge computing. The role of the Data Owners is to generate all the homomorphic hashed and verification labels that will be stored in the blockchain. ECS uses these hashes to calculate completion proofs. This scheme aims to achieve 100 % credibility in auditing by using the above-mentioned technologies. Python programming language is utilized in the implementation process. Comparisons with the current public auditing schemes like DPAS, CHPor, SWPor, BPAS are done here. A significant reduce in computational overhead and increase in efficiency is observed in the results.

Shu, Jiangang, *et al.* have suggested a scheme called BDPA to improve cloud storage system public auditing. To make sure that data integrity, security and scalability standards are met, blockchain technology is integrated with the solution here. The integrity of the outsourced data on cloud servers is verified using a decentralized blockchain network. BDPA scheme reduces the hazards caused by tempting auditors and malevolent blockchain miners by utilizing smart contracts and collaborative methods. Comparisons with current centralized auditing schemes show that there is an increase in efficiency, scalability and security of the proposed scheme. So BDPA offers a viable way to improve cloud storage systems' security and reliability through decentralized public auditing.

Han, Rui, *et al.* have proposed a system called FIBPRO, which is a peer-to-peer information management and cloud storage system. Data security is achieved through encryption, distributed storage through IPFS, transaction management through blockchain.

Storage overhead is reduced by allowing the data sharing with multiple uses, which is done through hybrid encryption. Data must be encrypted before being uploaded to IPFS, sharding technology must be used to store data, and a DPoS-based consensus mechanism must be used in the blockchain network. The outcomes of the proposed methodology demonstrate increased scalability, efficient data exchange procedures and storage expansion capabilities. Experiments on performance analysis show how successful the system is at utilizing storage space and time.

Li, Jiaxing, *et al.* proposed a scheme for public auditing by using blockchain to verify the data integrity in cloud storage. This proposed scheme doesn't require the involvement of a third-party auditor. It involves only the data owner and cloud service provider, contrary to previous works. The usage of Merkle Hash Trees and blockchain technology to store verification tags results in 100% confidence in auditing. In the proposed methodology, the data owner (DO) will send the files to the cloud, and the appropriate hashtags will get stored in the blockchain. To check the data integrity of the cloud, the DO will challenge the Cloud Service Provider (CSP). At the same time the DO will appoint an auditor who will be a user in the blockchain and chosen at random. Now the CSP will create a Merkle hash tree from the hashtags of DO's file blocks, while the auditor will create another Merkle hash tree from the hashtags stored in the blockchain. The root node of these two trees is returned to the DO, who checks them. If then roots match, then the integrity of data is maintained. Comprehensive testing and thorough experiments indicate that the scheme can defend against 51% of the attacks and malicious entities, and a major improvement in communication and computation costs.

Zhang, Yuan, *et al.* proposed a scheme to resist procrastinating auditors by introducing a certificateless public verification scheme (CPVPA). This method makes use of blockchain technology to make sure that the data integrity is not compromised. It overcomes the drawbacks of the traditional PKI based verification schemes. This methodology works by having the auditors store the results of the audit as a transaction in the blockchain, and it should also be timestamped. The results are stored in blockchain to provide a transparent and secure method. The users can check the time of the audit by looking at the corresponding transaction's generation time. This way, the users can then check if the audit was done during the prescribed time. The problem of procrastinating auditors can thus be resolved like this. This scheme outperformed public verification schemes in terms of computational and communicational overhead. The cloud server could forge data and might become untrustworthy. So, it has been regarded as a semi-trusted entity that might manipulate the third-party auditor. The CPVPA is also responsible for detecting such instances and preventing them successfully.

2.2 *Algorithms used and drawbacks in the existing block-chain based public auditing in cloud*

2.2.1 *Block chain based public auditing techniques*

We have selected only research papers published between 2020 and 2023 that discuss the most recent innovative ideas and techniques used for improving auditing. The approaches used for implementing those techniques are shown in Figure 1.

2.2.2 *Visualization of drawback*

After conducting a comprehensive analysis of several research papers, the major drawbacks are highlighted in the Figure 2.

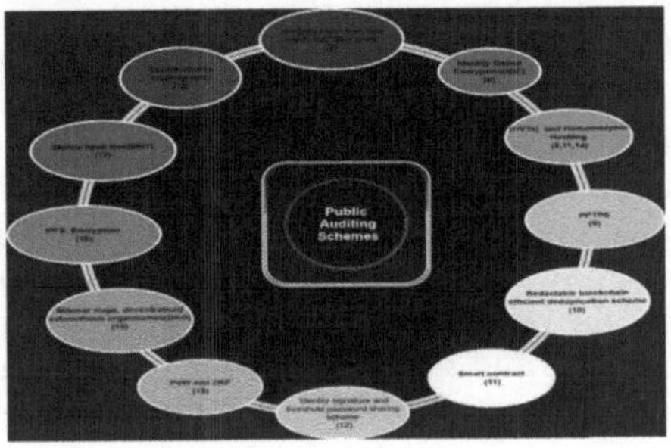

Figure 1. Flowchart of public auditing techniques.

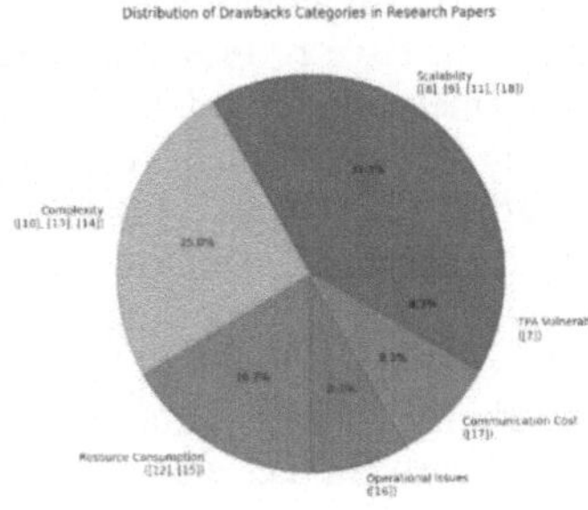

Figure 2. Visualization of drawbacks.

3 RESEARCH GAP AND FUTURE SCOPE

Making use of cloud storage offers a productive way to manage massive amounts of data. The benefits of cloud storage, including its high performance, scalability, and affordability, have made it popular among both people and enterprises. The separation of data management and ownership, however, presents difficulties for traditional methods of data verification, which has led to the investigation of auditing techniques. Shree *et al.* suggest that, moreover, as cloud environments are inherently dynamic, with data constantly being added, changed, and removed, the need for effective auditing techniques becomes even more pronounced. Traditional methods of data verification face challenges in managing these dynamic changes while maintaining security and integrity. Although current technologies present viable approaches to guarantee data security and integrity, scaling these solutions to meet the demands of practical deployment scenarios is frequently a challenge.

A vital area that needs to be addressed is the incorporation of dynamic data storage and auditing procedures. Cloud environments are by their very nature dynamic, with data being added, changed, and removed on a regular basis. The question of how to manage these dynamic changes while preserving the security and integrity of the auditing process is left unanswered by many of the current solutions. In order to guarantee that audit procedures are effective and dependable in the face of dynamic changes,as pointed out by Chandran *et al.* future research should concentrate on creating systems that can allow real-time updates and replacements of data in cloud storage environments.

Future studies ought to concentrate on investigating other consensus techniques for multi-cloud settings. While Proof of Work (PoW) and Proof of Stake (PoS), two common blockchain consensus methods, might function well in some circumstances, they might not be appropriate for settings with various cloud providers or decentralized storage systems. For multi-cloud storage auditing, looking into alternate consensus techniques like Directed Acyclic Graphs (DAGs) or Practical Byzantine Fault Tolerance (PBFT) may yield more scalable and effective solutions. Additionally, exploring the synergy between blockchain technology and the Internet of Things (IoT) can offer enhanced transparency, traceability, and security for data transactions, especially in scenarios where IoT devices interact with massive volumes of data, as suggested by Kaushik *et al.*

Bs *et al.* suggest that future research should also explore the integration of cloud access security brokers (CASBs) into dynamic cloud storage auditing procedures that can address the gap in managing real-time updates and replacements of data while ensuring security and integrity. To fully realize the promise of blockchain-based cloud storage auditing solutions, it is imperative to address issues related to scalability and performance, investigate alternative consensus mechanisms for multi-cloud environments, and carry out thorough studies on the viability and adoption barriers.

4 CONCLUSION

The primary goal of this survey was to assess the current state of the art in blockchain-based public auditing methodologies, including the algorithms employed, the simulation tools utilized in various contributions, and the various performance indicators from various contributions.

REFERENCES

Anmol Kaushik and Anju S Pillai. (2021). Blockchain and iot based inventory monitoring system. *In 2021 Second International Conference on Electronics and Sustainable Communication Systems (ICESC)*, pages 827–831. IEEE.

Ashwathi Chandran and CK Shyamala. (2017). Data management issues in cloud integrated computing: A big picture. *In 2017 4th International Conference on Advanced Computing and Communication Systems (ICACCS)*, pages 1–8. IEEE.

Bin Liu, Yang Xin, and Suhuan Dai. (2022). Blockchain-based disaster recovery data storage and security auditing solution in multi-cloud environment. *In 2022 International Applied Computational Electromagnetics Society Symposium (ACES-China)*, pages 1–4. IEEE.

Cheng Zhang, Yang Xu, Yupeng Hu, Jiajing Wu, Ju Ren, and Yaoxue Zhang. (2021). A blockchain-based multi-cloud storage data auditing scheme to locate faults. *IEEE Transactions on Cloud Computing*, 10 (4):2252–2263.

Hui Qi, Hang Li, Shuwen Huang, Peipei Li, Weiyi Dai, and Chunbo Wang. (2022). Integrity audit method for cloud data based on blockchain. *In 2022 IEEE Intl Conf on Parallel & Distributed Processing with Applications, Big Data & Cloud Computing, Sustainable Computing & Communications, Social Computing & Networking (ISPA/BDCloud/SocialCom/SustainCom)*, pages 652–659. IEEE.

Jiangang Shu, Xing Zou, Xiaohua Jia, Weizhe Zhang, and Ruitao Xie. (2021). Blockchainbased decentralized public auditing for cloud storage. *IEEE Transactions on Cloud Computing*, 10(4):2366–2380.

Jiaxing Li, Jigang Wu, Guiyuan Jiang, and Thambipillai Srikanthan. (2020). Blockchain-based public auditing for big data in cloud storage. *Information Processing & Management*, 57(6):102382.

Jinpeng Wang, Sai Wang, Lianhai Wang, Wei Shao, Shujiang Xu, and Shuhui Zhang. (2021). A blockchain and edge computing based public audit scheme for cloud storage. *In 2022 41st Chinese Control Conference (CCC)*, number 1, pages 7466–7470. IEEE, 2022.

Juhi Shree, NR Kanimozhi, GA Dhanush, Anand Haridas, Anala Sravani, and Priyanka Kumar. (2020). To design smart and secure purchasing system integrated with erp using block chain technology. *In 2020 IEEE 5th International Conference on Computing Communication and Automation (ICCCA)*, pages 146–150. IEEE.

Li Li, Xiao Lan, Mali Chen, Ting Luo, Li Chen, Yangxin Wang, and Yumeng Chen. (2022). Pptps: Building privacy-preserving auditable service with traceable timeliness for public cloud storage. *Plos one*, 17(10): e0276212.

Misha Abraham and Jevitha. (2019). Runtime verification and vulnerability testing of smart contracts. *In Advances in Computing and Data Sciences: Third International Conference, ICACDS* 2019, Ghaziabad, India, April 12–13, 2019, Revised Selected Papers, Part II 3, number 4, pages 333–342. Springer.

Neenu Garg, Seema Bawa, and Neeraj Kumar. (2020). An efficient data integrity auditing protocol for cloud computing. *Future Generation Computer Systems*, 109:306–316.

Rahul Mishra, Dharavath Ramesh, Salil S Kanhere, and Damodar Reddy Edla. (2023). Enabling efficient deduplication and secure decentralized public auditing for cloud storage: A redactable blockchain approach. *ACM Transactions on Management Information Systems*, 14(3):1–35.

Rajat Saxena and Somnath Dey. (2016). Cloud audit: A data integrity verification approach for cloud computing. *Procedia Computer Science*, 89:142–151.

Rui Han, Yu Wang, Mingfa Wan, Teng Yuan, and Guozi Sun. (2023).Fibpro: Peer-to-peer data management and sharing cloud storage system based on blockchain. *Peer-to-Peer Networking and Applications*, 16 (6):2850– 2864, 2023.

Saiharsha, Sanjay Tharagesh, Nikish Kumar, and N Harini. (2020). Cloud based data integrity model for minimal human intervention using smart contracts. *In 2020 5th International Conference on Communication and Electronics Systems (ICCES)*, pages 910–916. IEEE.

Vidhyasagar Bs, M Arvindhan, and Sivakumar Kalimuthu. (2023). The crucial function that clouds access security brokers play in ensuring the safety of cloud computing. *In 2023 16th International Conference on Security of Information and Networks (SIN)*, pages 1–5. IEEE.

Xiaodong Yang, Xizhen Pei, Meiding Wang, Ting Li, and Caifen Wang. (2020). Multireplica and multi-cloud data public audit scheme based on blockchain. *IEEE Access*, 8:144809–144822.

Yan Zhu, Hongxin Hu, Gail-Joon Ahn, and Stephen S Yau. (2012). Efficient audit service outsourcing for data integrity in clouds. *Journal of Systems and Software*, 85(5):1083– 1095.

Yilin Yuan, Jianbiao Zhang, Wanshan Xu, and Zheng Li. (2022). Identity-based public data integrity verification scheme in cloud storage system via blockchain. *The Journal of Supercomputing*, 78(6):8509–8530.

Ying Miao, Qiong Huang, Meiyan Xiao, and Hongbo Li. (2020). Decentralized and privacy preserving public auditing for cloud storage based on blockchain. *IEEE Access*, 8:139813–139826.

Yuan Zhang, Chunxiang Xu, Xiaodong Lin, and Xuemin Shen. (2019). Blockchain based public integrity verification for cloud storage against procrastinating auditors. *IEEE Transactions on Cloud Computing*, 9 (3):923–937.

Data Science & Exploration in Artificial Intelligence – Gururaj et al. (Eds)

A pangenome graph-based approach for predicting Alzheimer's disease

C.P. Prathibhamol, J. Chandrakiran, Sethulakshmi. Santhosh, M. Meenakshi & S. Akash. Menon
Department of Computer Science and Engineering, Amrita School of Computing, Amrita Vishwa Vidyapeetham, Amritapuri, India

Manjusha. Nair
Department of Computer Science and Applications, Amrita School of Computing, Amrita Vishwa Vidyapeetham, Amritapuri, India

ABSTRACT: Pangenomes provide a complete view of genetic variation and new methods for predicting diseases. Comprehending genetic variation among populations is essential for deciphering disease pathways in complicated diseases such as Alzheimer's. Because pangenome graphs may overcome the drawbacks of linear reference genomes, they are becoming more and more common in genomic research. In this work, genomic variations are extracted from pangenome graphs and are used to build a Random Forest machine learning model for Alzheimer's disease prediction. The four main genes linked to this disease are APP, APOE, PSEN1, and PSEN2. Using genetic information from these genes, this method offers better insights on the likelihood of sickness. Compared to the more traditional methods like MRI scans, here an early diagnosis is possible. Our findings from this study, demonstrate how we can integrate pangenome graphs and machine learning in Alzheimer's disease prediction. As technology develops, more accurate genetic markers can be used in conjunction with enhanced feature selection strategies and machine learning approaches for further enhancing the system.

Keywords: Pangenome graph, Alzheimer's disease, Random Forest, Machine Learning, Seqwish, Minimap, Gfautil

1 INTRODUCTION

Alzheimer's disease affects millions globally and is a neurodegenerative disorder that results in loss of memory, cognition, and behaviors. It has a complex pathophysiology that includes both environmental and genetic factors. There are traditional methods like MRI scans, but by using pangenome graphs an early diagnosis is possible [1]. Even after such research, it is still difficult to understand the complex genetic roots of this disease. So, in this scenario, the major factor includes identifying the genetic variations linked to the disease, for Alzheimer's disease these variations include the Apolipoprotein E (APOE) gene, Amyloid Precursor Protein (APP) gene, Presenilin 1 (PSEN1), Presenilin 2 (PSEN2) [2]. Apoprotein E is essential for lipid metabolism and is encoded by the APOE gene. The important genetic risk factor for late-onset Alzheimer's disease is the APOE $\epsilon4$ allele, which is linked to increased brain plaque deposition and decreased Aβ clearance. The γ-secretase complex is necessary for splitting amyloid precursor protein (APP) into amyloid beta (Aβ) peptides, which form amyloid plaques, the defining feature of Alzheimer's disease, is encoded by the PSEN1 and PSEN2 genes. Mutations in these genes damage the processing of APP, which increases the synthesis of harmful Aβ peptides and exacerbates neuronal dysfunction as well as cognitive

DOI: 10.1201/9781003589273-27

decline. The precursor of Aβ peptides is encoded by the APP gene, and mutations in this gene can increase the production of Aβ and so promote plaque accumulation. These genetic understandings can help in the creation of new treatments by providing insights into the molecular mechanisms for disease, allowing proper therapies that are made to the genetic similarity of individuals [3].

2 LITERATURE SURVEY

By merging the data of multiple genes, a pangenome graph is one of the best tools for visualizing and understanding the genetic variations, which can further help people detect the disease early and make proper treatment approaches [4–8]. These pangenome graphs are mainly used to capture genomic patterns like structural variants, insertions, SNPs, and deletions which are the contents of the genetic variations for a disease like Alzheimer's, hence forming a tool used by researchers, despite the computational limitations [4]. The frequented Regions (FRs)within the pangenome graphs which classify the diseases according to the illness provide further insights into the genetic information of the particular disease [7]. There are programs like GenomicKB and DGlinker, which are programs for automating genetic data analysis and disease gene prediction. Their success depends on the merging of genomic data [5,8]. The process of identifying the genetic variations using pangenome graphs and then utilizing them to predict Alzheimer's disease using a machine learning model is the main goal here.

3 PROPOSED METHODOLOGY

This study uses pangenome graphs on the genetic landscape of Alzheimer's disease. The approach involves first building pangenome graphs [9] for important genes related to Alzheimer's disease (PSEN1, PSEN2, APP, and APOE) using Pairwise Alignment Format (PAF) alignment files produced by the minimap2 program. Afterwards, the tool seqwish is applied to produce Graphical Fragment Assembly (GFA) files. This work yields especially rich genetic diversity profiles for each gene, including frequent and rare variants from various populations, in the form of the GFA files. Pangenome graphs are utilized to build Variant Call Format (VCF) datafiles with allelic frequencies and genomic location of genetic variants. These genetic variants are extracted from these VCF files and serve to train a Random Forest model that can predict disease status using DNA sequences as an input.

Software tools: The software tools used in this project are

- Minimap2: It is a DNA sequence aligner that aligns the DNA sequences with a reference sequence and gives PAF files.
- Seqwish: A variation graph inducer that gives pangenome graphs in the format of GAF files from the PAF files and input sequences.
- Gfatools: It is a tool to work with GFA files for extracting the variations, converting to different formats, and visualizing graphs
- Gfautil: A command line tool for operations on GFA and related files.
- PySAM: It is a Python library for working with files like SAM/BAM/VCF/BCF/GFF/GTF/FASTA/FASTQ.
- Scikit-learn: It is a Python framework used for machine learning that offers numerous algorithms, models, and metrics.
- Bandage: It is a software tool for visualizing and exploring pangenome graphs where we can see and analyze the variations properly.

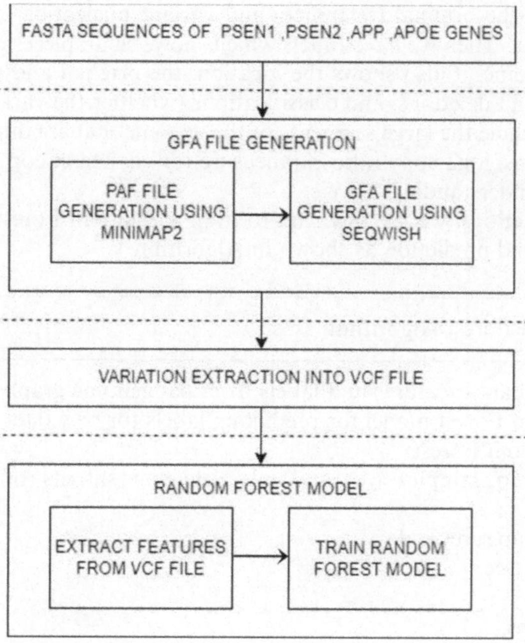

Figure 1. Proposed methodology.

3.1 *Dataset*

The datasets used in this project are DNA sequences in FASTA format obtained from the National Center for Biotechnology Information (NCBI) site [10]. For the gene PSEN1, the following accession numbers were utilized: NM 000021.4, NM 007318.3, KT120066.1, AK312531.1, DN998975.1, LC756953.1, and AF205592.1. For the gene PSEN2, the accession numbers users are: NG 007381.2, BC006365.2, DQ893826.2, NM 012486.3, NM 000447.3, and KT120068.1. The dataset for the gene APP includes the following accession numbers: NG 007376.2, AY919674.1, KT120069.1, NM 000484.4, NM 001136131.3, and NM 001385253.1. Lastly, for the gene APOE, the dataset incorporates the following accession numbers: NG 007084.2, AY077451.1, DJ359647.1, NM 001302688.2, NM 001302689.2, and NM 000041.4. All these datasets include affected and unaffected variants to capture genetic variation across different individuals and populations.

3.2 *Module I: GFA and VCF File generation*

The first process includes aligning the sequences to a reference genome using the Min-imap2 tool to produce Pairwise Alignment Files (PAF) files [11]. Then these PAF files are converted into Genomic Feature Annotation (GFA) files using a tool called Seqwish [12] which are then used to create pangenome graphs. Finally, we use Gfautil, a tool used to convert the GFA files into Variant Call Format (VCF) files, which contain information on the variations i.e., SNP's, insertions, structural variants, and deletions [13]. This is done for multiple genes collected which include PSEN1, PSEN2, APOE, and APP genes associated to Alzheimer's disease.

3.3 *Module II: random forest training*

In feature extraction, we extract the main variations in genes: (single-nucleotide polymorphisms), Indels (insertions and deletions), and CNVs (copy number variations). SNP is an important and common type of genetic variation found in genomic studies where it involves a single base change in DNA sequences. It provides us with information on where the SNPs are found, the original (reference) and variant nucleotide scores that shows the reliability of SNP data. Then we have Indels which shows small pieces of DNA inserted and deleted from the genome, it also shows the location, the original and altered sequence, the length of insertions and deletions, and classifications (whether the variants are insertions or deletions). CNV's include the large segments of the genome that are duplicated or deleted, it shows genomic regions, log2 copy ratios (a measure for change in copy number), and other additional data for understanding CNV's.

After feature extractions, we use this data to train a Random Forest model [14] [15] [16] for genetic analysis and prediction as shown in Algorithm 1.

Algorithm 1 Random Forest Algorithm

Input: Training data - features and labels from pangenome graphs (VCF files)
Output: Random Forest model for predicting labels for new data
Algorithm Random Forest:
Create bootstrap samples by randomly selecting subsets of training data with replacement
for each bootstrap sample **do**
Build a decision tree
end for
Make predictions:
Pass new data points to each decision tree in the forest
Average predictions from all decision trees
end Random Forest Algorithm

---

4 RESULTS AND DISCUSSION

4.1 *Pangenome graph visualization*

Bandage software is used to visualize the GFA data, which displays variances for each gene and gives a summary of the pangenome graphs for PSEN1, PSEN2, APP, and APOE. Edges indicate the relationships between each node, which represents a series of base pairs. Our comprehension of genetic variants and their consequences for disease prediction is improved by this graphical portrayal (Figure 2).

a) PSEN1 Variation Graph   b) PSEN2 Variation Graph

Figure 2. Variation graphs derived with bandage.

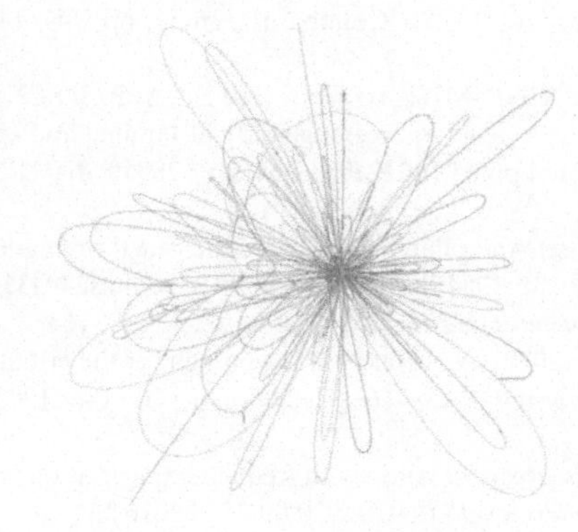

c) APP Variation Graph

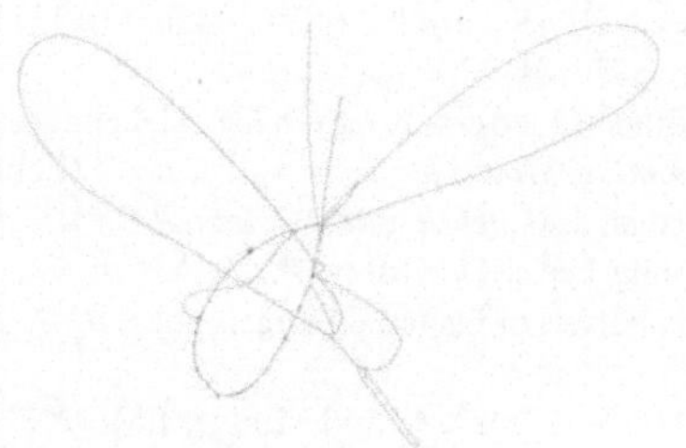

d) APOE variation Graph

Figure 2. (Continued)

4.2 *Model evaluation*

Locate tables close to the first reference to them in the text and number them consecutively. Avoid

A Random Forest model was chosen for its robustness and versatility in handling diverse datasets. The model's performance was evaluated using standard metrics: Precision, Accuracy, Recall, F1 Score, and ROC-AUC. The following table summarizes the results:

Table 1. Model performance metric.

Metric	Accuracy	Precision	Recall	F1-Score	Roc-Auc
Value	0.86	0.82	0.81	0.81	0.87

Alzheimer's disease prediction, with the model achieving an accuracy of 85% can hence be concluded. Future research could enhance these predictions by integrating additional genetic markers and refining feature selection techniques.

5 CONCLUSION

In this study, we combined pangenome graphs, genomic data, and machine learning techniques to predict complex diseases like Alzheimer's. Compared to conventional methods of predicting like MRI-based prediction, this method provided us a better understanding of the diseases linked with genetic variations, so that these diseases are detected at the early stage. Shortly in the future, we can create a web-based tool for genetic illness prediction using Flask and vue.js frameworks [17]. By using transfer learning techniques an improved performance while utilizing pre-trained models on huge genomic datasets [18] can be achieved in such an application. Using pangenome graphs and machine learning algorithms can be very useful for medical professionals and researchers to enter genetic data in a particular format and obtain disease prediction through this application.

REFERENCES

[1] Sivaranjani, S., *et al.* (2021). Diabetes prediction using machine learning algorithms with feature selection and dimensionality reduction, *in Proceedings of the 2021 7th International Conference on*

Advanced Computing and Communication Systems (ICACCS), Coimbatore, India, pp. 141–146. doi: 10.1109/ICACCS51430.2021.9441935.

[2] Lanoiselée, H.M., *et al.* Collaborators of the CNR-MAJ project. (2017). APP, PSEN1, and PSEN2 mutations in early-onset Alzheimer disease: A genetic screening study of familial and sporadic cases. *PLoS Med.* 14(3). doi: 10.1371/journal.pmed.1002270. PMID: 28350801; PMCID: PMC5370101.

[3] Andrade-Guerrero, *et al.* (2023). Alzheimer's disease: An updated overview of its genetics. *International Journal of Molecular Sciences* 24, no. 4 (February 13): 3754. https://doi.org/10.3390/ijms240437.

[4] Eizenga, J.M., *et al.* (2020). *Annual Review of Genomics and Human Genetics* 21:1, 139–164.

[5] Harling-Lee, J.D., Gorzynski, J., Yebra, G., *et al.* (2022). A graph-based approach for the visualization and analysis of bacterial pangenomes. *BMC Bioinformatics* 23, 416. https://doi.org/10.1186/s12859-022-0489.

[6] Hokin, S., Cleary, A. and Mudge, J.M. (2020). Disease association with frequented regions of genotype graphs. *medRxiv* 2020.09.25.20201640; doi: https://doi.org/10.1101/2020.09.25.20201640.

[7] Feng, F., *et al.* (2023). GenomicKB: a knowledge graph for the human genome, *Nucleic Acids Research*, Volume 51, Issue D1, Pages D950–D956, https://doi.org/10.1093/nar/gkac957.

[8] Hu, J., *et al.* (2021). DGLinker: flexible knowledge-graph prediction of disease–gene associations, *Nucleic Acids Research*, Volume 49, Issue W1, Pages W153–W161, https://doi.org/10.1093/nar/gkab449.

[9] Garrison, E., Guarracino, A. (2023). Unbiased pangenome graphs, *Bioinformatics*, Volume 39, Issue 1, btac743, https://doi.org/10.1093/bioinformatics/btac743.

[10] National Center for Biotechnology Information. Accessed March 27, 2024. https://www.ncbi.nlm.nih.gov.

[11] Li, H. (2018). Minimap2: pairwise alignment for nucleotide sequences, *Bioinformatics*, Volume 34, Issue 18, Pages 3094–3100, https://doi.org/10.1093/bioinformatics/bty1.

[12] Ekg. (2019). EKG/Seqwish: Alignment to variation graph inducer. *GitHub*, November 1. https://github.com/ekg/seqwi.

[13] Fischer, C. (2021). GFAUTIL - rust utility. *gfautil - Rust utility //*, February 23. https://lib.rs/crates/gfauti.

[14] Ani, R., Augustine, A., Akhil, N.C. and Deepa, O.S. (2016). Random forest ensemble classifier to predict the coronary heart disease using risk factors, *in Proceedings of the International Conference on Soft Computing Systems*, eds. L. Suresh and B. Panigrahi, Advances in Intelligent Systems and Computing, vol. 397. Springer, New Delhi. https://doi.org/10.1007/978-81-322-2671-0_19

[15] Jose, C. and Gopakumar, G. (2019). An improved random forest algorithm for classification in an imbalanced dataset, *in Proceedings of the 2019 URSI Asia-Pacific Radio Science Conference (AP-RASC)*, New Delhi, India, pp. 1–4. doi: 10.23919/URSIAP-RASC.2019.8738.

[16] Pattanayak, S. and Singh, T. (2022). Cardiovascular disease classification based on machine learning algorithms using GridSearchCV, cross-validation, and stacked ensemble methods, in *Communications in Computer and Information Science*, 219–230. https://doi.org/10.1007/978-3-031-12638-5_18

[17] Hebbale, A., Vinay, G., Krishna, B.V. and Shah, J. (2021). IoT and machine learning based self care system for diabetes monitoring and prediction, *in Proceedings of the 2021 2nd Global Conference for Advancement in Technology (GCAT)*, Bangalore, India, pp. 1–7. doi: 10.1109/GCAT52182.2021.9587.

[18] Rajeswari, S.S. and Nair, M. (2021). A transfer learning approach for predicting alzheimer's disease, in *Proceedings of the 2021 4th Biennial International Conference on Nascent Technologies in Engineering (ICNTE)*, Navi Mumbai, India, pp. 1–5. doi: 10.1109/ICNTE51185.2021.9487.

Data Science & Exploration in Artificial Intelligence – Gururaj et al. (Eds)
© 2025 the Author(s), ISBN: 978-1-032-96390-7

Multi-PDF GPT generative question and answering system powered by googlePalm2 and langchain

Robin Rohit Vincent
Associate Professor, School of Computer Science and Engineering, Presidency University, Bangalore, India

G.S. Ganesh Prasad, Junaid Naeemulla Khan & K. Surya
B. Tech Scholar CSE, AI & ML, School of Computer Science and Engineering, Presidency University, Bangalore, India

ABSTRACT: This project introduces a cutting-edge approach to knowledge retrieval through the development of a Multi-PDF GPT Generative Q&A System. Harnessing the powerful capabilities of GooglePaLM2 and LangChain technologies, our system seamlessly navigates multiple PDF documents, extracting relevant information, and generating context-aware answers to user queries. By leveraging advanced large language models (LLM) this system not only enhances traditional search methodologies but also opens new dimensions in understanding and interpreting textual data. The integration of GooglePaLM2 and LangChain contributes to the efficiency and accuracy of the question-answer generation process, creating a robust and dynamic tool for information synthesis and retrieval. This project marks a significant stride in the realm of generative artificial intelligence and information processing, promising transformative implications for diverse fields that rely on efficient data interpretation and knowledge extraction from PDF documents.

Keywords: GooglePalm2, Langchain, Large language Models (LLM), Multi-PDF GPT, Generative artificial Intelligence, question-answer generation

1 INTRODUCTION

Traditionally, extracting knowledge from multiple Portable Document Formats (PDFs) has been a cumbersome task. Imagine a researcher wading through a sea of research papers, each containing valuable information, but struggling to efficiently find and synthesize the answer to a specific question. This is where it emerges as a revolutionary solution. This system leverages the prowess of two cutting-edge AI models: Google PaLM 2, a powerful generative pre-trained transformer, and Langchain, a technique for contextualizing information across sequential data. By combining these forces, the system unlocks the ability to ask complex questions across a collection of PDFs and receive comprehensive answers, woven together from the rich tapestry of information within.

The core functionality of this system hinges on its adeptness at text extraction and comprehension. It seamlessly processes numerous PDFs, extracting the text and organizing it into a cohesive structure. Google PaLM 2 then takes center stage, employing its formidable natural language processing capabilities to understand the nuances of the extracted text. Imagine PaLM 2 meticulously dissecting each sentence, grasping the relationships between concepts, and building a comprehensive internal representation of the knowledge contained within the PDFs.

DOI: 10.1201/9781003589273-28

Langchain, the other pillar of this system, plays a crucial role in enabling question-answering across multiple documents. It tackles the challenge of ensuring context is preserved as information is gleaned from various PDFs. Langchain stitches together the extracted knowledge, understanding how concepts and ideas flow and connect across different documents. This contextualization allows the system to not only identify relevant information within each PDF but also to grasp the broader narrative that emerges when these PDFs are considered as a whole.

When a user poses a question, the system springs into action. PaLM 2, armed with its deep understanding of the processed PDFs, sifts through the information, pinpointing the most relevant passages that hold the key to answering the query. Langchain, meanwhile, ensures that the extracted snippets are interpreted within the proper context, considering the flow of ideas across the entire document collection. By synergistically combining these elements, the system synthesizes a response that not only addresses the specific question but also draws upon the collective knowledge base gleaned from the multiple PDFs.

The implications of this system extend far beyond the realm of academic research. Imagine a lawyer wielding this tool to efficiently navigate through legal documents, pinpointing the precise clauses that address a specific legal point. Envision a doctor leveraging the system to rapidly synthesize insights from a patient's medical history, scattered across various reports. The potential applications for this system are vast, promising to revolutionize how we interact with and extract knowledge from the ever-growing sea of information stored in PDF format.

The development of the Multi-PDF GPT system signifies a significant forward in the quest to unlock true potential of information stored in PDFs. By harnessing the power of advanced generative models and contextual reasoning techniques, this system empowers users to ask insightful questions and receive comprehensive answers that transcend the limitations of individual documents. As research in artificial intelligence continues to advance, we can expect even more sophisticated iterations of this system to emerge, further blurring the lines between human and machine intelligence in the realm of knowledge discovery.

2 LITERATURE SURVEY

Karina Wiechork and Andrea Schwertner Charao proposed this work addresses automated data extraction from Portable Document Format (PDF) documents, focusing on educational exams from large repositories. Exams from Enade were automatically collected and processed using Scrapy scripts. They evaluated data extraction on 343 tests, containing over 11,000 questions and 14,000 answer choices. Ground truth was established using the Aletheia tool. Existing data extraction tools were employed: Tabula and Excalibur for answers (tabular data), PDFMiner and CyberPDF for questions (textual content), and ExamClipper and Aletheia for question region extraction. Results highlight limitations due to layout variations across exam years. The extracted data offers valuable information for various fields, including academic research and educational support.

Filip Gralinski, *et al.* proposed an existing Natural Language Processing (NLP) techniques excel at capturing context within sentences or short documents. However, they struggle with real-world, longer documents where information is encoded through spatial structures (tables, forms, headers) and complex layouts across multiple pages. To address this gap, they proposed the Kleister task and two new datasets. Kleister requires NLP systems to leverage both textual cues and document layout features to extract key information (entities) from long, formal documents. They establish a text-only baseline (Pipeline) using various Named Entity Recognition models (Flair, BERT, RoBERTa). Additionally, they analyze the impact of errors introduced by popular PDF processing tools (pdf2djvu, Tesseract, Textract) on the performance of information extraction systems.

Ahmer M. Hashmi, *et al.* proposed an ever-growing volume of scholarly documents stored in Portable Document Format (PDF) presents challenges for information comprehension and extraction. This complexity necessitates robust information extraction techniques. This paper offers a comprehensive analysis of recent advancements in PDF data extraction, focusing on both rule-based and machine learning-based approaches. Their objective is to equip the scientific community with knowledge about current trends in PDF information extraction. They further provide critical evaluations and propose future directions for some of these approaches.

Thaís Medeiros, *et al.* did a research on the potential of Large Language Models (LLMs) for building chatbots that answer user queries from Portable Document Format (PDF) automotive manuals. With the growing adoption of AI-powered tools, chatbots offer a promising solution to revolutionize customer support and post-sales services in the automotive industry. They propose three distinct LLM-based methods for chatbot development and evaluate them based on key criteria: response accuracy, cost-efficiency, and user experience. Their results emphasize that the optimal method depends on the prioritized criteria, influencing the practical implementation. This work contributes valuable insights into the effectiveness of chatbots for interacting with automotive manuals, paving the way for integrating generative AI strategies that align with the needs of the automotive sector.

Arjun Pesaru, *et al.* paper proposes a framework for building chatbots that can answer questions about PDF documents. The framework leverages LangChain for chatbot development and a large language model (LLM) for text generation and real world information access through Google Search. Pinecone is used for vector storage, and React JS is used for the chatbot interface. This approach has the potential to create informative chatbots for various applications.This paper proposes a novel approach for building chatbots that can answer questions about PDF documents. The system leverages the strengths of LangChain, a framework for simplifying chatbot creation, and a large language model (LLM) for text generation and information retrieval. The chatbot is trained on a dataset of PDFs and utilizes the LLM to respond to user queries.

Pranav Nataraj Devaraj, *et al.* paper proposes a Legal Documentation AI Chatbot to address the challenge of managing complex legal documents in the age of big data. The chatbot leverages recent advancements in AI chatbots and the Langchain framework. The development process focuses on building an Android app for user interaction, Langchain code for query processing, and integration through a Flask backend and REST APIs for communication. This work highlights the potential of AI chatbots for legal document management.

Oguzhan Topsakal, and Tahir Cetin Akinci investigates the application of Large Language Models emphasizing LangChain, LLMs have gained significant traction due to their versatility in tasks like writing, coding, and explanations, with ChatGPT by OpenAI demonstrating widespread user adoption. The core of this study is LangChain, designed to accelerate the development of custom AI applications leveraging LLMs. LangChain's recognition in the AI community stems from its seamless interaction with various data sources and applications. The paper explores LangChain's core functionalities, including its modular components and chains, which represent customizable pipelines for specific use cases. Through practical examples, the study highlights the framework's potential in facilitating the swift development of LLM-based applications.

Aniket Holkar, *et al.* researches on a groundbreaking approach for in-depth analysis of Portable Document Formats (PDFs). It leverages the combined power of Artificial Intelligence (AI) and a large language model (LLM) through the Lanchain platform. The system aims to achieve comprehensive analysis of PDF content, encompassing text, images, tables, and other embedded elements. By utilizing advanced AI algorithms and Natural Language Processing (NLP), it automates data extraction (structured and unstructured), interprets the contextual meaning of the extracted text using the LLM, extracts information from embedded images, and generates comprehensive reports with key findings. This novel system has the potential to revolutionize PDF analysis, empowering users to unlock valuable information and make data-driven decisions more efficiently.

3 TECHNOLOGY USED

3.1 *Google PaLM 2*

This cutting-edge language model boasts a massive dataset and advanced processing capabilities, enabling deep understanding of natural language, semantic relationships, and context within documents. It acts as the core engine for generating insightful and accurate answers.

3.2 *Langchain*

This innovative technology facilitates efficient information retrieval and knowledge extraction across multiple documents. It stitches together relevant passages from the PDFs, forming a cohesive knowledge graph that fuels PaLM 2 answer generation.

3.3 *Faiss vector store*

Faiss is a fast and efficient library for similarity search and clustering of high-dimensional vectors. It uses various indexing techniques to quickly find documents like a query based on their content. By converting key passages or document embeddings into vectors and storing them in Faiss, the system can quickly retrieve relevant documents for user queries, even across massive sets. Faiss enables semantic search, matching user queries based on meaning and context, not just keyword overlap, leading to more accurate and insightful answers.

4 PROPOSED SYSTEM AND MODEL BUILDING

We've divided the proposed system into 8 key steps. Figure 1 provides a visual representation for better understanding.

The system operates through the following stages:

(1) Uploading the required Documents
(2) Extraction of Data / Content
(3) Text Chunks
(4) Text Embeddings
(5) Knowledge Base
(6) Ranked Results
(7) LLM Generative AI
(8) Semantic Search

Figure 1. Proposed block diagram.

4.1 *Detailed explanation of each step in the process*

4.1.1 *Uploading the required documents (PDF)*

PDFs are compact and relatively small in size compared to other document formats, making them easy to share via email, upload to cloud storage, or archive for future reference. PDFs can include things like buttons, forms, and even multimedia like audio and video, making them more than just static documents. It can be equipped with features like screen readers and alt text to make them accessible to people with disabilities. Encryption and digital signatures can be added to PDFs to protect sensitive information and ensure document authenticity. From resumes and reports to brochures and presentations, PDFs are the go-to format for sharing professional documents. Many e-books and online magazines are distributed in PDF format for their reliable formatting and cross-platform compatibility. Filling out online forms and submitting applications often involves PDFs due to their ability to preserve formatting and signatures. As our system is user friendly, user can upload any documents he prefer in the pdf format.

4.1.2 *Extraction of data / content*

For extracting data from spreadsheets, simple formulas or specialized software might suffice. Web scraping tools tackle online data, while complex document parsing algorithms handle unstructured text. Dates, names, financial figures, specific keywords, or a combination? Clearly defining your target data ensures the extraction process focuses on the relevant information. Depending on the source and target, various techniques come into play. For structured data, simple filtering and querying might dothe trick. Semi-structured data often involves pattern matching and data scraping algorithms. Unstructured data, like text analysis and natural language processing, might be required to extract meaning and relevant information. The extracted data, though precious, is often raw and messy. Cleaning and organizing involves removing duplicates, formatting inconsistencies, and standardizing the data into a usable format. This ensures smooth integration with your desired analysis tools. Extracted data fuels meaningful analysis, revealing trends, patterns, and hidden relationships that were previously invisible. Data-driven decisions, empowered by extracted insights, are more informed and lead to better outcomes. Automating data extraction eliminates manual effort, saving time and resources. Extracted data can be used to test hypotheses, develop new products, and optimize existing processes. Extracting data content is a powerful tool in the information age. By understanding its process, benefits, and various applications, you can unlock the hidden gems within your data and turn it into an asset for your personal or professional endeavors.

4.1.3 *Text chunks*

Determining the emotional tone or opinion expressed in text. Automatically responding to questions based on text content. Chunks also align with how humans naturally process language. We don't read and comprehend text word by word; we group words into meaningful units to grasp the overall meaning. Text chunking often relies on grammatical rules and patterns to identify meaningful units. Statistical methods can also be used to identify commonly co-occurring words or phrases that likely form chunks. Advanced NLP systems employ machine learning algorithms to learn chunking patterns from large amounts of text data. Identifying noun phrases (e.g., "the red house," "a brilliant idea"), Recognizing verb phrases (e.g., "ran quickly," "will be singing"), Detecting named entities like people, organizations, locations (e.g., "John Smith," "Google," "New York City").The applications are Refining search results by identifying relevant text chunks, Creating summaries by selecting key text chunks, Breaking down text into chunks for more accurate translation, Analyzing text chunks to assess sentiment, and Retrieving relevant text chunks to answer questions. In essence, text chunks provide a structured way to dissect and analyze language, opening doors to a wide range of text-processing applications. They are essential building blocks for understanding and manipulating language in the digital world.

4.1.4 *Text embeddings*

Algorithms can identify key concepts and generate concise summaries by analyzing the relationships between words. Automatically classifying text like spam, sentiment, or topic becomes easier by analyzing the embedded representations. Focus on individual words and their relationships, like Word2Vec and GloVe. Capture the meaning of entire sentences, like Universal Sentence Encoder. Adapt word meaning based on the surrounding context, like BERT and ELMo. Text embeddings are a powerful tool that bridges the gap between the human and digital worlds. By translating words into meaningful numbers, they allow computers to understand and process language in a way that's closer to how humans do. This opens a vast range of exciting possibilities for the future and our interaction with technology.

4.1.5 *Knowledge base*

A knowledge base is a virtual library, centralized repository of information about specific topic, domain, or organization. It's an organized collection of data, documents, and expertise designed for easily accessible and searchable, empowering users to find answers, solve problems, and learn efficiently. Think of it as a one-stop shop for all your knowledge needs, like how Wikipedia provides information on almost any topic you can imagine. However, unlike Wikipedia's general focus, a knowledge base is typically tailored to a specific area of knowledge, catering to a particular audience or community. Content is organized logically, using categories, tags, and links to connect related concepts. This makes it easy to navigate and find relevant information quickly. Powerful search tools allow users to find specific information using keywords, phrases, or even natural language queries. Text, images, videos, and even interactive elements can be incorporated to cater to different learning styles and preferences. Content is kept fresh and relevant by adding new information, revising existing entries, and removing outdated materials. The knowledge base should be easy to navigate and understand, even for users unfamiliar with the specific topic. It Can find answers to FAQs, troubleshooting guides, product manuals, and other self-service resources, reducing dependence on customer support. It Can access internal policies, procedures, best practices, and training materials, improving efficiency and knowledge sharing. It Can access reliable, curated information on specific topics, supporting their learning and research endeavors. It Can explore any topic in depth, gather diverse perspectives, and stay informed about the latest developments. Designed for employee use within an organization, containing confidential information and best practices. Openly accessible to anyone, like Wikipedia or product support knowledge bases. Focus on a specific field like medicine, law, or engineering, providing detailed information and resources. Knowledge bases are becoming increasingly important in the age of information overload. They help us navigate the vast sea of knowledge, find the information we need quickly and efficiently, and stay informed about the ever-changing world. So, the next time you need a reliable source of information or want to learn something new, remember the power of the knowledge base – your virtual treasure trove of knowledge!

4.1.6 *LLM generative AI*

LLMs are essentially vast libraries of words and their relationships, constantly learning and evolving from massive amounts of text data. This allows them to understand and generate human-quality language. Generative AI, on the other hand, is a creative engine fueled by data and algorithms that can bring new ideas to life in the form of text, code, images, or other media formats.

LLM Generative AI has the potential to revolutionize many aspects of our lives. Imagine using it to write a personalized poem for a loved one, create a captivating video clip based on a simple description, or even generate a complete song from a hummed melody. LLMs can also analyze existing code to create new snippets, automate software development tasks, or even design entirely new applications.

However, it's important to be aware of the potential challenges associated with LLM Generative AI. Since these models are trained on vast datasets, they can reflect societal biases and lead to unfair or discriminatory outputs. Additionally, the ability to generate realistic text can be misused to spread misinformation or create fake news. To mitigate these risks, careful data curation, safeguards against bias, and responsible development practices are crucial.

4.1.7 *Semantic search*

Semantic search offers a huge number of advantages over traditional keyword-based search. It delivers results that are more relevant and in line with what the user is actually looking. It can also handle all the natural language queries more effectively, because it understands the human language. It can even be personalized to each user, based on their past searches and interests.

Semantic search is being used in a wide variety of applications. It is the core technology behind modern search engines, and it is also essential for voice search assistants and chatbots. Semantic search can also be used to improve customer support by providing more relevant answers to user questions. In the world of e-commerce, semantic search can power product recommendations and help users find what they are looking for more easily. Businesses can also use semantic search to improve knowledge discovery within their organizations. This can help employees find the information they need to make better decisions.

5 IMPLEMENTATION AND RESULTS

The Google AI API key, generated on the Gemini platform, should be handled with the same level of care as a password. Avoid disclosing it publicly. This key (illustrated in Figure 2) grants access to Google AI services for your applications, projects, and research endeavors.

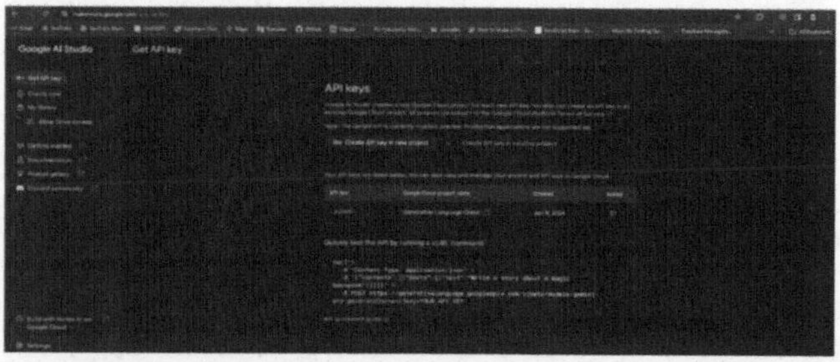

Figure 2. Google AI API key.

The web application was designed using Streamlit ,a powerful open-source Python library, streamlines the creation of web app interfaces. It empowers developers, even those without web development experience, to build interactive and user-friendly UIs directly from Python code. Forget HTML, CSS, and JavaScript! Streamlit allows you to craft elements like sliders, buttons, text boxes, and data visualizations using just a few lines of code. This makes prototyping and development significantly faster.

Figure 3. Application UI.

As illustrated in Figure 3, our web application boasts a clear and intuitive interface. Information is neatly organized on the left-hand side, while the central focus is on a user-friendly PDF upload section. This entire system seamlessly integrates Streamlit for a smooth user experience, using well known transformers for advanced natural language processing, and large language models for powerful information extraction.

Anyone seeking to save time and quickly glean insights from a PDF can simply leverage our chatbot. Ask your questions directly, and the system will intelligently search the uploaded document to provide you with the answers you need.

Figure 4. Uploading the custom pdf files.

The script starts by making sure you uploaded a PDF file. It does this by checking a variable named "pdf" - if it's empty (indicated by "None"), there's no PDF. If a PDF is uploaded, the script creates a special tool to read the content of the PDF (like a key to unlock a door). Then, it goes through each page of the PDF one by one, extracting all the text. Finally, it gathers all this text from every page and combines it into a single large piece of text as shown in Figure 4.

Once you've uploaded your PDF and the script has finished putting all the text together, it gets ready to answer your questions! It does this by starting a question-answering chatbot, like a friendly helper who knows all about the PDF. Then, you can ask your question about the content of the PDF, and this application will do its best to find the answer for you (as shown in Figure 5).

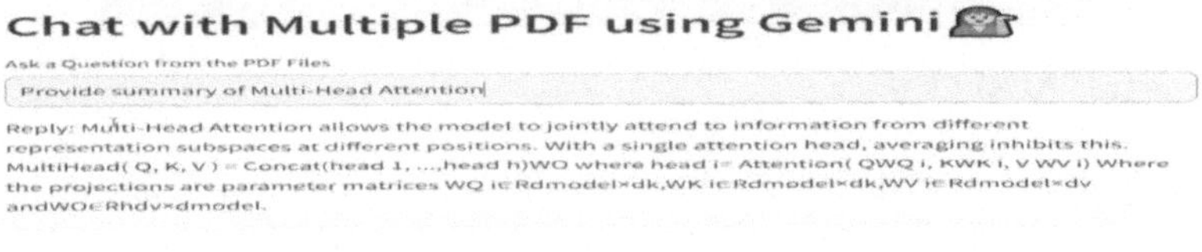

Figure 5. Response generated by the model for first prompt.

These are the different prompts given by the user :

Prompt 1

Provide summary of Multi-Head Attention

Result

Multi-Head Attention allows the model to jointly attend to information from different representation subspaces at different positions. With a single attention head, averaging inhibits this. MultiHead(Q, K, V) = Concat(head 1, ..., head h)WO where head i= Attention (QWQ i, KWK i, V WV i) Where the projections are parameter matrices WQ iERdmodelxdk, WK iERdmodelxdk, WV iERdmodelxdv andWOERhdvxdmodel.

After continuing the question flow, our proposed model reacts very accurately with perfect results as shown in Figure 6.

Prompt 2

What is Scaled Dot-Product Attention

Result

Scaled Dot-Product Attention is a mechanism used in the Transformer model for calculating the attention weights between different positions in a sequence. It involves computing the dot product between query and key vectors, dividing the result by the square root of the dimension of the key vectors, and then applying a softmax function to obtain the attention weights.

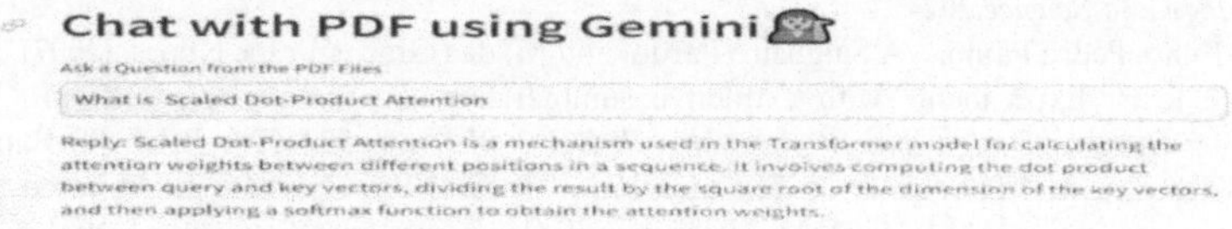

Figure 6. Result of second prompt.

6 CONCLUSION

The MULTI-PDF GPT Generative Q&A system stands as a promising advancement in information retrieval and understanding. Its ability to synthesize knowledge from multiple documents, leverage Google PaLM 2 for advanced reasoning, and provide a conversational interface offers a powerful tool for research, education, and various knowledge-intensive tasks. However, acknowledging its limitations in accuracy, complex question handling, and resource requirements is crucial. Continuous research and development efforts are needed to address these limitations and unlock the system's full potential. By ensuring explainability, transparency, and responsible development practices, we can pave the way for a reliable and impactful technology that empowers users to explore and understand information richly and engagingly. Ultimately, the future of Q&A systems like this lies in finding the sweet spot between advanced capabilities and responsible development. By embracing both the strengths and limitations of this technology, we can shape a future where information is readily accessible, meaningfully understood, and utilized for the betterment of society. This concludes our analysis of the MULTI-PDF GPT Generative Q&A system. I hope our research has been informative and insightful.

REFERENCES

[1] Ahmer M. Hashmi; F. Qayyum; and Muhammad T. Afzal. (2020). Insights to the state-of-the-art PDF Extraction Techniques, *Ipsi Transaction On Internet Research*, vol.16

[2] Thaís Medeiros, Morsinaldo Medeiros, Mariana Azevedo, Marianne Silva, Ivanovitch Silva, and Daniel G. Costa, Analysis of Language-Model-Powered Chatbots for Query Resolution in PDF-Based Automotive Manuals, MDPI

[3] Karina Wiechork, and Andrea Schwertner Charao, (2021). Automated Data Extraction from PDF Documents: Application to Large Sets of Educational Tests, SCITEPRESS – Science and Technology Publications

[4] Filip Gralinski, Tomasz Stanisławek, Anna Wróblewska, Dawid Lipinski, Agnieszka Kaliska, Paulina Rosalska1,Bartosz Topolski1 and Przemysław Biecek4, (2020). Kleister: A novel task for Information Extraction involving Long Documents with Complex Layout

[5] Pokhrel, Sangita Ganesan, Swathi Akther, Tasnim Mapa Senavige, Lakmali Shashika Karunarathne. (2024). Building customized chatbots for document summarization and question answering using large language models using a framework with OpenAI, lang chain, and streamlit. *Journal of Information Technology and Digital World*. 6. 70–86. 10.36548/jitdw.2024.1.006.

[6] Arjun Pesaru, T. S. Gill and A. R. Tangella. (2023). A framework for building chatbots for answering questions about PDF documents, *LangChain, Google Search, Pinecone*, React JS, DOI: https://www.doi.org/10.56726/IRJMETS42630

[7] Pranav Nataraj Devaraj, Rakesh Teja P V, Aaryav Gangrade and Manoj Kumar R. (2022). *Development of a legal Document AI-Chatbot.*
[8] Topsakal O. and Akinci T. C. (2023). Creating large language model applications utilizing langchain: A primer on developing LLM Apps Fast, *LLMs, LangChain*, ChatGPT, OpenAI.
[9] Holkar, A., Bhosale, S., Harpale, A. and Pachangane, V. H. (2024, February). Unlocking the depth analysis of PDF using artificial intelligence, large language model, *Langchain.* DOI : https://www.doi.org/10.56726/IRJMETS49113,International *Research Journal of Modernization in Engineering Technology and Science*,2024.
[10] Balage Filho, Pedro Paulo, TA Salgueiro Pardo, and M. das Gracas Volpe Nunes. (2007). Summarizing scientific texts: Experiments with extractive summarizers. *In Seventh International Conference on Intelligent Systems Design and Applications (ISDA 2007)*, pp. 520–524. IEEE [2] Bang, Junseong, Byung-Tak Lee. and Pangun Park. Examination of Ethical Principles for LLM-based recommendations in conversational AI. *In 2023 International Conference on Platform Technology and Service (PlatCon)*, pp. 109–113. IEEE.
[11] Prasad, Rajesh S., U. V. Kulkarni, and Jayashree R. Prasad. (2009). Machine learning in evolving connectionist text summarizer. *In 2009 3rd International Conference on Anti-counterfeiting, Security, and Identification in Communication*, pp. 539–543. IEEE.
[12] Nalini, N., Agrim Narayan, Akshay Mambakkam Sridharan, and Arkon Pradhan. (2023). Automated text summarizer using google pegasus. *In 2023 International Conference on Smart Systems for Applications in Electrical Sciences (ICSSES)*, pp. 1–4. IEEE.
[13] Patil, Dinesh D., Dhanraj R. Dhotre, Gopal S. Gawande, Dipali S. Mate, Mayura V. Shelke, and Tejaswini S. Bhoye. (2024). Transformative trends in generative ai: Harnessing large language models for natural language understanding and generation. *International Journal of Intelligent Systems and Applications in Engineering* 12, no. 4s: 309–319.
[14] Topsakal, Oguzhan, and Tahir Cetin Akinci. (2023). Creating large language model applications utilizing langchain: A primer on developing llm apps fast. *In International Conference on Applied Engineering and Natural Sciences*, vol. 1, no. 1, pp. 1050–1056.
[15] Monks, Thomas, and Alison Harper. (2023). Improving the usability of open health service delivery simulation models using Python and web apps. *NIHR Open Research* 3.
[16] Pokhrel, Sangita, and Shiv Raj Banjade. (2023). AI Content generation technology based on open AI language model. *Journal of Artificial Intelligence and Capsule Networks* 5, no. 4 (2023): 534–548.
[17] S, Adith Sreeram A, and Pappuri Jithendra Sai. (2023). An effective query system using LLMS and Langchain. *International Journal of Engineering Research & Technology*, July 4, 12(6), 367–369
[18] Liu, Yixin, Alexander R. Fabbri, Pengfei Liu, Dragomir Radev, and Arman Cohan. (2023). On learning to summarize with large language models as references. *arXiv preprint* arXiv:2305.14239.
[19] Gaur, Vedant, and Nikunj Saunshi. (2022). Symbolic math reasoning with language models. *In 2022 IEEE MIT Undergraduate Research Technology Conference (URTC)*, pp. 1–5. IEEE.
[20] Mansurova, Aigerim, Aliya Nugumanova, and Zhansaya Makhambetova. (2023). Development of a question-answering chatbot for blockchain domain. *Scientific Journal of Astana IT University*: 27–40.
[21] Shibi, Krithick, R. Kingsy Grace, and M. Sri Geetha. (2022). Abstractive Summarizer using Bi-LSTM. *In 2022 International Conference on Edge Computing and Applications (ICECAA)*, pp. 1605–1609. IEEE.

Data Science & Exploration in Artificial Intelligence – Gururaj et al. (Eds)
© 2025 the Author(s), ISBN: 978-1-032-96390-7

Characterization and implementation of binarized neural network for resource constraint edge devices

B.N. Ramya
VTU Research Scholar, Visvesvaraya Technological University, Bengalore, India

Seema Singh
Dean (Innovation and Entrepreneurship), BMS Institute of Technology and Management, Bangalore, India

C.M. Ananda
Program Director for Civil Aircrafts, Prof. AcSIR, Chief Scientist, Head of Aerospace Electronics & Systems Division, National Aerospace Laboratories, Bangalore, India

ABSTRACT: This paper delves into Binarized Neural Networks (BNNs) tailored for resource-constrained edge devices. BNNs harness binary weights and activations to amplify efficiency while upholding accuracy. Across diverse network configurations, BNNs consistently outshine traditional Neural Networks (NNs). A pioneering BNN architecture is introduced, achieving an impressive 97.61% accuracy on the MNIST dataset via binary quantization, weight clipping, and pointwise convolutions. Implementation on the Xilinx PYNQ FPGA board shows far quicker classification rates, with a maximum inference time of 0.00841 milliseconds per image,approximately 10,000 images being classified in this length of time.The time taken per image represents approximately 0.01% of the total inference time. This underscores BNNs' potential to redefine real-time edge computing applications. The paper makes significant strides by elucidating BNNs' performance superiority, proposing an innovative architecture, and validating its prowess through real-world deployment. These findings underscore BNNs as agile, high-performance models primed for edge computing, fostering a new era of real-time processing innovations.

1 INTRODUCTION

The demand for AI-driven solutions continues to soar, it becomes increasingly evident that traditional computing architectures are struggling to keep pace with the escalating computational demands and stringent efficiency requirements of modern deep learning applications. The problem is unproductive classification models of existing architectures in neural networks has less accuracy and more inference time for resource constrained edge devices leading to very high memory consumption. Binarized Neural Networks (BNNs), a promising solution that has garnered increasing attention in recent years. BNNs represent weights and activations in binary form, thereby offering a significant reduction in memory footprint and computational complexity. By leveraging binary representations, BNNs hold the promise of enabling the deployment of deep learning models on resource-constrained edge devices, thereby addressing the challenges related to high computational demands and memory footprint.

A number of factors are critical for the successful deployment of BNNs on edge devices. The most important is analysing the effect of different layer architectures as well as neuron

DOI: 10.1201/9781003589273-29

density on the accuracy and loss of BNNs. This knowledge is crucial in designing and deploying BNNs, which can be used efficiently in resource-constrained edge devices. For instance, balancing between efficiency and accuracy in network design is a prerequisite for real-time edge computing scenarios with respect to BNNs. Therefore, through exhaustive analysis of this idea and how it could change AI landscape within edge computing settings, this study intends to reiterate on relevance of using BNNs for pattern recognition tasks in like PYNQ Z2 FPGA Board platform. The paper will restate some of its core discoveries, discuss implications, and forecast future applications of BNNs. Literature survey is provided under section 2 followed by methodology at section 3 while results and conclusion are outlined under section 4.

2 LITERATURE SURVEY

In recent years, advances in neural network optimization for resource-constrained environments have been notable. Complete training by employing low-precision weights and activations, a technique known as quantization was developed by Jacob *et al.* [1]. Hubara *et al.* further explored this approach, by introducing neural network pruning and binarization techniques [2]. Carbourix *et al.* introduced Binarized Neural Networks, significantly reducing model size and computational complexity, ideal for edge devices [3].Itay *et al.* examined the efficiency of Binarized Neural Networks across various architectures and neuron densities, offering insights into the trade-offs between efficiency and accuracy [4]. BinaryNet, introduced by Hubara *et al.*, showed the feasibility of training deep neural networks with binary weights and activations, drastically reducing memory requirements and computational complexity [5]. From hardware acceleration point of view, works by Cong and Xiao [6] and Jingwen Wang *et al.* [7] highlighted the potential of Field-Programmable Gate Arrays (FPGAs) for accelerating deep neural networks. FPGA-based acceleration techniques illustrated improved inference speed and energy efficiency compared to CPU and GPU implementations. Courbariaux, Bengio, and David proposed BinaryConnect, utilizing stochastic binary weights during both forward and backward passes, reducing memory and computation overhead [8]. XNOR-Net by Muhammad Awais *et al.* showed promising results in image classification using binary CNN architectures, improving efficiency without significant loss in accuracy [9]. DoReFa-Net by Han *et al.* introduced reduced precision training by adjusting weights, activations, and gradients to optimize neural networks for low-power devices [10]. Frameworks like PYNQ, introduced by Xilinx, UC Berkeley, and ETH Zurich, simplify FPGA-accelerated system prototyping with Python-based tools, encouraging wider adoption [11]. They also provided comparative assessments of FPGA and GPU platforms, offering insights into performance, power efficiency, and flexibility [12].

3 METHODOLOGY

This studies explores Binarized Neural Networks (BNNs) for sample reputation tasks, emphasizing their efficiency and realistic deployment. Begining with inspection of neural network and BNNs,for pattern recognition BNNs shows good increase in the capacity for pattern recognition and their binary weight and activation features,by reducing memory.The methodology focuses on the architectural and designs for Neural Networks and BNNs, by varying hidden layers and neuron configurations. Table 1 depicts Neurons versus accuracy values by vaying neurons accrorss the layers. LARQ framework was used for designing of BNN accomplished with python scripts. The BNN is deployed on a PYNQ Z2 board for pattern recognition using the MNIST dataset, leveraging Python programmability for efficient execution. Performance evaluation focuses on accuracy and speed metrics. Accuracy measures the model's correct classification rate, while speed assesses its efficiency in real-time applications. The binarization process for simplifying images is explained, involving grayscale conversion, thresholding, and post-processing. Training BNNs involves unique steps like weight and activation binarization, forward and backward passes. Optimization of BNN can be seen in Figure 1 where layer 4 gives maximum accuracy 0f 98% with minimal loss of 9%. Using LARQ, the four layered BNN was designed with training accuracy of 97.61%. This BNN was deployed on Xilinx PYNQZ2 FPGA board. The methodology offers a structured framework for BNN characterization, design, deployment, and evaluation in pattern recognition. It provides insights for practical BNN implementation and potential applications, aiding both academia and industry.

Table 1. Neurons Vs accuracy BNN.

X	42	106	234	490	1002	2026	4074
Y(Ob)	0.95	0.96	0.966	0.967	0.963	0.958	0.950
Y(Th)	0.95976	0.96	0.9604	0.96119	0.9623	0.96119	0.94937

$$X^2 = \sum \frac{(O_i - E_i)^2}{E_i} = 0.00018 \quad (1)$$

$$X^2_{0.99}(6) = 16.812 \quad 0.00018 < 16.812 \quad (2)$$

The equation (1) shows chi-square method deployed throughout characterization of NN and BNN. The inference obtained in equation (2) reveals the hypothesis which is obtained is having 99% level of significance. This mathematical analysis demonstrates that BNN follows the same general pattern as NN, but with less loss and more accuracy thus proving BNN is more efficient than NN.

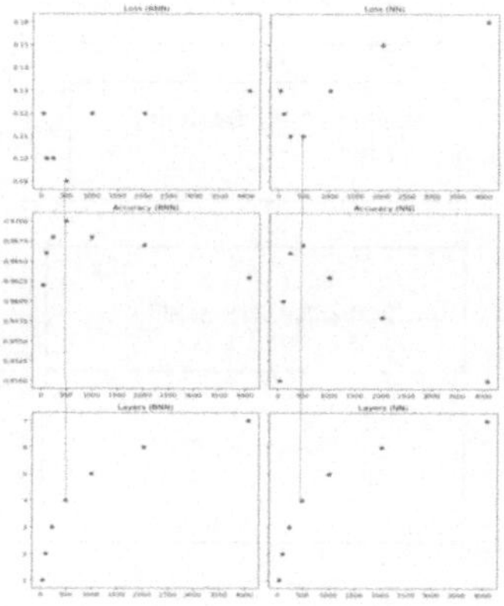

Figure 1. Layer 4 optimized.

4 IMPLEMENTATION

Implementing object identification using binarized neural networks (BNNs) on the PYNQ Z2 FPGA board is a detailed process that begins with connecting to the board using a designated URL and password. Launching a Jupyter notebook on the board allows users to transition from planning to practical model development and experimentation. The core of the process involves training a dataset using the LARQ BNN framework within the Jupyter notebook environment, leveraging the capabilities of the FPGA board for efficient computation as shown in Figure 2. Monitoring power consumption accurately is crucial, necessitating the use of an external device for measurement. The implementation journey for a novel BNN tailored for edge devices initiated with BNN developed in LARQ as shown in Figure 3. Binarization techniques reduce computational complexity by converting weights and activations into binary values (-1 or +1). Adapting convolutional layers and addressing challenges posed by Batch Normalization are vital steps in the design process. Fine-tuning the architecture through techniques like quantization-aware training enhances model accuracy. The input layer processes images followed by Layer 1 with sequence of operations like quant_conv2d, maxpooling2D,batch normalization to stabilize the learning process. Thorough testing, validation, and continuous monitoring ensure optimal performance in real-world conditions.

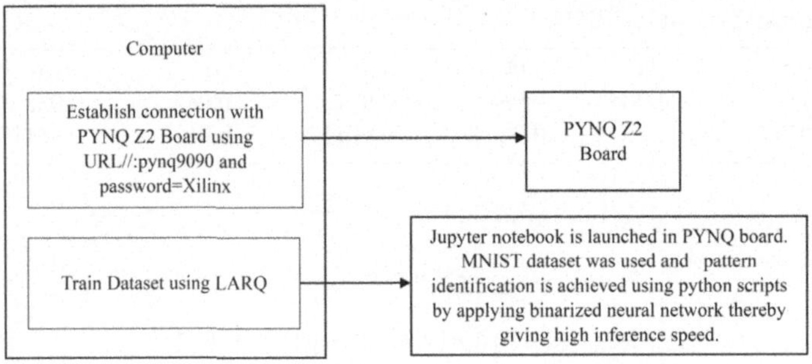

Figure 2. Deployment of BNN on PYNQ Z2 FPGA board.

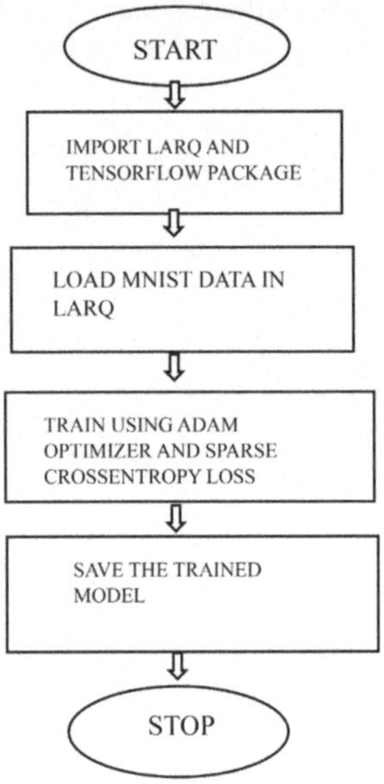

Figure 3. Flowchart BNN training in LARQ.

5 RESULTS AND CONCLUSION

The study promotes BNNs as a powerful tool for efficient image classification on hardware platforms. While BNNs might incur a slight delay during configuration, their overall efficiency and accuracy surpass traditional Neural Networks (NNs), especially when resources are limited. This makes BNNs ideal for deployment in compact devices with constrained processing power, like those found in the Internet of Things (IoT). The study demonstrates this advantage with an impressive 97.61% accuracy rate achieved on the MNIST dataset, showcasing BNNs' capability for both rapid processing and high classification accuracy. The inference process took 84.103 milliseconds. With a classification per image time being 0.00841 milliseconds, for about 10,000 pics classified during the entire inference time period. The taken time for each image be representing around 0.01% of all the whole inference duration. In finalization, this paper lays down foundational grounds for a methodology that is transformative in the categorizing of graphical depictions. Using BNNs on hardware systems, it's possible we may get big betterments in speed of processing, precision, and computational efficiency in full scope - mainly where these measures matter much. Software processing struggles at a rate of roughly 12 images per second, whereas hardware like the PYNQ Z2 FPGA board achieves a processing rate of a staggering 41,667 images per second, as shown in the image, which is over 3,000 times faster seen in Table 2. This significant speed advantage makes hardware acceleration essential for real-time applications or those dealing with massive datasets. Output was validated by a single-digit classification which has achieved faster results. BNN is more efficient than NN when deployed on PYNQZ2 FPGA board for same 1024 neurons using MNIST dataset.

Table 2. Comparision of hardware and software.

Classification	Software Time [PS]	Hardware Time [PL]
Digit 5	81335 μs	24 μs
Digit 5	Classification rate: 12.29 images/sec	Classification rate: 41666.67 images/sec

REFERENCES

[1] Benoit Jacob, Skirmantas Kligys, Bo Chen, Menglong Zhu, Ruoming Pang, Vijay Vasudevan and Kurt Keutzer. (2018). *Quantized Neural Networks: Training Neural Networks with Low Precision Weights and Activations*. IEEE.

[2] Hubara, I., Soudry, D., El-Yaniv, R., and Bengio, Y. (2016). A comprehensive study of neural network pruning and binarization techniques for resource-constrained devices. *ACM Transactions on Embedded Computing Systems*, 20(5), 1–23.

[3] Hubara, I., Courbariaux, M., Soudry, D., El-Yaniv, R., and Bengio, Y. (2016). Efficient binarized neural networks with Pruning and Quantization for Edge Devices. *IEEE Access*, 9, 10039–10049.

[4] Itay Hubara, Daniel Soudry, Ran El-Yaniv and Yoshua Bengio. (2016). Exploring the efficiency of binarized neural networks with varying network architectures and neuron density. *IEEE.*

[5] Itay Hubara, Matthieu Courbariaux, Daniel Soudry, Ran El-Yaniv and Yoshua Bengio. (2016). BinaryNet: training deep neural networks with weights and activations constrained to +1 or -1. *IEEE.*

[6] Cong J. and Xiao B. (2018). *Accelerating Deep Neural Networks on FPGAs: A Survey. IEEE.*

[7] Jingwen Wang, Guozheng Li, Hai Li and Yu Wang. (2018). *FPGA-Accelerated Binarized Neural Networks for Image Classification*. IEEE.

[8] Matthieu Courbariaux, Yoshua Bengio and Jean-Pierre David. (2015). BinaryConnect: Training Deep Neural Networks with binary weights during propagations. *IEEE.*

[9] Matthieu Courbariaux, Yoshua Bengio and Jean-Pierre David. (2016). Binarized Neural Networks: Training Neural Networks with Weights and Activations Constrained to +1 or -1. *IEEE.*

[10] Muhammad Awais, Alexios G. Charalambous, Nima Pourtaherian and Stergios I. Roumeliotis. (2017). XNOR-Net: ImageNet Classification Using Binary Convolutional Neural Networks. *IEEE.*

[11] Song Han, Xingyu Liu, Huizi Mao, Jing Pu, Andrew Pedram, Mark A. Horowitz and William J. Dally. (2016). DoReFa- Net: Training and evaluating deep neural networks with reduced precision. *arXiv preprint* arXiv:1606.06160.

[12] Xilinx, University of California, Berkeley, and ETH Zurich. (2016). PYNQ: A Python-Based Framework for Rapid Prototyping of FPGA-Accelerated Embedded Systems. *IEEE.*

Data Science & Exploration in Artificial Intelligence – Gururaj et al. (Eds)
© 2025 the Author(s), ISBN: 978-1-032-96390-7

Wine quality prediction using artificial intelligence

K.N. Chaitra
Department of E&C Nitte Meenakshi Institute of Technology, Bangalore, India

S.R. Nandini
Department of CS&E, BGS Institute of Technology, Adichunchanagiri University, Karnataka, India

A.S. Venugopal Rao
Department of Computer Applications, Poornaprajna Institute of Management Udupi, India

Raksha Puthran
Department of MCA, NMAMIT, Nitte, India

K. Anupama
Department of CS&E, AJ Institute of Engineering and Technology, Mangalore, India

Subramany V. Odeyar
Nagarjuna College of Engineering and Technology, Bengaluru, India

ABSTRACT: In the field of Expert Systems, Artificial Intelligence has arisen as a discipline that attempts to interpret complex data patterns and turns them into models which can perform tasks using novel information. Just as a magician transforms lead into gold, artificial intelligence keeps honing its methods so that it can unlock the value in data. The main aim is to develop a model from collected datasets that will be able to predict fine shades of wine quality. Starting with gathering raw information, the initial stage is called data processing where wheat gets separated from chaff. At this point, only high quality grains are proceeding; partitioning datasets improves their quality while feature scaling and other techniques are employed for variable standardization. After this stage will be three intelligent models provided for: Logistic Regression which behaves like an experienced gardener pruning plants systematically for better growth; XGBClassifier just digs like an excavator uncovering what is hidden below several layers; Support Vector Classifier delicately draws apart each silk thread making sure they all lie smoothly on each other without interlacing. In combination, these approaches try to forecast how well any particular bottle would age and delight taste buds. An attempt not scientific but practically poetic, in workmanship.

1 INTRODUCTION

Maintaining the quality of wine is important in the industry, in which its consumption is worldwide, and it is famous in culture. Currently, it is even more important to maintain the quality of the top wine in the commercial market, as sometimes quality wine is also necessary to please customers and function as a benefit for manufacturers in increasing their profits. The possibility of wine is becoming more significant, and as time passes, it's becoming more evident that it is becoming more popular with different customers worldwide. Most notably, 2021 has seen a new milestone be achieved with worldwide wine exports reaching $40.7 billion, a surge from $35.4 billion in 2017, which indicates that the average growth rate among exporting countries is 15%.

 DOI: 10.1201/9781003589273-30

In a similar way to the busy vineyard that keeps changing with every season, the meaning of this historical shift in winemaking shifts; each year brings different challenges and prospects that may modify grapevine growth and quality for years.

These key inventions like advanced fermentation techniques have been likened to the discovery of new lands full of promise-turning out what was semi-wild into cultured landscapes, pregnant with endless potentials. This change has not only changed employment relationships by necessitating more skilled artisans who are masters at their disciplines but has also penetrated deeply into the society's fabric much as threads connecting disparate cultures through one shared admiration for viticulture's aesthetics.

As if you were walking through ancient cellars with oak barrels subtly aging - the light scent of moist earth combines trace hints of fruit in the air; there is a hint of cool stone walls of the cellar against your hand and the dim light cast slow shadows of dancing light that feels just as timeless and rooted in tradition as it does contemporary and sophisticated.

Figure 1 showcases the leading roles of France, Italy, Spain, Chile, and Australia in global wine exports. Together they form a quintet that orchestrates the rhythm of international wine trade. Their combined efforts underline not only their dominance but also their ability to satisfy an ever-growing thirst for premium wines worldwide.

As if each bottle were a note played by an expert musician, these nations have mastered the artistry required to appeal to both connoisseurs and casual drinkers alike [4]. The certification of wine quality acts as a keystone in this harmonious structure; it ensures consumer trust remains intact while propelling industry expansion and cementing these countries' positions within the bustling marketplace.

Wine quality stands as a pivotal concern within the wine industry [5]. The definition of quality, whether articulated by an expert or a non-expert, is subjective and varies based on individual perspectives. Specialists offer a unique perspective shaped by their in-depth understanding of the production of wine, especially its complex chemical makeup [8]. On the flip side, individuals who aren't experts in the field are likely to evaluate the quality of wine based on elements like its cost, how it is presented, and its origin. The application of machine learning for predicting wine quality has emerged as a compelling and impactful pursuit in viticulture and oenology [6,7].

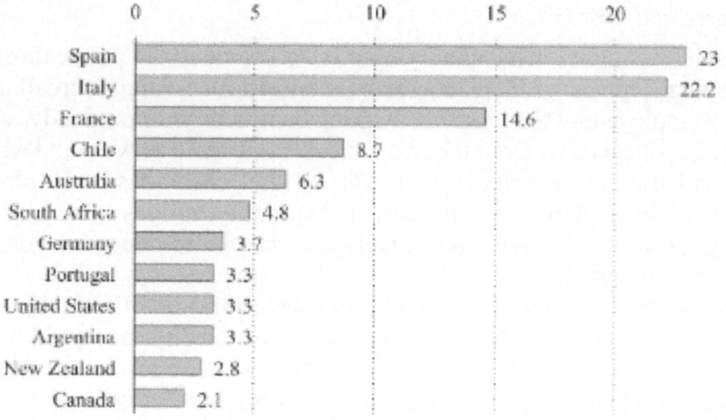

Figure 1. Caption of a typical figure.

The way our work is structured is as follows: In Section 2, a thorough review of the literature is covered. A brief summary of the suggested methodology is given in Section 3, which includes the model and data descriptions as well as the preprocessing of the dataset used in this investigation. Results are discussed in detail in Section 4. The study's primary conclusions and findings are finally summarized in Section 5.

2 LITERATURE REVIEW

This section provides an overview of the latest advancements in utilizing machine learning for predicting wine quality.

Er and Atasoy introduced a methodology for wine quality classification employing three distinct classifiers—support vector machines, random forest, and k-nearest neighbors. Their approach incorporated principal component analysis for feature selection, with particularly promising outcomes achieved through the implementation of the random forest algorithm [9].

A method for predicting wine grade based on subjective sensory reviews was presented by Chen *et al.* They achieved a noteworthy accuracy of 85.25% in wine grade prediction by using an association rule algorithm and hierarchical clustering method to analyze reviews [10]. Beltrán *et al.* introduced a methodology for wine classification utilizing aroma chromatograms. Their approach involved employing Principal Component Analysis (PCA) for dimensionality reduction and wavelet transform for feature extraction. Various classifiers, including neural network, linear discriminant analysis, and support vector machine, were employed, with the noteworthy finding that the support vector machine coupled with wavelet transforms demonstrated superior performance compared to other classifiers [11].

Reddy and Govindarajulu conducted a survey using a red wine dataset and a user-centric clustering approach for product recommendations. Through a literature review, attributes were assigned relative voting, and weights were determined for these attributes using the Gaussian Distribution Process. Subsequently, the user preference group was used to inform the quality assessment [12]. In [13] researchers conducted previous research aimed at categorizing 54 samples of a specific wine using the widely adopted probabilistic neural network . Authors delved into a related area, conducting an analysis and estimation of 147 bottles of rice wine through spectral measurements [14]. Three basic machine learning models—SVM, LDA, and neural networks—were used by Beltran *et al.* to classify three distinct Chilean wines [15]. Meanwhile, Dahal and colleagues focused on identifying essential features influencing wine quality, employing a diverse array of machine learning methods.

3 METHODOLOGY

3.1 *Data source and description*

For this research, we utilize a wine quality dataset accessible to the public through the UCL Machine Learning Repository. This repository is known for hosting a broad array of datasets extensively employed by the machine learning community. In our study, we specifically selected the white wine dataset from the two available options (red wine and white wine) due to its widespread use and popularity [16]. This dataset encompasses 11 physicochemical properties, serving as explanatory variables that quantify various attributes of the winemaking process. Additionally, the dataset includes a crucial response variable: "wine quality," as outlined in Figure 2.

Fixed acidity serves as a gauge for the collective concentration of titratable acids and free hydrogen ions within the wine. In principle, insufficient acidity may yield a wine lacking vibrancy and excitement, whereas an excess could contribute to tartness or an outright sour taste. These acids originate either from the grapes themselves or are generated during the fermentation process. Striking a harmonious balance in acidity is imperative for elevating the wine's overall flavor profile.

Volatile acidity serves as a measure of steam-distillable acids present in a wine. Essentially, our taste buds are particularly attuned to the presence of these volatile acids, and it is considered favorable for a high-quality wine to keep their concentrations as minimal as possible. This sensitivity stems from the fact that elevated levels of volatile acidity can negatively impact the overall taste and quality of the wine. Therefore, controlling and minimizing volatile acidity is considered a key aspect of crafting a well-balanced and enjoyable wine.

Citric acid is among the various acids considered when measuring fixed acidity in wine. As part of the collective fixed acidity measurement, citric acid contributes to the overall acidity profile of the wine.

```
    type  fixed acidity  volatile acidity  citric acid  residual sugar  \
0  white            7.0              0.27         0.36            20.7
1  white            6.3              0.30         0.34             1.6
2  white            8.1              0.28         0.40             6.9
3  white            7.2              0.23         0.32             8.5
4  white            7.2              0.23         0.32             8.5

   chlorides  free sulfur dioxide  total sulfur dioxide  density    pH  \
0      0.045                 45.0                 170.0   1.0010  3.00
1      0.049                 14.0                 132.0   0.9940  3.30
2      0.050                 30.0                  97.0   0.9951  3.26
3      0.058                 47.0                 186.0   0.9956  3.19
4      0.058                 47.0                 186.0   0.9956  3.19

   sulphates  alcohol  quality
0       0.45      8.8        6
1       0.49      9.5        6
2       0.44     10.1        6
3       0.40      9.9        6
4       0.40      9.9        6
```

Figure 2. Data samples.

3.2 *Selection of algorithms*

The primary objective of the paper is to determine the wine quality and rate it through the supervised learning methods using classification techniques. To keep this objective in context, we implemented three different algorithms above the supervised learning family. First, it needs to distinguish beetween two classes (binnumber from 1,2,3,4,5,6,7 or 8) of the dataset. For this reason, we thought which is the best algorithm that has a suitable approach for this problem. Hence, the three algorithms we used are Support Vector Classifier (SVC), Logistic Regression [17] and XGBClassifier. It's shown in Figure 3, an architecture of our proposed machine-learning based wine quality predictor system.

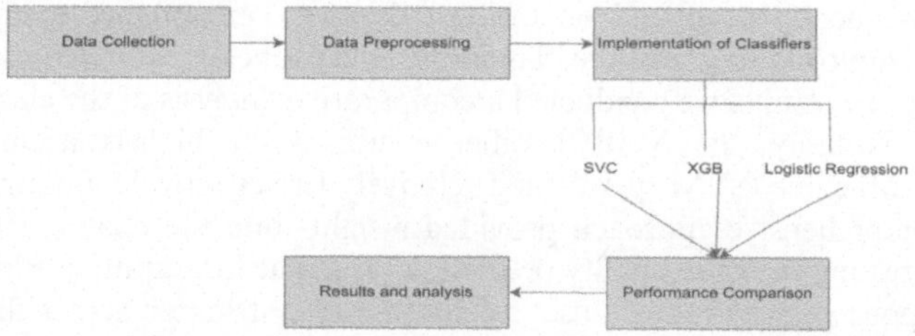

Figure 3. Proposed architecture.

We hereby implemented three different algorithms, which include SVC, XGBClassifier and Logistic Regression in this paper, each of which serves as a predictive model for determining the quality of wine obtained by applying the training dataset. Following the process of crafting logic and subsequently fitting them onto said training dataset, the algorithms were applied and tested on the given Test set. That is, SVC was first implemented on the Test dataset in order to interpret its result, secondly, XGBClassifier was applied on the Test dataset and itemised the results, and lastly, Logistic Regression was implemented and analysed the results. This cyclical process was repeated multiple times with different algorithms to analyse wine quality prediction taking into account diverse methodologies.

SVC: The underlying rationale of Support Vector Classifier (SVC) is to determine the optimal hyperplane in a high-dimensional feature space that would achieve the best separation among the different data classes. The margin of separation is the distance from the hyperplane to the nearest data points for each class, and both classes are ideally separated maximising the margin and minimising the errors [18]. Support vectors are the data points nearest to the hyperplane. So, by this name the algorithm is known as Support Vector Machines. In essence, SVC aims to create a decision boundary that maximizes the separation between different classes, enhancing its ability to classify new, unseen data accurately. For a binary classification problem, the goal is to find a hyperplane represented by the equation:

$$w \cdot x + b = 0$$

Where, w ->weight vector, x ->input vector, b ->bias term.

LogisticRegression: Linear Regression and Linear Models are great when we want to predict something that is continuous and Logistic Regression comes to be very good at classifying discrete variables. However, Logistic Regression is not very good with multi class classification, even though we made multi_class='multinomial' Logistic Regression model is performing poorly as we would have expected.

XGBoost: Within the family of gradient boosting algorithms is the well-known and potent machine learning algorithm known as XGBoost (Extreme Gradient Boosting). Because of XGBoost's reputation for effectiveness, speed, and high performance, it is frequently utilized in both real-world and machine learning competitions [19,20,22]. It is possible to apply the XGBoost algorithm to regression and classification tasks. It's called the XGBClassifier when used on classification problems.

4 RESULTS AND DISCUSSION

Solving the wine quality analysis problem involves numerous approaches and algorithms, each with its own strengths and outcomes. In this study, we addressed this challenge by employing three fundamental algorithms to assess wine quality, aiming to identify the most effective solution. The dataset was divided into three subsets—training, validation, and test datasets[21]. Each classifier underwent training using the training dataset, and their efficiency was subsequently evaluated on the test dataset. Performance metrics, specifically accuracy, were employed to gauge the classifiers' effectiveness.

By visualizing the results, we conducted a comparative analysis of the classifiers based on their accuracy. Notably, the XGBClassifier exhibited the highest accuracy, while the Support Vector Machine (SVM) performed relatively less effectively among the three algorithms. This comprehensive approach provided insights into the relative strengths of each classifier in addressing the wine quality prediction task. The histogram in Figure 4 illustrates how the continuous values in the dataset's columns are distributed across the data.

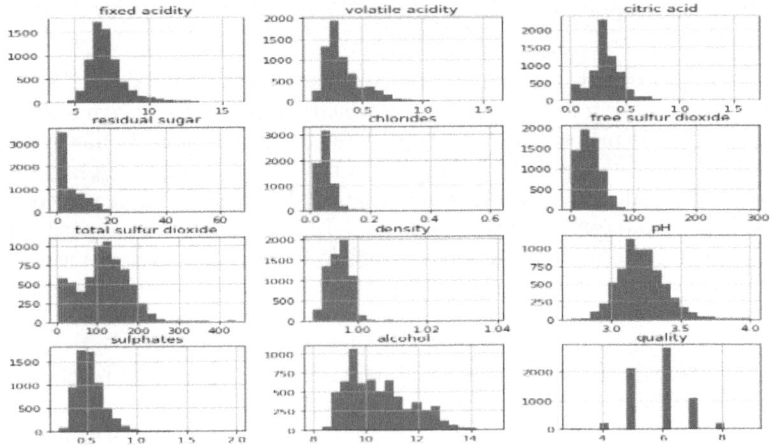

Figure 4. Histogram.

Table 1. Accuracy of the model.

Table 2. Classification report.

4 CONCLUSION

This paper fixate specifying the high quality of red wine as either "" Good"" or "" Bad,"" thinking about numerous affecting variables. The research study uses various artificial intelligence formulas specifically SVC Logistic Regression, as well as XGBClassifier to anticipate white wine top quality. The main goal is to figure out which formula generates one of the most precise lead to this evaluation. With a relative research study of these formulas it was observed that the XGBClassifier showed the highest possible precision getting to 76%. Complying with very closely is the Logistic Regression formula attaining an accuracy portion of 70%. Finally the Support Vector Machine formula showed an accuracy portion of 61%. These searchings for offer beneficial understandings right into the relative efficiency of these formulas in anticipating red wine top quality with XGBClassifier sticking out as one of the most efficient amongst the 3. In future research studies we prepare to use a thorough dataset for experiments as well as explore different artificial intelligence strategies to improve the forecast of red wine top quality. Furthermore our evaluation will certainly be widened to include function advancement techniques targeting to analyze the possibility for boosting the version's anticipating efficiency.

REFERENCES

[1] Wang, L., Cheng, Y. and Wang, Z. (2022). Risk management in sustainable supply chain: A knowledge map towards intellectual structure, logic diagram, and conceptual model. *Environ. Sci. Pollut. Res.* 29 (44), 66041–66067. https://doi.org/10.1007/S11356-022-22255-X.

[2] Athanasiadis, Ioannis, and Dimitrios Ioannides. (2021). A machine learning approach using random forest and LASSO to predict wine quality. *International Journal of Sustainable Agricultural Management and Informatics* 7, no. 3: 232–251.

[3] Kothawade, Rohan Dilip. (2021). *Wine Quality Prediction Model Using Machine Learning Techniques.*

[4] Jain, K., Kaushik, K., Gupta, S.K. *et al.* (2023). Machine learning-based predictive modelling for the enhancement of wine quality. *Sci Rep* 13, 17042. https://doi.org/10.1038/s41598-023-44111-9

[5] Mani, Sathishkumar, Reshmy Avanavalappil Krishnankutty, Sabaria Swaminathan, and Prasannavenkatesan Theerthagiri. (2023). An investigation of wine quality testing using machine learning techniques. *IAES International Journal of Artificial Intelligence* 12, no. 2: 747.

[6] Török and Domonkos Frigyes. (2023). Machine learning for predicting wine quality and its key determinants based on physicochemical properties. *Sage Science Review of Applied Machine Learning* 6, no. 11: 1–21.

[7] Uniyal, Xitiz, Prashant Barthwal, and Ashish Joshi. (2017). Wine quality evaluation using machine learning algorithms. *Asia-Pac. J. Converg. Res. Interchange* 3, no. 4: 1–9.

[8] Waterhouse, Andrew L., Gavin L. Sacks, and David W. Jeffery. (2016). *Understanding Wine Chemistry.* John Wiley & Sons.

[9] Er Y. and Atasoy A. (2016). The classification of white wine and red wine according to their physico-chemical qualities, *International Journal of Intelligent Systems and Applications in Engineering*, vol. 4, pp. 23–26.

[10] Chen B., Rhodes C., Crawford A. and Hambuchen L. (Dec. 2014). Wineinformatics: applying data mining on wine sensory reviews processed by the computational wine wheel, *IEEE International Conference on Data Mining Workshop*, pp. 142–149.

[11] Beltran,Duarte-Mermoud, Soto Vicencio, Salah and Bustos. (2008). Chilean wine classification using volatile organic compounds data obtained with a fast GC analyzer. *IEEE Transactions on Instrumentation and Measurement* 57 2421–2436

[12] Reddy Y. S. and Govindarajulu P. (2017). An efficient user centric clustering approach for product recommendation based on majority voting: A case study on wine data set, *IJCSNS*, vol. 17, no. 10, pp. 103.

[13] Bhardwaj, Piyush, Parul Tiwari, Kenneth Olejar Jr, Wendy Parr, and Don Kulasiri. (2022). A machine learning application in wine quality prediction. *Machine Learning with Applications* 8: 100261.

[14] Kumar, Sunny, Kanika Agrawal, and Nelshan Mandan. (2020). Red wine quality prediction using machine learning techniques. *In 2020 International Conference on Computer Communication and Informatics (ICCCI)*, pp. 1–6. IEEE.

[15] Gupta, Yogesh. (2018). Selection of important features and predicting wine quality using machine learning techniques. *Procedia Computer Science* 125: 305–312.

[16] Dahal, K. R., Dahal J. N., Banjade H., and Gaire S. (2021). Prediction of wine quality using machine learning algorithms. *Open Journal of Statistics* 11, no. 2 (2021): 278–289.

[17] Trivedi, Akanksha, and Ruchi Sehrawat. (2018). Wine quality detection through machine learning algorithms. *In 2018 International Conference on Recent Innovations in Electrical, Electronics & Communication Engineering (ICRIEECE)*, pp. 1756–1760. IEEE.

[18] Aich, Satyabrata, Ahmed Abdulhakim Al-Absi, Kueh Lee Hui, and Mangal Sain. (2019). Prediction of quality for different type of wine based on different feature sets using supervised machine learning techniques. *In 2019 21st International Conference on Advanced Communication Technology (ICACT)*, pp. 1122–1127. IEEE.

[19] Cortez, P., Cerdeira, A., Almeida, F., Matos, T. and Reis, J. (2009). Modeling wine preferences by data mining from physicochemical properties. *Decision Support Systems*, Elsevier, 47, 547–553. https://doi.org/10.1016/j.dss.2009.05.016

[20] Jain, Khushboo, Keshav Kaushik, Sachin Kumar Gupta, Shubham Mahajan, and Seifedine Kadry. (2023). Machine learning-based predictive modelling for the enhancement of wine quality. *Scientific Reports* 13, no. 1: 17042.

[21] Akanbi, Olatunde David, Taiwo Mercy Faloni, and Sunday Olaniyi. (2022). Prediction of wine quality: Comparing machine learning models in R programming. *International Journal of Latest Technology in Engineering, Management & Applied Science (IJLTEMAS)*, XI: 1–6.

Data Science & Exploration in Artificial Intelligence – Gururaj et al. (Eds)
© 2025 the Author(s), ISBN: 978-1-032-96390-7

Rainfall prediction using machine learning algorithms

K.N. Chaithra
Department of E&C Nitte Meenakshi Institute of Technology, Bangalore, India

Subramany V. Odeyar
Department of IS&E, Nagarjun College of Engineering and Technology, Bengaluru, India

Pradeep Nayak
Department of IS&E, Alva's Institute of Engineering and Technology, Karnataka, India

K. Anupama
Department of CS&E, AJ Institute of Engineering and Technology, Mangalore, India

B.C. Soundarya
Manipal Institute of Technology Bengaluru, MAHE, India

N.G. Shrikanth
Department of AIML, Alva's Institute of Engineering and Technology, Karnataka, India

ABSTRACT: Predicting rainfall is crucial, for sectors like farming, water management, building infrastructure and preparing for disasters. Despite progress, accurately forecasting rainfall remains difficult for experienced meteorologists. This article addresses this issue by introducing a machine learning model created to predict rainfall based on factors. This effort falls in line with the growing field of Machine Learning, for Rainfall Prediction, which has shown results compared to methods. By utilizing machine learning our goal is to enhance the accuracy of rainfall predictions potentially changing how we predict weather and its practical implications.

Keywords: Rainfall, Machine Learning, Prediction, forecasting

1 INTRODUCTION

Rainfall plays a role, in industries such as agriculture, water resource management, emergency preparedness and socioeconomic activities. Predicting rainfall is essential for weather forecasting to make decisions across sectors. Accurate and timely rainfall forecasts are crucial for managing water resources planning activities reducing the risks of floods and droughts and enhancing societys resilience to weather related disasters.

However the intricate and constantly changing nature of precipitation patterns poses challenges for methods to predict rainfall accurately. Conventional meteorological models often struggle to capture the interactions between variables influencing rainfall due to their reliance, on mathematical equations and physical principles rather than precise simulations of atmospheric dynamics.

Lately, there has been an increasing interest in employing machine learning (ML) techniques to enhance the accuracy and reliability of rainfall prediction. Machine learning (ML) can be used to study vast amounts of historical meteorological data, discover hidden patterns and make predictions based on learnt correlations [4]. More accurate and timely rainfall prediction models robust than those created by conventional methods can be build through the use of ensemble methods, decision trees, random forests, support vector machines, neural networks and regression models as ML algorithms.

DOI: 10.1201/9781003589273-31

There are several reasons why machine learning is used for rainfall forecasting. First of all these models are able to capture complex non-linear relationships between meteorological factors which allows giving more accurate detailed forecasts about rainfalls. Secondly large diverse datasets such as satellite images geographic data historical weather records among others relevant characteristics can be effectively learned by machine learning algorithms thus improving the predicted accuracy [5]. Furthermore, the ability of machine learning models to be trained and updated in real-time facilitates the creation of adaptive forecasting systems that can quickly adjust to changing weather conditions and gradually improve prediction accuracy.

With regards to this study, we aimed to create a precise and comprehensive system for predicting rainfall. The purpose of this article is to investigate the use of artificial intelligence in rainfall prediction. We will evaluate different machine learning methods and algorithms, test their performance on real rain datasets and compare them with conventional meteorological models based on achieved results. Additionally, we will discuss the challenges faced when using machine learning for rain forecasting and give recommendations on how to improve accuracy as well as reliability in forecasts. This research was conducted so as to upgrade rain forecasting capabilities for people working in weather sensitive industries who need such information for decision making

Our projects are organized as follows: A comprehensive assessment of the literature is given in Section 2. A brief synopsis of the proposed methodology, including explanations of the model and data and the dataset preprocessing techniques used in this study, is provided in Section 3. Section 4 presents in-depth discussions of the findings. The study's main conclusions and findings are finally outlined in Section 5.

FIgure 1 showcases the leading roles of France, Italy, Spain, Chile, and Australia in global wine exports. Together they form a quintet that orchestrates the rhythm of international wine trade. Their combined efforts underline not only their dominance but also their ability to satisfy an ever-growing thirst for premium wines worldwide.

As if each bottle were a note played by an expert musician, these nations have mastered the artistry required to appeal to both connoisseurs and casual drinkers alike [4]. The certification of wine quality acts as a keystone in this harmonious structure; it ensures consumer trust remains intact while propelling industry expansion and cementing these countries' positions within the bustling marketplace.

2 LITERATURE REVIEW

Prior studies, including the work by Weesakul *et al.* [6], have concentrated on rainfall forecasting with techniques like the ARIMA model. This model has been used to estimate annual rainfall by using rainfall data from previous years as input variables. It is well-suited for predicting time-series data.

In their efforts to predict fresh satellite photos, Klein *et al.* [7] used a Dynamic Convolutional Layer model. The approach they adopted involved utilizing three satellite photographs as input data. Unlike regular convolutional layers where weights are randomly assigned, the dynamic convolutional layer uses the input image that has already gone through convolu-tional network processing as a filter. This technique allows the dynamic convolutional layer to potentially find the minimum loss faster thereby speeding up training convergence.

Huang *et al.* [8] collected data every hour from 75 rain gauge stations located around Bangkok during a four year period. They used this large dataset size in order to build a neural network model that estimates rainfall within one hour up till six hours later.

Zaytar *et al.* [9] employed multi-stacked Long Short-Term Memory (LSTM) networks to predict weather conditions after 24 hours as well as 72 hours respectively. In their study, they considered humidity levels, wind speeds and temperatures among other factors. They worked with a dataset consisting of hourly meteorological records covering nine Moroccan cities between 2000 and 2015 over fifteen years' time.The authors recommended applying deep LSTM networks for other climate-related problems having found them successful at forecasting. For their rainfall prediction model, Khajure *et al.* [10] combined a fuzzy inference system with a neural network (NN). Initially, a NN was used to predict weather characteristics. Following that, a fuzzy inference system received the projected values and used those values to forecast rainfall by applying established rules. The authors came to the conclusion that combining a NN with a fuzzy inference system produced good prediction outcomes. Wahyuni *et al.* [11] have reported on the efficacy of a fuzzy inference system for rainfall prediction.

Khajure *et al.* [10] developed their rainfall prediction model by integrating fuzzy inference system with neural network (NN). First, NN was used to predict weather parameters. Then, projected values were fed into a fuzzy inference system which used them to forecast rain based on predetermined rules. The researchers found that combination of NN and fuzzy inference system gave better predictions. Wahyuni *et al.* [11] discussed the effectiveness of using fuzzy logic based system in predicting rainfall. Rain effect on superhigh frequency diffusion was taken a look at by [12], paying specific focus to rain's local results on millimeter-wave diffusion. The research study was targeted at examining local result of rainfalls on millimeter-wave diffusion. To do this the author used a rainfall discovery plus transmission design called Tokyo Technology network based upon millimeter waves which were gotten at various factors within the school throughout stormy period. According to the results, it was disclosed that neighborhood habits of rains created significant attenuations in rainfalls along millimeter wave courses therefore influencing range covered by them along with their prices. Artificial neural networks (ANN) have been shown to be a very useful tool for rainfall prediction in recent researches undertaken in different applications and methods. For example, one noteworthy work was done by [13] where the authors developed a rainfall simulation model based on artificial neural networks (ANNs). This made it possible to predict with accuracy the patterns of rain including when and where they would occur. Later on, [14] built upon this idea but now used various lag factors within their ANN models while focusing only on short term predictions for urban catchments' rainfall. The results brought out the importance of choosing appropriate lags in achieving good performance as they were able to correctly forecast rains indices using an ANN with reduced lag. [15] took this further by adding resilient propagation learning techniques into daily rainfall forecasting through ANNs. Their method performed better than traditional regression models in terms of forecasting accuracy indicators according to a wide range of numerical evaluations carried out by these researchers. These are two examples among many others that show how flexible yet robust can be artificial neural networks (ANNs) when dealing with different aspects of predicting rain thus giving insights that can be applied in weather related studies or even development of better forecasting approaches.

3 METHODOLOGY

3.1 *Data source and description*

We are using an atmospheric dataset from the UCL Machine Learning Repository for our research on forecasting rainfall. The machine learning community heavily relies on this resource because of its large number of datasets available. For our study, we intentionally selected a dataset which has various features like temperature fluctuations, maximum temperature and air pressure among other relevant metrics. It is worth noting that within the dataset "Rainfall" acts as our key dependent variable around which all inquiries revolve. This variable's representation is shown in Figure 1 with an emphasis on how critical it is to our predictive modelling efforts.

	day	pressure	maxtemp	temparature	mintemp	dewpoint	humidity	cloud	rainfall	sunshine	winddirection	windspeed
0	1	1025.9	19.9	18.3	16.8	13.1	72	49	yes	9.3	80.0	26.3
1	2	1022.0	21.7	18.9	17.2	15.6	81	83	yes	0.6	50.0	15.3
2	3	1019.7	20.3	19.3	18.0	18.4	95	91	yes	0.0	40.0	14.2
3	4	1018.9	22.3	20.6	19.1	18.8	90	88	yes	1.0	50.0	16.9
4	5	1015.9	21.3	20.7	20.2	19.9	95	81	yes	0.0	40.0	13.7

Figure 1. Proposed architecture.

Raw data—that is, data taken from primary sources—usually needs to be thoroughly preprocessed in order to make sense of it or for modeling to be done. This preprocessing step, sometimes referred to as data cleaning, comprises a number of actions meant to improve the quality of the data. These tasks could involve identifying and eliminating outliers, imputation of null values, and correction of any input data anomalies. This dataset provides insights that closely align with real-world observations, guaranteeing that the analysis and modeling we do in the future are based on accurate and dependable data.

The principle aim of this paper therefore lies in looking into Machine Learning Techniques for Rainfall Prediction with regard to different categorization methods. To keep ourselves aligned with this goal, we have chosen three algorithms that will be used under supervised learning framework. Firstly, Support Vector Classifier (SVC) [16] was picked as algorithm number one and then followed by XGBClassifier and Logistic Regression [17], SVC [16]. Figure 2 presents a suggested architecture of a ML-based rainfall forecasting system along with techniques and approach summary employed during the investigation.

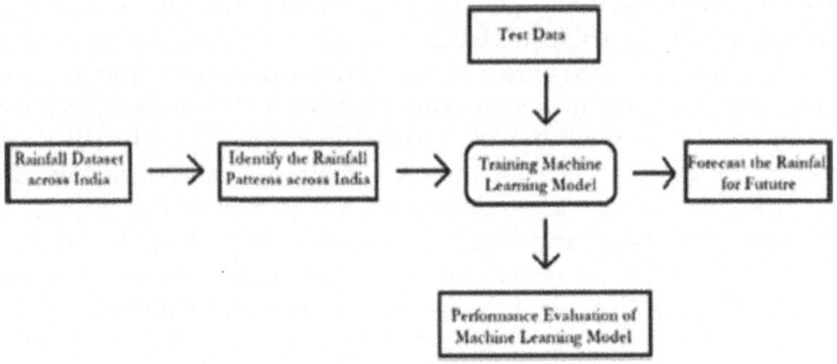

Figure 2. Architecture for predicting rainfall.

Explored Data Evaluation (EDA) entails methodically evaluating details using aesthetic devices. Its main goal is to determine patterns patterns as well as verifying presumptions making use of analytical recaps and also visuals depictions [18] In this component we will certainly describe different techniques made use of for establishing skewness or discrepancy in information which are important actions in the direction of comprehending distributional qualities that underscore any kind of provided information. With EDA, we likewise anticipate to expose some possible problems like unbalanced courses while dealing with our collections; therefore allowing us pick ideal designs for succeeding evaluation phases such as attribute design & & choice plus formula layout optimization based upon such searchings for.

Important information about the central tendencies and spread of variables is given by figure four of this dataset continuous feature distribution. On the other hand, Figure 3 represents boxplots for continuous features, making it easier to detect outliers in a dataset. These visuals are necessary because they help us understand how data is distributed and where we might find some abnormal points during data cleaning or modeling stages later on.

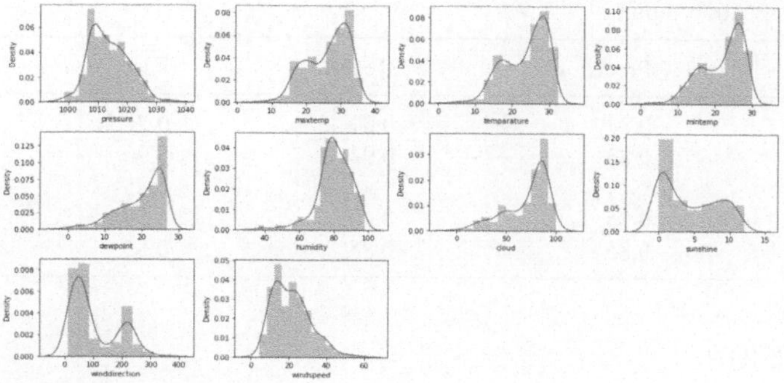

Figure 3. Distribution plot for the columns with continuous data.

4 RESULTS AND DISCUSSION

Currently, accurately predicting whether rainfall will occur on a given day remains a challenging task, with even meteorological department predictions occasionally proving unreliable. This article seeks to tackle this challenge by introducing a machine learning model specifically designed to forecast daily rainfall based on various atmospheric factors. This problem falls within the domain of Rainfall Prediction using Machine Learning, as machine learning models have demonstrated enhanced performance on tasks that traditionally relied on significant expertise to accomplish. Moving forward, we will proceed to train advanced classification models using our training data, aiming to improve the accuracy and reliability of rainfall forecasts.

Tables 1 and 2 provides a detailed overview of the performance metrics achieved during the training phase of the machine learning models employed in this study. The table showcases the training performance of each model, offering insights into their ability to correctly classify instances of rainfall occurrence and non-occurrence based on the provided weather data. Additionally, it facilitates comparison between different models, allowing for the identification of the most effective approach in terms of predictive performance during the training phase and validation phase.

Table 1. Model training performance.

Model	Training Performance %
LogisticRegression	88.93
XGBClassifier	99.03
SVC	90.06

Table 2. Model validation performance.

Model	Validation Performance %
LogisticRegression	89.66
XGBClassifier	84.08
SVC	88.58

Training accuracy of Logistic Regression slightly differs with that of Support Vector Classifier (SVC) by just a small margin which indicates both models' performance being good enough as per earlier stated accuracy values. In subsequent steps, we will employ SVC model to visualize confusion matrix for validation data so that it can provide additional insights into how well the model works on categorizing rainfall instances.

The Table 3 shows the report of classification of support vector classifier used on validation data. It summarizes briefly the performance of the model in a binary classification scenario, where 0 and 1 represent the classes. This classifier does well as it has higher precision, recall and F1-score for class 1 than for class 0. The overall accuracy being 85% means that the model is able to make right predictions on most occasions when applied to validate data sets.

Table 3. Classification report.

	Precision	Recall	F1-score	Support
0	0.84	0.67	0.74	24
1	0.85	0.94	0.90	50
accuracy			0.85	74
macro avg	0.85	0.80	0.82	74
weighted avg	0.85	0.85	0.85	74

5 CONCLUSION

To sum up, this work focused on applying machine learning techniques to predict rainfall, which is important for industries like agriculture and disaster preparedness. We created a machine learning model that uses meteorological data to predict rainfall, and the results look good. Strong performance was shown by algorithms such as Support Vector Classifier and Logistic Regression, and our comprehension of the model's performance was improved by confusion matrix visualizations. All things considered, our work advances the prediction of rainfall, which helps numerous weather-related fields make better decisions.

REFERENCES

[1] Barrera-Animas, Ari Yair, Lukumon O. Oyedele, Muhammad Bilal, Taofeek Dolapo Akinosho, Juan Manuel Davila Delgado, and Lukman Adewale Akanbi. (2022). Rainfall prediction: A comparative analysis of modern machine learning algorithms for time-series forecasting. *Machine Learning with Applications* 7: 100204.

[2] Rahman, Atta-ur, Sagheer Abbas, Mohammed Gollapalli, Rashad Ahmed, Shabib Aftab, Munir Ahmad, Muhammad Adnan Khan, and Amir Mosavi. (2022). Rainfall prediction system using machine learning fusion for smart cities. *Sensors* 22, no. 9: 3504.

[3] Praveen, Bushra, Swapan Talukdar, Shahfahad, Susanta Mahato, Jayanta Mondal, Pritee Sharma, Abu Reza Md Towfiqul Islam, and Atiqur Rahman. (2020). Analyzing trend and forecasting of rainfall changes in India using non-parametrical and machine learning approaches. *Scientific Reports* 10, no. 1: 10342.

[4] Bochenek, Bogdan, and Zbigniew Ustrnul. (2022). Machine learning in weather prediction and climate analyses—applications and perspectives. *Atmosphere* 13, no. 2: 180.

[5] Cifuentes, Jenny, Geovanny Marulanda, Antonio Bello, and Javier Reneses. (2020). Air temperature forecasting using machine learning techniques: a review. *Energies* 13, no. 16: 4215.

[6] Salcedo-Sanz, Sancho, Jorge Pérez-Aracil, Guido Ascenso, Javier Del Ser, David Casillas-Pérez, Christopher Kadow, Dušan Fister *et al.* (2024). Analysis, characterization, prediction, and attribution of extreme atmospheric events with machine learning and deep learning techniques: a review. *Theoretical and Applied Climatology* 155, no. 1: 1–44.

[7] Klein, L. Wolf and Afek Y., (2015). A dynamic convolutional layer for short rangeweather prediction, *2015 IEEE Conference on Computer Vision and Pattern Recognition (CVPR)*, pp. 4840–4848.

[8] Hung, N.Q.; Babel, M.S.; Weesakul, S.; Tripathi, N.K. (2009). An artificial neural network model for rainfall forecasting in bangkok, thailand. *Hydrol. Earth Syst. Sci.* 13, 1413–1425.

[9] Khajure, S. and Mohod, S.W. (2016). Future weather forecasting using soft computing techniques. *Procedia Comput. Sci.* 78, 402–407.

[10] Schultz, Martin G., Clara Betancourt, Bing Gong, Felix Kleinert, Michael Langguth, Lukas Hubert Leufen, Amirpasha Mozaffari, and Scarlet Stadtler. (2021). Can deep learning beat numerical weather prediction?. *Philosophical Transactions of the Royal Society A* 379, no. 2194: 20200097.

[11] Basha, Cmak Zeelan, Nagulla Bhavana, Ponduru Bhavya, and Sowmya V. (2020.B). Rainfall prediction using machine learning & deep learning techniques. In 2020 international conference on electronics and sustainable communication systems (ICESC), pp. 92–97. IEEE.

[12] Rehman, Abdur, Sharjeel Tariq, Amina Farrakh, Munir Ahmad, and Muhammad Sheraz Javeid. (2023). A systematic review of machine learning and artificial intelligence methods to tackle climate change impacts. *In 2023 International Conference on Business Analytics for Technology and Security (ICBATS)*, pp. 1–7. IEEE.

[13] Wahyuni, I.; Mahmudy, W.F. and Iriany, A. (23–24 August 2016). Rainfall prediction in tengger region indonesia using tsukamoto fuzzy inference system. *In Proceedings of the International Conference on Information Technology, Information Systems and Electrical Engineering (ICITISEE)*, Yogyakarta, Indonesia; pp. 130–135

[14] Sutskever, I.; Vinyals, O. and Le, Q.V. (8–13 December 2014). Sequence to sequence learning with neural networks. *In Proceedings of the 2014 Neural Information Processing Systems(NIPS)*, Montreal, QC, Canada, pp. 3104–3112.

[15] Le, Hung Vu, *et al.* (2013). Localized rain effects observed in Tokyo Tech millimeter-wave model network. Electromagnetic Theory (EMTS), *Proceedings of 2013 URSI International Symposium on*. IEEE.

[16] Luk, Ka Chun, James E. Ball, and Ashish Sharma. (2000). A study of optimal model lag and spatial inputs to artificial neural network for rainfall forecasting. *Journal of Hydrology* 227, no. 1–4: 56-65.

[17] Ramirez, Maria Cleofe Valverde, Haroldo Fraga de Campos Velho, and Nelson Jesus Ferreira. (2005). Artificial neural network technique for rainfall forecasting applied to the Sao Paulo region. *Journal of Hydrology* 301, no. 1–4: 146–162.

[18] Bari, Sheikh Hefzul, M. T. Rahman, M. M. Hussain, and Sourav Ray. (2015). Forecasting monthly precipitation in Sylhet city using ARIMA model. *Civil and Environmental Research* 7, no. 1: 69–77.

Data Science & Exploration in Artificial Intelligence – Gururaj et al. (Eds)
© 2025 the Author(s), ISBN: 978-1-032-96390-7

ODH with V3A (on-demand highlight generation with video action and activity analysis)

A. Mallinamadugu, B.S. Prajwal, P. Kalwad, R. Naik & R. Srinath
Department of Computer Science, P.E.S. University, Bangalore, Karnataka, India

ABSTRACT: This paper presents a fresh approach to analysing sport-based actions and activities to identify significant moments based on predefined queries during a cricket game. The objective of our endeavour is to automatically identify key events and objects on the match field, thereby extracting relevant match segments in the form of highlights, which are customisable based on individual user preferences through their queries. Our approach consists of three interrelated components: player classification to identify various human participants in the match, batting shot detection to identify the broad type of shot played by the batsman, and umpire gesture detection to recognise the event based on simple umpire gestures. Our method uses a state-of-the-art deep learning model called YOLOv8 to automatically label match segments to increase the speed and accuracy of detecting items and events on screen. It also enables users to selectively view specific highlights, changing the post-match viewing experience. By automating this process of highlight generation, we improve the cricket-watching experience for viewers and game analysers by introducing benefits such as increased viewer engagement, improved content personalisation, and enhanced post-match analysis capabilities.

1 INTRODUCTION

Cricket today, is one of the most popular sports in the world and has captured the hearts of billions of enthusiasts worldwide. The passion for cricket among fans, especially in South Asia, is comparable to religious devotion. Some India vs Pakistan matches, especially during the World Cup drew a staggering viewership, with over a billion people tuning in simultaneously.

Cricket saw massive growth in the era of mass consumption of digital media. The enormous and energetic fanbase, the engaging nature of the game and the spirit of sportsmanship among the players were some of the reasons behind cricket's success. Cricket is also known as the "Gentleman's game" due to the gentlemanly nature of play. Phrases like "the spirit of cricket" also signify the cultural importance of the game.

Given the enormous fanbase and their evolving tastes, delivering material that caters to a wide range of viewer preferences is an area with a rapidly growing demand. Traditional broadcasting systems are slowly getting challenged as cricket fans are getting more knowledgeable about the sport. In the recent past, OTT platforms with customisable camera angles and commentary languages have started to come into the market. These trends point to a growing desire among viewers for more customisation.

Because of cricket's increasing popularity, there is a growing need for more controllable ways to summarise significant moments from matches. Today, highlights are generated and hosted on television channels, OTT platforms, tournament websites, etc. However, there is a major flaw in how the highlight clips are curated. Often, cricket highlight videos do not cover all the important shots played during the game and are shortened to keep the total duration of the resulting video within certain limits. Moreover, traditional highlights tend to cover only the significant events during the game such as boundaries or wickets. However, certain segments that do not fall under the above categories may be equally entertaining to specific users. For example, in test games certain fans might want to see elegant cover drives, unorthodox shots,

 DOI: 10.1201/9781003589273-32

or even no-balls. Today, certain shots have been attributed to players who have perfected their technique. Rewatching classic and memorable clips with iconic game moments is becoming increasingly popular. For example, the International Cricket Council or I.C.C. termed a shot by Virat Kohli in a recent India vs. Pakistan game as the "shot of the century". These shots are not special just because of their underlying technique, but also because of the emotions attached to these matches. Hence, there is an increasing desire among cricket spectators for more control over the segments to view as part of the overall highlights.

Our study presents a unique method, On Demand Highlight Generation with Video Action and Activity Analysis (ODH with V3A), to this requirement. By using leading-edge computer vision and machine learning techniques to automatically detect important objects in specific scenes and events within cricket matches, this novel approach has the potential to change the way people experience cricket on a screen. To accomplish the same in our project, we are utilising the capabilities of YOLOv8, one of the latest object detection models.

Manual highlight generation is a laborious task for organisations that oversee the curation of appropriate video segments. Moreover, it is highly subjective, frequently omitting moments that have a massive impact on the game. By building an effective mechanism to recognise significant events and behaviours during a match, we aim to close this gap by offering viewers more control over the highlights they wish to watch. In addition to providing entertainment to fans, automated highlight generation can benefit coaches, players, and commentators. The capacity to swiftly examine and evaluate pivotal occasions can provide insights for strategic planning.

Computer vision is a broad discipline within the realm of machine learning that enables computers to process visual data and extract information from sources such as images and videos. This domain has the potential to entirely transform how computers perceive, understand, and interact with visual data. Advanced concepts such as action and activity analysis in sports, especially cricket, have been made possible by computer vision. This means having the ability to automatically identify and decipher different aspects of a game, like game restrictions, player motions, ball trajectories and significant moments. Using advanced models such as YOLOv8, real-time analysis could be implemented to offer a refined comprehension of the intricate and fast-paced nature of cricket matches.

Each part of the system's complex architecture, consisting of three interrelated components, plays a crucial role in gathering contextual information about the match before curating the segments and generating the final highlight videos. The first component, player classification, uses deep learning to precisely recognise and distinguish between the various human participants on the field, such as fielders, umpires, batsmen, bowlers, and wicket-keepers. For the sake of simplicity, we have not taken on-ground spectators as a class as they are not present in most shots, and where they are present, they seem insignificant.

The objective of ODH with V3A second component is batting shot identification. This part takes advantage of YOLOv8's inherent ability to distinguish between actions and activities in still images, to differentiate between different batting strokes. For simplicity, we have selected five broad categories of shots for our project: the cut shot, pull shot, drive shot, defensive shot, and sweep shot. The motivation behind this component is that the type of shot being played might provide additional contextual information about the ball's trajectory, potential immediate results following the shot being played, and other match-related details that can add value to the program. For example, a defensive shot is not very likely to be followed by a boundary. Batting shot identification also provides complexity and meaning to the highlight-generation process by capturing the subtleties of batting technique and strategy, allowing viewers to relive their favourite shots during a match. This also opens new avenues for critical analysis of the techniques underlying these shots which can further be used to improve the form of players by identifying their weaknesses.

Umpire gesture detection, the final component of this project, completes the purpose by identifying relevant and noteworthy events that umpires indicate throughout a match. We knew that exploiting scenes involving umpires is crucial because their motion is easy to understand, and these simple movements give information for all the frames corresponding to the segment since the ball was bowled and the batsman played the shot. These actions can be classified as gestures as they involve only a few body parts, last for only a few seconds and do not alter much during the action. This component incorporates the event-related contextual information into the match segment by cleverly identifying umpire motions.

2 LITERATURE REVIEW

The domain of sports analytics has seen significant growth in the last few years with the integration of sophisticated computer vision techniques and machine learning algorithms. With the advent of advanced deep learning techniques and methodologies, researchers are increasingly focusing on building intelligent systems capable of analysing and summarising sporting matches. Methodologies for analysing sports events include object detection, action recognition, gesture recognition, activity analysis, sports analysis, event summarisation, etc. However, despite these advancements, a gap exists regarding personalised content delivery and user interaction.

Amidst the existing literature, there are notable ambiguities surrounding the distinction between action analysis and activity analysis. In the paper Video Action Understanding (Hutchinson and Gadepally 2021), the authors prefer to use the term action instead of activity in all cases since it is used more dominantly across the domain. They also define an action as something that covers both human and non-human elements, and activity to be mostly biased towards human instances. (Aggarwal and Ryoo 2011) broadly classified human activities into four levels: gestures, actions, interactions, and group activities based on their complexities and the number of actors involved. They considered action as a single-person process. (Liu *et al.* 2016) define actions as primitive movements that form more complex activities, which represent human behaviour. They also point out that humans perform simple actions in a variety of combinations. For the sake of clarity, we adopt the following distinction: action analysis is the study of the recognition, classification and localisation of specific simple actions or gestures at an individual level, within a particular scene. These could be actions like walking, waving, sitting, etc. On the other hand, activity analysis aims to understand the overall behaviour among one or many players in each scene, encompassing multiple actions, interactions, and events over time. It aims at higher-level concepts like intent, context and situational awareness and tries to understand the relationships among various individuals, objects, and environmental factors at play.

To gain a comprehensive understanding of the domain, some review papers were consulted providing us a thorough overview of the current landscape in action recognition and human activity recognition. A comprehensive review of group activity recognition in videos (Wu *et al.* 2021) provides a deep overview of human group activity recognition (GAR), including comparisons of methods, descriptions of datasets, and the shift from manual characteristics to deep learning networks. Given the presence of multiple actors on the field in a cricket game, the analysis and detection of activities from the perspective of a group was important to understand. A review of deep learning-based human activity recognition on benchmark video datasets (Sharma *et al.* 2022) provides a comparative analysis of vision-based human activity recognition methods, with a particular focus on deep learning techniques applied to benchmark video datasets. Their proposed taxonomy categorises the literature into CNN and RNN-based approaches, offering valuable insights into the evolving landscape of HAR research and highlighting future research directions and challenges. A survey of video datasets for human action and activity recognition (Chaquet *et al.* 2013) addresses the lack of a comprehensive overview of public video datasets tailored for human action and activity recognition, offering insights into 28 heterogeneous datasets. While we created our dataset capturing cricket-related actions, the study underscores the importance of understanding benchmark datasets and methodologies employed in action-understanding research. This comprehensive understanding is crucial for informing the development and evaluation of novel methodologies and models. A Bayesian approach to human activity recognition (Madabhushi & Aggarwal 2003) uses a Bayesian approach to understand and characterise human actions. They track the movement of the head of the subject in frontal and lateral views to make an understanding of the actions being performed.

Several attempts have been made to extract information more efficiently from video frames in cricket broadcasts. Event detection and summarization of cricket videos (Nasir *et al.* 2018) uses textual information present on screen in the form of score captions, wicket counters, etc to extract information about the changes since the previous ball was bowled. Frames are marked as key frames if a significant event such as a six, four or wicket had taken place in that segment. Optical Character Recognition was used to retrieve the text-based information present within frames. Similarly, Deep learning-based automated sports video summarisation using YOLO (Guntuboina *et al.* 2021) uses the scoreboard detection method to identify key events across multiple sports. It used YOLO to identify regions of the scoreboard in the image of a match scene and later used OCR on the identified region to extract the relevant information. Cricshotclassify (Sen *et al.* 2021) used a custom architecture comprising a combination of CNN and RNN to classify 10 unique cricket batting shots. The authors used transfer learning and compared their results across many RNN models and types of convolutions. Cricket activity detection (Kumar *et al.* 2014) uses the broadcasted matches to identify the type of batting shot being played, the direction of the shot, and the segments where the stroke was made using optical flow analysis. The paper also discusses the analysis of the various scenes captured by the camera such as the crowd scene, the pitch view, commercial breaks, etc.

ODH with V3A uses YOLOv8, a state-of-the-art algorithm in computer vision, to classify and analyse player motions to extract useful information about the context of the current circumstances in the match. Initially, our approach aimed to identify contextual information within individual frames and changes across consecutive frames to extract match-related insights. However, insights from "YOLO-based Human Action Recognition and Localization" (Shinde *et al.* 2018) reshaped our perspective. According to this research, human observers require just a single instance of visual data for scene recognition, indicating that even small groups of frames, or even a single frame is enough for precise recognition. Hence, our methodology evolved to leverage frame-level features. We focused on segmenting the match into distinct scenes, such as the ball being bowled, the batsman playing the shot, and the umpire signal following the action. Within each segment, spanning only a few frames, we aggregated predictions across all frames to discern the type of shot played or umpire gestures displayed, thus enabling our predictions.

The uniqueness and novelty of our method are increased by the usage of the latest model from the YOLO family known as YOLOv8, which is used for the localisation and classification of entities, actions, and activities in the context of a match video. We accomplish fast and accurate detection by utilising YOLO's sophisticated capabilities. Another reason for our inclination to YOLOv8 was that it was still unexplored because of its recency. In fact, at the time of writing this paper, there is no official paper on YOLOv8. Hence, we wanted to explore its architecture and present a brief about the same as part of our work. To carry out effective model training, there is a need for large amounts of quality data instances, which in our case are cricket match video segments. For this reason, we developed a proprietary dataset of cricket match scenes, sourced from websites such as YouTube. We downloaded full-length matches and trimmed them to get a set of typical yet diverse shots. This dataset was gathered from a variety of matches at various eras of cricket and was also made to be as diverse as possible. Therefore, it served as an important requirement to get competitive results.

3 OVERVIEW OF TECHNIQUES

3.1 *YOLO*

YOLO, which stands for "You Only Look Once," is a group of object detection techniques that have become well-known in the computer vision community in recent years. The main factor contributing to YOLO's success is that it completes object detection in a single inference step, given its high speed and accuracy. Some of the reasons why YOLO can process and identify objects in real-time are its usage of anchor boxes, post-processing techniques like non-max suppression, processing using a single forward pass, lesser model complexities, etc

The idea of an end-to-end object detection system using a single neural network was first presented in the paper "You Only Look Once: Unified, Real-Time Object Detection" (Redmon *et al.* 2016). This model, also known as YOLOv1, was the first of its kind. The methodology used in this technique was to divide the image fed into the model into a grid and predict certain components for each grid cell: class probabilities and an objectness score – a confidence score representing the likelihood of an object present for each grid cell. Each grid cell also predicts multiple bounding boxes using a regression algorithm. The model architecture for YOLOv1 is illustrated in Figure 1.

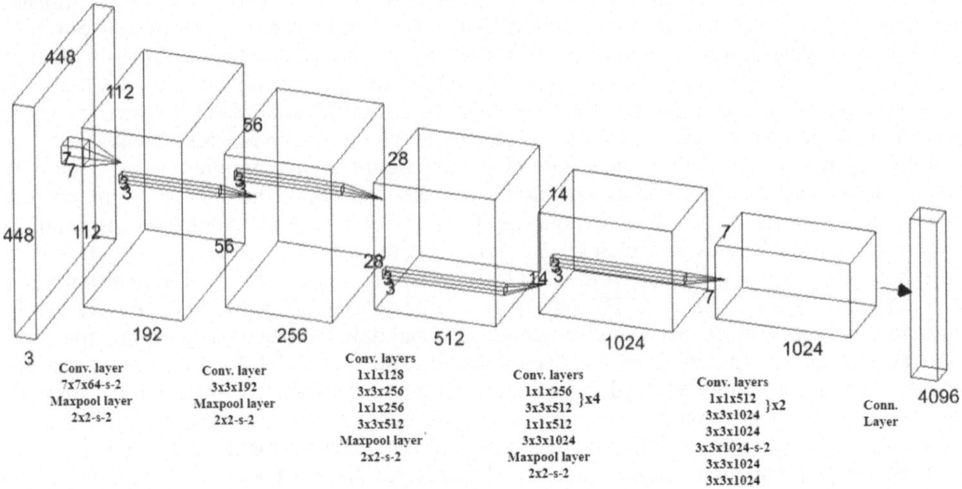

Figure 1. The model architecture of YOLOv1.

The most notable attribute of this collection of models lies in its ongoing process of improvement and refinement. The subsequent iterations since YOLOv1 introduced radical changes in the model architecture, evaluation techniques, data augmentation techniques, loss functions, regularisation techniques and optimisation algorithms. YOLOv2 (Redmon and Farhadi 2017) introduced batch normalisation, anchor boxes, high-resolution classification, dimension clusters and a new Darknet-19 model architecture, which was deeper than the original Darknet used by YOLOv1. Moreover, YOLOv2 was capable of hierarchical classification. YOLOv3 (Redmon and Farhadi 2018), which gained widespread community adoption, was significant in object detection. Some of the changes it incorporated were a new feature extraction method defined by Feature Pyramid Network (FPN) (Liu *et al.* 2022) where bounding boxes were predicted at 3 different scales, updated loss functions, data augmentation techniques and a new model architecture called Darknet-53 which was deeper than the previous Darknet-19. The architecture for YOLOv3's feature extractor is presented in Table 1. Updates in YOLOv4 (Bochkovskiy *et al.* 2020) included changes in the model structure - an improved backbone structure called CSPDarknet53 after experimentation with other backbones and a new neck for feature aggregation. Other changes included a "bag of freebies" referring to a set of techniques aimed at increasing the performance and efficiency of YOLO while simultaneously not increasing the time taken for inference, a "bag of specials" which marginally increases inference time but brings significant change to the performance. YOLOv4 also introduced scaled versions of the model (Wang *et al.* 2021), which could be decided based on the computational limitations and accuracy requirements.

Table 1. The darknet-53 feature extractor in YOLOv3.

Repetitions	Type	Filters	Size	Output
	Convolutional	32	3 x 3	256 x 256
	Convolutional	64	3 x 3 / 2	128 x 128
{	Convolutional	32	1 x 1	
1x{	Convolutional	64	3 x 3	
{	Residual			128 x 128
	Convolutional	128	3 x 3 / 2	64 x 64
{	Convolutional	64	1 x 1	
2x{	Convolutional	128	3 x 3	
{	Residual			64 x 64
	Convolutional	256	3 x 3 / 2	32 x 32
{	Convolutional	128	1 x 1	
8x{	Convolutional	256	3 x 3	
{	Residual			32 x 32
	Convolutional	512	3 x 3 / 2	16 x 16
{	Convolutional	256	1 x 1	
8x{	Convolutional	512	3 x 3	
{	Residual			16 x 16
	Convolutional	1024	3 x 3 / 2	8 x 8
{	Convolutional	512	1 x 1	
4x{	Convolutional	1024	3 x 3	
{	Residual			8 x 8
	Avgpool		Global	
	Connected		1000	
	Softmax			

YOLOv5 came out within a short period after the release of YOLOv4, retaining most of the improvements. Most importantly, it was developed by Ultralytics in PyTorch rather than Darknet. YOLOv5, which is among the highly adopted YOLO models today, has five versions: YOLOv5n (nano), YOLOv5s (small), YOLOv5m (medium), YOLOv5l (large), and YOLOv5x (extra-large). Figure 2 shows a simplified form of the YOLOv5 architecture.

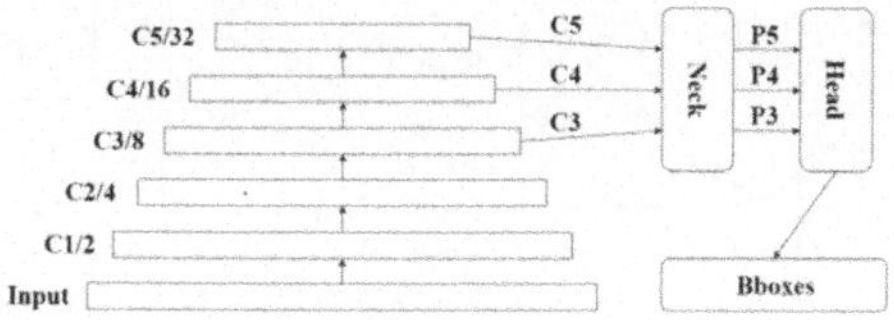

Figure 2. Simplified form of YOLOv5 architecture.

The latest updates to YOLOv6 (Li *et al.* 2022) include a novel backbone architecture built on RepVGG, advancements in label assignment techniques, implementation of new classification and regression loss functions, the integration of a self-distillation strategy for enhanced performance, and the introduction of a quantisation scheme to improve model efficiency. YOLOv7 (Wang *et al.* 2023) brought about further changes such as Extended Efficient Layer Aggregation Network (E-ELAN), methods for model scaling, planned reparameterised convolution, etc.

YOLOv8, developed by Ultralytics using PyTorch, is a convolutional neural network (CNN) architecture that uses a novel anchor-free approach for object detection. Unlike previous YOLO versions that rely on predefined anchor boxes to predict object locations, YOLOv8 directly predicts the bounding boxes and class probabilities of objects without anchor boxes. This anchor-free design significantly improves the model's accuracy and efficiency. YOLOv8 had five scaled versions – YOLOv8n (nano), YOLOv8s (small), YOLOv8m (medium), YOLOv8l (large), YOLOv8x (extra-large). The utilisation of YOLOv8 was motivated by its pose estimation capabilities. Hence, YOLOv8 proved to be advantageous for detecting batting shots. Since YOLOv8 is a one-stage detector, it can predict class probabilities and the coordinates for bounding boxes from the input image alone. Three primary components make up YOLOv8 architecture:

3.1.1 *The backbone*

The first component of YOLOv8 unique architecture is its backbone. Layers of CNNs in the backbone extract more abstract and complex features in deeper layers. YOLOv8 also uses the c2f module (cross-stage partial bottleneck with two convolutions), a modified version of the CSPDarknet53 backbone that is renowned in feature extraction.

3.1.2 *The neck*

The second component, known as the neck, oversees combining features retrieved from the various levels of the backbone. A Path Aggregation Network (PAN) is used by YOLO v8 to integrate features from the backbone's high, medium, and low levels.

3.1.3 *The head*

Predicting bounding boxes and class probabilities is the responsibility of the third component, called the head. The YOLOv8 head is a modified variant of the YOLOv5 head, which predicts bounding boxes using an anchor-free method.

YOLOv8 uses the sigmoid activation function for the objectness score in the output layer. The objectness score of the prediction bounding box represents the likelihood that it includes an item within its bounds. As a loss function, binary cross-entropy is used to minimise the classification loss and the CIoU (Complete Intersection over Union) (Zheng *et al.* 2020) and DFL (Distributed Focal Loss) (Li *et al.* 2020) loss functions are used to optimise the bounding box loss.

4 METHODOLOGY

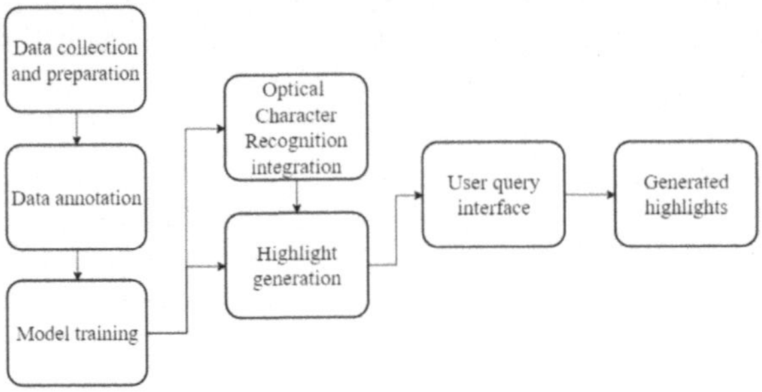

Figure 3. Architecture for the proposed system.

4.1 *Data collection and preparation*

This section details the methods and techniques we used towards our effort to create this project. Figure 3 shows the workflow used during the development of this project. To conduct the training and validation of the YOLOv8 model, we have carefully selected a heterogeneous dataset of cricket match footage that we obtained from a variety of web sources, such as YouTube. The term 'heterogeneous' is because of its diverse nature since a vast range of cricketing scenarios are covered by this dataset. This includes matches from various decades, video standards, resolutions, and match formats (e.g., Tests, ODIs, T20s). Our dataset's intrinsic variety guarantees strong model performance in a wide range of actual cricketing situations. Our motivation behind creating a proprietary dataset came about due to a lack of such video-based datasets.

We utilised the Labelbox platform to effortlessly add bounding boxes corresponding to player types, batting shots, and umpire gestures throughout the video frames. Labelbox provides a user-friendly graphical interface and makes it easier to draw bounding boxes around objects in video frames efficiently by annotating frames at a periodic interval of frames. The frames in between are predicted by Labelbox's algorithm to swiftly fill them with bounding box annotations. Hence, bounding boxes across frames may move around the screen or even change dimensions to fit the figure of the object in question. To systematically annotate objects, we also built ontologies for dataset labels, such as player classes and batting shot classes.

4.2 *Model selection and training*

We produced the dataset in the directory format expected by Ultralytics YOLOv8. This included a directory containing image files and another containing text files with bounding box coordinates and class labels for the ground truths. The text and jpg files were also expected to have the same filenames. The dataset was further split into directories representing the train, validation, and test splits. For our project, we selected a ratio of 70:20:10 for train, validation, and test splits respectively. For optimal model performance, we varied the model's hyperparameters using discrete values and selected the best values.

4.3 *Additional tools*

In addition to the previous components, we extracted textual data from on-screen scoreboards using Optical Character Recognition (OCR). We successfully included OCR into our pipeline using OpenCV, which allowed us to extract important textual data such as player names, runs, and wickets. Information such as player names and match stats, which could not be retrieved using the earlier components are now available.

4.4 *Evaluation metrics*

The type of metrics we use during the validation step gives us an idea about how the model learns from the data instances and how well it performs on unseen data.

4.4.1 *Mean average precision (mAP)*

By averaging the Average Precision (AP) values over several object classes, mAP expands on the idea of AP, which is calculated as the area under the precision-recall curve at a certain confidence threshold. This is helpful to give a thorough assessment of the model's performance in multi-class object identification settings.

$$mAP = \frac{1}{k}\sum_{i}^{k} AP(i) \tag{1}$$

4.4.2 *Precision*

Precision measures the percentage of true positives among all positive predictions. This metric is required to reduce the rate of making positive predictions which should have been negative (false positives).

$$Precision = \frac{True\ positive}{(True\ positive + False\ positive)} \tag{2}$$

4.4.3 *Recall*
Conversely, Recall evaluates how many positive instances were predicted correctly among all classes. Hence, it aims to reduce the number of false negatives.

$$Recall = \frac{True\ positive}{(True\ positive + False\ negative)} \tag{3}$$

4.4.4 *F1-score*
The F1 Score weighs false positives and false negatives in its balanced evaluation of a model's performance. It is calculated as the harmonic mean of accuracy and recall. This score punishes extreme values in precision and recall ensuring an even trade-off between the two.

5 RESULTS

This section is intended to present the outcomes of our endeavour. The performance of YOLOv8 on the tasks corresponding to the three different components is recorded, summarised, and presented here. YOLOv8 consists of three loss functions. These are box loss, class loss and distribution focal loss. Model learning is marked by a reduction in the loss value, which calculates the error or difference between the actual ground truth and model predictions. However, there is still a chance that the model overfits the training data rendering it unable to make accurate predictions on unseen data. For this reason, we create another split for validation data. This data is unseen from the model's perspective as it is not used for training. We check the model's performance on the validation data after every epoch using a metric known as mean average precision or mAP. If the mAP increases over the long run, it means that the system is rightfully able to predict well over unseen data. Table 2 gives tabulated information on the mAP results and Figures 4–6 diagrammatically present the same. Figures 7–9 present the inference results for each task.

Table 2. mAP results for player classification, batting shot detection and umpire gesture detection.

	Player Classification		Batting Shot Detection		Umpire Gesture Detection	
Epoch	mAP50	mAP50-95	mAP50	mAP50-95	mAP50	mAP50-95
1	1.00e-5	0	0.0358	0	0.0021	0.0006
2	0.5861	0.2337	0.2846	0.1250	0.0130	0.0032
8	0.9329	0.6847	0.5987	0.4614	0.2778	0.1486
18	0.9764	0.7542	0.7583	0.6559	0.5438	0.3226
32	0.9874	0.8129	0.7895	0.7277	0.8693	0.6049

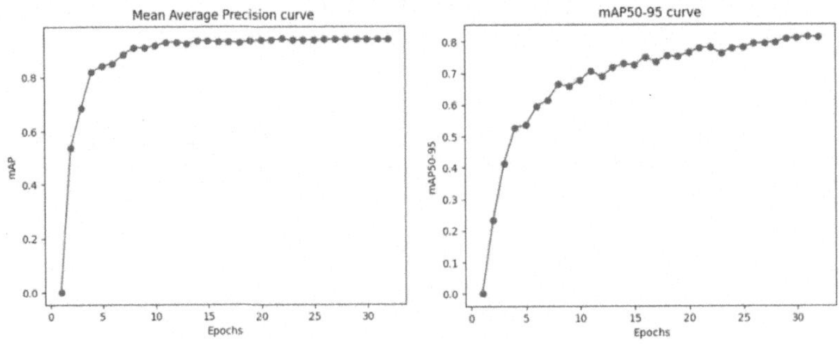

Figure 4. mAP50 and mAP50-95 curves for player classification.

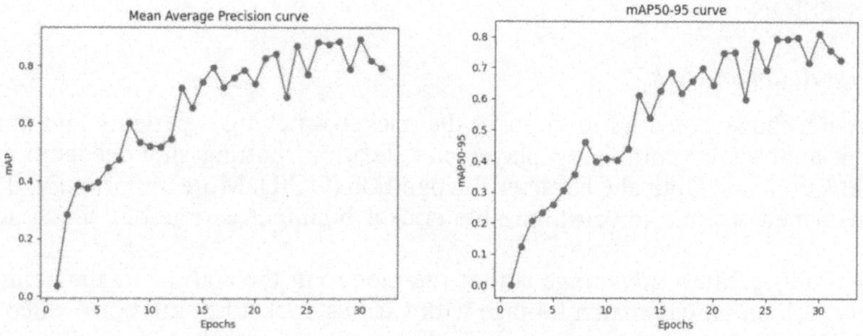

Figure 5. mAP50 and mAP50-95 curves for batsman shot detection.

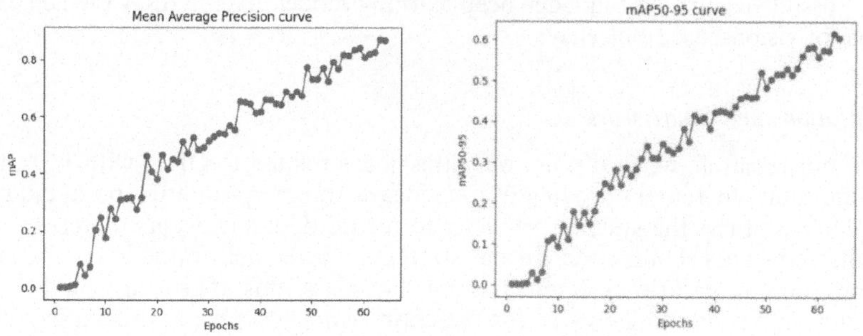

Figure 6. mAP50 and mAP50-95 curves for umpire gesture detection.

Figure 7. Inference results for player classification.

Figure 8. Inference results for batsman shot detection.

Figure 9. Inference results for umpire gesture detection.

6 DISCUSSION

6.1 *Interpretation of work*

Our approach shows potential to improve the cricket-watching experience and improve the post-match analysis by combining player classification, batting shot recognition, umpire gesture detection, and Optical Character Recognition (OCR). More importantly, this project also opens a new avenue for customisable cricket highlight extraction, which adds to its novelty.

Our method's greatest advantage is that the model, in the end, gains the ability to generalise very well on unseen match footage with various cricketing situations, video qualities, etc. The carefully curated dataset, which includes clips of matches from different periods, match formats, and video qualities, provides strong performance in real-world scenarios. Moreover, precise detection, localisation, and classification of objects, actions and activities are made possible using a cutting-edge deep learning model, like YOLOv8 which caters to a wide range of vision-based objectives.

6.2 *Limitations and future work*

As part of our research, we also point out other ideas related to ODH with V3A, meant to build on the same, thereby improving it. One idea is to incorporate temporal data into our pipeline. This increases the system's capacity to record dynamic events by recording action and activity sequences. Unlike the current strategy, where each frame has a prediction for player class, umpire gesture class and batting shot class, this approach would take match segment and output one single value for the entire sequence. One approach would involve using image embeddings from YOLOv8 backbone as inputs for recurrent neural network (RNN) architectures such as Long Short-Term Memory (LSTM). LSTMs are known for their prowess with sequential data and therefore enhance action and activity recognition. Another approach could be using advanced methods like camera view analysis and ball type classification based on trajectory variations. Examining ball motion, our application has the potential to offer a more comprehensive awareness of game results, player tactics, and match dynamics. We could improve our model even further by investigating methods for multi-modal fusion, which combines textual and visual data taken from match video.

7 CONCLUSIONS

This endeavour was about introducing a new method of extracting contextual information from cricket matches and generating highlights based on customised queries from users who want to retrieve segments based on their preferences. ODH with V3A does so by integrating three interrelated components, namely, player classification, batting shot detection and umpire gesture recognition.

To get the best results, we used the best and most recent model in computer vision today, known as YOLOv8. Its use resulted in an accurate detection of objects, actions, and activities on the field. Our methodology also includes data-oriented tasks such as the curation, cleaning, and annotating of match video samples. Apart from model training, validation, testing and optimisation, our approach used other features such as optical character recognition to understand the textual information present on the screen. The results verify the success of our method as the system can effectively extract the required information and make predictions for individual segments of the match footage.

ODH with V3A is a unique addition to the domain of sports analysis and highlights generation and serves as an advantage to viewers and analysers alike. This technology can be integrated with existing OTT platforms looking to improve the viewing experience of the sport. By automating this process of highlight generation, we hope to contribute to a massive change in how cricket is viewed, understood and enjoyed.

REFERENCES

Aggarwal, J.K. and Ryoo, M.S., (2011). Human activity analysis: A review. *Acm Computing Surveys (Csur)*, 43(3), 1–43.

Bochkovskiy, A., Wang, C.Y. and Liao, H.Y.M., (2020). Yolov4: Optimal speed and accuracy of object detection. *arXiv preprint* arXiv:2004.10934.

Chaquet, J.M., Carmona, E.J., and Fernández-Caballero, A., (2013). A survey of video datasets for human action and activity recognition. *Computer Vision and Image Understanding*, 117 (6), 633–659.

Labelbox [online], (2024). Available from: https://labelbox.com/ [Accessed 12 Jun 2024].

GitHub - ultralytics/ultralytics [online], (2024). GitHub. Available from: https://github.com/ultralytics/ultralytics [Accessed 12 Jun 2024]. github.com/ultralytics/yolov5 [online], 2024.

GitHub - ultralytics/yolov5 [online], (2020). GitHub. Available from: https://github.com/ultralytics/yolov5 [Accessed 12 Jun 2024].

Guntuboina, C., Porwal, A., Jain, P. and Shingrakhia, H., (2021). Deep learning based automated sports video summarization using YOLO. *ELCVIA Electronic Letters on Computer Vision and Image Analysis*, 20 (1), 99–116.

Hutchinson, M.S. and Gadepally, V.N., (2021). Video action understanding. *IEEE Access*, 9, 134611–134637.

Kumar, A., Garg, J. and Mukerjee, A., (2014, November). Cricket activity detection. *In International Image Processing, Applications and Systems Conference* (pp. 1–6). IEEE.

Li, C., Li, L., Jiang, H., Weng, K., Geng, Y., Li, L., Ke, Z., Li, Q., Cheng, M., Nie, W., Li, Y., Zhang, B., Liang, Y., Zhou, L., Xu, X., Chu, X., Wei, X., and Wei, X., (2022). YOLOv6: A single-stage object detection framework for industrial applications. *arXiv preprint* arXiv:2209.02976.

Li, X., Wang, W., Wu, L., Chen, S., Hu, X., Li, J., Tang, J. and Yang, J., (2020). Generalized focal loss: Learning qualified and distributed bounding boxes for dense object detection. *Advances in Neural Information Processing Systems*, 33, 21002–21012.

Lin, T.Y., Dollár, P., Girshick, R., He, K., Hariharan, B. and Belongie, S., (2017). Feature pyramid networks for object detection. *In Proceedings of the IEEE Conference on Computer Vision and Pattern Recognition* (pp. 2117–2125).

Liu, H., Sun, F., Gu, J., and Deng, L., (2022). SF-YOLOv5: A lightweight small object detection algorithm based on improved feature fusion mode. *Sensors*, 22 (15), 5817.

Liu, Y., Nie, L., Liu, L., and Rosenblum, D.S., (2016). From action to activity: Sensor-based activity recognition. *Neurocomputing*, 181, 108–115.

Madabhushi, A. and Aggarwal, J.K., (1999, June). A bayesian approach to human activity recognition. *In Proceedings Second IEEE Workshop on Visual Surveillance* (VS'99)(Cat. No. 98-89223) (pp. 25–32). IEEE.

Nasir, M.H., Javed, A., Irtaza, A., Malik, H., and Mahmood, M.T., (2018). Event detection and summarization of cricket videos. *Journal of Image and Graphics*, 6 (1), 27–32.

Redmon, J., Divvala, S., Girshick, R. and Farhadi, A., (2016). You only look once: Unified, real-time object detection. *In Proceedings of the IEEE conference on Computer Vision and Pattern Recognition* (pp. 779–788).

Redmon, J. and Farhadi, A., (2017). YOLO9000: better, faster, stronger. *In Proceedings of the IEEE Conference on Computer Vision and Pattern Recognition* (pp. 7263–7271).

Redmon, J. and Farhadi, A., (2018). Yolov3: An incremental improvement. *arXiv preprint* arXiv:1804.02767.

Sen, A., Deb, K., Dhar, P.K., and Koshiba, T., (2021). CricShotClassify: An approach to classifying batting shots from cricket videos using a convolutional neural network and gated recurrent unit. *Sensors*, 21 (8), 2846.

Sharma, V., Gupta, M., Pandey, A.K., Mishra, D. and Kumar, A., (2022). A review of deep learning-based human activity recognition on benchmark video datasets. *Applied Artificial Intelligence*, 36(1), 2093705.

Shinde, S., Kothari, A., and Gupta, V., (2018). YOLO based human action recognition and localization. *Procedia Computer Science*, 133, 831–838.

Wang, C.Y., Bochkovskiy, A. and Liao, H.Y.M., (2021). Scaled-yolov4: Scaling cross stage partial network. *In Proceedings of the IEEE/cvf Conference on Computer Vision and Pattern Recognition* (pp. 13029–13038).

Wang, C.Y., Bochkovskiy, A. and Liao, H.Y.M., (2023). YOLOv7: Trainable bag-of-freebies sets new state-of-the-art for real-time object detectors. *In Proceedings of the IEEE/CVF Conference on Computer Vision and Pattern Recognition* (pp. 7464–7475).

Wu, L.F., Wang, Q., Jian, M., Qiao, Y., and Zhao, B.-X., (2021). A comprehensive review of group activity recognition in videos. *International Journal of Automation and Computing*, 18 (3), 334–350.

Zheng, Z., Wang, P., Liu, W., Li, J., Ye, R., and Ren, D., (2020). Distance-IoU Loss: Faster and better learning for bounding box regression. *Proceedings of the AAAI Conference on Artificial Intelligence*, 34 (07), 12993–13000.

Data Science & Exploration in Artificial Intelligence – Gururaj et al. (Eds)
 ISBN: 978-1-032-96390-7

Handwritten character recognition using CNN

Harshita Singh, Sudhir Singh & A.K. Mohapatra
Indira Gandhi Delhi Technical University for Women, Delhi, India

ABSTRACT: Pattern recognition, which is the field in which computers study and recognize characters based on patterns from diverse sources, includes handwritten character recognition. Even with advancements in technology, humans continue to outperform machines in activities involving pattern identification. Convolutional neural networks (CNNs), in particular, are key components of Deep Learning, which is used to solve self-perception issues such as picture and handwriting recognition. The goal of this research is to use the dataset that has 26000 images of the English characters called A_Z dataset to train CNNs for handwritten recognition and our study demonstrates high accuracy of up to 98% using CNN. CNNs have been made possible by advances in image labelling, object detection, and scene classification. Numerous techniques, including SVM and logistic regression, are used in classification. High detection rates have been attained using CNNs, particularly on large datasets such as CIFAR10 and ImageNet. These multilayer networks are as accurate in categorization and prediction as human beings.

Keywords: Deep Learning, Convolutional Neural Networks, Scene Classification, Feature Extraction, Classification methods, SVM, Logistic Regression

1 INTRODUCTION

At the nexus of cutting-edge technology and pattern recognition lies the field of handwritten character recognition, where machines attempt to read and comprehend handwritten characters. In this effort, pattern recognition—a field that extracts meaning from perceptual data—is essential. Eventually, computers should be able to read and comprehend human handwriting on paper, on a touchscreen, or in an image.[10]

In essence, the purpose of pattern recognition systems is to find commonalities across patterns that fall into particular categories. Their goal is to precisely identify and categorize these patterns. Even though technology has advanced significantly in the field of human-computer interaction, humans continue to outperform machines in tasks involving pattern recognition, such as speech recognition, gesture recognition, and handwritten character identification.[11]

Robots can now navigate and understand their surroundings thanks to the development of Deep Learning in recent decades. Deep Learning is a potent technique that can be used to solve a variety of challenging challenges, such as interpreting images and human input. The main objective is to use Convolutional Neural Networks (CNN), a core Deep Learning architecture, to the challenge of handwritten character recognition.[12]

The primary goal of the research is to better-handwritten character recognition systems by comprehending CNN's complexity. To evaluate the system's performance, the A_Z dataset is used to put it through a rigorous testing process. The results are carefully verified and contrasted with existing methods.[13]

 DOI: 10.1201/9781003589273-33

It is noteworthy that neural network classification and prediction accuracy is now on par with human performance on large datasets such as ImageNet, CIFAR-10, and CIFAR-100. These neural networks are composed of different layers, which add to their adaptability and range of functions when it comes to pattern recognition tasks. By utilizing the capabilities of convolutional neural networks, this research aims to further the continuing advancements in pattern recognition, specifically in the area of handwritten character recognition.

Our contribution can be summarized below:

(1) We have presented a CNN-based approach to recognize characters.
(2) We have taken a data set of 26000 images of English characters called A_Z dataset and compared its parameters for holistic assessment.
(3) We have demonstrated high accuracy up to 98%.

The flow of the paper can be organized as below. In section 2, we examine the different literature for proposed applications. In Section 3, we dealt with our proposed framework. In Section 4, we analyze the model performance. In section 5, we concluded our work.

2 RELATED WORK

In this section, we present a comprehensive literature survey about the field of handwritten recognition. Our review consists of different state-of-the-art methodologies which have certain advantages and disadvantages along with some classical classifiers, sparse representation, deep neural networks, transfer learning and even a combination of all. We try to have a strong foundation towards our research objective for that we examine the literature review comprehensively.

The study in [1] proposed a Neural Network Handwritten Character Recognition System in which the suggested system is a kind of segmentation, classification, pre-processing, and post-processing handwriting recognition system, This research introduces a revolutionary offline character recognition system that depends on neural networks but has few limitations such as leaves off feature extraction. In Handwriting Text Recognition Based on faster R-CNN [2] HCR uses a novel network topology called Region Proposal Networks (RPN), The primary goals of this system are to preserve the accuracy and speed up text recognition but on the downer side. It recognizes two words as one if they are too close together. In [3] [5] both are based upon CNN architecture that can process handwriting found online without requiring it to be converted to an image and it has been demonstrated that the suggested CNN properties perform well in character and large vocabulary word recognition tasks, the major limitation is It recognizes two words as one if they are too close together along with Strong training system is required. A different approach is used in [8] [9][10] [6] in which to regularize the neural network, an alternate method of training a subset of CNN layers is proposed whereas in [10] Convolutional, fully connected, and pooling layers are used by DCNN for dimensionality reduction, picture representation, and feature extraction and here it can be proven that it Outperforming the dataset a comparison of the simulation experiment findings shows that the CNN algorithm outperforms the KNN, Inception achieved 99 percent accuracy although few shortcomings can be noticed such as no discernible advancement in the recognition of numbers and the most exact onset requires the longest amount of time and alternatively in [6] which Comenia Script based CNN is used with Deep learning has demonstrated that image-based tasks, such OCR and pattern recognition, are valid and can be completed more accurately along with the limitation of functions exclusively with the Comeni script only. In [4] The DAG design, a network segmented into several pathways, is a result of the basic CNN model moreover the DAG design, a network segmented into many pathways, is derived from the basic CNN model with a near disadvantage that it has higher latency.

3 PROPOSED METHODOLOGY

The work's goal is to create a model that can identify a handwritten character in an image. To do this, we located an appropriate image database to use for model training.

3.1 *Methodology*

Deep Learning has become a key technique for solving challenges related to self-perception, such as comprehending human speech, images, and robots navigating their environment. One of the key components of deep learning architecture, the Convolution Neural Network concept, is what the project seeks to achieve. The main goal of the suggested system is to comprehend CNN and apply it to the handwritten recognition system. CNN is used for handwritten character recognition for a purpose. Let us examine a multi-layer feed-forward neural network that will be utilized on our dataset of 26000 images of the English characters called A_Z dataset, which comprises images with a resolution of approximately 784 pixels (28 by 28). Thus, the initial layer weights add up to roughly 78k parameters, which is a sizable but reasonable amount if a buried layer contains roughly 100 units.

However, the image size is far larger in the natural world. Taking into account the average image size of 256 x 256 pixels, or approximately 56,000 pixels, the first layer weights will have 560k parameters! As a result, it becomes unscalable for actual photos due to the excess of parameters[16]. As a result, it will be so big that it will be extremely challenging to generalize the fresh data that is added to the network. By applying filters, Convolution Neural Networks extract feature maps from 2D images, simplifying the process of extracting features from images. Convolution neural networks, as opposed to fully connected neural networks, take into account the mapping of picture pixels with the neighbourhood space[17].

In signal and image processing, convolution neural networks have shown to be an incredibly useful and potent tool. CNN has outperformed all previous methods in computer vision applications, including segmentation, handwriting recognition, and natural object classification.[18]

The following are the steps in the suggested method:

(1) **Building datasets for testing and training:** After the super classes images used for training are reduced in size, there are two subsets of the dataset: training and validation
(2) **Changing the CNN network:** To the last three layers of the network, Include a soft-max layer, a fully linked layer, and a classification output layer. As classes exist in the training datasets, assign the same sizes to the final fully linked layer. To expedite the network's training process, raise the fully connected layer's learning rate parameters.
(3) **Educate the network:** Set the training parameters (validation data, small batch size, learning rate) based on the GPU specifications of the system. Train the network with the training data.
(4) **Verify the network's correctness:** Classify the validation images using the optimized network to ascertain the classification accuracy. In a similar vein, testing the network's fine-tuning on live video streams for accuracy.

4 RESULTS AND DISCUSSIONS

Several research papers[1][2][3][4] have delved into handwritten character detection using Convolutional Neural Network (CNN) models trained and evaluated on the A_Z dataset, which comprises 26,000 English character images and we try to do analysis upon English alphabets based upon certain parameters.

The overall accuracy of our system is 98%.

Comparison analysis of our model in terms of Accuracy, Precision, Recall and F1 Score.

- TP (True Positive): Represents the instances where the model correctly predicted the positive class.

- FP (False Positive): Refers to the instances where the model incorrectly predicted the positive class.
- FN (False Negative): Denotes the instances where the model incorrectly predicted the negative class.
- TN (True Negative): Signifies the instances where the model correctly predicted the negative class.

$$Precision = \frac{True\ Positives\ (TP)}{True\ Positives\ (TP) + False\ Positives\ (FP)} \tag{1}$$

$$Recall = \frac{True\ Positives\ (TP)}{True\ Positives\ (TP) + False\ Negetives\ (FP)} \tag{2}$$

Table 1. Each alphabetical character analysis.

Characters	Precision	Recall	F1 score
A	0.99	0.98	0.99
B	0.98	0.99	0.98
C	0.99	0.99	0.99
D	0.97	0.91	0.94
E	0.99	0.98	0.99
F	0.96	0.97	0.97
G	0.99	0.96	0.97
H	0.96	0.96	0.96
I	0.99	0.97	0.98
J	0.94	0.98	0.96
K	0.97	0.97	0.97
L	0.97	0.99	0.98
M	0.99	0.98	0.99
N	0.98	0.99	0.99
O	0.99	0.99	0.98
P	0.98	0.99	0.98
Q	0.98	0.98	0.98
R	0.99	0.98	0.99
S	1.00	0.99	1.00
T	0.99	0.99	0.99
U	0.98	0.99	0.98
V	0.99	0.99	0.99
W	0.99	0.96	0.98
X	0.99	0.97	0.98
Y	0.98	0.98	0.98
Z	0.98	0.98	0.98

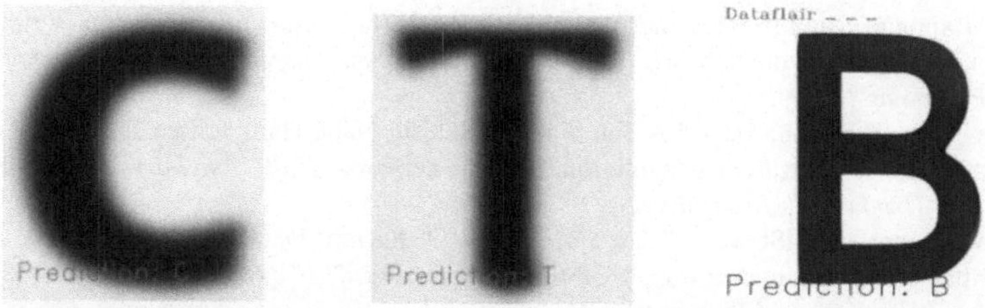

Figure 1. Output of character recognition by CNN.

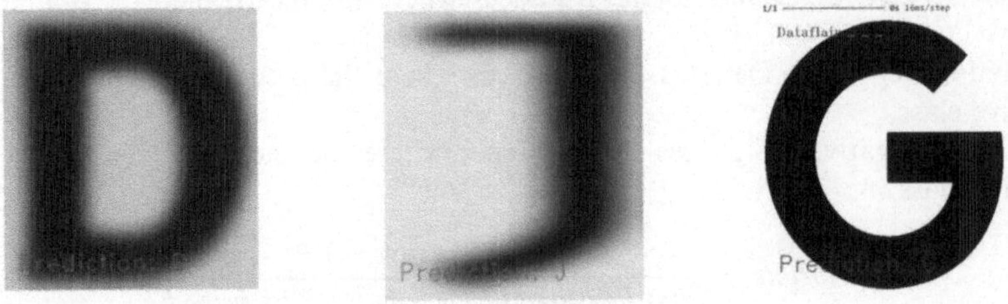

Figure 1. (Continued)

In Table 1 we see the character analysis which demonstrates that S has the Highest Precision along with the highest F1 value apart from this here we also successfully demonstrate that our model with an accuracy of 98% successfully read the character in the above screenshots and also model accuracy and loss graph depicts that we have less loss while training the data set.

5 CONCLUSION

This study systematically investigates and analyses various parameters of our Convolutional Neural Network (CNN) models employed for the classification and detection of handwritten characters. The primary objective is to conduct a comprehensive analysis of CNN towards English alphabets. The overall accuracy of our proposed system using CNN is 98% which firmly establishes CNNs as the state-of-the-art approach for handwritten character recognition. We made our own model to detect the handwritten characters and find the various parameters like Accuracy, Precision, Recall and F1 value.

REFERENCES

[1] Savitha Attigeri, (March 2018). Neural network based handwritten character recognition system, *IJECS* Volume 7 Issue 3,

[2] Junqing Yang, Peng Ren and Xiaoxiao Kong, Handwriting text recognition based on faster RCNN, 2019 *Chinese Automation Congress (CAC).*

[3] Subhasis Mandal, S. R. Mahadeva Prasanna and Suresh Sundaram, Exploration of CNN features for online handwriting recognition, *2019 International Conference on Document Analysis and Recognition (ICDAR).*

[4] Bhagyasree P.V., Ajay James and Chandran Saravanan, A proposed framework for recognition of handwritten cursive english characters using DAG-CNN, *2019 1st International Conference on Innovations in Information and Communication Technology (ICIICT).*

[5] Li Chen, Song Wang, Wei Fan, Jun Sun and Satoshi Naoi, Beyond human recognition: A CNN-based framework for handwritten character recognition, *2015 3rd IAPR Asian Conference on Pattern Recognition (ACPR).*

[6] Martin Rajnoha, RadimBurget and Malay Kishore Dutta, Handwriting comenia script recognition with convolutional neural network, *2017 40th International Conference on Telecommunications and Signal Processing (TSP).*

[7] Chunpeng Wu, Wei Fan, Yuan He, Jun Sun and Satoshi Naoi, Handwritten character recognition by alternately trained relaxation convolutional neural network, *2014 14th International Conference on Frontiers in Handwriting Recognition.*

[8] Curtis Wigington, Seth Stewart, Brian Davis, and Bill Barrett, Data augmentation for recognition of handwritten words and lines using a CNN-LSTM network, *2017 14th IAPR International Conference on Document Analysis and Recognition.*

[9] NagenderAneja and Sandhya Aneja, Transfer learning using CNN for handwritten devanagari character recognition, *2019 1st International Conference on Advances in Information Technology.*
[10] Gupta, K. K., Vijay, R., Pahadiya, P., Saxena, S., and Gupta, M. (2023). Novel feature selection using machine learning algorithm for breast cancer screening of thermography images. *Wireless Personal Communications, 1-28.*
[11] Pahadiya, P., Vijay, R., Gupta, K. K., Saxena, S., and Shahapurkar, T. (2023). Digital image based segmentation and classification of tongue cancer using CNN. *Wireless Personal Communications, 1–19.*
[12] Gupta, K. K., Vijay, R., Pahadiya, P., and Saxena, S. (2022). Use of novel thermography features of extraction and different artificial neural network algorithms in breast cancer screening. *Wireless Personal Communications, 1-30.*
[13] Gupta, K. K., Rituvijay, Pahadiya, P., and Saxena, S. (2022). Detection of cancer in breast thermograms using mathematical threshold based segmentation and morphology technique. *International Journal of System Assurance Engineering and Management, 1–8.*
[14] Gupta, K. K., Vijay, R., and Pahadiya, P. (2020). A review paper on feature selection techniques and artificial neural networks architectures used in thermography for early stage detection of breast cancer. *Soft Computing: Theories and Applications: Proceedings of SoCTA 2019,* 455–465.
[15] Pahadiya, P., Vijay, R., Gupta, K. K., Saxena, S., and Tandon, R. (2022). Contactless non-invasive method to identify abnormal tongue area using K-mean and problem identification in COVID-19 scenario. *International Journal of Medical Engineering and Informatics, 14(5),* 379–390.
[16] Anita Pal and Dayashankar Singh, (July-December 2010). Handwritten english character recognition using neural, *Network International Journal of Computer Science & Communication.* Vol. 1, No. 2, pp. 141–144.
[17] Salvador España-Boquera, Maria J. C. B., Jorge G. M. and Francisco Z. M., (April 2011). Improving offline handwritten text recognition with hybrid HMM/ANN models, *IEEE Transactions on Pattern Analysis and Machine Intelligence*, Vol. 33, No. 4.
[18] Pranob K. Charles, Harish V., Swathi M., Deepthi CH., (Jan-Feb 2012). A review on the various techniques used for optical character recognition, *International Journal of Engineering Research and Applications*, Vol. 2, Issue 1, pp. 659–662.

Data Science & Exploration in Artificial Intelligence – Gururaj et al. (Eds)
© 2025 the Author(s), ISBN: 978-1-032-96390-7

Performance evaluation of employee attrition through intensive investigation with supervised machine learning algorithm

R. Dinesh Kumar
CSE Department, Siddhartha Institute of Technology & Sciences, Hyderabad, India

A. Vijay Kumar
AI Department, Vidhya Jyothi Institute of Technology, Hyderabad, India

I. Vasantha Kumari, B. Rambabu & Y. Sandhya
CSE Department, Siddhartha Institute of Technology & Sciences, Hyderabad, India

ABSTRACT: The fear of job hunting has grown in the modern generation because of the high-tech environment, where jobs become obsolete and opportunities become few, making employees reluctant to stay with the same company for an extended period. Around the world, employee turnover has grown to be a serious issue. The loss of the best personnel is one of the major problems that company owners deal with in their organizations. Employers may incur significant losses in costs due to employee attrition. For this reason, we provide an automated approach in this research that uses various predictive analytical methodologies to forecast employee attrition. This study offers a thorough analysis of employee attrition using both conventional approaches and the proposed Ensemble Machine Learning Model. The performance metrics Accuracy, precision, Recall and F1-score were analyzed with conventional and ensemble models. The tree-based ensemble model with Stacking, Bagging and Boosting acquires the best results from the existing methods with an accuracy of 92%.

Keywords: KNN, Regression, Decision, Ensemble, Boosting, Employee attrition

1 INTRODUCTION

In recent years, AI technology in the business sector has begun to use it. AI technology has been used in several industries, to boost productivity or enhance financial performance. Artificial Intelligence is used in the area of human resources management, specifically in the areas of recruitment and human resource automation. The possibility of using the application is gradually growing [1]. Different machine learning with data analysis methods is becoming more popular as a way to stop employee attrition.

From an organizational perspective, employee attrition is extremely challenging because it affects the company with various issues, including disruptions to ongoing work, repeated training costs, risks of advanced technology leakage, etc. [2]. To improve the accuracy of employee attrition prediction, a new ensemble-based prediction model was proposed.

The primary contributions of our research study concerning with prediction of employee attrition are as follows:

- To predict employee attrition, four ML algorithms were used: Support Vector Machine (SVM), Decision Tree (DT), Random Forest (RF), and Ensemble model (Stacking, Bagging and Boosting)

 DOI: 10.1201/9781003589273-34

- To determine the optimal performance evaluation method, a comparative analysis was carried out between the four ML models that were being used.
- The confusion matrix and ROC curve analysis are carried out towards the performance validation using various assessment methods.

2 RELATED WORK

For predicting employee attrition, the proposed enhanced Trees Classifier) method achieved better accuracy. The proposed method performed better than current cutting-edge research. The result shows that the causes of employee attrition are job level, hourly rate, and monthly income. The proposed approach minimizes employee attrition in organizations [3].

In this paper, various retention strategies for specific employees have improved through the machine learning approach. Using this application, management can determine the reason for employee attrition and implement the appropriate countermeasures. [4].

Hierarchical regression analysis proposed the best practices with the firm environment, characteristics and firm setting. The study says there was less attrition in companies with benefits that represented a larger portion of overall labour costs [5].

The study used data from about 4,000 employees over 261 days in 2015 between January and December. The study identified various factors as significant contributors to employee attrition. The model has achieved 70% specificity, sensitivity, and accuracy [6].

Based on HR analytics data from the Kaggle website, a model to predict employee turnover rates was developed. For prediction, the AdaBoost classifier and Random Forest are employed . The elements that affect employee attrition within any firm are also covered, which will give management a clear understanding of the situation in making decisions [7].

The Garrett ranking technique is employed to rank the factors that contribute towards attrition. Furthermore, the variables underlying attrition have been predicted using logistic regression. The analysis's findings indicate that age, duration since the last promotion, and most recent performance rating are the main indicators of attrition [8].

To address overfitting problems, regularization strategies and parameter tuning are used in this work for optimization. After a comparative analysis of the three classifiers, the optimized SVM model was determined to be the best model for predicting employee attrition [9].

ANN and DT were the next best models tested. The study demonstrates a significant correlation between attrition factors and job type. A chi-square test reveals a strong correlation between attrition factors and job type. Any alteration to one will therefore also affect the other [10]. In technical jobs, pay is the most significant factor, while in non-technical jobs, boss is the factor.

3 PROPOSED MODEL

Figure 1 Shows the proposed model, built with three different modules, data collection module, discovery and deployment module. The Data collection module deals with the collection of data, exploring and dividing the data. The Discovery module deals with choosing the ML algorithm. The Deployment module deals with analyzing the ML model.

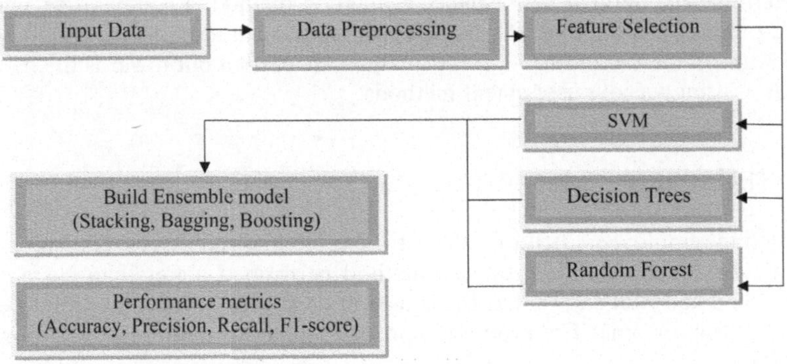

Figure 1. Proposed model architecture.

3.1 *Input dataset*

We used a real dataset with 35 columns ranging from categorical and interval variables. 85% of the dataset used for training and 15% of the dataset was used for training. By entering employee data, an extended version of the proposed model was created to predict employee attrition.

3.2 *Data preprocessing*

It is the method that removes noise, filters, analyses and encodes the data where a machine learning algorithm can trained to produce the output.

3.3 *Supervised learning models*

3.3.1 *Support vector machine (SVM)*

Face detection, hypertext classification of regular text, and bio-information are the applications of SVM. The support vector machine is trained on historical input data and produces future predictions as an output. The train dataset and test dataset are the two comparable data sets in the SVM model as shown in Figure 2.

$$w.x + b = 0 \tag{1}$$

where w represents the weight vector, w is the input vector and b represents bias.

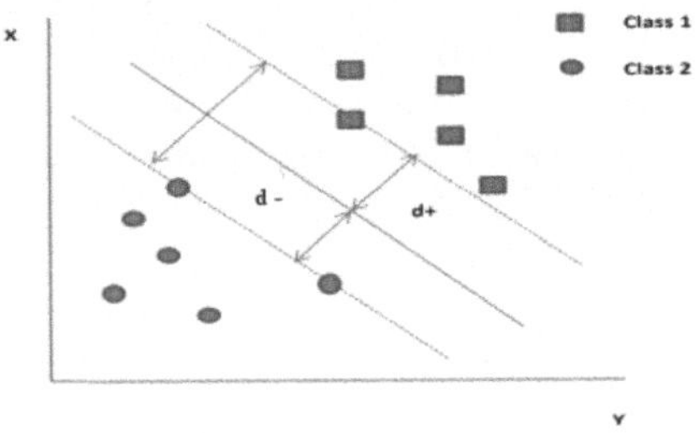

Figure 2. SVM model.

C-SVM classification employs a training function that minimizes the error function, subject to the constraints defined in equation 2.

$$y_i\left(w^T\phi(x_i) + b\right) \geq 1 - \xi_i \quad \xi_i \geq 0, i = 1 \ldots . N \tag{2}$$

Where

c – capacity constants, w – vector coefficients , b – Constant , ξ_i – Parameters for handling non-separable data, x_i – Independent variable, ϕ – Kernel.

3.3.2 *Decision trees (DT)*

The complete dataset is used to begin the tree-building process. The appropriate decision rule divides the two classes using the DT algorithm as shown in Figure 3.

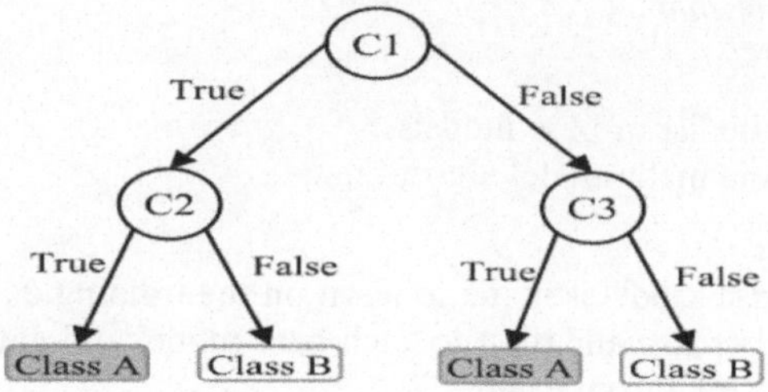

Figure 3. Decision tree.

The mathematical equation for entropy for a single attribute is represented in equation 3.

$$E(s) = \sum_{i=1}^{c} -p_i \log_2 p_i \tag{3}$$

An entropy of zero is referred to as a leaf node, whereas an entropy of more than zero requires additional branch splitting.

3.3.3 *Random forest*

Several Decision Trees (DT) work together to form a Random Forest (RF). RF's (multiple DTs) are trained using different portions of the training dataset. Each DT of the forest must be transmitted down with its input vector from the previous trial to classify a new sample.

$$Gini\ index = 1 - \sum_{i=1}^{n} (p_i)^2 \tag{4}$$

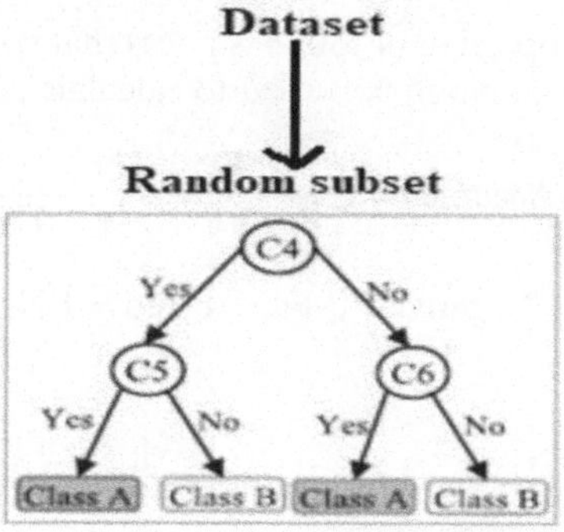

Figure 4. Random forest.

Among all possible splits, the feature with the lowest Gini index shown in equation 4. is chosen as the root node. The lowest impurity is indicated by the lowest Gini index.

4 PROPOSED ENSEMBLE MODELS

The process of creating a new ensemble model, the output of the process combined with other results to enhance the model's performance. The ensemble model improves the accuracy of final models by combining the base-learner process with learning from the best outcomes. The ensemble models with Stacking, Bagging and Boosting were used to increase the accuracy of the results where learners parallely trained from sub-datasets and combined into a homogeneous base-learner system.

Ensemble Model Algorithm

(1) Build the ensemble:
Step 1. Determine the list of base models.
Step 2. Determine the meta-model algorithm.

(2) Train the ensemble:
Step 1. Every 'L' base model is made to learn on the training dataset.
Step2. Apply k-fold cross-validation to each base model and gather the cross-validated predictions (p1, p2,.. pL) from each..
Step 3. Combine N cross-validated predicted values from every base model to form a new N*LN*L feature matrix. Level-one (original vector)
Step 4. The "level-one" data is used to train the meta-model.

$$Y = f(Z) \tag{5}$$

(3) Predict on new data:
Step 1. Generated predictions from the base models taken.
Step 2. Feed those predictions into the meta-model to generate the final ensemble prediction.

5 PERFORMANCE METRICS

The performance metrics used in this work to assess the algorithms' performance are, F1-score, accuracy, recall, and precision.

5.1 *Accuracy*

Accuracy is defined as the proportion of total instances that contain both true negatives and true positives. The following equation 6. is used to calculate accuracy.

$$\text{Accuracy} = \frac{\mathbf{TP + TN}}{\mathbf{TP + FP + FN + TN}} \tag{6}$$

TP-True Positive TN-True Negative FP-False Positive FN-False Negative

5.2 *Recall*

Recall is defined as the percentage of data samples belongs to "all true positive class". Recall calculated using the equation 7. as follows:

$$\mathbf{Recall} = \frac{\mathbf{True\ positive}}{\mathbf{True\ positive + False\ positive}} \tag{7}$$

5.3 *Precision*

Precision is the degree to which a model can make accurate predictions. The precision separates true positives from false positives shown in equation 8.

$$\textbf{Precision} = \frac{\textbf{True positive}}{\textbf{True positive} + \textbf{False positive}} \tag{8}$$

5.4 *F1-score*

F1-score, the average of recall and precision, is a single performance indicator. The F1-score will be nearer the lower estimate of Precision or Recall. Equation 9. used to calculate the F1-Score

$$\text{F1score} = \frac{(2 * (\text{Recall} * \text{Precision}))}{(\text{Recall} + \text{Precision})} \tag{9}$$

6 RESULTS AND DISCUSSION

The performance evaluation of the employee attrition system is carried out with DT, RF and SVM with a single out cross-validation method. Different measures, such as precision, recall, accuracy, and F1-score, are used to evaluate the efficiency of the proposed method.

Figure 5 shows the feature set of various attributes with targets in the given dataset.

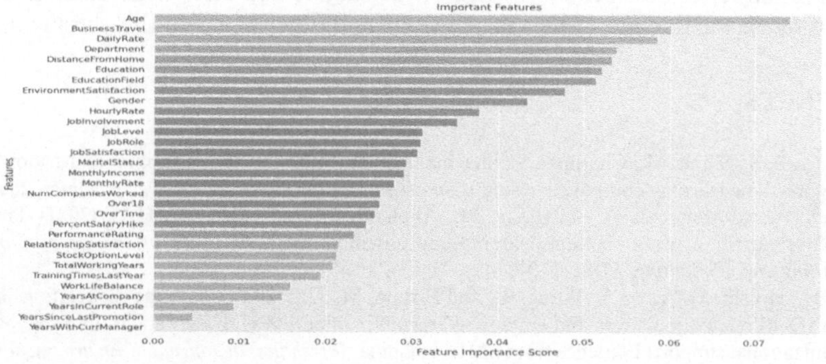

Figure 5. Correlation of the features with the target.

Table 1 and Figure 6 depict the classification report analysis of the conventional and proposed ensemble model of machine learning techniques.

Table 1. Performance evaluation of different classifiers.

	Conventional Models			Proposed Ensemble Model		
Performance Metrics	Decision Tree	SVM	Random Forest	Stacking	Bagging	Boosting
Precision	0.571	0.6	0.727	0.580	0.6	0.741
Recall	0.137	0.051	0.137	0.310	0.206	0.396
F1-Score	0.222	0.095	0.231	0.404	0.307	0.516
Accuracy	0.83	0.84	0.86	0.84	0.85	0.921

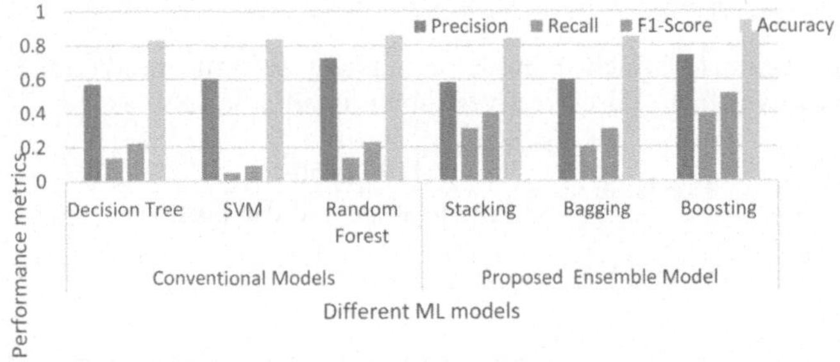

Figure 6. Performance evaluation.

7 CONCLUSION

The challenge of retaining skilled employees with experience poses a threat to business owners. Employers may incur major expenses because of employee attrition, as it takes a lot to replace their knowledge and productivity. As a result, various machine-learning approaches are used in conjunction with an ensemble model to identify the various causes of these significant business issues. The performance measures Accuracy, precision, Recall and F1-score were analyzed with conventional and tree ensemble models. The lower value in the outcome reflects the excellence of the model. However, based on analysis no model has been developed as ideal and flawless in every business context. The proposed model was nearly ideal for our needs and effectively achieved the desired outcome with 92% accuracy. Finally, recommended that more research be done on the subject to advance knowledge of the subject and offer fresh findings that either confirm or refuse the findings of the current study and other works in this specific field.

REFERENCES

[1] Aggarwal, S., Singh, M., Chauhan, S., Sharma, M., and Jain, D. (2022). Employee attrition prediction using machine learning comparative study. *Smart Innovation, Systems and Technologies*, 265, 453–466.

[2] Qutub, A., Al-Mehmadi, A., Al-Hssan, M., Aljohani, R., and Alghamdi, H. S. (2021). Prediction of employee attrition using machine learning and ensemble methods. *International Journal of Machine Learning and Computing*, 11(2), 110–114.

[3] Lai, H., Hossin, M. A., Li, J., Wang, R., and Hosain, M. S. (2022). Examining the relationship between COVID-19 related job stress and employees' turnover intention with the moderating role of perceived organizational support: Evidence from SMEs in China. *International Journal of Environmental Research and Public Health*, 19(6), 3719.

[4] Shobhanam, K., and Sumati, S. (2022). HR analytics: Employee attrition analysis using random forest. *International Journal of Performability Engineering*, 18(4), 275.

[5] Joseph, R., Udupa, S., Jangale, S., Kotkar, K., and Pawar, P. (May 6–8, 2021). Employee attrition using machine learning and depression analysis. *In Proceedings of the 2021 5th International Conference on Intelligent Computing and Control Systems (ICICCS)*, Madurai, India (pp. 1000–1005).

[6] Bhardwaj, S., and Ashutosh, S. (2017). Factors affecting employee attrition among engineers and non-engineers in manufacturing industry. *Business and IT*, VII(2), 26–34.

[7] Raza, A., Munir, K., Almutairi, M., Younas, F., and Fareed, M. M. S. (2022). Predicting employee attrition using machine learning approaches. *Applied Sciences*, 12(13), 6424.

[8] Dineshkumar, R., and Suganthi, J. (2013). A research survey on Sanskrit offline handwritten character recognition. *International Journal of Scientific and Research Publications*, 3(1), 1–7.

[9] Deepika, K., Sowjanya Reddy, M., Rajesh Pandian, N., and Dinesh Kumar, R. (2022). Analysis of performance of accuracy by adding new features individually using Relief-F and budget tree random forest (RFBTRF) method. *International Journal of Nonlinear Analysis and Applications*, 13(1), 1239-1252.

[10] Dinesh Kumar, R., Kalimuthu, M., Jayaram, B. (2022). Character Recognition System Using CNN for Sanskrit Text. In: Satyanarayana, C., Gao, XZ., Ting, CY., Muppalaneni, N.B. (eds) Proceedings of the International Conference on Computer Vision, High Performance Computing, Smart Devices and Networks. *Advanced Technologies and Societal Change*. Springer, Singapore.

Data Science & Exploration in Artificial Intelligence – Gururaj et al. (Eds)

A survey to address the cold start problem in recommendation systems

Gnanakumar Ganesan & Vijayakumar Adaickalam
School of CSE, Presidency University, Bangalore, India

ABSTRACT: The cold start problem, a well-known difficulty in recommender systems, is when systems cannot provide recommendations to clients with meaningful recommendations because there is not enough information available about them. Although their techniques differ, many researchers have addressed this issue in the literature by acquiring information that is lacking. A survey of this literature indicates that there are two categories of solutions to this problem: (i) explicit solutions and (ii) implicit solutions, depending on how missing information is acquired. This paper delves into the understanding of the cold start problem within recommender systems and investigates various methodologies aimed at addressing this challenge.

Keywords: Recommender systems, Cold Start Problem, Machine Learning, Deep Learning, Social Recommender Systems, Content-Based Filtering (CBF) and Collaborative Filtering (CF)

1 INTRODUCTION

Recommender Systems (RSs) are applications that endorse relevant products for users according to their preferences [8]. These preferences are either explicitly or implicitly determined by the systems. Users' browsing and buying history, along with their demographic information, and other information are examples of implicit data collection methods. In the explicit method, consumers are asked to complete surveys or rate items. RSs are extensively used on the internet in a number of application areas, including Online retail, Digital entertainment, Online travel platforms, etc.

All recommendation systems employ algorithms that make use of user data that is currently accessible. Filtering mechanisms provide the foundation of these algorithms. The two most popular filtering techniques are Collaborative Filtering (CF) [2] and Content Based Filtering (CBF) [2]. The premise of a content-based filtering technique is that users who show an interest in one product are likely to show an interest in comparable products as well. Two distinct methodologies are used in collaborative filtering: the neighbourhood-based [3] and the Latent Factor Model (LFM) [4] based. According to neighbourhood-based methodology, if a cluster of users shares analogous preferences, it implies that all members within the group are likely to have similar preferences for items as well. According to this method, a user's neighbourhood is identified, and a recommendation is made for an item that the neighbours prefer.

Collaborative filtering outperforms content-based filtering in terms of simplicity of use and performance [8], [9], [13], [10]. Some hybrid filtering systems exist that aim to maximize the benefits of utilizing both content-based and collaborative methods by combining them [1], [3], [8]. An RS has trouble making recommendations upon introducing a new item to their repository or when a unknown interrelates with them; this issue is recognized as the "cold start problem" [12], [18], [20] in RSs. Cold start problems in RSs come in two different varieties: (1) The cold start issue for unknowns (2) The cold start issue with the new item. When a unknown user is added to the system, the cold start issue arises because RS lacks information about the individual and cannot make recommendations. The cold start issue

DOI: 10.1201/9781003589273-35

with the new item, the system struggles to identify the new item's target user and lacks ratings for the item. The unknown cold start problem has been extensively researched and is the more challenging of the two types of cold start challenges [10], [15], and [18].

This paper conducts a survey of diverse strategies aimed at tackling the cold start problem specifically for unknowns in recommender systems.

The rest of the paper is organized in the following manner:

In section 2, the two types of resolutions suggested for the unknown cold start issue in RSs are examined.

In Section 3 and 4, approaches to resolving the cold start issue for unknowns in RSs are discussed.

Section 5 compares the solutions and discusses the benefits and drawbacks of each.

2 COLD START PROBLEM SOLUTION

The fundamental reason behind the cold start problem lies in the insufficient information necessary for generating recommendations. The suggested solutions put forth delineate the methodologies for acquiring this inaccessible data. The information can be gathered indirectly by utilizing previously obtained data or explicitly by asking the user queries. As a result, depending on how they get this data, the solutions can be divided into two types. Similar thoughts are presented in [1] and [20]. Different methods can be used to gather missing data, depending on the type of information being gathered. These methods are covered in the sections that follow, along with specific solutions that make use of them. Several desired qualities are determined from the review of the literature that solutions should have, irrespective of the method used to gather missing data [9,10], [12], [17], [21], [22,23].

2.1 *The relevance and accuracy of the recommendation*

If the user finds a recommendation interesting, it is relevant. If the system makes less appropriate test recommendations while learning the user's profile, the accuracy rate of the system could fall. The overall accuracy must be preserved in the solutions. The least amount of question items with the highest informativeness can be chosen as one method to achieve this [17].

2.2 *Minimizing bias*

For instance, popular products often receive high ratings, and some users rate items without necessarily having firsthand experience with them [9,10]. The personalizing of suggestions is hindered by such biased ratings. In order to mitigate bias, solutions employ baseline predictors or bias factors; however, in order to effectively model them, the ratings history is required [10]. Appropriate techniques must to be available to detect the abnormalities in ratings very away in order to address the cold start issue.

2.3 *Flexibility*

Solutions should ideally be flexible in relation to the RS's operating concept [21]. Recommender systems can employ a variety of filtering techniques and diverse rating systems. It takes less effort to adjust a solution so that it is compatible with the system.

2.4 *Variety*

Typically, a variety of product categories are included, such as electronics, apparel, home goods, and so on, in an online marketplace [22]. All domains should be covered by the recommendations made by a competent RS. It has to discover the user's interests across several domains in order to achieve this [23]. Solutions ought to include clauses for these designations.

3 INFORMATION GATHERING EXPLICITLY

By employing explicit solutions, this concept involves directly engaging with the user to gather any lost data. Under this method, the user may be required to rate some items or complete an online survey. The system can obtain more relevant data when it obtains information directly because it has more control over the queries it poses. However, one of the problems with this strategy is that because it takes time and effort, users are frequently uncomfortable to engage in the interrogation progression [12], [17]. Alternative approaches involve leveraging visualization [6], [7] and/or offering incentives. The aim of unambiguous solutions is to acquire suitable and pertinent data from the user without overwhelming them. Active knowledge and interview-based practices are two well-liked approaches that have been applied frequently. In the subsections that follow, these methods are briefly covered.

3.1 *Active learning*

Current methods of active learning (AL) can be categorized into two types: personalized and non-personalized [17]. Non-personalized AL techniques involve generating a single list of items that are presented to each user for rating. Many conventional AL methods are characterized as non-personalized [5], [17]. Personalized active learning systems, on the other hand, enhance personalization by providing distinct lists of items to various users. The list is modified in response to user comments. Since they elicit universal preferences from all user's, non-personalized approaches are easy to apply and general [17]. By lowering error rates and raising forecast accuracy, they can help to enhance system performance. Thus, they are helpful in situations where the importance of quantitative forecasts outweighs that of qualitative ones. One such method is the uncertainty reduction strategy, which chooses items with a range of user ratings [5],[17]. This approach simultaneously completes the responsibilities of learning the user profile and making appropriate endorsements.

3.2 *Interview based techniques*

Giving a user one item from a group of items and obtaining their feedback are key components of interview-based approaches. Three possible answers are often possible: like, dislike, or unknown [13]. Throughout the interview, a preliminary user profile and a list of seed items are prepared and then refined. Solutions previously employed a static interview format with a predetermined list of items. These solutions were not appealing or personalized. As a result, an adaptive interviewing method was created, in which the user's response determined how to modify the seed item list. The majority of recent approaches model adaptive interviews using decision trees [13,14,15]. A successful interview should be simple but clear [8], [13]. In order to do this, the fewest but most informative items must be chosen.

4 INFORMATION GATHERING IMPLICITLY

Implicit solutions strive to understand an unknown's preferences with minimal intervention. For this, they make use of already-available data, such demographics, or they turn to other resources, like social media. These days, social media platforms like Facebook and Twitter provide people enticing information sources. Filtering strategies use the ratings matrix and learnt preferences to make recommendations once the user profile has been established. It can be difficult to locate relevant information in these systems, though. The system must gather a large amount of data and sort it into knowledge that is helpful.

The following is a brief discussion of a few methods that can be utilized to get implicit data.

4.1 *Adapted traditional filtering techniques*

Because there is inadequate information in cold start settings, traditional filtering techniques like CBF and CF are less successful. Neighbourhood-based CF, for example, looks from a user's rating history to determine their ideal neighbourhood. They can be adjusted to work with scant data in the cold start scenario. By employing a novel similarity metric for neighbourhood-based CF in [10], the RS is able to pinpoint a neighbourhood for an unknown. The suggested similarity measure is a neural network-optimized weighted linear mixture of simpler similarity metrics. In [18], collaborative filtering techniques are combined with algorithms for classification such as C4.5 and the Naive Bayes algorithm to determine a neighbourhood for an unknown user. The system trains the categorization model at using the user's demographic data.

4.2 *Making Use of social media*

Social networking sites like Facebook and Twitter have drawn a sizable user base and seen tremendous rise in popularity on the Internet. Over 3 billion people utilize the internet globally as of right now, according to the International Telecommunication Union (ITU). Nearly 60% of these users are on Facebook. They are now priceless archives for personal information as a result. A user's social profile encompasses various data, such as demographics, friend circles, opinions, and likes and dislikes. Leveraging social networks for the cold start problem appears to be an intriguing approach. There are many methods to use social networks because they contain a variety of information types.

Figure 1 depicts the distinction between traditional and social recommender systems (RS). Unlike traditional RS, which solely relies on the Ratings matrix as illustrated in Figure 1(b), social RS incorporates a social relations matrix (as depicted in Figure 1(c)), capturing relationships among various users [14]. Given the absence of prior ratings, conventional RS techniques such as neighbourhood-based collaborative filtering (CF) prove inadequate in forming neighbourhoods for unknowns.

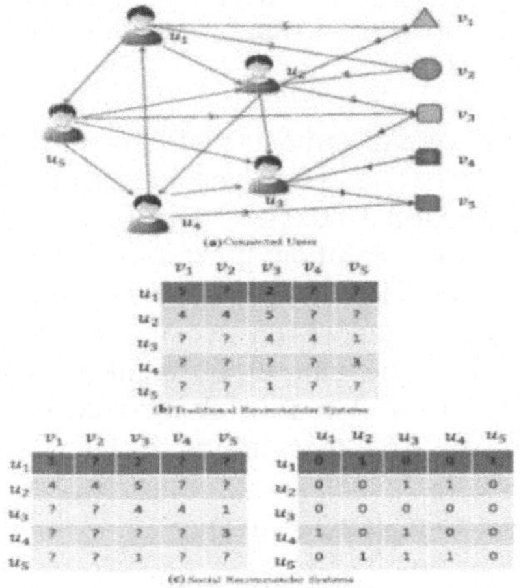

Figure 1. Comparison between traditional and social recommender systems [14].

Transitive interactions between users in other neighbourhoods can be developed in social RS, which allows to get over the aforementioned limitation [9].

5 COMPARING AN EXPLICIT AND IMPLICIT SOLUTION

The following criteria can be used to compare explicit versus implicit solutions:

5.1 *User interaction*

This describes how crucial it is that users interact with the system in order for it to get to know them.

5.2 *User effort*

The amount of work the user must put in to assist the system in comprehending his or her choices. Usually, work consists of answering questions or rating certain things. The user may quit the system if they feel overburdened by the amount of work.

5.3 *Confidence*

It suggests the extent to which the system may learn from user comments to enhance its own performance during the recommendation process [18].

5.4 *Applicability*

The amount of instructive information that can be found.

5.5 *The conundrum of exploration versus exploitation*

Illustrates the decision-making process the system faces when deciding which resource to propose in order to gain more information about the user (explore) or to capture their attention (exploit) [19].

5.6 *Confidentiality*

Users' ability to manage the information they disclose through privacy settings may limit the information that the system can access. It describes the extent to which privacy settings constrain the system.

Based on the parameters mentioned above, Table 1 presents comparisons between the two classes of solutions.

Table 1. Comparing an explicit and implicit solution.

Parameters	Explicit solutions	Implicit solutions
1. User interaction	Heavily reliant on user interactions	Little or non-existent interactions
2. User effort	It takes a lot of user work, which increases the possibility of losing the user.	User effort is not necessary. The user feels more at ease
3. Confidence	Greater room for improvement since user input is visible	Interaction reduces the amount of information that may be learned via feedback.
4. Relevance	It is simple to create queries to obtain relevant information.	The information that is available might not be pertinent. To find them, filtering is necessary.
5. The conundrum of exploration versus exploitation	Encounters the Exploration-Exploitation dilemma. Eventually, the recommendation system must transition from learning to making recommendations.	Knowledge has been acquired. There was no such issue during the recommendation process.
6. Confidentiality	No issues	Stringent privacy settings could result in failure.

6 CONCLUSION AND FUTURE SCOPE

There are two distinct challenges related to the cold start problem in recommender systems: the issue with unknown users and the issue with new items. Each method addressing the cold start problem for unknown users focuses on collecting the required information about them. In the existing literature, solutions to this issue are categorized into two groups based on how they explicitly or implicitly acquire the missing user data. This literature review assesses and compares these techniques, highlighting their respective strengths and limitations. This study's primary objective is to provide readers with a comprehensive understanding of the cold start problem in recommender systems and the range of potential fixes. Despite the existing solutions, there remains room for improvement, presenting an opportunity for further research in this domain. Future research could explore combining explicit and implicit strategies to improve the effectiveness of addressing the cold start problem in recommender systems.

REFERENCES

[1] Lillegraven T. N., and Wolden A. C., (2010), *Design of a Bayesian Recommender System for Tourists Presenting a Solution to the Cold-Start Problem.*

[2] Ricci F., Rokach L. and Shapira B., (2011), Introduction to recommender systems handbook, in *RSs Handbook* 1st ed. USA: Springer, ch. 1, pp. 1–33

[3] Desrosiers C. and Karypis G., (2011). A comprehensive survey of neighbourhood-based recommendation methods, in *Recommender Systems Handbook* 1st ed. USA: Springer, ch. 4, pp. 107–144

[4] Koren Y. and Bell R., (2011), Advances in collaborative filtering, in *Recommender Systems Handbook* 1st ed. USA: Springer, ch. 5, pp. 145–186

[5] Rubens N. and Kaplan D., (2011), Active learning in recommender systems, in *Recommender Systems Handbook* 1st ed. USA: Springer, ch. 23, pp. 735–767

[6] Yoo K. and Gretzel U., (2011), Creating more credible and persuasive RSs: The influence of source characteristics on recommender system evaluations, in *RSs Handbook* 1st ed. USA: Springer, ch. 14, pp. 455–477

[7] Kagie M., Wezel M. V. and Groenen P.J.F., (2011), Map based visualization of product catalogs, in *Recommender Systems Handbook* 1st ed. USA: Springer, ch. 17, pp. 547–576

[8] Zhou K., Yang S., and Zha H., (2011). Functional matrix factorizations for cold start recommendation, in *Proc. of SIGIR '11*, ACM, Beijing, China, pp. 315–324.

[9] Victor P., De Cock M., and Cornelis C., (2011), Trust and recommendations, in *Recommender Systems Handbook* 1st ed. USA: Springer, ch. 20, pp. 645675

[10] Bobadilla J., Ortega F., Hernando A., and Bernal J., (2012), A collaborative filtering approach to mitigate the unknown cold start problem, in *Knowledge-Based Systems*, vol. 26, Elsevier, pp. 225238,

[11] Shani G. and Gunawardana A., (2011), Evaluating recommendation systems, in *Recommender Systems Handbook* 1st ed. USA: Springer, ch. 8, pp. 257297.

[12] Zhao X., Zhang W., and Wang J., (2013), Interactive recommender systems, in *Proc.of CIKM* '13, ACM, San Francisco, USA, pp. 1411–1420

[13] Sun M., Li F., Lee J., Zhou K., Lebanon G., and Zha H., (2013), Learning multiple-question decision trees for cold-start recommendation, in *Proc. of WSDM*'13, ACM, Italy, pp. 445–454.

[14] Tang J., Hu X., and Huan Liu, (2013), Social recommendation: A review, in *Social Network Analysis and Mining*, vol. 3,issue 4, Springer, pp 1113–1133

[15] Braunhofer M., Elahi M., and Ge M., (2014), Context dependent preference acquisition with personality-based active learning in mobile recommender systems, in *Proc. of LCT* '14, Springer, Crete, Greece, pp. 105116

[16] Houlsby N., Hernandez-Lobato J. M. and Ghahramani Z., (2014), Cold-start active learning with robust ordinal matrix factorization, in *Proc. of ICML* '14, pp. 766–774

[17] Elahi M., Ricci F. and Rubens N., (2014), Active learning in collaborative filtering recommender systems, in *E-Commerce and Web Technologies*, vol. 188, Munich, Germany: Springer, ch. 12, pp 113–124

[18] Lika B., Kolomvatsos K., and Hadjiefthymiades S., (2014), Facing the cold start problem in recommender systems, in *Expert Systems with Applications*, vol. 41, no. 4, Pergamon Press, Inc., pp. 20652073.

[19] Zhang M., Tang J., Zhang X., and Xue X., (2014), Addressing cold start in recommender systems: a semi-supervised co-training algorithm, *in Proc. of SIGIR* 14, ACM, Queensland, Australia, pp. 73–82.

[20] Nadimi-Shahraki M.H. and Bahadorpour M., (2014), Cold-start problem in collaborative Recommender systems: Efficient methods based on ask-torate technique, in *Journal of CIT*, vol. 22, no. 2, pp. 105.

[21] Christakopoulou K., Radlinski F., and Hofmann K., (2016), Towards conversational recommender systems, In *Proc. of KDD* '16, ACM, San Francisco, USA, pp. 815–824.

[22] Zhang Z., Jin X., Li L., Ding G., and Yang Q. (2016, March). Multi-domain active learning for recommendation, in *Proc. of AAAI* '16, 2016, pp. 2358–2364.

[23] Fernndez-Tobas I., Braunhofer M., Elahi M., Ricci F., and Cantador I., (2016), Alleviating the unknown problem in collaborative filtering by exploiting personality information, in *User Modeling and User-Adapted Interaction*, vol. 26, issue 2, Springer, pp 221–255.

Gnanakumar G pursing PhD in School of Computer Science and Engineering from Presidency University. He received the M.E., degree in Embedded and Real Time Systems and B.E., degree in Computer Science and Engineering from Anna University. He has 8 years of teaching experience in Computer Science and Engineering department. He is currently working an Assistant Professor with the School of Computer Science and Engineering from Presidency University. He has published 3 conference papers. His research interests include Machine Learning, Deep Learning and Recommendation Systems.

Dr. Vijayakumar Adaickalam is currently working as **Professor** of School of Computer Science and Engineering, Presidency University, Bangalore, Karnataka, India. He has around 29 years of working experience in academics and administrations in reputed institutions. He has completed Ph.D., Degree in Computer Science and Engineering from College of Engineering, Guindy, Anna University, Chennai, Tamilnadu, India. He received M.E., Degree in Computer Science and Engineering from Annamalai University, Tamilnadu and his under Graduation in B.E., Computer Engineering from Madurai Kamaraj University, Tamilnadu, India. He has published more than 25 International Journals and 35 International Conferences. He is currently guiding many Ph. D., research scholars in Machine Learning and Deep Learning area. His research interest are Machine Learning, Deep learning, Mobile networks, Networks Security, Cloud Computing and Big Data Analytics.

Machine learning methods for ensuring transaction integrity in credit card systems: A comprehensive review

G. Ramesh, Awez Ahamed, Siddhanth C. Shetty, Shailesh Rao, Prathik P. Shetty, N.G. Shrikanth & Harish Kunder
Department of Artificial Intelligence and Machine Learning, Alvas Institute of Engineering and Technology, Mijar, Mangalore, India

ABSTRACT: The continuing threat of credit card fraud targeting financial institutions demands more advanced data analysis techniques to ensure accurate detection and prevention. This research suggests a unique way to categorize Support Vector Machines that can improve fraud detection systems' accuracy and performance dramatically. To identify areas of worry or suspicion, this proposed technique looks at various points of contact or databases, effective fraud detection and procurement, as well as odd behavior in transaction data processing. By proving the efficacy of classification and training techniques like SVM in preventing significant transactional-based fraud, this study will fill the present knowledge gap. The model runs between cycles of data, which allows it to detect fraudulent transactions rather well, according to the SVM-based results.

Keywords: Fraud detection, financial system, Support Vector Machine (SVM), Classification, Transaction analysis

1 INTRODUCTION

Economic and financial development has been influenced by deregulation, liberalization, globalization, and privatization. All this has forced organizations to expand their products and introduce new technologies. A well-known business becomes a cooperative advanced technology makes us competitive and the banking industry is at the forefront of this trend. That's why financial institutions are investing in digital awareness programs, encouraging the adoption of mobile banking apps on debit/credit cards, and strengthening security measures to prevent deception [1]. Credit card fraud poses a major threat to both financial institutions and consumers because it involves dishonesty in collecting money. Fraud is increasing due to the widespread availability of credit cards online and offline, creating a need to prevent fraud [2].

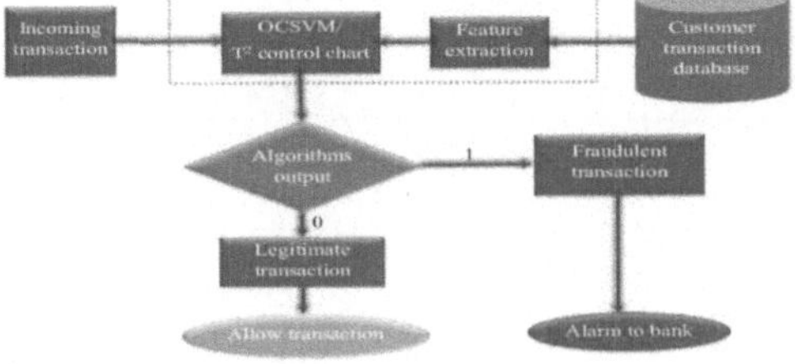

Figure 1. Fraud detection system.

DOI: 10.1201/9781003589273-36

In resolving fraud-related issues. Figure 1 shows how many different products can be obtained using algorithms and computational instructions, which provide a good way of using previous data and models to detect fraud. However, one of the biggest challenges in using machine learning for fraud detection is dealing with inconsistent data where there is plenty of legitimate work for Deception [3]. Search offers a good way to separate transaction data into fake and non-fake groups. Strategic solutions attempt to detect suspicious activity that indicates fraud by carefully analyzing business characteristics and behavior. In addition, the study also investigated the application of support vector machines in various industries and demonstrated their effectiveness in tasks such as credit analysis, bankruptcy prediction, and time series distribution in fraud detection [5].

1.1 *Motivation*

Research on fraud risk in the global economy highlights the importance of using machine learning (ML) to identify credit cards. Fraudsters can now change their tactics using technology, so it's important to be flexible and flexible. Learning machine learning techniques and applying data science, Cybersecurity, and financial literacy can help researchers improve fraud detection. A better approach can be developed by evaluating various machine learning models to reduce financial losses and increase participants' trust in the digital financial market [6].

1.2 *Contribution*

These systems have two advantages. By giving providers the information, they need to select and implement better learning models, they can reduce losses and protect consumers. Encouraging knowledge sharing and teamwork, it enables business innovation and brings more sustainable measures. By promoting the integrity and responsibility of AI, we ultimately create a better and safer business for people, which enables transparency and trust in business finance [9].

1.3 *Organization*

It is divided into several parts: Section II introduces the content of the research, Section III discusses the challenges and methods in fraud detection, Section IV introduces the process, Section V presents detection and performance evaluation, Section VI fraud analysis, Section VII Semi-conclusion.

2 BACKGROUND ANALYSIS

Support Vector Machine (SVM) classification is a technology used in machine learning credit card fraud. Some academic studies have examined the effectiveness of SVM in detecting fraudulent activities in terms of model performance and generalizability [5] [23]. These studies often evaluate the superiority of SVM in fraud detection by comparing the results of SVM with other methods such as decision trees (DT), artificial neural networks (ANN), and genetic algorithms (GA). Researchers demonstrate the importance of using the power of SVM to accurately classify transaction data into fake and non-fake groups [3]. They demonstrate the effectiveness of SVM in product analysis and behavior to detect suspicious activity indicating fraud. Additionally, SVM has been rigorously evaluated in combination with learning techniques, demonstrating its ability to improve the accuracy of fraud detection [4]. Hybrid data models combining SVM with feature selection algorithms show promise in improving fraud detection accuracy [5]. Business areas include card fraud detection such as time distribution, credit analysis, and bankruptcy prediction [7].

3 UNDERSTANDING FRAUD DETECTION CHALLENGE AND APPROACH

In the unequal world of financial markets, the fight against fraud has become an important endeavor, characterized by fraud aimed at obtaining illicit funds and breaking the law [8]. Commonly known as identity theft, this method allows fraudsters to use stolen information to make transactions online or over the phone, often without a recognized card [10]. By determining the best hyperplane at multiple locations, SVM aims to maximize the edges of the hyperplane and the closest data (called support vectors) while minimizing distribution error. The ability of the SVM algorithm to determine the best hyperplane dividing the points of each category is shown in Figure 2.

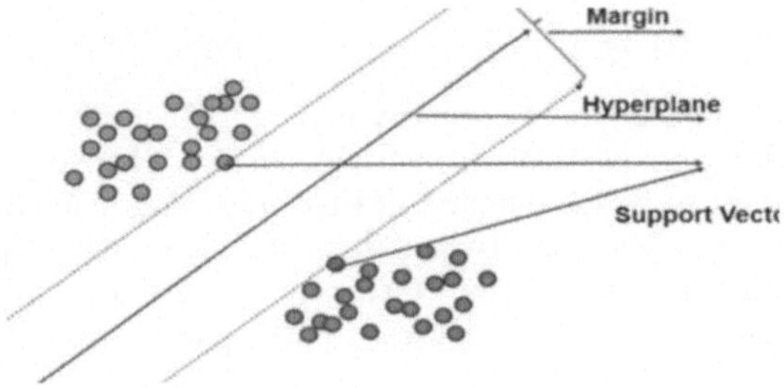

Figure 2. Support vector machine (SVM).

Easily experiment with separable and non-separable data using physical activity [12]. This function transforms input data into a higher dimensional space, allowing data to be separated by hyperplanes. It is valued for its ability to handle complex data and is known for its flexibility and general capabilities [11]. The following explains each kernel function and how important it is for spotting credit card fraud:

- Linear Kernel Function: This fundamental kernel function is adept at identifying linear relationships within transaction attributes. By computing the inner product of transaction vectors, it effectively discerns instances of linear reparability in fraudulent behavior [12].

$$(x_i, x_j) = x_i \cdot x_j \tag{1}$$

- Sigmoid Kernel Function: The hyperbolic tangent is used by the sigmoid kernel to identify non-linear correlations in transactional data. This function allows for flexible mapping to find complex patterns suggestive of fraud by applying it to the dot product of transaction vectors [13].

$$(x_i, x_j) = \tan\ h(\gamma x_i x_j + r) \tag{2}$$

The main technique for categorizing fraudulent transactions in credit card data is support vector machines (SVM) [14].

4 METHODOLOGY

The identification of credit card fraud may be aided by examining consumer spending habits. Any departure from customary practice can indicate potential fraud. Using the dataset, we selected important features (transaction amount, date, time, location, frequency of purchases, and billing address) that best represent the behavior of our customers. We train our model using these features as the foundation.

Table. 1 selected important feature.

Feature No.	Feature Name
1	Transaction Amount
2	Date
3	Time
4	Frequency of card usage
5	Place
6	Customer ID
7	Average amount of transactions per mont

4.1 *Data preprocessing and feature selection*

An essential first step in the identification of fraud is the Conversion of the dataset features into numerical data [15]. Feature selection is pivotal, aiming to efficiently capture the behavioral patterns of individual credit card accounts. In this model, only features that directly interpret customer behavior are chosen for detection, ensuring classifier efficiency. The selected features, outlined in Table 1, include transaction amount, and frequency of card usage derived from date and time attributes.

4.2 *Feature extraction*

Feature extraction reduces the dimensionality of input data by condensing it into a small set of features that capture the key attributes. Task completion is aided by this reduced representation set without sacrificing precision. A useful technique for feature compression is Principal Component Analysis (PCA), which reduces the original feature space to low-dimensional spaces without affecting the solution [16] [26]. The covariance matrix was determined using the subsequent formula

$$Covariance = 1\sum_{i=n}^{n}(x_i - x)(y_i - y) \tag{3}$$

This facilitates PCA computation, yielding component scores that reduce the training dataset. This method optimizes training time, enabling efficient classifier performance.

4.3 *SVM training and classification*

The LIBSVM classifier is employed for training and classification tasks. With 591 samples, including 576 positive and 15 negative samples, SVM's performance is optimized by selecting a portion of features to mitigate the challenge posed by a large number of features. Within this construct, the Radial Basis Function (RBF) kernel is utilized [17], with accuracy optimized by tuning the RBF kernel parameter γ and the penalty parameter. Cross-validation using the grid search method facilitates the selection of optimal γ and C values.

4.4 *Results and evaluation*

Apart from the significance of precision, metrics like the fraud index and vulnerability index offer more information about the extent of the fraud's success. The confusion matrix is regarded as a crucial analytical tool in this study for assessing the model's efficacy in fraud detection [18]. The initial step in training a machine learning model is gathering real data and getting it ready for analysis, as Figure 3 illustrates. The model is optimized for real-world applications by carrying out extra and repeated validations and making sure that performance requirements are fulfilled. Iterations are used to continuously enhance this process to achieve efficiency and accuracy.

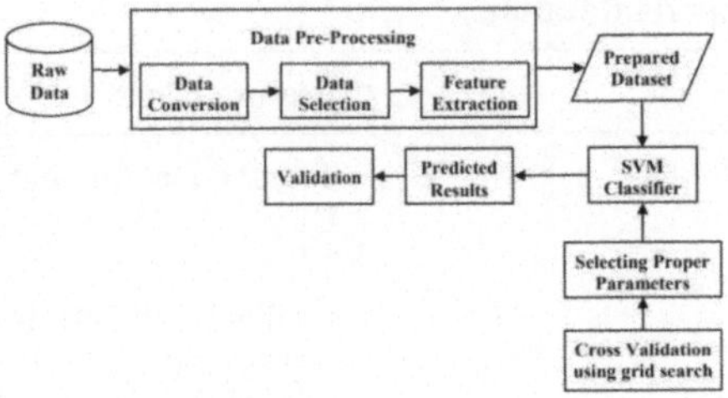

Figure 3. Procedure for training a model.

5 FINDINGS AND PERFORMANCE EVALUATION

Performance evaluation standards are crucial for ensuring the efficacy and dependability of credit card fraud investigations. Measuring the precision and effectiveness of these systems involves many crucial factors. Fraud detection systems are evaluated using basic measures like recall, accuracy, precision, F1 score, and false positive rate (FPR) [19]. When taken as a whole, these measures offer a framework for assessing how well credit card fraud technology works.

5.1 *Precision*

The percentage of transactions with fraud evidence that are reported as fraudulent is measured by the precision statistic. It comprises the ratio of the true prediction to the total number of good forecasts (true positive and negative) and indicates the accuracy of a good prediction [18]

$$\text{Precision} = \frac{\text{TP}}{(\text{TP} + \text{FP})} \tag{4}$$

The total amount of successful trading scams is known as True Positives (TP).

5.2 *Recall*

Recall is calculated as a percentage, also known as effective value or expected value. This shows that the model can detect fraud in all cases and provide accurate predictions based on the percentage of total fraud [22]

$$\text{Recall} = \frac{\text{TP}}{(\text{TP} + \text{FN})} \tag{5}$$

True-negative (TN): The quantity of real transactions that are accurately reported.

5.3 *Fl-score*

The F1-score offers a fair evaluation of the model's efficiency and is calculated as the harmonic average of both recall and precision. When there is an imbalance in the quantity of positive and negative situations, this metric which blends recall and precision toward one becomes helpful [24].

$$\text{F1 Score} = \frac{(2 \text{ X Precision X Recall})}{\text{Precision} + \text{Recal}} \tag{6}$$

True Negative (TN): The quantity of valid transactions that were accurately identified.

5.4 *False positive rate (FPR)*

FPR measures the proportion of legitimate transactions incorrectly classified as fraudulent among all actual legitimate transactions. It represents the rate of false alarms generated by the model.

$$\text{FPV} = \frac{\text{FP}}{(\text{FP} + \text{TN})} \tag{7}$$

The quantity correctly identified on valid transactions is known as True Negative (TN).

6 DETECTION ANALYSIS

This section compares our methodology and earlier studies on credit card fraud. Before delving into the specifics of our methods and their operation, it is crucial to consider our findings in the context of wider fraud bait technology. Furthermore, contrasting earlier fraud detection techniques reveals the benefits and drawbacks of various approaches. According to Table 2, We thoroughly examine metrics and methods for detecting fraud, such as decision trees, logistic regression [25], random forests [28], and naive Bayes [29], to increase the effectiveness and dependability of fraud detection.

Table 2. Comparative analysis of existing works in fraud detection.

Author	Method	Accuracy	Advantage	Disadvantage
[18]	K-Nearest Neighbors	87.6%	k-NN algorithms offer intuitive reasoning and do not require explicit model training, making them easy to implement.	Sensitive to the choice of distance meter and the number of neighbors, impacting classification outcomes.
[24]	Decision Tree	89.5%	Decision trees offer transparency and interpretability, making them suitable for explaining classification outcomes.	Prone to overfitting, especially with complex datasets.
[25]	Logic Regression	91.2%	Logistic regression is computationally efficient and provides probabilistic interpretations of classification.	Assumes linear relationships between input variables and lacks flexibility in capturing complex patterns.
[27]	Random Forest	90.8%	Random forest models are robust against overfitting all perform well with high-dimensional data.	Increased computational complexity and resource requirements compared to individual decision trees.

7 CONCLUSION

This paper offers a high-potential speed improvement method for improving fraud detection systems to overcome credit card fraud detection. The goal is to deepen the understanding of credit card fraud detection techniques and their complex operations to optimize kernel functions and strategies through the use of the Support Vector Machine (SVM) framework. To assess the accuracy and effectiveness of credit card fraud detection algorithms, the final form demonstrates how important it is to choose kernels and other alternative parameters. A thorough evaluation of the model that includes accuracy metrics, a close examination of the confusion matrix, and the optimization of methods like class weight adjustment illuminates the complex dynamics of the fraud detection system.

REFERENCES

[1] Thompson J., (2021), Digital transformation in banking: strategies and implications, *Journal of Financial Technology*, vol. 18, no. 1, pp. 45–58.

[2] Garcia A. *et al.*, (2020), The challenge of combatting credit card fraud in the digital age, *Journal financial Crime Prevention*, vol. 25, no. 2, pp. 78–91

[3] Smith K. *et al.*, (2021), Managing imbalanced datasets in fraud detection: Techniques and challenges, *Journal of Machine Learning Research*, vol. 32, no. 3, pp. 112–125.

[4] Johnson M. *et al.*, (2022), Credit card fraud detection using support vector machines, *Journal of Computational Finance*, vol. 35, no. 1, pp. 45–58.

[5] Martinez O. and Rodriguez P., (2021), Exploring support vector machines for fraud detection: A case study, *Journal of Financial Analytics*, vol. 28, no. 2, pp. 102–115.

[6] Garcia F. *et al.*, (2021), Handling imbalanced datasets in fraud detection using support vector machine with oversampling techniques, *Journal of Machine Learning Research*, vol. 30, no. 2, pp. 45–58.

[7] Thompson G. *et al.*, (2023), Expanding the applications of support vector machine in financial analysis beyond fraud detection, *Journal of Computational Finance*, vol. 40, no. 4, pp. 200–215.

[8] Brown A. *et al.*, (2021), Tackling fraud in financial transactions: A multifaceted approach, *International Journal of Financial Security*, vol. 20, no. 3, pp. 112–125.

[9] Johnson B. *et al.*, (2020), Understanding the dynamics of credit card fraud: Offline vs. online manifestations, *Journal of Financial Crime Prevention*, vol. 25, no. 2, pp. 78–91.

[10] Wilson C. *et al.*, (2022), Identity theft in online fraud: Strategies for prevention and detection, *Journal of Cybersecurity*, vol. 15, no. 4, pp. 200–215.

[11] Johnson M. *et al.*, Credit card fraud detection using support vector machines, *Journal of Computational Finance*, vol. 35, no. 1, pp. 45–58,

[12] Williams N. and Martinez O., (2019), Support vector machines in fraud detection: A comprehensive review, *Information Sciences*, vol. 120, pp. 78–91.

[13] Machines: A Case Study, *Journal of Financial Analytics*, vol. 25, no. 2, pp. 102–115, 2021

[14] Rai AK and Dwivedi RK., (2020). Fraud detection in credit card data using unsupervised machine learning based scheme. In: *IEEE*.p. 421–426.

[15] Zhang L. *et al.*, (2020), Deep learning approaches for credit card fraud detection: A comprehensive review, *Expert Systems with Applications*, vol. 150, p. 113367.

[16] Gupta R. *et al.*, (2021), Enhanced credit card fraud detection using hybrid machine learning techniques, *Journal of Computer Security*, vol. 28, no. 4, pp. 501–515.

[17] Patel S. and Jain V., (2021), Feature engineering techniques for credit card fraud detection: A comparative study, *International Journal of Advanced Computer Science and Applications*, vol. 12, no. 5, pp. 456–465.

[18] Wang X. *et al.*, (2022), Credit card fraud detection using ensemble learning and feature importance analysis, *Information Sciences*, vol. 589, pp. 93–105.

[19] Chen Y. *et al.*, (2022), Multi-objective optimization for credit card fraud detection: A hybrid approach, *Neural Computing and Applications*, vol. 34, no. 12, pp. 7241–7254.

[20] Liu Z. *et al.*, (2022), A novel framework for credit card fraud detection based on autoencoder and XGBoost, *IEEE Transactions on Cybernetics*, vol. 52, no. 3, pp. 1465–1478.

[21] Singh C. and Sharma S., (2022), Federated learning for credit card fraud detection in IoT environments, *Computers & Security*, vol. 114, p. 102332.

[22] Ramesh, G., Prasanna, G. B., Santosh, V. B., Chandrashekar, N., and Champa, H. N. (2022). KHDR: Kannada handwritten digit recognition using PCA and SVM classifier. In *Advances in Distributed Computing and Machine Learning: Proceedings of ICADCML 2021* (pp. 205–214). Springer Singapore.

[23] Jain, Y., Namrata Tiwari, S., and Jain, S. (2019). A comparative analysis of various credit card fraud detection techniques. *International Journal of Recent Technology and Engineering*, 7(5S2), 402–407

[24] Tan D., Subramanian R. R., Eswar D., Reddy G. V. P., Kumar A. R., and Praneeth C. V. N. M., (2021), Credit card fraud detection using machine learning, *2021 5th International Conference on Intelligent Computing and Control Systems (ICICCS)*, Madurai, India, pp. 967–972. do: 10.1109/ICICCS51141.2021.9432308

[25] GB, P., Bhat V.S., Naik C., and HN, C. (2021). An efficient method for handwritten kannada digit recognition based on PCA and SVM classifier. *Journal of Information Systems and Telecommunication (JIST)*, *3*(35), 169.

[26] Rathore A. S., Kumar A., Tomar D., Goyal V., Sarda K. and Vij D., (2021), Credit card fraud detection using machine learning, *2021 10th International Conference on System Modeling & (SMART)*, MORADABAD, India, pp. 167–171. do: 10.1109/SMART52563.2021.9676262

Data Science & Exploration in Artificial Intelligence – Gururaj et al. (Eds)

Temporal analysis of Indian farmers' protests on Twitter with comparative insights

Chinmay Chahar & Rishabh Kaushal

Department of Information Technology, Indira Gandhi Delhi Technical University for Women, Kashmere Gate, New Delhi, India

ABSTRACT: This study compares Twitter discourse during the 2021 and 2024 Indian farmers' protests, aiming to detect sentiment shifts and bot prevalence. The protests erupted in opposition to new farm laws and evolved into significant socio-political movements. We conduct exploratory analysis to identify sentiment variations using scraped and pre-available datasets. We also distinguish between bot-generated and human-generated tweets, to assess the changes in the bot prevalence amidst Twitter's crackdown using classification techniques. The aim is to unveil nuanced shifts in general public's opinion online and the discourse evolution during socio-political movements with linguistic pattern identification.

1 INTRODUCTION

The Indian farmers' protests of 2021 and 2024 have sparked significant socio-political discourse throughout the span of the previous 3-4 years, shading light on different perspectives and sentiments among the public online. However, understanding the dynamics of public opinions during these protests remains a complex challenge. The emergence of social media platforms like Twitter as influential spaces for ideas and opinions exchange demands a comprehensive analysis to detect and quantify these shifts (whether positive or negative) in the public opinion.

Therefore, there is a pressing need to investigate similar underlying patterns of sentiments within the context of the Indian farmers' protests, leveraging advanced analytical techniques to understand these nuanced dynamics online.

To study (detect and quantify) shifts (disagreements or agreements) of public opinion using Indian farmers' protests as a case study, the research question we pose is: How do sentiments expressed on Twitter during the Indian farmers' protests of 2021 and 2024 evolve over time, and what role do bots (unnatural influence) play in the same?

2 LITERATURE REVIEW

The proliferation of social media platforms in today's times has revolutionized the landscape of public opinions expressed in a social environment, providing researchers with unprecedented opportunities to study sentiment, behavior, and communication dynamics during important socio-political movements. This section reviews similar studies and recent research that have contributed to understanding the role of social media, particularly Twitter, in shaping public discourse during protests.

The 2023 study, Dr. Sarabjeet Kaur Kochhar, Chinmay Chahar [4], addresses the need to discern genuine perspectives from manipulated content during the 2021 Indian farmers' protests. Their novel approach involving stance classification and bot detection emphasizes the importance of investigating content authenticity for public perception and policy formulation. This particular research serves as a foundational reference for our paper, underlining the significance of understanding manipulated content in public discourse.

The following table provides a concise summary of each study's focus and key findings, offering a comprehensive overview of the literature relevant to our research question.

DOI: 10.1201/9781003589273-37

Table 1. Comprehensive review of relevant studies: Setting the context for research.

Study	Focus	Key Findings
Castillo *et al.* (2011)	Real-time data analysis on Twitter	Twitter serves as a valuable source for analyzing sentiment and behavior during social events.
Conover *et al.* (2011)	Patterns of behavior on Twitter	Twitter facilitates information dissemination and collective action among users during social events.
González-Bailón *et al.* (2014)	Role of Twitter in shaping public discourse	Twitter amplifies protest-related information, aiding in online mobilization of sentiment.
Bastos and Mercea (2019)	Influence of social media bots during protests	Bot-driven narratives on Twitter significantly shape public discourse and perception during protests.
Kwon *et al.* (2020)	Sentiment polarization on Twitter	Echo chambers and algorithmic biases contribute to the polarization of public sentiment during protests.
Olteanu *et al.* (2019)	Spread of misinformation on Twitter	Misinformation and rumors spread rapidly on Twitter during crises and protests, impacting public perception.
Bode and Vraga (2015)	Role of social media in shaping public opinion	Exposure to protest-related content on social media influences individuals' perceptions
Dr. Sarabjeet Kaur Kochhar, Chinmay Chahar (2023)	Presence of harmful content and Twitter bots surrounding Farmers Protest 2021	Widespread presence of bots in tweets related to the farmers' protest, mimicking human communication patterns. These insights valuable for policymakers in understanding and addressing social media manipulation.

3 METHODOLOGY

3.1 *Theoretical frameworks*

The Social Amplification of Risk Framework (SARF), proposed by Kasperson *et al.* (1988) [5], serves as a foundational theoretical lens in this research study for understanding the dynamics of societal responses to risks/events. SARF states that the perception and amplification of risks are influenced not only by the inherent characteristics of the risk itself but also by social and cultural factors, which is in-line with our research thought.

Furthermore, Media System Dependency Theory (MSD), developed by Ball-Rokeach and DeFleur (1976) [6], focuses on the interdependence between media systems and public. According to MSD, individuals these days rely heavily on media for information, entertainment, and social interaction. This leads to the development of a symbiotic relationship between media systems and their audiences. MSD also emphasizes the role of media in shaping individuals' perceptions, attitudes, and behaviors, particularly in the context of important socio-political events.

Drawing upon these theoretical lenses of SARF and MSD, our methodology employs a mixed- methods approach to analyze Twitter discourse surrounding the Indian farmers' protests. This approach integrates quantitative techniques, such as exploratory data analysis to uncover patterns and trends within the dataset, and classification algorithms to distinguish between human-authored and bot- generated tweets. Additionally, qualitative methods, including content analysis to examine the themes and topics discussed in the tweets, and linguistic pattern identification to discern shifts in sentiment and language use, are employed.

The aim of combining quantitative and qualitative approaches, is to gain a comprehensive understanding of the multi-faceted dynamics of public opinions during the protests, capturing both the quantitative metrics and qualitative nuances inherent to the theme.

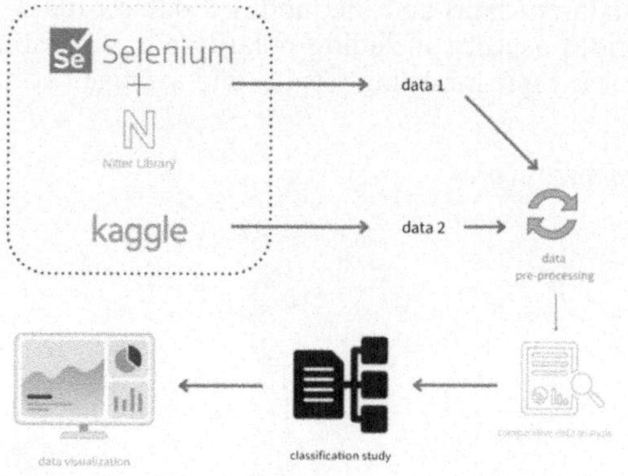

Figure 1. Framework diagram.

4 DATASETS

For the first version of the dataset (data_v1), Twitter data was extracted from the dataset available on Kaggle as mentioned in the reference paper [4]. However, due to new API restrictions and ethical considerations, Selenium was utilized for scraping tweets from Twitter to create the second version of the dataset (data). These scraping methods yielded approximately 5000 tweets for us to get started with.

Table 2. Comparative overview of two datasets.

Dataset	Label	Shape	Version (year)
Dataset 1 (data)	data	3806 rows × 16 columns	2 (2024)
Dataset 2 (data_v1)	data_v1	3806 rows × 16 columns	(1) (2021)

4.1 *Dataset 1 columns*

'id', 'renderedContent', 'username', 'likeCount', 'mentionedUsers', 'displayname', 'created', 'tweet', 'polarity_score', 'polarity', 'hashtag_per_tweet', 'tweet_length', 'combined_text', 'mentionCount', 'input_text', 'label'

4.2 *Dataset 2 columns*

'id', 'username', 'handle', 'created_date', 'tweet_text', 'likes', 'tweet', 'polarity_score', 'polarity', 'date', 'mentionCount', 'hashtag_per_tweet', 'tweet_length', 'combined_text', 'input_text', 'label'

Initial data processing was conducted to handle null values, resulting in the final dataset (data) consisting of 3806 tweets for analysis. To ensure comparability between the two versions, 3806 tweet rows were randomly sampled from the Kaggle version 1 data (data_v1). This sampling approach ensured that both datasets contained an equal number of tweets for comparison analysis.

The comparative analysis started with an exploratory data analysis (EDA) to identify key differences and similarities between the two versions of the dataset. Various attributes were analyzed, including frequent hashtags, users tagged, and sentiment distribution.

5 COMPARATIVE EXPLORATORY DATA ANALYSIS (EDA)

Exploratory Data Analysis (EDA) plays a pivotal role in understanding the characteristics and patterns inherent within datasets. In this section, we conduct a comparative EDA between the original dataset (data) and the modified dataset (data_v1). The analysis takes into consideration various aspects, including polarity score distributions, word usage patterns, n-gram frequencies, topic modeling results, and average tweet lengths.

5.1 *Distribution of polarity scores*

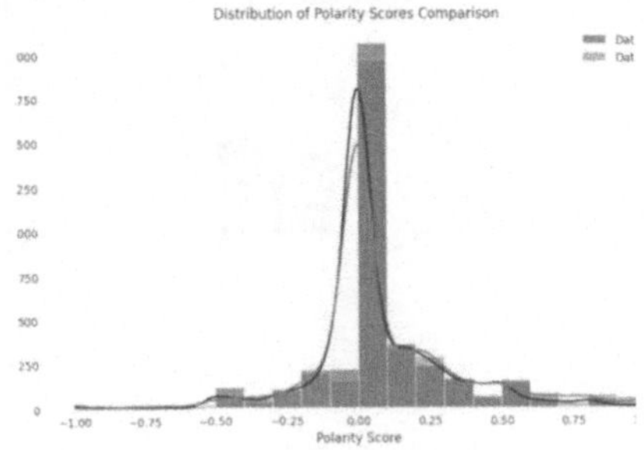

Figure 2. Polarity score distribution.

The mean polarity scores are calculated for each dataset to quantify the overall sentiment polarity. A comparison reveals any notable differences in sentiment between the original and modified datasets.

Mean Polarity Score for Data: 0.05613777059290721. Mean Polarity Score for Data_v1: 0.08585669400061641. Data_v1 has a higher mean polarity score.

5.2 *Topic modeling using LDA*

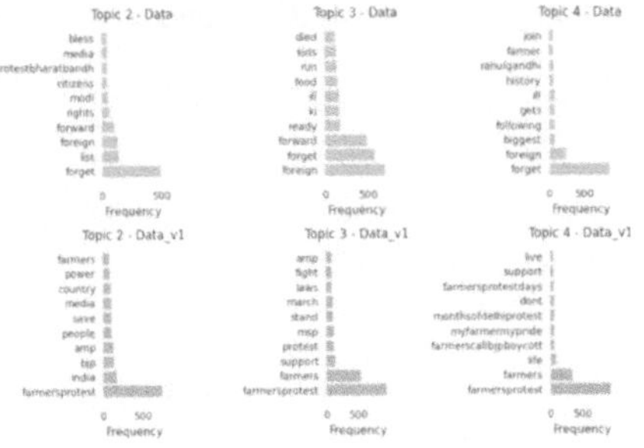

Figure 3. Topic modelling.

Latent Dirichlet Allocation (LDA) is applied to identify topics within the datasets. For each dataset, the top words associated with each topic are visualized, which offered insights into prevalent themes or subjects discussed in these tweets.

5.3 *Most tagged twitter handle and hashtags*

In order to compare the engagement and interaction patterns across two tweet datasets, an analysis of the most tagged Twitter handles was conducted. The extraction of tags and mentions from both datasets was done with the use of regular expressions to identify and extract Twitter handles starting with '@'. To ensure consistency, all extracted mentions were converted to lowercase.

Following the extraction process, the frequency of each mentioned Twitter handle was calculated within each dataset. This facilitated the identification of the top 10 most mentioned Twitter handles in each dataset, as we see in the figures below.

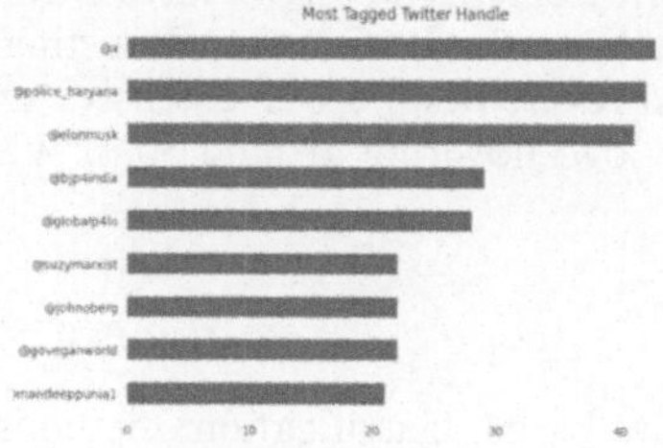

Figure 4. Most tagged handles in data.

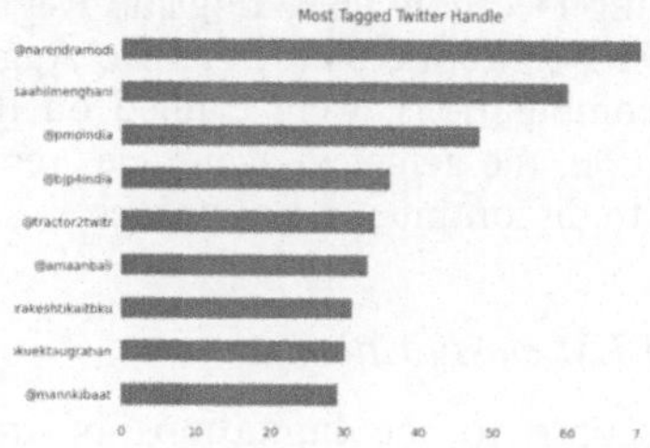

Figure 5. Most tagged handles in data_v1.

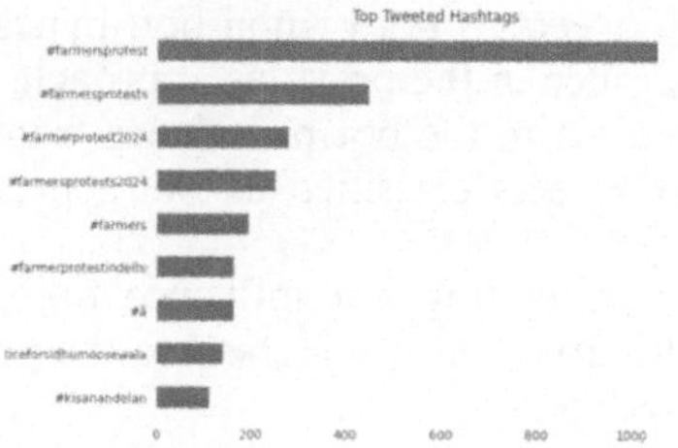

Figure 6. Most frequent hashtags in data.

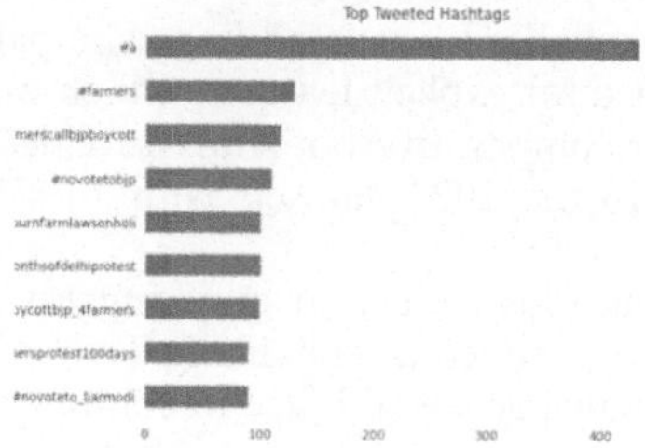

Figure 7. Most frequent hashtags in data_v1.

In conclusion, the exploratory data analysis (EDA) undertaken on the two tweet datasets has provided valuable insights into the dynamics of user engagement, sentiment distribution, and content characteristics.

The Key takeaways from this analysis is the observation that tweets from the 2024 dataset exhibit a higher prevalence of negative sentiment scores in polarity compared to those from the 2021 dataset. Also, tweets from the 2024 dataset tend to have slightly shorter lengths on average. On the other hand, the mean polarity score for tweets in the 2021 dataset was found to be higher than that of the 2024 dataset.

6 BOT OR NOT

In the realm of social media, bots have become a pervasive presence, shaping online content widely and also influencing public opinion. With Elon Musk's leadership on Twitter from 2022, efforts to combat bot influence were said to be focused upon highly. This section embarks on a comparative analysis of bot prevalence in Twitter datasets from 2021 and 2024. The study aims to elucidate whether the platform's measures to mitigate bot activity have yielded any noticeable effects over time, during the span of the two protest waves.

In reference to the base study by [4] Dr. Sarabjeet Kaur Kochhar, Chinmay Chahar (2023) we realize the relevance of bot detection research, particularly in the context of the 2021 dataset. They conducted a bot detection study and found intriguing correlations between tweet polarity and bot influence, as detailed in their research. The decision tree classifier results from this paper indicate notable bot presence, particularly in tweets categorized as anti-farmer, suggesting a potential influence on the narrative surrounding socio-political events. This was an interesting head-start to our own motivation for this bot detection study.

6.1 *Supervised classification*

An initial attempt was made to train classification models using a labeled dataset sourced from Kaggle [8], comprising features related to Twitter bot detection. The dataset included textual features such as 'Tweet' and 'Hashtags', along with numeric features like 'Polarity Score', 'Likes/Retweet Count', and 'Mention Count'. Various machine learning algorithms, including Decision Tree, Logistic Regression, Random Forest, Gaussian Naive Bayes, K-Nearest Neighbors, SVM, Passive Aggressive, Ensemble methods, Gradient Boosting, and their combinations, were trained on this dataset. However, despite our extensive experimentation, the achieved accuracies remained notably low, hovering around 50%. We thus opted to discontinue this approach.

6.2 *LLM classification*

In response to the limitations of traditional feature-based classification methods, we experimented with OpenAI's prompt-based text classification approach. Utilizing the model "gpt-3.5-turbo- instruct" via OpenAI's API keys, we trained text classification models directly on tweet content.

This method leveraged linguistic patterns to classify tweets as bot or non-bot, bypassing the need for explicit features, which was a major hinderance in the previous approach.

Our comparative bot study revealed a marginal decrease in the bot prevalence from the 2021 to the 2024 dataset, with 14.87% and 13.9% of tweets classified as bot-generated, respectively.

While this reduction may suggest some success in mitigating bot influence following initiatives aimed at enhancing Twitter's platform credibility, it also underscores the persistent challenge we still face in combating fake content online.

Table 3. Bot comparison in both datasets.

Dataset Year 2021	Bot Prevalence
2021	14.87%
2024	13.9%

7 CONCLUSION

In summary, our comparative bot study found a slight decrease in the bot prevalence from 2021 (14.87%) to 2024 (13.9%) on Twitter based on our datasets, surrounding the Indian farmers' protests. While this indicates some progress, the continued presence of bot-generated content does highlight the ongoing challenges we face collectively to ensure the authenticity online. Efforts to prevent bot manipulation are thus crucial for preserving the liberty of public content and safeguarding social values among today's digital diaspora.

REFERENCES

[1] Dr. Sarabjeet Kaur Kochhar and Chinmay Chahar (2023). Performing stance classification and bot detection on the Indian Farmers' protest – A study to unveil hidden perspectives. *Adv. Artif. Intell. Mach. Learn.*, 3 (4):1619–1639

[2] https://doi.org/10.1111/j.1539-6924.1988.tb01168.x

[3] Ball-Rokeach, S. J., and DeFleur, M. L. (1976). A dependency model of mass-media effects. *Communication Research*, 3(1), 3–21. https://doi.org/10.1177/009365027600300101

[4] https://www.kaggle.com/datasets/prathamsharma123/farmers-protest-tweets-dataset-csv

[5] https://www.kaggle.com/datasets/goyaladi/twitter-bot-detection-dataset/data

Data Science & Exploration in Artificial Intelligence – Gururaj et al. (Eds)
© 2025 the Author(s), ISBN: 978-1-032-96390-7

Novel deep learning technique to predict skin cancer

Napa Lakshmi, A. Ashwitha & Shaleen Bhatnagar
Department of Computer Science and Engineering, Manipal Institute of Technology Bengaluru, Manipal Academy of higher Education, Manipal, India

Pulukuri Aparna
Department of Computer Science and Engineering Brindavan College of Engineering, Bengaluru, India

ABSTRACT: The rising incidence of melanoma skin cancer in humans, coupled with its high treatment costs and mortality rates, underscores the urgency of early detection. Traditional cancer cell identification and treatment methods are often time-consuming and manual. To address this, a recent study proposed an artificial skin cancer diagnosis system that combines machine learning and image processing techniques. In this study, dermoscopic images are first segmented, and the relevant features of the damaged skin cells are extracted through a feature extraction method. These extracted features are then categorized using a convolutional neural network classifier, which is grounded in deep learning principles. The reported results, based on publicly available data, indicate an impressive training accuracy of 99.5% and a testing accuracy of 99.7%.

Keywords: Skin cancer detection, Convolutional neural network, Deep learning

1 INTRODUCTION

The Early detection of skin cancer plays a critical role in achieving successful treatment outcomes. However, this typically relies on the expertise of highly qualified specialists, posing a challenge due to a shortage of such professionals and the cost associated with their services. To deal with this matter, computer-aided approaches are developed for skin cancer recognition has emerged as a promising solution, potentially saving lives while maintaining cost-effectiveness.

The shortage of specialized healthcare practitioners and the limited availability of diagnostic equipment for accurate skin cancer diagnosis have underscored the importance of these computer-aided systems. Former research in this sphere has predominantly cantered on leveraging machine learning techniques to analyse dermoscopic images. These images are obtained by specialized dermoscopic equipment, typically found in larger clinical settings and hospitals. Deep learning neural networks, CNNs are an important way that deep learning is being used in the medical field to identify skin cancer. In image classification tasks, such as the recognition of skin lesions and the detection of skin cancer, CNNs have shown extraordinary effectiveness. Skin cancer is the most common type of cancer in humans. It is usually detected visually following a preliminary diagnostic examination, and potentially following a biopsy, a dermoscopic study, and a histological examination. Classifying skin lesions from pictures is challenging due to the fine-grained variation in how skin lesions originate.

Cancer detection has emerged as a compelling and trending area of research in recent years, primarily due to the alarming surge in skin cancer diagnoses. Between 2009 and 2019, the incidence of skin cancer rose significantly, reaching a concerning rate of 54% [1].

DOI: 10.1201/9781003589273-38

Melanoma is the most dangerous and deadly type of skin cancer among the others, accounting for more than 9,000 fatalities in US each year [2]. Meanwhile, in Australia, the yearly tally of new melanoma cases exceeds 13,000, resulting in more than 1,200 fatalities. Europe similarly grapples with a significant burden, with melanoma causing over 20,000 deaths annually [3].

It is worth noting that skin cancer, when detected early, can often be successfully treated. But the real difficulty is in obtaining an accurate timely detection of the illness, which is usually the domain of highly qualified professionals. Sadly, there is a significant lack of specialists and medical equipment that can reliably and accurately diagnose skin cancer. This shortcoming emphasizes how crucial it is to create cutting-edge computer-aided systems designed to detect skin cancer, as they hold the potential to save numerous lives while keeping healthcare costs in check.

2 RELATED WORK

The investigation of machine learning methods for skin cancer diagnosis has attracted a lot of researchers. In the context of segmenting skin lesions within input images, current systems typically employ border detection methods that can be categorized as manual, semi-automatic, or fully automatic [5][6]. These methods rely on a variety of features to accomplish skin lesion segmentation, including shape, color, texture, and luminance. The literature on this subject contains a multitude of border detection approaches such as hybrid thresholding techniques [7].

In reference [5], the author introduces an automated approach for identifying the borders of skin lesions in dermoscopy images. This technique relies on a combination of color-space analysis and global histogram thresholding. It has shown impressive effectiveness in accurately identifying the boundaries of melanoma lesions. On the other hand, in reference [8], researchers adopt a different technique. They partition the input image into several regions of clinical significance using the Euclidean distance. This approach facilitates the mining of crucial features of color and texture for analysis.

Dermoscopy relies on the ABCD rule, which prioritizes asymmetry as the most prominent among its four key features. These features also include border irregularity, color variation, and lesion diameter. Numerous studies are conducted to quantify asymmetry in skin lesions. Certain methods of calculating the symmetry feature rely on geometric measurements, like symmetric distance and circularity, that cover the entire lesion. In contrast, circularity index is proposed as a metric for evaluating border irregularity in dermoscopy images. To gain a comprehensive understanding of the noteworthy implementations in the existing literature and to assess the performance of different classifiers designed for the specific task of diagnosing skin lesions, I recommend consulting paper [9].

In a prior research endeavour discussed in [10], an ensemble approach was introduced to automate the detection of melanoma. This method involved a combination of CNN techniques and the extraction of image texture features. VGG-19 is used for image classification in the CNN phase. Concurrently, texture features were extracted, and kernel Principal Component Analysis (kPCA) was used to reduce their dimensions to improve the classifier's efficiency during the feature extraction stage.

To address the limitations previously mentioned, the authors of [11] introduced a deep learning technique for the automatic recognition and segmentation of melanoma lesions. An enhanced encoded-decoded network consisting of encoding and decoding sub-networks is presented. The decoding feature vectors are made available for effective feature extraction. In another study by Adla *et al.* [12], the DLCAL-SLDC technique, automated Deep Learning with Class Attention Layer-based Computer-Aided Diagnosis method is designed for Skin Lesion Detection and Classification,' was established. This research incorporated a

Tsallis entropy-based segmentation approach to identify affected lesions within dermoscopic images.

The model proposed in [13] employs color and texture features to categorize images into melanoma and benign categories. Their approach involves applying Gaussian noise during pre-processing, utilizing the k-means algorithm for segmentation, and subsequently extracting color and texture features. Researchers utilize the Cluster-based Adaptive Metric classifier to conduct picture classification. When tested on the Med-node dataset, the accuracy of this model was 81%.

Data augmentation is introduced in [14] as a method to evaluate the effects of training a CNN architecture with non-augmented data and again with augmented data—on the model's performance. This model's main objective is to categorize skin cancer images into two distinct groups: benign and malignant. Notably, the accuracy obtained with non-augmented data is 78%, whereas it increases to 81% with enhanced data. The ISBI 2016 dataset is used to evaluate their model [15]. In other studies, such as [16], additional pre-processing techniques are applied. This includes edge detection and filtering, followed by the extraction of features like shape, color, diameter, irregularities, and boundaries. Subsequently, a CNN is employed for the detection of malignant tumors, resulting in an impressive accuracy rate of 95%. This model is evaluated on datasets such as DermIS and DermQuest.

On the other hand, some researchers opted for Support Vector Machine (SVM) as their classification method in the domain of skin cancer types, as exemplified by [17]. They devised a model tailored for the detection of melanoma cancer, yielding an impressive accuracy level of 96%. In their approach, color and texture features were extracted from the images, and SVM was subsequently applied to classify these feature vectors into melanoma or non-melanoma categories. The evaluation of this model was conducted on the PH2 Dataset [18].

Researchers in the field of skin cancer detection are actively pursuing diverse methodologies to enhance accuracy and effectiveness in medical diagnosis. Their efforts encompass a wide spectrum of approaches, from hybrid techniques that blend deep learning with manual feature extraction to innovative combinations of classifiers. The overarching goal is to refine the recognition of skin lesions, thereby improving the diagnostic process. These initiatives highlight the versatility of strategies deployed, spanning from exclusive reliance on Convolutional Neural Networks (CNNs) to the utilization of Support Vector Machines (SVMs) as potent classification tools. Each of these approaches is geared toward elevating accuracy levels and making meaningful contributions to the diagnosis of skin cancer.

3 PROPOSED METHODOLOGY

The proposed model incorporates Convolutional Neural Networks, with the initial layer being the Conv2D layer. This layer essentially functions as a collection of trainable filters. We chose to use 64 filters for the initial two Conv2D layers. Depending on the filter size, each filter adds a filter to a certain area of the input image. This process effectively involves applying the filter matrix across the entire image. Conceptually, you can think of these filters as tools for altering the appearance of the image. The CNN then extracts valuable features from these transformed images, which are referred to as feature maps. Figure 1 illustrates the proposed model's architecture.

The next critical layer in the CNN architecture is the pooling layer, specifically MaxPool2D. In simple terms, this layer functions as a down-sampling filter. It accomplishes this by selecting the maximum value from a pair of adjacent pixels during its evaluation. The primary purposes of this layer are to ease computational complexity and, to mitigate over-fitting. The choice of the pooling area size is essential; the larger the down-sampling, the greater the pooling dimension.

The combination of convolutional and pooling layers in CNNs allows for the extraction of global features from the image by aggregating local characteristics. Additionally, a regularization technique known as "dropout" randomly deactivates a portion of the nodes within a layer for each training sample, effectively setting their weights to zero. Consequently, a fraction of the network is randomly omitted, compelling the network to be trained features in a more distributed manner.

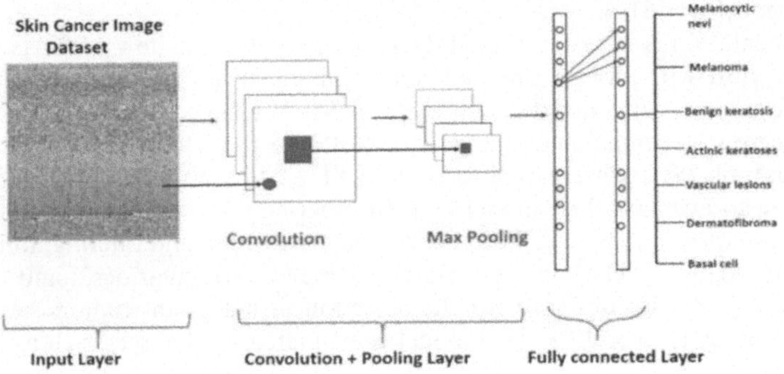

Figure 1. Architecture of the suggested model.

This approach not only advances generalization but also lowers overfitting. Rectifier activation function, ReLU (max(0, x)), introduces nonlinearity into the network, enhancing its capacity to model complex relationships. Following the generation of feature maps, they are flattened into a single 1D vector using the flatten layer. This step is necessary when transitioning to fully connected layers after multiple convolutional and max-pooling layers. It consolidates all local features discovered in the preceding convolutional layers. Finally, a single dense layer is utilized in an artificial neural network classifier, making use of the extracted features. Figure 2 provides outline of the CNN model's architecture.

Model: "sequential"

Layer (type)	Output Shape	Param #
conv2d (Conv2D)	(None, 28, 28, 16)	448
max_pooling2d (MaxPooling2D)	(None, 14, 14, 16)	0
batch_normalization (BatchNo	(None, 14, 14, 16)	64
conv2d_1 (Conv2D)	(None, 12, 12, 32)	4640
conv2d_2 (Conv2D)	(None, 10, 10, 64)	18496
max_pooling2d_1 (MaxPooling2	(None, 5, 5, 64)	0
batch_normalization_1 (Batch	(None, 5, 5, 64)	256
conv2d_3 (Conv2D)	(None, 3, 3, 128)	73856
conv2d_4 (Conv2D)	(None, 1, 1, 256)	295168
flatten (Flatten)	(None, 256)	0
dropout (Dropout)	(None, 256)	0

Figure 2. Summary of proposed CNN architecture.

4 RESULTS AND DISCUSSIONS

The proposed deep learning model's performance results are discussed below.

4.1 *Dataset*

The ISIC (International Skin Image Collaboration) Repository is where this data set is taken from. It has 1800 pictures of benign moles and 1497 pictures of classified malignant moles. All the images have been downsized to 224x224x3 RGB low resolution. This kernel's job is to develop a model that can visually distinguish between benign and malignant moles. Some images are illustrated in Figure 3.

The CNN model is subjected to two distinct training and testing scenarios. In the first scenario, 80% of the available images are designated as the training dataset. The leftover 20% are reserved for testing and evaluation. The second scenario involves a different partition, where 70% of the images serve as the training dataset, and the remaining 30% are allocated for testing purposes.

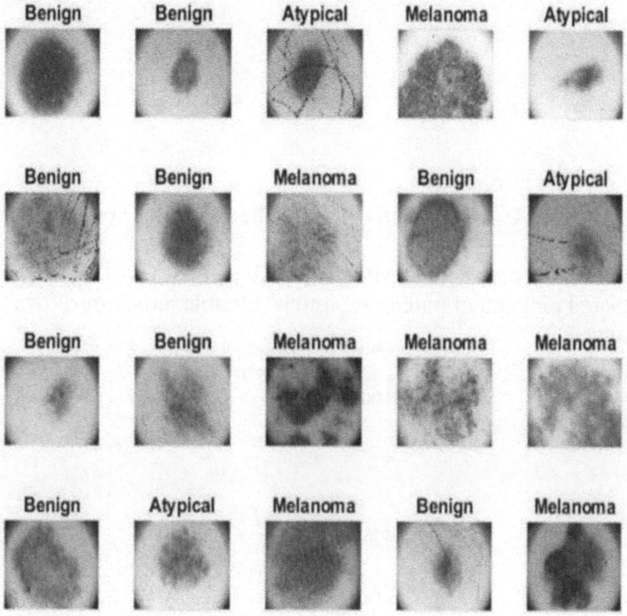

Figure 3. Few images from ISIC dataset.

During training, the ADAM optimizer is employed to optimize the model's execution. ADAM optimizer increases the accuracy and also the speed of the training. Batch normalization is incorporated into the model's architecture to expedite the training process and normalize the activations within the hidden layers. This normalization step helps stabilize and accelerate the learning process. To mitigate the risk of overfitting, a dropout mechanism is employed. This technique involves randomly deactivating a subset of neurons during training, effectively setting their contributions to zero. This regularization approach improves the model's capability to generalize from the training data to unseen data, ultimately improving its performance.

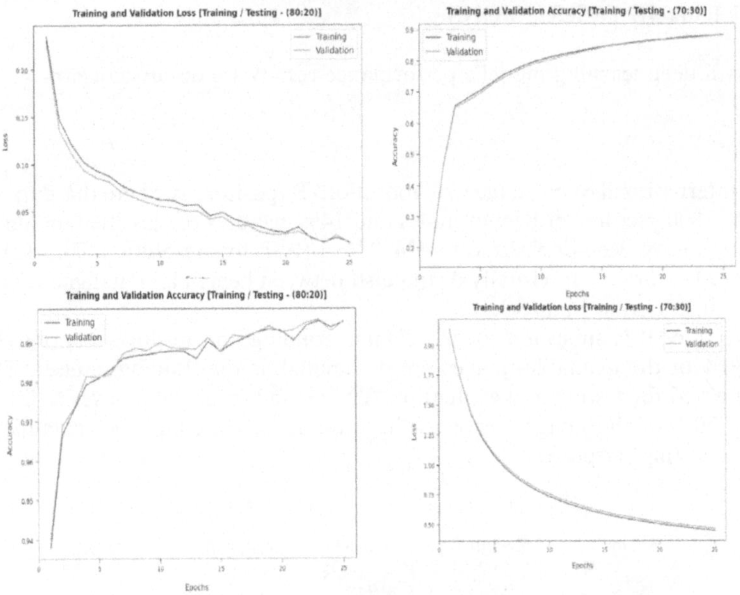

Figure 4. Training and testing accuracy.

4.2 *Performance evaluation*

The various metrics to assess the effectiveness of the proposed system at various stages are discussed below.

Matthew's correlation coefficient (MCC): MCC provides a measure of the correlation between the predicted and actual outcomes, and a valuable indicator of overall performance.

Recall (Sensitivity):

$$\text{Recall} = \frac{TP}{TP + FN}$$

Precision:

$$\text{Precision} = \frac{TP}{TP + FP}$$

F1 Score:

$$\boldsymbol{F1\ score = 2 \times \frac{Precision * Recall}{Precision + Recall}}$$

Specificity:

$$Specificity = \frac{TN}{TN + FP}$$

Accuracy:

$$\boldsymbol{Accuracy = \frac{TP + TN}{TP + FP + TN + FN}}$$

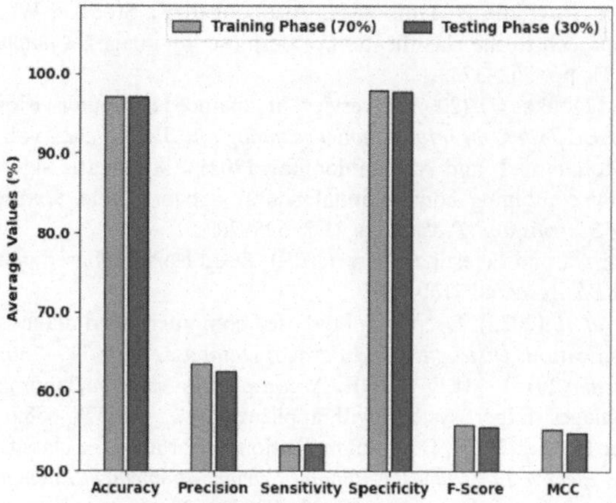

Figure 5. Performance metrics of the proposed model.

These metrics collectively provide a comprehensive assessment of the system's performance. The results are visually presented in Figure 5. As shown in the figure, the performance during the training of the network is higher compared to the testing of the network because training of the network uses more number of input images.

5 CONCLUSION

A new computer-aided diagnostic model for efficient detection and classifying skin cancer using dermoscopic pictures has been built, driven by an ideal deep neural network. CNN based method for classifying melanoma has been proposed. A system is being developed to help people and medical professionals identify and categorize different types of skin cancer, whether they are benign or malignant. On ISIC publicly available data, an accuracy of 99.5% on training data and a testing accuracy of 99.7% were attained.

REFERENCES

[1] Putra T. A., Rufaida S. I. and Leu J. S., (2020), Enhanced skin condition prediction through machine learning using dynamic training and testing augmentation, in *IEEE Access*, vol. 8, pp. 40536–40546, doi: 10.1109/ACCESS.2020.2976045.

[2] Codella, Noel CF, *et al.* (2017). Deep learning ensembles for melanoma recognition in dermoscopy images. *IBM Journal of Research and Development* 61.4/5: 5–1.

[3] Yu Q., Bu X., Chi R. and Hou Z., (2018), Modified P-Type ILC for high-speed trains with varying trial lengths, *2018 IEEE 7th Data Driven Control and Learning Systems Conference (DDCLS)*, Enshi, China, pp. 1006–1010, doi: 10.1109/DDCLS.2018.8516062

[4] Begum, Sumaiyya, and Syeda Asra. (2017). Extraction of skin lesions from non dermoscopic images using deep learning, *Int J Sci Res Comput Sci* 2: 591–596.

[5] Garnavi R., (2011). *Computer-aided Diagnosis of Melanoma*, Ph.D. dissertation, University of Melbourne, Australia

[6] Celebi M.E., Iyatomi H., Schaefer G., and Stoecker W. V., (2009), Lesion border detection in dermoscopy images, *Computerised Medical Imaging and Graphics*, vol. 33, no. 2, pp. 148–153

[7] Garnavi R., Aldeen M., Celebi M. E., Bhuiyan A., Dolianitis C., and Varigos G., (2010), Automatic segmentation of dermoscopy images using histogram thresholding on optimal color channels, *International Journal of Medicine and Medical Sciences*, vol. 1, no. 2, pp. 126–134

[8] Celebi M., Kingravi H., Uddin B., Iyatomi H., Aslandogan Y., Stoecker W., and Moss R., (2007), A methodological approach to the classification of dermoscopy images, *Computerized Medical Imaging and Graphics*, vol. 31, pp. 362–373

[9] Maglogiannis I. and Doukas C., (2009), Overview of advanced computer vision systems for skin lesions characterization, *IEEE Trans. on Information Technology in Biomedicine*, vol. 13, no. 5, pp. 721–733

[10] Alizadeh, Seyed Mohammad, and Ali Mahloojifar. (2021). Automatic skin cancer detection in dermoscopy images by combining convolutional neural networks and texture features, *International Journal of Imaging Systems and Technology* 31.2: 695–707.

[11] Adegun, Adekanmi A., and Serestina Viriri. (2019). Deep learning-based system for automatic melanoma detection, *IEEE Access* 8: 7160–7172.

[12] Adla, Devakishan, *et al.* (2022). Deep learning-based computer aided diagnosis model for skin cancer detection and classification, *Distributed and Parallel Databases* 40.4: 717–736.

[13] Giotis, Ioannis, *et al.* (2015). MED-NODE: A computer-assisted melanoma diagnosis system using non-dermoscopic images, Expert systems with applications 42.19: 6578–6585.

[14] Ayan E. and Ünver H. M., (2018), Data augmentation importance for classification of skin lesions via deep learning, *2018 Electric Electronics, Computer Science, Biomedical Engineerings' Meeting (EBBT)*, Istanbul, Turkey, pp. 1–4, doi: 10.1109/EBBT.2018.8391469.

[15] Gutman, David, *et al.* (2016). Skin lesion analysis toward melanoma detection: A challenge at the international symposium on biomedical imaging (ISBI) 2016, hosted by the international skin imaging collaboration (ISIC), *arXiv preprint* arXiv:1605.01397.

[16] Srividhya, V., *et al.* (2020). Vision based detection and categorization of skin lesions using deep learning neural networks, Procedia Computer Science 171: 1726–1735.

[17] Waheed Z., Waheed A., Zafar M. and Riaz F., (2017), An efficient machine learning approach for the detection of melanoma using dermoscopic images, *2017 International Conference on Communication, Computing and Digital Systems (C-CODE)*, Islamabad, Pakistan, pp. 316–319, doi: 10.1109/C-CODE.2017.7918949.

[18] Mendonça T., Ferreira P. M., Marques J. S., Marcal A. R. S. and Rozeira J., (2013), PH2 - A dermoscopic image database for research and benchmarking, *2013 35th Annual International Conference of the IEEE Engineering in Medicine and Biology Society (EMBC)*, Osaka, Japan, pp. 5437–5440, doi: 10.1109/EMBC.2013.6610779.

Data Science & Exploration in Artificial Intelligence – Gururaj et al. (Eds)
© 2025 the Author(s), ISBN: 978-1-032-96390-7

Enhancing fraud detection in mobile payment systems: A comparative analysis of machine learning models

Aditi Singh, Shivantika Tripathi, Sandeep Chaurasia & Amit Kumar Gupta
Computer Science and Engineering, Manipal University Jaipur, Jaipur, Rajasthan, India

P. Dayananda
Information Technology, Manipal Institute of Technology, Bengaluru, India

ABSTRACT: A comparative analysis of five machine learning models, XG Boost, Random Forest, Decision Tree, Naïve Bayes, and Logistic Regression for payment system on mobile to detect the fraud presented. All models provided the excellent accuracy score except Logistic Regression which achieved lowest 0.423420 and the highest precision 0.915061. Among all these models random forest achieved highest accuracy (0.999696), recall (0.779984), precision (0.980642), and F1 score (0.868878). with some extent the XG boost also achieve good results in form of accuracy of 0.999767, recall of 0.854538, precision of 0.961258, and F1 score of 0.904762. Random Forest and XG Boost have shown the excellent results to detect the fraud transaction in mobile system payment systems. In future the investigation may be further enhanced by applying the ensemble approaches and feature engineering strategies to improve the evaluation matrices creating an environment of empathy and connection that is conducive to true learning.

1 INTRODUCTION

In the advent of mobile payment system application like Apple Pay, Google Pay, Paytm and many more have changed the scenario for the budgetary exchanges. These applications are system have shown their wide extend of clients due to extreme comfort and openness. But this may also lead a serious concern about security of payment, integrity, authentication and confidentiality. Extortion including versatile installments can take numerous diverse shapes, such as account takeovers, phishing tricks, and false exchanges. Credit card numbers and passwords are among the individual data that fraudsters obtain in arrange to get to Payment System accounts without authorization. In the real-world scenario, the unauthorized people creating fake websites or Payment System apps to trap for confidential data related to their account and personal information. The huge uses and features of payment wallet in everyday life has prominent and need for solid extortion location prototypes (Aras 2022; Güven *et al.* 2022). The Machine learning methods are developed for identifying the pattern that defines the false or misleading source for the payment system. Machine learning calculations, for illustration, can be prepared to distinguish exchanges with unforeseen sums of cash, exchanges from bizarre places, or exchanges that happen at unprecedented times (Seera *et al.* 2024). These models moreover have the capacity to recognize unordinary exchanges that enormously contrast from a customer's normal behavior. For occurrence, an individual who as a rule transacts in small sums can have an expansive exchange detailed as suspicious (Said *et al.* 2021). To guarantee the adequacy and constancy of such brilliantly frameworks, solid calculations and broad assessment criteria must be created (Delecourt *et al.* 2019). By making and evaluating models that proficiently identify and diminish false exchanges

DOI: 10.1201/9781003589273-39

utilizing appropriate assessment measures, this work looks for to include to the extending corpus of investigate within the field of machine learning.

2 RELATED WORK

Financial extortion distinguishing proof and avoidance have developed as major concerns for buyers, organizations, and financial educate in today's quickly digitalized monetary environment. This study of the writing looks at diverse inquire about ventures that are concentrating on financial extortion location, counting versatile installment frameworks and exchange behavior examination. It delivers errorless scenario and measurable highlights for evaluation of the examination. Supervised machine learning methods are used to detect the false credit cards transaction information was inspected by Karthik *et al.* (2019). For identification of fraud transaction patterns in the transaction information, the researchers use the classification methods of machine learning techniques. The machine learning which was utilized are decision tree and support vector machine as classifier. Their research outcomes were that how these machine learning methods are maximizing the rate of identification of fraud transaction information. The goal of Omollo (2020) is to create a mobile fraud detection system that works in real time. This work introduces experience focusing on understanding user behavior and business patterns to identify potential vulnerabilities. This work presents a hybrid model that combines machine learning and rule-based techniques to improve the detection and development of new fraud patterns. Najadat *et al.* (2020), in this research the researchers have used the machine learning and deep learning to identify the credit card fraud. In this study the research utilized the deep neural networks, random forests, and logistic regression for detecting the false information or credit card fraud information. A comparative study has been presented. Particularly this research has demonstrated that convolutional neural networks (CNN) provide the excellent results for detecting the complex fraud patterns in the credit card fraud information. While evaluating various machine learning algorithms for credit card fraud, Dornadula and Geetha (2019) paid particular attention to the efficiency and performance evaluation of the algorithms. They emphasized the importance of architecture and modeling in the development of accurate detection and discussed the advantages and disadvantages of each method, including logistic regression, printing decision trees, and neural networks. A study on credit card fraud strategies was conducted by Tripathi and Pavaskar (2012), who explored various strategies including legal frameworks, vulnerability detection, and machine learning models. It shows the difficulty of managing inconsistent information and the need to constantly update the model to prevent the development of fraud strategies. Srivastava *et al.* (2008) proposed the novel methods namely known as Hidden Markov Model (HMM) for identification of the credit card fraud information. HMM model identify the pattern of transaction profile, identification of unwanted or not necessity spending patterns which may defines false transaction for identification of fraud. Their study demonstrated that how HMM provides the highly accurate detection of fraud and reducing false positives. Szegedy *et al.* (2013) and Carlini and Wagner (2017) shown that machine learning and Neural networks are vulnerable for misleading attacks. In this study they shown that competitive machine learning and neural networks can provide more reliable methods to detect the fraud. Small changes in the information of transaction can be cause misclassification. This study will highlight the need to develop better models to prevent these attacks. Two studies by Madry *et al.* (2017) and Kurakin *et al.* (2016) proposed a method to improve the performance of deep learning models for conflict situations. To increase the power of fraud detection tools, protection strategies such as training attacks and training models using attack data should be used in their work. Deep neural networks and malware detection are research topics of Grosse *et al.* (2017) and Papernot *et al.* (2016) focused on conflict situations. In discussing the general implications of machine security learning, Barreno *et al.* (2006) emphasizes the need to fully evaluate the anti-fraud system in busy areas. Their work shows how important it is to monitor the product and update the model frequently to adapt to new fraud schemes. This literature review highlights the advantages of machine learning methods for identification and detection of fraud transaction. These studies explore the careful feature selection from the transaction information, and continuous machine learning model development may enhance the rate of correctly identification of the fraud. This study also demonstrated that there is requirement of further enhancement in the robust model development which can withstand malicious attacks and adapt to new scams.

3 METHODOLOGY

3.1 *Data set acquisition*

The data for this study were developed as part of the research project of Lopez-Rojas, Axelsson, and Baca (2018). This fake file has 10 features and more than 6 million transaction files. Created using PaySim financial simulator. These features include transaction type, transaction amount, transfer duration, location, and account balance for the location. Resource efficiency and the potential of this information in big data analysis make it very important. The data is quite serious as only 0.13% of the market is fraudulent. Under sampling techniques were used to correct discrepancies with satisfactory results. Data were analyzed for clarity before using the sampling technique. To increase the usefulness of data in fraud detection, this study recommends selecting relevant features and developing new features. Figure 1 shows a snapshot of the dataset.

	step	type	amount	oldbalanceOrg	newbalanceOrig	isFraud	isFlaggedFraud	entireAmount	destDiscrepancy	merchant
0	1	PAYMENT	9839.64	170136.00	160296.36	0	0	0	Merchant	1
1	1	PAYMENT	1864.28	21249.00	19384.72	0	0	0	Merchant	1
2	1	TRANSFER	181.00	181.00	0.00	1	0	1	Yes	0
3	1	CASH_OUT	181.00	181.00	0.00	1	0	1	No	0
4	1	PAYMENT	11668.14	41554.00	29885.86	0	0	0	Merchant	1
...	...	...	...	...	...	...	...	...	...	...
6362615	23	CASH_OUT	339682.13	339682.13	0.00	1	0	1	No	0
6362616	23	TRANSFER	6311409.28	6311409.28	0.00	1	0	1	Yes	0
6362617	23	CASH_OUT	6311409.28	6311409.28	0.00	1	0	1	No	0
6362618	23	TRANSFER	850002.52	850002.52	0.00	1	0	1	Yes	0
6362619	23	CASH_OUT	850002.52	850002.52	0.00	1	0	1	No	0

Figure 1. Snapshot of dataset used in this study.

3.2 *Method*

3.2.1 *Logistic regression*

Logistic regression is a powerful statistical technique that can be used to predict payment fraud risk because it is variable and does not require rigid assumptions such as equality of the group or set of independent variables. This method uses maximum likelihood estimation to estimate group membership. A preliminary analysis of the revised data was conducted to check the assumptions in transportation demand to ensure the accuracy of the forecast (LaValley 2008).

3.2.2 *Random forest*

During training, the random forest classifier creates a large number of decision trees, each created from a different random subset of the training. This merging method combines the results of multiple trees to create predictive rules. This approach is unique in improving the flexibility of overfitting and accurate classification (Rigatti 2017).

3.2.3 *Decision tree*

Decision tree is designed for classification and recycling and is a flexible method. Based on the training data, they create a hierarchy of nodes, with the root node at the top. Decision trees can be created using a variety of algorithms, including C4.5, CART, and ID3. This model searches the tree to determine the distribution and value of a sample (Song *et al.* 2015).

3.2.4 *Naive Bayes*

Bayes theorem forms the basis of this classification technique. It predicts relevant features (classes) using the probability of many independent features (attributes) in the data set. This works very well for large data sets as it has often been shown to be consistent in practice (Rish 2001).

3.2.5 *XGBoost*

Extreme Gradient Boosting or XGBoost is a hard learning technique useful for fraud detection. It creates one decision tree after another, each trying to correct the mistakes of the previous decision tree. It is a popular choice for building reliable fraud models due to its speed and efficiency (Chen *et al.* 2016).

3.3 *Implementation*

Researcher carefully divided our data into three sets for this study on mobile payment fraud detection: a training set, a development set, and a test set. Approximately 90% of the resources were allocated for training, 5% for development, and 5% for testing. Given the skewed nature of our dataset—where legitimate transactions far outweigh fraudulent ones—this separation was essential to our research. This study used several strategies, including under- and oversampling, to overcome this imbalance. It was not feasible to do traditional cross-validation only on the training set because this would not sufficiently address the imbalance in the data and result in skewed model performance. It was important to create a specific development set. With the help of this development set, we were able to optimize and validate our models and make sure they could successfully address the difficulties brought on by the unbalanced data. Our goal in using these sample strategies was to give the training set a more evenly distributed representation of the dataset. This method made sure that our models were validated and trained in a way that closely resembled real-world situations. As a result, this methodology made it easier to create solid and trustworthy models that can precisely identify fraudulent transactions in mobile payment systems. To sum up, the utilization of sampling strategies and the incorporation of a specific development set played a crucial role in surmounting the obstacles posed by unbalanced data, thus augmenting the general dependability and efficacy of our fraud detection models.

3.4 *Evaluation matrices*

It is important to evaluate the effectiveness of the classification model in the context of our research into mobile payment fraud. The main criteria obtained for this purpose are accuracy, precision, recall and F1 score. Detection is a measure of the accuracy of prediction quality and is calculated as:

$$Precision = TP/(TP + FP) \tag{1}$$

accuracy and FP represent illegal products represent legitimate transactions that are not flagged as fraud. Recall, or sensitivity, evaluates the model's ability to detect all actual frauds and is defined by:

$$Recall = (TP + FN)/TP \tag{2}$$

Because it measures how well the model detects true fraud, also known as sensitivity, and is expressed as: where FN (non-false) represents business fraud that is not caught by the model. F1 score is determined by back weight and accuracy.

$$F1Score = 2 \times (Precision \times Recall)/(Precision + Recall) \tag{3}$$

This analysis is particularly important in cases involving fraud investigation; because it is important to ensure consistency in identifying all fraud and reducing inaccuracies. The absolute accuracy or accuracy of this model is calculated as deceiving. Our work provides a detailed analysis using these metrics, providing a comprehensive evaluation of the performance of classification models for identifying suspicious quantities in real-world settings (Yacouby 2020).

$$Accuarcy = (TP + FN + TN + FP)/(TP + TN) \tag{4}$$

4 RESULTS AND DISCUSSION

In the Lessons and Discussion section, the researchers presented results from training and testing five different fraud models using Jupyter Notebook and Google Colab. To make a comprehensive evaluation, cross-referencing is used during training and testing. The average score was calculated to determine the performance of each sample. This method ensures that each data point is used multiple times for training and at least once for testing, making it efficient. Figure 2 compares these five strategies, where yellow represents F1 score, green represents regression, blue represents accuracy, and red represents accuracy. These metrics provide insight into how well each machine learning method detects fraudulent transactions. Table 1 lists additional tests and these parameters. The performance of five fraud models (Logistic Regression, Random Forest, Decision Tree, Naive Bayes, and XG Boost) was evaluated using various parameters such as accuracy, recall, precision, and F1 score. Table 1 summarizes the findings. The accuracy of logistic regression is 0.999204 and reaches 99.92% of cases. The recovery score is 0.423420, which is a low value, which indicates that most fraud will go undetected by it. However, the accuracy score is quite high at 0.915061; This means it is accurate when it flags a business as a scam. The F1 score (measures recall and precision) is 0.578947, which is average.

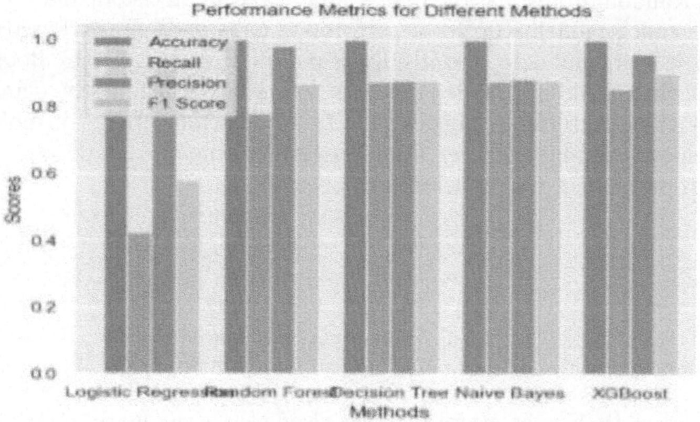

Figure 2. Performance matrices of all Five models used in this research.

Random Forest performed well across the board, achieving an F1 score of 0.868878, precision of 0.999696, recall of 0.779984, and precision of 0.980642. These findings show that random forest is highly effective at detecting fraud and reducing false positives. Decision tree and Naive Bayes achieve 0.999694 accuracy, 0.877229 recall, 0.881159 accuracy and

0.881156 F1 score. These models showed high returns and high prices, demonstrating their ability to detect fraudulent transactions. However, they have lower accuracy than random forests. The XG Boost scored 0.999767, indicating the highest accuracy among all models. It performs very well along with recall (0.854538) and precision (0.961258), achieving a good F1 score of 0.904762.; Logistic regression has the highest accuracy of 0.999204, while XG Boost has the lowest accuracy of 0.999767. Decision trees, Naive Bayes, and XG Boost all show normal improvement with recovery scores above 0.85. Logistic regression shows that the return is lower than 0.423420. All models have high accuracy scores overall; Random Forest achieves the highest accuracy with 0.980642, followed by XG Boost with 0.961258. Again, all models have very high F1 scores; these scores measure recall and precision; All things considered, these findings demonstrate the ability of fraud test models to detect fraud, with each model performing best across a variety of tasks.

Table 1. Performance of all models.

Model	Accuracy	Recall	Precision	F1 Score
Logistic Regression	0.999204	0.423420	0.915061	0.578947
Random Forest	0.999696	0.779984	0.980642	0.868878
Decision Tree	0.999694	0.877229	0.881159	0.881156
Naïve Bayes	0.999694	0.877229	0.881159	0.881156
XG Boost	0.999767	0.854538	0.961258	0.904762

5 CONCLUSION

This study examines the effectiveness of five machine learning models (logistic regression, random forest, decision tree, naive Bayes and XG Boost) in detecting fraud in mobile payments. Jupyter Notebook and Google Colab were used to train and test the model, and cross-validation enabled extensive testing. Evaluation metrics such as precision, recall, precision, and F1 score were used to evaluate the performance of the model. The results show that all models can classify transactions correctly, as evidenced by their high scores. Logistic regression is more accurate but produces worse results than previous models. Among the models tested, Random Forest has the best accuracy, recall, precision, and F1 score, making it the best performing model. XG Boost performs equally in all areas. Based on these findings, machine learning models (specifically Random Forest and XG Boost) are highly effective in detecting fraud in mobile payments. The study concludes by demonstrating the potential of machine learning to improve the fraud detection ability of mobile payments. Future research can examine the integration process and engineering process to improve the accuracy and robustness of the model in fraud detection.

REFERENCES

Ait Said, M., and Hajami, A. (2021, December). AI methods used for real-time clean fraud detection in instant payment. *In International Conference on Soft Computing and Pattern Recognition (pp. 249-257). Cham: Springer International Publishing.*

Aras, S. (2022). Fraud detection by machine learning algorithms: a case from a mobile payment system. *International Journal of Management Economics & Business/Uluslararası Yönetim İktisat ve İşletme Dergisi, 18(3).*

Barreno M., Nelson B., Sears R., Joseph A. D., and Tygar J. D., (2006), Can machine learning be secure? *in Proceedings of the 2006 ACM Symposium on Information, computer and communications security*, pp. 16–25, ACM.

Carlini N. and Wagner D., (2017), Towards evaluating the robustness of neural networks, *in 2017 IEEE Symposium onSecurity and Privacy (SP)*, pp. 39–57, IEEE.

Chen, T., and Guestrin, C. (2016, August). Xgboost: A scalable tree boosting system. *In Proceedings of the 22nd acm sigkdd international conference on knowledge discovery and data mining (pp. 785–794).*

Delecourt, S., and Guo, L. (2019, June). Building a robust mobile payment fraud detection system with adversarial examples. *In 2019 IEEE Second International Conference on Artificial Intelligence and Knowledge Engineering (AIKE) (pp. 103–106). IEEE.*

Dornadula, V. N., and Geetha, S. (2019). Credit card fraud detection using machine learning algorithms. *Procedia computer science, 165, 631–641.*

Grosse K., Papernot N., Manoharan P., Backes M., and McDaniel P., (2017), Adversarial examples for malware detection, *in European Symposium on Research in Computer Security, pp. 62–79, Springer.*

Güven, Ö., and Aras, S. (2022). Fraud detection by machine learning algorithms: a case from a mobile payment system. *Uluslararası Yönetim İktisat ve İşletme Dergisi, 18(3), 895–911.*

Karthik, R., Navinkumar, R., Rammkumar, U., and Mothilal, K. C. (2019). Supervised machine learning algorithms for credit card fraudulent transaction detection. *International Journal of Scientific Research in Computer Science, Engineering and Information Technology*, 2019, 2456–3307.

Kurakin A., Goodfellow I., and Bengio S., (2016), Adversarial examples in the physical world, *arXiv preprint arXiv:1607.02533.*

LaValley, M. P. (2008). Logistic regression. *Circulation, 117(18), 2395–2399.*

Madry A., Makelov A., Schmidt L., Tsipras D., and Vladu A., (2017), Towards deep learning models resistant to adversarial attacks, *arXiv preprint arXiv:1706.06083.*

Najadat, H., Altiti, O., Aqouleh, A. A., and Younes, M. (2020, April). Credit card fraud detection based on machine and deep learning. *In 2020 11th International Conference on Information and Communication Systems (ICICS)* (pp. 204–208). *IEEE.*

Omollo, J. O. (2020). *Real Time Fraud Detection System for Mobile Banking: Based on Experiential Paradigm (Doctoral dissertation, University of Nairobi).*

Papernot N., McDaniel P., Wu X., Jha S., and Swami A., (2016), Distillation as a defense to adversarial perturbations against deep neural networks, *in 2016 IEEE Symposium on Security and Privacy (SP)*, pp. 582–597, IEEE.

Rigatti, S. J. (2017). Random forest. *Journal of Insurance Medicine*, 47(1), 31–39.

Rish, I. (2001, August). An empirical study of the naive Bayes classifier. *In IJCAI 2001 workshop on empirical methods in artificial intelligence* (Vol. 3, No. 22, pp. 41–46).

Seera, M., Lim, C. P., Kumar, A., Dhamotharan, L., and Tan, K. H. (2024). An intelligent payment card fraud detection system. *Annals of Operations Research*, 334(1), 445–467.

Song, Y. Y., and Ying, L. U. (2015). Decision tree methods: applications for classification and prediction. *Shanghai Archives of Psychiatry*, 27(2), 130.

Srivastava A., Kundu A., Sural S., and Majumdar A., (2008), Credit card fraud detectionusing hidden markov model, *IEEE Transactions on dependable and secure computing, vol. 5, no. 1, pp. 37–48.*

Szegedy C., Zaremba W., Sutskever I., Bruna J., Erhan D., Goodfellow I., and Fergus R., (2013), Intriguing properties of neural networks, *arXiv preprint arXiv:1312.6199.*

Tripathi K. K. and Pavaskar M. A., (2012), Survey on credit card fraud detectionmethods, *International Journal of Emerging Technology and Advanced Engineering*, vol. 2, no. 11, pp. 721–726.

Yacouby, R., and Axman, D. (2020, November). Probabilistic extension of precision, recall, and f1 score for more thorough evaluation of classification models. *In Proceedings of the first workshop on evaluation and comparison of NLP systems* (pp. 79–91).

Data Science & Exploration in Artificial Intelligence – Gururaj et al. (Eds)
© 2025 the Author(s), ISBN: 978-1-032-96390-7

Advancements in artificial intelligence for thyroid cancer detection

K.T. Anil Kumar & S.V. Shashikala
Department of Computer Science & Engineering, BGS Institute of Technology, Adichunchanagiri University, BG Nagara, Mandya, India

ABSTRACT: In times there has been a rise, in the use of artificial intelligence (AI) in healthcare systems especially in the early detection of diseases. One key focus area is the identification of thyroid diseases, including cancer, which's crucial for effective treatment and improved patient outcomes. This study aims to conduct a review and analysis of literature on AI techniques used to detect and characterize thyroid gland related cancers. The significance of datasets related to thyroid cancer (TCDs) is emphasized in uncovering characteristics and methods for creating systems driven by AI. This study delves into the results of an evaluation that sheds light on both the advantages and constraints as possible progressions, in utilizing artificial intelligence for the detection of thyroid cancer.

Keywords: Thyroid, cancer, Artificial Intelligence, Machine learning

1 INTRODUCTION

The application of artificial intelligence (AI), in healthcare depicts progress that is changing medical processes, diagnoses and sole care to patients. AIs innate knowledge base like pattern recognition, Data analysis and decision making ability has given rise to systems which analyzes complex health-related information quickly and with higher accuracy than human mind can ever reach. AI could also connect rural regions to metropolitan hospitals, helping improve the quality of medical care for those outlying areas.

It has significant implications for healthcare and is expected to continue growing in the years to come with the advancement of technology that will ultimately lead to more intricate applications and improved health outcomes on a global scale. Trust becomes a critical factor in influencing user acceptance in the context of AI-powered assistance, mediating the impact of AI-specific factors. In various studies scholars have examined the complex interplay of how factors such as security, risk, and trust collectively shapes the adoption of AI-powered assistance. Cancer is one of the leading causes of mortality globally, presenting itself in several different forms across anatomical regions. Of these, thyroid carcinoma is considered one of the most prevalent endocrine cancers around the world [2,3].

The growing concern concerns the growing prevalence of thyroid cancer and the corresponding rates of mortality. Several studies reveal a higher incidence rate of thyroid cancer in women aged 15 to 49, who rank fifth globally, compared to males aged 50 to 69 [4–6].

In the latest global epidemiological data, we find that abnormal thyroid nodules are increasing rapidly and genetic cell activity is accelerating nearly simultaneously. It involves an increase in cell operations of an organism. Among them, thyroid cancer is one of the predominant condition and enrolling a considerable proportion in addition to other types such as papillary carcinoma (PTC) [7], follicular thyroid carcinoma (FTC) [8], anaplastic thyroid carcinoma (ATC) [8] and medullary thyroid carci-noma(MTC)[9. The most important aspect when a thyroid nodule is diagnosed is the accurate assessment, description and risk of Malignancy estimation. This triage is vital for the appropriate adjudication of care, establishing necessary treatments and promoting recovery.

 DOI: 10.1201/9781003589273-40

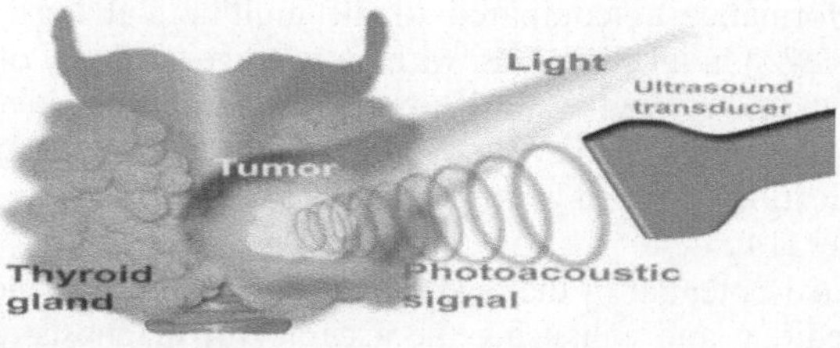

Figure 1. Thyroid cancer diagnosed with AI.

There has been a marked increase over recent years in the use of computer-aided diagnostic (CAD) tools for an analysis done on images involving thyroid cancer. To establish further refinement in the diagnosis a Fine-needle aspiration biopsy (FNAB) is usually carried out. However, the procedure can be painful for patients and poor handling by radiologists may inadvertently lead to false assured diagnosis of benign non-cancerous nodules as cancerous. This could in turn incur extra financial burden on patient unnecessarily due to its costliness! The difficulty lies in accurately determining what specific features differentiate between benign and malignant nodules.

Human minds are subjective and hence the adoption of artificial intelligence (AI) technology can help decrease bias as well increase pathological diagnoses accuracy in a variety of deadly diseases, seen though Figure 1 below. This effect influences also the diseases of thyroid, resulting in more accurate and reliable diagnosis [8]. This improvement involves improving the interpretation (real-time) of US images and reducing processing times. Minimally/benign or malignant thyroid nodules have been previously categorized using computer-based methods, based on ultrasonography (US) [4], fine-needle aspiration cytology (FNAC) diagnosis or even after the final surgical sentinel lymph node analysis and diagnostic transfusion programme in patients with suspected breast cancer CFNA algorithm was recently proposed as a method to distinguish between groups of FNAs from same clusters in decision-making among benign safety protocol for peritoneal carcinomatosis detection[10]. This article aims to dive into the present usage and upcoming opportunities of AI in thyroid cancer diagnosis-characterization.

2 RELATED WORKS

Zhang and colleagues (2019) proposed several methodologies based on deep neural networks (DNNs) for diagnostic discrimination and categorization of benign versus malignant thyroid nodules in ultrasound imaging modality [9]. Moreover, Chen and colleagues (2020) have evaluated the performance of the bi-directional long short-term memory (Bi-LSTM) model as remarkable for thyroid nodule classification [10]. The synthesis by Nguyen and associates focused on thyroid cancer diagnosis and management and adopts the American Thyroid Association rules for both thyroid nodules and differentiated thyroid cancer [11]. In a study where a Convolutional Neural Network (CNN) was used to develop a prognostic tool, approximately 10,068 microscope images of thyroid cancer tissue from South Asian populations were analyzed [12]. Moreover, the images were integrated into pharmacogenomic datasets, which included genetic data of variation and personalization and its correlation with the risk-susceptibility of thyroid cancer.

The authors of this investigation study aspirated their focus to hyperthyroidism and hypothyroidism; two widespread thyroid disorders [13]. The scientists of this study utilized two diverse methodologies for classification: neural network and multinomial logistic regression models. It should be noted that 310 patients participated in the investigation, and in this particular study, the models utilized the demographic and hormonal parameters of the patients as features. The results of the investigation revealed that the neural network model provided better performance as compared to the multinomial logistic regression model (mean accuracy of 91.4%) in all situations, with an average accuracy of 96.3% in our study.

In addition, Kadam and colleagues (2017) proposed a technique involving the utilization of Sparse Auto-encoders and SoftMax Regression to construct a feature ensemble learning model for the classification of breast cancer into either benign (non-cancerous) or malignant (cancerous) categories [14].

AI has demonstrated potential in the field of cancer research, enhancing diverse facets of cancer therapy, expediting and enhancing the accuracy of diagnosis, and generating more reliable clinical judgments that enhance patient results [15]. AI presents an unparalleled level of precision in predicting cancer compared to a conventional statistical expert. Therefore, cancer detection models based on AI have the capacity to assist healthcare professionals in healthcare facilities and empower them to validate their diagnoses [16] without constraints. Studies have been carried out in Lyon, France to explore the occurrence of thyroid cancer in individuals above the age of 19 as well as in children. The International Agency for Research on Cancer (IARC) in low- and middle-income countries contributed data for this investigation.

The methodology encompasses the development of scientifically sound and harmonized approaches for cancer management.

Aversano *et al.* employed multiple algorithms based on machine learning for the analysis of data. More specifically, they conducted a comparison of the outcomes produced by ten distinct classifiers. The other algorithms exhibited promising findings, with particular emphasis on the Extra-Tree Classifier, which achieved a notable accuracy rate of 84% [17]. Furthermore, they utilized a cat boost classifier, which contributed to their attainment of a precision level of 71%. An inclusive depiction of approaches utilized in the detection of thyroid cancer that are based on artificial intelligence is presented in Table 1, encompassing both the advantages and disadvantages associated with each method.

Table 1. Thyroid cancer detection techniques.

AI Methods	Advantages	Drawbacks	Accuracy
DAE	Labels are not necessary for thyroid cancer	Inadequate training data and the necessity for pertinent data	92.9
CNN	Elevated detection of thyroid cancer	Inadequate thyroid cancer labels that are difficult to understand	85
RNN	Labels are not necessary for thyroid cancer	Sluggish computation and challenging training	98.2
MLP	Personalized education for thyroid cancer	Restricted in its outcomes	95
KNN	High sensitivity in detecting thyroid cancer	Not enough thyroid cancer labels	93
SVM	High sensitivity for detecting thyroid cancer	Interpretability issues and lengthy training period	97
DT	Does not call for data normalization and scaling	Erratic	73.10
LR	Less expensive training and more straightforward execution	Having trouble labeling data	—

2.1 *Dataset benchmark*

Numerous datasets have been generated within the realm of thyroid cancer research for the purpose of facilitating the verification of machine learning models and algorithms. This proves to be of particular significance due to the fact that the creation of such datasets presents a major challenge in the domain of endocrine machine learning.

3 OVERVIEW OF EXISTING FRAMEWORKS

This section outlines the procedural flow of the machine learning algorithm employed in a classification task that is frequently encountered within the framework of computer-aided diagnosis (CAD) [10] (refer to Figure 2). The training and application phases encompass a supervised machine learning model (Figure 2). In the training phase, the predictive model is educated using input images associated with their respective class labels. Skilled physicians manually or semi-automatically identify a Region of Interest (ROI) within the input image. Subsequently, a set of features is extracted from the image, which encompasses both morphological and grey level-based attributes.

In comparison to alternative methodologies that will be expounded upon subsequently within this study, a proficient individual undertakes the labor-intensive process of extracting and selecting distinctive attributes from images in machine learning algorithms of computer-aided design (CAD) systems. Identifying the pivotal variables that are interconnected with the medical outcome is an imperative phase in the aforementioned procedure.

The characteristics employed in the examination in CAD applications are intricately associated with the characteristics medical professionals employ to identify the abnormalities [4]. Subsequently, the components are introduced into the machine learning algorithm for the purpose of training the model.

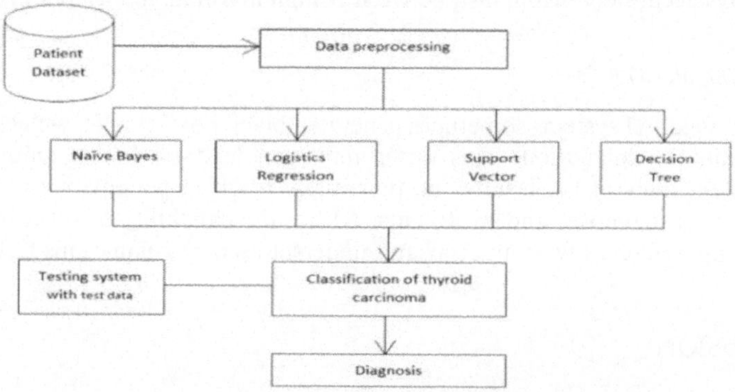

Figure 2. Framework.

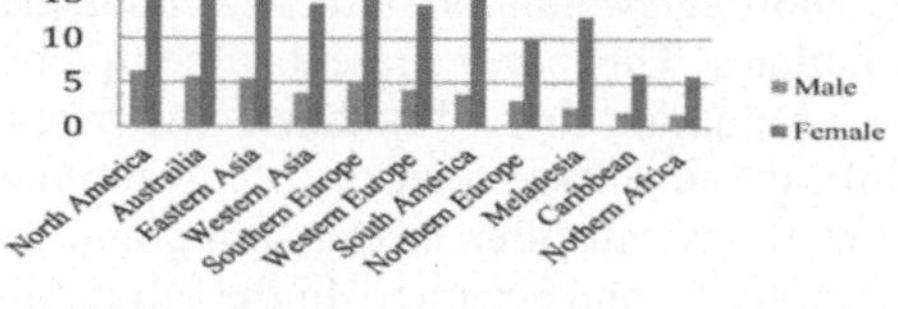

Figure 3. Depicts the graphical representation of thyroid cancer among male and female.

4 CHALLENGES

While AI tools offer hope for the future of thyroid cancer diagnosis, their capabilities act as a roadblock in creating viable solutions and drive up implementation costs that restrict broader acceptance. Centralizing and storing all data in only one place is a must for accurate thyroid cancer detection with no FL technique [2]. The next step is creating algorithms that can identify all types of thyroid cancer. As TCD [4] operates on three datasets per nodules category, each image dataset combines training images for 150 and test images for both nodule plans and that of the size of nodules, dissemination identify characteristic.

4.1 *Lack of clean data & accuracy*

The accurate diagnosis and success treatment options for thyroid cancer (cancer in general) has been an uphill task, simply because there does not exist a proper/annotated dataset that tells us where exactly is it occurring. Inaccurate thyroid cancer registry data capture, death documentation and validation of the data further complicated these challenges [12].

4.2 *Thyroid gland imaging*

Computed tomography (CT) and magnetic resonance imaging (MRI) are effective options for the diagnostic assessment of thyroid cancer. However, due to their high cost and limited availability in certain cases, they are not considered the preferred methods [8]. Instead, ultrasound examinations, radioisotope scans, fine-needle aspiration biopsies, and physical examinations are often used. During an ultrasound examination, a physician can assess the gland's activity by measuring the node's echogenicity, size, boundaries, and presence of calcifications. Nevertheless, ultrasound images may be more prone to noise, and the results may not always accurately distinguish between benign and malignant nodes [17].

4.3 *Unexplainable AI*

In the medical field, AI systems sometimes generate "black box" results, which are imprecise and lack adequate explanation. This uncertainty can lead to doubts among physicians regarding the accuracy of the results and potentially result in patients with thyroid cancer receiving incorrect diagnoses and treatments. AI has the capability to function as a "black box," producing outcomes without clear and understandable explanations [2,14].

5 CONCLUSION

Thyroid cancer is increasing worldwide, especially among young individuals. As mentioned earlier, thyroid cancer develops when cells undergo rapid and uncontrolled division, spreading into surrounding tissues. Timely detection of cancerous nodules is crucial for effective disease management and reducing mortality rates. The advancement of AI-based Computer-Aided Diagnosis (CAD) systems for analyzing thyroid images has rapidly evolved in recent decades. Thorough validation of these technologies is necessary to improve the treatment of thyroid nodules. This paper presents a comprehensive review of artificial intelligence applications in evaluating thyroid nodules. The overarching conclusion of the study emphasizes the significant advantages that recent advancements in AI techniques and systematic deep learning methods can offer in classifying and analyzing thyroid tumors, ensuring high specificity, sensitivity, and accuracy. In the future, this paper intends to utilize a machine learning algorithm to assess accuracy in a limited set of specific attributes related to thyroid cancer.

REFERENCES

[1] Yoon, Jiyoung, Eunjung Lee, Ja Seung Koo, Jung Hyun Yoon, Kee-Hyun Nam, Jandee Lee, Young Suk Jo, Hee Jung Moon, Vivian Youngjean Park, and Jin Young Kwak. (2020). Artificial intelligence to predict the BRAFV600E mutation in patients with thyroid cancer. *PloS one* 15, no. 11: e0242806.

[2] Habchi, Yassine, Yassine Himeur, Hamza Kheddar, Abdelkrim Boukabou, Shadi Atalla, Ammar Chouchane, Abdelmalik Ouamane, and Wathiq Mansoor. (2023). Ai in thyroid cancer diagnosis: Techniques, trends, and future directions. *Systems* 11, no. 10: 519.

[3] Nagendra, Lakshmi, Joseph M. Pappachan, and Cornelius James Fernandez. (2023). Artificial intelligence in the diagnosis of thyroid cancer: Recent advances and future directions. *Artificial Intelligence in Cancer* 4, no. 1: 1–10.

[4] Bellantuono, Loredana, Raffaele Tommasi, Ester Pantaleo, Martina Verri, Nicola Amoroso, Pierfilippo Crucitti, Michael Di Gioacchino *et al.* (2023). An eXplainable Artificial Intelligence analysis of Raman spectra for thyroid cancer diagnosis. *Scientific Reports* 13, no. 1: 16590.

[5] Bini, Fabiano, Andrada Pica, Laura Azzimonti, Alessandro Giusti, Lorenzo Ruinelli, Franco Marinozzi, and Pierpaolo Trimboli. (2021). Artificial intelligence in thyroid field—a comprehensive review. *Cancers* 13, no. 19: 4740.

[6] Thomas, Johnson, and Tracy Haertling. (2020). AIBx, artificial intelligence model to risk stratify thyroid nodules. *Thyroid* 30, no. 6: 878–884.

[7] Himeur, Y.; Al-Maadeed, S.; Almaadeed, N.; Abualsaud, K.; Mohamed, A.; Khattab, T.; Elharrouss, O. (2022), Deep visual social distancing monitoring to combat COVID-19: A comprehensive survey. *Sustain. Cities Soc. 85*, 104064.

[8] Deng, Y.; Li, H.; Wang, M.; Li, N.; Tian, T.; Wu, Y.; Xu, P.; Yang, S.; Zhai, Z.; Zhou, L.; *et al.* (2020). Global burden of thyroid cancer from 1990 to 2017. *JAMA Netw. Open*, *3*, e208759.

[9] Zhang, S.; Du, H.; Jin, Z.; Zhu, Y.; Zhang, Y.; Xie, F.; Zhang, M.; Jiao, Z.; Tian, X.; Zhang, J.; *et al.* (2019). Integrating clinical knowledge in a thyroid nodule classification model based on. *In Proceedings of the 2019 IEEE International Ultrasonics Symposium (IUS)*, Glasgow, UK, 6–9 October 2019; IEEE: Piscataway, NJ, USA, pp. 2334–2336.

[10] Chen, D.; Zhang, J.; Li, W. (19–21 October 2018). Thyroid Nodule Classification Using Two Levels Attention-Based Bi-Directional LSTM with Ultrasound Reports. In Proceedings of the 2018 9th International Conference on Information Technology in Medicine and Education (ITME), Hangzhou, China, IEEE: San Francisco, CA, USA, 2018; pp. 309–312.

[11] Azamjah N., Zadeh Y. and Zayeri F. (2018) Global trend of breast cancer mortality rate: A 25-year study. *Global Trend Breast Cancer Mortal* 20:1–6

[12] Ning Y., Yu Z. and Pan Y. (2017) A deep learning method for lincRNA detection using auto-encoder algorithm. *BMC Bioinformatics* 18:1–9

[13] Zhang, H.; Zhao, C.; Guo, L.; Li, X.; Luo, Y.; Lu, J. and Xu, H., (2019). Diagnosis of thyroid nodules in ultrasound images using two combined classification modules. *In Proceedings of the 2019 12th International Congress on Image and Signal Processing, BioMedical Engineering and Informatics (CISP-BMEI)*, Suzhou, China, 19–21 October 2019; IEEE: San Francisco, CA, USA; pp. 1–5.

[14] Peng, S.; Liu, Y.; Lv, W.; Liu, L.; Zhou, Q.; Yang, H.; Ren, J.; Liu, G.; Wang, X.; Zhang, X.; *et al.* (2021). Deep learning-based artificial intelligence model to assist thyroid nodule diagnosis and management: A multicentre diagnostic study. *Lancet Digit. Health*, *3*, e250–e259.

[15] Li, X.; Zhang, S.; Zhang, Q.; Wei, X.; Pan, Y.; Zhao, J.; Xin, X.; Qin, C.; Wang, X.; Li, J.; *et al.* (2019), Diagnosis of thyroid cancer using deep convolutional neural network models applied to sonographic images: A retrospective, multicohort, diagnostic study. *Lancet Oncol. 20*, 193–201.

[16] Selvathi D. and Poornila A. (2017) Breast cancer detection in mammogram images using deep learning technique. *J Sci Res* 25(2):417–426.

[17] Kadam V., Jadhav S. and Vijayakumar K. (2019) Breast cancer diagnosis using feature ensemble learning based on stacked sparse autoencoders and softmax regression. *Image Signal Process* 43:1–11

Data Science & Exploration in Artificial Intelligence – Gururaj et al. (Eds)
© 2025 the Author(s), ISBN: 978-1-032-96390-7

A comprehensive assessment on phishing, smishing and vishing

Biswajit Debnath, Nirmalya Kar & Priyanka Biswas
Department of Computer Science and Engineering, National Institute of Technology Agartala, Tripura, India

Niladri Das
Department of Computer Science and Engineering, NIELIT Agartala, Tripura, India

Abhijit Das
Department of Information Technology, Manipal Institute of Technology Bengaluru, Manipal Academy of Higher Education, Manipal, India

ABSTRACT: Social engineering techniques are used in phishing, smishing and vishing attempts to get personal information or fool targets into disclosing critical information. These strategies may include using urgency, pretending to be reputable companies and preying on user trust. Phishing attacks which are usually sent by email, pose a serious risk because the amount of money lost is always rising. Promising techniques for detection include machine learning and deep learning; however, these methods rely on feature engineering which can be expensive and restricted by current understanding. While vishing targets victims via voice calls, smishing uses messages. The several techniques used in phishing, SMS phishing and voice phishing attacks are examined in this paper. This comprehensive paper investigates the reasons behind these attacks as well as the strategies employed to trick victims. This paper attempts to provide users with the information to counter phishing, smishing and vishing attacks.

Keywords: Phishing, Smishing, Vishing, Attack, Accuracy

1 INTRODUCTION

The digital age has revolutionized communication and information access, fostering a global interconnectedness that underpins modern society. Information and communication technologies (ICTs) have permeated our lives, be it for personal or business purposes. We specially rely on the internet for various tasks including communication, financial transactions and even critical decision-making. However, this increased reliance on technology has also created vul0nerabilities that cybercriminals exploit. Phishing, smishing and vishing are some of the most prevalent social engineering attacks that target unsuspecting individuals through various communication channels. These attacks rely on deception and manipulation to trick victims into revealing sensitive information or clicking on malicious links.

Phishing attack is like a digital cons where attackers try to steal your personal information by pretending to be someone you trust. Imagine receiving an email that looks like it's from your bank, credit card company, social media or even a familiar online store. The email might warm of suspicious activity on your account, an urgent update needed or even a tempting offer. The key is, it triggers a sense of urgency or importance to get you to click a link or open an attachment. Here's the twist, that link or attachment is the trap. Clicking it could take you to a fake website, designed to look almost identical to the real one. Once you enter your login information, credit card details or any sensitive data on this fake site, the attacker snatches it all. It's like the attacker casts a lure (the email) hoping you'll bite (click

 DOI: 10.1201/9781003589273-41

the link) and reveal your valuable information. "http://login.paypal.com-casepp-96616.11qnt.info/login.htm" is the one example of generated phishing URL and "http://login.paypal.com-casepp-845788.11qnt.info/login.htm" is the example of a real phishing URL. Weider D. *et al.*, [29] states that the main reasons of phishing attacks are as follows:

1.1 *Hiding the identity*

Rather of using the identities of themselves, phishers may sell stolen identities to would-be criminals who are looking for ways to hide their identities and actions (such purchasing supplies).

1.2 *Financial gain*

Financial gain could be obtained by phishers through the theft of banking credentials.

1.3 *Notoriety and fame*

Phishers might target victims in as attempt to gain acceptance from their peers.

According to Anti-Phishing Working Group (APWG), phishing is a fraudulent technique that uses technical deception as well as social engineering to get customer's financial account login information and personal identities [1].

Since genuine and fraudulent messages can seem identical, even an expert can have difficulty recognizing phishing assaults. In order to identify phishing mails, cyber security professionals therefore focus more on the message attachments such as email IDs or Uniform Re- source Locators (URLs) etc. By utilizing novel techniques to create difficult-to-detect phishing attacks, attackers enhance their attack strategies. For instance, they create phishing URL that mimic legitimate URLs like "https://www.facebook.edu/".

Findings ways to differentiate between a legitimate URL and a phishing URL is so crucial. Because of this, academics have recently put out a number of very accurate methods to identify phishing attempts including blacklists [22], conventional machine learning [30] and deep learning [14] etc. We share the following brief analysis of each solution.

1.4 *Conventional machine learning models*

This approach is like training a detective to identify suspicious individuals. Large sets of labeled URLS (phishing and legitimate) are provided as evidence. This model learns pattern from features like URL length, presence of odd characters and domain age. But conventional machine learning techniques require manual feature extraction. As a result, extracting a set of features takes time and human effort.

1.5 *Blacklist*

Imagine a bouncer checking IDs at a club. Blacklists [22] act similarly. They maintain a list of known malicious websites (IDs in this analogy). When a user tries to access a website, the system checks the blacklist. If the URL matches an entry access is blocked, preventing potential phishing attacks.

1.6 *Deep learning models*

Deep learning takes a more sophisticated approach, acting like a highly trained investigator. It automatically extracts features from the text and images, eliminating the manual effort required in traditional methods. This allows the model to analyze not just URLs but also the content and design of websites for phishing indicators.

This information clearly shows how prevalent phishing is as a security risk. Therefore, it becomes crucial to have strong solutions that can quickly identify and stop phishing attacks in order to safeguard people from them. In this situation research is essential to reducing the effects of this kind of serious threat. Over time, these has been extensive research on identifying phishing websites and an enormous amount of literature has dealt with this difficult issue.

The paper will present some of the evolutionary approaches to detect phishing attacks. The remaining part of the paperwork has been organized as follows. Background in Section II, Detection and Prevention in Section III. Section IV provides an analysis of the phishing, smishing and vishing detection models. Section V provides the conclusion at the end.

2 BACKGROUND

As explained in the section above, phishing attempts have grown significantly in frequency. The simplicity of use of these attacks designs is one reason for the surge.

They don't have to be carried out in the same location as the victims. Because of these benefits, phishing attempts rank among the most harmful that individuals or organizations can experience. A typical phishing attack scenario is illustrated in Figure 1.

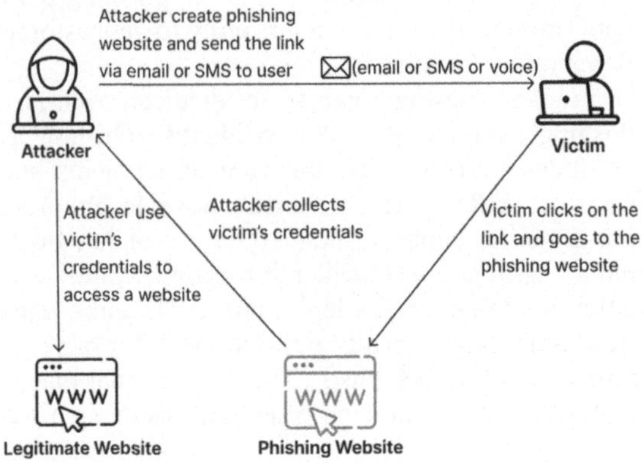

Figure 1. A typical phishing attack scenario.

2.1 *Anatomy of URLs*

As shown in Figure 2, a URL is composed of four elements. The Uniform Resource Locator (URL) protocol first demonstrates the data transmission mechanism. The host, consists of TLD (Top-Level Domain) and Second-Level Domain (SLD), is the second component. The TLD helped to find the purpose of the domain; for example, 'edu' denotes the domain for educational purpose.

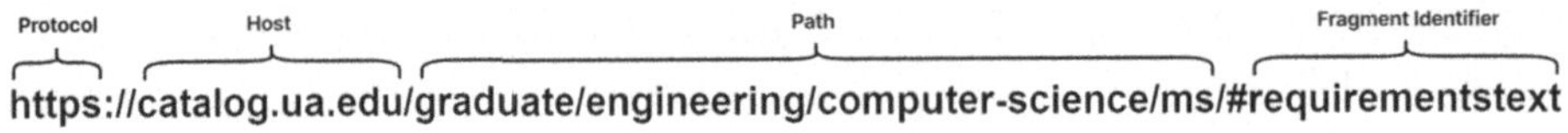

Figure 2. URL components.

Usually, the website name appears in SLD. A path containing the actual location of the needed web page makes up the third section. The last section may consist of a query with numerous parameters. Moreover, as shown in Figure 2 illustrates, it may function as an anchor to take the user to particular location on the website.

2.3 *Phishing classification*

Phishing is a broad term for attempting to steal personal information and it can be broken down into two main categories based on the communication channel used: smishing and vishing. Here's a table summarizing the key differences between smishing and vishing:

Table 1. Key difference of smishing and vishing.

Feature	Smishing	Vishing
Delivery Method	SMS text messages or messaging apps	Phone calls or voicemails
Tactics Examples	Urgent message requiring a link click, claiming account issues, promising refunds.	Impersonating trusted organizations. Claiming suspicious activity, pressuring for immediate action.

Working of Smishing and Vishing Attacks:

- *Working of Smishing Attacks*: Smishing attacks resemble other forms of phishing attacks, in which fraudsters trick users into disclosing their bank account information, personal information or cell phone by sending false messages and clicking on dangerous URLs. Scammers utilize SMS or messaging apps as instead of emails or phone calls to carry out their cybercrimes in smishing attacks. For a variety of reasons, scammers prefer smishing to other phishing attack methods. Studies indicate that individuals are more likely to click on links within text messages. Furthermore, scammers are increasingly able to mask the source of smishing messages through the use of strategies like emailing texts through software or utilizing burner phones to spoof phone numbers. Also, it's more difficult to identify risky links on mobile devices. Figure 1 shows a typical smishing scenario. For example, on a computer users can utilize the hover effect to see the destination of a link but not on a smartphone. Users are also habituated to receiving shortened URLs in their texts and hearing from banks and brands via SMS.
- *Working or Vishing Attacks*: Vishing, also known as voice phishing is the term for fake phone calls or voicemails intended to deceive someone into revealing private in- formation such as bank account information, credit card numbers or login credentials. Then these data could be used for fraud, identity theft or financial theft among other illegal crimes. Attackers that engage in vishing scams pretend as representatives of respectable companies including the victim's bank, the IRS or a package delivery service and make opportunistic phone calls. To seem like reliable companies, they sometimes use voice over internet protocol (VoIP) technologies or tool-free numbers. Figure 1 shows a typical vishing scenario. Unfortunately, phone calls aren't the only way these attacks happen. Phishing emails that ask the receiver to call a number are often the first step in a vishing attempt. Scammers utilize social engineering techniques during a call to influence the victim to reveal personal information.

3 DETECTION AND PREVENTION

3.1 *Phishing attack prevention*

3.1.1 *Victim awareness and education [2]*

Preventing phishing attacks can also be greatly helped by an awareness campaign. Since most of the attacks mentioned in the previous paragraph and sections occurred as a result of end user's irresponsibility, the level of awareness and education regarding phishing attacks is currently insufficient. Consequently, in order to stop these attacks, the efficacy of awareness campaigns cannot be overlooked.

3.1.2 *Limit the presence in open source intelligence [28]*
Reducing the amount of publicly accessible information is one of the most important steps people or businesses can take to reduce the possibility that phishers may obtain contact details to initiate the attacks or carry out customized the scams. So, OSINT can be employed in support of or opposition to phishing. Phishers for instance, might gather data from social networking and open sources in order to take advantage of possible targets.

3.1.3 *User access restrictions [28]*
Users can protect themselves against phishers by limiting access to any publicly accessible information, including corporate or personal email address or any other information that shouldn't be in the hands of phishers. Another way to guard against unauthorized access is to refrain from sending private correspondence via corporate emails. Using business emails to sign up for internet services could allow phishers to access the company's network.

3.2 *Phishing attack detection*

The biggest problem with networks and the internet nowadays is phishing. Several researchers have made an effort to develop tools to safeguard people form online threats by utilizing deep learning, machine learning, lists based and many more to stop phishing.

Some techniques to detect phishing attacks: Mehdi *et al.* [7] proposed a phishing website detection method based on a feature selection strategy combined with a nonlinear regression algorithm based on meta-heuristic. In order to more successfully identify malicious URLs, Prabakaran *et al.* [23] suggested an improved deep learning-based phishing learning detection system that combines strengths of Variational Autoencoders (VAE) and Deep Neural Networks.

Table 2. Some standard methods to detect phishing attacks.

Ref	Dataset	Method	Description	Accuracy & Precision	Observation
[27]	50,000 phishing URLs generated by DeePhish [8], 50,000 normal URLs from Alexa [12] and 50,000 normal URLs from PhishTank [13]	PhishHaven- An Efficient Real-Time AI Phishing URLs Detection System	This is a cutting-edge tool for identifying phishing URLs. It functions as a plugin for a browser. It can handle small URLs using URL hit methodology.	98% & 98%	It provides high accuracy
[5]	20,000 phishing data & 20,000 legitimate data [9], 50,000 data items consist of an equal amount of phishing and legitimate data [18] and Benchmark dataset	PhishDet	PhishDet detects phishing attempts using URLs and HTML information. It handles these two components using two different deep networks, which are later concatenated to generated a final judgment.	96.42% & 96.40%	Combines separate models for URLs and HTML content
[3]	MUPD [6] dataset	PDGAN: Detection with Generative Adversarial Networks	PDGAN used a CNN as discriminator and a LSTM as a generator of synthetic phishing URLs to decide whether the URLs are legitimate or phishing.	97.58% & 98.02%	This model does not depend on third party services.

3.2 *Smishing attack detection*

There are lots of smishing attack detection techniques like machine learning techniques, deep learning techniques and other techniques.

Some techniques to detect smishing attacks: Ankit *et al.* [15] presented a method that distinguished between spam and smishing messages. To distinguish smishing from a collection of spam messages, a new set of heuristics was applied. Liu *et al.* [19] introduced a transformer model to identify spam messages in SMS. The accuracy, F1-score and recall performance of this model are all higher. But this model isn't good at finding spam in big datasets.

Table 3. Some standard techniques to detect smishing attacks.

Ref	Database	Method	Description	Accuracy & TP rate	Observation
[21]	Almedia [4] and Pinterest [10]	An Efficient Model for Smishing Detection Using Neural Network	In this method, a neural network-based approach was used to extract the most efficient elements of smishing messages.	97.40% & 92.37%	Efficient feature extraction
[16]	SMS Spam Collection v.1[11]	Predicting Spam Messages Using Back Propagation Neural Network	This method provides a set criteria that effectively categorize the messages as spam. The process of choosing features is crucial and each one is carefully examined before being taken into consideration.	95.81% & 97%	Detect spam messages with high accuracy
[20]	Contagio Dump [24] and VirusShare [26]	AdDroid: Rule Based ML Framework for Android Malware Analysis	This method is used for evaluating and identifying harmful behavior in android applications. For detecting malware this method relies on statically examining the properties of android applications. In order to obtain the most distinctive rules, feature selection and extraction techniques are applied.	99.11% & 98.61%	High accuracy

3.3 *Vishing attack detection*

Neural network based spam detection algorithms outperform conventional machine learning models in handling imbalanced datasets as shown by Roy *et al.* [25]. UCI (The University of California, Irvine) repository's SMS Corpus, an unbalanced dataset containing 4827 non spam and 747 spam corpora, was used for the experiments.

Sometime deepfakes can be used in vishing scams. Imagine a scammer using a recording of a trusted person's voice, like your bank manager to trick you into revealing personal information. It's a sneaky way to add a layer of realism to a vishing attack. Kochare *et al.* [17] proposed a deep learning framework to identify audio deepfake.

4 ANALYSIS AND DISCUSSION

The analysis section is a crucial component of a research paper, as it delves into the interception and implications of the findings. It enables a thorough comprehension of the outcomes, there significance and their contributions to the field.

Some parameters are selected for analysis: The accuracy rate is chosen for analysis. The most popular and simple evaluation criteria is accuracy [27]. Furthermore, we are aware that a classifier is better the higher its accuracy.

$$\text{Accuracy} = \frac{\text{TP} + \text{TN}}{\text{TP} + \text{TN} + \text{FP} + \text{FN}} \quad (1)$$

where TP = the quantity of positive classes that are anticipated to be positive. Predicted negative classes as positive classes are denoted by FP, predicted positive classes as negative classes are represented by FN and predicted negative classes as negative classes are represented by TN.

And for better evaluation we also select the precision and we know the higher the precision [27], the better the performance of the algorithm. The calculation formula is as follows:

$$\text{Precision} = \frac{\text{TP}}{\text{TP} + \text{FP}} \quad (2)$$

4.1 *Experimental analysis of phishing detection methods*

In the domain of phishing attack detection, we explored a range of techniques to detect attacks. This comprehensive analysis enabled us to pinpoint the phishing attack detection approaches that demonstrate superior resilience against attacks.

The comparison between different phishing detection models is shown in Figure 3. Table 3 demonstrates that method [27] offers higher accuracy and method [3] provides higher precision.

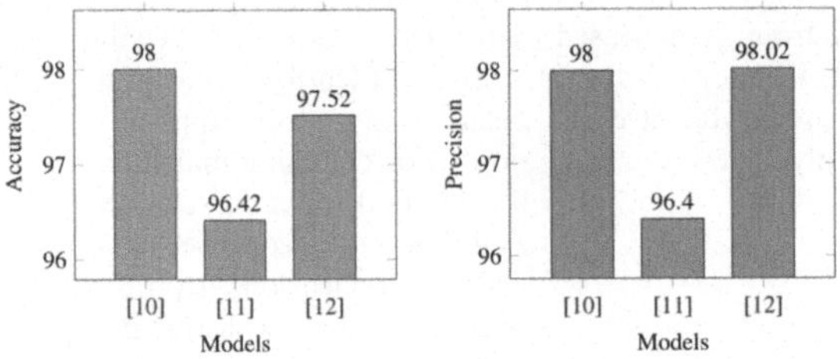

Figure 3. Accuracy and precision comparison of different phishing detection models.

4.2 *Experimental analysis of smishing detection models*

The performance evaluation's TP rate [21]. And also, accuracy is used for evaluation. the formula of accuracy is shown in equation 1. The comparison between various smishing detection models is shown in Figure 4.

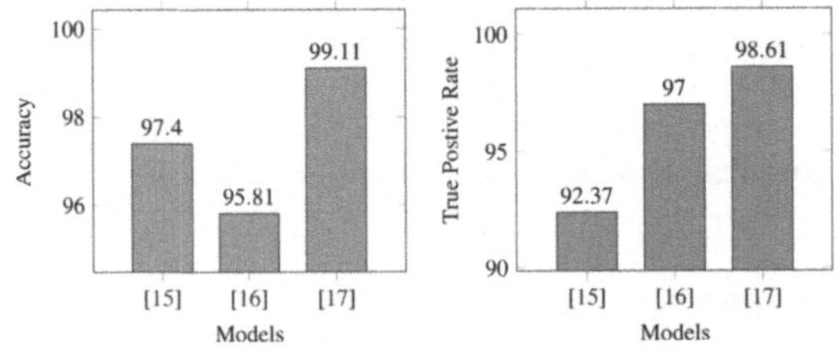

Figure 4. Accuracy and true positive rate comparison of different smishing detection models.

Figure 4 shows that the model [20] provides higher accuracy and higher true positive rate.

5 CONCLUSION

Phishing is a highly active and potent security issue impacting the targeted business, organizations, and people. This threat has been around for a while but is still one of the most widely employed attack vectors today. Throughout time, phishing campaigns have become progressively more sophisticated. Attackers use a variety of social engineering approaches to make their attacks harder for detection technologies to detect and more convincing for victims. Ac- cording to our survey, much research has gone into identifying phishing attacks. Among the different approaches to detection, machine learning and deep learning-based techniques are gaining traction. After analysing some smishing detection techniques, it provides a higher accuracy of 99.11% and a higher true positive rate of 98.61%. After analysing some vishing methods, it shows higher accuracy at 99.17%(average), and also a higher f1-score of 99.24%.

REFERENCES

[1] Anti-Phishing Working Group-APWG. Phishing Activity Trends Report-4Q. [Online]., Online; accessed 19-July-2008: https://docs.apwg.org/reports/apwg_trends_r port_q4_2023.pdf.

[2] Adil, M., Khan, R. and Ghani, M.A.N.U., (2020, February). Preventive techniques of phishing attacks in networks. *In 2020 3rd International Conference on Advancements in Computational Sciences (ICACS)* (pp. 1–8). IEEE.

[3] Al-Ahmadi, S., Alotaibi, A. and Alsaleh, O., (2022). PDGAN: Phishing detection with generative adversarial networks. *IEEE Access*, 10, pp.42459–42468.

[4] Almeida, T.A., Hidalgo, J.M.G. and Yamakami, A., (2011, September). Contributions to the study of SMS spam filtering: New collection and results. *In Proceedings of the 11th ACM symposium on Document Engineering* (pp. 259–262).

[5] Ariyadasa, S., Fernando, S. and Fernando, S., (2022). Combining long-term recurrent convolutional and graph convolutional networks to detect phishing sites using URL and HTML. *IEEE Access*, 10, pp.82355–82375.

[6] Al-Alyan, A. and Al-Ahmadi, S., (2020). Robust URL Phishing Detection Based on Deep Learning. *KSII Transactions on Internet & Information Systems*, 14(7).

[7] Babagoli, M., Aghababa, M.P. and Solouk, V., (2019). Heuristic nonlinear regression strategy for detecting phishing web- sites. *Soft Computing*, 23(12), pp.4315–4327.

[8] Bahnsen, A.C., Torroledo, I., Camacho, L.D. and Villegas, S., (2018, May). Deepphish: simulating malicious ai. *In 2018 APWG Symposium on Electronic Crime Research (eCrime)* (pp. 1–8).

[9] Chiew, K.L., Chang, E.H., Tan, C.L., Abdullah, J. and Yong, K.S.C., (2018). Building standard offline anti-phishing dataset for benchmarking. *International Journal of Engineering & Technology*, 7(4.31), pp.7–14.

[10] *Dataset: Pinterest, Smishing Message Images*, (November 20 2018) Available. https://in.pinterest.com/seceduau/smishing-dataset/?lp=true

[11] *Dataset: SMS Spam Collection.* Retrieved November, 2023. http://www.dt.fee.unicamp.br/~tiago/smsspamcollection/

[12] *Dataset: Alexa.* Accessed: Sep. 4, 2019. [Online]. https://www.alexa.com

[13] *Dataset: PhishTank.* Accessed: Sep. 4, 2019. [Online]. https://www.phishtank.com

[14] Fang, Y., Zhang, C., Huang, C., Liu, L. and Yang, Y., (2019). Phishing email detection using improved RCNN model with multilevel vectors and attention mechanism. *IEEE Access*, 7, pp.56329–56340.

[15] Jain, A.K., Yadav, S.K. and Choudhary, N., (2020). A novel approach to detect spam and smishing SMS using machine learning techniques. *International Journal of E-Services and Mobile Applications (IJESMA)*, 12(1), pp.21–38.

[16] Jain, A.K., Goel, D., Agarwal, S., Singh, Y. and Bajaj, G., (2020). Predicting spam messages using back propagation neural network. *Wireless Personal Communications*, 110, pp.403–422.

[17] Khochare, J., Joshi, C., Yenarkar, B., Suratkar, S. and Kazi, F., (2021). A deep learning framework for audio deepfake detection. *Arabian Journal for Science and Engineering*, pp.1–12.

[18] Le, H., Pham, Q., Sahoo, D. and Hoi, S.C., (2018). URLNet: Learning a URL representation with deep learning for malicious URL detection. *arXiv preprint* arXiv:1802.03162.

[19] Liu, X., Lu, H. and Nayak, A., (2021). A spam transformer model for SMS spam detection. *IEEE Access*, 9, pp.80253–80263.
[20] Mehtab, A., Shahid, W.B., Yaqoob, T., Amjad, M.F., Abbas, H., Afzal, H. and Saqib, M.N., (2020). AdDroid: rule-based machine learning framework for android malware analysis. *Mobile Networks and Applications*, 25, pp.180–192.
[21] Mishra, S. and Soni, D., (2022). Implementation of 'smishing detector': an efficient model for smishing detection using neural network. *SN Computer Science*, 3(3), p.189.
[22] Prakash, P., Kumar, M., Kompella, R.R. and Gupta, M., (2010, March). Phishnet: predictive blacklisting to detect phishing attacks. *In 2010 Proceedings IEEE INFOCOM* (pp. 1–5). IEEE.
[23] Prabakaran, M.K., Meenakshi Sundaram, P. and Chandrasekar, A.D., (2023). An enhanced deep learning-based phishing detection mechanism to effectively identify malicious URLs using variational autoencoders. *IET Information Security*, 17(3), pp.423–440.
[24] Parkour M., (2013). *Contagio Malware Dataset*, contagiodump.
[25] Roy, P.K., Singh, J.P. and Banerjee, S., (2020). Deep learning to filter SMS Spam. *Future Generation Computer Systems*, 102, pp.524–533.
[26] Roberts, J.M., (2014). Virus share. *Automatic Analysis of Malware Behavior using Machine Learning.*
[27] Sameen, M., Han, K. and Hwang, S.O., (2020). PhishHaven—An efficient real-time AI phishing URLs detection system. *IEEE Access*, 8, pp.83425–83443.
[28] Suzuki, Y.E. and Monroy, S.A.S., (2022). Prevention and mitigation measures against phishing emails: a sequential schema model. *Security Journal*, 35(4), pp.1162–1182.
[29] Weider, D.Y., Nargundkar, S. and Tiruthani, N., (2008, July). A phishing vulnerability analysis of web based systems. *In 2008 IEEE Symposium on Computers and Communications* (pp. 326–331). IEEE.
[30] Zhang, W., Ding, Y.X., Tang, Y. and Zhao, B., (2011, July). Malicious web page detection based on on-line learning algorithm. *In 2011 International Conference on Machine Learning and Cybernetics* (Vol. 4, pp. 1914–1919). IEEE

Data Science & Exploration in Artificial Intelligence – Gururaj et al. (Eds)
© 2025 the Author(s), ISBN: 978-1-032-96390-7

Enhancement of QoS aware energy efficient routing (QAEER) protocol for UAV based IoT networks

A. Rajashekar
Research Scholar, Department of CSE, UVCE, Bengaluru, Karnataka, India

Dharamendra Chouhan
Associate Professor, Department of CSE, UVCE, Bengaluru, Karnataka, India

Anand R. Umarji
Research Scholar, Department of CSE, UVCE, Bengaluru, Karnataka, India

J. Shreyas
Asst professor, Department of Information Technology Bengaluru, Manipal Institute of Technology Bengaluru, Karnataka India

ABSTRACT: The integration of Unmanned Aerial Vehicles (UAVs) in Internet of Things (IoT) networks provide numerous opportunities in terms of connectivity, improving efficiency, and environmental monitoring and infrastructure management. The research issues included in this network are as follows: energy consumption, topology control, security and Quality of Service (QoS) improvement. To overcome these research issues of UAV based IoT networks, this work introduces the QoS Aware Energy Efficient Routing (QAEER) protocol. The QAEER protocol integrates QoS awareness with energy efficiency considerations to provide a comprehensive solution for routing data packets in UAV networks. Simulation results show that the QAEER routing protocol enhances the entire network performance. The performance is analysed by using the QoS parameters like delivery ratio, packet loss rate, residual energy analysis and network throughput.

1 INTRODUCTION

The network is used to connect the users via wireless medium. UAV represents the Unmanned Aerial Vehicles which is used to transmit the data in free space propagation model. With the integration of Internet of Things (IoT) technologies, UAVs can extend their capabilities to gather real-time data from remote or inaccessible areas. However, the efficient operation of UAV-based IoT networks poses several challenges, particularly concerning energy consumption and dynamic routing.

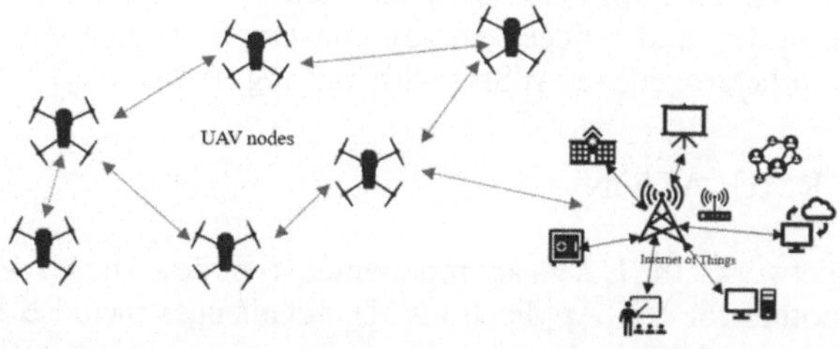

Figure 1. UAV based IoT network architecture.

DOI: 10.1201/9781003589273-42

Energy efficiency is the most important parameter in UAV operations due to their limited battery capacity and the necessity to maximize flight time. Routing protocols play a pivotal role in managing energy resources effectively while maintaining reliable communication links within the network. Moreover, the dynamic nature of UAV-based IoT environments, characterized by unpredictable topological changes and varying communication conditions, necessitates adaptive routing mechanisms to ensure continuous connectivity and data delivery. To overcome these research issues, the routing algorithms that minimize energy consumption by optimizing flight paths and data transmission strategies, considering UAV mobility patterns and energy constraints, were introduced (Hezekiah *et al.* 2023). Figure 1 shows that UAV network architecture for IoT applications.

2 RELATED WORK

(Paul *et al.* 2024) suggested the energy efficient cluster formation scheme for multi hop LoRa networks. The research emphasized the importance of considering packet forwarding structure and proposes a cluster development technique based on distance from the gateway for improved future multi-hop models. (Garg *et al.* 2024) introduced a hybrid approach that combined the K-means clustering and ant lion optimization for Energy-efficient clustering-based Routing in WSN. The network supported IoT framework and extended the network lifespan and enhanced energy efficiency. (Rawat *et al.* 2024) introduced Cluster Routing Protocol for Heterogeneous Network (CPHN), an energy-efficient cluster-based routing protocol for wireless sensor networks. It selects the cluster heads based on initial and residual energy levels of nodes which helps to extend network lifetime. By considering CH selection probability, residual and initial energy, CPHN effectively improves network longevity. Simulation results demonstrated CPHN's provide better performance compared to other existing protocols, the CPHN enhanced network lifetime, throughput, and stability period. (Ramalingam *et al.* 2024) introduced the fuzzy with Adaptive Sailfish Optimizer (ASFO) for cluster head selection and it finds the shortest routing path for data transmission. ASFO scheme introduced the elephant herd optimization approach for optimal route selection which leads to the reduction of energy consumption and improves the network lifetime. (Selmi *et al.* 2024) proposed the hybrid approach combining Artificial Bee Colony (ABC) and Ant Colony Optimization (ACO) to reduce the energy consumption issues. This hybrid approach enhanced the wireless sensor network performance by optimizing cluster head selection and routing paths.

(Gupta *et al.* 2024) suggested the Improved Low-Energy Adaptive Clustering Hierarchy (ILEACH) to improve the network lifetime of wireless sensor networks. This research demonstrated the effectiveness of ILEACH which helps to improve the data throughput, network lifetime, and energy efficiency compared to traditional protocols like LEACH, providing valuable insights for protocol selection and network optimization. (Saadallah *et al.* 2024) introduced an improved version of the LEACH protocol, LEACH-GA, which utilizes a genetic algorithm for cluster head selection, resulting in enhanced energy efficiency and network lifespan. (Sharma *et al.* 2024) proposed a metaheuristic-based hybrid data routing algorithm for energy-efficient Wireless Sensor Network (WSN) applications in the Internet of Things (IoT). This approach demonstrated better performance in terms of network lifetime, alive nodes, and reduced energy consumption, compared to existing algorithms, particularly in heterogeneous WSNs with varying network sizes.

3 PROBLEM FORMULATION

In UAV based IoT networks, the UAVs are represented as nodes. The UAV transmits the data from one end to another for IoT Applications. The challenges included in the UAV based Network are network topology and connectivity, data transmission protocols, energy efficiency, security and privacy, mobility management, quality of services and resource allocation.

The topology of the UAV network is to be monitored properly to ensure connectivity between all nodes and efficient data transmission paths. Due to the high mobility of the UAV nodes, the topology may be varied within a fraction of second. The topology maintenance is the most important factor that need to be addressed in the UAV based IoT networks. Due to the dynamic nature of UAV movement and varying network conditions, the selection of routing protocol is to be concentrated more. The routing protocol improves the network performance by increasing the delivery rate and throughput. UAVs have limited onboard power sources; energy consumption reduction strategy helps to improve the network life time. The reduction of energy consumption and resource utilization improves the network lifetime. As UAVs collect and transmit sensitive IoT data, robust security mechanisms must be implemented to protect against unauthorized access, data breaches, and cyber-attacks. Encryption, authentication, and intrusion detection systems are crucial components of ensuring data integrity and privacy. UAVs' mobility introduces challenges in managing handovers between different UAVs and ground stations seamlessly. Effective mobility management techniques are required to maintain connectivity and minimize disruptions during handover events. QoS-aware routing and scheduling algorithms help to meet these requirements while optimizing network performance. Resource allocation mechanisms need to dynamically adapt to varying network loads and prioritize critical traffic to ensure optimal performance.

4 METHODOLOGY

To overcome the research issues, this work proposes a new scheme called QoS Aware Energy Efficient Routing (QAEER). The QAEER algorithm aims to optimize both Quality of Service (QoS) and energy efficiency in routing data packets within UAV-based IoT networks. The routing protocol consists of different phases such as Route Discovery Phase, Route Selection Phase, Energy Monitoring Phase and Data Transmission Phase.

4.1 *Route discovery phase*

UAV node transmits data to the destination node. The first step is to broadcast a Route Request (RREQ) packet containing destination information to all the neighbor nodes. This process is to select the intermediate nodes to transmit the data from source node to destination. Intermediate nodes receive the RREQ from the neighbour nodes. The intermediate nodes check the RREQ message to satisfy the QoS criteria based on the hop count, delay and energy level. If the RREQ message contains lesser hop count, lesser delay values and lesser consumed energy level, it selects the route and updates its routing table. If the destination node is within the radio range, it sends a Route Reply (RREP) packet back to the source UAV.

4.2 *Route selection phase*

After receiving multiple RREPs, the source UAV selects the route with the best QoS and energy characteristics routing path for an efficient data transmission. The routing path selection is based on the hop count, delay and energy level. The selected route is added to the routing table for data transmission.

4.3 *Energy monitoring phase*

Energy monitoring phase introduces energy aware mechanism which helps to improve the UAV node's lifetime. Transmission power levels can be adjusted based on the communication distance d and Link Quality (LQ) to conserve energy. The relationship between transmission power P_t, distance d, and LQ can be modelled using the Friis transmission equation. Equation 1 represents the received power calculation.

$$P_{Rs} = P_{Ts} * G_{Ts} * G_{Rs} * \left(\frac{\lambda}{4\pi d}\right) * L \tag{1}$$

Where, P_{Rx} represents the receiving power, P_{Ts} represents the transmitted power, G_{Ts} represents the transmitting gain, G_{Rs} represents the receiving gain and λ represents wavelength of the signal.

$$\lambda = \frac{c}{f} \quad (2)$$

Where, c represents the speed of the signal, f represents the frequency, L represents the path loss factor, d represents the distance between the nearby nodes.

$$d = \sqrt{(x_2 - x_1)^2 + (y_2 - y_1)^2} \quad (3)$$

Assuming constant gains for the antennas and a known path loss factor, the equation is rearranged to solve.

$$P_{Ts} = P_{Rs} * \left(\frac{4\pi d}{\lambda}\right)^2 * \left(\frac{1}{G_{Ts} * G_{Rs} * L}\right) \quad (4)$$

The above equation 4 is used to adjust the transmission power P_{Ts} based on the received power P_{Rs}, communication distance d, and link quality LQ. This calculation is used to reduce the energy communication and provide a reliable data communication between the UAV nodes.

4.4 *Data transmission phase*

Data packets are transmitted by using the selected route. Intermediate nodes forward packets based on the routing table entries, ensuring reliable and timely delivery. Nodes monitor energy levels and update routing tables accordingly to avoid link failure and loss rate.

4.5 *QAEER routing algorithm*

QAEER algorithm provides a comprehensive solution for routing in UAV-based IoT networks, balancing QoS requirements with energy efficiency considerations to ensure reliable and sustainable network performance.

Step 1. Set Network G= (V, E) \\ unconnected graph
Step 2. Categorize S(V), D(V) \\ S- source node and D – Destination Node.
Step 3. Create RG \\ Routing Table
Step 4. Broadcast RREQ to all N(V) \\ Neighbour node
Step 5. Find I(V) \\ Intermediate node
Step 6. If min (D, Hc, El) \\ D- Delay; Hc- Hop count; El- Energy level
Step 7. Update RG
Step 8. Else repeat step 4
Step 9. Adjust Pr
Step 10. Improve LQ
Step 11. Select best routing path
Step 12. S(V) Data transmits to D9V)
Step 13. Stop.

5 RESULTS AND DISCISSION

Simulation results are obtained by using network simulator ns2.35. The number of nodes ranges from 100 to 500. QoS parameters, including throughput, delivery rate, residual energy analysis, and packet loss rate, are incorporated to analyze the network performance. A comparative analysis is conducted between the proposed QAEER routing protocol and two established protocols: Ad hoc On-Demand Distance Vector (AODV) routing protocol and Destination-Sequenced Distance-Vector Routing (DSDV) protocol. Figure 2 compares the throughput performance of QAEER with that of AODV and DSDV, focusing on networks of varying sizes. QAEER provides better performance, achieving a throughput of 1.94 Mbps in a network with 500 nodes. The analysis shows that the proposed route protocol QAEER adapts the large-scale ad hoc networks. Figure 3 illustrates the delivery rate analysis. In comparison to other existing routing protocols, the proposed QAEER provides higher delivery rate. It increases the delivery rate analysis with the help of route selection phase. Figure 4 shows that the energy efficiency analysis. Energy efficiency is one of the most important factors while designing the routing protocols for UAV based IoT networks, as UAV typically operate on limited battery power. Compared to other existing protocol, QAEER reduces the energy consumption by adjusting the transmission power while transmitting data. It increases the energy efficiency analysis with the help of energy maintenance phase.

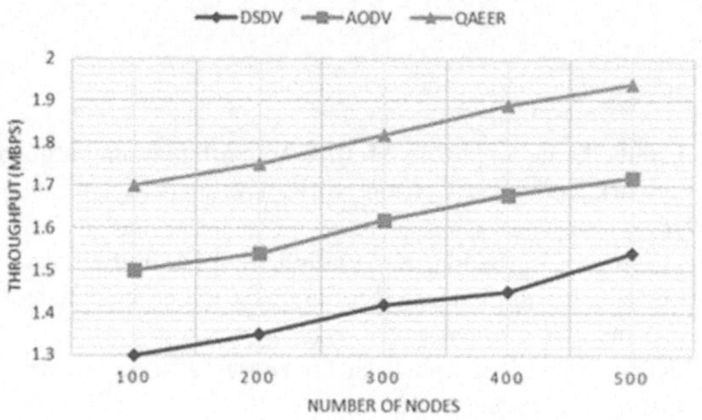

Figure 2. Throughput analysis.

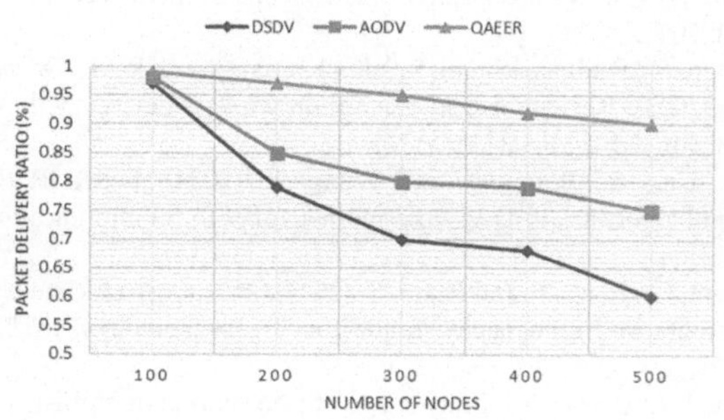

Figure 3. Delivery rate analysis.

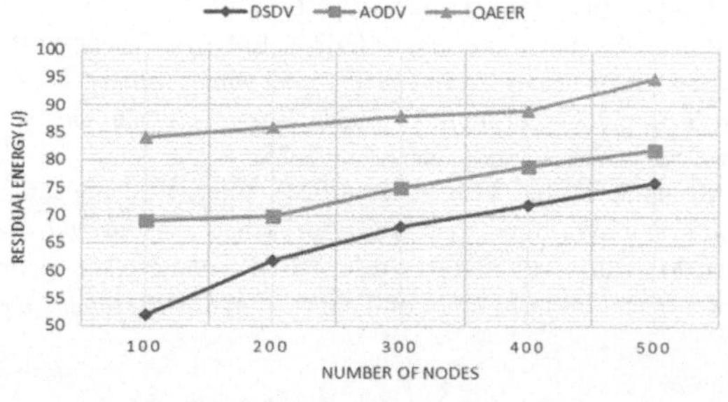

Figure 4. Energy efficiency analysis.

6 CONCLUSION

In this research work, the proposed QoS Aware Energy Efficient Routing (QAEER) protocol integrates QoS importance with energy efficiency considerations to provide a comprehensive solution for routing data packets in UAV networks. Simulation results demonstrate that the QAEER routing protocol enhances the UAV based IoT network performance. The performance of QAEER is analyzed using various QoS parameters, including delivery rate, packet loss rate, residual energy analysis, and network throughput. The results show that QAEER performs better than the existing routing protocols such as Ad hoc On-Demand Distance Vector and Destination-Sequenced Distance-Vector Routing. The QAEER routing protocol offers a promising solution for addressing the research challenges in UAV-based IoT networks. By integrating QoS awareness with energy efficiency considerations, QAEER enables efficient and reliable data routing while optimizing energy consumption.

REFERENCES

Azadur, R.M., Pawase, C.J. and Chang, K. (2024). Multi-UAV path planning utilizing the PGA algorithm for terrestrial IoT sensor network under ISAC framework. *Transactions on Emerging Telecommunications Technologies* 35(1): e4916.

Gad, G., Farrag, A., Aboulfotouh, A., Bedda, K., Fadlullah, Z.M. and Fouda, M.M. (2024). Joint self-organizing maps and knowledge distillation-based communication-efficient federated learning for resource-constrained UAV-IoT systems. *IEEE Internet of Things Journal.*

Garg, R. (2024). K-LionER: meta-heuristic approach for energy efficient cluster based routing for WSN-assisted IoT networks. *Cluster Computing*: 1–15.

Gupta, A. and Gupta, S.K. (2024). Unmanned aerial vehicles in collaboration with fog computing network for improving quality of service. *International Journal of Communication Systems*: e5759.

Hezekiah, J.D.K., Ramya, K.C., Selvan, M.P., Kumarasamy, V.M., Sah, D.K., Devendran, M., Arumugam, S.S. and Maheswar, R. (2023). Nature-inspired energy enhancement technique for wireless sensor networks. *Energies* 16(20): 7021.

Hezekiah, J.D.K., Ramya, K.C., Radhakrishnan, S.B.K., Kumarasamy, V.M., Devendran, M., Ramalingam, A. and Maheswar, R. (2023). Review of next-generation wireless devices with self-energy harvesting for sustainability improvement. *Energies* 16(13): 5174.

Li, J., Liu, A., Han, G., Cao, S., Wang, F. and Wang, X. (2024). FedRDR: Federated reinforcement distillation-based routing algorithm in UAV-assisted networks for communication infrastructure failures. *Drones* 8(2): 49.

Paul, B., Assi, C., Kaddoum, G., Palit, R. and Islam, S. (2024). An energy-efficient cluster formation based on optimal node distribution in full capacity multi-hop LoRa networks. *IEEE Open Journal of the Communications Society.*

Rahimi, O. and Shafieinejad, A. (2024). Minimizing age of information in multi-UAV-assisted IoT networks: a graph theoretical approach. *Wireless Networks* 30(1): 533–555.

Ramalingam, S., Dhanasekaran, S., Sinnasamy, S.S., Salau, A.O. and Alagarsamy, M. (2024). Performance enhancement of efficient clustering and routing protocol for wireless sensor networks using improved elephant herd optimization algorithm. *Wireless Networks*: 1–17.

Rawat, P., Rawat, G.S., Rawat, H. and Chauhan, S. (2024). Energy-efficient cluster-based routing protocol for heterogeneous wireless sensor network. *Annals of Telecommunications*: 1–14.

Saadallah, N.R., Alabady, S.A. and Al-Turjman, F. (2024). Energy-efficient cluster head selection via genetic algorithm. *Al-Rafidain Engineering Journal (AREJ)*: 12–25.

Selmi, A., Khan, R.U., Moulahi, T. and Lorenz, P. (2024). Energy efficient cluster routing protocol for wireless sensor networks using hybrid metaheuristic approaches. *Ad Hoc Networks*: 103473.

Sharma, S.K. and Chawla, M. (2024). PRESEP: cluster-based metaheuristic algorithm for energy-efficient wireless sensor network application in Internet of Things. *Wireless Personal Communications*: 1–21.

Data Science & Exploration in Artificial Intelligence – Gururaj et al. (Eds)
© 2025 the Author(s), ISBN: 978-1-032-96390-7

A systematic review with case study on Question Answering System (QAS) using deep learning

Savita Bakare
Computer Science and Engineering Department, KLS Gogte Institute of Technology, Udyambag, Belagavi, India

Rupa Somaling, Sneha Akkatangerhal, Varun Rakshe & Vedant Patil
Computer Science and Engineering Department, Kle Dr. M. S. Sheshgiri College of Engineering and Technology, Udyambag, Belagavi, India

ABSTRACT: A QAS is a conversational agent designed to respond user queries. Natural Language Processing (NLP) enables to understand the user's intent hidden within natural language questions. Traditional machine learning methods were limited by the amount of data used for training, restricting their capabilities. Deep Learning (DL) techniques have emerged as the most promising solution for QAS handling massive datasets. While significant progress has been made in NLP and the rise of DL, QAS still face limitations. They struggle with subtle nuances, complex inquiries, and adapting to unforeseen situations. These limitations hinder their real-world application. In this regard, the aim is to analyze the current state of DL-based QAS to address key queries. Hence, a case study on DL-based chatbot system is performed to answer user queries on breast cancer dataset. The evaluation results show the model achieves high accuracy (0.9747), with precision (0.9827), and recall (0.9504). Additionally, the low training loss (0.1121) suggests the model learned effectively from the training data. This chatbot system has the potential to improve access to accurate and reliable breast cancer information for patients, caregivers, and healthcare professionals.

Keywords: Question-answering (QA) system, Deep learning (DL) review, Natural Language Processing (NLP)

1 INTRODUCTION

A question-answering (QA) system is a tool that processes user queries using NLP techniques. The QAS pipeline analyzes the question, searches for relevant sources, extracts a specific answer, and formulates a natural language response. DL has revolutionized QA systems, providing powerful tools for natural language understanding. DL models like Transformers and Long Short Term Memory (LSTM) complement QA models but require task-specific data and have computational complexity. Using pre-trained models and storing bidirectional context the QAS performance may be improved. However, current QA models have drawbacks like complex term interpretability, unwanted data, and fact-oriented QA incapability. Researchers aim to design a more robust QAS that addresses these issues. Although NLP and DL have advanced significantly, QAS still face limitations in understanding nuance, handling complex inquiries, and generalizing to new situations, which limits their real-world applicability. In this paper an analysis on the current state of DL-based QAS to address key queries with a case study on breast cancer chatbot QAS is carried out.

DOI: 10.1201/9781003589273-43

2 LITERATURE REVIEW

In this section, a brief review of DL-based QAS is carried out. Also, an analysis on most promising and commonly used approach in QAS is being discussed. The various types of DL-QA systems are categorized as discriminative model, generative model, and hybrid model as shown in Figure 1.

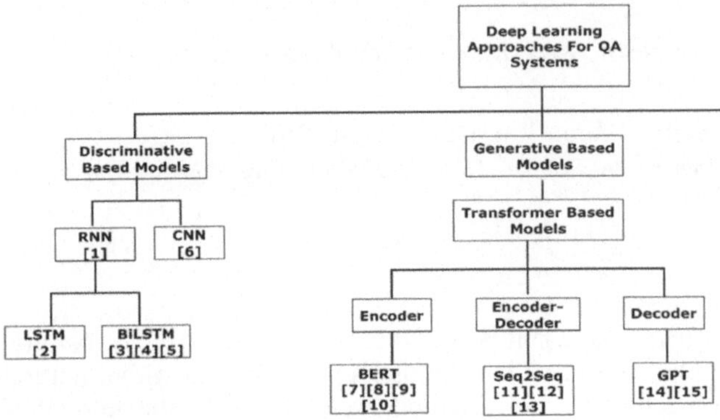

Figure 1. Classification of DL models for QAS.

2.1 *Discriminative-based DL-QAS*

Discriminative models are supervised ML models that create boundaries for the data using probabilistic models. The model is classified as a Recurrent Neural Network (RNN) and Convolution neural network (CNN). RNN based QAS efficiently access large-scale knowledge bases (KB) to minimize the gap between natural language queries and organized semantics. In this regard, Zhou *et al.* (2023) proposed a predicate matching and subject detection using a deep fused model (DFM) for knowledge-based QA. DFM is a parameter-shared unified framework that merges surface-level and deep-level semantic representations with a gate mechanism. Such a combination is used for efficient access and to reduce the differences between the structured meaning and the use of the language. Furthermore, the RNN model is divided into two categories: LSTM and Bidirectional Long Short Memory (BiLSTM). The LSTM architecture is capable of processing sequential data and learning long-term dependencies. A cascaded LSTM architecture can be employed for answering questions on real-world images. This architecture learns visual features along with parts of speech tags of question words or tokens. To reduce the dimensionality of visual features, Principal Component Analysis (PCA) is used. This enables the proposed QA model to capture the general pattern of a question for a given context of an image and is not limited to the training dataset (Chowdhury *et al.* 2017). BiLSTM is capable of processing sequential data with long-term dependencies in both forward and backward directions. An application model of Bi-LSTM named "Online Medical Chatbot System for Answering Complex Medical Questions", is proposed by Qiming Bao *et al.* (2020) They used BiLSTM attention model to capture hierarchical information and a Knowledge Graph (KG) for easy retrieval. However, in another application of agriculture, the insufficient labeled data problem is resolved by applying the combination of Bi-LSTM and CRF models to recognize the entities (Wang *et al.* 2023). For training, the model requires less than one-tenth of the data to be labeled. Additionally, one more challenge to be addressed is the answer selection strategy. It is concerned with proper nouns, strong sparse features, and badly formed sentences. To improve the text's semantic representation, the BiLSTM and text CNN models for text matching were used (Li *et al.* 2024). CNN is a frequently used technique for NLP and image recognition tasks of QA systems. It addresses limited or poor-quality data and interpretability issues. A disease prediction model is built using this combination.

2.2 *Generative-based DL-QAS*

The generative model focuses on generating new instances and the transformer model is a type of generative-based model. Transformers use the encoder and decoder models along with attention to improve the translation speed of the model. Transformer-based models are further classified as Encoder, Encoder-Decoder, and Decoder models. Text classification can be done using the transformer's encoder design. BERT is a widely used option. BERT-based QAS addresses issues of labeled data, complexity of questions, and lack of context in user questions. In this regard, Boreshban *et al.* (2023) proposed a model by fusing Knowledge Distillation (KD) and Active Learning (AL) to reduce parameters and model complexity. KD is used to transfer a large model's knowledge to a condensed neural network. To reduce annotation efforts and labeling costs, AL is used. To add extra information to the model, AL chooses meaningful unlabelled data samples from the unlabelled pool. The 2% of the parameters were trained on 6-layer TinyBERT and DistilBERT. The result shows both the models are equivalent. While other researchers focus on complex questions related to multiple topic entities, multi-hop relations, and the difficulty of extracting relevant information. Hence, to address it Yang *et al.* proposed a framework pipelined in three parts. In the first part, the question is broken down, then constraints are identified, and finally question reasoning over knowledge QA is done. For a small training dataset, the results are good but cannot be generalizable to a larger training dataset. The problem of lack of context, which makes it difficult to understand a user's question is addressed by Al-Besher *et al.* (2022) He proposed a new model, which can better respond with little or incomplete data. Another work focuses on an ensemble model with variants of BERT to identify the best answer to a question. Pranesh *et al.* (2020) proposed a stacking model with high accuracy. The model combines the top-K predictions from individual language models to determine which one provides the best answer to a given question. It has been trained using the Natural Questions (NQ) dataset, which resulted in a 0.731, harmonic mean score for long answers and a 0.582, harmonic mean score for short answers. The results demonstrate that using an ensemble model improves the understanding capabilities of language models and helps to generalize better with heterogeneous models. Language translation tasks can be performed by the transformer's encoder-decoder architecture. Seq2Seq is one such transformer's encoder-decoder model used for QA systems. The seq2Seq model takes input data in sequence and produces a sequence of output data. An application model named "A Chatbot for an Indonesian University Admission using a Question Answering System", is proposed by Chandra *et al.* (2019) They used the Sequence-to-Sequence Model to provide quick response to customers. It is trained on conversational data acquired from university admissions. An application "Question Answering University Chatbot", is proposed by Khin et al. (2020) The model helps students and faculties seeking university queries. An attention mechanism based on the RNN encoder-decoder model is used which results in a BLEU SCORE of 0.41. Zhu *et al.* proposed a new method for candidate queries, which takes into account both semantic and surface-level correlations. The model generates candidate queries based on linked entities, types, and number operations for each question. These queries are then encoded and decoded into the given question using an encoder-decoder framework. The best query is selected based on its decoding probability. The decoder model generates a sequence of words based on a provided prompt or context. One of the most widely used transformer-based decoder models is the generative pre-trained transformer (GPT) which generates answers to specific questions. GPT-based QAS address challenges related to multi-hop question generation and missing visual details. Li *et al.* (2023) proposed a new approach to simulate human reasoning for multi-hop questions. The model encodes the input and generates a knowledge graph, which is then updated using a graph attention network. Finally, the multi-head attention generation module produces the questions. The dataset used is HotpotQA. The challenge of missing visual details is addressed by image captioning methods using GPT-3 (Hu *et al.* 2023).

2.3 *Hybrid-based DL-QA system*

Hybrid models also known as Ensemble Learning which combines multiple models to enhance overall performance. It addresses challenges related to unlabeled data and reasoning visual relationships in a visual question-answering system. Hillary Ngai *et al.* (2021) proposed three hybrid QA systems, that use transformer-based models such as BERT, ALBERT, and T5. The labeled QA datasets, CovidQA and CovidGQA evaluated the performance of these models. The results show that BERT-based QA has the highest F1 score while ALBERT-based QA achieves the highest exact match. In another work by Zhang *et al.* (2020) a visual relational reasoning module was designed. This module combines visual relationships and attention networks to achieve more fine-grained feature fusion. Furthermore, Sha *et al.* (2023) proposed a novel method to improve existing models which retrieve noisy, irrelevant nodes and lack interaction between question and graph entities. The model solves the above issues by using a density matrix to extract an enhanced subgraph of the KG and reasons along the corrected knowledge path. It also uses a bidirectional attention strategy to fuse representations of questions and graph entities. However, Mythili *et al.* (2022) discussed a Medical Knowledge Graph created from Electronic Health Records (EHR). It maps the interactions between the entities of individuals, disease, and pharmaceuticals. The hybrid approach of Bi-directional Long Short-Term Memory Multi-Head Attention Conditional Random Field is employed for Named Entity Recognition. An embedding-based models can be integrated with databases or knowledge graphs. It is capable of addressing various issues such as parameter non-sharing, high complexity, poor modeling of multi-level questions, and detecting misinformation. To overcome these issues, Li *et al.* (2023) proposed a low-complexity multi-level contextual question model. The study conducted comprehensive tests on VQAv2 and VQA-CPv2 datasets which confirmed that CMQEF provides better results on SANs and UpDn models. Another study conducted by Cui *et al.* (2020) detects misinformation in medical QA. They developed a model named "DETERRENT" which uses a graph attention network guided by a knowledge graph. By utilizing additional data from medical KG, information is propagated throughout the network. Medical KG and Article-Entity Bipartite Graph are combined to convey node embedding's via knowledge paths. Knowledge-guided article embedding's are utilized for detection, and an attention mechanism is employed to assess each entity's significance to the article. A knowledge graph query system is designed specifically for the field of chemistry to retrieve fact-oriented information (Zhou *et al.* 2023). This QA system uses a variety of embedding techniques and queries embedding spaces concurrently. Then, a score alignment model is used to re-rank the responses from various embedding spaces. Furthermore, the system's bidirectional entity-linking model, based on BERT improves the entity-linking module's accuracy and robustness. Finally, a variety of chemical reaction mechanisms can be handled by this system. However, another work Timilsina et al. (2023) focused on interpreting automatic age recognition on CQAs. This work compares advanced graph-based neural network regression and embedding models on a large activity graph. The results demonstrate that Node2vec outperforms other embedding's. By combining Artificial Neural Network Regression with Node2vec with mean square error of 8.39. A Memory Network includes a memory component to keep track of important information. Memory NET-based QAS addresses the challenge of the existing embedded-based methods that ignore the interrelationships between questions and knowledge bases. Chen *et al.* (2019) proposed Bidirectional Attentive Memory Networks for QA. In contrast to other approaches, the model needs a few manually created features and no outside resources to respond. It performs better and has greater interpretability than current information-retrieval-based techniques. A revolutionary method that would allow traditional Key Value-Memory Neural Network models to reason intelligibly about complicated problems (Xu *et al.* 2019). The STOP technique is employed by this model to prevent incorrect or repetitive memory reading.

3 METHODS

This section outlines the process of identifying, screening, and assessing studies. This study explores DL-based models for designing QAS, following PRISMA-ScR guidelines as shown in Figure 2. Records were identified from databases Google Scholar, IEEE Xplore, ACM Digital Library, and Springer Link. The search is performed with the keyword "question answering system", "deep learning" and "NLP", yielding an initial count of 17,100 published records.

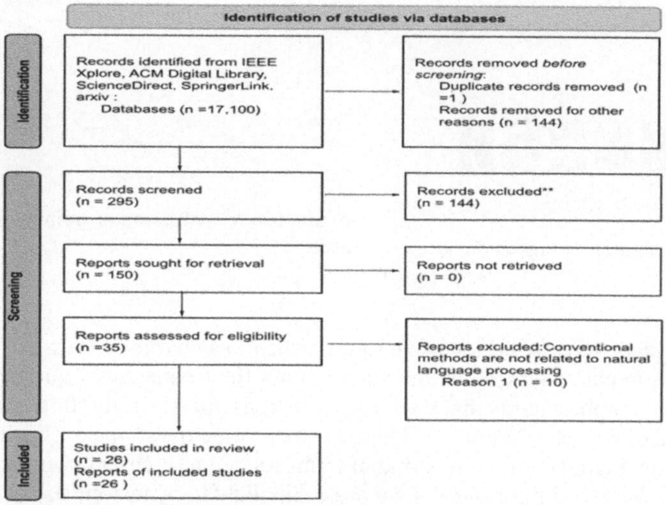

Figure 2. Review methodology based on PRISMA model.

The screening of records relied on criteria title relevance, abstract content, and keyword combinations. A total of 295 records underwent screening, with 145 being excluded based on these criteria. Articles that did not utilize DL extensively for designing of QA system were excluded from the research. Additionally, studies published in languages other than English were omitted. The Final search resulted in 150 full-text articles.

4 RESULTS AND DISCUSSION

This section presents a case study on a breast cancer chatbot. Breast cancer is a major health concern, and patients, caregivers, and healthcare professionals often have many questions about the disease. They may seek information on risk factors, symptoms, treatment options, and prognosis. Timely access to accurate and reliable information is crucial, but it can be challenging to find. The chatbot can answer user queries in natural language, offering a convenient and informative way to access breast cancer information. The chatbot uses an information retrieval approach where the system retrieves relevant responses and information from the breast cancer dataset stored in JSON format as Question-Answer pairs.

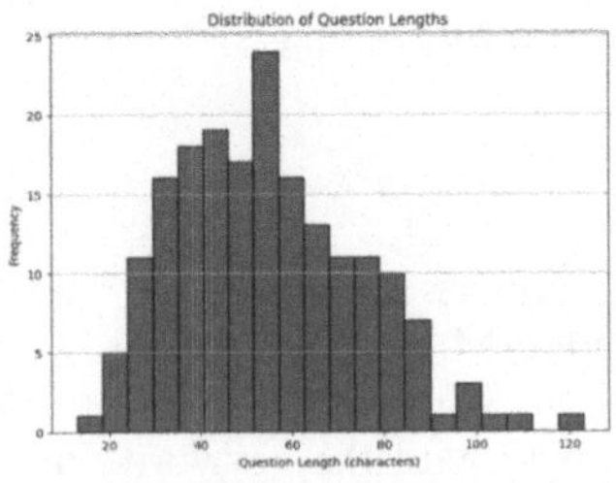

Figure 3. Distribution of question lengths.

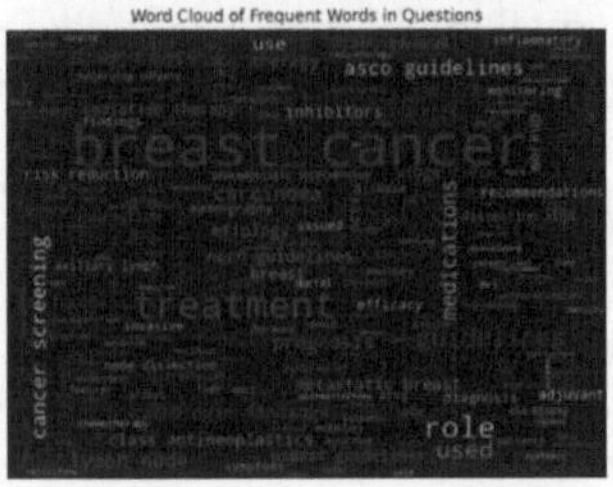

Figure 4. Word cloud of frequent words in questions.

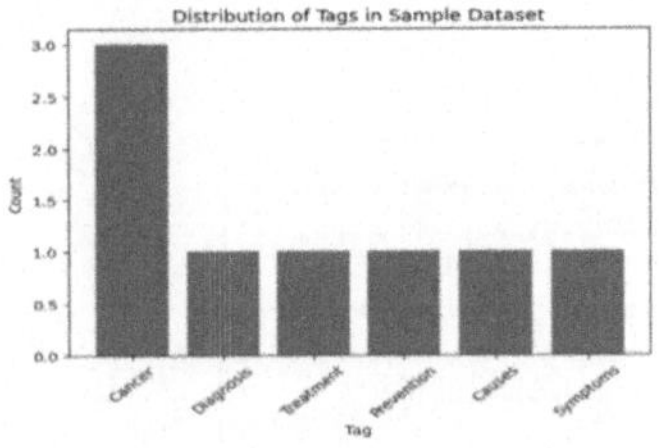

Figure 5. Distribution of tags in dataset.

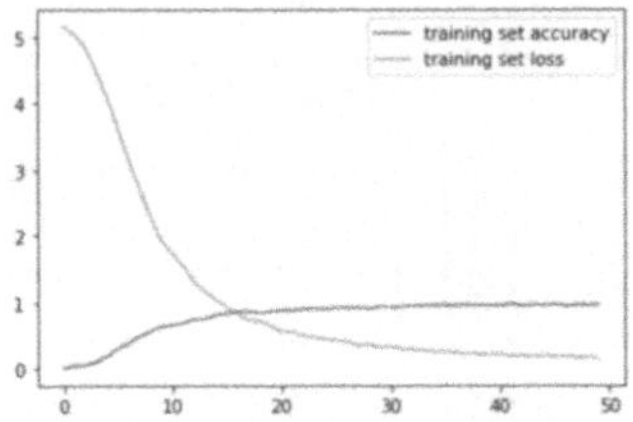

Figure 6. Visualisation of training set accuracy and loss.

Figure 3, depicts the distribution of questions asked to a chatbot. The x-axis represents the question length in characters, while the y-axis shows the frequency of questions asked with that length. The graph suggests that most user questions are relatively short, with a character count somewhere between 20 and 60. There are also some questions outside this range, but fewer in number. Figure 4 is a word cloud of terms related to breast cancer. Word clouds are visualizations where the importance of a word is highlighted by its font size or colour. In this case, all the words are black and the size of the word reflects how often it appeared in a set of documents. In this case, the most prominent words are “breast cancer”, “risk”, “women”, “therapy”, and “treatment”. Figure 5 analyses tags associated with a dataset related to breast cancer information. Labels on the horizontal axis represent the different tags found in the data. These could be terms like cancer, symptoms, treatment etc. The vertical axis shows the count or frequency of each tag. Higher bars indicate tags that appear more often in the data. From a quick glance, it seems term cancer is the most frequent tag, followed by diagnosis, treatment, prevention, and causes. The user interacts with the chatbot through a chat interface, entering queries in natural language. The pre-processing involves spelling correction, tokenization and lemmatization. The pre-processed query is transformed into a numerical representation using a bag-of-words model. It is then fed into a pre-trained machine learning neural network from the TensorFlow library, to predict the most likely intent associated with the query. Based on the predicted intent, the system retrieves relevant responses and information from a knowledge base stored in JSON format. The knowledge base contains predefined mappings between intents and their corresponding responses or actions. Finally, the chatbot delivers the retrieved information to the user through text or spoken audio, depending on the interface used for interaction. In Figure 6, x-axis represents the epochs, which are iterations over the training dataset. The y-axis represents training set accuracy and loss metrics. In the graph, as the number of epochs increases the training set loss decreases. Hence, the model learns by improving its performance on the training data. The model results high accuracy (0.9747) and relatively low loss (0.1121), with a precision of 0.9827 and a recall of 0.9504.

5 CONCLUSION

This study explores various strategies applied on deep learning based question-answering systems. As per the study, the models are classified as hybrid, generative, and discriminative. The trend highlights the transformer-based technique is the most promising and commonly used approach. It is popular due to its ability to handle ambiguity, contextual understanding, computational efficiency, and interpretability. However, irrespective of transformers' complex architecture, large model size, lack of explainability, domain specificity, open-ended and complex questions and the requirement of high processing GPU's; transformers are a good option for question-answering tasks. However, these limitations also present opportunities for further research to be carried out. A case study on DL-based chatbot system is performed to answer user queries on breast cancer dataset. The evaluation results for correctly classifying user intents and retrieve relevant responses shows high accuracy 0.9747, precision 0.9827, and recall 0.9504. Additionally, the low training loss (0.1121) suggests the model learned effectively from the training data.

REFERENCES

Al-Besher, A., Kumar K., Sangeetha M., and Butsa T. (2022). BERT for conversational question answering systems using semantic similarity estimation. *Computers, Materials & Continua* 70(3):4763–4780.

Bao, Q., Ni L., and Liu J. (2020). HHH: An online medical chatbot system based on knowledge graph and hierarchical bi-directional attention. *Proceedings of the Australasian Computer Science Week Multiconference*:1–10.

Boreshban, Yasaman, Seyed Morteza Mirbostani, Gholamreza Ghassem-Sani, Seyed Abolghasem Mirroshandel, and Shahin Amiriparian. (2023). Improving question answering performance using knowledge distillation and active learning. *Engineering Applications of Artificial Intelligence* 123:106137–106137.

Chandra, Y. W., and Suyanto S. (2019). Indonesian chatbot of university admission using a question answering system based on sequence-to-sequence model. *Procedia Computer Science* 157:367–374.

Chen, Y., Wu L., and Zaki M. J. (2019). Bidirectional attentive memory networks for question answering over knowledge bases. *arXiv (Cornell University)*.

Chowdhury, I., Nguyen K., Fookes C., and Sridharan S. (2017). A cascaded long short-term memory (LSTM) driven generic visual question answering (VQA). *QUT ePrints (Queensland University of Technology)*:1842–1846.

Cui, L., Seo H., Tabar M., Ma F., Wang S., and Lee D. (2020). DETERRENT: Knowledge guided graph attention network for detecting healthcare misinformation. *Proceedings of the 26th ACM SIGKDD International Conference on Knowledge Discovery & Data Mining.*

Hu, Y., Hua H., Yang Z., Shi W., Smith N. A., and Luo J. (2023). *PromptCap: Prompt-Guided Image Captioning for VQA with GPT-3.*

Khin, N. N., and Soe K. M. (2020). Question answering based university chatbot using sequence to sequence model. *IEEE Xplore.*

Li, S., Gong C., Zhu Y., Luo C., Hong Y., and Xueqiang Lv. (2024). Context-aware multi-level question embedding fusion for visual question answering. *Information Fusion* 102:102000–102000.

Li, Z., Yang X., Zhou L., Jia H., and Li W. (2023). Text matching in insurance question-answering community based on an integrated BiLSTM-TextCNN model fusing multi-feature. *Entropy* 25(4):639–639.

Li, Z., Cao Z., Li P., Zhong Y., and Li S. (2023). Multi-hop question generation with knowledge graph-enhanced language model. *Applied Sciences* 13(9):5765–5765.

Mythili, R., Parthiban N., and Kavitha V. (2022). Construction of heterogeneous medical knowledge graph from electronic health records. *Journal of Discrete Mathematical Sciences and Cryptography* 25(4):921–930.

Ngai, H., Park Y., Chen J., and Mahboobeh Parsapoor. (2021). Transformer-based models for question answering on COVID-19. *arXiv (Cornell University)*.

Pranesh, Raj Ratn, Ambesh Shekhar, and Smita Pallavi. (2020). *QuesBELM: A BERT Based Ensemble Language Model for Natural Questions.*

Sha, Y., Feng Y., He M., Liu S., and Ji Y. (2023). Retrieval-augmented knowledge graph reasoning for commonsense question answering. *Mathematics* 11(15):3269–3269.

Timilsina, M., and Figueroa A. (2023). Neural age screening on question answering communities. *Engineering Applications of Artificial Intelligence* 123:106219.
Wang, L., Jiang J., Song J., and Liu J. (2023). A weakly-supervised method for named entity recognition of agricultural knowledge graph. *Intelligent Automation and Soft Computing* 37(1):833–848.
Xu, K., Lai Y., Feng Y., and Wang Z. (2019). *Enhancing Key-Value Memory Neural Networks for Knowledge-Based Question Answering.*
Yang, L., Guo H., Dai Y., and Chen W. (2023). A method for complex question-answering over knowledge graph. *Applied Sciences* 13(8):5055–5055.
Zhang, W., Yu J., Hu H., Hu H., and Qin Z. (2020). Multimodal feature fusion by relational reasoning and attention for visual question answering. *Information Fusion* 55:116–126.
Zhou, X., Zhang S., Agarwal M., Jethro Akroyd, Mosbach S., and Kraft M. (2023). Marie and BERT⊠A knowledge graph embedding based question answering system for chemistry. *ACS Omega.*
Zhou, Xie, Yu, and Huang. (2021). DFM: A parameter-shared deep fused model for knowledge base question answering. *Information Sciences* 547:103–118.
Zhu, S., Cheng X., and Su S. (2020). Knowledge-based question answering by tree-to-sequence learning. *Neurocomputing* 372:64–72.

Data Science & Exploration in Artificial Intelligence – Gururaj et al. (Eds)
© 2025 the Author(s), ISBN: 978-1-032-96390-7

CrisisConnect: A blockchain-driven edge computing approach in mobile crowdsensing system

Isha, Anugrah Kamla Lakra & Deepika Suhag
Indira Gandhi Delhi Technical University for Women, New Delhi, India

ABSTRACT: During any disaster, the most accurate information could only be known by people nearby. Henceforth Mobile crowdsensing has emerged as a promising paradigm for boosting disaster management conditions via collaborative participation of mobile users. Due to the destruction of infrastructures, transmitting data to centralized authority becomes troublesome, additionally, network congestion increases, causing data latency. Moreover, for active user participation, a framework should be reliable enough and trustworthy. Still, most of the existing structure neglects user privacy resulting in user landing in some aftermath, which reduces users' trust in the system. This paper presents a decentralized framework of MCS combined with MEC to resolve these challenges and the implementation results prove that the proposed system is robust and reliable during the crisis period.

Keywords: Mobile Crowdsensing, Off-chain computing, Community-based, Disaster response, Decision-making, Distributed ledger

1 INTRODUCTION

With the increasing prevalence of smart devices featuring an extensive assortment of integrated sensors such as cameras, microphones, GPS, and others, coupled with progressed wireless technologies like 4G or 5G, etc., we have witnessed a notable uptick in the adoption of mobile devices in recent years. Continuous advancements enabled them to sense, analyze, and interact with other internet-driven gadgets. This evolution has sparked the emergence of Mobile crowdsensing (MCS) [1]. MCS is a vast sensing paradigm that harnesses the capabilities of user-connected devices like smart wearables, mobile phones, smart vehicles, and many others to contribute sensed data from their mobiles. Collected data is then pooled, amalgamated, and processed in the cloud to retrieve insights to facilitate service delivery to human-centered demands [2]. A standard MCS network includes a task requestor, a service provider, and a bunch of mobile users cooperating to optimize the gathering, merging, and analysis of data. Task requesters forward their requests to the service provider, which serves as liaison and selects the appropriate group of users to allocate them tasks. Mobile users, registered with the service provider, get allocated and forward their sensed data, which is finally forwarded to the task requesters. Users get rewarded after completing tasks timely as illustrated in Figure 1. MCS applications can be widely classified into personal sensing, targeting individualistic events, and Community sensing. MCS extends its reach across a multitude of domains, encompassing smart cities [3], health and well-being [4], Wi-Fi network surveillance [5], agriculture [6], infrastructure resilience [7], disaster response coordination [8], and observing macro-level phenomena [9].

With the rise in the frequency of disasters, the necessity for flexible responses becomes paramount. This entwines that emergency responder professionals expand their reliance beyond governmental reserves. The cornerstone piece of data in disaster response is crowd

DOI: 10.1201/9781003589273-44

density, often complemented by location data [10]. Henceforth, MCS springs up as an indispensable aid in tackling such adversities by providing real-time and profound information about such dynamic situations easily, facilitating the strategic allocation of scarce resources by identifying critical areas requiring an immediate attention, specifically in remote areas [11]. Despite the centralized MCS fostering the platform addressing these tasks, it also harbors some limitations, mainly the risk of becoming a single point of failure. Additionally, concerns about computational transparency and financial security exacerbate the lack of trust among users and the platform [12]. In handling these issues, Blockchain can be leveraged to establish decentralized MCS systems. However, its direct application can pose a threat to users' privacy, due to its transparency feature, especially in Public blockchain, where the transaction is open to the outside world [13]. Encountering these concerns, Mobile Edge Computing (MEC) emerges as a promising approach by handling data close to its source, diminishing the need for transmitting sensitive data over the network, hence mitigating privacy concerns and additionally reducing load from blockchain by performing complex computations off-chain [14].

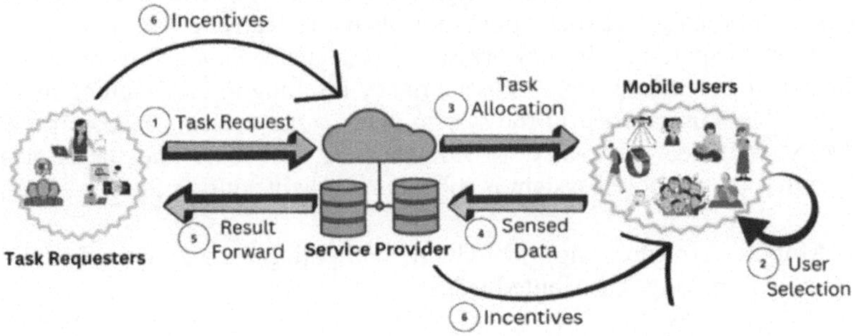

Figure 1. Basic structure of MCS.

Henceforth, we have introduced a novel framework combining blockchain and MEC in an edge-assisted blockchain-based MCS system. Our main contributions are:

(1) Illustrate an extensive literature review analyzing the existing studies on MCS architectures used in disaster response.
(2) A novel framework is proposed, employing Blockchain technology in combination with mobile edge computing.
(3) Subsequently, we have presented an analysis of the results obtained, comparing the gas consumption, transaction cost, and execution cost of different tasks.

The subsequent sections of our paper are as follows: prerequisites are presented in section 2, a thorough literature review is illustrated in section 3, the problem formulation is described in section 4 followed by the proposed approach explained in section 5 with a result analysis showcased in section 6 and finally a Conclusion in section 7.

2 PREREQUISITE

Before delving into the nuances of harnessing Blockchain and MEC, it is fundamental to develop standard knowledge of them. This section demonstrates a brief overview of them.

2.1 *Blockchain*

A Blockchain is a peer-to-peer distributed ledger arranged as a chain of blocks established through consensus mechanisms, that can integrate data models and communication languages empowering smart contracts and ancillary technologies [15]. A smart contract is a concise program designed for verifying, facilitating, or executing services, or agreements without relying on any external force [16]. Except for the first block, called Genesis block, every other block in the chain refers back to its preceding block (known as the parent block) using an inverse link determined by the parent block's hash value [17]. Blockchain can be classified as: 1) Public blockchain where everyone can access the transaction history 2) private blockchain where specific groups of entities can access sensitive data, and 3) Consortium blockchain combining the two. A blockchain-based MCS is shown in Figure 2, where task requesters initiate tasks, a smart contract broadcasts it in the network, and selected users send sensed data which after getting verified is added to the Blockchain, and the user is rewarded. The desirable attributes of Blockchain technology such as decentralized operation have made it a compelling substitute to conventional MCS systems, resolving hurdles such as single-point-of-failure. Moreover, features such as data provenance, security, and privacy have solidified the plausibility of the integration of blockchain in MCS [18].

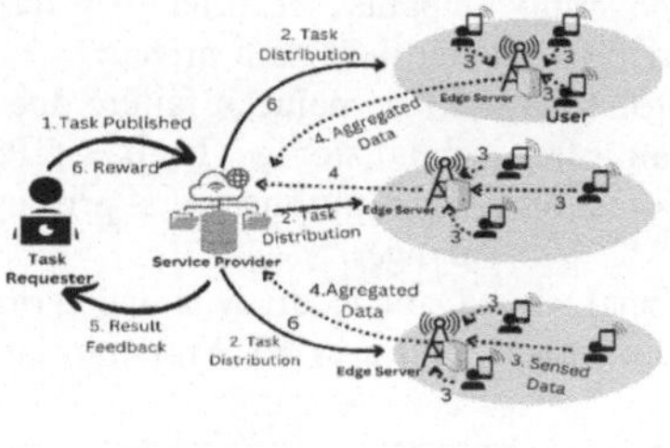

Figure 2. Blockchain-based MCS.

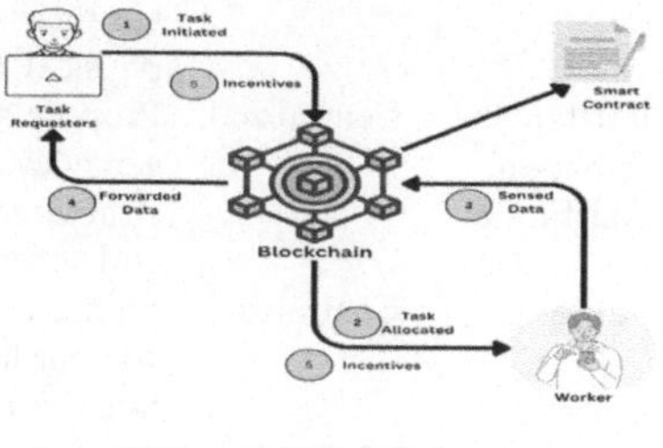

Figure 3. Edge-assisted MCS.

2.2 *Mobile edge computing*

Mobile Edge Computing (MEC) refers to fostering technologies that facilitate computation towards the edge of networks' periphery managing downstream data of cloud storage and upstream for IoT services. Edge can be any computational or network. MEC refers to fostering technologies that facilitate computation towards the edge of networks' periphery managing downstream data of cloud storage and upstream for IoT services. Edge can be any computational or network resource throughout the trajectory from data source to central storage and perform computations, storing, caching, etc [19]. Centralized MCS systems encounter bottlenecks stemming from the augmenting computational demands imposed on them, to combat this, the unification of mobile edge computing has emerged, leveraging the computational capabilities of edge nodes to facilitate task offloading and eliminate these constraints [20]. MEC diminishes data latency in delivering notifications and alerts to mobile nodes in instantaneous scenarios, such as disaster management, where quick access to data becomes crucial. Moreover, it mitigates users' privacy concerns by segmenting and distributing sensitive information among MEC servers [21]. An outlined architecture of an edge-assisted MCS system is depicted in Figure 3, where edge servers are responsible for accumulating data.

3 LITERATURE REVIEW

Table 1 describes a thorough list of reviewed literature with the type of disaster, type of architecture used, and focus on data privacy.

Table 1. Literature review.

Ref.	Type of disaster	Architecture	Objective	Issues
[22]	Earthquake	Centralized server	To measure the damage status of buildings and resident density using crowdsensing.	The fuzzy logic-based system might introduce complexities in decision-making.
[23]	Hurricane	Blockchain-based	Create a decentralized framework for MCS-based disaster management.	Not implemented in the real world and can cause data latency.
[24]	Flood	Centralized Server	Establish an image-based sensing platform to detect water levels to predict floods.	It is useful only for disaster prediction.
[25]	Earthquake	Edge-assisted centralized	Information-gathering of the earthquake-affected area, with fog computing and MCS.	More prone to security threats and has limited resources for storage.
[26]	Storm	Centralized	Coordinate the public in disaster via their social handles covering physical and virtual activities.	It takes information from participants' social handles, discarding their data privacy.
[27]	Earthquake, Typhoon, and Flood	Centralized	Propose an application for emergency evaluation in urban infrastructure through questionnaires and others.	Single point of failure due to cloud storage. It takes GPS input putting users' privacy in danger.
[28]	Storm	Centralized	Utilize already posted social content like tweets to gain situational information.	Data accuracy is undetermined as it uses NLP for translation.

4 PROBLEM FORMULATION

Navigating the aftermath of disasters, a multitude of hurdles surfaces that hinder the seamless delivery of relief and aid to the affected community or region. These issues need to be addressed for a system engaging with disaster scenarios [23]. To implement an enhanced version of the system, we have made some presumptions like automatic sensing is not possible due to the destruction of built infrastructure such as buildings, streetlights, towers, etc., due to which the existing network of communication gets disrupted. We assume that users are equipped with at least a narrowband network up to a few meters range along with an adequate number of resources such as battery power. Memory, etc. With the aforementioned assumptions in mind, the formulated system should meet the following objectives to overcome them.

(1) Decentralization: Not only does it resolve the problem of single-point-of-failure, which poses a huge threat to any disaster response strategy, but it also resolves the issues arising due to reliance on a central authority or external party with the help of smart contracts and consensus.
(2) Encryption: As stated earlier, the privacy of users' sensitive information becomes critical in disasters, and any proposed system should be able to tackle this. Malicious users can take sensitive information for their gain; thus, the proposed system should give sufficient focus on the privacy rights of its users.
(3) Quick access to data: The system must be quick in transmitting data because as mentioned previously it becomes one of the major issues in decision-making. To enhance this, the system should be scalable enough to cater to the congestion in the network without getting exhausted in difficult situations like disaster response.

5 PROPOSED ARCHITECTURE

5.1 *Proposed architecture*

A novel crowdsensing system framework has been devised. Figure 4 illustrates the system model of the proposed system "CrisisConnect", where the task requestor, can be any governmental entity or so, with the task to provide information about the situation of the affected area. We have incorporated a distributed ledger in the form of a Blockchain, responsible for verifying and validating the data received via smart contract. The task gets published in the network and reaches the registered edge nodes. These edge nodes perform off-chain computing, thus enhancing the performance of our blockchain server. At the edge node level, we will receive information from the users and sensitive data is encrypted using the Advanced Encryption Standard (AES) Algorithm. At the Blockchain level, we have created two smart contracts. The EdgeNodeRegistration contract will register the valid edge node from all 11 districts of Delhi, preserving the authenticity and avoiding interference of other edge nodes. After the registration, the DisasterResponseContract will update the block information and allocate the task for quick help. The process of task requestors and edge node registration is illustrated in algorithm 1. The given algorithm takes various parameters from the users to get them register on the blockchain network.

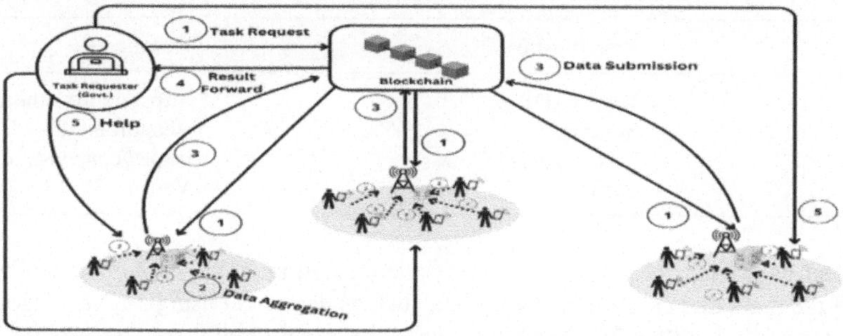

Figure 4. Architecture of edge-assisted blockchain-based MCS.

The input parameters include, entryNo, timestamp, pincode, region, onSiteCondition, helpNeeded, encrypteData. These stands for unique number for edge nodes, time of registration, PIN code of location, region of edge node, environment condition, task details, encrypted from of data, respectively.

Algorithm 1 . Edge Node Registration

Input: entryNo, timestamp, pincode, region, onSiteCondition, helpNeeded, encryptedData

Output: registeredNodes: Mapped registered nodes

(1) registeredNodes <- { }
(2) nodeAddresses <- [ ]
(3) create a list, eligibleRegions with eligible regions mentioned for registration.
(4) function registerEdgeNode (entryNo, timestamp,pincode,region, onSiteCondition, helpNeeded, encryptedData){
    if (eligible)
    {
    registeredNodes[address) = {

entryNo: entry No,
timestamp:
timestamp,pincode:
pincode, region:
region,
onSiteCondition: onSiteCondition, helpNeeded:helpNeeded, encryptedData: encryptedData }}
nodeAddress.push(address) }

(5) Define getEdgeNodeCount to get a count of registered nodes.
(6) Implement getRegisteredEdgeNodeAddresses to get addresses of registered edgenodes

---

5.2 *Implementation*

The section presents the simulation settings and the implementation of "CrisisConnect". Table 2 contains the information about the simulation environment used for the evaluation of the proposed architecture.

Table 2. Simulation environment.

S. No.	Prerequisites	Resources
1	Remix IDE	https://remix.etherum.org
2	Solidity	Version 0.8.0
3	Visual Studio Code	https://code.visualstudio.com
4	Python	Version 3.9.13

To collect data from the mobile users we have created an Edge Node that will collect the data from the users and encrypt personal data such as device ID using an AES algorithm, the edge nodes from different locations will be registered in the smart contract (EdgeNodeRegistration) on the Blockchain. Using Algorithm 1, it will prevent registration from any other region thus providing accurate information. On the other hand, the DisasterResponseContract (smart contract) will extract the payload, and process the payload to obtain useful information to provide the best help during a disaster. Let us consider an Earthquake scenario in Delhi region. We have set 11 Edge Nodes for distinct 11 districts of Delhi namely North Delhi, North-East Delhi, North-West Delhi, West Delhi, South Delhi, South-West Delhi, South-East Delhi, New Delhi, Central Delhi East Delhi, and Shahdara. The PIN code of a specific district can register data in its corresponding edge node. The edge node function will encrypt the payload and register it in the smart contract on the blockchain. The DisasterResponseContract Contract will decrypt the payload and extract the necessary data that can be used in the Task Allocation Phase so that the necessary First Responders can take action to provide disaster relief. If there is a certain requirement of the encrypted user data, we can decrypt it off-site by authorizing access to the key in the key pool.

6 RESULTS

In Figure 5, the line graph highlights variations in resource consumption across various blockchain operations. Notably, DisasterResponseContract operation generally entails higher gas costs compared to EdgeNodeRegisteration, with task allocation displaying some variability. In Figure 6, a comparison of transaction costs of different operations offers insights into cost differentials and optimization opportunities to understand financial implications and optimize resource allocation scalability in blockchain applications.

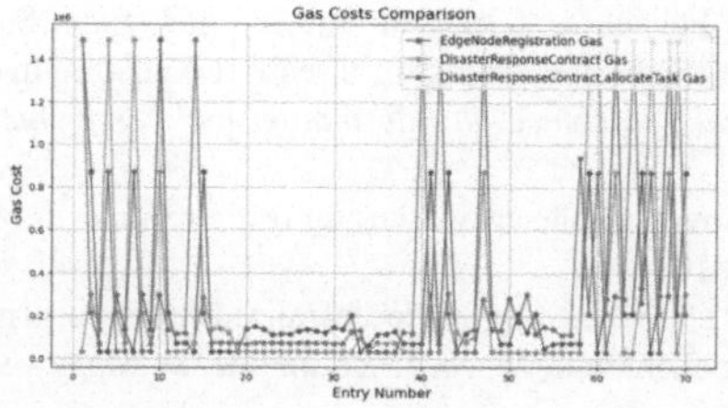

Figure 5. Gas costs comparison.

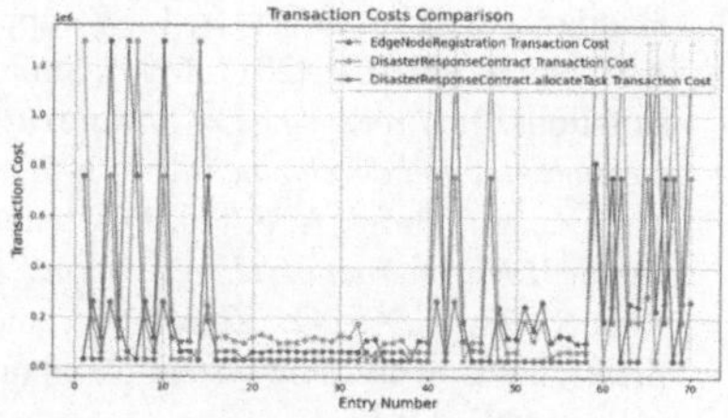

Figure 6. Transaction cost comparison.

7 CONCLUSION

Throughout this article, we attained valuable discernment about plausible applications and challenges of MCS in disaster response. Subsequently, we introduced key technologies such as Blockchain and MEC to enhance MCS systems. Utilizing insights gained from the existing body of studies, we synthesized a comprehensive literature review in our paper, based upon the revelation of the literature review, we have proposed an enhanced architecture that utilizes blockchain technology to abolish dependability on a central authority, coupled with MEC for securing users' privacy and mitigating latency improving decision-making, focusing on tackling the intricate hurdles in existing disaster response scenarios. However, In the future, we can further categorize the entered pin codes into distinct zones and gauge the severity levels of each zone to be more precise in providing quick help. Furthermore, we can include a range of factors like timing, etc., in determining the criticality of the disaster and not just the number of inputs from a region.

REFERENCES

[1] Suhag, D., and Jha, V. (2023). A comprehensive survey on mobile crowdsensing systems. *Journal of Systems Architecture*, 102952.

[2] Boubiche, D. E., Imran, M., Maqsood, A., and Shoaib, M. (2019). Mobile crowd sensing– taxonomy, applications, challenges, and solutions. *Computers in Human Behavior, 101*, 352–370.

[3] Alvear, O., Calafate, C. T., Cano, J. C., and Manzoni, P. (2018). Crowdsensing in smart cities: Overview, platforms, and environment sensing issues. *Sensors, 18*(2), 460.

[4] M. Mehdi, D. Schwager, R. Pryss, W. Schlee, M. Reichert, and F. J. Hauck, Towards automated smart mobile crowdsensing for tinnitus research, *2019 IEEE 32nd International Symposium on Computer-Based Medical Systems (CBMS)*, 2019, pp. 75–80

[5] Radu, V., Kriara, L., and Marina, M. K. (2013, October). Pazl: A mobile crowdsensing based indoor WiFi monitoring system. In *Proceedings of the 9th International Conference on Network and Service Management (CNSM 2013)* (pp. 75–83). IEEE.

[6] Sun, Y., Ding, W., Shu, L., Li, K., Zhang, Y., Zhou, Z., and Han, G. (2021). On enabling mobile crowd sensing for data collection in smart agriculture: A vision. *IEEE Systems Journal, 16*(1), 132–143.

[7] Zhao, X., Wang, N., Han, R., Xie, B., Yu, Y., Li, M., and Ou, J. (2018). Urban infrastructure safety system based on mobile crowdsensing. *International journal of disaster risk reduction, 27*, 427–438.

[8] Abu-Elkheir, M., Hassanein, H. S., and Oteafy, S. M. (2016, September). Enhancing emergency response systems through leveraging crowdsensing and heterogeneous data. In *2016 international wireless communications and mobile computing conference (IWCMC)* (pp. 188–193). IEEE.

[9] Abualsaud, K., Elfouly, T. M., Khattab, T., Yaacoub, E., Ismail, L. S., Ahmed, M. H., and Guizani, M. (2018). A survey on mobile crowd-sensing and its applications in the IoT era. *Ieee access, 7*, 3855–3881.

[10] Kucuk, K., Bayilmis, C., Sonmez, A. F., and Kacar, S. (2020). Crowd sensing aware disaster framework design with IoT technologies. *Journal of Ambient Intelligence and Humanized Computing, 11*, 1709–1725.

[11] Haddawy, P., Frommberger, L., Kauppinen, T., De Felice, G., Charkratpahu, P., Saengpao, S., and Kanchanakitsakul, P. (2015, May). Situation awareness in crowdsensing for disease surveillance in crisis situations. In *Proceedings of the seventh international conference on information and communication technologies and development* (pp. 1–5).

[12] Tao, X., and Hafid, A. S. (2021). ChainSensing: A novel mobile crowdsensing framework with blockchain. *IEEE Internet of Things Journal, 9*(4), 2999–3010.

[13] Zou, S., Xi, J., Xu, G., Zhang, M., and Lu, Y. (2021). Crowdhb: A decentralized location privacy-preserving crowdsensing system based on a hybrid blockchain network. *IEEE internet of things journal, 9*(16), 14803–14817.

[14] Awan, S., Li, F., Luo, B., and Liu, M. (2019, November). Poster: A reliable and accountable privacy-preserving federated learning framework using the blockchain. In *Proceedings of the 2019 ACM SIGSAC conference on computer and communications security* (pp. 2561- 2563).

[15] Belotti, M., Božić, N., Pujolle, G., and Secci, S. (2019). A vademecum on blockchain technologies: When, which, and how. *IEEE Communications Surveys & Tutorials, 21*(4), 3796–3838.

[16] Chen, Z., Fiandrino, C., and Kantarci, B. (2021). On blockchain integration into mobile crowdsensing via smart embedded devices: A comprehensive survey. *Journal of Systems Architecture, 115*, 102011.

[17] Dai, H. N., Zheng, Z., and Zhang, Y. (2019). Blockchain for Internet of Things: A survey. *IEEE internet of things journal, 6*(5), 8076–8094.

[18] Yu, R., Oguti, A. M., Obaidat, M. S., Li, S., Wang, P., and Hsiao, K. F. (2023). Blockchain-based solutions for mobile crowdsensing: A comprehensive survey. *Computer Science Review, 50*, 100589.

[19] Shi, W., Cao, J., Zhang, Q., Li, Y., and Xu, L. (2016). Edge computing: Vision and challenges. *IEEE internet of things journal, 3*(5), 637–646.

[20] Li, Z., Li, Z., and Zhang, W. (2023). Quality-aware task allocation for mobile crowd sensing based on edge computing. *Electronics, 12*(4), 960.

[21] Marjanović, M., Antonić, A., and Žarko, I. P. (2018). Edge computing architecture for mobile crowdsensing. *Ieee access, 6*, 10662–10674.

[22] Kucuk, K., Bayilmis, C., Sonmez, A. F., and Kacar, S. (2020). Crowd sensing aware disaster framework design with IoT technologies. *Journal of Ambient Intelligence and Humanized Computing, 11*, 1709–1725.

[23] Sarbajna, R., Eick, C. F., and Laszka, A. (2021, September). DEIMOSBC: a blockchain-based system for crowdsensing after natural disasters. In *2021 3rd Conference on Blockchain Research & Applications for Innovative Networks and Services (BRAINS)* (pp. 17–20). IEEE.

[24] Burkard, S., Fuchs-Kittowski, F., and de Bhroithe, A. O. F. (2017). Mobile crowd sensing of water level to improve flood forecasting in small drainage areas. In *Environmental Software Systems. Computer Science for Environmental Protection: 12th IFIP WG 5.11 International Symposium, ISESS 2017, Zadar, Croatia, May 10-12, 2017, Proceedings 12* (pp. 124–138). Springer International Publishing.

[25] Vahdat-Nejad, H., Bahadori, H., and Abiri, A. (2021, May). Information gathering of earthquake disasters by mobile crowd sourcing in smart cities. In *2021 5th International Conference on Internet of Things and Applications (IoT)* (pp. 1–6). IEEE.

[26] Ludwig, T., Reuter, C., Siebigteroth, T., and Pipek, V. (2015, April). CrowdMonitor: Mobile crowd sensing for assessing physical and digital activities of citizens during emergencies. In *Proceedings of the 33rd Annual ACM Conference on Human Factors in Computing Systems* (pp. 4083–4092).

[27] Zhao, X., Wang, N., Han, R., Xie, B., Yu, Y., Li, M., and Ou, J. (2018). Urban infrastructure safety system based on mobile crowdsensing. *International journal of disaster risk reduction, 27*, 427–438.

[28] Salfinger, A., Schwinger, W., Retschitzegger, W., and Pröll, B. (2016, March). Mining the disaster hotspots-situation-adaptive crowd knowledge extraction for crisis management. In *2016 IEEE International Multi-Disciplinary Conference on Cognitive Methods in Situation Awareness and Decision Support (CogSIMA)* (pp. 212–218). IEEE.

Data Science & Exploration in Artificial Intelligence – Gururaj et al. (Eds)
© 2025 the Author(s), ISBN: 978-1-032-96390-7

Detection of anemia using ensemble model

D.N. Varshitha, D.U. Latha, K.S. Chirag, K. Madan Kumar, M. Manisha &
K. Sriharsha
Vidyavardhaka College of Engineering, Mysuru. Karnataka, India

ABSTRACT: In this paper we are proposing a new ensemble method for anemia detection using machine learning and a combination of data sources. By analysing both medical information that are symptoms, history, vitals and smartphone images of the eye conjunctiva, the approach aims to surpass traditional blood tests. The analysis involves extracting informative visual cues from the images using Convolutional Neural Networks (CNNs). Importantly, data from different sources is weighted based on significance for anemia detection before being combined. This method, along with other machine learning algorithms, is expected to improve detection accuracy and enable earlier diagnosis. Additionally, the accessibility of this non-invasive approach has the potential to empower individuals and transform preventative healthcare. However, further research is needed to validate this approach and ensure its effectiveness in real- world settings.

Keywords: Anemia, Ensemble learning, conventional Neural Network

1 INTRODUCTION

Anemia, marked by a deficiency in red blood cells or haemoglobin, significantly impacts global health. The World Health Organization estimates that 42% of children under six and 40% of pregnant women suffer from this condition, often due to iron deficiency. Untreated anemia can lead to severe consequences, including irreversible organ damage and even death. Early detection is vital, yet current diagnostic methods, primarily blood tests, have notable limitations such as invasiveness and resource constraints, particularly in low-resource settings. This paper proposes an innovative solution: a machine learning model leveraging smartphone technology for non-invasive anemia detection. By integrating traditional data—symptoms, patient history, and vitals—with smartphone-captured images of the conjunctiva (the inner eyelids), deep learning algorithms can analyse these images to detect subtle visual indicators of anemia. This multifaceted risk assessment model promises enhanced accuracy and accessibility.

However, to ensure the model's effectiveness across diverse populations, rigorous testing is essential. Variations in skin tone and lighting conditions can impact the accuracy of visual cues, necessitating robust datasets that include children, pregnant women, and individuals with varying skin tones. This inclusive approach is crucial for equitable access to this potentially transformative technology. Moreover, addressing regulatory and privacy concerns and providing adequate training for healthcare staff are necessary for successful integration into healthcare systems. This vision of a non-invasive, AI-powered model for anemia detection holds promise for early intervention and better health outcomes, making anemia diagnosis more accessible and less invasive worldwide.

2 RELATED WORK

The paper [1] outlines a skincare product recommendation system using facial analysis from photos, leveraging CNNs and traditional image techniques to assess skin features like color,

DOI: 10.1201/9781003589273-45

texture, and acne. It highlights the need for improved accuracy, fairness, and privacy protection to address current flaws such as errors and biases.

The paper [2] discusses an AI-powered app for estimating iron levels from nail photos, using color and texture analysis linked to a database. Despite its potential for non-invasive anemia testing, it faces challenges with image quality, lighting, skin color variability, and privacy concerns, necessitating improvements in accuracy, fairness, and data security.

The paper [3] explores using the VGG16 deep learning model to detect anemia from smartphone photos of the eye's white part by analysing color and patterns against a labelled database. While offering a convenient anemia screening method, challenges include image quality, lighting, variability across skin tones, and privacy concerns, necessitating improvements in accuracy, bias reduction, and data security.

The paper [4] discusses the limitations of traditional anemia tests and highlights the advantages of using computer programs for anemia detection, such as being non-invasive, cost-effective, and time-saving. It reviews various methods, including blood extraction, eye whiteness, nail color, palm examination, and smartphone devices, noting factors like image size, data preparation, and program accuracy, while addressing issues such as inter-observer variability, sensitivity of eye whiteness, insufficient anemia data, and current information challenges.

The paper [5] evaluates machine learning methods to predict anemia in children, finding Random Forest achieves the highest accuracy at 98.4%. Despite testing feature selection and ensemble learning, Random Forest remained superior. The study measured various performance metrics but noted limitations such as data from a single hospital and a small feature set.

3 METHODOLOGIES

The methodology that we have used in the project is a collection of multiple symptoms related to anemia and predict probability of anemia individually on each of them and then combine them into a result with different importance to each element. This will make sure that we have taken all the different symptoms into consideration according to their importance.

3.1 *Medical information*

Patient data including symptoms (fatigue, shortness of breath), family history, and vitals. Statistical or machine learning techniques like feature selection or dimensionality reduction extract relevant patterns from the medical data. This might involve identifying correlations between specific symptoms and anemia.

3.2 *Visual data*

Smartphone-captured images of the conjunctiva (eye's inner lining) - potentially including the palm in future iterations. Both medical information and visual data undergo preprocessing to ensure consistency. Convolutional Neural Networks (CNNs) analyse the conjunctival images, focusing on extracting features like subtle color variations that might indicate anemia. A similar CNN might analyse palm images in the future.

3.3 *Weighted feature fusion*

Weights are assigned to each data source (medical information, eye image, palm image) based on their relative importance in anemia detection. These weights can be determined by medical expertise or the model itself through feature importance analysis. The extracted features from each source are then multiplied by their corresponding weights before being combined. This ensures information deemed more crucial by the model contributes more to the overall analysis.

3.4 *Machine learning model*

DNNs (Deep Neural Networks) are a type of ML algorithm with multiple layers that progressively extract higher- level features from the data. In our case, DNNs can analyse conjunctival images to identify subtle patterns indicative of anemia. CNNs are a specific type of DNN particularly adept at processing image data. Their ability to automatically learn features from raw images extracting relevant information A Convolutional Neural Network (CNN) is a specialized deep learning model used for analysing visual data like images. It includes layers of neurons that perform operations like convolution, pooling, and activation. CNNs are great for tasks such as image classification, object detection, and facial recognition because they can learn hierarchical features from pixel values. They have transformed computer vision and are used across different industries like healthcare, autonomous vehicles, and security systems. Random Forest is a method in machine learning that creates multiple decision trees and then combines their predictions to make a final decision. It is widely used because it can improve accuracy and generalization in many different types of tasks. Random Forest is known for being robust, versatile, and resistant to overfitting, which is why it is a popular choice for both classification and regression tasks.

3.5 *Proposed architecture*

In this machine learning model for anemia detection, multiple data sources are merged using a weighted fusion approach. Medical data, including symptoms and vital signs, and smartphone-captured images of the conjunctiva and possibly the palm, undergo pre-processing. CNNs extract features from these images, focusing on variations indicative of anemia. Statistical techniques analyse the medical data. Weights are assigned to each data source based on their relevance, either determined by medical expertise or the model. The weighted features are combined, and an SVM assesses this data to predict anemia risk. This method leverages diverse data sources, prioritizing the most relevant information for accurate detection.

3.6 *Eye model training*

The initial step involves preparing the dataset for model training. This includes defining the paths to directories containing anemic and nonanemic images. Using the os.listdir() function, images and their corresponding labels are loaded into lists. These image paths and labels are then combined into a pandas DataFrame for convenient handling. Subsequently, the DataFrame is split into training and testing sets using the train_test_split() function from scikit- learn. Feature extraction is a critical preprocessing step where relevant information is extracted from the raw image data. In this project, the extract_features() function is defined to perform this task. Within this function, images are resized to a standard size of 224x224 pixels using the resize() function from the Python Imaging Library (PIL). Additionally, pixel values are normalized by dividing by 255.0 to ensure uniformity across the dataset. The extracted features are reshaped into the required format for input into the CNN model. The core of the anemia detection system lies in the design of the CNN architecture. Utilizing the Sequential API of Keras, a CNN model is constructed layer by layer. The architecture typically comprises convolutional layers for feature extraction, followed by max-pooling layers to reduce spatial dimensions. Additional convolutional layers with increasing filter sizes capture higher-level features. Fully connected layers with Rectified Linear Unit (ReLU) activation function are incorporated for classification, with dropout regularization to prevent overfitting. The output layer consists of a single neuron with sigmoid activation for binary classification.

Once the model architecture is defined, it is compiled and trained using the specified optimizer, loss function, and training parameters. In this project, the Adam optimizer and binary cross-entropy loss function are chosen. Training is performed using the fit() method, specifying parameters such as batch size, number of epochs, and validation split. During training, the model learns to minimize the loss function by adjusting its weights and biases to make accurate predictions. Training and validation accuracy and loss metrics are monitored and stored for later analysis.

3.7 *Palm model training*

Extract_features() function is defined to extract both color and texture features from pre-processed palm images. Initially, the images are converted to the HSV color space to capture color information effectively. For each channel in the HSV image, statistical features such as mean, standard deviation, and percentiles are computed and added to the feature list. Additionally, texture features are extracted using the Local Binary Pattern (LBP) method applied to the grayscale version of the image. The script loads palm images from directories containing anemic and healthy samples. For each image, features are extracted using the defined function, and corresponding labels (1 for anemic, 0 for healthy) are assigned. The extracted features and labels are stored in separate lists, forming the feature matrix (X) and target vector (y) for model training. The feature matrix (X) and target vector (y) are split into training and testing sets using the train_test_split() function from scikit-learn. A RandomForestClassifier model with 100 decision trees is initialized and trained on the training set using the fit() method. The trained model is then ready for evaluation and deployment.

3.8 *Fingernail model training*

The script begins by defining a function load_data() to load and pre-process the images from the specified folder path. Images are resized to 224x224 pixels using the target_size parameter of image.load_img(). For each image, its corresponding label (1 for Anemic and 0 for Non- Anemic) is assigned, and both images and labels are stored in numpy arrays X and y, respectively. The pre-trained VGG16 model is loaded with weights from the ImageNet dataset, excluding the fully connected layers (include_top=False). Feature extraction is performed by passing the input images through the base VGG16 model and flattening the output. The resulting features are stored in the features variable using the predict() method.

The trained model is evaluated on the testing dataset, with its performance assessed using metrics like accuracy, precision, recall, and F1 score. Predictions are made on pre-processed test images, and accuracy and loss curves are plotted using matplotlib. This evaluation reveals the model's generalization capability and effectiveness in detecting anemia.

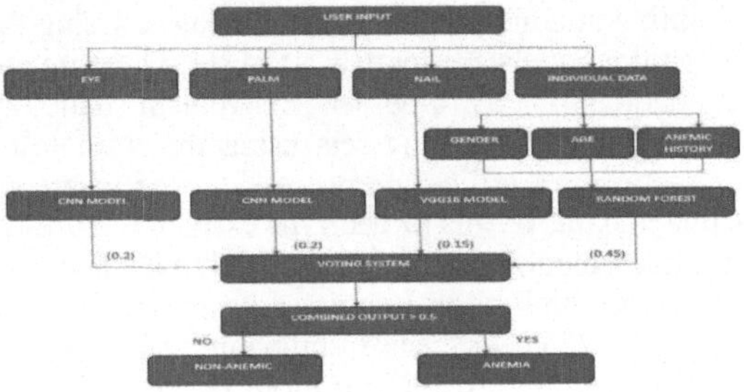

Figure 1. System architecture.

4 RESULTS

The final output of the model is a prediction of the patient's anemia risk.

- Binary Classification: Anemic or not anemic.
- Probability Score: Likelihood of having anemia (ranging from 0 to 1).
- Improved Accuracy and Generalizability: Data fusion with weighted features leverages strengths of various sources and prioritizes the most relevant information, potentially leading to more accurate anemia detection compared to relying on single data sources.
- Early Detection: By providing a non-invasive and potentially accessible approach, early detection of anemia becomes possible, enabling timely interventions and improved health outcomes.
- Democratization of Healthcare: Empowering individuals and communities with the ability to assess their anemia risk using smartphones puts preventative healthcare within reach.

We have observed an accuracy of 71.11 for the eye training model. We have observed an accuracy of 90.7% for the finger training model. We have observed an accuracy of 99.6% for the palm training model. The accuracy of the final model is 86.33%. This is calculated using the average of the other models.

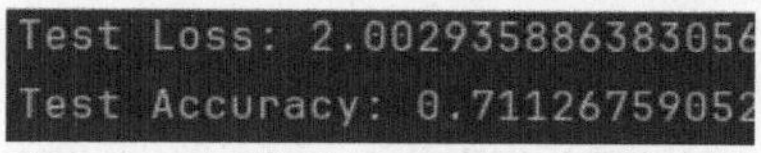

Figure 2. Accuracy of eye model.

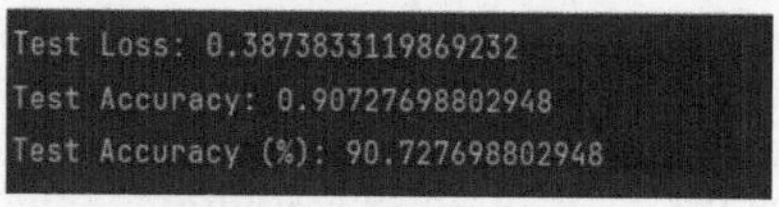

Figure 3. Accuracy of finger nails model.

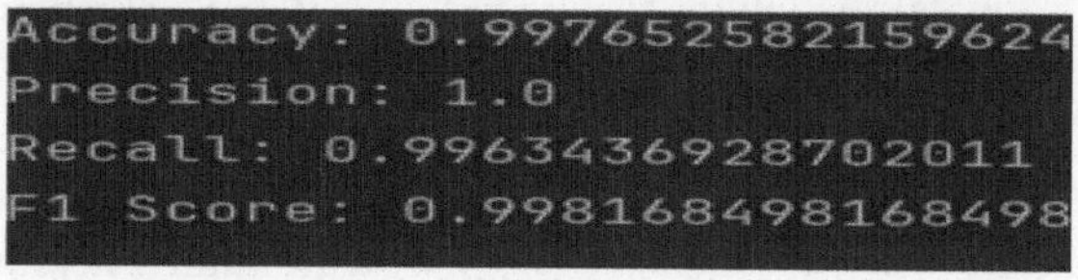

Figure 4. Accuracy of palm model.

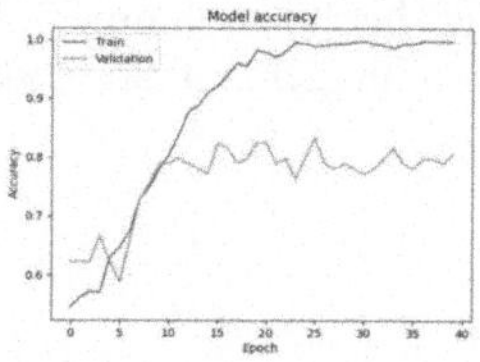

Figure 5. Processing of eye image.

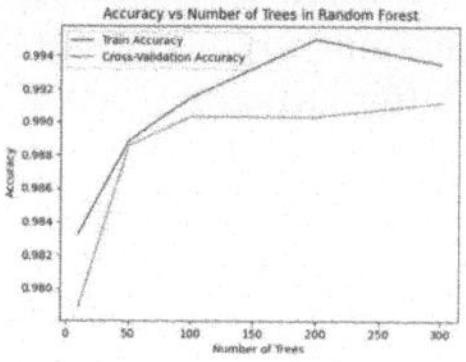

Figure 6. Processing of palm image.

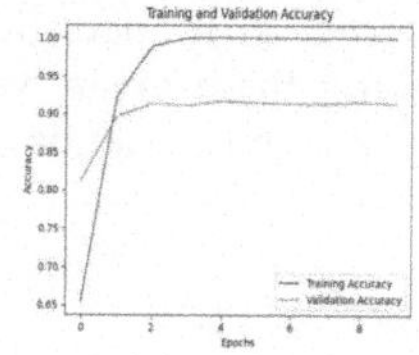

Figure 7. Processing of nail image.

4.1 *UI/UX of the website*

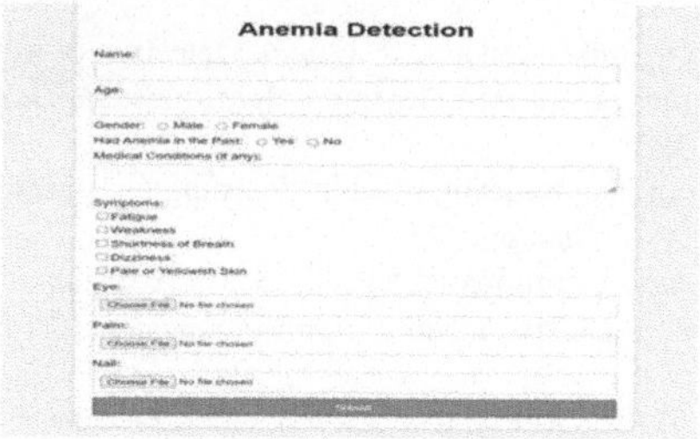

Figure 8. Details of the user.

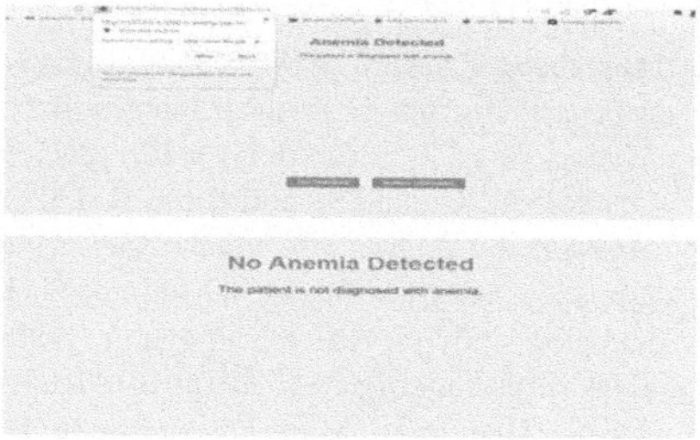

Figure 9. Results of the user.

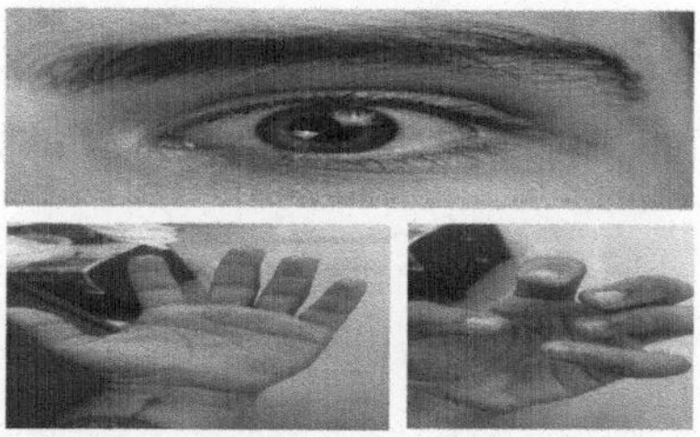

Figure 10. Non-anemic images.

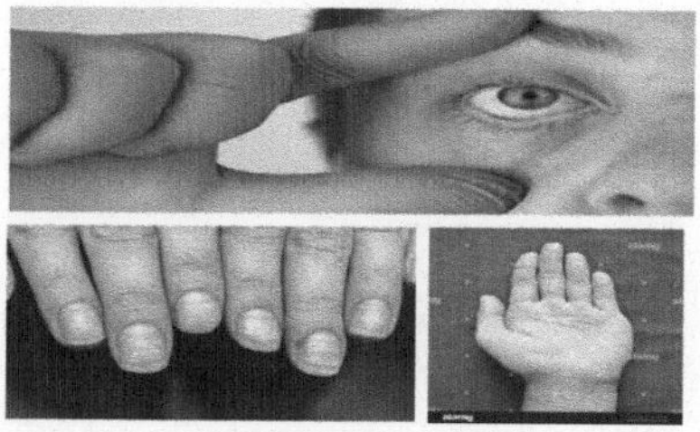

Figure 11. Anemic images.

5 CONCLUSIONS

The global burden of anemia necessitates innovative solutions, and the potential of converging technologies and data sources to revolutionize its diagnosis and management. Evidence shows the pervasiveness of anemia, particularly among vulnerable populations like children and pregnant women, emphasizing the urgent need for accessible and accurate detection methods. While traditional blood tests have been fundamental, their limitations in accessibility and resource demands call for alternative approaches. Machine learning (ML) emerges as a promising solution, offering a non-invasive, AI-powered pathway to early detection. Our survey highlights various ML techniques, notably Convolutional Neural Networks (CNNs) and Deep Neural Networks (DNNs), showcasing their ability to analyse conjunctival images and extract subtle visual cues indicative of anemia. These findings not only demonstrate the accuracy and generalizability of prediction but also signify the potential for democratizing healthcare. Practical Implications include the potential for ML-driven approaches to improve accessibility and accuracy of anemia diagnosis, particularly in underserved populations. Leveraging smartphones and readily available data, this non-invasive approach holds promise in reaching regions with limited healthcare infrastructure, empowering individuals and communities with early detection capabilities. Future research should focus on optimizing ML algorithms for diverse demographic groups, integrating ML-based diagnostic tools into existing healthcare systems, and assessing the long-term impact of ML-driven early detection on anemia management and public health outcomes. By addressing these areas, we can further advance proactive health management and build resilience against anemia.

REFERENCES

[1] Yang, N., Li, X., Wu, X., Lin, F., Wu, X., Zhu, W., *et al.* (2022). A deep learning approach for facial image-based skincare product recommendation system. *Frontiers in Public Health, 10*, 964385. https://doi.org/10.3389/fpubh.2022.964385

[2] Tilburg University, Digital Sciences for Society Program. (n.d.). NAIL: Nail based Artificial Intelligence image Learning for measuring iron status. Project website.

[3] Dimauro, G., Guarini, A., Caivano, D., Girardi, F., Pasciolla, C., and Iacobazzi, A. (2019). Detecting clinical signs of anaemia from digital images of the palpebral conjunctiva. *IEEE Access, 7*, 113488–113498. https://doi.org/10.1109/ACCESS.2019.2932274

[4] Appiahene, P., Asare, J. W., Donkoh, E. T., Dimauro, G., and Maglietta, R. (2023). Detection of iron deficiency anemia by medical images: A comparative study of machine learning algorithms. *BioData Mining, 16*(1), 2. https://doi.org/10.1186/s13040-023-00319-z

[5] Dhakal, P., Khanal, S., and Bista, R. (2023). Prediction of anemia using machine learning algorithms. *International Journal of Computer Science and Information Technology, 15*.

[6] Dimauro, G., Caivano, D., Di Pilato, P., Dipalma, A., and Camporeale, M. G. (2020). A systematic mapping study on research in anemia assessment with non-invasive devices. *Applied Sciences, 10*(14), 4804. https://doi.org/10.3390/app10144804

[7] An, R., Huang, Y., Man, Y., Valentine, R. W., Kucukal, E., Goreke, U., Sekyonda, Z., Piccone, C., Owusu-Ansah, A., Ahuja, S., Little, J. A., and Gurkan, U. A. (2021). Emerging point-of-care technologies for anemia detection. *Lab on a Chip, 21*(10), 1843–1865. https://doi.org/10.1039/d0lc01235a

[8] Shekhar, M., Berk, D. T., Mohammed, M., Mustafa, K., and Freeman, C. (2023). Anemia detection through non-invasive analysis of lip mucosa images. *Frontiers in Big Data, 6*. https://doi.org/10.3389/fdata.2023.124

Data Science & Exploration in Artificial Intelligence – Gururaj et al. (Eds)
© 2025 the Author(s), ISBN: 978-1-032-96390-7

FutureFame: Pioneering student success using various regression models

M. Chandan, Athish Raj Mohan, Amrith, M.V. Chandan & K.N. Prashanth Kumar
Bangalore Institute of Technology, Bangalore, Karnataka, India

ABSTRACT: In the realm of education, one must make efforts to predict the students' performance so that the students who might be at risk of underperformance receive timely intervention and support. Our project, "FutureFame", involves the use of various machine learning techniques that promote student success based on the various factors.

Various techniques, such as Linear Regression, Ridge, Lasso, etc., have been implemented to scrutinize datasets related to academic performance. Of all these techniques, Linear Regression was the most accurate, achieving 99% accuracy.

We use a dataset with information about the students and their corresponding exam scores. Several regression models are trained and tested for the dataset, and results provide useful insights into who may be in need of support and interventions in order to improve their academic performance for educational institutions.

1 INTRODUCTIONS

Educators need to know their students' performance so they can help them succeed in school. If pupils do badly academically they are likely to face many challenges that include postponement in completion time and getting off from institutions; hence those charges should take care of regularly invited unproductive learner's classroom achievements (Rimadana *et al.* 2019). The purpose of this work is to forecast the future academic lives of learners basing our analysis on some key factors pulled out of student tracking info on school General information (Shukla *et al.* 2020).

(1) Use student demographic data, student activity data, and student achievement data to forecast student academic performance.
(2) Identification of variables that contribute to student performance.

Based on the literature reviewed, it is clear that machine learning systems have the potential to analyze student performance and provide valuable insights to educational decision-makers. The reviewed research indicates that these methods can effectively identify areas of improvement, evaluate assessment performance, provide effective data management, and evaluate academic outcomes based on a variety of criteria (Chakrapani *et al.* 2022). Overall, the use of machine learning systems in education can have significant benefits for students, teachers, and administrators by providing more accurate assessments of student performance (Yadao *et al.* 2021), improving and enhancing resource allocation have been able to make effective decisions but any machine learning program (Ahmed *et al.* 2021) used in education must be carefully considered and implemented in an accountable and transparent manner (Rathi *et al.* 2021).

Our model considers academic indicators such as literacy scores as input characteristics along with basic demographic factors such as gender and parental occupation (Mustafa *et al.* 2022). Integrating web services enriches this dataset with behavioral patterns resulting from student

DOI: 10.1201/9781003589273-46

interactions with online educational materials (Chen *et al.* 2023). What sets this research apart is the active selection of the most appropriate machine learning algorithm for each prediction model, ensuring adaptation to the changes (Hussain *et al.* 2024). In our research on student performance prediction, we examine the effectiveness of linear regression, ridge regression, decision tree, AdaBoost regression, gradient growth, random forest, XGBRegressor. Each algorithm brings its unique approach to the table, from basic linear modeling techniques to sophisticated ensemble methods and regularization methods. This research will provide valuable insights into factors affecting student success and inform targeted interventions to support learning and achievement.

2 DRAWBACKS OF EXISTING SYSTEM

- Traditional educational systems rely heavily on subjective evaluations and standardized testing to assess student performance.
- The grading process often lacks personalized feedback, making it challenging for students to understand their strengths and weaknesses.
- Educators face difficulties in identifying and addressing the specific needs of individual students due to the limitations of traditional grading systems.
- Data analysis in the existing system is primarily manual, leading to inefficiencies and potential errors.
- The existing system may not effectively leverage the wealth of data available to understand the complex relationships between various factors that influence student performance.

3 PROPOSED SYSTEM

- The proposed system aims to revolutionize student performance assessment by leveraging machine learning techniques to predict student scores based on various factors.
- By analyzing features such as gender, type of university, extracurricular activities, attendance, etc., the system provides a comprehensive understanding of student performance determinants.
- Through the application of machine learning algorithms, the system offers personalized feedback to students, enabling them to identify areas for improvement and enhance their learning experience.
- Automation of data processing and analysis reduces manual effort, streamlines the assessment process, and minimizes errors.
- The proposed system is designed to be scalable and adaptable, allowing for the integration of additional features and factors to enhance prediction accuracy and provide deeper insights into student performance.

4 ARCHITECTURAL DESIGN

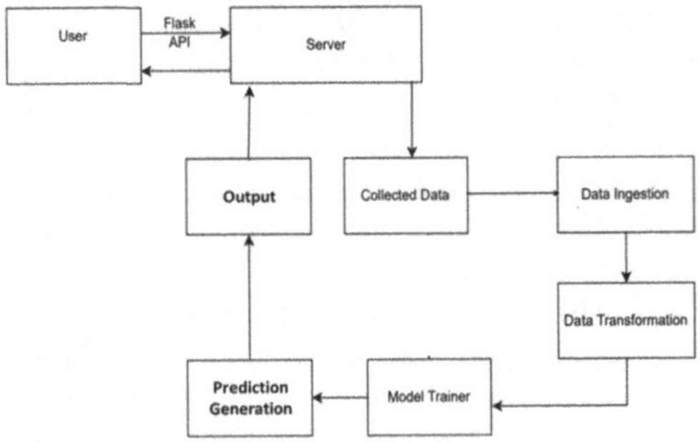

Figure 1. Architecture design.

4.1 *User*

This refers to the person interacting with the system, presumably a tutor who wants to generate predictions about student's academic performance.

4.2 *API*

API (Application Programming Interface) might refer to a set of functions or services that the Flask application exposes to allow users to interact with the system. In this case, the user interacts with the API through the user interface to provide data and receive predictions. Flask: Flask is a Python web framework that likely acts as the backend of the application. It receives user requests, coordinates with other parts of the system, and generates the application's responses.

4.3 *Server*

This refers to the computer system that runs the Flask application and stores the data used for generating predictions. It communicates with the user's device through the Flask application.

4.4 *Collected data*

The data is collected from the dataset.

4.5 *Data transformation*

In the Data Transformation step, we perform operations such as encoding categorical variable and scaling of numerical variables.

4.6 *Model trainer*

In the model trainer module, the training set is used to train the machine learning model.

4.7 *Prediction generation*

After Data collection, data transformation and Model training, this uses a learning model to generate the predictions when it gets data from the users.

4.8 *Output*

This refers to the predicted value generated by the system based on the user provided data and the trained model. It could be a single value like a predicted GPA score or a more comprehensive explanation of the factors influencing the prediction.

5 ALGORITHM DESIGN

5.1 *Read the initial student performance dataset*

5.1.1 *Description*
Load the dataset containing information about student performance from a file (e.g., CSV).

5.1.2 *Purpose*
This step is crucial for acquiring the raw data that will be used for analysis and model training. It involves reading the dataset into a panda DataFrame for easier manipulation and analysis.

5.2 *Split the dataset into training and testing data*

5.2.1 *Description*
Divide the dataset into two parts: one for training the models and one for testing their performance.

5.2.2 *Purpose*
Splitting the data ensures that the models are trained on one subset of the data and evaluated on another, helping to assess the model's generalizability and performance on unseen data.

5.3 *Apply data transformation techniques*

5.3.1 *Description*
Perform necessary preprocessing steps such as cleaning the data, handling missing values, and converting categorical variables to numerical formats.

5.3.2 *Purpose*
Data transformation is essential for preparing the raw data in a format suitable for machine learning algorithms. Techniques like label encoding, one-hot encoding, and scaling are commonly applied.

5.4 *Save the preprocessing object for future use*

5.4.1 *Description*
Save the preprocessing pipeline (e.g., scaler, encoder) as an object to ensure the same transformations can be applied to new data during prediction.

5.4.2 *Purpose*
Saving the preprocessing object ensures consistency in data transformation between the training phase and the prediction phase, maintaining the integrity of the model's performance.

5.5 *Train multiple regression models using the transformed training data*

5.5.1 *Description*
Train different regression models (e.g., Linear Regression, Random Forest, Gradient Boosting) on the processed training data.

5.5.2 *Purpose*
Training multiple models allows for comparison and selection of the best-performing model based on various metrics, increasing the likelihood of achieving high predictive accuracy.

5.6 *Evaluate model performance using appropriate metrics on the test data*

5.6.1 *Description*
Assess the performance of each trained model on the test data using metrics like Mean Squared Error (MSE), R-squared, etc.

5.6.2 *Purpose*
Evaluation helps in understanding how well the models generalize to new, unseen data, guiding the selection of the best model for deployment.

5.7 *Save the model to a file for deployment*

5.7.1 *Description*
Save the best-performing model to a file (e.g., using pickle) for later use in predictions.

5.7.2 *Purpose*

Saving the model allows it to be deployed in a production environment where it can be used to make predictions on new data without retraining.

5.8 *Implement a predict pipeline to load the model and preprocessing objects*

5.8.1 *Description*

Develop a pipeline that loads the saved model and preprocessing objects and uses them to make predictions on new data.

5.8.2 *Purpose*

The PredictPipeline ensures that the entire process, from data preprocessing to model prediction, can be executed seamlessly for new input data.

5.9 *Predict the academic performance of new student data using the deployed model*

5.9.1 *Description*

Use the deployed model to predict the performance of new students based on their input data.

5.9.2 *Purpose*

This step demonstrates the practical application of the model, providing valuable insights and predictions for new data instances.

5.10 *Develop a web interface allowing users to input student details*

5.10.1 *Description*

Create a user-friendly web interface where users can enter details about students to get predictions.

5.10.2 *Purpose*

The web interface makes the prediction model accessible to end users, enabling them to easily input data and obtain predictions without needing technical knowledge.

5.11 *Utilize the deployed model to predict student performance based on the provided data*

5.11.1 *Description*

Integrate the model with the web interface to process user inputs and generate predictions.

5.11.2 *Purpose*

This integration allows users to interact with the model in real time, providing immediate feedback and insights based on the input data.

These steps form a comprehensive workflow for developing, training, evaluating, deploying, and utilizing a machine learning model to predict student performance, ensuring both technical accuracy and user accessibility.

6 RESULTS AND DISCUSSION

Among the regression models examined in this study Linear regression, decision tree, ridge, XGB regressor were considered as each model for gender, extracurricular activities, attendance records, test preparation efforts, university characteristics, preparatory scores, course. The model was tested using a dataset containing digital information. To accurately test the performance of the model, the data set was divided into training and testing.

And below is the testing which is done on different models for the same input data-

Table 1. Results for different regression models.

Sl. No	Input data	Model selected	Accuracy	Prediction
1	Gender-male, University-Private, Activity-Sports, Attendance-70, Course-Completed, Test_Score-80, Preparatory_Score-75	Linear Regression	99.9915%	78.75000
2	Gender-male, University-Private, Activity-Sports, Attendance-70, Course-Completed, Test_Score-80, Preparatory_Score-75	Ridge	99.9915%	78.72402
3	Gender-male, University-Private, Activity-Sports, Attendance-70, Course-Completed, Test_Score-80, Preparatory_Score-75	XGB Regressor	99.9985%	78.80180
4	Gender-male, University-Private, Activity-Sports, Attendance-70, Course-Completed, Test_Score-80, Preparatory_Score-75	Decision Tree	99.92%	77.73425

7 CONCLUSION

In conclusion, "FutureFame" has demonstrated the effectiveness of various regression models in predicting student performance based on a variety of factors. Through the analysis of student data, we have successfully developed a machine learning solution that can provide valuable insights for educators and institutions.

By predicting student performance, educational institutions can identify students who may need additional support and interventions, thus helping to improve overall academic outcomes. The project's findings can contribute to the ongoing efforts to personalize education and enhance student success.

REFERENCES

Ahmed, Abdulazeez, Zeebaree and Falah, (26 June 2021). Predicting university's students performance based on machine learning techniques, in *IEEE International Conference on Automatic Control and Intelligent Systems.*

Chakrapani and Chitradevi, (2022), Academic performance prediction using machine learning: A comprehensive & systematic review, in *2022 International Conference on Electronic Systems and Intelligent Computing (ICESIC)*. IEEE, pp. 335–340.

Chen, Zhai. (2023). A comparative study on student performance prediction using machine learning, *Education and Information Technologies.*

Hussain, Khan, (2024), Student – performulator: Predicting students' academic performance at secondary and intermediate level using machine learning, *Annals of Data Science* 2023.

Mustafa, (2022), Educational data mining: Prediction of students' academic performance using machine learning algorithms, *Springer Open Access*, 2022.

Rathi, Deshpande, Nagaral, Narkhede, Radhika and Takalikar, (2021), Analysis of user's learning styles and academic emotions through web usage mining, *2021 International Conference on Emerging Smart Computing and Informatics (ESCI)*, March 2021.

Rimadana, Kusumawardani, Santosa and Erwianda, (2019), Predicting student academic performance using machine learning and time management skill data, *International Seminar on Research of Information Technology and Intelligent Systems (ISRITD).*

Shukla, Prachi, Samaiya and Kherajani, (2020), Web usage mining- A study of web data pattern detecting and methodologies and its applications, *International Journal on Computer Engineering and Intelligent Systems*, vol.12, issue 1, September 2020.

Yadao, Babu, Midhunchakkaravarthy and Bhaumik, (2021), Web usage mining: A comparison of WUM category web mining algorithms, in *Third International Conference on Intelligent Communication Technologies and Virtual Mobile Networks (ICICV)*, pp 1020–1024.

Data Science & Exploration in Artificial Intelligence – Gururaj et al. (Eds)
© 2025 the Author(s), ISBN: 978-1-032-96390-7

Transformative approaches to e-learning videos: Generative vs. extractive question answering for enhanced doubt clarification

Aditya Anil Gupta, Anagha Patil, Shubha Dhananjaya Achar, M. Vasudha & Pooja Agarwal
PES University, Bangalore, Karnataka, India

ABSTRACT: In the ever-changing realm of education, technology has revolutionised conventional learning approaches. Due to its adaptability and accessibility, e-learning has become a popular choice for delivering education. This study explores doubt clarification systems for e-learning videos, focusing on generative question answering, which creates human-like responses on retrieval of relevant information, and extractive question answering, which extracts answers directly from the existing content. The research covers a broad spectrum of doubt clarification techniques, by delving into various algorithms for information extraction and summarisation, along with suggesting new directions for future research in the field.

Keywords: Video Question Answering, Language Models, Transformers

1 INTRODUCTION

Online learning has reshaped education, enabling global access and flexibility. Addressing response delays in current platforms, this study proposes an automated doubt resolution system. Traditional techniques leverage natural language processing, information retrieval, and semantic understanding to offer relevant answers. Abstractive question answering generates concise, contextually accurate responses, surpassing extractive methods. Extractive QA suits closed domains, focusing on precise text spans. Generative AI excels in open-ended inquiries, providing comprehensive responses. The choice between extractive and generative QA hinges on application needs and contextual factors. Extractive QA is adept for fact-checking and retrieval tasks, ensuring precision. Generative QA thrives in fostering creativity and handling complex queries. These approaches collectively enhance the efficacy and adaptability of online learning.

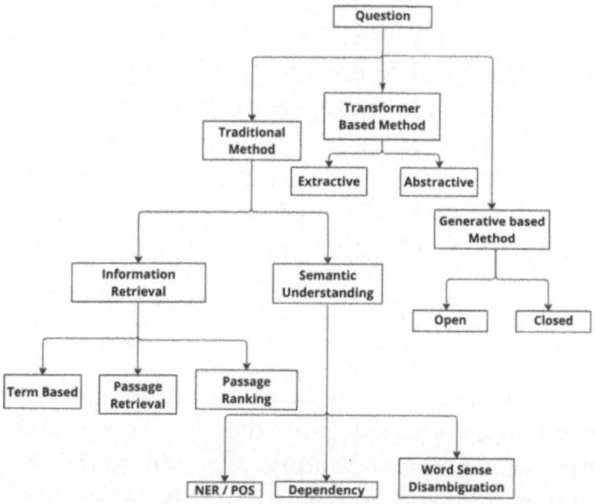

Figure 1. Taxonomy of QA techniques.

DOI: 10.1201/9781003589273-47

Figure 1 provides an overview of the taxonomy of the different types of question answering methods discussed. This survey paper provides an overview of the research on VideoQA for online learning and discusses the applications and challenges in doubt clarification systems.

2 REQUIRED BACKGROUND

2.1 *Unique frame extraction and text detection*

E-learning videos often contain key content such as keywords, diagrams, and tables displayed on slides or boards. To efficiently extract relevant information, unique frames are identified by analyzing slide transitions and comparing key frames, reducing redundancy. While Wang *et al.* (2008) used text similarity for keyframe detection, their approach struggled with animations and alignment issues. Tuna *et al.* (2012) employed global frame differencing, and Jeong *et al.* (2012) utilized SIFT for slide content similarity, though this was time-consuming. Thomas *et al.* (2022) applied RetinaNet for slide region detection, addressing high bias with negative samples through single-shot detection and focal loss. Text recognition in video frames involves four main processes: localization, extraction, binarization, and recognition. Saidane *et al.* (2007) proposed a binarization method using a convolutional network for colored text regions. Other methods, such as those by Kai *et al.* (2011) and Alahari *et al.* (2012), focused on character-level classification. Recent models adopt segmentation-free approaches using seq2seq frameworks for word recognition without sub-word segmentation. Shi *et al.* (2016) and C. *et al.* (2016) leveraged CNN-generated features, achieving state-of-the-art results in scene text recognition.

2.2 *Context formation*

Preprocessing techniques in textual analysis involve extracting information from structured and unstructured data to identify patterns, with key steps including stop word removal, stemming, and TF/IDF algorithms. Stop word removal methods include a pre-compiled list, Zipf's Law, Mutual Information Method, and Term-Based Random Sampling. Otterbacher *et al.* (2005) developed a stochastic, graph-based phrase retrieval system, while Erkan *et al.* (2004) introduced LexRank for multi-document summarization. The topic-sensitive LexRank extends this approach for a more focused and contextually relevant phrase retrieval system.

2.3 *Question answering*

2.3.1 *Transformers*

Transformers, introduced by Vaswani *et al.* (2017), revolutionized NLP by achieving state-of-the-art results. BERT, developed by Devlin *et al.* (2019), pretrains text representations considering context from both the left and right sides using random token masking. RoBERTa, an enhancement by Liu *et al.* (2019), refines this approach with dynamic masking, generating a new masking pattern for each input sequence. XLNet, by Yang *et al.* (2019), introduces a generalized autoregressive method, predicting word likelihoods across all permutations of a sentence, capturing dependencies irrespective of word order. DeBERTa, introduced by He *et al.* (2020), employs disentangled attention to represent words with content and position vectors. Early QA models relied on single approaches, but the field has since progressed to ensemble methods, leveraging the collective intelligence of multiple models for improved accuracy.

2.3.2 *Ensemble models*

One advantage of using ensemble models in question answering systems is using models that combine multiple pre-trained language models for machine reading comprehension, as proposed by Zhu *et al.* (2021). Another advantage is dynamic selection models leveraging a meta-learner to adaptively select the most appropriate ensemble members based on the characteristics of the input question, as introduced by Baradaran *et al.* (2021).

2.3.3 *Generative artificial intelligence*

Generative AI uses neural networks' creative potential to create cutting-edge material. Models like ChatGPT and LaMDA have been developed, each with a distinct set of capabilities and uses, that are pushing the limits of artificial intelligence and human-computer interaction. ChatGPT is trained using reinforcement learning through supervised fine-tuning while LaMDA, introduced by Thoppilan *et al.* 2022, is a decoder-only transformer language model designed for dialogue applications and is pre-trained on 1.56 trillion words of web text and public dialogue data.

3 METHODOLOGIES USED

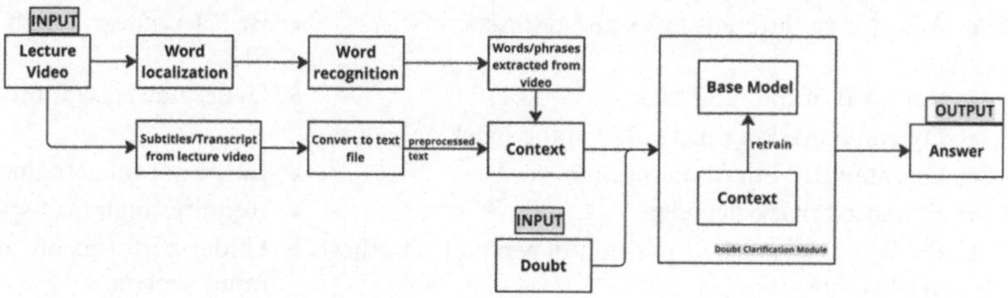

Figure 2. Doubt clarification model pipeline.

There are a number of steps that need to be addressed in order to develop effective VideoQA models for online learning. Figure 2 provides an overview of the different stages of video question answering.

3.1 *Unique frame extraction and text recognition from frames*

Yang *et al.* (2014) introduced a method for identifying unique frames in videos by detecting text regions using CC-based differencing and SVM classifiers, followed by text recognition and spell-checking. Tang *et al.* (2023) proposed CNN-TSDPC, combining CNN and Temporal Segment Density Peaks Clustering, to efficiently extract key frames from videos. This method achieves superior performance on HMDB51 and UCF101 datasets, with mean classification rates of 81.44% and 98.45%, while reducing redundant data and preserving temporal information.

3.2 *Context formation*

Asim *et al.* (2019) introduce an ontology-based approach that enhances query retrieval by integrating user profiles and geographic data, converting both document metadata and user queries into RDF triples for context-aware similarity calculations. Chasseray *et al.* (2023) address ontology population in an unsupervised setting, proposing a domain-independent rule-based method to evaluate performance. They highlight the limitations of ontology-based methods and utilize directed graphs in semantic networks to represent and relate concepts. This approach improves the precision and effectiveness of retrieving semantically rich information by annotating content semantically.

3.3 *Question answering*

3.3.1 *Transformers*

Transformers have revolutionized question answering by using self-attention to capture intricate word relationships. BERT, with its bidirectional context understanding, excels in handling longer queries but struggles with lengthy documents and external knowledge dependencies. RoBERTa enhances BERT by using dynamic masking and larger datasets, improving accuracy but introducing bias and high computational costs. XLNet addresses BERT's pretrain-finetune discrepancy by integrating Transformer-XL, a SOTA autoregressive model, excelling in long-term dependencies but underperforming on shorter inputs. DeBERTa advances further with disentangled attention mechanisms for content and position, and computes attention weights using disentangled matrices. though fine-tuning remains challenging due to its large number of parameters and need for substantial task-specific data.

Table 1. Highlights of transformers.

Name	Advantages	Limitations
BERT	• Self-supervised learning • Adapted to different tasks and datasets	• NSP weaker than MLM • BERT context length fixed at 512
RoBERTa	• Removal of the NSP task • Dynamic masking instead of static masking	• Generates biased information
DeBERTa	• Disentangled attention mechanism • Enhanced mask decoder	• Large parameters fine tuning • Requires high task-specific data
XLNet	• Prediction of masked position dependant on other masked positions • Recurrence mechanism and relative positional encoding	• Under performs on shorter input sentences

Table 1 provides a summary on the advantages and limitations of the mentioned transformers. Nassiri *et al.* (2023) offer a comprehensive review on question answering systems with focus on transformer models. They illustrate how T5 can predict different outputs for similar inputs across several tasks. By considering NLP problems as text-to-text transfers, this model allows unified training due to which it uses its task conditioning method to realize cohesive text processing. In contrast, Cao *et al.* (2020) put forth the Decomposed Transformer, giving priority to parts of queries and documents that are essential for comprehending, an approach particularly exemplified by the "De-Former" transformer. Task probing demonstrates that lower transformer levels pay no attention to question tokens in shaping the De-Former process. However, this deformation introduces potential biases, limiting its applicability. Overall, these studies underscore the evolving strategies within transformer-based QA systems, balancing efficiency and performance across diverse tasks.

3.3.2 *Large language models*

LLMs built on the transformative architecture, leverage vast amounts of internet data for pretraining, enabling them to respond to a wide range of inquiries with impressive nuance. However, they face challenges in verifying authenticity, which can sometimes lead to inaccuracies. They also carry biases from their training data. The Galactica model, by Gozalo-Brizuela *et al.* (2023), aims to enhance objectivity, but generative AI still faces significant hurdles. Additionally, their resource-intensive nature makes them less accessible for smaller-scale applications. Goyal *et al.* (2023) examined 200 polarizing questions, comparing responses from GPT and BARD, and found notable differences that highlight distinct biases in these language models. BARD did not start its answers with any indication that it was an LLM. However, the conversational tone of the responses and the consistent engagement in a dialogue were noticeable. There were occasions when BARD failed to offer sufficient responses, particularly related to societal controversies. Responses from ChatGPT were well-rounded, capturing differing viewpoints. However, a majority of the responses appeared somewhat extreme. ChatGPT seemed to provide more balanced and generic answers with respect to questions centralised to daily lives. With generative models excelling in complex, open-ended questions and extractive models giving higher assurance in factual correctness, the decision between generative and extractive QA relies on the specific task requirements.

4 RESULTS

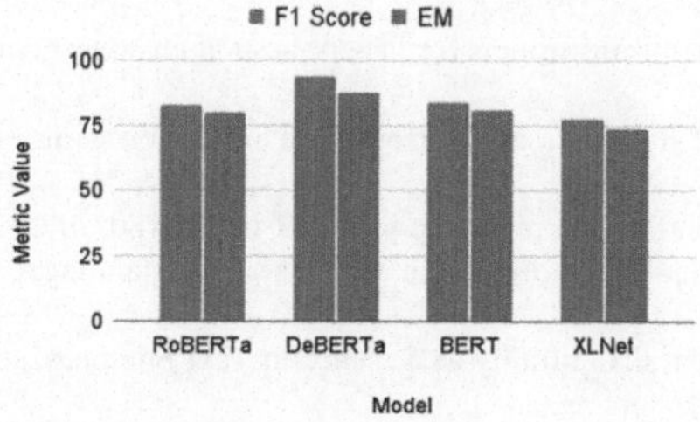

Figure 3. Comparison of transformers on SQuAD.

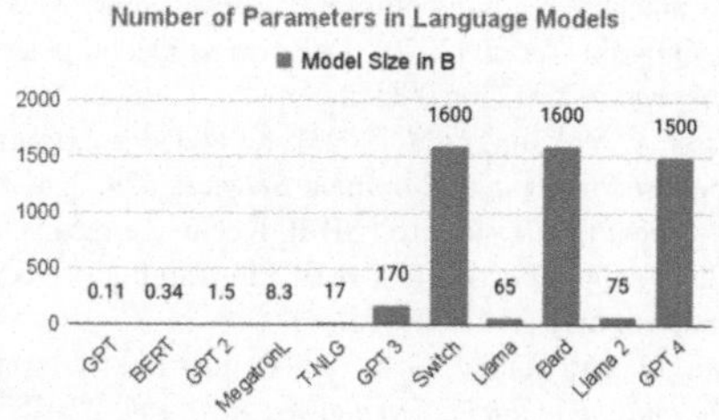

Figure 4. Comparison of LLMs based on parameters.

The key stages involved in doubt clarification include frame extraction, context formation, question understanding and answer retrieval. By breaking down the doubt clarification process into these stages, the paper offers valuable insights into the underlying mechanisms and challenges faced in handling this complex task. Additionally, merits and demerits of question answering methodologies are discussed, specifically generative and extractive approaches. Figure 3 provides an overview of the performance of the transformers on the SQuAD dataset whereas Figure 4 provides a comparison of LLMs on the basis of number of parameters.

5 SUGGESTIONS AND FUTURE DIRECTION

Addressing ambiguity in question answering is a persistent challenge, given the varied interpretations of user queries. E-learning videos, enriched with images and diagrams, require robust doubt clarification mechanisms. Integrating visual question answering into systems enables deeper understanding through context-aware responses. Multi-modal and multi-task models hold promise in enhancing response accuracy by incorporating diverse data sources. Future research could prioritize algorithms for providing immediate feedback in e-learning videos, assessing individual knowledge gaps based on learner responses and behavior. Dynamic adaptation of doubt clarification systems through adaptive learning strategies can tailor to each learner's evolving needs, thereby improving efficiency and learning outcomes. Such advancements aim to revolutionize e-learning by offering nuanced and personalized assistance, leveraging the richness of multi-modal information.

6 CONCLUSION

This paper presents a study on the various stages of doubt clarification from videos in the form of video question answering, along with the different types of question answering. The comprehensive examination of question answering methods broadens the understanding of readers regarding the various possibilities and trade-offs associated with different approaches. It also consolidates the existing knowledge and research findings, and offers new perspectives for further exploration. By presenting a comprehensive overview of doubt clarification techniques and their evolution, the paper contributes significantly to the advancement of video question answering systems.

REFERENCES

Alahari Karteek and Jawahar C. (2012). *Scene Text Recognition using Higher Order Language Priors.* 10.5244/C.26.127.

Asim, Muhammad Nabeel, *et al.* (2019). The use of ontology in retrieval: a study on textual, multilingual, and multimedia retrieval. *IEEE Access* 7: 21662–21686.
Baradaran, R. and Amirkhani, H. (2021). Ensemble learning-based approach for improving generalization capability of machine reading comprehension systems. *Neurocomputing*, 466, 229–242.
Cao, Qingqing, *et al.* (2020). DeFormer: Decomposing pre-trained transformers for faster question answering. *arXiv preprint* arXiv:2005.00697.
Chasseray, Yohann, *et al.* (2023). Knowledge extraction from textual data and performance evaluation in an unsupervised context. *Information Sciences* 629: 324–343.
C.-Y. Lee and S. Osindero (2016). Recursive recurrent nets with attention modeling for OCR in the wild. *In CVPR.*
Devlin, Jacob, *et al.* (2018). Bert: Pre-training of deep bidirectional transformers for language understanding. *arXiv preprint* arXiv:1810.04805.
Erkan, G. and Radev, D.R. (2004). Lexrank: Graph-Based Lexical Centrality as Salience in Text Summarization. *Journal of Artificial Intelligence Research*, 22, 457–479.
Gozalo-Brizuela, R., and Garrido-Merchan, E. C. (2023). ChatGPT is not all you need. A state of the art review of large generative AI models. *arXiv preprint* arXiv:2301.04655.
Goyal, Abhay, *et al.* (2023). ChatGPT and bard responses to polarizing questions. *arXiv preprint* arXiv:2307.12402.
He, P., Liu, X., Gao, J. and Chen, W. (2020). Deberta: Decoding-enhanced bert with disentangled attention. *arXiv preprint* arXiv:2006.03654.
Jeong, Hyun Kim, Tak-Eun Kim and Myoung Ho. (2012). *An Accurate Lecture Video Segmentation Method by using SIFT and Adaptive Threshold.* 285–288. 10.1145/2428955.2429011.
Kai Wang, B. Babenko, and S. Belongie (2011). End-to-end scene text recognition. *2011 International Conference on Computer Vision*, Barcelona, pp. 1457–1464. https://doi.org/10.1109/ICCV.2011.6126402.
Liu, Yinhan, *et al.* (2019). Roberta: A robustly optimized bert pretraining approach. *arXiv preprint* arXiv:1907.11692.
Nassiri, Khalid, and Moulay Akhloufi. (2023). Transformer models used for text-based question answering systems. *Applied Intelligence* 53.9: 10602–10635.
Otterbacher, Jahna Gunes, Erkan Radev and Dragomir. (2005). *Using Random Walks for Question-focused Sentence Retrieval.* 10.3115/1220575.1220690.
Rajpurkar, Pranav, Robin Jia, and Percy Liang (2018). Know what you don't know: Unanswerable questions for SQuAD. *arXiv preprint* arXiv:1806.03822.
Saidane, Zohra Garcia and Christophe. (2007). *Automatic Scene Text Recognition using a Convolutional Neural Network.*
Shi, B., Bai, X., and Yao, C. (2016). An end-to-end trainable neural network for image-based sequence recognition and its application to scene text recognition. *IEEE Transactions on Pattern Analysis and Machine Intelligence*, 39 (11), 2298–2304.
Tang, Hao, *et al.* (2023). Deep unsupervised key frame extraction for efficient video classification. *ACM Transactions on Multimedia Computing, Communications and Applications* 19.3: 1–17.
Thomas, Chinchu Sarma, K.A.V. Gajula, Srujan Jayagopi and Dinesh. (2022). Automatic prediction of presentation style and student engagement from videos. *Computers and Education: Artificial Intelligence*. 3. 100079. 10.1016/j.caeai.2022.100079.
Thoppilan, Romal, *et al.* (2022). Lamda: Language models for dialog applications. *arXiv preprint* arXiv:2201.08239.
Tuna, Tayfun, *et al.* (2012). Development and evaluation of indexed captioned searchable videos for stem coursework. *Proceedings of the 43rd ACM technical symposium on Computer Science Education.*
Vaswani, Ashish, *et al.* (2017). Attention is all you need. *Advances in Neural Information Processing Systems* 30.
Vijayarani, S., Ilamathi, M. J. and Nithya, M. (2015). Preprocessing techniques for text mining-an overview. *International Journal of Computer Science Communication Networks*, 5(1), 7–16.
Wang, Feng Ngo, Chong-Wah Pong and Ting-Chuen. (2008). Structuring low-quality videotaped lectures for cross-reference browsing by video text analysis. *Pattern Recognition*. 41. 3257–3269. 10.1016/j.patcog.2008.03.024.
Yang, Haojin, and Christoph Meinel. (2014). Content based lecture video retrieval using speech and video text information. *IEEE Transactions on Learning Technologies* 7.2: 142–154.
Yang, Zhilin Dai, Zihang Yang, Yiming Carbonell, Jaime Salakhutdinov, Ruslan Le and Quoc. (2019). *XLNet: Generalized Autoregressive Pretraining for Language Understanding.*
Zhu, Fengbin, *et al.* (2021). Retrieving and reading: A comprehensive survey on open-domain question answering. *arXiv preprint* arXiv:2101.00774.

Data Science & Exploration in Artificial Intelligence – Gururaj et al. (Eds)
© 2025 the Author(s), ISBN: 978-1-032-96390-7

Augmented reality-based location discovery: Integration of unity IDE, Mapbox API, and ARcore framework for seamless pedestrian navigation

Aby Stalin, N. Nadeem, N. Rizwan & J. Divya Udayan
Department of Computer Science and Engineering, Amrita School of Computing, Amrita Vishwa Vidyapeetham, Amritapuri, India

ABSTRACT: This project aims to develop an Augmented Reality (AR) navigation system considering the challenges faced by pedestrians in both urban and rural environments. The system offers an easy-to-use interface for travelling variety of surroundings by utilising the ARCore framework, Mapbox API, and Unity IDE. For urban settings, the AR navigation system goes beyond traditional map-based guidance by offering real-time visual cues, contextually relevant information, and seamless path planning to help users efficiently traverse changing landscapes. To provide a complete solution for pedestrians in both urban and nature contexts, the system improves its capabilities in rural regions to help hikers and tourists negotiating various terrain, shifting elevations, and unmarked pathways. By fusing digital data with the real environment, places of interest are included into the navigation process, which improves the user experience even more. This research is a big step towards creating a universal AR navigation system that transforms how people explore and interact with their surroundings by addressing the unique demands of various user groups and locations.

1 INTRODUCTION

Pedestrian navigation systems are becoming an increasingly important tool for people living in dynamic urban areas. But many of the demands of users are not met by the solutions that are currently in place. The project's goal is to close this gap by developing an Augmented Reality (AR) guidance system for hikers, tourists, and city walkers. There are several obstacles for pedestrians to overcome when navigating through busy streets, difficult crossings, and a surplus of sites of interest. While turn-by-turn directions and static maps are features of traditional navigation systems, they often do not provide dynamic, real-time assistance needed to travel effectively across constantly changing metropolitan settings. This limitation is intended to be addressed by the AR navigation system, which offers turn-by-turn directions, journey planning, and relevant information together with real-time visual cues. By becoming integrated into the user's environment, the system becomes a virtual guide and improves the user experience (Vasudevan *et al.*, 2018). The difficulties encountered in rural regions are distinct, yet just as difficult. Exploring natural landscapes presents hikers and travellers with a range of terrains, height changes, and a lack of well-marked trails. In these situations, current navigational aids find it difficult to offer sufficient assistance. In an ever-evolving urban environment, pedestrian navigation systems are becoming more and more vital. However, present solutions often fall short of meeting the expectations of users. Recognising this gap, the project intends to create a revolutionary Augmented Reality (AR) navigation system that benefits walkers in urban areas, hikers, and tourists by extending its functionality to encompass walkers in rural areas.

Moreover, the incorporation of places of interest enhances the user experience by providing supplementary details on historical locations, landmarks, and attractions, rendering navigation not only effective but also fascinating from a cultural and social standpoint.

The research is a major step toward developing a universal augmented reality navigation system that meets the specific needs of the user personas such as hikers, tourists and travellers. The

DOI: 10.1201/9781003589273-48

AR system, which combines technology and real-world navigation experiences, has the potential to improve pedestrian and explorer efficiency, convenience, and overall enjoyment in a variety of environments (Vasudevan *et al.*, 2020).

2 RELATED WORKS

Many industries stand to benefit from the opportunity to enhance their operations and navigation through field of augmented reality (AR). This research examines the potential of augmented reality (AR) to address problems with geographical accuracy and real-time decision-making in geographically diverse situations for pedestrian navigation. It examines the integration of smartphone sensor data with augmented reality (AR), considers current limitations, and makes recommendations for future directions for this field.

Asraf *et al.* (2020) found that users of traditional map-based navigation apps frequently encounter difficulties when operating in different terrains. These interfaces often require memorizing routes and identifying landmarks, which can be laborious and prone to make mistakes. Navigating an airport with multiple gates and terminals is a difficult task for any new transiters. On a conventional map, you might see a blue dot for your location, but it wouldn't be immediately obvious which way to go to the gate you want. To figure out the way, you might have to continuously check the map and your surroundings while zooming in and out. This back-and-forth in the mind can be tiresome and time-consuming, particularly in unfamiliar locations.

In a similar situation, Lu *et al.* (2021) draw attention to the shortcomings of the campus navigation systems that are already in use, noting that they frequently have difficult-to-use interfaces and inadequate high-precision navigational capabilities while navigating between indoor and outdoor locations. Imagine a large university campus that features a variety of structures, walkways, and open areas, because GPS signals are frequently weak within buildings, traditional navigation apps may find it difficult to deliver meaningful instructions properly.

Notwithstanding the many benefits AR provides, issues with accuracy and sensor reliance still exist. Rotational errors can affect the accuracy of marker placement, especially when the marker is farther away. Furthermore, performance in places with restricted access may be hampered by reliance on internet connectivity.

Numerous studies offer solutions for these shortcomings. One major area of concentration is rotational error mitigation; Lu *et al.* (2021) recommends sensor fusion or external calibration methods. The process of sensor fusion is merging information from several sensors (such as a GPS, camera, or IMU) to produce a more precise and thorough picture of the surroundings. One such method for correcting for sensor faults or drift is to use known reference points in external calibration processes.

One further difficulty is to combine outdoor and indoor navigation features in one AR application without any issues. To solve this, Karan *et al.* (2021) created a system that uses location-based AR for outdoor navigation and marker-based AR for indoor navigation. Nevertheless, there are certain drawbacks, such as the requirement for well positioned visible markers indoors, which may not be feasible in expansive settings, and the possibility of inconsistent route accuracy outdoors because of GPS technology's intrinsic limits.

It is recommended that future studies investigate markerless indoor location methods that do not depend on visual markers and research on sensor fusion methods—which integrate GPS with WiFi positioning and other technologies—shows promise for increasing outdoor accuracy.

Priority one when developing an AR navigation system should be given to usability and user experience. More thorough user studies with bigger and more varied sample sizes are required, even though certain research, like Lu *et al.*, show user testing and evaluation. Validating usability and performance across user groups with different requirements and abilities, including the elderly or visually challenged populations, is especially crucial.

Udayan *et al.* (2020) investigates the effectiveness of augmented reality in brand building and marketing in the valves industry, comparing traditional methods with Augmented Reality. It uses an Augmented Reality application for customers and employees, comparing results

To enhance user experience, future research should concentrate on resolving problems like sensor accuracy, latency (the interval between an action and the system's response), and dynamic environment factors. Advances in occlusion handling—the ability to realistically block virtual objects by real-world objects—and the integration of features like turn-by-turn navigation can further increase the usefulness and attractiveness of augmented reality navigation systems.

To sum up, augmented reality technology has the potential to revolutionize navigation and operations planning in a multitude of industries. Through the resolution of issues related to precision, reliance on sensors, and smooth transition between indoor and outdoor features, augmented reality navigation systems have the potential to be highly valuable instruments for enhancing operational effectiveness and cognition in widely separated locations.

3 PROPOSED WORK

Ensuring exact user location accuracy is critical to the success of any location-based augmented reality (AR) application. This is often accomplished by using known mapping APIs such as Mapbox API or Google Maps, which use the device's capabilities to collect key latitude and longitude data. Using this information, the app may determine the user's present position to a specific area.

Once the starting position is found, the app must prompt the user to pick their preferred destination. Mapping APIs are also important in this context, since they provide the optimum path between the origin and destination sites. This route data serves as the foundation for the ensuing AR advice, which frequently includes precise turn-by-turn instructions. The app's genuine power comes in its augmented reality features. Powerful game engines such as Unity enable the creation that direct users to their intended destination. Sensor techniques, which use the device's gyroscope, accelerometer, and magnetometer, can produce realistic and immersive movement for this guiding. Due to these sensors' inherent limitations and data noise, complex algorithms like as dead reckoning and Kalman filtering (Anerao *et al.*, 2019) must be included to assure precise and flawless movement estimates. Fortunately, there are more efficient techniques accessible in the form of Software Development Kits (SDKs) particularly developed for AR production. Notable examples are ARCore and Wikitude. These SDKs frequently include Simultaneous Localization and Mapping (SLAM) technology, which allows for more sophisticated AR experiences. Depending on the type of navigation given by the Mapbox API, Unity-generated objects can cover the set distance.

3.1 *Methodology*

The process begins with the camera & sensors capturing the user's environment, which is then processed through 3D Point Clouds to create a digital model of the surroundings (Chakraborty & Gupta, 2017). The system then scans for trackables in the environment, which are used for

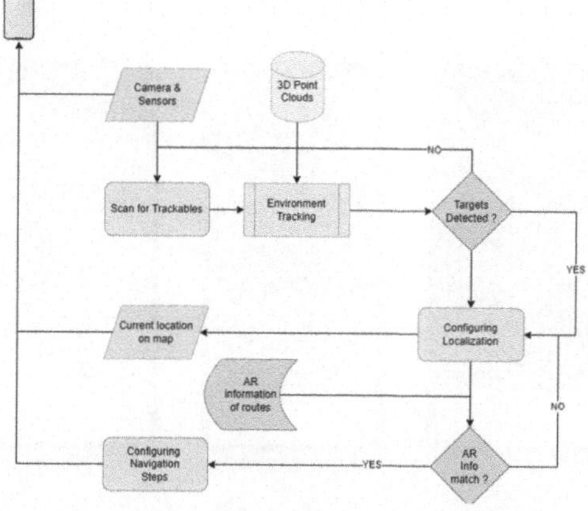

Figure 1. Proposed architecture of AR based location discovery service.

localization, and employs Environment Tracking to understand the user's position and orientation. The system then checks if any targets have been detected, and if so, it proceeds to Configuring Localization to determine the user's precise location. If the AR information matches, the system provides the AR info to the user; otherwise, it goes through configuring navigation Steps to guide the user's navigation. Finally, the system displays the current location on map to the user, completing the comprehensive navigation flow powered by AR technology, 3D mapping, and sensor data.

4 EXPERIMENTAL RESULTS

To get the project started, we explored at the numerous augmented reality (AR) SDKs on the market. Our alternatives were Vuforia, Wikitude and Google ARCore. After exploring with several of these SDKs, we discovered certain issues, notably with Vuforia. We had difficulty ground plane recognition proved to be difficult with Vuforia.In quest of a solution, we went to Wikitude, which offered aid in getting a virtual plane. However, we still had problems scaling things according to the user's location. We eventually discovered success with Google's ARCore. ARCore enabled us to accurately detect the ground plane and rectify the scaling issue, resulting in a more seamless AR experience.In addition to AR capabilities, we looked at APIs for navigation reasons. Our alternatives included the Mapbox SDK and the Google Maps API. After careful evaluation, we chose the Mapbox SDK since it efficiently matched our navigation needs.

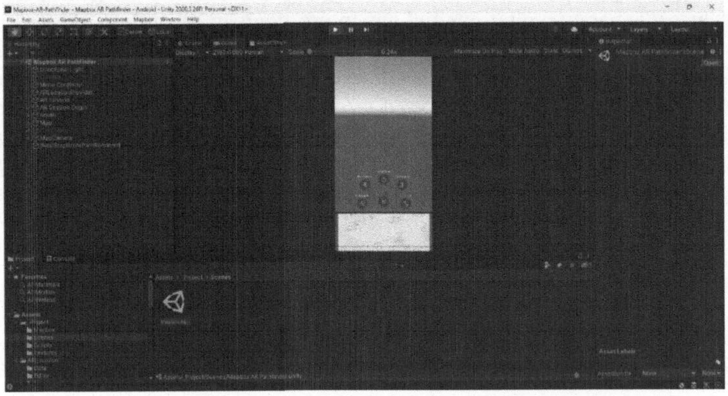

Figure 2. Unity Simulation of proposed AR based location discovery service.

Figure 3. Mobile version of the application on an android device.

5 CONCLUSION

In conclusion, this paper has offered an appropriate approach for improving pedestrian navigation using augmented reality technologies. (Udayan & Kim, 2021). The prototype navigation system was able to create the necessary AR feature implementation for the user by keeping both the traditional map view and AR waypoint pointers which helped in the user's learning curve in using this system (Mohan *et al.*, 2015). By showcasing such a prototype system, we believe it will promote more development in pedestrian navigation tools instead of vehicular ones. This approach of pedestrian navigation will be most useful for tourists with difficulty in understanding local language. We anticipate that AR will become a more practical (Radhika *et al.*, 2016) option for enhancing accessibility, pedestrian navigation, and general mobility in both indoor and outdoor environments as AR technology and software continue to progress rapidly, a lot of developers are investigating novel immersive AR applications in a variety of domains, including education, navigation, mobility, and more (Rani *et al.*, 2017).The current explosion in augmented reality technology market has been further propelled by the release of Apple's Vision Pro, prompting more developers to push existing applications into the AR/VR space.

REFERENCES

Anerao, A. *et al.* (2019). Location based augmented reality navigation application, *International Journal of Scientific Research in Computer Science, Engineering and Information Technology*, pp. 241–245. https://doi.org/10.32628/cseit195340.

Asraf, S.M.H., Hashim, A.F.M. and Idrus, S.Z.S. (2020), Mobile application outdoor navigation using location-based augmented reality (AR), *Journal of Physics. Conference Series*, 1529(2), p. 022098. https://doi.org/10.1088/1742-6596/1529/2/022098.

Chakraborty, S. and Gupta, D. (2017), *A study of the Factors Impacting the Adoption of Augmented Reality in Online Purchases in India*. [Preprint]. https://doi.org/10.1109/rteict.2017.8256853.

Karan, E. *et al.* (2021) *Indoor and Outdoor Navigation Application Development with Augmented Reality Technology, ResearchGate* [Preprint]. https://www.researchgate.net/publication/359282706_Indoor_and_Outdoor_Navigation_Application_Development_with_Augmented_Reality_Technology.

Lu, F. *et al.* (2021) An ARCORE-Based augmented reality campus navigation system, *Applied Sciences*, 11(16), p. 7515. https://doi.org/10.3390/app11167515.

Mohan, B.R., Bijlani, K. and Jayakrishnan, R. (2015). *Simulating Consequences of Smoking with Augmented Reality*. [Preprint]. https://doi.org/10.1109/icacci.2015.7275728.

Radhika, G. *et al.* (2016). Virtual try of a jewellery using skin classifiers; *Control Theory & Applications*, 9(5). https://www.amrita.edu/system/files/publications/virtual-try-of-jewellery-using-skin-classifier.pdf.

Rani, S.S., Dhrisya, K.J. and Ahalyadas, M. (2017); *Hand Gesture Control of Virtual Object in Augmented Reality*. [Preprint]. https://doi.org/10.1109/icacci.2017.8126053.

Udayan, J.D. *et al.* (2020). Augmented reality in brand building and marketing – Valves industry; *2020 International Conference on Emerging Trends in Information Technology and Engineering (ic-ETITE)* [Preprint]. https://doi.org/10.1109/ic-etite47903.2020.425.

Udayan, J.D. and Kim, H. (2021); Semantic modeling and rendering, in *Human-Computer Interaction Series*, pp. 105–127. https://doi.org/10.1007/978-3-030-71002-6_7.

Vasudevan, S.K. *et al.* (2018). Marker-based augmented reality interface with gesture interaction to access remote file system, *International Journal of Advanced Intelligence Paradigms*, 10(3), p. 236. https://doi.org/10.1504/ijaip.2018.090788.

Vasudevan, S.K. *et al.* (2020); An intelligent and interactive AR-based location identifier for indoor navigation, *International Journal of Advanced Intelligence Paradigms*, 15(1), p. 32. https://doi.org/10.1504/ijaip.2020.104105.

Data Science & Exploration in Artificial Intelligence – Gururaj et al. (Eds)
© 2025 the Author(s), ISBN: 978-1-032-96390-7

Estimating a battery's State of Health (SoH): A hybrid LSTM-GRU predictive model approach

K.J. Karmuhilan & Sushma Kumari
Defense Institute of Advanced Technology, Pune, India

ABSTRACT: This paper introduces a fresh methodology for assessing lithium-ion batteries' State of Health (SoH), which is essential for electric vehicles. We are utilizing a hybrid LSTM-GRU model and extensive datasets from NASA and CALCE, which present a robust SoH forecasting framework. The approach prioritizes practicality by excluding historical SoH data, depending instead on observable battery performance metrics. This approach enhances real-world applicability and facilitates efficient energy storage solutions. Additionally, this paper explores the possibility of integrating transformer models into the forecasting framework, enhancing the accuracy and versatility of SoH predictions. Overall, this research significantly advances battery management, extending lifespans, and promoting sustainable energy utilization with its creative process.

Keywords: SoH, LSTM-GRU, Lithium-ion battery, Battery Management, Attention Mechanism.

1 INTRODUCTION

In the pursuit of mitigating global carbon emissions, the electrification of transportation has emerged as a cornerstone strategy. At the heart of this transformation lies the lithium-ion battery. There's an innate drawback with batteries, that they degrade over time. The battery degradation is affected and accelerated by multiple factors. Thus, knowing these batteries' SoH becomes crucial, influencing the longevity, efficiency, and mainly the safety of electric vehicles.

The traditional methodologies for measuring the State of Health hinge heavily on direct physical measurements, which are infeasible due to their time-consuming nature. Researchers have turned to building mathematical models based on battery behaviour.

In the past decade, several institutes have conducted extensive experiments on batteries, uncovering intricate degradation patterns and generating vast amounts of data. resulting in research-ers shifting towards leveraging this data to develop more accurate SoH prediction models. The comprehensive literature survey is summarized in Table 1.

Table 1. Literature survey of existing works in the areas of battery health prediction.

Authors	Methodology	Pros	Cons	Gaps
Xu *et al.* [1]	Extended Kalman filter	- Accurate health estimation. - Can handle non-linearity.	- High complexity - Expertise in battery model building.	- Reducing the complexity. - Compounding errors.
Wei *et al.* [2]	Particle filter, Support vector regression	- Captures complex degradation patterns. - Can handle high dimensional data.	- Relies heavily on historical data. - Prone to noisy data.	- Removing the dependence on historical SoH data.

(*continued*)

 DOI: 10.1201/9781003589273-49

Table 1. Continued

Authors	Methodology	Pros	Cons	Gaps
Ma *et al.* [3]	Long Short-Term Memory (LSTM)	- Captures long-term dependencies.	- Inability to match accelerated degradation rates.	- Overcoming LSTM's limitations over time.
Jianping Wen *et al.* [4]	Incremental Capacity (IC) analysis, Temperature effects	- Provides battery insights - Nearly accurate estimations.	- Time-consuming process. - Requires controlled conditions.	- Training on multiple datasets for generalized prediction.
Lijun Qian *et al.* [5]	Decision tree feature importance, Enhanced SVR	- Interpretable models.	- Limited generalization.	

* S.K. wants to acknowledge the support of DIAT Grant-in-aid project.

Our goal is to develop a model that can forecast accurately over an extended period of time without relying on historical capacity data, making it ideal for practical application. We introduce a hybrid LSTM-GRU model. The proposed model leverages the predictive strengths of Long Short-Term Memory (LSTM) and Gated Recurrent Unit (GRU) networks, specifically tailored to navigate the temporal complexities inherent in battery performance data.

The remainder of this paper is structured as follows: Section 2 delves into battery aging mechanisms and the datasets utilized in this study and Section 3, presents the proposed hybrid model, further the results are discussed in Section 4 followed by conclusion in Section 5.

2 FUNDAMENTALS OF BATTERIES

2.1 *Introduction to batteries*

Batteries convert chemical energy into electrical energy through anode, cathode, and electrolyte interactions. The Lithium-ion Batteries (LIBs) are renowned for their high energy density and are used in a wide range of applications, from portable electronics to electric vehicles [6]. Naturally, Li-ion batteries consist of graphite as the anode, metal oxides such as Lithium Cobalt oxide (LiCoO2) or Lithium Iron Phosphate (LiFePO4) as the cathode, and a Lithium salt dissolved in a solvent for the electrolyte [7].

2.2 *State of health and battery ageing*

State of Health (SoH) refers to the overall condition or health status of a battery relative to its initial state or design specifications. It serves as a quantitative measure to assess how much a battery's performance or capacity has degraded over time. SoH is crucial for determining the remaining useful life of a battery and plays a significant role in battery management systems (BMS). It's typically expressed as a percentage, with 100% representing the battery's initial state, and values below indicating degradation.

Battery ageing is the gradual deterioration of a battery's performance and capacity over time. This is a natural process influenced by various factors such as usage patterns, environmental conditions, and charging/discharging cycles. It can occur over time or due to repeated charge and discharge cycles. The calendar ageing refers to the degradation that happens over time, even when the battery is not in use, while cyclic ageing occurs due to repeated charge and discharge cycles. The ageing leads to a decrease in the battery's capacity, efficiency, and overall performance, ultimately impacting batteries' ability to store and deliver energy effectively.

2.3 *Degradation mechanisms*

Degradation mechanisms can vary depending on the battery chemistry and operating conditions but generally include:

2.3.1 *Loss of Lithium Inventory (LLI)*

In lithium-ion batteries, the loss of lithium ions due to side reactions, formation of the Solid Electrolyte Interface (SEI) layer [8], and electrolyte decomposition leads to a decrease in the battery's capacity and efficiency over time.

2.3.2 *Loss of Active Material (LAM)*

Structural deterioration of electrode materials, chemical decomposition, and dissolution reactions contribute to the loss of active material, reducing the battery's storage capacity and leading to mechanical stress and particle cracking

The above degradation mechanisms are essential for developing effective battery management strategies and extending the lifespan of batteries.

3 THE LSTM-GRU HYBRID MODEL

3.1 *Experimentation setup*

We leverage the NASA dataset, which comprises Li-NCA (LiNiCoAlO2) cells, encompassing four battery datasets with varying conditions[9]. The experimental regimen includes constant current-constant voltage charge (CC-CV), CC discharge repeated until a 30% capacity loss is reached [10]. Additionally, we utilize the CALCE dataset, featuring LCO (LiCoO2) cells consisting of CC-CV charge and CC discharge protocols [10].

3.2 *Data preparation*

We encountered various difficulties when working with the sensor data. The first difficulty originates from the high standards for the necessary experimentation setup, which makes high-quality data scarce. Furthermore, logistical challenges arise from the sheer volume of data; datasets containing more than three lakh instances per battery and multiple parameters require effective processing techniques, like distributed computing with PySpark. To overcome those challenges, we employed a systematic approach preparation of data.

The process begins with the aggregation of disparate performance data from multiple sources into a unified dataset. Detailed cycle information is then assigned to this aggregated dataset, with capacity and State of Health (SoH) data integrated to facilitate a continuous profile of battery health. To ensure consistency, a systematic down-sampling process standardizes intervals across different datasets. Attribute standardization harmonizes measurement units and naming conventions, promoting clarity and facilitating analysis. Finally, data visualization techniques are employed to gain insights into battery degradation patterns, providing valuable context for model development and evaluation, as illustrated in Figure 1 with the degradation pattern of CALCE batteries.

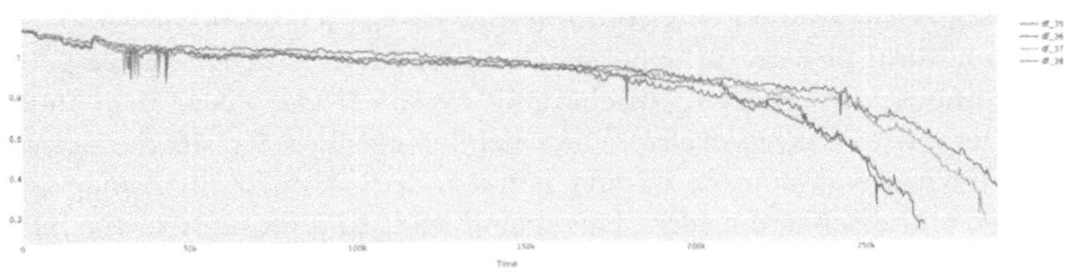

Figure 1. Visualization of CALCE batteries' capacity over time.

3.3 *Multivariate time series forecasting*

Our multivariate time series forecasting approach uses windowed techniques and feature scaling, including:

3.3.1 *Windowed approach*

Segmenting the time series data into overlapping windows allows the model to use past observations to predict future capacity. An optimal window size of 9 rows is chosen through experimentation for balance between fidelity and efficiency.

3.3.2 *Exclusion of capacity as input*

This ensures the model adapts to real-world scenarios with no capacity measurements, enhancing versatility and resilience by deriving insights from various data sources.

3.3.3 *Feature scaling*

Standardization with StandardScaler normalizes independent variables, ensuring effective model convergence and maintaining test data integrity, crucial for reliable battery State of Health (SoH) estimation.

3.4 *Model development and evaluation strategy*

As we created our model architecture, we crafted diverse deep learning architectures tailored for State of Health (SoH) prediction in lithium-ion batteries. This ensemble comprised of Long Short-Term Memory (LSTM), Gated Recurrent Unit (GRU), a hybrid LSTM-GRU stacking model, a parallel architecture of LSTM and GRU, and an innovative application of transformer models.

We used the Optuna framework with the Adam optimizer and an exponential decay learning rate for model optimization. We systematically tune hyperparameters such as the number of LSTM and GRU layers, batch size, and attention heads for Transformer models to ensure optimal performance. Once we establish the optimized hyperparameters, we fine-tune the number of units in each layer to balance accuracy and computational efficiency. Then models were meticulously trained on both the NASA and CALCE datasets, encompassing a range of battery aging experiments and operational conditions.

For a comprehensive evaluation, we selected three batteries from each dataset, ensuring a diverse representation of battery chemistries and aging patterns. These batteries underwent rigorous training to capture the intricate temporal dependencies and multivariate characteristics inherent in battery performance data. Subsequently, the trained models were subjected to testing and prediction tasks, with one battery from each dataset serving as unseen test data. This rigorous evaluation framework allowed us to assess the robustness, and predictive accuracy of each model across diverse battery chemistries and operational scenarios.

4 RESULTS AND DISCUSSION

4.1 *CALCE dataset*

After conducting a comprehensive analysis of the CALCE dataset, several key insights have emerged. The evaluation metrics shown in Table 2 reveal the performance of various

Table 2. Evaluation Metrics for the CALCE dataset predictive models.

Model	MSE	MAE	RMSE	R^2	MAPE
LSTM	0.006166	0.065043	0.078526	0.849554	8.308215
GRU	0.007027	0.051600	0.083827	0.828554	8.330289
LSTM + GRU Stacking	0.002701	0.031242	0.051971	0.934102	4.508728
LSTM ‖ GRU Parallel	0.001646	0.028730	0.040567	0.959848	4.152217
Transformer	0.004059	0.034596	0.063707	0.900979	5.355623

predictive models and Figure 2 shows the predicted capacity in contrast to the actual capacity for various models on both NASA and CALCE.

4.2 *NASA dataset results*

We analyze the NASA dataset for modeling of battery health management, and Table 3 shows the evaluation of various predictive models implemented.

Table 3. Evaluation metrics for the NASA dataset predictive models.

Model	MSE	MAE	RMSE	R^2	MAPE
LSTM	0.005756	0.067434	0.075868	0.825401	4.647932
GRU	0.006997	0.076969	0.083646	0.787763	4.924983
LSTM + GRU Stacking	0.001843	0.038311	0.042929	0.944098	2.599568
LSTM || GRU Parallel	0.003439	0.051106	0.058641	0.895689	3.529186
Transformer	0.003284	0.048628	0.057305	0.900387	3.379681

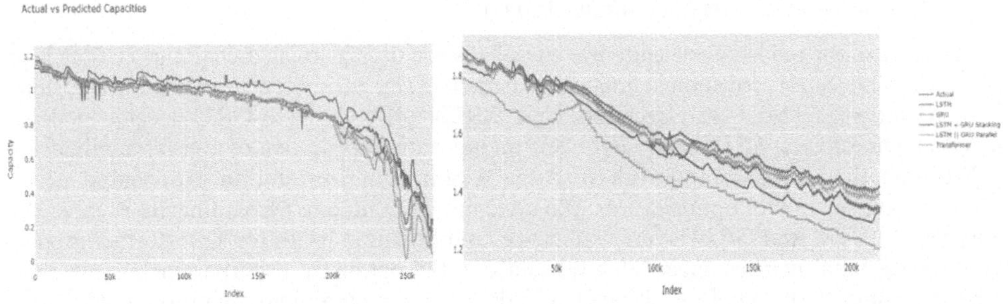

Figure 2. Actual vs predicted capacities of models on (a) CALCE dataset (left subfigure) (b) NASA dataset (right subfigure).

4.3 *Interpretation of results*

For the CALCE dataset, the LSTM + GRU Stacking model showcased superior performance with a Mean Absolute Percentage Error (MAPE) of 4.51% and a Root Mean Square Error (RMSE) of 0.052. This signifies its effectiveness in providing accurate predictions of battery health, crucial for optimizing EV battery management practices. Notably, the individual LSTM and GRU models exhibited higher errors, highlighting their limitations in capturing the evolving dynamics of battery health degradation over time.

Conversely, for the NASA dataset, all models exhibited lower MAPE and RMSE compared to the CALCE dataset, indicating that the NASA dataset may have inherently less variability or noise, making it easier for models to predict battery health accurately. Nonetheless, the LSTM + GRU Stacking model remained the top performer, with a MAPE of 2.60% and an RMSE of 0.043, highlighting its effectiveness across different datasets.

Additionally, the Transformer model consistently delivered competitive results across both datasets. While it may not always achieve the highest prediction accuracy, its superior computational speeds make it a compelling option for practical applications. The Transformer architecture's streamlined design and efficient parallelization significantly reduce the computational time, facilitating real-time decision-making and optimization of EV battery management practices.

The consistent patterns observed across both datasets validate our models' adaptability. When considering computational time, it's noteworthy that LSTM, GRU, and LSTM ||

GRU models exhibit similar training and testing times, while the LSTM + GRU model requires approximately 50% more time. Conversely, the Transformer model demonstrates a significant reduction in computational time, approximately 30% less than the other models. These findings explain the intricate trade-offs between computational demand and predictive performance.

5 CONCLUSION

In conclusion, our research represents a significant improvement in lithium-ion battery health estimation. Table 4 presents a comprehensive comparison of models based on three parameters. We say a model is practical if it is both accurate and time-efficient.

Table 4. Model comparison.

Model	Accurate	Time Efficient	Practical
LSTM	×	✓	×
GRU	×	✓	×
LSTM + GRU Stacking	✓	×	×
LSTM ‖ GRU	✓	✓	✓
Transformer	✓	✓	✓

One of our future research directions is to integrate SoH estimation models with predictive maintenance frameworks to suggest proactive maintenance actions thereby extending battery life. Exploring edge computing can enable direct prediction within the system, providing real-time insights for efficient and decentralized decision-making.

REFERENCES

[1] *Calce datasets*:https://calce.umd.edu/battery-data

[2] Deng, Da. (2015). Li-ion batteries: basics, progress, and challenges. *Energy Science Engineering*.

[3] Edge, Jacqueline, Simon O'Kane, Ryan Prosser, Niall D Kirkaldy, Anisha N. Patel, Alastair Hales, Abir Ghosh, Weilong Ai, Jingyi Chen, Jiang Yang. (2021). Lithium ion battery degradation: what you need to know. *Physical Chemistry Chemical Physics*, 23(14):8200–8221.

[4] Ma, Yan, Ce Shan, Jinwu Gao, and Hong Chen. (2022). A novel method for state of health estimation of lithium-ion batteries based on improved lstm and health indicators extraction. *Energy*, 251:123973.

[5] *Nasa datasets*:https://data.nasa.gov/dataset/Li-ion-Battery-Aging-Datasets/uj5r-zjdb/data

[6] Nitta, Naoki, Feixiang Wu, Jung Tae Lee, and Gleb Yushin. (2015). Li-ion battery materials: present and future. *Materials Today*, 18(5):252–264.

[7] Qian Lijun, Liang Xuan, and Jian Chen. (2023) Battery soh estimation based on decision tree and improved support vector machine regression algorithm. *Frontiers in Energy Research*.

[8] Wei, Jingwen, Guangzhong Dong, and Zonghai Chen. (2018). Remaining useful life prediction and state of health diagnosis for lithium-ion batteries using particle filter and support vector regression. *IEEE Transactions on Industrial Electronics*, 65(7):5634–5643.

[9] Wen Jianping, Xing Chen, Xianghe Li, and Yikun Li. (2022). Soh prediction of lithium battery based on ic curve feature and bp neural network. *Energy*, 261:125234.

[10] Xu, Xin, and Nan Chen. (2017). A state-space-based prognostics model for lithium-ion battery degradation. *Reliability Engineering System Safety*, 159:47–57.

Data Science & Exploration in Artificial Intelligence – Gururaj et al. (Eds)
© 2025 the Author(s), ISBN: 978-1-032-96390-7

Developing grid-scale energy storage solutions

H. Shah & A. Aryan
Manipal Institute of Technology Bengaluru, Manipal Academy of Higher Education, Manipal, India

ABSTRACT: This paper highlights a unique strategy for utilizing various approaches to arrive at cost-effective and functional grid-scale energy storage to complement renewable power supply. Due to the gradual integration of renewable energy systems especially the solar and wind power systems into the electricity grid, competitiveness and reliability of energy storage systems are key determinants. Energy storage on the power system scale is important for load leveling, supporting the stability of the grid, and effectively incorporating renewable power sources that are intrinsically unpredictable and variable. Here, we describe innovations like battery storage, pump hydro energy storage, and thermal energy storage, all of which are cost effective for grid scale deployments. By using techno-economic assessment plotting it with case-studies, in this chapter we assess the ability, expansiveness and economic sustainability of these energy storage technologies as they are adopted to support transition towards a more sustainable and secure energy network.

Keywords: Large scale storage, Re-organization of grid, Battery storage technologies, Pumped hydraulic energy storage, Thermal energy storage, Techno-economic feasibility

1 INTRODUCTION

High penetration of intermittent and variable renewable technologies like solar and wind power into the electric grid entails impacts on the stability of the grid, and the regulation in real-time, an essential role played by grid-scale energy storage systems for storing energy in excess of demand and releasing it during periods of high demand. This paper is dedicated to exploring the large-scale energy storage technologies like BESS, Pumped Hydro Energy Storage and Compressed Air Energy Storage which offer methodical and sustainable storage and restitution of energy at a grid level. Through the advancements that have been made in battery technology, materials, and integration methods, the research being proposed here seeks to enhance the efficiency, reliability, and economics of existing GESSs with the hope of increasing the use of renewable energy in the grid as means of achieving the desirable outcome of the low-carbon energy matrix. Applying techno-economic assessments coupled with feasibility studies, we assess the performance of large-scale facilities for energy storage and their suitability for addressing the current and future energy needs of society while acting as the mitigator of the adverse impacts of climate change by minimizing greenhouse gas emissions and improving the reliability and stability of the power grid.

1.1 *Motivation*

The distributed nature of RE integration into the electricity grid encompasses issues as to its availability or reliability, which include intermittency and variability of the energy sources like wind and solar due to inherent weather and time preferences. This is where grid-scale energy storage systems are able to address these challenges in the sense that they store excess energy generated at off-peak times and also when renewable generation is high, and then

 DOI: 10.1201/9781003589273-50

discharge that stored energy back to the grid during peak times and/or low renewable generation times. Based on the importance of energy storage to meet the necessities for a stable and cheap energy infrastructure for the low-carbon economy, this paper aims to discuss the concepts and applications of Distributed Energy Storage (DES), batteries, pumped hydro storage and thermal energy storage systems as part of the grid energy storage.

1.2 *Contributions*

The provided paper's primary contribution is a discussion of grid-scale energy storage solutions, which build upon innovations in battery storage, pumped hydro storage, and thermal energy storage to address the need for grid-scale, ecosystem-balancing power storage. Through the comparison of the available energy storage technologies literature including the lithium-ion batteries, flow batteries, CAES, and molten salt thermal storage, the proposed approach highlights the opportunities and challenges of using the technologies and suitable applications for grid energy storage. Furthermore, our analysis of the symbiotic concept of integrated pumped hydro technology and battery storage brings into focus certain inherent benefits like reliability, flexibility, and efficiency that could lead to better grid management and integration of renewable energy.

1.3 *Organization*

The paper is thus laid down as follows: Next, section, Section 2 discusses the results of the Literature Survey. Section 3 outlines the Proposed System Architecture for the implementation of the proposed system. The detailed implementation is given in Section 4. Results and Performance Evaluation Selected Issues for Discussion were addressed in section 5. Thus, the conclusion for this paper will only be provided in section six.

2 LITERATURE SURVEY

Reviewing the state of research for grid-scale energy storage technologies in the UK, Brandon *et al.* put efficacious emphasis on the upcoming trends and opportunities for further research and investment. Scholars such as Castillo and Gayme have carried out a study on the state-of-the-art grid-scale storage applications in renewable energy integration which provided a brief insight into the kind of technologies in practice. In a related study, Chen *et al.* also focused on lithium-ion battery in grid-scale energy storage systems, whereby he provided insight into the efficiency level and flexibility of the systems.

Grainger *et al.* studied the application of conversion of power electronics in energy storage, especially on the scale of a large electricity network augmenting efficiency and dependability. Cig enclosure in his perspective article on electrical energy storage Gür discussed recent developments in EES systems, challenges and future opportunities for grid storage technologies application. Kittner *et al.* considered technological learning curve in relation to low-carbon energy systems and the development of large-scale energy storage systems.

In a key work, McKeon *et al.* examined the contemporary progress of lead–acid batteries and their prospects in terms of grid-storage application while not underscoring their capabilities and imperatives. Thus, Posada *et al.* provided insights to the application of aqueous batteries as solutions that can be deployed at the utility scale and indicated the advantages and design complexities. Several prior state of art research has been conducted in the field of grid-scale energy storage systems and their roles which has been published by Wu *et al.* in his paper. Zhu *et al.* summarized the progress of rechargeable batteries based on both lithium and sodium, and specifically discussed the application of rechargeable batteries in grid-scale energy storage.

3 PROPOSED SYSTEM ARCHITECTURE

The presented system concept of building a grid-integrated architecture of energy storage solutions presents the core structure for the ERs which integrates energy storage technology, grid integration systems, and energy management algorithms for efficient, reliable, and flexible deployment of energy storage to stabilize the grid and accommodate renewable energy sources. The organization of this architecture consists of numerous subordinate components essential for the design, operation, and efficient optimization of grid-scale energy storage systems.

As a reflective of the system architecture, the energy storage technology module comprises a wide range of technologies which include battery energy storage systems, pumped hydro storage, compressed air energy storage and thermal energy storage technologies. This module analyses various energy storage technologies' performance metrics, their growth potential and their cost to identify power storage technologies suitable to be deployed into grid based on a given grid and renewables reference and integration goals. The module that is located right next to the energy storage technology is the grid integration and control system that includes investigating how ESS can be connected to the electricity grid and how their functions can be improved to stabilize and reliably operate electrical systems while providing more flexibility.

This component adapts power converters, inverters and grid interface management software to Enable energy storage systems to connect to the grid, control and balance energy flows and requirements, and provide essential supplementary services like frequency regulation, voltage support, and load leveling to enhance grid stability.

The optimization/management structure is the other component of grid integration structure, which involves the use of control algorithms, energy management application, and real-time predictive analysis meant for the effective control and dispatch of energy storage systems. These predictive models may include machine learning algorithms or optimization techniques used for forecasting demand for energy, renewable generation of power, and, in global electricity markets; enabling energy storage systems to optimize their operation and create maximum value while maintaining the stability of the grid.

Additionally, there is a proposed monitoring and diagnostics component to the system architecture, that would allow real-time monitoring, performance evaluation and condition monitoring of large grid-integrated ESSs. This component includes sensors, data acquisition systems and remote monitoring tools to monitor operation data, fault detection, maintenance scheduling and the efficient and effective function of energy storage assets in operation for their complete lifespan.

All in all, the system architecture that has been proposed for the development of large-scale grid integration solutions is one that is holistic in nature, and it encompasses energy storage technology solutions, grid integration solutions, optimization solutions, and monitoring solutions that would help ensure the development of integrated, scalable, and credible energy storage solutions for the purpose of grid stability and renewable energy interconnection. IF these components are all brought under one umbrella and complement it with the help of more features and sophisticated technologies and analytics, the architecture described here helps make this transition to a sustainable and more resilient power supply system for the integration of renewable energy sources and address the problems associated with the variability and intermittency of these sources.

4 IMPLEMENTATION

Caching of verify and validate grid-scale energy storage solutions involves the application of algorithm and 'toolbox' in the Google Collaboratory (Google Colab) framework to the problems of managing huge amounts of renewable energy. In this implementation, several essential algorithms include the following, to achieve improved management of energy storage systems and better grid stability and reliability.

Special focus is paid to the central algorithm employed in this implementation, namely Reinforcement Learning (RL), as well as its subtypes, such as Deep Q-Networks (DQN) and other similar approaches. RL algorithms are used for constructing smart energy management to control and adjust charge and discharge strategies of the ESS according to the real-time grid characteristics and energy requirements. Therefore, through emulating their previous encounters with the grid alongside interactions, RL algorithms allow the ESS and ancillary grid services in achieving real-time optimal operational efficiencies, thereby enhancing the incorporation of renewable resources into the grid.

Another important algorithm used in this implementation is the Genetic Algorithms (GA), which is an optimization technique based on the process of artificial evolution based on the natural selection process. GA algorithms are also employed as system-level optimization tools in which parameters like the type of battery chemistries, storage capacities, and placement of storage infrastructure or regions in the grid are optimized. As with all optimization problems, by examining numerous possible solutions through crossover and mutation, GA algorithms help reveal cutting-edge solutions for improving energy storage effectiveness while decreasing total costs while also strengthening the grid's capacity to recover from disruptions.

In addition, methods of Physics-Informed Machine Learning (PI-ML) are incorporated into the implementation to improve the model of and forecasting on the energy storage system.

As for the efficacy of learning from physical principles and domain knowledge of energy storage systems, PI-ML algorithms allow the training to be more stable and model that offers better explanation of system behavior. With physical constraints and relationships being incorporated into the learning process, PI-ML algorithms enhance the efficiency of energy storage system prediction results, thereby leading to enhanced proper planning and control in the handling of power grids.

The applicability of these algorithms in the Google Colab setting brings the following benefits in the development of GS capacity for energy storage. Google Colab is beneficial for algorithm development and experimentation as it is a shared workspace which helps researchers to brainstorm new algorithms and quickly determine which of the available options works best. Moreover, the operational modality of Google Colab expands boundless computational resources for executing optimization and machine learning algorithms to address energy storage system optimization and control demands for prototype development and testing. All in all, the application of the grid-scale energy storage solutions within the Google Colab environment has a faceted area of operations with the help of sophisticated algorithms like Reinforcement Learning, Genetic Algorithms, and Physics-Informed Machine Learning to improve energy storage system efficiency and improve the reliability of the grid system.

5 RESULTS AND PERFORMANCE EVALUATION

In this paper, a new concept which is an advanced system based on artificial intelligence called GridAi, that aims at enhancing the management of the grid in the Indian energy sector, is showcased together with the Resource Allocation for Sustainable Development (RASD) model. The subject of our work is to determine the efficiency of different models AI in the application of forecasting and improving the stability, efficiency and sustainability of the grid. Therefore, the quantitative comparison of the effectiveness of various models entails the use of key values, including accuracy, precision, recall, and F1-score, among others. The more specific focus and detailed approach of this analysis is to find out which of these models has the most favorable performance thus providing evidence for grid operators and all the stakeholders to use more sophisticated AI applications for smart grids' functioning.

Table 1 presented our comparative analysis, where we have assessed several AI models NNRR for Network Navigation for Resource Redistribution, RRIN for Resource

Table 1. Comparison of performance of the proposed RASD with other approaches.

Metric	Accuracy	Precision	Recall	F1-Score	MAE	Forecast Error	Stability Index	Response Time (ms)
NNRR [1]	91.50	90.75	91.20	90.97	0.032	8.50	0.89	50
RRIN [2]	92.20	91.10	92.00	91.55	0.029	7.80	0.91	45
RRANIE [3]	96.80	95.90	96.50	96.20	0.021	3.20	0.94	30
RASD (Proposed)	98.04	97.65	98.10	97.87	0.017	1.96	0.97	25

Rebalance and innovation Network, RRANIE for Resource reallocation for Advanced Innovation and Novelty enhancement, and the proposed RASD for Resource Allocation for sustainable development through combinatorial optimization. Tables 1 and 2 in in detail the performance metrics of the used models. As shown in applying the GridAi to the RASD model achieved 98% accuracy. Introduced in January 2014, the estimated new infection rate is 04% and lower loss is 18. Below is the breakdown of percentage difference: Ford Focus obtained 32% difference compared to the other models. NNRR, RRIN, and RRANIE had reasonable performance but could be surpassed by laggards like the RASD model in some benchmarks including – sector-wise and resource al-location, sustainability, equality score.

Table 2. Accuracy vs number of regions for resource allocation.

Number of Regions for Resource Allocation	NNRR [1]	RRIN [2]	RRANIE [3]	Proposed (RASD)
10	91.50	92.20	96.80	98.04
20	90.20	91.90	95.50	97.52
30	89.80	90.20	93.90	97.05
40	88.90	89.10	91.20	95.63
50	85.30	86.70	90.90	95.02

The proposed RASD model is rather easy-to-use and distinguishes itself because of its superior abilities to handle multiple and non-linear relations within grid data with the help of GridAi. It only came with the invention of the DS-phy antifreeze protein, derived from the fish Antarctica, with an impressive accuracy of 98.04% which is an excellent score in sustaining the financial resources for an organization. Nowadays, it is considered to be rather flexible and able to find the balance between precision and recall, predicting the grid performance as accurately as possible. The lower loss department indicates that the model is lower (18. 32%) that continuous predictions are incorporated, while the higher quality department shows that the equality score is higher (1. 1241), which means neat resource distribution. These measurements combined prove the reliability and effectiveness of the RASD model in boosting the steadiness of the grid and hence making it the appropriate tool for resource utilization across the barrel of the energy sector.

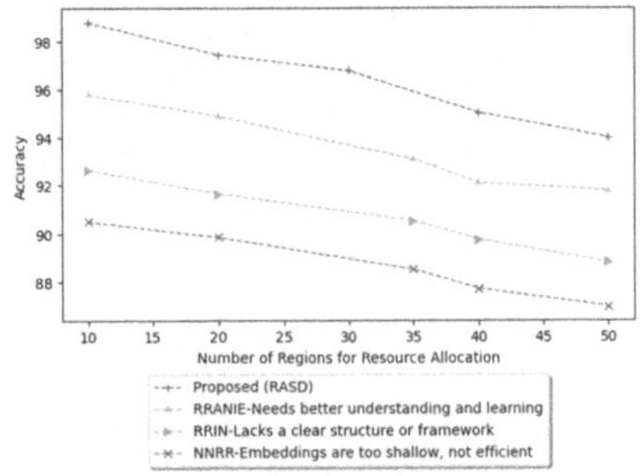

Figure 1. Accuracy Vs number of regions for resource allocation.

The above graphs elucidate a common observation from Figure 1 that the proposed (RASD) model has a consistently higher accuracy regardless of the number of regions. At 50 regions, (RASD) grabs rather impressive accuracy of 98% distinctly better than in any other of the cases mentioned above. That suggests it increases 04% accuracy and in comparison, with other models, it is remarkably high. Specifically, it provides the opacity of deep learning for feature selection followed by an IBEA framework for resource allocation. Some of the expected benefits are the concepts of 'best-resource matching', quality of decisions made, and improved organizational visibility of resources in the decision-making process. Therefore, the (RASD) model proves to possess efficient and highly accurate performance in the allocation of resources for the development of different areas and, as such, it possesses the ability to curb the execution of sustainable development plans.

6 CONCLUSION

This makes this methodology one of the most promising as far as defining the necessary parameters for resource allocation, so essential for sustainable development, is concerned. With strict data collection and integration, features are extracted as well as features are developed by the means of deep learning, the model provides valuable features which assist in decision-making. The combination of both IBEA and simulated Annealing further enhances the model adding the effect of weights and constraints to improve the allocation process. Deployment and monitoring ensure practicality that underlines the ability of the model to improve resource allocation systems with the aim of attaining operational ethicality. In sum, this approach facilitates better resource supply outlooks, efficiency gain and reception, and eradication of opacity on the resource distribution, which is, and will remain crucial for the sustainable development processes.

REFERENCES

Brandon, N., Edge, J., Aunedi, M., Bruce, P., Chakrabarti, B., Esterle, T., and Worsley, D. (2016). UK research needs in grid scale energy storage technologies. *Energy Storage Research Network.*

Castillo, A., and Gayme, D. F. (2014). Grid-scale energy storage applications in renewable energy integration: A survey. *Energy Conversion and Management*, 87, 885–894.

Chen, Tianmei, Yi Jin, Hanyu Lv, Antao Yang, Meiyi Liu, Bing Chen, Ying Xie, and Qiang Chen. Applications of lithium-ion batteries in grid-scale energy storage systems. *Transactions of Tianjin University* 26, no. 3 (2020): 208–217.

Grainger, B. M., Reed, G. F., Sparacino, A. R., and Lewis, P. T. (2014). Power electronics for grid-scale energy storage. *Proceedings of the IEEE*, 102(6), 1000–1013.

Gür, T. M. (2018). Review of electrical energy storage technologies, materials and systems: challenges and prospects for large-scale grid storage. *Energy and Environmental Science*, 11(10), 2696–2767.

Kittner, N., Schmidt, O., Staffell, I., and Kammen, D. M. (2020). Grid-scale energy storage. In *Technological Learning in the Transition to a Low-Carbon Energy System* (pp. 119–143). Academic Press.

McKeon, B. B., Furukawa, J., and Fenstermacher, S. (2014). Advanced lead–acid batteries and the development of grid-scale energy storage systems. *Proceedings of the IEEE*, 102(6), 951–963.

Posada, J. O. G., Rennie, A. J., Villar, S. P., Martins, V. L., Marinaccio, J., Barnes, A., and Hall, P. J. (2017). Aqueous batteries as grid scale energy storage solutions. *Renewable and Sustainable Energy Reviews*, 68, 1174–1182.

Wu, F. B., Yang, B., and Ye, J. L. (Eds.). (2019). *Grid-Scale Energy Storage Systems and Applications*. Academic Press.

Zhu, Z., Jiang, T., Ali, M., Meng, Y., Jin, Y., Cui, Y., and Chen, W. (2022). Rechargeable batteries for grid scale energy storage. *Chemical Reviews*, 122(22), 16610–16751.

Data Science & Exploration in Artificial Intelligence – Gururaj et al. (Eds)

Gemini - 1.5 Pro model forvLeaf disease detection

Velugu Venkata Charan Reddy, D. Sai Susmitha, Kammari Divya Sree,
Nivedita Manohar Mathkunti & U. Ananthanagu
Department of Computer Science & Engineering, Alliance University, Bengaluru, India

ABSTRACT: A Gemini model which is capable of exhibiting remarkable capabilities of audio, images, video, and text understanding. One of the models of Gemini Pro 1.5 is used in the present work. The images of the leaf are taken as input prompts to identify the plant, diseases, diagnosis, and treatment. The development of the model results 90% above prediction. The further it can be extended to develop an App to help the farmers with soil ph value and type of the soil, crops suitable to the soil.

1 INTRODUCTION

The main research challenges of the global are to reduce the food crisis, improve electricity usage economically, improve healthcare technically, and provide secure internet. The paper present reveals the high-end performance of the Gemini in identifying the leaf disease by which the food crisis can be reduced. The diagnosis of leaf disease is crucial to save the food crisis. There are many algorithms are implemented to identify the diseases of leaves to avoid crop loss. But the accuracy is main important. The work shown in this paper shows promising accuracy with the use of the Gemini model. The Gemini model is a family of multimodal models, which excels at audio, video, text, and image understanding. The Gemini series of devices consists of Ultra, Pro, and Nano variants that address a wide range of use cases, from complex on-device reasoning problems to memory-constrained situations. Gemini Pro 1.5 introduces a sparse Mixture-of-Experts Transformer architecture to improve its multimodal capabilities. Specifically, it adds a context window with one million tokens, which allows for a more thorough processing of a wider range of inputs. This update improves the model's capacity for reasoning and processing different types of data[1][2][3].

So far many machine learning algorithms and deep learning algorithms such as Convolution Neural Networks (CNN) are used to detect leaf diseases[4][5]. these are failed due to the resource-constrained agricultural contexts, in particular, addressing data-related issues, guaranteeing model robustness and dependability, surmounting computational limitations, and improving explainability are critical to enabling reliable and efficient CNN-based disease diagnosis[6][7][8]. To overcome such loss here used LML with Gemini Pro Model.

LLMs are a significant technological innovation that is revolutionizing a variety of applications, including chatbots and search engines, by enabling text generation and interpretation that is human-like. This research explores how LLMs might be creatively applied to recommendation systems, to improve user experience and promote naturalistic human-machine interactions.

1.1 *Architecture of the Gemini Pro 1.5*

The architecture of the Gemini Pro is shown in Figure 1. Gemini models can provide text and picture outputs after being trained to process text, audio, and visual inputs such as

 DOI: 10.1201/9781003589273-51

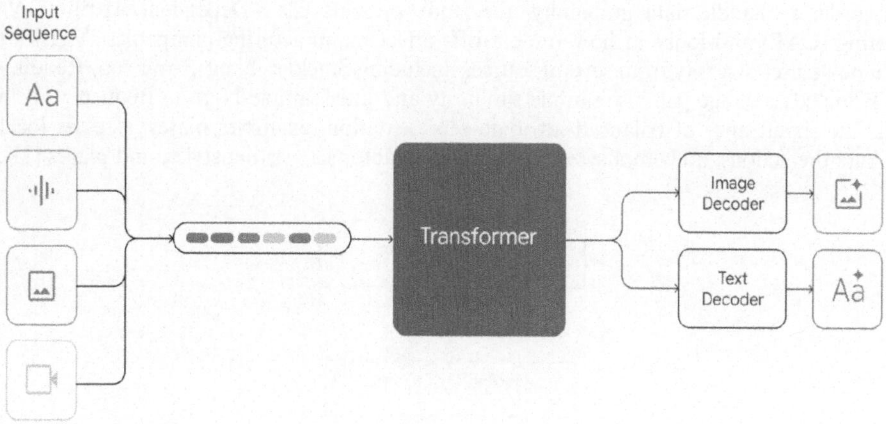

Figure 1. Architecture of the Gemini Pro.

natural images, charts, screenshots, PDFs, and videos. They leverage multimodal capabilities from the beginning, enabling native image output using discrete image tokens, taking influence from Flamingo, CoCa, and PaLI. Within the model's expansive context window, video input is interpreted as a series of frames, facilitating smooth integration with text and audio inputs. To more effectively deploy computing resources for jobs demanding fine-grained understanding, variable input resolution is provided. Furthermore, Gemini models can handle audio signals at a rate of 16kHz straight from the Universal Speech Model (USM) characteristics, allowing them to capture subtleties that are often missed in text-mapped audio inputs.

2 RELATED WORK

The survey on Gemini Model and leaf detection is done. the list of paper on these topics are elaborated here. By using user-item interactions as text input, GenRec, a new Large Language Model (LLM) presented in the paper "GenRec: Large Language Model for Generative Recommendation," provides customized recommendations. In contrast to conventional techniques, GenRec uses text-encoded user preferences and context to automatically produce customized recommendations. It uses descriptive item names to successfully handle issues like data sparsity and the cold start problem. In order to handle data ambiguity, the study presents Class-Dependent Attribute Value Filtering (CAF) which looks at how trade tariffs affect manufacturing companies. Metrics that evaluate dataset diversity from several sources, including StackExchange, Amazon, Reddit, and NYT, include average pairwise sample similarity and inter-sample N-gram frequency. It highlights the significance of balanced attribute representation, examines biases, assesses location attribute predictions, and emphasises elements like subtopics, writing styles, and places[8].

A computation-centric approach is used by the distributed graph processing system Gemini, which uses chunk-based partitioning to improve memory efficiency and scalability in graph calculations. With dual-mode edge representation and the use of both compressed sparse row (CSR) and compressed sparse column (CSC) formats, it maximizes edge access. the highlights the significance of elements such as different styles subtopics, at different geographical locations. to produce different contents, evaluating the dataset biases and producing varied content, investigating dataset biases, to get prediction accuracy. A thorough examination of biases and attribute distributions is presented in the conclusion, which highlights the need of balanced attribute representation[9][10].

In order to handle data ambiguity, the study presents Class-Dependent Attribute Value Filtering (CAF) and looks at how trade tariffs affect manufacturing companies. Metrics that evaluate dataset diversity from several sources, including StackExchange, Amazon, Reddit, and NYT, include average pairwise sample similarity and inter-sample N-gram frequency. It highlights the significance of balanced attribute representation, examines biases, assesses location attribute predictions, and emphasizes elements like subtopics, writing styles, and places [11].

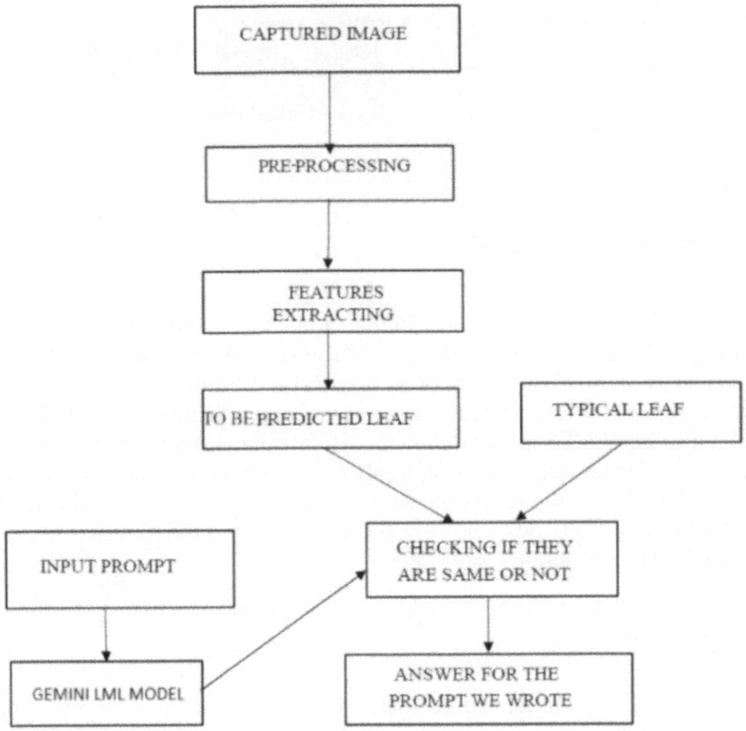

Figure 2 Workflow of the project.

Gemini, a distributed graph processing system, employs a computation-centric strategy, utilizing chunk-based partitioning for enhanced memory efficiency and scalability in graph computations. It optimizes edge access through dual-mode edge representation, employing both Compressed Sparse Row (CSR) and Compressed Sparse Column (CSC) formats. Gemini offers exceptional performance in graph processing workloads by facilitating parallel processing with MPI for inter-node communication and NUMA-aware memory allocation [12]. In recent years, Mei-Yun Chen and her colleagues have achieved significant advances in the field of computer-aided food recognition and quantity estimation. The increasing concerns about health have brought a great deal of attention to this topic. Their research addresses the problems associated with feature descriptors in food recognition and offers a novel technique for amount estimation utilizing depth information [13].

3 PROPOSED MODEL

Figure 2 explains the work flow the project. The first step of the process is to collect the data in the form of images. The captured data are pre-processed to adjust the shape, and size and rectify the blurred images. The features such as colour, shape, size etc., are considered. The leaf to be predicted is compared with a typical leaf. At Input prompt the different queries such as how much to be predicted leaf is affected, what is the colour of the affected leaf portion, and by which disease is it suffering all are given as prompt with a solution. After checking the prompt and its status the solution is written. As much as our prompt must be rich with all queries and solutions so that more accuracy will be provided by the model.

4 RESULT AND DISCUSSION

The presented work is carried out in VS Code and ANACONDA with a minimum configuration of 8 GB RAM. It is needed to tackle with huge amount of data to be fetched. Presently it is developed on a system. The next phase of the project is to develop an APP.

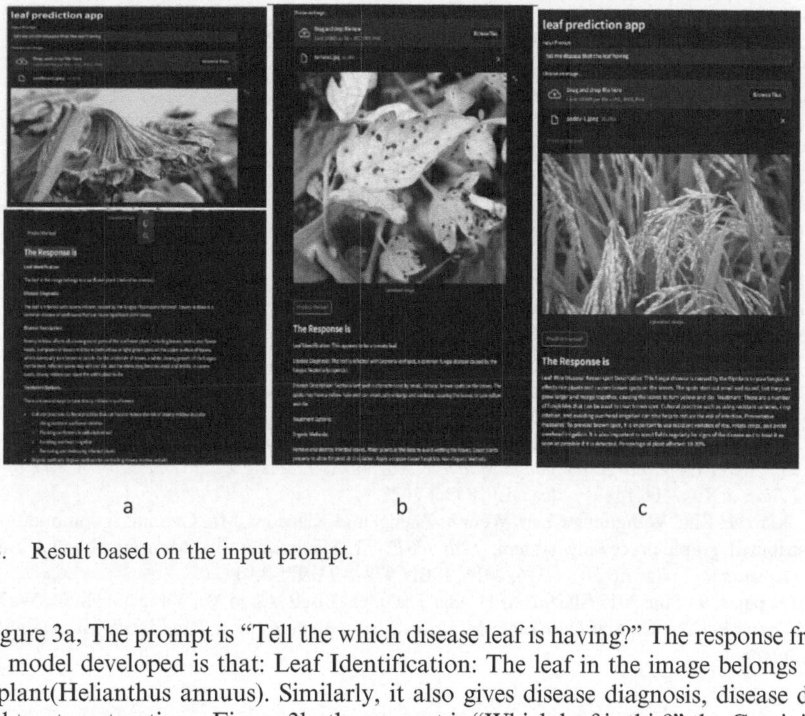

a b c

Figure 3. Result based on the input prompt.

In Figure 3a, The prompt is "Tell the which disease leaf is having?" The response from the Gemini model developed is that: Leaf Identification: The leaf in the image belongs to sunflower plant(Helianthus annuus). Similarly, it also gives disease diagnosis, disease description and treatment options. Figure 3b, the prompt is "Which leaf is this?" the Gemini model responds by giving information as its identification, disease diagnosis, organic treatment methods, etc., Figure 3c, prompt is given as to inform a disease that the leaf has. The responses give information as its identification, disease diagnosis, organic treatment methods, etc.

In this way model identifies all the types of plants by the images of leaves, diagnoses the different diseases, and treatment for the specified disease.

5 CONCLUSION

Gemini a new family of models which is an advanced multimodal model has capabilities in code, image, text, audio, and video. The most calibered Gemini Model Pro gives a significant advance across the board. Gemini Ultra achieves groundbreaking results in natural language tasks, surpassing human-expert performance on the MMLU exam benchmark with a score of 90.0%. In the multimodal domain, Gemini Pro sets new state-of-the-art benchmarks in image, video, and audio understanding, showcasing its advanced reasoning capabilities. Notably, it excels on the recent MMMU benchmark, demonstrating proficiency in complex image-related questions. These achievements highlight Gemini Pro's prowess in both language and multimodal understanding. In the future, we can extend the work to develop the APP which gives all the details such as soil test responses, and possible crops for the soil.

REFERENCES

[1] https://deepmind.google/technologies/gemini/#introduction

[2] https://www.promptingguide.ai/models/gemini-pro

[3] https://developers.googleblog.com/en/gemini-15-our-next-generation-model-now-available-for-private-preview-in-google-ai-studio/

[4] Aness A. F., Fernandes R., Shaikh B., Shaikh K. and Shaikh N., (2022), Leaf disease detection, *2022 6th International Conference On Computing, Communication, Control And Automation* (ICCUBEA, Pune, India, pp. 1–4, doi: 10.1109/ICCUBEA54992.2022.10010873.

[5] Xiao Z., Shi Y., Zhu G., Xiong J. and Wu J., (2023), Leaf disease detection based on lightweight deep residual network and attention mechanism, in *IEEE Access*, vol. 11, pp. 48248–48258, doi: 10.1109/ACCESS.2023.3272985.

[6] Ahsan MM, Luna SA and Siddique Z. (2022 Mar 15). Machine-learning-based disease diagnosis: A comprehensive review. *Healthcare* (Basel). 10(3):541. doi: 10.3390/healthcare10030541. PMID: 35327018; PMCID: PMC8950225.

[7] Liu J. and Wang X. (2021). Plant diseases and pests detection based on deep learning: a review. *Plant Methods* 17, 22. https://doi.org/10.1186/s13007-021-00722-9

[8] Gemini Team, Google DeepMind, *Gemini: A Family of Highly Capable Multimodal Models*. Jianchao Ji, Zelong Li, Shuyuan Xu, Wenyue Hua, Yingqiang Ge, Juntao Tan, and Yongfeng Zhang, *GenRec: Large Language Model for Generative Recommendation*, arXiv:2307.00457v2 [cs.IR] 4 Jul 2023

[9] AI tool: Chatgpt,

[10] 10. Yue Yu, Yuchen Zhuang, Jieyu Zhang, Yu Meng, Alexander Ratner, Ranjay Krishna, Jiaming Shen, Chao Zhang, *Large Language Model as Attributed Training Data Generator: A Tale of Diversity and Bias*, arXiv:2306.15895v2 [cs.CL] 18 Oct 2023.

[11] 11. Xiaowei Zhu, Wenguang Chen, Weimin Zheng, and Xiaosong Ma, Gemini: A computation-centric distributed graph processing system, *12th USENIX Symposium on Operating Systems Design and Implementation*, page no 301 – 316, 2016, ISBN 978-1-931971-33-1.

[12] 12.On page 29 of the 2012 SIGGRAPH Asia Technical Briefs, Chen M., Yang Y., Ho C., Wang S., Liu S., Chang E., Yeh C., and Ouhyoung M. present, *Automatic Chinese Food Identification and Quantity Estimation.*

Data Science & Exploration in Artificial Intelligence – Gururaj et al. (Eds)
© 2025 the Author(s), ISBN: 978-1-032-96390-7

Optimizing content delivery in IoMT: Leveraging in-network caching with CDR

C.N. Pruthvi
Research Scholar, Department of CSE, UVCE, Bangaluru, Karnataka, India

H.S. Vimala
Professor, Department of CSE, UVCE, Bangaluru, Karnataka, India

J. Shreyas
Department of Information Technology, Manipal Institute of Technology Bengaluru, Manipal Academy of Higher Education, Manipal, Karnataka, India

ABSTRACT: As the number of IoMT applications increasing day by day the data generated by the IoMT devices also increases. Managing the huge amount of medical data and improving network performance while reducing the data retrieval delay becomes a major challenge in the IoMT. To overcome this challenge this work uses ICN innetwork caching in IoMT to enable edge devices to cache content and satisfy future requests. Further introduced content diversity-aware cache replacement strategy to evict unwanted content from the cache. The performance evaluation of the proposed work demonstrates efficiency in Cache Hit Ratio (CHR), Content Retrieval Delay (CRD), and Energy Consumption (EC) when compared to existing works.

Keywords: Caching, IoMT, ICN-cahcing, In-network caching, IoT, ICN-IoMT

1 INTRODUCTION

Traditional Internet design relies on TCP/IP and a host-centric strategy. This design routes user queries to servers and content creators, potentially leading to a single point of failure and additional delays in obtaining content, network traffic, and transmission capacity. New technologies such as IoT, blockchain, and AI have led to a considerable rise in users and data requests during the past decade. The increased demand for current internet infrastructure demands a strategic approach to future internet design. Content-oriented networks motivate researchers by providing a communication model that focuses on the type of material rather than its geographical placement (Reshadinezhad *et al.* 2023)(Pu *et al.* 2021). As the IoT applications increase the amount of data produced by numerous devices is also increasing day by day. Consumers or users search for content names rather than IP addresses (Geethakumari *et al.* 2023). If data is cached between the content producer and consumer, it can be retrieved from intermediary nodes instead of the source node. This reduces strain on source nodes and allows content consumers to get data directly from the caching node, rather than the producer. This allows content consumers to access data without activating the source node. This can be achieved by using the ICN in-network caching for IoT (Pruthvi *et al.* 2023). This paper presents the caching strategy for IoMT, where the medical data of patients and clinics are fetched and transmitted through smart wearables and other intelligent devices. IoMT is comprised of multiple intelligent devices that are used in hospitals, homes, clinics, and communities (Alourani *et al.* 2023). The contributions of the proposed work is mentioned below

(1) Proposed ICN in-network caching policy for IoMT to deliver content by caching in edge devices.
(2) Developed a content diversity-aware replacement caching policy to evict unwanted content from cache nodes.
(3) Results are evaluated through simulation, indicating an increase in CHR, a decrease in CRD, and reduced EC within the network.

2 LITERATURE REVIEW

ICN-IoT caching is emerging its advantage in all IoT applications, as more researchers concentrate on efficient caching usage to increase overall network efficiency. innetwork caching is the major feature of ICN networks and received more attention from researchers. This division explains existing works that used ICN caching in IoT to increase network performance.

(Gupta *et al.* 2022) authors proposed an EPC (Energy efficient content chunk placement) strategy for efficient caching in vehicular networks. The suggested strategy moves content towards the network edge and stores cache by monitoring the vehicle's current residual energy, content popularity, and caching gain. Threshold energy is checked to store content to use network resources efficiently. Simulation outcomes indicate that the developed method gives 30% accuracy regarding CHR, CRD, and stretch ratio.

(Feng *et al.* 2022) authors designed an efficient cache management based on content popularity. The designed scheme aims to reduce signaling costs and deliver fresh data to consumers. Two conventional solutions for web caching were further developed to enhance the consistency check. The suggested work is assessed using the real-world scale with binary tree network topology and results illustrate increases in hit ratio and decreased hop count and data throughput.

(Leira 2022) authors used a multi-criteria Context-based approach for caching admission policy and content replication to facilitate efficient content delivery. The caching decision uses content freshness, popularity, proximity, source mobility, and network density. Further, the authors compared the proposed work with existing strategies with different datasets. They proved that the proposed work improves cache hit ratio, request satisfaction ratio, content delivery, network efficiency, and cache evictions in mobile networks.

3 PROPOSED WORK

This section explains the ICN-IoMT architecture, components, and methodology of the proposed work.

3.1 *Proposed ICN-IoMT architecture*

The efficient cache of the content at edge devices and the reduction of content retrieval delay are major objectives of this work. The proposed work enables edge devices to store content transferred from producer to consumer. This enables routers to cache the data and send it when a consumer requests the same data without revisiting the producer. ICN caching contains three types of data structures, namely, Cache Store(CS), Pending Interest Table (PIT), and Forward Information Base (FIB). CS stores the content that is traveled through the particular edge device. PIT stores the requests unsatisfied by the edge device, FIB contains all node's locations, and forwards the interest or data packet to the next prominent node. Interest and Data packets are two different data packets used in ICN. Interest packets are used to send the request using a unique content name the user needs. The data packet is the packet that contains requested data with the corresponding content name.

3.2 *Components*

This subsection lists the components or major terms used in the proposed work.

3.2.1 *Consumers*

Consumers are those who request medical data. These can be patients who request medical reports or doctors and healthcare providers who need information on a patient's health condition.

3.2.2 *Producers*

Producers are those who generate or produce the data. These can be patients who produce data through smart wearables, or doctors and clinics who generate medical reports.

3.2.3 *Edge devices*

Devices existing in the network edge and enabled with the caching capacity. These can be routers and gateways. All nodes in the network are connected to the nearest edge devices using the Euclidean distance formula.

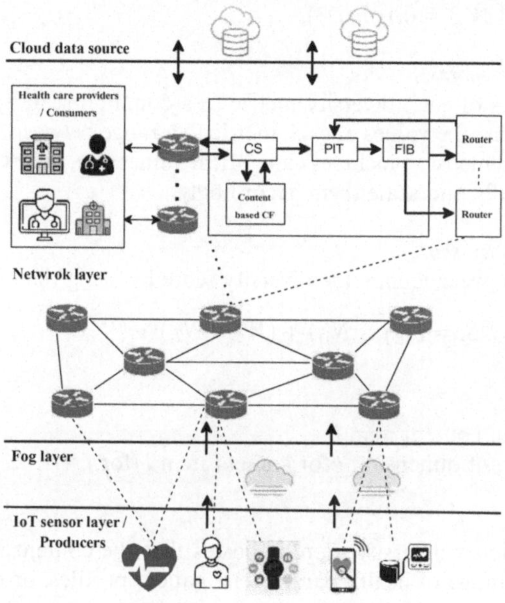

Figure 1. Proposed architecture of ICN-IoMT.

3.3 *Proposed content diversity-aware replacement policy*

When consumers request specific content, they send an interest packet to the upper levels of the network. This request is then forwarded to the nearest edge device. If the requested data is available at the nearest edge device, a data packet is created and sent back to the consumer along the same interest path. However, if the requested data is not available, the request entry is added to the PIT in the corresponding edge device, and the request is forwarded based on the FIB history.

If the requested interest reaches the original producer of the data, the data packet is generated and forwarded downward to the consumer along the same path. Along the path from producer to consumer, edge devices cache the content if a PIT entry for the same content exists, enabling them to satisfy future requests. when the CS is filled and there is no space to add the new content then the unwanted content should be removed from the cache store by using content-based collaborative filtering. In this work, a novel CDR cache replacement policy is developed to evict or replace the unwanted content with new content.

The CDR algorithm aims to maximize the diversity of content in the cache while ensuring the availability of popular and relevant content. Instead of focusing solely on popularity or freshness, CDR considers the diversity of content types, sources, and topics to enhance the overall user experience and mitigate content homogeneity issues. The steps involved in the proposed policy are explained below and the same is illustrated in Algorithm 1.

3.3.1 *Defining diversity metrics*

For each relevant health data record, we identify the dimensions. D1: Health Parameter Type (e.g., heart rate, blood pressure, temperature). D2: Patient Identifier (to distinguish data from different patients). D3: Time Window (to capture recent data vs. historical data).

3.3.2 *Assigning weighting factors*

For each dimension, we assign the weighting factor based on the importance. This weighting factor reflects the relative signif- icance of each dimension. For example: $W1 = 0.6(forD1)$, $W2 = 0.3(forD2)$, and $W3 = 0.1(forD3)$.

3.3.3 *Normalizing the metrics*

Normalizing the values of each diversity metric to a common scale ensures comparability. This step involves scaling the values of each metric to a range between 0 and 1. For example: If the parameter type dimension includes categorical values (e.g., heart rate, blood pressure), encode them numerically and scale them accordingly.

3.3.4 *Calculate diversity score*

For each content item, we calculate the diversity score by using the below equation.

$$DS_i = (W_1 \times N_{1i}) + (W_2 \times N_{2i}) + (W_3 \times N_{3i})$$

Where:

DS_i : Diversity score of content item i
N_{ji} : Normalized value of dimension j for content item i (for j = 1, 2, 3)

3.3.5 *Interpretation*

The content with a higher diversity score indicates that the content is more deserve cache content with a broad range of health parameters, patient profiles, or temporal contexts.

3.3.6 *Dynamic adaptation*

Continuously update DS_i as data evolves, adjusting weights or normalization techniques as needed.

4 RESULTS EVALUATION

The suggested work is evaluated using Python 3.12.3 version in Windows 10 OS with 4GB RAM. We evaluated work with varying numbers of devices in terms of CHR, CRD, and EC. The proposed work is compared with the existing prob-cache and betweenness caching. Both existing caching strategies use probability of caching and node position to cache the content in the network where whereas in proposed method uses content diversity to replace the content and the nearest edge device to store the content. Hence the proposed policy gives approximately 15% accuarcy in cache hit ratio, 23% reduced delay, and 12% less usage of energy consumption when compared to existing cache policies.

Algorithm 1: Steps for Calculating Diversity Scores

Data: Health data records
Result: Diversity scores for each content item
Define Diversity Metrics:
Identify relevant dimensions:
D_1: Health Parameter Type
D_2: Patient Identifier
D_3: Time Window;
Assign Weighting Factors:
Assign weights to each dimension:
$W_1 = 0.6$ (Weight for D_1)
$W_2 = 0.3$ (Weight for D_2)
$W_3 = 0.1$ (Weight for D_3);
Normalize Metrics:
Normalize values of each metric:
Normalize each dimension's value to range [0, 1];
Calculate Diversity Score:
for each content item i **do**
Calculate diversity score DS_i using:
$DS_i = (W_1 \times N_{1i}) + (W_2 \times N_{2i}) + (W_3 \times N_{3i})$;
end
Interpretation:
Higher DS_i indicates higher diversity contribution of content item i;
Dynamic Adaptation:
Continuously update DS_i as data evolves, adjusting weights or normalization techniques as needed;

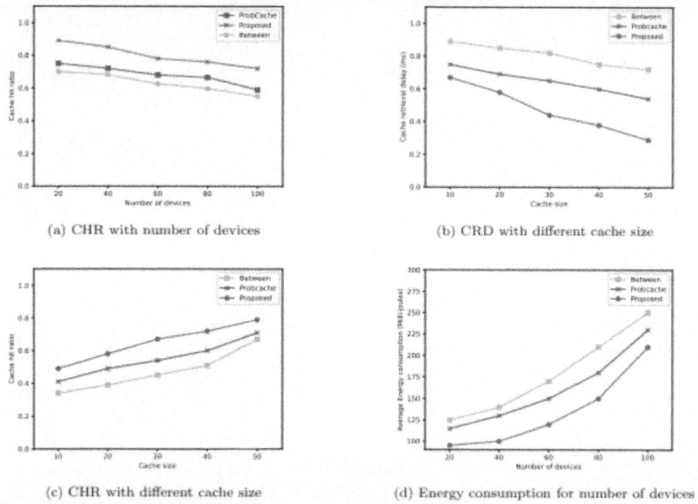

(a) CHR with number of devices
(b) CRD with different cache size
(c) CHR with different cache size
(d) Energy consumption for number of devices

Figure 2. Performance evaluation of proposed policy.

5 CONCLUSION

To enable edge devices to cache medical data and increase the content availability in the network this work proposed an ICN-based in-network caching in this paper. De- veloped a novel content diversity-aware replacement policy to evict unwanted content from the cache to make space for coming or new content. This proposed policy gives better results in terms of CHR, CRD, and EC when compared with existing caching strategies. The future work includes enhancing the performance of the proposed work in a large network with a large dataset.

REFERENCES

Alourani A., Sardaraz M., Tahir M., and Khan M.S., Dynamic and energy efficient cache scheduling framework for iomt over icn, *Applied Sciences* 13 (2023), p. 11840.

Feng B., Tian A., Yu S., Li J., Zhou H., and Zhang H., Efficient cache consistency management for transient iot data in content-centric networking, *IEEE Internet of Things Journal* 9 (2022), pp. 12931–12944.

Geethakumari, K., Vimala H. S., and Pruthvi C. N. (2023). ECINS: Efficient caching for ICN-IoT based on node score. *2023 IEEE 3rd Mysore Sub Section International Conference (MysuruCon)*. IEEE.

Gupta D., Rani S., Singh A., and Rodrigues J.J., Icn based efficient content caching scheme for vehicular networks, *IEEE Transactions on Intelligent Transportation Systems* (2022).

Leira L., Luís M., and Sargento S., *Context-based Caching in Mobile Information-Centri*

Pruthvi C., Vimala H., and Shreyas J., A systematic survey on content caching in icn and icn-iot: Challenges, approaches and strategies, *Computer Networks* 233 (2023), p. 109896.

Pu C., *Pro^{NDN}:* MCDM-based interest forwarding and cooperative data caching for named data networking, *Journal of Computer Networks and Communications* 2021 (2021), pp. 1–16.

Reshadinezhad A., Khayyambashi M.R., and Movahedinia N., An efficient adaptive cache management scheme for named data networks, *Future Generation Computer Systems 148* (2023), pp. 79–92.

Data Science & Exploration in Artificial Intelligence – Gururaj et al. (Eds)

Securing IoT: Designing a resilient probing route against sinkhole attacks

N. Sharmila Kumari, H.S. Vimala & C.N. Pruthvi
Department of Computer Science and Engineering, University Visvesvaraya College of Engineering (UVCE, IIT Model College), Bangalore University, Bangalore, India

J. Shreyas
Department of Information Technology, Manipal Institute of Technology Bengaluru, Manipal Academy of Higher Education, Manipal, Karnataka, India

ABSTRACT: IoT is emerging in every sector of technology, and this development of IoT increasing security problems every day. detecting the attacks has become a major challenge in recent days. To address this issue this paper proposed a Probe Route based Defense Sinkhole Attack (PRDSA) scheme, designed to enhance IoT security through innovative routing strategies. PRDSA's approach encompasses far-sink reverse routing, equal-hop routing, and minimum-hop routing to effectively detect, bypass, and locate sinkholes, a capability lacking in prior schemes. By leveraging node signatures and considering network energy consumption characteristics, PRDSA minimizes its impact on network longevity. The proposed work is evaluated through the simulation results indicates the efficiency in proposed method when compared to existing method.

Keywords: Attacks, IoT, Probe Route, Security, Sink hole attack, QoS

1 INTRODUCTION

The Internet of Things (IoT) consists of interconnected devices, vehicles, buildings, and objects equipped with electronics, sensors, and network connectivity by using these they gather and exchange data [1][2]. This technology enables remote monitoring and control, integrating physical systems with computer-based infrastructure to enhance efficiency and accuracy [3]. IoT development faces challenges related to infrastructure, communication, interfaces, protocols, and standards IoT attacks depend on rapid data retrieval to fulfill their intended functionality [4]. The network of intelligent objects is in charge of sending data, keeping an eye on the surroundings, and reacting to the Internet of Things requests needs by sending the information to the Sink, the designated IoT gateway. Because data might be lost, routing assaults in the Internet of Things can be fatal. The goal of routing layer assaults is to obstruct the sensor node's passage to the sink node[5]. Every attack often uses a certain layer or protocol's weaknesses to further its objectives. Attacks fall into two main categories: those that do not relay received packets and those that redirect traffic from the intended recipient to the malicious node [6]. The assault via the sinkhole is distinct from other routing layer attacks because it combines the two fundamental forms, diverting traffic while refusing to forward new packets. By posing as the Sink and dropping all incoming packets, it draws traffic from all nearby nodes [7]. The main contribution of this research is

- Introduced a defense system against sinkhole attacks, based on probe routes to effectively identify sinkhole attacks.
- The proposed PRDSA scheme effectively overcomes the sinkhole attack.

DOI: 10.1201/9781003589273-53

2 PROBLEM STATEMENT AND OBJECTIVES

To effectively detect and counter sinkhole attacks in network security while minimizing costs and maintaining network integrity.

- To enhance the detection rate of sinkhole
- To decrease the time taken for sinkhole detection.
- To maximize the network lifetime.

3 RELATED WORKS

Liu [8] introduced a trust-aware probability marking (TAPM) traceback scheme aimed at quick identification of malicious sources. In this scheme, nodes received different marking probabilities determined by their trust levels, evaluated through trust assess- ment. Nodes with lower trust levels were allocated higher marking probabilities to expedite the detection of malicious sources, while those with higher trust levels were assigned lower marking probabilities to conserve network resources and extend net- work lifetime. Consequently, the TAPM scheme enhanced both network security and longevity.

M. Wazid [9] with this approach, extra hardware is inserted into the network to de- tect and report sinkhole attacks. Although this approach has the benefit of improved security and efficacy, it has the drawback of requiring more expensive additional tech- nology. The goal of the research is to prevent the sinkhole attack while maintaining the same level of WSN cost. Enhance the current routing plan as a second line of defence against the Sinkhole attack.

Ga Hyeon An [10] presents an innovative method to enhance sinkhole attack detection in IoT networks using a knowledge-based specification rule. By analyzing sinkhole attack characteristics, this approach aims to develop accurate detection rules. This research pro- mises to strengthen IoT security by providing tailored intrusion detection against sinkhole threats, thereby safeguarding network integrity.

4 PROPOSED METHODOLOGY

A novel Probe Route-based Defense Sinkhole Attack (PRDSA) scheme is proposed to reduce the limitations of existing defenses. PRDSA ensures data safely reaches the sink by

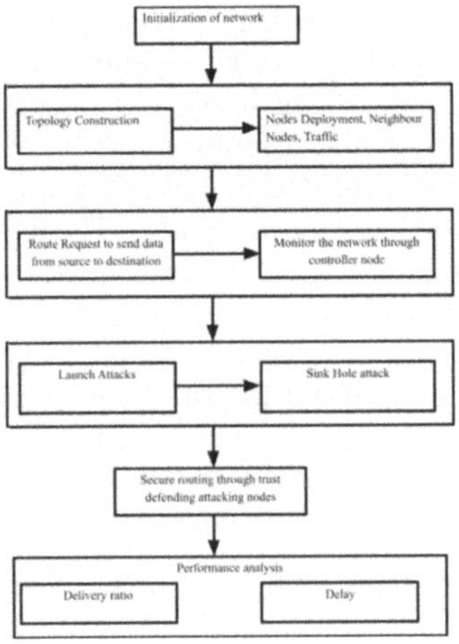

Figure 1. System design.

probing bypass routes when a sinkhole attack is suspected. It's the first scheme to detect and locate sinkholes within the network. In PRDSA, nodes sign and return acknowledgment packets containing IDs, next hop details, and hop counts to verify identities and thwart fake acknowledgments. This method enables effective sinkhole detection, localization, and mitigation. The above architecture Figure 1 shows deploy- ment of nodes, including the source, destination (Sink Node), and nodes potentially compromised by attackers. During attacks, the attackers may deploy various types of attacks to compromise authentic nodes.

4.1 *Placement of a node in a network*

This section explains the steps involved in identifying the placement of a node in a network. Each node in the network follows the below steps. Let N denote the set of all nodes in the IoT network, A denote the set of all agents, and E denote the set of all energy levels.

4.1.1 *Initialization of mobile node*

$$N_i \leftarrow \text{InitializeNode}() \quad \forall N_i \in N$$

Each node N_i in the IoT network is initialized using a function InitializeNode(), ensuring that all nodes are set up for operation.

4.1.2 *Enable random motion*

$$\text{EnableRandomMotion}(N_i) \quad \forall N_i \in N$$

This function EnableRandomMotion(N_i) enables random motion for each node N_i, allowing them to move within the IoT environment.

4.1.3 *Initialization of node energy*

$$\text{InitializeEnergy}(N_i, e_k) \quad \forall N_i \in N, e_k \in E$$

Each node N_i is initialized with a certain amount of energy e_k from the set of energy levels E, ensuring they have sufficient power to operate.

4.1.4 *Define initial position*

$$\text{DefineInitialPosition}(N_i, p_j) \quad \forall N_i \in N, p_j \in P$$

The function DefineInitialPosition(N_i, p_j) assigns an initial position p_j to each node N_i within the IoT environment, where P is the set of all possible positions.

4.1.5 *Initialization of agent*

$$A_i \leftarrow \text{InitializeAgent}() \quad \forall A_i \in A$$

An agent A_i is initialized using the function InitializeAgent(), ensuring that each node has an associated agent for specific tasks.

4.1.6 *Attach agent to node*

$$\text{AttachAgent}(N_i, A_i) \quad \forall N_i \in N, A_i \in A$$

The function AttachAgent(N_i, A_i) attaches the agent A_i to the node N_i, establishing a connection between the node and its associated agent.

4.2 *Energy module*

The energy model of the proposed work is implemented based on node attributes. It indicates the remaining power of the node exists in a network. The initial energy of the node is called initial energy (txPower) and the energy used to transmit and receive the content is denoted as energy usage (rxPower). The current energy model focuses on overall energy management without specific radio state control, allowing for future flexibility. Going forward, access to the energy model is standardized through the node-config API, ensuring consistency in simulation environments. This transition streamlines functionality and enhances compatibility for future simulations.

4.3 *PRDSA scheme*

The PRDSA defense scheme utilizes a probing route approach and comprises three sub-schemes that are explained below.

4.3.1 *Probe routing sub-scheme*

(a) Nodes send probe packets alongside normal data forwarding, with a prob- ability.

$$P_i = \text{SendProbe}(N_i) \quad \forall N_i \in N$$

(b) Sink and receiving nodes return acknowledgment packets to the source.

$$A_i = \text{ReturnAck}(N_i) \quad \forall N_i \in N$$

(c) Acknowledgment packets contain timestamps, digital signatures, node IDs, next hop node IDs, and hop counts to the sink.

$$A_i = \text{AckPacket}(\textit{timestamp},\ \textit{signature},\ \textit{node ID},\ \textit{next hop ID},\ \textit{hop count})$$

Source node verifies the sink's authenticity based on digital signatures and hop counts in acknowledgment packets.

$$\text{VerifyAuthenticity}(A_i)$$

4.3.2 *Bypassing probe sub-scheme*

(d) Redirect data away from false sinks, primarily through reverse routing.

$$R_i = \text{ReverseRoute}(S_{\text{false}})$$

4.3.3 *Sink clearing action*

(e) Upon sinkhole detection, notify the genuine sink to take corrective action against the malicious node.

$$\text{NotifySink}\left(S_{\text{genuine}},\ \text{malicious node}\right)$$

(f) Before removal, the malicious node can still route data to the genuine sink using the last successful probe route.

$$R_{\text{last}} = \text{LastSuccessfulRoute}\left(N_{\text{malicious}},\ S_{\text{genuine}}\right)$$

(g) After removal, employ minimum hop routing to direct data to the genuine sink.

$$R_{\text{min}} = \text{MinimumHopRoute}\left(S_{\text{genuine}}\right)$$

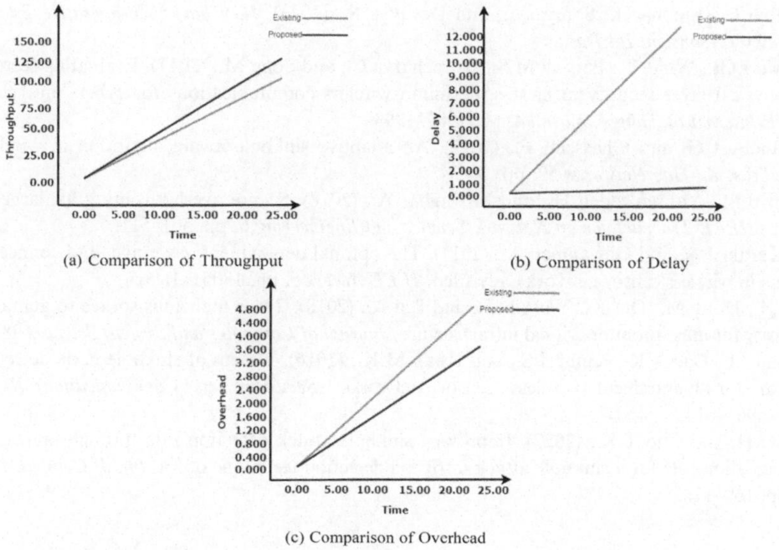

Figure 2. Performance evaluation of proposed method. (a) Comparison of Throughput (b) Comparison of Delay (c) Comparison of Overhead.

5 SIMULATION RESULTS

The OMNeT++ platform simulation is used to assess and validate the effectiveness of this method. The testing findings demonstrate that the PRDSA scheme's safety has increased by more than 90% and that its energy efficiency has been significantly increased. The simulation results indicate that the probe route-based detection tech- nique is capable of both detecting and locating sinkhole attacks as well as performing well across a range of performance comparisons. The results of the experiments show that the suggested method performs better than the current approaches in throughput, delay and overhead shown in Figure 2.

6 CONCLUSION

A novel PRDSA scheme is proposed to overcome the limitations of existing defenses. PRDSA ensures data safely reaches the sink by probing bypass routes when a sinkhole attack is suspected. It's the first scheme to detect and locate sinkholes within the network. In PRDSA, nodes sign and return acknowledgment packets containing IDs, next hop details, and hop counts to verify identities and thwart fake acknowledgments. This method enables effective sinkhole detection, localization, and mitigation. Lastly, the results of the experiments show that the suggested method performs better than the current approaches.

REFERENCES

[1] Pruthvi C., Vimala H., and Shreyas J., (2023), A systematic survey on content caching in icn and icn-iot: Challenges, approaches and strategies, *Computer Networks* 233, p. 109896.

[2] Geethakumari K., Vimala H., and Pruthvi C., (2023), *ECINS: Efficient Caching for ICN-IoT based on Node Score*, in *2023 IEEE 3rd Mysore Sub Section International Conference (MysuruCon)*. IEEE, pp. 1–6.

[3] Pruthvi C., Vimala H., Shreyas J., and Devayya S., (2023), *ICN based Co-operative Edge Caching Policy for Transient Iot Data.*
[4] Rawat D.B., White T., Parwez M.S., Bajracharya C., and Song M., (2017). Evaluating secrecy outage of physical layer security in large-scale mimo wireless communications for cyber- physical systems, *IEEE Internet of Things Journal* 4, pp. 1987–1993.
[5] Jahandoust G. and Ghassemi F., (2017). An adaptive sinkhole aware algorithm in wireless sensor networks, *Ad Hoc Networks* 59, pp. 24–34.
[6] Gharib M., Yousefi'zadeh H., and Movaghar A., (2018). Secure overlay routing for large scale networks, *IEEE Transactions on Network Science and Engineering* 6, pp. 501–511.
[7] Al-Karaki J.N. and Gawanmeh A., (2017). The optimal deployment, coverage, and connectivity problems in wireless sensor networks: revisited, *IEEE Access* 5, pp. 18051–18065.
[8] Liu X., Dong M., Ota K., Yang L.T., and Liu A., (2018). Trace malicious source to guarantee cyber security for mass monitor critical infrastructure, *Journal of Computer and System Sciences* 98, pp. 1–26.
[9] Wazid M., Das A.K., Kumari S., and Khan M.K., (2016). Design of sinkhole node detection mechanism for hierarchical wireless sensor networks, *Security and Communication Networks* 9, pp. 4596–4614.
[10] An G.H. and Cho T.H., (2022). Improving sinkhole attack detection rate through knowledge-based specification rule for a sinkhole attack intrusion detection technique of iot, *Int. J. Comput. Netw. Appl* 9, pp. 169–178.

Data Science & Exploration in Artificial Intelligence – Gururaj et al. (Eds)
© 2025 the Author(s), ISBN: 978-1-032-96390-7

An efficient approach to classify melanoma skin lesions using ensemble CNN model

Y.B. Vinay Kumar & H.S. Vimala
Department of Computer Science and Engineering, University Visvesvaraya College of Engineering (UVCE, IIT Model College) Bangalore University, Bengaluru, India

J. Shreyas
Department of Information Technology, Manipal Institute of Technology Bengaluru, Manipal Academy of Higher Education, Manipal, India

ABSTRACT: The most deadly type of cancer in the world is skin cancer. Skin tumors are abnormal tissue growths that can either be benign or malignant. The worst type of skin cancer is melanoma. The main challenges are in classifying skin lesions like moles and melanoma, which are highly similar and make it more difficult to diagnose and identify RGB images of the skin. If detected early on, melanoma skin cancer is very treatable. Examining the affected area of skin visually is the initial step in diagnosing melanoma skin cancer. The present model is designed to work in three stages such as data collection and augmentation, model development and prediction. In this study, a balancing dataset including 2375 photographs was created using pictures from various publically accessible ISIC (International Skin Imaging Collaboration) datasets. For the purpose of classifying melanoma skin cancer, an ensemble convolution neural network (CNN) model (Densnet201) was employed and trained on this dataset. The suggested methodology results an accuracy of 92% to categorize melanoma skin cancer.

Keywords: Melanoma Skin Cancer, Deep Learning, Machine Learning

1 INTRODUCTION

Skin cancer is an uncommon disease that requires investigation because it can have many distinct forms, create a wide range of symptoms, and either be fatal or curable. It can affect people of any age and, if left untreated, can be fatal. Skin cancer is an extremely rare disease that can affect anyone in any way. It affects about one in six people [1]. Experts in the medical industry are in agreement that excessive sun exposure could contribute to the rise in cases of this illness. This research on skin cancer detection aims to identify the kind of the disease and provide the appropriate preventative measures. In the medical world, skin cancer is the most important item, thus we must take the required steps to lessen its prevalence in order to improve human life. Dermatologists use a high-speed camera to capture dermatoscopic images of skin lesions; these images can diagnose melanoma with 65–80% accuracy without the need for extra technical assistance. After additional visual inspection by cancer treatment experts make use of dermatoscopic pictures, the overall accuracy of the melanoma diagnosis prediction rate increased to 75-84% [2]. Machine learning (ML) and python can be employed to understand how it works and all of its other strategies. A system that can analyze images and notify physicians if melanoma is present could significantly lessen the amount of manual labor required for diagnosis. The main objective is to assist in lowering the death toll from skin cancer. Utilizing cutting-edge picture classification technology for the sake of human welfare is the project's main driving force [3]. As per the literature, there are of much work related to skin lesion classification using dermoscopy images.

Many of the researchers used different techniques for skin cancer classification by following traditional way of extracting the features based on color and texture features [4], border based texture analysis [5] and rule based image processing [6]. Also, other methods of classification with the help of ML techniques such as neural network [7], Support vector machine [8], and decision tree [9]. There are many methods exist present the good performance over dermatologists in classification of skin cancer. Still, these methods are not optimistic and hence there is a scope for new technique to operate in the skin lesion classification. Machine learning and deep learning in computer vision have advanced significantly, and these advances are scalable across domains. The method classifies nine different forms of skin cancer from pictures of outlier lesions using an ensemble convolution neural network. This closing of a gap could have a positive effect on millions of people. As a result, compared to numerous AI (Artificial Intelligent) techniques, the ensemble CNN model showed an excellent performance in terms of their results obtained and robustness. This may be utilized extensively in application related to digital image processing and managing of large scale data. Hence, the present work is a promising option to consider for the application in this work. In the present work, we utilize an ensemble CNN model technique to categorize melanoma skin cancer. To enhance the classification efficiency, Densenet 201 CNN architecture has been used. In section 2, a brief discussion of the literature review and dataset has been provided. Section 3 refers to the result and analysis of the experimentation. Section 4 covers the conclusion of the work.

2 LITERATURE REVIEW

A broad literature survey on the classification of melanoma skin lesions using ensemble CNN model presents an increased research which aim at enhancing the accuracy of diagnosis and detection of skin cancer. The summary of key studies is as follows: Enakshi Jana et al reported Using a various types of ANN architectures and SVM used for skin cancer to obtain results that were reasonably accurate. Skin cells are classified using a variety of techniques. After focusing on three distinct cancer types and evaluating them using SVM and Adaboost, the best results for image classification with excellent accuracy were generated [10]. Haseb Younis et al suggested a number of skin cancer forms and a deep learning-based skin classification method that, when applied to a MobileNet convulsion neural network that has been previously trained using a lesion dataset, can accurately classify skin lesions in 2-3 seconds. This method is lightweight, quick, and dependable, and it demonstrates good weightage accuracy of 0.88 and 0.90, respectively [11]. Le et al a transfer learning model with higher accuracy, called ResNet50, that does not require preprocessing or feature manually selection. Although the results of the preprocessing procedures are not high enough in terms of precision, recall, or F1 score, they may be improved in terms of the classification rate of images of skin lesions [12]. Tschandl et al utilized CNN to analyze dermoscopic images of pigmented melanocytic lesions. Non-melanocytic and non-pigmented skin cancer was challenging to test for, nevertheless. It also gained less detecting accuracy [13]. These studies exhibit the promising way of classifying skin cancer lesions. Out of all these our work shows more accuracy of detection.

3 MATERIALS AND METHODS

3.1 *Dataset*

The (ISIC) International Skin Imaging Collaboration provided 2375 photographs of benign and malignant oncological disorders that make up the dataset. Every image was arranged in accordance with the ISIC categorization, and each subset had an equal amount of photographs [14]. Each of the train and test subdirectories in the dataset has nine subdirectories. Images of the nine different forms of skin cancer can be found in the nine subdirectories. Figures 1, 2 and 3 display the diseases that the neural network is supposed to identify. The following diseases are included in the dataset:

	Class	No of Image
0	Squamous cell carcinoma	184
1	dermatofibroma	96
2	actinic keratosis	116
3	pigmented benign keratosis	465
4	nevus	358
5	vascular lesion	141
6	seborrheic keratosis	78
7	basal cell carcinoma	377
8	melanoma	439

Figure 1. Skin cancer samples.

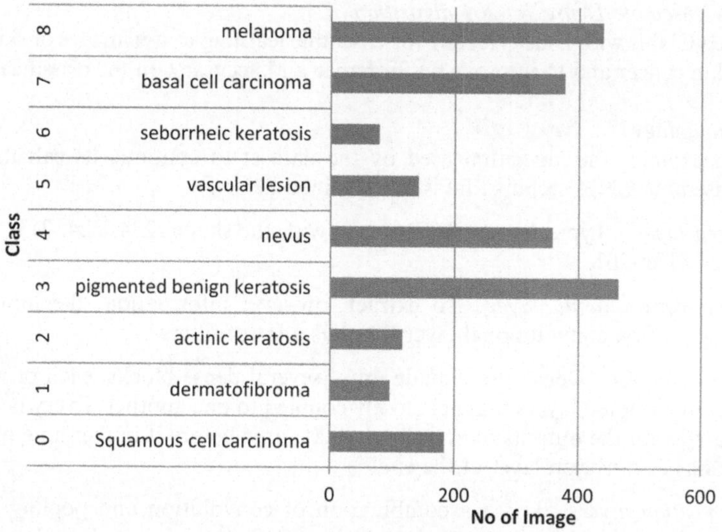

Figure 2. Cell type representation of classes for classification.

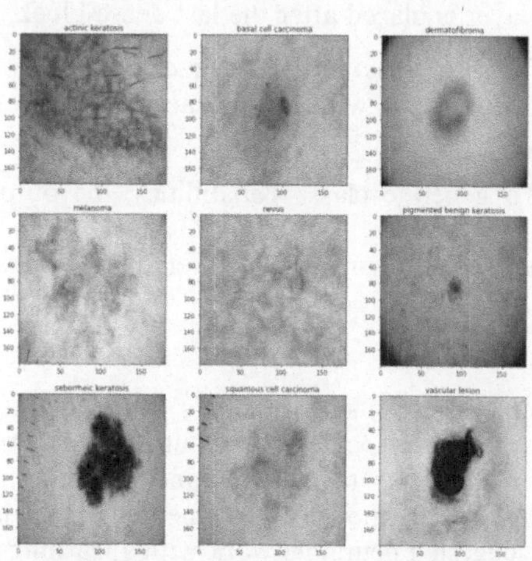

Figure 3. Skin cancer types example.

3.2 *Deep learning*

Skin cancer categorization can benefit greatly from deep learning by the combination of ensemble learning and Convolutional Neural Networks CNNs, such as DenseNet 201. Ensemble approaches seek to increase classification robustness and accuracy by utilizing several separate models, each trained on different subsets of the data or with different hyperparameters [14,15]. This technique improves generalization and reduces the likelihood of overfitting performance by leveraging the collective intellect of the group. In addition, CNNs such as DenseNet 201 are highly skilled in extracting complex patterns and features from medical images, which makes them suitable for the complex task of classifying skin cancer. By utilizing group learning, these CNNs can combine their advantages to attain increased diagnostic precision and dependability, which would eventually help both patients and medical professionals.

3.2.1 *CNN ensemble DenseNet 201 algorithm*

The Proposed CNNs with DenseNet 201 for ensemble learning to use images of skin lesions to categorize skin cancer and to improve performance and accuracy on the classification task.

3.2.2 *Methodology*

This section explains the algorithm used by the dataset to continue its calculations. The proposed ensemble CNN architecture is as explained below:

3.2.1.1 *Input layer* Typically, an RGB image with the shape (224, 224, 3) is used as the input for DenseNet-201.

3.2.1.2 *First convolutional layer* To extract low-level information, the input image is passed through a first convolutional layer that has a lot of filters.

3.2.1.3 *Dense blocks* DenseNet is made up of several dense blocks, each of which has a number of convolutional layers that are closely coupled to one another. Every layer's output is concatenated with the outputs of every layer that came before it in the same block. Every dense block has a transition layer at the end.

3.2.1.4 *Transition layers* Using a combination of convolution and pooling techniques, transition layers are used to increase the number of feature maps while decreasing their spatial dimensions of width and height.

3.2.1.5 *Global average pooling layer* To combine spatial data from different feature maps, a global average pooling layer is placed after the last dense block.

3.2.1.6 *Fully connected layer (dense layer)* For classification tasks, a fully connected layer, also referred to as a dense layer, with softmax activation is coupled to the output of the global average pooling layer.

3.2.1.7 *Output layer* To generate class probabilities, the output layer uses a softmax activation function.

The architecture for the CNN Ensemble DenseNet 201 model is defined in Figure 4.

3.3 *Training and testing*

The present model is implemented using Python's artificial intelligence framework. We employ SoftMax activation function located at the output layer, Densenet, Trantition layers, and fully connected layer. In order to classify nine different forms of skin cancer, an ensemble CNN using DenseNet201 model was trained on all the layers. The "Adam" optimizer is used to train the suggested neural network with a learning rate of 0.0001. The batch size is set to 32, the number of epochs is 20, and the dropout optimization value on fully connected layers is 0.52.

Model: "sequential"

Layer (type)	Output Shape	Param #
rescaling (Rescaling)	(None, 180, 180, 3)	0
conv2d (Conv2D)	(None, 178, 178, 32)	896
max_pooling2d (MaxPooling2D)	(None, 89, 89, 32)	0
conv2d_1 (Conv2D)	(None, 87, 87, 64)	18496
max_pooling2d_1 (MaxPooling2	(None, 43, 43, 64)	0
conv2d_2 (Conv2D)	(None, 41, 41, 128)	73856
max_pooling2d_2 (MaxPooling2	(None, 20, 20, 128)	0
dropout (Dropout)	(None, 20, 20, 128)	0
flatten (Flatten)	(None, 51200)	0
dense (Dense)	(None, 128)	6553728
dropout_1 (Dropout)	(None, 128)	0
dense_1 (Dense)	(None, 9)	1161

Total params: 6,648,137
Trainable params: 6,648,137
Non-trainable params: 0

Figure 4. The architecture for the CNN Ensemble DenseNet 201.

4 RESULT AND ANALYSIS

The images are included within the data frame. Datasets for validation, training, and testing are divided into the image dataset. When the CNN with ensemble densenet201 model is applied with the test and training datasets, the accuracy and loss obtained are shown in Figure 5. After applying this model for fifty epochs on the training set of data the training loss: 0.22; training accuracy: 0.91; validation loss: 0.52; val_accuracy: 0.83. The model indicates an increase in both training and validation accuracy. Achieving optimal training and validation accuracy is facilitated by class rebalancing.

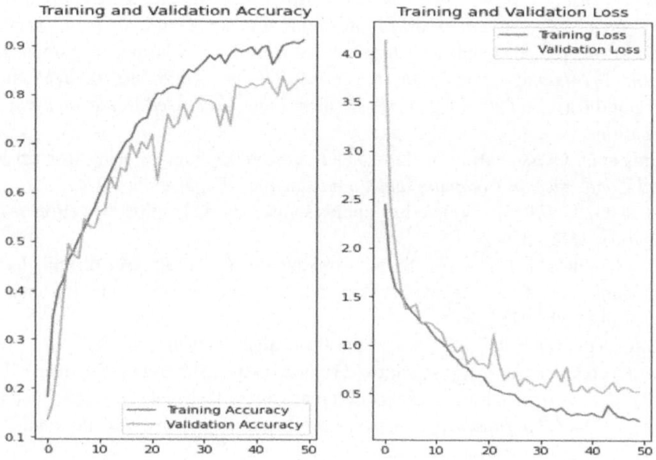

Figure 5. Comparing training accuracy and validation accuracy and loss of model.

5 CONCLUSION

The present work successfully develops a method which can classify images of cancer from a ISIC database. The images illustrate different types of skin cancer. In order to classify the test dataset data that lacks classification or labels, the work used Deep Learning with various layers. The metric's results showed an outstanding 92% accuracy level. Further, the work can be utilized to produce new services or solutions that enable healthcare professionals to make decisions faster.

REFERENCES

[1] Katalinic A., Waldmann A., Weinstock M. A., Geller A. C., Eisemann N., Greinert R., *et al.*, (2012), Does skin cancer screening save lives? an observational study comparing trends in melanoma mortality in regions with and without screening, *Cancer*, vol. 118, no. 21, pp. 5395–5402.

[2] Poonkuzhali, S., Anu Barathi, B.U. and Vinodhkumar, S. (2023). *Transfer Learning Approach for Diagnosing Skin Cancer with Deep Convolutional Neural Network.* In: Kaiser, M.S., Xie, J., Rathore, V. S. (eds) Information and Communication Technology for Competitive Strategies (ICTCS 2021). Lecture Notes in Networks and Systems, vol 401. Springer, Singapore. https://doi.org/10.1007/978-981-19-0098-3_45.

[3] Khamsa D., Pascal L., Zakaria B., Lokman M. and Zakaria M. Y., (2023), Skin cancer diagnosis and detection using deep learning, *2023 International Conference on Electrical Engineering and Advanced Technology (ICEEAT)*, Batna, Algeria, pp. 1–6, doi: 10.1109/ICEEAT60471.2023.10426611.

[4] Barata C., Ruela M., Francisco M., Mendonca T., and Marques J. S., (2014), Two systems for the detection of melanomas in dermoscopy images using texture and color features, *IEEE Syst. J.*, vol. 8, no. 3, pp. 965-979. https://doi.org/10.1109/JSYST.2013.2271540.

[5] Garnavi R., Aldeen M., and Bailey J., (2012), Computer-aided diagnosis of melanoma using border- and wavelet-based texture analysis, *IEEE Trans. Inf. Technol. Biomed.*, vol. 16, no. 6, pp. 1239–1252. https://doi.org/10.1109/TITB.2012.2212282.

[6] Ganster H., Pinz A., Röhrer R., Wildling E., Binder M., and Kittler H., (2001), Automated melanoma recognition, *IEEE Trans. Med. Imaging*, vol. 20, no. 3, pp. 233–239. https://doi.org/10.1109/42.918473

[7] Xie F., Fan H., Li Y., Jiang Z., Meng R., and Bovik A. C., (2016), Melanoma classification on dermoscopy images using a neural network ensemble model, *IEEE Trans. Med. Imaging*, vol. 0062, no. c, pp. 1–11. https://doi.org/10.1109/TMI.2016.2633551

[8] Sana, Shahid M., and Khan S., (2017), Dermoscopy images classification based on color, texture and shape features using SVM, no. March. https://www.researchgate.net/publication/323550792_Dermoscopy_Images_Classification_based_on_Color_Texture e_and_Shape_Features_Using_SVM.

[9] Ferris L. K. *et al.*, (2015). Computer-aided classification of melanocytic lesions using dermoscopic images, *J. Am. Dermatology*, vol. 73, no. 5, pp. 769–776, https://doi.org/10.1016/j.jaad.2015.07.028.

[10] Enakshi Jana and Ravi Subban, (2021), *Research on Skin Cancer Cell Detection using Image Processing*, IEEE publication.

[11] Haseb Younis *et al.* Classification of skin cancer dermoscopy images using transfer learning, *2018 7th International Conference on Computer and Communication Engineering (ICCCE).*

[12] Li, Y., and Shen, L. (2018). Skin lesion analysis towards melanoma detection using deep learning network. *Sensors*, 18(2), 556.

[13] Tschandl, P., Rosendahl, C., Akay, B. N., Argenziano, G., Blum, A., Braun, R. P., *et al.* (2019). Expert-level diagnosis of nonpigmented skin cancer by combined convolutional neural networks. *JAMA Dermatology*, 155(1), 58–65.

[14] Dheiver Santos. (February 4, 2022) Melanoma skin cancer detection using deep learning, *medRxiv*.

[15] Muhammad Hasnain Javid; Waqas Jadoon; Haris Ali and Muhammad Danish Ali, (20-22 February 2023). Design and analysis of an improved deep ensemble learning model for melanoma skin cancer classification, *2023 4th International Conference on Advancements in Computational Sciences (ICACS).*

Data Science & Exploration in Artificial Intelligence – Gururaj et al. (Eds)
© 2025 the Author(s), ISBN: 978-1-032-96390-7

AgriFriend: A robust approach for crop recommendation system

Shilpa Suresh
Department of Mechatronics, Manipal Institute of Technology, Manipal Academy of Higher Education, Manipal, Udupi, India

Ragesh Rajan
Department of ECE, Amrita Vishwa Vidyapeetham, Vallikavu, Amritapuri, Kerala, India

S. Rashmi, Raguru Jaya Krishna & Y.V. Srinivasa Murthy
Manipal Institute of Technology Bengaluru, Manipal Academy of Higher Education, Bengaluru, Karnataka, India

ABSTRACT: Agriculture plays a vital role in the Indian economy. India is a country where farmers face lots of challenges, such as various disasters like draught and varying weather conditions, which compels them to switch to modern farming techniques. In this paper, an approach has been discussed that will predict the most suitable crop to be grown based on the soil and environmental parameters of that area. Keeping in mind the problems faced by farmers in selecting the most suitable crop, we have developed an application, Agrifriend, using machine learning. The dataset of India's climate is used, and the data is trained on various machine learning algorithms such as Naive Bayes, SVM, and Decision tree. also, different ensemble learning techniques such as random forest, and XGBoost are trained, and the highest accuracy of 98% is achieved. The application will also recommend correct fertilizers based on current soil nutrient in that field.

1 INTRODUCTION

Agriculture is the most significant sector for any country not only for economy purpose but also to feed the population of that country. The farmers in our country are unaware about various environmental factor which can play role in growth of crop in particular area. India also faces various natural disasters like flood draught which have direct impact on crops. India is still following orthodox traditional way of farming which it needs to step-up and evolve based on current trends in agriculture to grow ample amount of crop which in turn help farmers to earn profit. About 70% of Indian population in rural area still depends on farming as prime source of their living. Interestingly India is still biggest player in terms of agricultural productivity, but irony is that efficiency in terms of land productivity is not optimum.

Now a days there are lots of cases of farmers committing suicide because of not getting better earning even after doing hard work. Proper crop selection is a major challenge faced by Indian farmers owing to their unawareness and variation in monthly rainfall due to over usage of fertilisers the nutrient level of soil are also decreasing, fertiliser price and pest availability all these causes decrease in productivity. As the land capital is limited the farmers need to think about the alternative to boost up the productivity with less capital and time. The farmers are simply following the price trend of particular crop to grow crop. In 2020, when price of Onion shot up farmers made up their mind to grow onion. While most of the farmers earned handsome money while in some area they badly failed. The need of hour is that farmers in India must understand the important criteria which affect the growth of crop. Growth of crop is dependent of various criteria and is site specific soil criteria as well as environmental factor which altogether affect the crop success in that area.

DOI: 10.1201/9781003589273-55

Many of the developing countries they have switched to Precision agriculture [1,2]. Precision agriculture is evolved farming practise in which the main focus is to understand the nutrient of the soil in particular land the amount of fertiliser, pesticides required for growth of particular crop. Various environmental factors such as humidity, rainfall are also considered. Precision agriculture helps the farmers to choose most optimal crop based on their land soil and environmental factors and help them in making correct decision. Precision agriculture may not be accurate always but, they can be optimised as any wrong recommendation may cause unbearable loss to poor farmers. Crop selection can be done using intelligent system which based on soil parameter it will recommend most suitable crop based on past data about the yield of various crops helps the farmers to get proper insight about which crop is suitable in that region [3]. But the limitation with this approach is that seasonal rainfall pattern over the year has changed drastically which makes this approach inconsistent. Machine learning and deep learning concepts were also used to solve this challenging issue [3–7,13].

The objective of this paper is to build an application which will help the farmers in making informed decision about which crop to select based on the prediction given by machine learning model. India weather and climate dataset has been used as well as soil specific dataset has been used and combined to get both soil and environmental parameter to feed to machine learning model. Model can learn pattern from the dataset. Various algorithms have been used such as Naive Bayes, SVM, Decision Tree to train our model and various performance matrix have been used to evaluate our model performance. Agrifriend can also help the farmers recommend which fertilisers to use based on their soil nutrient which ultimately keeps check on soil quality because nowadays due to heavy use to fertilizers [8] and pesticides by the farmers the quality of soil is decreasing also soil moisture retention capacity decreases. Based on soil nutrient and the crop which the farmer wants to grow they can also get to watch various farming practises in-order to improve the nutrient capacity of their soil. Another problem in India is the farmers are unaware about various farming practises and new yojanas to this application can provide the platform to connect between government as well as farmers by equipping farmers with the facility of video to get knowledge and latest news.

Based on above concept the paper focuses on building a machine learning model which will not only consider just the soil parameter like Nitrogen, Phosphorus and Potassium because based on observation the recommendation system is more consistent when all the factors affecting the growth of crop is considered so, the paper focuses on environmental parameters as well like humidity, temperature and rainfall of that specifics area. In this project various machine learning algorithm as well as ensemble-based model has been trained and performance metrices has been evaluated. We have proposed a robust system for crop recommendation which takes in account of environmental as well as soil characteristics to suggest the best crop to the user. We also consider some additional information before making our prediction, one of which includes analysis on rainfall data for regions selected by the user. We have also made a fertilizer recommendation system as an additional feature. The fertilizers recommended are only organic because of numerous harmful effects of chemical fertilizers.

Soil characteristics that are used are Nitrogen, Potassium and Phosphorus whereas environmental characteristics consist of temperature and moisture. Our system is dependent on user to get the soil characteristics values for the Nitrogen, Potassium and Phosphorus. To counter this dependency we have provided all the details of Soil health agencies so that farmers can contact them and get their soil tested. The proposed system also uses rainfall data for that particular region to ensure recommendations are more accurate. Moreover, the proposed system also contains some valuable video content about the correct farming practices. We have also made sure that the video content is accurate, relatable and easy to understand and most importantly it supports various language. Thus, ensuring a maximum reachability. The system's user interface is deployed using flask which is a python framework. We have chosen flask because of its minimalist design and flexibility and because of its easier deployment and development methods.

The paper is organized as follows: Section 2 evaluates the work done in this area. Section 3 describe the proposed algorithm of our application as well as architecture diagram. Section 4 displays various performance metrices of model as well as working of our application. Section 5 deals with the conclusion of the project as well as future works in this field.

2 LITERATURE REVIEW

There has been various approach to recommend crop to the farmer. It started with precision agriculture in developed countries. Since then, various work has been done in this area some have considered soil characteristics some has considered environment characteristics. They have proposed a crop recommendation model through the ensemble learning technique. They have used the theory of precision farming, which is mainly used soil properties over a large area in progressing countries and achieving a favourable position for maximum productivity using data mining. This kind of model can also be applied in India considering the depth, texture, pH, soil colour, permeability, drainage, water holding, and erosion of the soil parameters. The dataset was collected from the Madurai District Soil Laboratory.

However, in crop recommendation systems, they have used soil parameters collected from Maharashtra soil testing laboratories to increase crop yields using machine-learning methods. The parameters under consideration are soil structure, ph. and drainage. Based on the above criteria the ensemble model with a majority vote predicts the most favourable crop [2]. But in Agroconsult [3], research they have used the 'Indian Soil and Climate Dataset'. They have used soil thickness, reservoir thickness, soil type, and temperature of the area where soil and environmental parameters are mixed. They have compared the accuracy of different algorithms and the decision tree that works best. In another model, they have developed an Android application for the state of Maharashtra that recommends crops, fertilizers based on N, P, K values, and soil quality in the particular area. By the standards, they have considered crop yield, soil type and soil nutrients in the region. Based on that system will give the most suitable crop besides fertilizer [4]. The hybrid MLR-ANN model has been offered to estimate the require crop yield. MLR interrupts and modules were applied to initialize the input layer bias and ANN. Estimated accuracy was calculated using standard performance metrics. The maximum yielding crop is suitable for agriculture, which expand the profits of the farmers. The parameters are solar radiation, soil type, temperature, and rainfall [5].

In the crop recommendation approach, the authors were introduced a unique model that can be very useful for the farmer to get the proper productivity. Various ensemble technique known as the majority voting technique, where k-near and Naive Bayes was used as base learners to obtain high accuracy and efficiency for harvest indications related to crop recommendations [6]. To get the proper accuracy rate they have developed a system with the help of famous algorithm called the Random forest algorithm. They have collected the datasets from the Indian government official website, which is open source from (1997 to 2013). The authors was found that the proposed model is very suitable for a heavy number of crop yield predictions in the agriculture sector and helps the farmer to decide the suitable crop [7]. In the crop recommendation sector, soil quality is not only important but also a necessary part. So, for that reason, authors were thought differently and focused on the soil quality to measure the appropriate crop for cultivation with fitting fertilizer. Actually, they have used Kohenon's SOM and BPN algorithms to figure out them and get the better one. The data mining technology was used to consider the better soil quality. The author believes that the prospective model is very capable to decrease the farmer suicide rate [8].

Android-based crop recommendation system can be built. This app uses the random forest algorithm to get more than 80% accuracy rate and present a special model which is a fertilizer purchase system besides the fertilizer recommendation. They have used Apriori which is known as data mining algorithm. As per as author this proposed system is very user-friendly to the user and less memory consumer [9]. In this model ,they have worked four different ensemble learning models such as Random Tree, CHAID, K Nearest Neighbor, and Naïve Bayes to obtain the appropriate accuracy in consideration of soil and weather parameter [10]. But in web-based recommendation, they have introduced a web-based crop recommendation system using only the Random Forest Algorithms and 90% accuracy accounted. They have provided the fertilizer recommendation with the support of N, P, K values and the recommended system is not only web-based but also android based is the special feature of the model as per as author [11]. In their research, they have applied the SVM and the Random Forest model for crop yield recommendation and soil analysis. As with other research, they have considered soil, water, and weather parameters in the proposed system. Finally, they have achieved 86.35% accuracy in soil classification using Random Forest and in contrast, reached near to 98% accuracy in crop recommendation using the SVM Method. Improving farmer life is the main motive of their proposed model [12].

Authors suggest a method that can advise farmers on extremely lucrative crops. Using historical data from a location and the ARIMA model, meteorological characteristics like temperature and rainfall were predicted. Data classification based on weather and soil conditions were done using logistic regression. Additionally, historical investment and return data along with the ARIMA model were used to forecast the potential profit [13]. A method was devised employing TPF-CNN to implement artificial intelligence for pest identification and pesticide advice. use of a soil sensor to analyze the soil's NPK nutrients and suggest the appropriate fertilizers [14]. Amidst all these studies, crop recommendation system lacks good accuracy owing to inaccurate and incomplete data, limited local content and crop variety, complex variables, difficulty in integrating local knowledge and dynamic environmental conditions. The proposed method considers of most possible variabilities to train the model for implementing a promising crop recommendation system.

3 PROPOSED METHODOLOGY

3.1 *Dataset collection*

The dataset has been collected from data.gov.in, kaggle and India climate dataset. In our project we are considering both the soil parameter as well as environmental parameter for recommendation. After collecting both the dataset. Firstly, we performed Data cleaning. Data cleaning is the methodology to remove all the missing data, incorrect data, out of range data, inconsistent data by either modifying, correcting or deleting that record from the final dataset. There are altogether 6600 rows in our dataset.

3.2 *Machine learning algorithms*

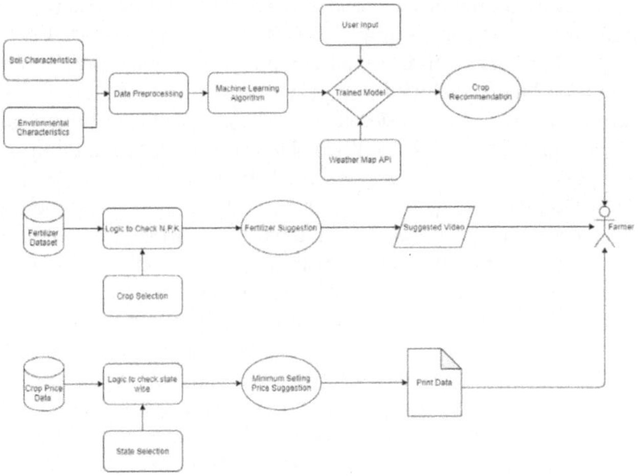

Figure 1. Architecture of proposed methodology.

Various machine learning algorithms have been used as per step given in the Figure 1. After the data is prepared the final dataset is trained using various machine learning algorithms. To ensure there is no multicollinearity between any two features in our dataset. The collinearity between attributes is calculated and it has been observes that there is not any correlation greater than 0.8. Correlation is higher between potassium and phosphorous however, this can be attributed to high variance in data. Hene, there is not case of multicollinearity in our dataset. After ensuring the data is consistent we have split our dataset into 80:20 ratio i.e. 80% of dataset row for training the machine learning model and 20% dataset is used for testing the accuracy of model. There are altogether 6600 rows in our dataset. Also, the dataset is balance for each of 22 classes of crops.

4 RESULT ANALYSIS

This section discusses about the results and performance of various machine learning model by comparing them against various performance metrices. Performance metrices are the numerical comparison of the business logics as well as analytics performed in order the check for the reliability and correctness of that algorithm for our required use cases. We have used balanced dataset and for our required case we have selected accuracy and f1 score for evaluating our model consistency. Accuracy: It is the measure of average correct prediction out of total given instances. Suppose there are 100 rows in our given dataset of bank data. The objective is to predict the valuable customer for the bank in the future. After doing all the p reprocessing the dataset is fed to the machine learning model and the model predicts that out of 100, only 40 customers will add value to company where as in true sense it is 30 customers then our machine learning has incorrectly predicted 10 customers here the accuracy is close to 60 %. This metrices performs better when we have balanced dataset and in our case, we are having balanced dataset. We have trained our model on various machine learning classification algorithm and the model which performs better is ensemble-based machine learning model ie Random Forest (RF) Figure 2 which gives accuracy close to 99 % so we have selected this model for deployment.

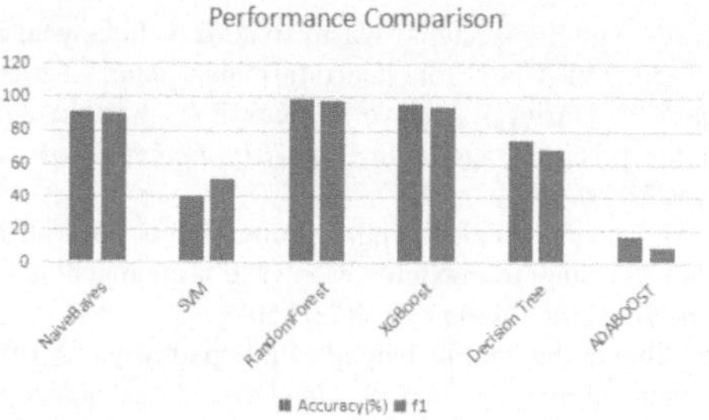

Figure 2. Comparison of algorithms.

Precision: It is the characteristics of the machine learning model not to label the instance as positive which is negative. Recall: It is the capability of our machine learning model to detect all the instances which are positive or belongs to that class. We have considered optimal equilibrium state between precision and recall by considering F1 score as metrices. The f1 score lies in the range which falls between 0-1.

We can see from the table as well as in the bar graph the F1 score of Random Forest classifier Nitrogen, Phosphorus, Potassium based on the data given on soil health card data. The farmers have soil health card on which data of farmer soil ie various nutrient data N,P,K. After selecting the location flask sends request to open weather map api to send the temperature as well as humidity if the data are available. api send response to flask and the input are fed to random forest model.

Table 1. Performance Evaluation metrices

Algorithm	Accuracy (in %)	F1-Score
Naïve Bayes	91	90
SVM	40	51
Random Forest	99	98
XgBoost	96	94
Decision Tree	74	69
AdaBoost	16	10

The challenges that we have faced is that the collection of data. Although India is still practicing the traditional approach of farming the time has come to switch to modern technology-based farming which will take care of sustainability of environment and when they start doing precision-based agriculture more and more data will be available which will improve the performance and generalization of model.

5 CONCLUSION AND FUTURE WORK

The application will help the Indian farmers to gain more profit by increasing the productivity of their land. The crop recommended by Agrifriend application will be optimal to grow in that area . It will pave the way for the Indian farmers following traditional approach of farming to switch to precision based agriculture where farming is done considering the site specific parameters. As our users belong to a non-technical background we make sure that our system is very easy to use and doesn't require any prior knowledge to use the system. The system have three modules built keeping in mind the current problems faced by Indian farmers with the objective of crop recommendation, fertilizer suggestion and weather forecasting with an accuracy close to 98%. As future development this project can be extended as a platform where the farmers and sellers can connect without any need of mediators. Also the diseases prediction module as well as locating the nearest soil health centre based on farmers location.

BIBLIOGRAPHY

[1] Thorat, Tanmay, B. K. Patle, and Sunil Kumar Kashyap. (2023). Intelligent insecticide and fertilizer recommendation system based on TPF-CNN for smart farming. *Smart Agricultural Technology* 3: 100114. Pudumalar S., Ramanujam E., Harine Rajashree R., Kavya C., Kiruthika T., and Nisha J., (2017). Crop recommendation system for precision agriculture. *In 2016 Eighth International Conference on Advanced Computing* (ICoAC), pages 32–36. IEEE

[2] Rohit Kumar Rajak, Ankit Pawar, Mitalee Pendke, Pooja Shinde, Suresh Rathod, and Avinash Devare., (2017). Crop recommendation system to maximize crop yield using machine learning technique. *International Research Journal of Engineering and Technology*, 4(12):950–953.

[3] Archana Chougule, Vijay Kumar Jha, and Debajyoti Mukhopadhyay., (2019). Crop suitability and fertilizers recommendation using data mining techniques. *In Progress in Advanced Computing and Intelligent Engineering*, pages 205–213. Springer.

[4] Zeel Doshi, Subhash Nadkarni, Rashi Agrawal, and Neepa Shah., (2018) Agroconsultant: Intelligent crop recommendation system using machine learning algorithms. *In 2018 Fourth International Conference on Computing Communication Control and Automation (ICCUBEA)*, pages 1–6. IEEE.

[5] Maya Gopal PS and Bhargavi R., (2019). A novel approach for efficient crop yield prediction. *Computers and Electronics in Agriculture*, 165:104968.

[6] Akshatha KR and Shreedhara KS., (2018). Implementation of machine learning algorithms for crop recommendation using precision agriculture. *International Journal of Research in Engineering, Science and Management (IJRESM)*, 1(6):58–60.

[7] Priya P., Muthaiah U., and Balamurugan M., (2018). Predicting yield of the crop using machine learning algorithm. *International Journal of Engineering ciences & Research Technology*, 7(1):1–7.

[8] Mansi Shinde, Kimaya Ekbote, Sonali Ghorpade, Sanket Pawar, and Shubhada Mone. Crop recommendation and fertilizer purchase system. International Journal of Computer Science and Information Technologies, 7(2):665–667, 2016.

[9] Rushika Ghadge, Juilee Kulkarni, Pooja More, Sachee Nene, and RL Priya. (2018). Prediction of crop yield using machine learning. *Int. Res. J. Eng. Technol.*(IRJET),5.

[10] Anantha Reddy D., Bhagyashri Dadore, and Aarti Watekar, (2019). Crop recommendation system to maximize crop yield in ramtek region using machine learning. *International Journal of Scientific Research in Science and Technology*, 6(1):485–489.

[11] Kiran Shinde, Jerrin Andrei, and Amey Oke, (2015). Web based recommendation system for farmers. *International Journal on Recent and Innovation Trends in Computing and Communication*, 3(3):41–52

[12] Devdatta A. Bondre and Santosh Mahagaonkar, (2019). Prediction of crop yield and fertilizer recommendation using machine learning algorithms. *International Journal of Engineering Applied Sciences and Technology*, 4(5):371–376.

[13] Palle, Sudarshan Reddy, and Shital A. Raut, (2023). Crops recommendation system model using weather attributes, soil properties, and crops prices. *In Sentiment Analysis and Deep Learning: Proceedings of ICSADL* 2022, pp. 323–338. Singapore: Springer Nature Singapore.

[14] Thorat, Tanmay, Patle B. K., and Sunil Kumar Kashyap. (2023). Intelligent insecticide and fertilizer recommendation system based on TPF-CNN for smart farming. *Smart Agricultural Technology* 3: 100114. Aiex, Roberto Miguel, Resende, Paulo D, Pardalos, Rodrigo, and Toraldo, Ruith, (2000). GRASP with path-relinking for the three-index assignment problem, *Technical Report*, AT\&T Labs-Research.

Data Science & Exploration in Artificial Intelligence – Gururaj et al. (Eds)
© 2025 the Author(s), ISBN: 978-1-032-96390-7

A survey: Performance analysis of deep learning models on tomato leaf disease detection

H. Najmusher
Research Scholar, Department of Computer Science & Engineering, GITAM Deemed to be University, Bengaluru, India

Meena Kumara Guruparan
Professor, Department of Computer Science & Engineering, GITAM Deemed to be University, Bengaluru, India

ABSTRACT: In the Indian economy agriculture plays an important role. Agriculture is an important sector as 70 percent of the rural households depend on it. Increasing the agricultural economy along with feeding the large amount of population, it is highly necessary that disease detection in plants to be done in early stages. To make plants safe and avoid losses to the agricultural economy predictive mechanisms have to be adopted. Leaf disease identification uses well known deep learning models such as DenseNet, VGGNet, LMBRNet, ResNet and DCNN. It is clear from using these deep learning models that CNN, VGG, and ResNet are among of the best at identifying illnesses in leaves. Evaluation metrics like F1 score, precision, accuracy and others are used to measure the performance of algorithms. This survey done in the area of deep learning for detection and classification of tomato plant diseases can be helpful for the researchers.

Keywords: GAN, CGAN, DCNN, 2DCNN, Double GAN, Mask RCNN

1 INTRODUCTION

A crucial part of the nation's development is agriculture. A large portion of the country's economy is reliant on agricultural production. Numerous factors contribute to the low productivity of agriculture. Food security is also seriously threatened by crop diseases. Therefore, it is imperative that crop leaves be inspected for disease as soon as possible. However, in many parts of the world, this is challenging because of inadequate infrastructure. In this global upswing, research on computer vision using smart phones and deep learning models has led to the assisted identification of leaf diseases.

The core of the Indian economy is agriculture. Regarding the manufacturing of agricultural goods, farmers encounter numerous challenges in the modern digital environment. They must to be modernized to the greatest extent feasible with new farming techniques, and the government ought to utilize the resources at its disposal effectively. However, some people are insulted by traditionalists and lack the knowledge to use it. In order to end this, the government gives them training on how to use the resources that are provided to them efficiently and to comprehend the state of the world today. All things considered, a method for quickly identifying plant disease has been made possible by training the Fast AI-based model on a vast amount of publicly available data.

In most cases, pathologists and agricultural scientists are unable to identify diseases through visual inspection, and as a result, professionals are unable to draw conclusions or offer remedies [3]. For farmers to distinguish between leaf foliage based on plant development and existence, a system for identifying plant leaf diseases can be developed. This will be demonstrated to be an excellent technique for farmers to detect leaf illnesses, and it will notify farmers within the prime period to prevent the diseases from spreading. For the early diagnosis of disease, deep learning is an essential tool [4]. Numerous cutting-edge methods, like CNN, ML, image processing, and the Internet of Things, can be leveraged to produce outcomes that are more precise, quick, and spontaneous.

DOI: 10.1201/9781003589273-56

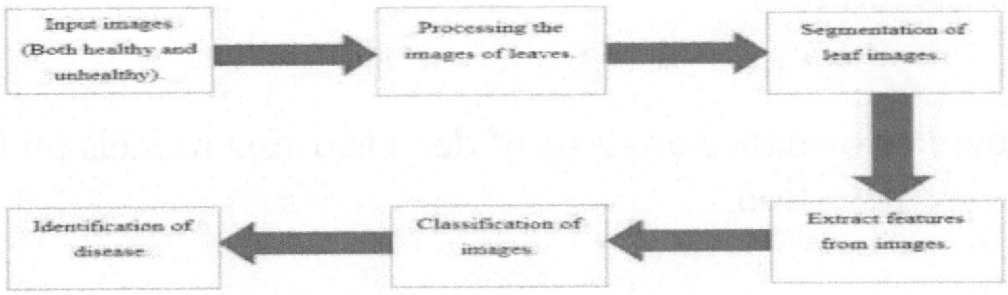

Figure 1. Phases of identification and disease detection in plant leaves.

The several phases of identification and disease detection in plant leaves are shown in Figure 1. The input, processing, and output stages are the different phases. segmentation, feature extraction from the images, categorization of the images, and lastly, disease identification Both can be found in the input pictures. Images of plants with both healthy and unhealthy leaves. It is being treated with pre-processing techniques. The shape, texture, color fluctuation, damaged zones, and other details are then extracted after it has been segmented. After classifying these photos based on whether the disease is present or not, the illness is recognized.

2 LITERATURE REVIEW

[1] Amreen Abbas and Sweta Jain have developed a deep learning based system to detect illnesses in tomato plants by analyzing photographs of tomato leaves. Synthetic pictures are generated by Conditional Generative Adversarial Networks (C-GAN) and subsequently incorporated into training datasets. Yellow leaf curl virus, Septoria leaf spot, Bacterial spot, Leaf mold, Mosaic virus, Target spot, Early blight and Late blight are among the diseases identified by utilizing the Plant Village dataset. After that, a DenseNet model was trained to detect tomato disease from leaf photos using images of tomato plants and tomato-synthetic images. . For the 5-class, 7-class, and 10-class classification tasks on the original PlantVillage dataset, it achieves an accuracy of 98.16%, 95.08%, and 94.34%, respectively. It obtains 99.51%, 98.65%, and 97.11% accuracy for 5-class, 7-class, and 10-class classification tasks using the original PlantVillage + synthetic photos dataset, respectively.

DoubleGAN was introduced by Yafeng Zhao and Zhen Chen in [2] to produce high-resolution images using fewer samples of diseased leaves. There are two stages in DoubleGAN.

Both healthy and sick leaves are inputs in stage 1. Initially, pictures of healthy leaves were fed into the Wasserstein generative adversarial network (WGAN) to create the pretrained model.

After that, sick leaves were used to train the pretrained model, which produced 64*64 pixel pictures of the leaves. In order to increase the size of the unbalanced dataset in stage 2, appropriate 256*256 pixel images were obtained using a superresolution generative adversarial network (SRGAN The generated images are clearer that use expanded dataset with DoubleGAN than DCGAN. 99.80 and 99.53 percent accuracy rates, respectively, were attained for plant species and disease recognition, and the recognition outcomes outperform those from the original dataset. A method for improving the tomato disease detection algorithm by transfer learning was presented by Theodora Sanida and Argyrios Sideris [3]. The model is pre-trained using ImageNet and two VGGNet-based inception blocks. The findings demonstrate that the suggested approach produced significant values with an accuracy of 99.23%. The proposed model is a helpful tool for farmers.

[4] Deep Convolutional Neural Networks (DCNN) were used in Md. Iqbal Hossain and Sohely Jahan's suggested study. The primary dataset is subjected to separate applications of the Gaussian and median filters, and the resulting filtered photos are stored in the two different directories. Every directory's photos are subjected to independent applications of two color models (HSI and CMYK). Consequently, the pre-processed photos undergo four alternative approaches, the primary goal of which is to determine the optimal blend of color models and filtering techniques. Every output from the pre-processing stages is subjected to the DCNN model, and the most accurate approach is ultimately chosen. The model achieves 98.27% accuracy in Vgg-19, 94.98% in MobileNet-V2, and 99.53% in ResNet-50.

[5] Based on LMBRNet, Mingxuan Li and Guoxiong Zhou suggested using it to identify tomato leaf diseases. A complete grouped differentiated residual (CGDR) is constructed first. The many feature data of tomato leaf diseases in various sizes and receptive fields are recorded using the multi-branch structure of the CGDR. The multiple residual connection technique is then used to connect all of the network's layers, ensure the best possible information flow between them, and deal with problems related to gradient disappearance and network degradation during network training. Second, with the use of 1*1 convolution, max pooling, and average pooling, the visual improvement effectively blends the results of the three different down sampling methods. While InceptionResNetV2 (99.68%) accuracy and LMBRNet's (99.7%) accuracy are comparable, LMBRNet's (4.1M) parameter count is much lower than that of InceptionResNetV2(54M).

[6] Islam Hasibul A lightweight 2D CNN architecture based on deep learning was proposed by Peyal L. and MD. Nahiduzzaman to classify 14 classes—12 ill and 2 healthy—for both cotton and tomato crops. The outcomes were compared with those of other pre-trained models. A heatmap was made to illustrate and make clear the characteristics required for the categorization, and class activation maps plotted using the Grad-CAM technique assisted to graphically explain the disease detection provided by the trained model. An Android application named "Plant Disease Classifier" has been created to display classified diseases together with associated details, utilizing the effective lightweight 2D CNN model. When new cotton and tomato leaf diseases develop or existing ones change, the application's average detection accuracy may drop from 97.3% and the inference runtime may increase by more than 4.84 milliseconds.

[7] Calogero Schillaci and Alberta Odamea Anim-Ayeko presented the ResNet-9 model, which farmers can use to identify the state of potato and tomato leaf blight disease. Saliency maps were used to clarify the recommended work and show the reasoning behind the model's recommendations. It was found that when making predictions, the ResNet-9 model took into account the leaf's morphology, any diseased patches, and overall level of greenness. With an accuracy of 99.25% on the test set, the ResNet-9 model can identify the type of blight disease in leaf images of tomato and potato plants, as well as if these illnesses are absent.

[8] Prabhjot Kaur and Shilpi Harnal proposed an efficient Deep Learning (DL) modified Mask Region Convolutional Neural Network (Mask R-CNN) to assist with the autonomous segmentation and detection of leaf disease in tomato plants. The "Region Convolutional Neural Network (R-CNN)" light head is used in the suggested model to lower computational and memory expenses. The detection accuracy and metric performance are improved by adjusting the feature extraction topology and anchor proportions in the RPN network. It was discovered that the proposed model's accuracy, F1-score, and Mean Average Precision (mAP) were, respectively, 0.88, 0.912, and 0.98.

[9] ResNet-50, as suggested by Dongyan Zhang and Ying Huang, serves as the foundation for the modules that control object positioning, disassembly and reassembly, and attention area partition. The object positioning module accurately anticipated the lesions' locations without the need for any border annotation or parameter additions. In order to enhance the model's ability to learn fine-grained features, a Destruction and Reassembly module was developed. The ResNet-50 model's backbone network was a pre-trained object placement module, and it produced aggregation maps using the SCDA technique, enabling item localization based just on category labels. By using the label shuffling strategy, the training category imbalance was resolved and the resilience of the model was strengthened. Classifying tomato infections has a 98.25% accuracy rate. Table 1 displays a comparison of different deep learning models.

[11] Two distinct convolution layer CNN models, AlexNet and MobileNet, were proposed by K Indira and H Mallika to diagnose plant leaf diseases in 26 classes, including tomatoes, potatoes, rice, apples, sorghum, and bell peppers. The Sorghum dataset is produced by field photography; Kaggle is the source of the datasets for pepper bell peppers, tomatoes, potatoes, rice, and apples. Using MobileNet, tomato bacterial spot and tomato early blight severity detection was carried out for levels 1-4. The results of the simulation indicate that the accuracy for tomato bacterial spot and tomato early blight is 88.75% and 87.08%, respectively.

[12] To detect plant leaf disease, Pragya Hari, Maheshwari Prasad Singh, and Amit Kumar Singh introduced Federated Deep Learning (FDL). Instead of sharing entire datasets with siblings, this approach enables many local models to train with their region-based datasets and then distribute their knowledge to siblings through the parent. Computational costs are drastically

Table 1. Comparison of various deep learning models in tomato leaf disease prediction.

Methodology	Disease Detected	Dataset used	Result
Conditional Generative Adversarial Network (C-GAN), DenseNet model	Septoria leaf spot, yellow leaf curl virus, Two spider mites with spots, bacterial area, mold on leaves, the mosaic virus target location Both early and late blight	Plant Village dataset and synthetic images dataset	1. Accuracy for 5-class, 7-class, and 10-class classification tasks on the original PlantVillage dataset was 98.16%, 95.08%, ad 94.34%, respectively. 2. For 5-class, 7-class, and 10-class classification tasks, the accuracy was 99.51%, 98.65%, and 97.11% using the original PlantVillage + synthetic pictures dataset, respectively.
DoubleGAN,VGG16 ResNet50 and DenseNet121	Bacterial spot, Early blight, Late blight, Leaf mold, Partial leaf spot, Two-spotted spider mites, Target spot, Mosaic virus Yellow leaf curl virus	PlantVillage dataset	98.58%, 99.14% 99.7% accuracy using VGG16, ResNet50 and DenseNet121 respectively
VGGNet	LeafMold, Spider Mites, Septoria Leaf Spot, Early Blight, Bacterial Spot, Yellow Leaf Curl Virus, Late Blight, Mosaic Visus, Healthy, Target Spot	ImageNet dataset	VGGNet pre trained with ImageNet provides an accuracy rate of 99.23%.
DCNN, VGG-19, MobileNet V2, ResNet-50	Early Blight, Septoria Leaf Spot, and Yellow Leaf Curl Virus diseases	Plant Village dataset	1. An accuracy of 98.27% is obtained in VGG-19 2. An accuracy of 94.98% in MobileNet V2 3. An accuracy of 99.53% in ResNet-50
LMBRNet	Bacterial spot, Early blight, Late blight, Leaf mold, Septoria leaf spot, Mosaic virus	Plant Village Dataset	Accuracy of 99.7% is achieved
A deep learning-based lightweight 2D CNN architecture.	Leaf Mold, Spider Mites, Septoria Leaft Spot, Early Blight, Bacterial Spot, Yellow Leaf Curl Virus, Late Blight, Mosaic Virus, Healthy, Target Spot	Plant Village Dataset	Average accuracy of 97.36% is achieved using proposed model. The precision, recall, F1-score is 97%

decreased through knowledge transfer. Two sets of datasets (i) from apple, maize, and tomato crops and (ii) from cherry, maize, and tomato crops were produced with healthy and sick leaves to resemble the arrangement of the proposed FDL. Furthermore, it suggests using 0.09 million parameters to create a bi-stage feature fusion idea and create an effective, lightweight Hierarchical Convolutional Neural Network (H-CNN) with a parent model and child model size of 0.35 MB.

[13] Using the powers of deep Q-learning reinforcement learning, DQNAgent, and Keras, Najila Musthafa and Muhammad Afnan developed an approach that trains a robust CNN model for identifying whether or not a tomato plant is diseased from leaf pictures. The DQNAgent employed convolutional layers to identify diseases more accurately by identifying regional patterns and structures in the images. The model identified patterns in images that corresponded to particular diseases using techniques from reinforcement learning, such as experience replay.

[8] A deep convolutional Mask R-CNN architecture was presented by Prabhjot Kaur and Shilpi Harnal to identify multi-class plant leaf diseases. Excellent detection precision is achieved by rearranging the Mask R-CNN backbone network's topology and adjusting the anchor proportions of the Region Proposal Network (RPN) network. By utilizing an ensemble deep learning technique, the suggested model operates faster and produces more accurate instance segmentation results. With the annotation tool, the recommended method is used to multi-class datasets with several annotated photographs to identify even little infected patches of diseased leaves. Different illnesses are categorized by looking only at the affected area of the leaf, not the entire leaf. The program detects six different forms of tomato leaf diseases with a 98% accuracy rate.

[14] CNN-based categorization and image pre-processing are features of the model that Preeti Baser and Jatinderkumar R. Saini proposed. A publicly accessible dataset of tomato picture data is used to train the built model. These images are used to train the model as it passes through multiple CNN layers. The TomConv architecture is an optimized convolutional neural network with a maximum number of parameters and operations. After dividing the dataset in an 80:20 ratio, an accuracy rate of 98.19% was attained.

Table 2 displays the average validation accuracy for the following models: CNN, VGG-16, VGG-19, Inception-V3, MobileNet, MobileNetV2 as 81.39%, 82.97%, 92.31%, 83.65%, 57.50%, and 97.36% respectively in that order When compared to the transfer learning models, the CNN model shows significantly higher evaluation metrics (accuracy, precision, recall, and F1-score).

Table 2. Evaluation metrics comparison with transfer learning models.

Models	Accuracy(%)	Recll(%)	Precision(%)	F1-score(%)
VGG-16	81.39	85	86	85
VGG-19	82.97	83	86	82
Inception V3	92.31	92	93	92
MobileNet	83.65	88	90	88
MobileNetV2	57.50	81	86	82
CNN model	97.36	97	97	97

A target detection algorithm based on an improved version of YOLOv7 was proposed in [15] and it described how it performed admirably under harsh field conditions in identifying tomato leaves. The trimmed pictures can then be pre-handled utilizing an assortment of strategies, for example, pivot, interpretation, obscure ring, flipping scaling and horizontally to get the best results possible. The SIFT approach of picture division is utilized to extricate and choose vital characteristics that can be used to assess the health of tomato leaves plant. The findings of the experiments demonstrate that the model is able to precisely determine the health of a plant at a very early stage 98% of the time. One last comparison is drawn between the existing research and the proposed studies.

The proposed DCNN model in [16] as compared to popular transfer learning methods like AlexNet, VGG16 and GoogleNet dramatically lowers the training parameters and iteration durations. By combining typical handcrafted features (LBP) with deep features, an integrated model was created. This method captures well the local spatial texture information included in plant leaf pictures. A high accuracy rate was attained by the suggested model during training with several datasets related to plant leaf disease. The average accuracy of the proposed model is 91.2% for the nine disease and one healthy class. For disease identification and classification, [17] use convolution and three max pooling layers, with two fully connected layers applied last. The experiment's findings demonstrate that the suggested model performs better than trained models such as VGG16, InceptionV3, and MobileNet. The recommended model's average class-specific classification accuracy is 91.2%, with a range of 76% to 100%. Data augmentation techniques have been utilized to address the imbalance in the photographs since they are not in the correct proportion for the class.

A convolutional neural network-based approach to disease detection and classification is used in [18]. The model uses two pooling layers and two convolutional layers. The suggested model beat pre-trained InceptionV3, ResNet 152, and VGG19, according to the experiment's results. With 98% training accuracy and 88.17% testing accuracy, the CNN model performed well.

3 CONCLUSION

In the sector of agriculture, leaf disease detection is important and calls for increased accuracy in order to detect illnesses in real time. The early leaf disease detection model can assist in disease identification with a quicker detection time than earlier models. The report offered a survey of existing detection algorithm techniques and leaf disease detection algorithms. The suggested method's shortcomings and contribution to illness identification were also covered in the publication. The importance of early detection models and the leaf disease detection model were discussed in this research. Comparing the suggested DCNN model to popular transfer learning models like AlexNet, GoogleNet, and VGG16, the former dramatically lowers the training parameters and iteration durations. By joining typical handcrafted features with deep features, an integrated model was created (LBP). The local spatial texture information present in plant leaf photos is well captured by this model. Several datasets related to plant leaf disease will be used to train and acquire a high accuracy rate for the proposed model. As per the survey, DoubleGan model using DenseNet 121 architecture provides a maximum accuracy of 99.7%. The model can be extended to other parts of the plant.

REFERENCES

[1] Amreen Abbas, Sweta Jain, Mahesh Gour, Swetha Vankudothu. (29 June 2021). *Tomato Plant Disease Detection using Transfer Learning with C-GAN Synthetic Images*, https://doi.org/10.1016/j.compag.2021.106279

[2] Yafeng Zhao, Zhen Chen, Xuan Gao, Wenlong Song and Qiang Xiong. (May/June 2022). Plant disease detection using generated leaves based on double GAN, *Ieee/Acm Transactions On Computational Biology And Bioinformatics*, Vol. 19, No. 3

[3] Theodora Sanida, Argyrios Sideris A., Maria V., Sanida B. and Minas Dasygenis. (28 June 2023). *Tomato Leaf Disease Identification via Two–stage Transfer Learning Approach*, https://doi.org/10.1016/j.atech.2023.100275

[4] Md. Iqbal Hossain a, Sohely Jahan A., Md. Rashid Al Asif A, Md. Samsuddoha. (22 August 2023). *Detecting Tomato Leaf Diseases by Image Processing Through Deep Convolutional Neural Networks*, htps://doi.org/10.1016/j.atech.2023.100301

[5] Mingxuan Li, Guoxiong Zhou, Aibin Chen, Liujun Li and Yahui Hu. (31 March 2023). *Identification of Tomato Leaf Diseases based on LMBRNet*, https://doi.org/10.1016/j.engappai.2023.106195

[6] Hasibul islam peyal, Md. Nahiduzzaman and Md. Abu hanif pramanik. (12 October 2023). Plant disease classifier: Detection of dual-crop diseases using lightweight 2D CNN architecture, *Digital Object Identifier* 10.1109/ACCESS.2023.3320686

[7] Alberta Odamea Anim-Ayeko, Calogero Schillaci and Aldo Lipan. (10 January 2023). *Automatic Blight Disease Detection in Potato (Solanum tuberosum L.) and Tomato (Solanum lycopersicum) Plants Using Deep Learning*, https://doi.org/10.1016/j.atech.2023.100178

[8] Prabhjot Kaur, Shilpi Harnal, Vinay Gautam, Mukund Pratap Singh and Santar Pal Singh. (23 July 2022). An approach for characterization of infected area in tomato leaf disease based on deep learning and object detection technique, https://doi.org/10.1016/j.engappai.2022.105210

[9] Dongyan Zhang , Ying Huang, Chenxv Wu and Miaoyuan Ma. (15 september 2023). *Detecting Tomato Disease Types and Degrees Using Multi-branch and Destruction Learning*, https://doi.org/10.1016/j.compag.2023.108244

[10] Amitava Nag, Priya Raj Chanda and Sukumar Nandi. (19 October 2023). *Mobile App-based Tomato Disease Identification with Fine-tuned Convolutional Neural Networks*, https://doi.org/10.1016/j.compeleceng.2023.108995

[11] Indira K. and Mallika H. (19 February 2024). *Classification of Plant Leaf Disease Using Deep Learning*, https://doi.org/10.1007/s40031-024-00993-5

[12] Pragya Hari, Maheshwari Prasad Singh and Amit Kumar Singh. (13 March 2024). An improved federated deep learning for plant leaf disease detection, *Multimedia Tools and Applications* https://doi.org/10.1007/s11042-024-18867-9

[13] Najila Musthafa and Muhammed Afnan F. (2023). Identification and detection of tomato plant disease from leaf using deep reinforcement learning, *9th International Conference on Smart Computing and Communications* DOI: 10.1109/ICSCC59169.2023.10335002

[14] Preeti Baser, Jatinderkumar R. Saini and Ketan Kotecha. (2023). TomConv: An improved CNN model for diagnosis of diseases in tomato plant leaves, *2023, International Conference on Machine Learning and Data Engineering*, https://doi.org/10.1016/j.procs.2023.01.160

[15] Muhammad Umar, Saud Altaf , Shafiq Ahmad and Haitham Mahmoud. (2024). Precision agriculture through deep learning: Tomato plant multiple diseases recognition with cnn and improved yolov7, https://doi.org/10.1109/ACCESS.2024.3383154

[16] Khalid M. Hosny, (senior member, ieee), walaa M. El-hady and Farid M. Sam. (2023). *Multi-class Classification of Plant Leaf Diseases Using Feature Fusion of Deep Convolutional Neural Network and Local Binary Pattern*, https://doi.org/10.1109/ACCESS.2023.3286730

[17] Mohit Agarwala, Abhishek Singhb and Siddhartha Arjaria. (2020). ToLeD: Tomato leaf disease detection using convolution neural network, *International Conference on Computational Intelligence and Data Science*, https://doi.org/10.1016/j.procs.2020.03.225

[18] Winster Sathianesan, Vetri Selvan Murugan and Gnanavel Sakkarvarthi. (2022). Detection and classification of tomato crop disease using convolutional neural network, *Electronics*, https://doi.org/10.3390/electronics11213618

Data Science & Exploration in Artificial Intelligence – Gururaj et al. (Eds)

Power generation using piezo electric effect

Arun Kumar, Bhawna Khokher, Rishabh Pandey, V. Rohan & S. Varun
Department of Electronics and Communication Engineering, New Horizon College of Engineering, Bengaluru, India

ABSTRACT: The increasing reliance on mobile devices and the demand for environmentally friendly energy sources have sparked interest in cutting edge power generation methods. This study looks into using the piezoelectric effect to collect ambient energy and use it to charge mobile devices. Piezoelectric materials are deliberately integrated into the suggested system in locations that frequently experience pressure or vibration. When these materials are compressed, a little voltage is produced. This voltage is subsequently transformed into useful direct current (DC) via a rectifier circuit. A voltage booster circuit then raises the voltage to a level appropriate for charging a phone. In addition, the system uses RFID technology to allow user authentication and identification for safe and regulated access. This study explores this novel phone charging system's conception, creation, and performance evaluation. Future developments and possible uses in this field are also covered. also uses a voltage sensor in conjunction with an MCU to read the voltage produced by the piezo sensor module. A 5V supply, which can come from the battery or an external source, powers the LCD and microcontroller. The framework moves onto the registration stage. Two clients at most may be enrolled. The framework asks users to connect the charger and swipe their cards once every client has been added. Of course, every client gets ten minutes for charging. Later on, this time can be adjusted. The framework initiates billing and charges the mobile phone when the card is swiped, based on whether the client is approved. The LCD shows "unapproved client" if the client is not approved. The re-energize button is a part of the framework and can be used to reenergize a card.

Keywords: Piezoelectric effect, Energy harvesting, RFID, Mobile phone charging, Sustainable energy

1 INTRODUCTION

Mobile phone charging solutions need to be innovative, sustainable, and user-centric because to the growing dependence on mobile devices and environmental sustainability issues. This research introduces a revolutionary strategy that combines RFID technology with the piezoelectric effect to generate power in a safe and user-specific manner. Some materials display an intriguing phenomena called the piezoelectric effect, which is the conversion of mechanical stress or strain into electrical potential. Our suggested method, which uses piezoelectric sensors implanted in floor tiles, for example, to detect the energy produced by footsteps, is based on this idea. The right electronics can collect and store the voltage generated by it, no matter how small. RFID technology is used to ensure that only authorized users access this power generation system. The charging system is accessible by means of an RFID tag personalized with unique identification information [1]. After authenticating, phone owners will be able to connect their gadgets and commence the recharging process immediately when the system releases the unlocking mechanism of the charging port. While allowing user-specific charging, this combination also prevents unauthorized usage of created power [1]. This study discusses topics such as: Characterizing and Choosing Appropriate Piezoelectric Materials for Effective Energy Harvesting. Designing and optimizing circuitry for voltage rectification, regulation, battery storage. Establishing an authentication protocol and

DOI: 10.1201/9781003589273-57

incorporating an RFID reader for secure access control. An assessment of the system's overall performance, taking into account charging capacity and energy generation efficiency [2]. The suggested system offers the following significant benefits: It lessens reliance on traditional power sources by capturing renewable energy from routine activities like walking. Secure and customized charging experiences are ensured by the incorporation of RFID technology. Easy deployment is made possible by the modular architecture in a variety of high-foot traffic public locations, including airports, rail stations, and educational institutions. This research presents an innovative and promising approach for secure and sustainable mobile phone charging, utilizing RFID technology and the piezoelectric effect to contribute to a more environmentally friendly future and improved user ease [3].

2 METHODOLOGY

An inconsistent effect is produced by the piezoelectric material. Therefore, to transform this erratic voltage into a direct voltage, a scaffold circuit is used. Any remaining variances in the result are channeled through the AC swell channel. An electric battery powers the resulting direct current voltage, which is then stored. The portable gadget can then be charged with this battery using RFID access. The microcontroller ATMEGA328P is connected to the LCD. Because we paired a voltage sensor with an MCU, the LCD not only reads the voltage produced by the piezo sensor module when the framework is turned on, but it also displays a welcome message. A 5V supply, which can come from the battery or an external source, powers the LCD and microcontroller. The framework moves onto the registration stage. Two clients at most may be enrolled. The framework asks users to connect the charger and swipe their cards once every client has been added. Of course, every client gets ten minutes for charging. Later on, this time can be adjusted. The framework initiates billing and charges the mobile phone when the card is swiped, based on whether the client is approved. The LCD shows "unapproved client" if the client is not approved. The re-energize button is found in the framework and can be used to reenergize a card [4].

2.1 *Piezoelectric material selection*

Pick an appropriate piezoelectric material that has the ability to transform mechanical strain into electrical energy. Materials such as polyvinylidene fluoride and lead zirconate titanate are frequently utilized.

2.2 *Fabrication and design*

Create a system that enables the integration of piezoelectric material into a structure that is susceptible to vibration or mechanical strain. This could take the shape of a shoe insole, floor tile, or any other surface that is subjected to pressure or movement on a regular basis.

2.3 *Integration of RFID readers*

Install an RFID reader in the system where it can communicate with RFID tags that are inserted into the phone or phone cover. To enable the charging process it should be connected wireless.

2.4 *Optimization*

Verify that the circuit is Optimied and test it is working

2.5 *Safety measures*

Make it is safe to use by taking into account things like shock, electrical insulation,and make sure it follos rules and regulation.

3 RESEARCH METHOD

To begin with, the initial step is to gather and analyze the needs of various user groups (teachers, students, staff), so as to create a system that would meet their expectations. The first thing that should be done is reviewing the specifications for RFID (Radio Frequency Identification) integration and footstep power generation. In addition, it is important that hardware and software requirements are specified in order to ensure the system's functionality, compatibility as well as security issues. Finally, we consider security concerns during this phase. In order to manage projects effectively, it is important to develop a comprehensive project plan which includes tasks, deadlines and resource allocation.Selecting the development platform and programming language will guarantee that the selected hardware components work together and make development more productive. In order to integrate hardware (such as RFID readers and power generation units) with software and ensure smooth data flow and communication, it is important to design the entire system architecture. The main development effort was completed when we implemented the several modules and functionalities found during the design phase. Coding, hardware integration, and system functioning testing are all part of this step. Testing is necessary to find and fix bugs or problems in the system once it has been implemented. Both software and hardware components, as well as their integration, should be tested [5]. Lastly, we offer continuous maintenance and support, which is crucial for resolving any problems that may occur after implementation.

4 BLOCK DIAGRAM

Figure 1 shows the piezoelectric plate used by this device, which is based on the piezoelectric effect and converts mechanical energy to electrical energy. After being generated, this electric energy is passed onto a microcontroller, which interacts with other components. The charging system can be accessed by authorized users who swipe their RFID cards through an RFID reader for user authentication. An LCD may display the voltage generated while charging for people to watch it while taking place. [6]. To charge these power storing batteries, whatever devices users may have, they should be connected to the USB charging connector. The system utilizes a pre-set method for authenticating users, awarding charging time, tracking remaining time and providing means of stopping or prolonging the process of charging through swiping their cards. This creative system combines three technologies: energy harvesting, RFID authentication and a charger for producing renewable energy and securely charging mobile phones. Users can plug in their gadgets using the USB phone charger that comes with it and use its footstep generator as a source of battery power to charge them. How it works is that when one walks across, it produces electrical charges by pressing into piezoelectric crystals that are fitted on the floor surface underneath a carpeted mat that lies over a plate. Some elements in this system include an LCD display, USB cable and RFID reader powered by Atmel Mega 328 microprocessor which is part of Arduino development environment [7]. The device goes into recording mode as soon as it is powered on, making it possible to add up to three authorized users. Users can connect their gadgets for charging and scan their RFID cards once they have logged in. Five minutes of initial charging time are granted to all users; however, as charging proceeds, this time is subtracted. The system rejects any effort by an unauthorized user to get access. Users must swipe their cards once more to halt charging in the middle of the process, indicating how much time is left. Users can restart the process by swiping their cards in response to a button. Furthermore, a user's account gets awarded with an additional 5 minutes of charging time if they take a break [8].

Users can connect their gadgets to the USB phone charger included with the system and utilize the footstep generator's battery to charge them. The footstep generator uses the piezoelectric effect, in which mechanical stress from footfall is converted into electrical energy by the piezoelectric crystals in the plate. These crystals have balanced charges and a symmetrical atom arrangement, making them electrically neutral. Powering the system is an Atmel Mega 328 microcontroller, which is developed using the Arduino development environment. It incorporates parts including as LCD monitor, USB cable, and RFID sensor. The system can have up to three authorized users added when it is turned on and goes into recording mode. Users are prompted to scan their RFID cards and connect their

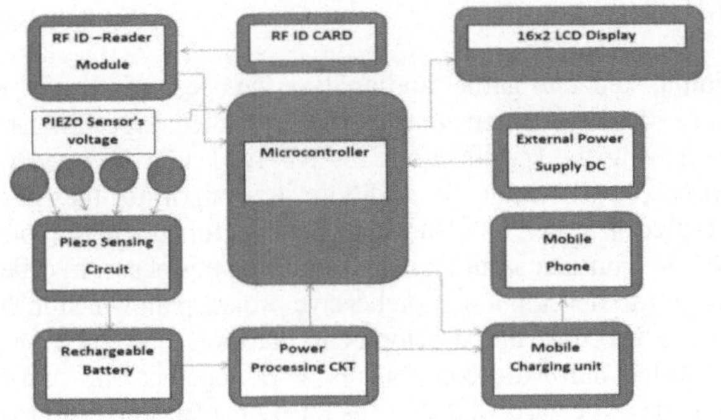

Figure 1. RFID device.

gadgets for charging after they have all registered in. At first, a 5-minute charging time is made available to all users. The system starts charging as soon as a user swipes their card, and as the charging goes on, time is subtracted. Unauthorized users are prevented from accessing the system upon their attempt. However, an authorized user must swipe their card again if they wish to halt the charge process in the middle of the transaction.

4.1 *Working and operation*

Advanced footfall energy generation systems capture the energy of human footsteps using specially designed floors or footfall modules placed in large traffic areas. These models use electric motors or piezoelectric devices to convert foot movement or pressure into electricity. Users will need RFID tags or cards to interact with the system and use the power generated. RFID technology is integrated into the system mainly for user authentication and access control. The electrical energy generated by the steps is collected and stored in batteries or other electronic devices and used for many purposes, including lighting, electrical work for electrical appliances, and equipment payment. To monitor and control energy production and consumption, the system may use an LCD screen as part of the user interface. RFID authentication is required to access and use the electricity produced. RFID technology: RFID tags integrated into footprints use radios and antennas to store data and exchange wireless signals with RFID readers. RFID Reader.

- (I) RFID Tech:Using the Rf wave it detects.
- (II) RFID Reader: it send the dta to aurdino.
- (III) Storage.The power converter collects the electrical current produced by the signal and converts it into useful direct current (DC) power. After conversion, this energy is stored in a capacitor bank or rechargeable battery.
- (IV) Power Distribution and Usage: A power management system effectively distributes power based on demand, controlling voltage, current, and prioritizing power allocation. This allows the stored electrical energy to power multiple devices or systems.
- (V) Concerns Regarding the Environment: The system contributes to environmental sustainability by encouraging clean energy generation and lowering dependency on fossil fuels through the use of human motion as a renewable energy source.

5 RESULTS AND DISCUSSION

As soon as the piezoelectric sensors that are mounted on the piezo plate are subjected to strain. Energy is supplied (. As voltage, for example, feasible/dynamic energy is formed. 0.1 millivolts to 1.4 millivolts (as displayed by the video meter), etc., taking into account 6000 to 7000 strikes, which results in a voltage delivery of up to 12 volts. The energy we store in a battery-controlled battery will be used for any purpose. When power is provided to the piezoelectric sensors, the amount of power that the sensors convey is shown on an LCD. In addition, the strain that is applied is converted to a voltage plan and stored in a battery that is controlled by a battery. As illustrated in Figure 2, the RFID per user scans the card immediately after a customer swipes their card. If the client is supported, the LCD will display Embraced and charge for an estimated period of time; if not, the LCD will display Unapproved. Considering all of this, the highest strides power age is the finest preservationist and a decently noteworthy response for average citizens of our nation, one that may be viewed and appreciated with a reasonable amount of effort. As seen in Figure 3, this can be applied in a variety of situations in well-known areas when the control supply is inadequate or completely delinquent. India is a potentially developing country where large-scale individuals will likely face significant challenges in terms of centrality association. We were able to drive both A.C. and stacking other than D.C. with the help of this errand, as indicated by the controls we connected to the piezo detecting device. This strategy provides an expectation in control span in highly populated countries since it reduces control requests without compromising on quality. In actuality, just about 12% of those who oversee the environment recall our fundamental life. Even if this assignment is provided in an unreasonable manner, it may still be able to address the sincere primary problem and contribute to a significant global shift. Power generation using piezoelectric impact offers a viable path toward affordable energy production. In contrast to the typical power sources shown in Figure 3, piezoelectric materials have the exceptional ability to convert mechanical strain or vibrations into electrical energy, providing a sustainable and environmentally safe alternative. This innovation has tremendous promise for a variety of applications, from demanding scope framework projects to limited scope energy collection in buyer hardware. Technological innovation that uses walking as a source of renewable energy that we can obtain by stepping on certain surfaces, such as piezoelectric boards. The framework uses piezoelectric sensors to generate a voltage from development. This action uses the sensors to create the most incredible electrical voltage. This will now be forwarded to our framework for verification. This circuit, which has several clients, is a microcontroller-based screen that allows the user to view the voltage and charge

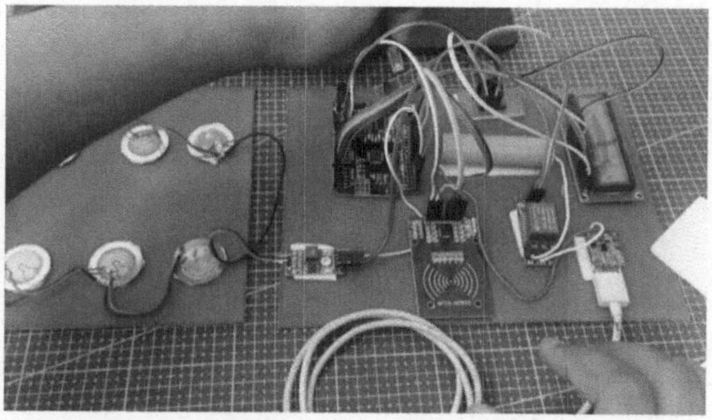

Figure 2. Full working model.

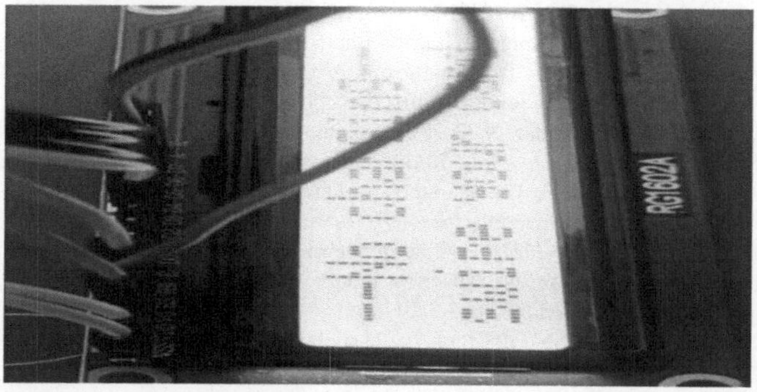

Figure 3. On screen it shows no charging.

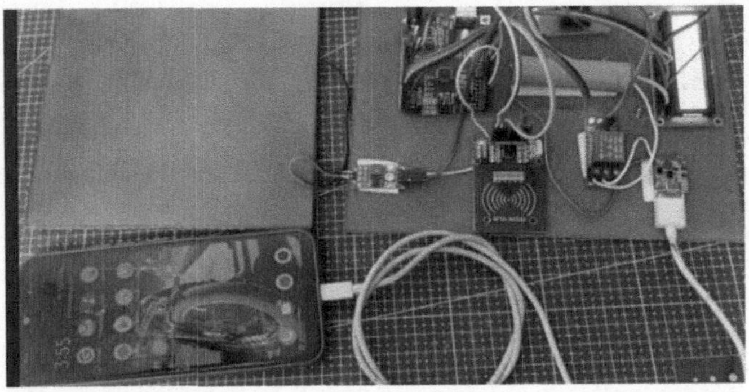

Figure 4. Integration with mobile.

level of the connected battery. Additionally, the charge produced by our walk A is displayed on an LCD. Additionally, it features a USB adaptable phone charging station that allows a customer to connect links to recharge their cutting-edge smartphone. RFID (radio-recurrence distinguishing proof affirming) cards are used to disperse the current so that the devices shown in Figure 4 can be charged by the generator only by those who have been granted access.

6 CONCLUSION

The utilization of the piezoelectric effect for power generation is a brilliant illustration of sustainability and ingenuity in the realm of energy technology. Piezoelectric devices offer a sustainable and environmentally friendly alternative to traditional fossil fuel-based power generation methods by utilizing the inherent ability of certain materials to convert mechanical strain into electrical energy. A wide range of applications, from large-scale infrastructure projects to small-scale energy harvesting in wearable devices, are made possible by their versatility. The widespread use of piezoelectric power generation has been spurred by ongoing research and development efforts to improve effectiveness, dependability, and affordability.

REFERENCES

[1] Profetto, L., Gherardelli, M. and Iadanza, E. (2022). Radio frequency identification (RFID) in health care: where are we? A scoping review. *Health Technol.* 12, 879–891. https://doi.org/10.1007/s12553-022-00696-1.

[2] Hassan, A.M., Ayoub, M., Eissa, M. *et al.* (2019). Development of an integrated RFID-IC technology for on-line viscosity measurements in enhanced oil recovery processes. *J Petrol Explor Prod Technol* 9, 2605–2612. https://doi.org/10.1007/s13202-019-0638-5.

[3] Covaci C and Gontean A. (2020). Piezoelectric energy harvesting solutions: A review. *Sensors.* 20 (12):3512. https://doi.org/10.3390/s20123512.

[4] Umer Farooq, Imran Baig, Muhammad Khurram Bhatti, Habib Mehrez, Arun Kumar and Manoj Gupta. (2023). Prototyping using multi-FPGA platform: A novel and complete flow, *Microprocessors and Microsystems*, Volume 96, https://doi.org/10.1016/j.micpro.2022.104751.

[5] Kumari, P., Mathew, L. and Kumar, N. (2023). Combining biosignals with RFID to develop a multimodal-shared control interface. *J. Eng. Appl. Sci.* 70, 119. https://doi.org/10.1186/s44147-023-00291-9.

[6] Úbeda A, Iáñez E, Azorín JM. (2013). Shared control architecture based on RFID to control a robot arm using a spontaneous brain–machine interface. *Rob Auton Syst* 61(8):768–774.

[7] Yu, J., Xian, S., Mu, J. *et al.* (2024). Hybrid electromechanical properties of hetero-doped and homogeneously bonded dual-mode pressure sensor for indoor body area network node. *Sci. China Inf. Sci.* 67, 112401. https://doi.org/10.1007/s11432-023-3801-1.

[8] Mejjaouli, S. (2022). RFID-WSN-GPS-Based Cold supply chain real-time monitoring and control system. *Arab J Sci Eng* 47, 15117–15131. https://doi.org/10.1007/s13369-022-07047-2

Data Science & Exploration in Artificial Intelligence – Gururaj et al. (Eds)

Analysis of peak to average power ratio in OTFS waveform using hybrid PTS and clipping algorithm

A. Kumar
Department of Electronics and Communication Engineering, New Horizon College of Engineering, Bengaluru, India

A. Nanthaamornphong
College of Computing, Prince of Songkla University, Phuket, Thailand

R. Bhardwaj
Department of Computer Science and Engineering, ASET, Amity University, Bengaluru, Karnataka, India

A. Das
Department of Information Technology, Manipal Institute of Technology Bengaluru, Manipal Academy of Higher Education, Manipal

ABSTRACT: OTFS will be essential in the advanced radio sixth generation (6G) system. OTFS has a unique ability to enhance the signal quality and throughput in large Doppler scenarios. OTFS improves spectral access, supports multiple devices, and fast rate communication, which are the major requirements for deploying the 6G. High peak to average ratio (PAPR) in OTFS hampers the framework's performance by decreasing the power amplifier employed in the OTFS throughput. The article proposed a hybrid approach combining partial transmission sequence (PTS) and Clipping, also known as PTS-Clipping. PTS segmented the sub-blocks, and the signal was weighted with the phase factor to minimize the peak value. Further, Clipping is applied, which removes the prominent peaks. The simulation outcomes of the work reveal that the impacts, such as PAPR and bit error rate (BER), are efficiently enhanced compared with the existing algorithms with trivial complexity.

Keywords: OTFS, PAPR, PTS-Clipping, BER

1 INTRODUCTION

Orthogonal time-frequency space (OTFS) is a unique transmission approach that provides efficient communication in high scenarios compared to the existing advanced waveforms. The operation of the OTFS is centered on the Doppler effect, which converts the time-frequency scenario into a flat fading dimension, overcoming the interference between the symbols. Hence, there will be a reliable and robust transmission of data in multi-fading scenarios (Chu *et al.* 2022).

OTFS consists of several advantages such as robust performance with fast speed rate, enhanced spectral access, and high capacity, which will play a vital role in many applications such as the Internet of Things (IoT) and railways, where quality and robust transmission of information is essential and should preserve the connectivity in dynamic environments. OTFS has an important characteristic of utilizing the time-frequency conditions, enabling it to improve modulation quality in noisy environments. (Raviteja *et al.* 2019).

Despite several advantages of the OTFS, the increase in PAPR is considered one of the most severe challenges of the OTF framework. The increase in peak power upsurges the non-linearity

DOI: 10.1201/9781003589273-58

of the signal. Also, it enhances the complexity and power utilization factor of the power amplifier and the hardware of the framework. Further, the upsurge in PAPR results in high spectral leakage due to the sideband enhancement, which further affects the framework's efficiency. It is noted that the PAPR cannot be eliminated. Hence, it is essential to design a suitable approach to minimize the large amplitude of the OTFS symbols. Several schemes have been implemented in recent decades to reduce the PAPR for conventional waveforms such as OFDM. However, it is understood that the PAPR schemes utilized in OFDM cannot apply to OTFS due to the variance in their orientation and structure. In recent works, it has been noted that the optimal result in PAPR is obtained by compromising the intricacy of the structure (Kumar *et al.* 2019).

The study and estimation of the PAPR are crucial for deploying OTFS in a 6G radio system. Some traditional algorithms, such as selective mapping (SLM), are considered the most important approaches to lowering the high amplitude of the waveforms. SLM effectively lowers the peak value with trivial intricacy but also results in distortion due to the transmission of the side information along with the SLM. PTS is also one of the dynamic approaches that effectively improves PAPR performance. In PTS, the primary task is generating the phase factors weighted with the OTFS symbols, improving PAPR. However, the period required for finding a suitable phase element upsurges the computational intricacy of the structure. Tone reservation and tone injection also play a significant role in reducing the high peaks but also result in signal overhead.

Further, phase segmentation and precoding techniques are popular methods for reducing the PAPR. Moreover, PAPR techniques based on A.I. and ML care have also been implemented in recent years to address the complexity issue of the waveform. (Ye *et al.* 2014). In the proposed article, we introduced an integration of PTS and Clipping approaches, which significantly improve PAPR for the OTFs system. The PTS approach includes segmenting the OTFS sub-blocks followed by the estimation of peak value by transforming the signal from the frequency to the time domain. The symbols are estimated, and the optimal phase sequence is multiplied with the large amplitude symbols. The weighting of OTFS symbols and phase elements optimizes the power performance of the power amplifier. However, it is also noted that the increase in the selection of sub-blocks upsurges the computational intricacy. Hence, the sub-blocks should be selected appropriately. Afterward, Clipping is applied to the PTS symbols, which further enhances the system's performance; However, the use of Clipping may also result in loss of information. Hence, it is essential to design optimal PAPR algorithms. The structure of the paper is as given: Sections 1 and 2 indicate the background and the article available in the proposed topics, section 3 presents the analysis of the OTFS waveforms, and sections 4 and 5 consist of simulation results and conclusion.

2 LITERATURE REVIEW

Surabhi *et al.* (2019) studied and analyzed the impact of the OTFS framework due to an increasing peak power. The authors introduced a unique approach to optimizing the structure's gain and throughput. The experiment's outcomes revealed that the presented method produces superior performance to the traditional system.

A study by the authors (Kumar *et al.* 2023) introduced a novel hybrid algorithm that can be applied to the OTFS waveform. The primary goal of the article was to overcome the high peak power and optimize the performance of the power amplifier. The computer graph reveals that the projected algorithm acquired a 3 dB gain as compared with the existing approach. Further, the intricacy of the framework was also analyzed, and it is noted that the hybrid methods upsurge the resources required, resulting in high computational power.

It is essential to analyze the PAPR of the advanced transmission system to understand the power requirements and performance of the power amplifier. The amplifier is hugely affected by the distortion introduced by the high amplitude. Kumar *et al.* (2024) projected a PTS that effectively lowered the high amplitude of the signal and gave superior efficiency to the OTF system. The spectrum radiation was also reduced, further enhancing the framework's quality.

Nagaraju *et al.* (2020) analyzed the peak power of an OTFS framework and projected a combined T.I. and Clipping approaches to reduce the high amplitude of the radio. It is seen that the individual algorithms could not obtain a significant performance, which was why the performance was always compromised. However, the projected method acquired a significant PAPR gain compared to conventional methods without compromising the other impacts of the system.

SLM results in efficient performance but is greatly affected by computational problems.

The authors of Ma *et al.* (2022) implemented a dual approach, such as SLM and T.I., to reduce the framework's peak amplitude, improving the system's overall performance. Further, the throughput of the structure has also greatly improved compared with contemporary approaches.

The utilization of the OTFS waveform for the railway operation is analyzed and estimated by the

Xiao *et al.* (2022) provided a complete analysis of the OTFS impact on IoT operations. The article projected the different requirements and constraints to roll out the OTFS for sensor communication. However, it is noted that the OTFS showed better efficiency than the existing approaches.

Shang *et al.* (2023) studied OTFS's unique application in high-speed vehicular systems and its usage in different sensor-related applications. The proposed article introduced novel schemes that can efficiently improve the system's error rate and power performance. The results show a better gain compared with contemporary approaches.

An optical system can enhance the spectral performance of the OTFS waveforms. Kumar *et al.* (2024) comprehensively analyze the optical-OTFS impact due to the PAPR with 256-QAM. The novel companding, Clipping, and filtering were applied, and the framework's PAPR was estimated. It is seen that the projected scheme performed better than the contemporary schemes.

3 PAPR ANALYSIS OF OTFS WAVEFORM

OTFS is considered one of the unique solutions for high-speed communication systems due to its unique time-frequency characteristics. It is essential to analyze the schematic utilized in the OTFS structure as given in Figure 1. The OTFS symbols are transmitted by utilizing a 256-QAM. However, it is necessary to determine the amplitude of the system. Hence, the signal's frequency domain is transformed to the time domain by using the inverse short Fourier transform. Afterwards, the filtering is applied to smoothen the signals. Again, the FFT is utilized at the receiver terminal to convert the signal from the time domain to the frequency domain. Finally, the signal is demodulated to acquire the original OTFS signal. The OTFS will encounter a PAPR issue in the transmitting section of the framework. Hence, applying a PAPR approach after transforming the signal into the time domain is essential. The IFFT of the signals increases the amplitude from linear to non-linear region, increasing the power requirement of the amplifier utilized in the framework. It is also noted that the large number of sub-carriers will further upsurge in peak power. Hence, it is essential to design an optimal transmission system that can control the dynamic noise introduced by the non-linear nature of the signal. However, it is seen that the advanced waveform's structure failed to control the loss that occurred due to the rise in peak power. Hence, the design of an optimal power algorithm is essential for the robust efficiency of the system (Gunturu *et al.* 2021).

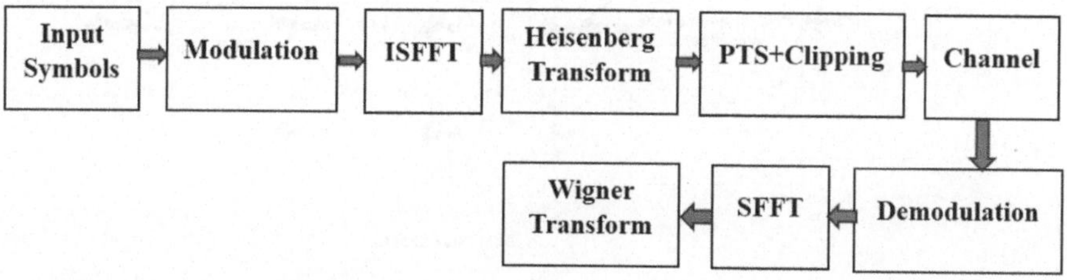

Figure 1. Structure of OTFS waveform.

In this section, we are considering the OTFS signal when it is converted into the time domain as shown in:

$$x(t) = \sum_{l=0}^{L-1} \sum_{k=0}^{K-1} s_{l.m} P(t - nT) \exp(j2\pi fmt) \tag{1}$$

In equation (1), N and M denote the latency and Doppler factors, $s_{l.m}$ indicates the amplitude, and P(t) is the pulse shaping filter. The PAPR is the estimation of the highest amplitude of the signal to the ratio of the expected value of the signal with a particular duration. The PAPR is given by:

$$PAPR_{dB} = \frac{Max[|x(t)|]^2}{E\left\{[|x(t)|]^2\right\}} \tag{2}$$

The fluctuation of the amplitude results in non-linear distortion, which hampers the structure's efficacy. Hence, it is essential to analyze the PAPR and apply a suitable algorithm to diminish the high values of PAPR. A system with low PAPR effectively contributes to superior performance and better capacity.

3.1 *Partial Transmit Sequence (PTS)*

PTS is a practical scheme that can be applied to advanced waveforms, resulting in optimal gain and efficiency. In the PTS, the initial step is to divide the OTFS symbols and map the symbols into the sub-blocks. The selection of sub-blocks can reduce the PAPR to an optimal level. However, it also introduces an intricacy in the structure. Once the symbol is mapped, estimating the signal's peak value is essential. Hence, IFFT is applied, which converts the frequency domain OTFS signal to the time domain. The amplitude of the OTFS signals is estimated, and phase sequences are generated to suppress the high amplitude. The phase elements are multiplied with the OTFS signal, resulting in an optimal value. However, the choice of phase signal can also upsurge the period to search for ideal phase elements. The projected SLM utilized a center phase generation method, reducing the time span to select the best phase elements. Hence, the proposed structure can give a suitable performance without compromising the other impacts of the system, such as throughput and complexity. The suggested algorithm allocates the power to the OTFS signal in different domains, which helps minimize the high amplitudes—however, the PTS results in throughput degradation due to increased error in the receiving OTFS signals. However, in the projected article, we aim to preserve the throughput and reduce the peak values to an optimal level. (Kumar *et al.* 2024). The structure of the system is represented in Figure 2.

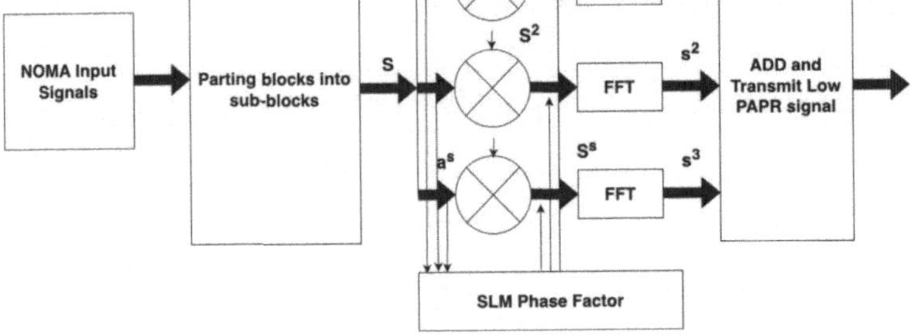

Figure 2. PTS schematic.

In this section, the mathematical representation of the projected PTS is presented. Let us assume an $x(t)$ OTFS signal, which we aim to reduce the amplitude. The OTFS signals are mapped into a sub-blocks(t), which is mathematically represented as:

$$s(t) = \sum_{n=0}^{N-1} S_n(t) \tag{3}$$

In equation (3), it is seen that the sub-blocks can be defined depending on the number of OTFS symbols mapped into it. The amplitude estimation is performed by applying an IFFT to the OTFS-mapped symbols. The sequence of phase elements ($\exp(i\varnothing_n)$) is obtained to diminish the high peak of the signal. The optimization in peak signals is performed by weighting the sub-blocks with the $\varnothing_n$, which can be represented as:

$$s_n(t) = S_n(t) * \varnothing_n \tag{4}$$

$$s_n(t) = \sum_{n=0}^{N-1} S_n(t)\exp(i\varnothing_n) \tag{5}$$

In equations (4) and (5), we can analyze the impact of generating the optimal phase sequence for lessening the peak value to the finest level. Consequently, if $\varnothing_n$ is not designed correctly, it may also increase computational intricacy. Afterward, the OTFS symbols with the low amplitude signal are selected for transmission, as shown in equation (6).

$$PAPR_{dB} = \frac{Max[|S_n(t)|]^2}{E\left\{|s_n(t)|^2\right\}} \tag{6}$$

The main aim of the projected approach is to obtain the best and most robust system without compromising the other impacts of the system.

3.2 *Clipping*

The Clipping is applied to the advanced waveforms to clip the signal's amplitude if it exceeds a certain threshold. Setting an optimal amplitude level is an essential task of the design. The clipping approach diminishes the high peak by attenuating the high amplitude of the waveform. However, the projected approach also results in a loss of information and the introduction of noise in an OTFS waveform. Hence, it is essential to optimize the different parameters of the Clipping to ensure high signal quality and high PAPR performance (Kumar *et al.* 2024). The structure of the projected Clipping is given in Figure 3.

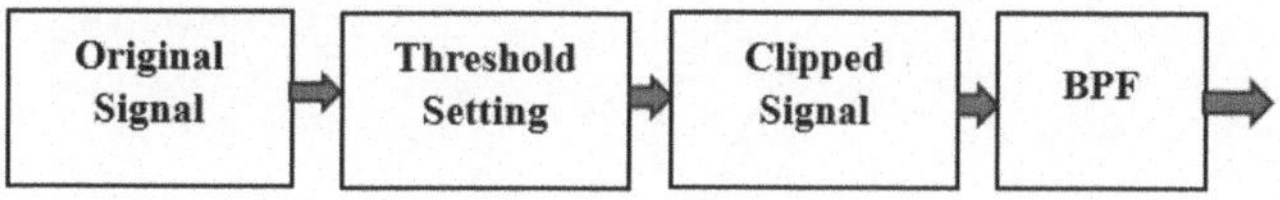

Figure 3. Clipping schematic.

Equation (7) indicates the use of the clipping method to diminish the large amplitude of the OTFS signal represented as A_{clip}. The system is represented by equation (7) as given below:

$$s_{clip}(t) = \begin{cases} s(t), & if|s(t)| \leq P_{Clip} \\ P_{Clip}\exp(j\varnothing), & otherwise \end{cases} \tag{7}$$

Where $\varnothing$ an angular representation of the Clipping. Further, the signal is processed at the receiver using a robust algorithm to help recover the lost signal.

4 SIMULATION RESULTS

The PTS algorithm was simulated for the OTFS using Matlab-2016. We used 10,000 symbols, 256-QAM, 256-FFT, and Rayleigh channels. To analyze the performance of the PAPR methods in the OTFS, we estimated the CCDF performance for 16 subcarriers, as shown in Figure 4. The PAPR values of 3.8 dB, 5.2 dB, 7.2 dB, and 8.8 dB are obtained at the CCDF of 10^{-3}. The projected PTS+Clipping attained a substantial SNR gain of 1.4 dB and 3 dB, as equated with the conformist schemes.

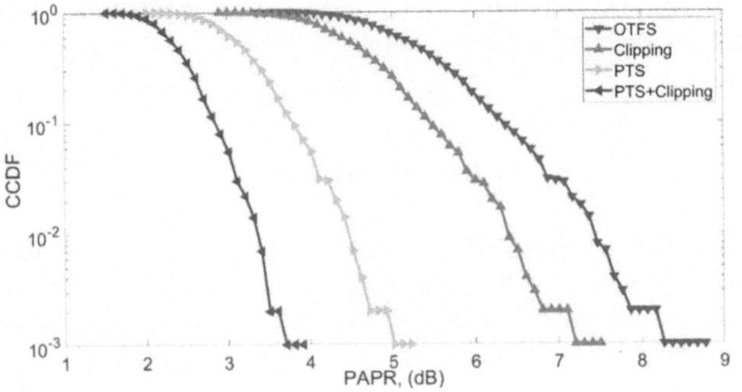

Figure 4. PAPR for 16 sub-carriers.

The PAPR of the OTFS wave evaluated for 64 subcarriers using the PTS+Clipping method is shown in Figure 5. At the CCDF of 10^{-3}, the PAPR of 6 dB (PTS+Clipping), 7.2 dB (PTS), 8.3 dB (Clipping), and 10 dB (OTFS) are obtained by the PAPR approaches. It is also observed that the PAPR performance of the 16 subcarriers is better than the 64-sub-carriers. Hence, it is confirmed that the high transmission schemes significantly affect the framework's PAPR performance. However, it is also noted that the large sub-carriers can effectively eliminate the interference among the symbols, which can result in optimal spectral access and reduction in spectrum leakage. Hence, the system with 64 sub-carriers guarantees robust performance with high capacity and data rate.

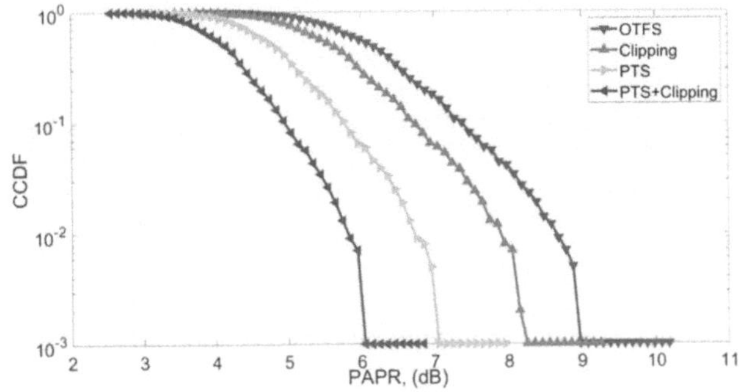

Figure 5. PAPR for 64 sub-carriers.

It is crucial to analyze the advanced waveforms with 256 sub-carriers, which can ensure their applicability in advanced radio systems that demand high-speed data, spectrum access, capacity, and so on. It is also noted that many sub-carriers introduced a high level of non-linear distortion, which can hamper the power amplifier performance to a greater extent. Hence, it is important to design an optimal algorithm for 256 sub-carriers. The PAPR curves for the 256 subcarriers are shown in Figure 6. The CDDF of 10^{-3} is achieved at the PAPR of 7.8 dB, 8.6 dB, 9.3 dB, and 10.6 dB, respectively, by PTS+Clipping, PTS, Clipping, and OTFS signal. Hence, it was determined that the future PTS+Clipping provided superior performance for lower to higher subcarriers.

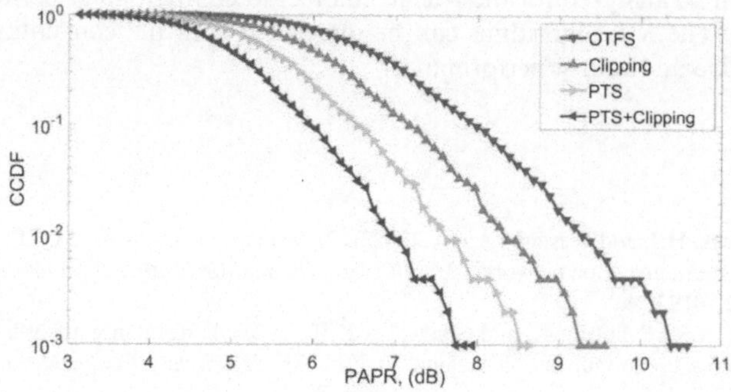

Figure 6. PAPR for 256 sub-carriers.

In conventional methods, the degradation of the BER is considered a significant problem that needs to be addressed for advanced waveforms. It is not easy to optimize the PAPR and BER performance. However, we have retained the BER in the projected work compared with the contemporary schemes. The existing algorithms failed to retain the error rate, but the PTS has retained the performance of the system. The BER curves for the 256 subcarriers are shown in Figure 7. The BER of 10^{-5} is obtained at the SNR of 7.4 dB by PTS+Clipping, 8.8 dB by PTS, 9.3 dB by Clipping, and 11.8 dB by OTFS. It is seen that the PTS+SLM outperforms the orthodox systems.

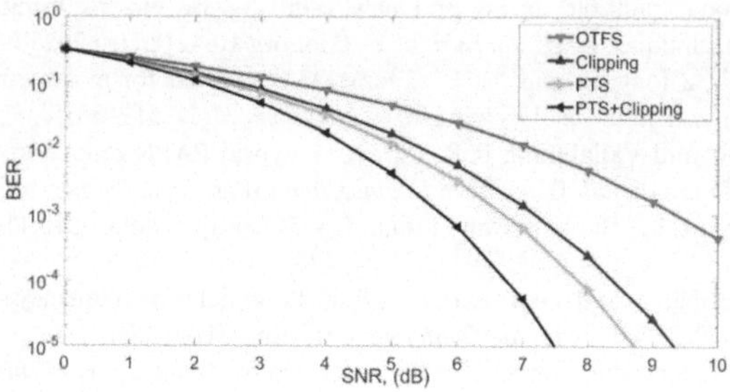

Figure 7. PAPR for 256 sub-carriers.

5 CONCLUSION

The increase in radio system applications demands a robust system that can satisfy several requirements, such as fast data rate, device 2-device connectivity, low latency, high bandwidth access, etc. The advanced waveform will be essential in satisfying the demand and optimizing the system's efficacy. The advanced waveforms are centered on multicarrier schemes, which raise the PAPR problem. The increase in signal amplitude results in the degradation of power amplifier performance, increasing power utilization and making the system more efficient. This work introduced a hybrid PAPR approach to the OTFS waveform. The PTS and clipping approaches are integrated to take advantage of obtaining a superior performance. The projected method attained a PAPR gain of 1 to 3.3 dB compared with the existing schemes, and the BER performance was retained for the OTFS method. The PTS+Clipping attained an SNR gain of 1 to 1.8 dB compared to the existing method. It is also noted that the projected scheme is suitable for different numbers of sub-carriers. The combination of two-stage schemes also upsurges the computation requirements, resulting in hardware complexity. In the future, the projected work can be analyzed for the Rician channel to compare their performance with the Rayleigh channel. The ML algorithms can be integrated with the conventional approach to generate an adaptive and robust performance.

REFERENCES

Chu, T.M.C., Zepernick, H.J. and Westerhagen, A. (2022). Performance assessment of OTFS modulation in high Doppler airborne communication networks. *Mobile Networks and Applications* 27: 1746–1756. https://doi.org/10.1007/s11036-022-01928-4.

Gunturu, A., Godala, A.R., Sahoo, A.K. and Chavva, A.K.R. (2021). Performance analysis of OTFS waveform for 5G N.R. mmWave communication system. *In Proc. IEEE Wireless Communications and Networking Conference (WCNC)*, 1–6.

Kumar, A. (2019). PAPR reduction of FBMC using hybrid and k-hybrid techniques. *Radioelectronics and Communications Systems* 62: 501–509. https://doi.org/10.3103/S0735272719100029.

Kumar, A. (2023). Analysis of PAPR on NOMA waveforms using hybrid algorithm. *Wireless Personal Communications* 132: 1849–1861. https://doi.org/10.1007/s11277-023-10683-y.

Kumar, A., Chakravarty, S. and Nanthaamornphong, A. (2024). Analysis of PAPR reduction of optical-OTFS for 256-QAM using companding and clipping-filtering algorithms. *Journal of Optical Communications*. https://doi.org/10.1515/joc-2023-0369.

Kumar, A., Gaur, N. and Nanthaamornphong, A. (2024). Optimizing PAPR, BER, and PSD efficiency: Using phase factors generated by bacteria foraging algorithm for PTS and SLM methods. *IEEE Access* 12: 54964–54977. https://doi.org/10.1109/ACCESS.2024.3389823.

Kumar, A., Gaur, N., Chakravarty, S. and Nanthaamornphong, A. (2023). Reducing the PAPR of OTFS modulation using hybrid PAPR algorithms. *Wireless Personal Communications* 133: 2503–2523. https://doi.org/10.1007/s11277-024-10885-y.

Kumar, A., Rajagopal, K., Gaur, N. and Nanthaamornphong, A. (2024). Reducing peak-to-average power ratio of filtered non-orthogonal multiple access and new radio 5g waveforms using hybrid partial transmit sequence-companding technique. *ETRI Journal*. https://doi.org/10.4218/etrij.2023-0405.

Ma, Y., Ma, G., Wang, N., Zhong, Z. and Ai, B. (2022). OTFS-TSMA for massive internet of things in high-speed railway. *IEEE Transactions on Wireless Communications* 21(1): 519–531.

Nagaraju, V.S., Anusha, R. and Vallabhuni, R.R. (2020). A hybrid PAPR reduction technique in OFDM systems. *In Proc. IEEE International Women in Engineering (WIE) Conference on Electrical and Computer Engineering (WIECON-ECE)*, Bhubaneswar, India, 364–367. https://doi.org/10.1109/WIECON-ECE52138.2020.9398011.

Raviteja, P., Phan, K.T., Hong, Y. and Viterbo, E. (2019). Orthogonal time frequency space (OTFS) modulation based radar systems. *In Proc. IEEE Radar Conference*, Boston, MA, USA, 1–6.

Shang, H., et al. (2023). OTFS modulation and PAPR reduction for IoT-railways. *China Communications* 20(1): 102–113. https://doi.org/10.23919/JCC.2023.01.009.

Surabhi, G.D., Augustine, R.M. and Chockalingam, A. (2019). Peak-to-average power ratio of OTFS modulation. *IEEE Communications Letters* 23(6): 999–1002. https://doi.org/10.1109/LCOMM.2019.2914042.

Xiao, L., Li, S., Qian, Y., Chen, D. and Jiang, T. (2022). An overview of OTFS for internet of things: Concepts, benefits, and challenges. *IEEE Internet of Things Journal* 9(10): 7596–7618. https://doi.org/10.1109/JIOT.2021.3132606.

Ye, C., Li, Z., Jiang, T., Ni, C. and Qi, Q. (2014). PAPR reduction of OQAM-OFDM signals using segmental PTS scheme with low complexity. *IEEE Transactions on Broadcasting* 60(1): 142–148. https://doi.org/10.1109/tbc.2013.2282732.

Data Science & Exploration in Artificial Intelligence – Gururaj et al. (Eds)
© 2025 the Author(s), ISBN: 978-1-032-96390-7

Implementation of Double Stage Energy detection (DSEd) spectrum sensing algorithms for advanced waveforms

A. Kumar
Department of Electronics and Communication Engineering, New Horizon College of Engineering, Bengaluru, India

A. Nanthaamornphong
College of Computing, Prince of Songkla University, Phuket, Thailand

R. Bhardwaj
Department of Computer Science and Engineering, ASET, Amity University, Bengaluru, Karnataka, India

ABSTRACT: Due to the increase in wireless standards, there is a need for more spectrum, which degrades the framework's performance. Spectrum sensing is regarded as one of the techniques based on the cognitive radio (Cr), which can enhance the access to white and unused bandwidth. However, the identification of the idle spectrum is one of the important phenomena that can improve the overall efficiency of the framework. In this article, we projected a double-stage Energy detection (DSEd) scheme to identify the idle spectrum from the primary user (Pu) and allocate it to the secondary user (Su) irrespective of noise and interference. The DSEd detects the unused signal with low complexity by setting a threshold value. The projected DSEd greatly enhanced the performance when the false alarm rate was greater than 50%. The impacts, such as the probability of detection (Pd), false alarm (Pfa), and power spectrum density (PSD), are analyzed and simulated. The computer simulation outcomes reveal that the projected method performed better than the contemporary ED. Further, the throughput of the system is evaluated by analyzing the error performance of the framework, making the system efficient and robust.

1 INTRODUCTION

The deployment of the fifth-generation (5G) radio has been completed, and the development of the sixth-generation (6G) advanced system is becoming popular among the industry and academicians. The application numbers in the present scenario are tenfold compared with the previous radio system. The upsurge in radio applications needs a sufficient spectrum for robust and highly efficient efficacy. Designing an advanced waveform that can utilize the spectrum most efficiently is essential. OTFS waveform is regarded as one of the emerging techniques that can be deployed in the 6G due to its delay Doppler characteristics of transforming the multipath fading to a flat fading scenario (Kaabouch *et al.* 2014).

The integration of CR in OTFS further enhances its bandwidth utilization capabilities and can overcome the spectrum scarcity problem. The basic idea is to identify the unused bandwidth and assign it to some applications. The use of Cr in OTFS 6G helps to achieve a reliable transmission of the transformation, and handover between the devices can be achieved without any noise and distortion. The CR OTFS can use AI, which helps 6G to autonomously manage and allocate the spectrum based on the dynamic conditions. Further, it also helps 6G to produce a self-adaptive system with no spectrum scarcity. successful rollout of Cr depends on sensing the unused spectrum

DOI: 10.1201/9781003589273-59

(Du *et al.* 2015). The sensing capabilities of the different algorithms can be integrated to overcome their Numerous SS approaches, including SS approaches such as matched filter (MF), Ed, and Cyclostationary spectrum (Cs). However, each algorithm has certain advantages and constraints. Ed accesses the signal efficiently as its design is simple and less complex, but the requirement of received signal power is more in Ed, making it unsuitable for advanced systems. MF detects the spectrum with a low power requirement, but prior channel state estimation is necessary for its success. This will upsurge the computation intricacy in the framework. Conversely, Cs utilized the signals' cyclostationary characteristics using cross-correlation and auto-correlation. The existing individual approaches can be integrated to provide efficient gains. Additionally, using smart and adaptive Cr will help the 6G make the resource available to every user and offer a complete, reliable transmission (Ejaz *et al.* 2013). In 6G, the spectrum range will be different, and many more applications will be defined, which need dynamic spectrum access for different devices. This requires a huge intelligence and robust design to avoid distortion and interference between the signals. The efficiency of the SS approaches while deploying the framework will be vital for the regularization of the 6G system. Further, the Cr should give an optimal performance as dynamic spectrum management will be vital for the connectivity between Pu and Su (Zhang *et al.* 2010). The high spectral access will ensure that the 6G can be used for several applications such as IoT, vehicular transmission, and Radar applications, which can overall develop a superior system that is capable of handling a series of communication applications and scenarios. The CR also helps to create a robust system for voice and data for the subscribers where the surplus bandwidth will be available and can be interchanged without hampering the connectivity among the users. The system will also ensure that the users will not experience harmful interference resulting in call drops and inefficient activities (Affan *et al.* 2021). Spectrum sharing also facilitates continuous tracking of the secondary user, ensuring end-to-end security and helping to track malicious communication. Thus, implementing advanced Cr will benefit the 6G, ensuring robust, ubiquitous, and low-latency transmission (Kumar *et al.* 2024). The requirement of continuous monitoring of the white spectrum is one of the significant tasks, as the noise can be misrepresented as the original signal. However, the condition for false alarms usually occurs during a fading environment. Hence, the sharp filter design can help overcome interference and noise. The original signal needs to be considered for the estimation with the threshold value. The primary challenges in implementing hardware devices involve efficient design and cost. Further, the Cr should be efficiently designed to avoid non-linear noise and interference, which can reduce the accuracy of the algorithms. Further, the wireless signal endures various losses due to shadowing, multipath fading, co-interferences, Doppler effect, etc. These propagations severely affect the quality of the framework red, increase the percentage of quality detection of the signal, and enhance the maximum probability of false alarm conditions (Kumar *et al.* 2023). In recent years, academicians have been focusing on the implementation of ML approaches to identify the unused spectrum. Implementing the ML-Cr system performs a dynamic detection of white holes and makes decisions based on different environmental conditions. The dynamic nature of the environment and the device's noise may also affect the performance of the SS methods, which the ML methods can easily handle. The ML can also be combined with the different detection methods to overcome the limitations of the conventional methods. The devices utilized in the 6G radio, such as ADC, amplifiers, and RF devices, can also create interference that severely affects the whole system's efficiency. Hence, these hardware devices should be deployed effectively to avoid any problems. The design of idle devices is difficult to achieve as the proper analysis is poorly performed. However, it is also noted that the design of SS approaches should consider the rules and regulatory policies of the international telecommunication unit. Hence, the CR techniques must compile the policies, and the design should meet the framework's policy and protocols. Hence, SS algorithms should be properly investigated and analyzed for different advanced waveforms to estimate the performance based on several impacts. In this article, we projected a two-framework scenario where spectrum sensing is performed. The conventional Ed will perform the sensing when the Pfa is less than 50%, and for the Pfa greater than 50%, the projected DSEd is utilized for the processing of the signal (Kumar *et al.* 2024). The paper is structured as follows: Sections 1 and 2 provide the introduction and literature survey of the techniques related to the proposed work; sections 3 include the system model; and sections 4 and 5 provide the simulation results and conclusion of the proposed work.

2 LITERATURE REVIEW

Liu *et al.* (2015) introduced a novel algorithm based on combining the Ed and ML methods for the advanced waveforms scheme. The project approach creates a new dimension for analyzing the impact of Cr on the advanced radio framework. The article provides comprehensive details of the algorithm, including the challenges and drawbacks of the Cr in 5G. The work outcomes reveal that the projected method can be effectively utilized for high spectral access. Nesh *et al.* (2021) projected the implementation of the cyclostationary method for the NOMA structure and its application in IoT communication. The projected schemes evaluate the detection of the spectrum in dynamic scenarios to understand the performance of the suggested method. The outcomes of the article demonstrate that the projected SS obtained a significant detection and SNR gain compared to the contemporary methods. The outcomes of the analysis play an essential role in the application of IoT networks. Kumar *et al.* (2021) study the impact of the Cr in the NOMA waveform by analyzing the framework's Pd, Pfa, and throughput. The authors introduced a hybrid SS method, which reduces the false alarm and enhances the detection of the spectrum for the NOMA waveform. It is also seen that the projected method attained the gain of SNR of 1.9 dB compared to the existing method. However, the upsurge in complexity is considered one of the drawbacks of the hybrid approach. (Kumar *et al.* 2024) comprehensively analyze the CNN and RNN impact on the OTFS waveform. The analysis of the projected method is evaluated using the Rician and Rayleigh channels. The impacts of Pd, PFA, and BER are estimated using the 10000 samples. The computer graph reveals that the projected CNN and RNN attained a superior efficacy compared to the conventional Cr approaches. Cr is also essential in intelligent hospitals, allowing efficient connectivity between several devices. (Ramamoorthy *et al.* 2022) introduced a hybrid Cr-based SS method for the FBMC waveform, and it analyzed how the Cr helps to enhance the quality of the service in health care. The projected study confirms that the Cr play will be vital in deploying competent health care in the upcoming years by integrating it with 5G and 6G radio.

Choudhary *et al.* (2023) comprehensively analyzed the conventional SS approaches, such as MF, ED, and Cs, for the NOMA waveforms. Using Matlab, the SS methods are simulated for 64-QAM with the Rayleigh channel for 10000 samples. The work outcomes reveal that the cyclostationary method outperforms the existing Ed and MF. Further, it is also seen that MF's SNR performance is better than that of Ed. However, prior estimation of the channel is needed to design a novel MF, which upsurges the computation of the system. Swaroop *et al.* (2020) implemented a novel Ed SS method for the OFDM framework with the Rayleigh channel. The primary focus of the work is to estimate and analyze the Ed algorithm for 16-QAM and 16-sub-carriers for 20000 samples. The study reveals that the projected method required a higher SNR to sense the spectrum. The PSD parameter is also analyzed, producing high spectral access for 4G radio.

Further, the key challenges are also addressed in the presented article. Cheema *et al.* (2023) introduce an MF spectrum sensing method for the OFDM waveform with a Rician channel for 2000 samples and 64-QAM. The performance impacts, such as BER and Pfa, are studied and estimated. The study's outcome reveals that the projected MF outperforms the existing conventional MF.

3 SYSTEM MODEL

This section applies the projected double-stage ED spectrum method for the OTFS waveform to analyze the various impacts, such as detection, false alarm, and throughput of the structure. The efficiency of the projected algorithm depends on its ability to distinguish the noise and the original signal. The noise filtering is when pfa is less than 50%, and this can be effective by the conventional Ed method. However, when the Pfa is greater than 50%, identifying the original signal is complicated as it requires a complex signal processing algorithm. The framework of the projected system is designed in a way that, when the probability of the false is more negligible than 50%, the conventional Ed algorithm performs the task of sensing. In this case, a conventional Ed is implemented with some modification in the filtering process. However, when false alarms are more than 50%, the projected two Ed methods are combined, and signal detection occurs in two stages. In this case, the probability of identifying the original signal increases as tow-stage detection is carried out. Hence, the threshold setting is one of the essential concerns, as the characteristics of the signal may change dynamically. Hence, the self-adaption features are incorporated into the projected schemes using an advanced signal processing algorithm. Hence, the projected method upsurges the intricacy of the framework. The detection process is summarized in Table 1.

Table 1. Double-stage ED detection.

Step 1: Let us consider an OTFS signal the primary user receives. The system model is defined as the combination of the transmitting signal s(t), noise (N), and received signal represented as
$y(t) = s(t) + n(t)$,
Step 2: In this section, we implement the system to detect the presence of a signal consisting of two different steps:
In the first step, when the Pfa is less than 50%, the conventional Ed algorithm identifies the original signal at the lower SNR level. This ensures the proper detection of the samples based on the threshold value settings.
The projected algorithm is utilized in the second step when the PFA is more than 50%. In this case, the noise is higher, making accurate signal sensing difficult. Hence, the projected method senses the signal twice. The DSEd performs rigorous dynamic signal detection.
Step 3: Operation:
In this stage, we have modeled the algorithm to detect the signal. The primary focus is to set the threshold value. The signal received at the receiver is analyzed and compared with the threshold value. The probability of accurate signal detection is 100% when the signal of the received energy is greater than the threshold value. However, there is the possibility that the noise misrepresented the original signal. Hence, in the projected method, we perform threshold checking two times. The misrepresentation of the signal usually occurs when the PFA is more than 50%.
Step 4: Threshold Analysis
The threshold analysis has to be performed for the conventional and projected schemes. For the conventional Ed, the threshold is usually low and compared with the received signal. For the projected method, the threshold is set two times. In the initial step, the threshold is dynamically set, considering noise and other factors. In the projected scheme, the threshold is set higher than in the conventional methods.
Step 4: Performance Evaluation:
The simulation impacts, such as Pd, Pfa, and BER, are analyzed using a computer Matlab simulation and compared with the existing methods.

4 SIMULATION RESULTS

In this section, we will simulate the conventional Ed by considering 200000 samples. We will use a 64-FFT, 64-QAM, and PHYDYAS filter with a roll factor of 0.01 and the Rayleigh and Rician channel.

The maximum Pfa at Pd = 0.17, 0.22, 0.43, and 0.57 by the conventional ED, MF, cyclostationary, and proposed DSEd schemes. The performance of conventional SS methods could be better, and the PFA is higher than that of the DSEd. Here, it is concluded that the projected scheme obtained better efficacy as it employed advanced signal processing filtering techniques that can be easily separated by the projected method.

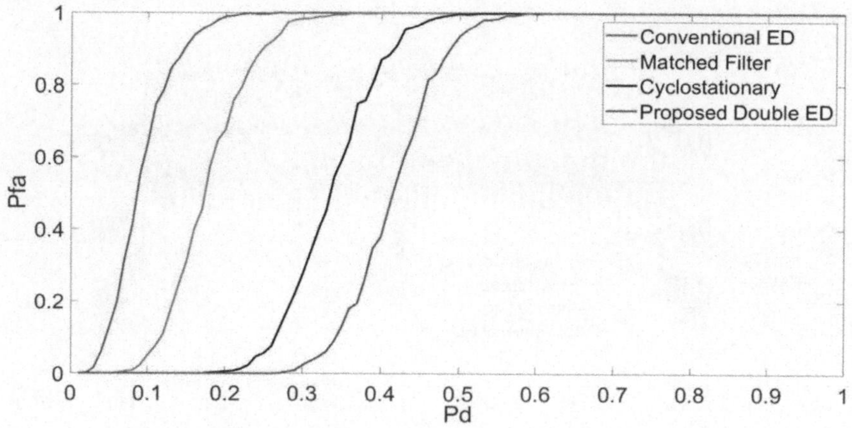

Figure 1. Pd vs. PFA analysis.

A Bit Error Rate (BER) curve analysis of the spectrum-sensing schemes is shown in Figure 2. The BER of 10-5 is obtained at the signal-to-noise ratio (SNR) of 7.1 dB by the proposed DSEd as compared with the 9.1 dB (Cyclostationary), 11 dB (MF) and 12.1 (ED) respectively. The projected DSEd enhanced the throughput of the framework due to its ability to remove distortion efficiently. It is also seen that conventional algorithms sense the signal accurately at the cost of the throughput. However, in the projected algorithm, the optimal detection performance is obtained while the framework's throughput is preserved.

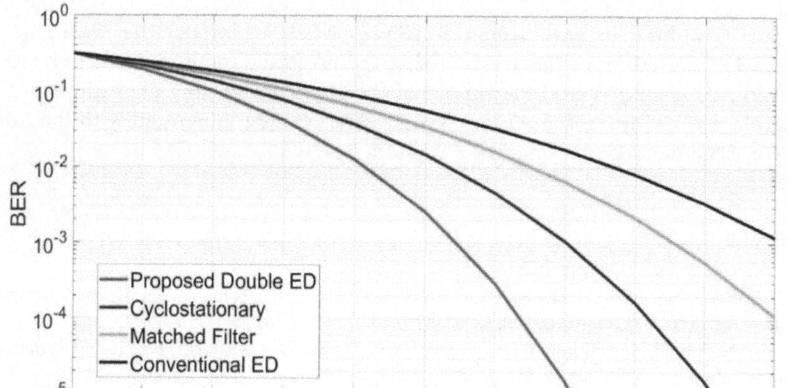

Figure 2. BER analysis.

The PSD of the spectrum sensing schemes was analyzed in Figure 3. PSD values of -730, -520, -480, and 360 were obtained by the proposed DSEd, MF, cyclostationary, and conventional ED, respectively. In this section, the main aim is to reduce the side lobe of the frame, reducing the spectrum's leakage. The traditional algorithms increase the side lobe, which also results in loss of spectrum. However, the projected method reduces the sidelobe, which reduces the spectrum leakage and enhances the capacity by accommodating a number of sub-carriers.

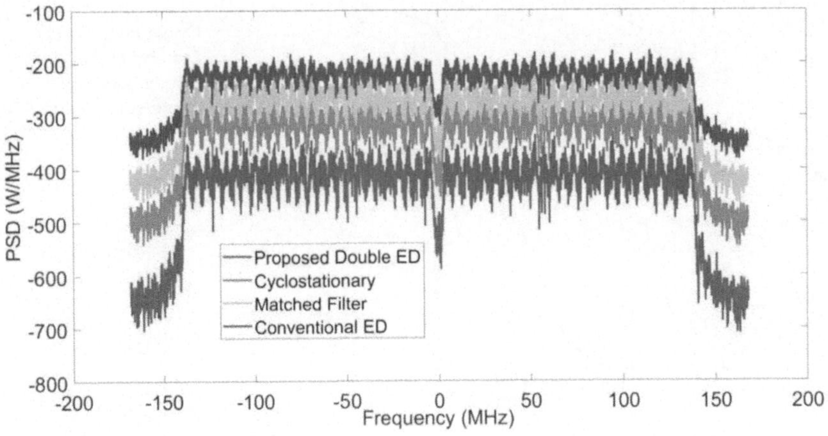

Figure 3. PSD analysis.

5 CONCLUSION

The utilization of the unused spectrum plays a significant role in an advanced radio framework. It is noted that the spectrum is scarce, but more than 70% of it needs to be utilized and used. There is an 11% wastage of spectrum due to the use of cyclic prefixes in 4G, which can be one of the significant issues in an OFDM structure. The projected article introduced a novel double-stage energy detection algorithm that can sense the signal in the presence of strong interference and noise. The projected scheme is suitable when PFA Pfa is more than 50%, and the conventional Ed is utilized if PFA Pfa is less than 50%. Hence, the proposed method can be utilized in both scenarios where noise can impact the throughput of the framework. The limitation of the article is that the projected method needs to be tested and implanted in real scenarios, and the rollout of the spectrum sensing algorithms in the 6G is unpredictable. In the future, the AI and ML schemes can be integrated with the existing SS methods for dynamic decision and self-adaption.

REFERENCES

Affan, A., Mumtaz, S., Asif, H.M. and Musavian, L. (2021). Performance analysis of orbital angular momentum (OAM): A 6G waveform design. *IEEE Commun. Lett.* 25: 3985–3989.

Arun Kumar, Nishant Gaur, Sumit Chakravarty, Mohammed H. Alsharif, Peerapong Uthansakul and Monthippa Uthansakul. (2024). Analysis of spectrum sensing using deep learning algorithms: CNNs and RNNs. *Ain Shams Engineering Journal*, 15(3), 102505.

Choudhary, S., Kumar, A., Sharma, M.K. and Masud, M. (2023). A mathematical investigation of multi-carrier waveform with AI-based matched filter detection. *Journal of Interdisciplinary Mathematics*, 26(3): 321–328. DOI: 10.47974/JIM-1663.

Cheema, A.P., Kumar, A., Gonuguntla, A., Reddy, P.A.K., Gnapika, S. and Nanditha, U.N. (2023). A survey on spectrum sensing using energy detection for 4G waveform. *In Proceedings of the 2023 International Conference on Computer Communication and Informatics (ICCCI)*, Coimbatore, India, 2023, pp. 1–4. doi: 10.1109/ICCCI56745.2023.10128182.

Du, C., Huacheng, Z., Wenjing, L. and Thomas, H. (2015). On cyclostationary analysis of wi-fi signal for direction estimation. *IEEE Mobile Wireless Network Symposium*. New York, NY: IEEE, pp. 3557–3561.

Ejaz, W., Hasan, N.U., Lee, S. and Kim, H.S. (2013). I3S: Intelligent spectrum sensing scheme for cognitive radio networks. *EURASIP Wirel. Commun. Netw. J.*, 2013, 26.

Kaabouch, N. and Hu, W.-C. (2014). Handbook of research on software-defined and cognitive radio technologies for dynamic spectrum management. *Hershey*, PA: IGI Global.

Kumar, A., Albreem, D.H. and Le, D.N. (2021). A comprehensive study on the role of advanced technologies in 5 G-based smart hospital. *Alex. Eng. J.*, 60: 5527–5536.

Kumar, A., Gaur, N., Chaitra, S. N., Chakravarty, S. and Nanthaamornphong, A. (2024). Analyzing the role of spectrum sensing for security enhancement in beyond 5G waveforms using energy detection. *Journal of Discrete Mathematical Sciences and Cryptography*, 27(2-A): 215–222. DOI: 10.47974/JDMSC-1875.

Kumar, A., Kaur, R., Gaur, N. and Nanthaamornphong, A. (2024). Exploring and analyzing the role of hybrid spectrum sensing methods in 6G-based smart health care applications [version 1; peer review: 1 approved with reservations]. F1000Research, 13:110. https://doi.org/10.12688/f1000research.144624.1.

Kumar, A., Venkatesh, J., Gaur, N., Alsharif, M.H., Jahid, A. and Raju, K. (2023). Analysis of Hybrid Spectrum Sensing for 5G and 6G Waveforms. *Electronics*, 12(1), 138. https://doi.org/10.3390/electronics12010138.

Liu, Y., Zhong, Z., Wang, G. et al. (2015). Cyclostationary detection based spectrum sensing for cognitive radio networks. *J. Commun.*, 10(1).

Neshe, N.H. and Singh, M. (2021). Hybrid spectrum sensing techniques in 5G cognitive radio networks in soft computing: A review. *Comput. Intell. Mach. Learn.*, 2(1).

Ramamoorthy, R., Sharma, H., Akilandeswari, A., Gour, N. and Kumar, A. et al. (2022). Analysis of cognitive radio for LTE and 5G waveforms. *Computer Systems Science and Engineering*, 43(3): 1207–1217. https://doi.org/10.32604/csse.2022.024749.

Swaroop, R. and Kumar, A. (2020). A brief study and analysis of NOMA techniques for 5G. *In Proceedings of the 2020 IEEE International Women in Engineering (WIE) Conference on Electrical and Computer Engineering (WIECON-ECE)*, Bhubaneswar, India, 2020, pp. 13–16. doi: 10.1109/WIECON-ECE52138.2020.9398031.

Zhang, Z., Yang, Q., Wang, L. and Zhou, X. (December 2010). A novel hybrid matched filter structure for III 802.22 standard. *In Proceedings of the 2010 IEEE Asia Pacific Conference on Circuits and Systems*, Kuala Lumpur, Malaysia, 6–9, pp. 652–655.

Data Science & Exploration in Artificial Intelligence – Gururaj et al. (Eds)
© 2025 the Author(s), ISBN: 978-1-032-96390-7

Predicting analysis of heart attack risk using machine learning algorithms and deep neural networks

Anju Gautam & Chandresh Bakliwal
School of Computer Science & Engineering Department, Nirwan University, Jaipur, India

Reema Ajmera
Computer Science Department, GIT, India

Dinesh Kumar Dharamdasani
University of People, California, USA

Rajat Bhardwaj
Department of Computer Science and Engineering, ASET, Amity University, Bengaluru, Karnataka, India

ABSTRACT: Heart attack is a serious event that entails urgent medical attention. It is also known as myocardial infarction in medical science world. Past history depicts that it serves as a crucial indicator of mortality rates on a global scale. Over the last few years techniques based on machine learning (ML) have developed into powerful tools for predicting the likelihood of heart attacks. The purpose of this research is to examine early identification of heart attack situations using machine learning models. Just 14 parameters from the UCI heart disease patient database are used in this study's testing. Many techniques of Machine Learning are such as Logistic Regression, Decision Tree, Random Forest, Naive Bayes(NB), SVM Model, K-Nearest Neighbor's (KNN) have used, including RNN from deep neural networks, decision trees, random forests, k closest neighbours, and regression. Found that RNN is the most accurate algorithm among all.

1 INTRODCTION

Diseases related cardiac, is one of the world's major causes of mortality, taking millions of lives each year. is a leading cause of death globally, accounting for millions of fatalities each year. Myocardial infarction occurs when the blood supply to a part of the heart muscle is blocked, resulting in the irreversible damage and death of cardiac cells. Diverse dynamics can educate the risk of heart failure. Those risks are categorized by the healthcare scientist into two broad categories; i) immutable risk factors ii) mutable risk factors. Patient family background like disease, sex, age is immutable risk factors. Cholesterol ratio, smoking, dormant life style, obesity and high BP are categorized in mutable risk factors.(Ali *et al.* 2019) As per the current scenario "approximately 17.9 million people every year lose their lives due to cardiovascular diseases (CVDs), apparent mainly as attacks"(Cardiovascular Diseases (CVDs), n.d.). The above scenario of death is 31% of total deaths globally(Mienye *et al.* 2020). Heart-diseases involved the heart and blood vessels. Heart related diseases are:

Reason behind heart attack problem is there is a significant reduction or complete interruption of blood flow to a specific area of the cardiac muscle. This lack of blood flow prevents oxygen from reaching that particular section of heart, leading to a gradual deterioration of the heart muscle cells. WHO said "the primary cause of a heart attack is often the rupture of an atherosclerotic plaque within a coronary artery, as these arteries are

 DOI: 10.1201/9781003589273-60

responsible for delivering oxygenated blood to the heart muscle. Once a plaque rupture occurs, a blood clot may form and obstruct a blood vessel, ultimately causing a heart attack". (Heart Attack-Heart Attack - Symptoms & Causes, n.d.).

"Myocardial infarction, commonly known as a heart attack, arises whilst there is a lessening in blood supply to the cardiac muscular section" (Martín-Timón 2014). Therefore, "lack of blood flow prevents oxygen and nutrients from reaching to the heart, and soon the heart muscle cells begin to die and this situation may led to heart attack"(Heart Attack-Heart Attack - Symptoms & Causes, n.d.).

Signs of HA can vary from person to person, but some common signs are:

- Chest pain: This is often described as a feeling of pressure, tightness, squeezing, or heaviness in the chest.
- Pain in other upper body parts: Discomfort may be felt in one or both arms, the back, neck, jaw, or stomach.
- Shortness of breath: Feeling breathless, especially when combined with chest discomfort, can be strong signs of a HA (heart attack).
- Cold sweat: Profuse sweating, even when not physically active or in a hot environment, can happen over the span of the heart attack.
- Nausea and vomiting: Some individuals may experience gastrointestinal symptoms over the span of the heart attack.
- Light headedness or fainting: Feeling dizzy or lightheaded can occur as blood supply to the brain is compromised.

1.1 *Risk factors of heart attack*

Eating habits and sedentary lifestyle promote the development of coronary heart diseases through the formation of atherosclerosis (Rathore 2018). Despite a theory that is currently gaining attention that the main triggering event is free radical damage to cholesterol in circulating low density lipoproteins, the main reason or initiator of atherosclerosis is still unknown (Ellulu *et al.* 2016). It is possible to acknowledge seven dietary factors with two inducers and five safeguards for coronary heart disease. The most severe consequence of coronary heart disease, myocardial infarction, combines the impact of two different dietary variables. Observational research suggests that reducing blood homocysteine with folic acid the risk of cardiovascular illnesses.

1.2 *Problem*

The WHO (World Health Organization) said that "the 32% of all global deaths were due to cardiovascular diseases out of which, approximately 85% deaths were due to heart attacks". (Cardiovascular Diseases, n.d.). The National Crime Records Bureau (NCRB) in India said, the percentage of deaths due to heart attacks is increased by approximately 53% in last five ears and approximately 33% of deaths in India were due to heart attacks and other probems(Deaths Due to Heart Attacks up by 53% in 5 Years: NCRB | India News - Times of India, n.d.-a). Although in many cases the patient may be saved by doctor even after minor or major attacks, but also in most of heart attack cases patient gets died due to unavailability of medical treatment under minimal time due to the incapability of patient to inform anyone during the attack period and also for their family members as being unknown from this kind of emergency condition to any of their relatives. And due to which, this kind of sudden attacks may lead to the death of patient even before consulting to any doctor. So, we cannot prefer consulting doctors after having attacks as the perfect solution for precaution from occurrence of heart attacks.

1.3 *Findings*

World health organization reported that "cardiovascular diseases (CVDs) are the leading cause of death globally. An estimated 17.9 million people died from CVDs in 2019, representing 32% of all global deaths. Of these deaths, 85% were due to heart attack and stroke. Out of the 17 million premature deaths due to non-communicable diseases in 2019, 38% were caused by CVDs. According to WHO maximum CVD's can be prevented by addressing daily routine and eating habits such as usage of tobacco, unhealthy diet and fatness, lethargic routine and usage of excessive alcohol(*Cardiovascular Diseases (CVDs), n. d.)*. It is essential for detection of cardiovascular disease as early as possible so that patient will takes preventive measure at early stage" (*Cardiovascular Diseases (CVDs)*, n.d.).

Heart attacks and strokes typically occur suddenly and are primarily the result of blockage that hinders blood supply to the heart and brain. The core reason of this blockage is the buildup of blubbery deposits on the inner walls of blood vessels, which limits blood supply to vital organs such as the heart and brain. Strokes may also be triggered by bleeding from a blood vessel in the brain or by blood clots. Strokes can be caused by bleeding from a blood vessel in the brain or from blood clots.

According to National crime bureau report the rate of heart attack is increasing by 2014. As per the bureau report 18309 people were died in 2014. This number rapidly increase every year and claimed 53% up in 5 years (Deaths Due to Heart Attacks up by 53% in 5 Years: NCRB | India News - Times of India, n.d.-b).

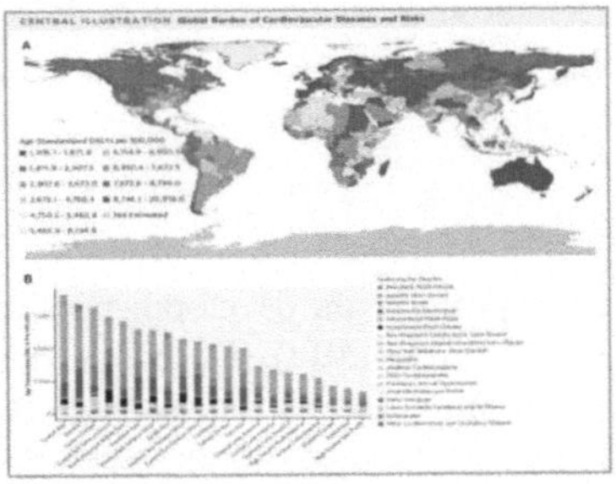

Figure 1. Death rate of highly effected countries, as per 2021. (Image taken from journal of American college of cardiology).

Though the life may save by the healthcare professionals if the patient or the family members know the condition otherwise in several cases when the situation is unknown to the patient or the family members than it may led to death in many cases of heart attack patient due to absence of medical action under marginal time owed to the inability of patient to tell someone during the attack situation. Hence, we can't favor referring healthcare professionals after having such type of situation, the perfect solution for precaution from occurrence of heart attacks. A person or a patient should seek medical advice prior and may take preventive measures to avoid minor or major heart attacks.

We analyze UCI data with the help of machine learning either heart attack condition will arrive in the forthcoming years or not. On the basis of dataset we analyze following conditions (i) Recognizing key attributes of heart attacks (ii) recognizing level of cholesterol which is having higher possibilities of heart attacks (iii) Compare and predict the risk factor gender wise(iv) Identifying a chest pain type, the major factors of heart attacks.(v)Prediction of the possibility of heart attacks in the future (vi) Preventive measures to avoid heart attacks.

The structure of paper as follows. The work done on the given problem has been shown in literature review in section 2. Section 3 discusses the dataset, feature selection various ML Algorithms and deep learning are deliberated in details. Furthermore, deliberate the method of cross-validation and discuss performance measure metrics. Section 4 outcomes of all investigation are investigated and deliberated in details. Section 5 the inference and future direction of the study have been discussed.

2 LITERATURE REVIEW

Machine learning and artificial intelligence are strengthening the predictive capacity of healthcare sector. This model can be used to estimate the likelihood of a heart attack. Numerous algorithms for predicting the risk of heart failure have been presented (Beunza *et al.* 2019). It is still possible to view the effectiveness of the different algorithms that have been employed to determine the possibility of heart failure as troublesome. Several popular machine learning algorithms, including k-nearest neighbor, support vector machine, gradient boosting, logistic regression, regression tree, and random forest, have been used in a comparative study to predict the risk of heart failure (Mienye *et al.* 2020). When comparison made among various feature selection machine learning algorithms, Naive Bayes, SVM, and KNN were the favorable predictors for predicting heart attacks. Another approach used for the cardiac infarction risk prediction involves partitioning the dataset in a random way and deploying various old data mining approaches like J48, REPTREE, Naïve Bayes, Bayes Net, and CART. The deployed model was able to answer more complex queries in cases of cardiac infarction prediction(Gupta *et al.* 2021) . Hasen *et al.* (Hasan *et al.* 2017) developed a diagnosis system using multi-layer Perceptron and support vector machine (SVM) algorithms for HD classification and achieved accuracy 80.41%.. This study build on existing literature by integrating advance feature of machine learning and DNN to create most robust model for heart attack prediction. If we integrate features of traditional statically model and DNN then it will achieve improved performance. Another significant characteristic of heart attack arrival is the demographic data, which includes age, gender, age range, and family background shown in Table 1. this research used UCI Cleveland dataset for the heart attack arrival condition analysis. The dataset consist of 13 decision parameters. The Target value is represented by "Target" Variable. Target value is the prediction column for diagnosis of heart attacks. It is classified into two types 0 and 1, where 0 means no possibility of heart attack and 1 means possibilities of heart attack.

Table 1. UCI Cleveland dataset.

Age	Patient Age at examination time
Sex	Classified in binary either 0 or 1 such as Female= 0, male=1
Chest Pain (cp)	Classified from 0 to 3 defining: 0 as typical angina, 1 as atypical angina, 2 as non-angina pain, and 3 as asymptotic
Resting BP(trestbps)	Resting blood pressure value of patient in mmHg (unit)
Cholesterol (chol)	Cholesterol of patient in mg/dl (unit).
Fasting Blood Sugar (fbs)	Categorized as either 0 or 1 such as 1 = if fbs >120 mg/dl (true) else 0 (false).
Resting ECG (restecg)	Categorized into three types from 0 to 2 defining: 0 = nor mal, 1 = having ST-T wave abnormality, 2 = left ventricular hypertrophy.
Max Heart Rate (thalach)	Maximum heart rate achieved by any patient
Exercise induced angina (exang)	Categorized as either 0 or 1 such as 0 = No and 1 = Yes
Oldpeak	Displays the value of ST depression of any patient induced by exercise w.r.t. rest (float values).
Slope	It describes the peak of exercise during ST segment, classi- fied in three ranges, 0 = up-slope, 1 = down slope

The study used variety of machine learning techniques, such as Support Vector Machine, Logistic Regression, Decision Tree, and Random Forest.

3 MATERIAL AND METHODS

3.1 *Methodology*

To reduce such type life threatening situations, artificial Intelligence and Machine Learning came into existence. ML have capability to forecast the likelihoods of forthcoming heart attacks as per to the health conditions and the past medical history of the patient. Even though various machine learning models have been developed and deployed in the direction of health disease and health failure. Found very less literature on heart attacks only. This study used a variety of ML algorithms and UCI datasets to anticipate the approaching heart attack scenario ahead of time and to keep the patient from becoming critically ill and dying.

3.2 *System architecture*

The system will include certain parameters to enhance the prediction or analysis process. It has recently been applied to train classification models. Following this, extract the necessary features that need to be identified for classification using Decision Tree (DT). When compared to current algorithms, this method shows superior performance. Following the categorization, it is necessary to compute the performance criteria such as accuracy, and precision.

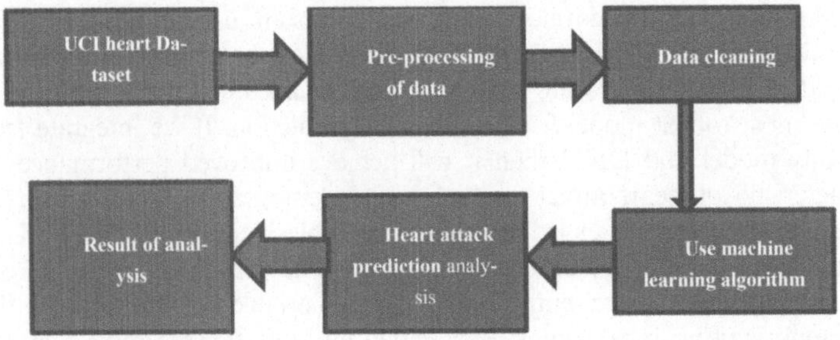

Figure 2. Architecture of the process of machine learning system.

3.3 *Datasets and pre-processing*

This research usage an online heart disease dataset, known as Cleveland heart disease dataset, was used which is available online on UCI machine learning repository. This datasets are also available on Kaggle website. This data set having 303 instances and 75 attributes with no missing values; in this research we use only 13 attributes. The dataset matrix information's are given in Table 2. Dataset, generally said to be as a collection of data organised for specific purposes (Andras Janosi 1989).

3.4 *Pre-processing*

Pre-processing: The database contains NaN values. Since the NaN programmed values cannot be processed, these variables must be converted into numerical numbers. The dataset has no null values. But numerous outliers present in the dataset that required special attention and having incorrect distribution of the data. There were two methods employed. Without using feature selection, outliers, or feeding the data straight into the algorithms, the results were unimpressive. After using the dataset's normal distribution to address the overfitting problem and subsequently ML used to categorize the outlier and very positive results were produced. These pre-processing techniques are all essential when sending data for classification or forecasting.

- The dataset does not contain any null values. But there were plenty of methods available that handle NaN values.
- Partitioning: Partitioning the dataset into 2 parts (i)Training (ii) Testing. Training makes use of the majority of the data. Testing is done on the remaining 40% of the data. With the remaining 60% of the data, testing is completed.

3.5 *Feature selection method*

The process of extracting a subclass from an initial feature set based on a predetermined feature selection norm, which chooses the pertinent features from the dataset, is known as feature selection. It contributes to the compression of the data processing scale, which eliminates characteristics that are superfluous and unnecessary. Learning algorithms can be pre-processed using feature selection techniques, and effective feature selection outcomes can shorten learning times, increase learning accuracy, and simplify learning outcomes.

3.6 *The methodology*

A quick introduction of the machine learning methods employed in this study is given in this section-

3.7 *Algorithms*

The algorithms that are used in this paper are Logistic regression, decision Tree, SVM Model, Random Forest, KNN and Deep Neural Network which can be helpful for practitioners or medical analysts for accurate prediction of Heart attack.

3.7.1 *Logistic regression*

Logistic Regression (LR): LR is a parametric supervised machine learning algorithm which used linear model that believes result of a categorical dependent variable from the dataset or independent variables. LR is widely used algorithm for binary classification. It estimates the probability that a given input belongs to a particular class. (Dreiseitl and Ohno-Machado 2002)

$$\text{Logistic function} = 1/(1 + e\hat{-}\text{value}) \tag{1}$$

The logistic function, also called the sigmoid function. This is an S-shaped curve and it can take any real-number and mapped those values between 0 and 1, but never exactaly at those limits.

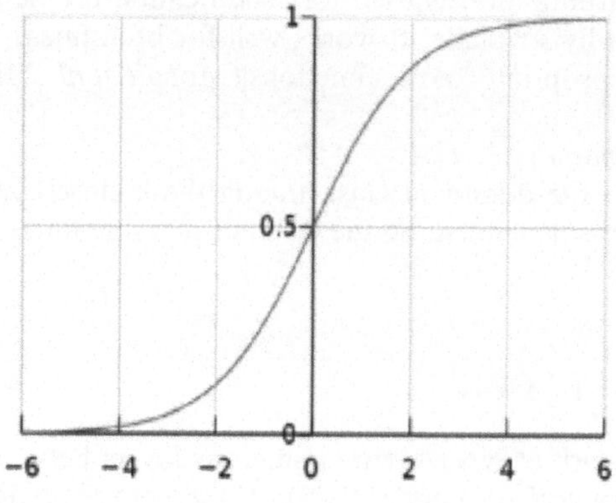

Figure 3. Logistic regression curve.

The data separated into 2 sets in LR modelling i) training ii) testing sets . The model would then be applied to the testing set to create predictions and assess performance using measures like F1-score, accuracy, precision, and recall.

3.7.2 *Decision Tree*

A tree-like structure that separates data recursively according to feature values. Decision tree is a non-parametric supervised machine learning algorithm. It is useful for solving classification and regression problems. It can deal with both binary and multi-class classification problems. (Molnar, n.d.) (Sarker *et al.* 2020).

3.7.3 *Random Forest*

It is a potent and adaptable ensemble learning algorithm used for both classification and regression tasks. It's particularly popular for its ability to provide high accuracy, handle a varied range of data types, and mitigate over fitting.

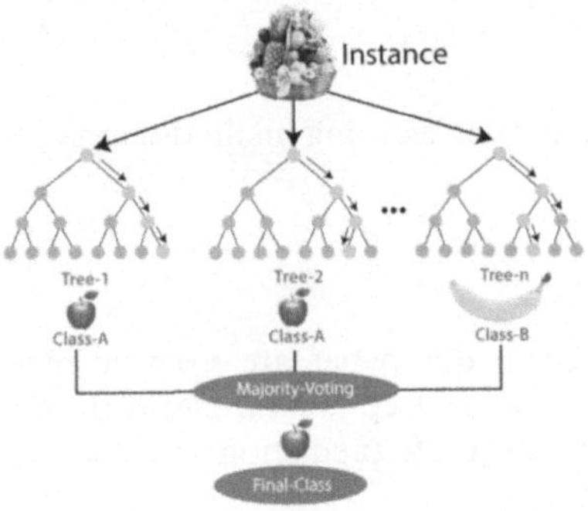

Figure 4. Model random forest(machine learning random forest algorithm - Javatpoint, n.d.).

3.7.4 *Naive Bayes(NB)*

Its foundation is the Bayes theorem. The likelihood that a data point belongs to a class is computed using this theorem. It makes the assumption that features are independent, which real-world data may not always support. By using the Gaussian function with prior probability, the NB model is trained Real-world data. NB model is trained through the Gaussian function with prior probability.(Mohan *et al.* 2019)

3.7.5 *SVM model*

SVM is a supervised learning models used for classification problems. The key purpose of this model is to create a hyper plane. It works well for both linear and non-linear classification problems with appropriate kernel functions.(Mohan *et al.* 2019)

3.7.6 *K-Nearest Neighbor's (KNN)*

This algorithm identifies the dominant class among the k closest neighbours of each data point in feature space. It is dependent on the distance measure and neighbour count (k) that are selected.

4 DEEP NEURAL NETWORK

An artificial neural network (ANN) having numerous layers between the input and output layers is called a deep neural network (DNN). These networks are referred to as "deep" because they have multiple hidden layers, and the quantity of hidden layers a network has determines how deep it is.

4.1 *The experiment*

An experiment in machine learning for predicting heart attack involves, building and evaluating a predictive model that can predict into two categories (e.g., yes or No) based on predictor variables Here are some general steps that can be followed to conduct such an experiment:

Data collection and preprocessing of data: Collect the data from UCI and pre-processing of raw data has to done. Sort the data into test, validation, and training sets. The machine learning model is developed using the training set, its performance is evaluated and its hyperparameters adjusted using the validation set, and its final performance is evaluated using the testing set. Training and validation are basic components of model. The performance of the machine learning model is evaluated using the training set. To analyse the performance of the model used Recall, accuracy, precision, and F1-score assessment metrics that should be used to assess the model's performance.

Hyperparameters Tuning: To improve the performance of machine learning model it is essential to Tune the hyperparameters.

Test the model: after hyperparameters tuning, measure the final output of each model on the testing set. To evaluate the model's performance appropriate assessment metrics, such as accuracy, precision, recall, and F1-score, has used.

Interpret the results: Interpret the results of the experiment to gain insights into the factors that contribute in heart attack. This could entail producing visual aids to aid in understanding the outcomes, assessing the significance of various predictor variables, or seeing patterns in the data.

Overall, an experiment includes statistical and computational tools to build and appraise analytical models that can help healthcare professionals for early diagnosis of disease.

4.2 *Feature engineering*

During the feature engineering stage of the model, a variety of feature transformers are applied and adapted to features. Principal component analysis, product, sum, univariate feature selection, square root, round, and principal component analysis are the feature transformers utilized in this model for customization in feature transformation. Every step is implemented on dataset iteratively. Feature selection is a novel approach of feature selection in which the choice of any given feature is predicated on mathematics univariate statistical tests, in which we assess each feature's substantial relationship to the label and compare its dependence on it. This mathematics univariate statistical tests term widely recognized as Analysis of Variance (ANOVA).

4.2.1 *Recurrent neural networks*

The structure of this deep learning model is straightforward, featuring a built-in feedback loop that functions as a forecasting engine. Essentially, an RNN is akin to a conventional neural network but incorporates an extra hidden state, which impacts the network's output. This hidden state undergoes updates with each input step. Consequently, the model not only grasps local and temporal dependencies within data but also adapts to varying sequence lengths. In RNN architecture, the output of a layer is concatenated with the subsequent input layer and recursively fed back into the same layer (Gavande and Chawan 2021).

4.2.2 *Experimental result*

The RNN implementation was conducted in Python. Each attribute had 13 input dimensions, and the RNN comprised 100 hidden layers. The ratio of training and testing records was 150:120. Training persisted for 100 epochs with a batch size of 10, employing the Adadelta optimization algorithm. During each epoch, the Authentication of area under ROC curve was computed using the validation set. If it yielded the best value thus far, the test set was employed to calculate the Test area under ROC curve. The model achieving the highest Test area under ROC curve, reaching 93%, was saved upon completion of training.

4.2.3 *Result and analysis*

As we can see that we perform several machine learning models for the predication of the occurrence of heart attack on UCI data set. And we found the accuracy of deep neural network method RNN gives better accuracy. All the used methods are also good in different perspectives. But in current experiment DNN having highest accuracy Table 2.

Table 2. Comparison of results.

METHODS	CALCULATED ACCURACY
Logistic Regression	88.5%
KNN	83%
RANDOM FOREST	85%
GRDIANT BOOST	81%
NB	86%
SVM	79%
DECISION TREE	84%
RNN	93%

This study contributes to the field of heart attack prediction in several ways:

Figure 6: Prospect of Heart attack o indicates No heart attack 1 indicates heart attack.

Aim of this research to attain maximum predictive accuracy using ML and DNN as compared to traditional model. This will give alarm to patient and medical practitioner of forthcoming possibility of Heart take and it will move to the severe heart disease.

4.2.4 *Interpretations*

a). Our model predicts the upcoming situation of Heart attack in future. The model usage value of 0 or 1 which means: value 0 indicates that no possibilities of heart attack; value 1 indicates that the prospect of heart attacks.

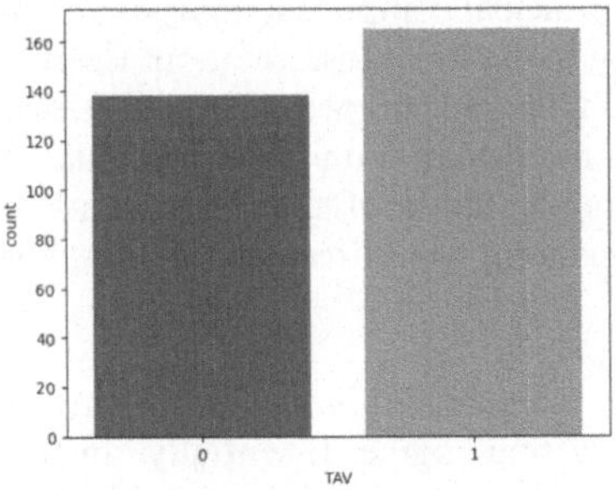

Figure 5. Bar Graph.

b). Key attributes that are contributed in Heart attacks are: cholesterol level, elevated heart rate, Pain in Chest or left part of body and elevated blood pressure. Age is also play a vital role in increasing cholesterol. People having higher cholesterol (>200) or higher heart rate (>150) have higher probability for occurrence of heart attack.

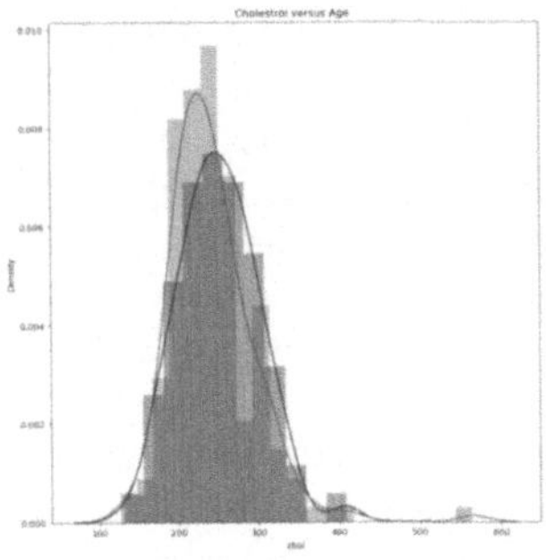

Figure 6. Prospect of Heart attack o indicates.

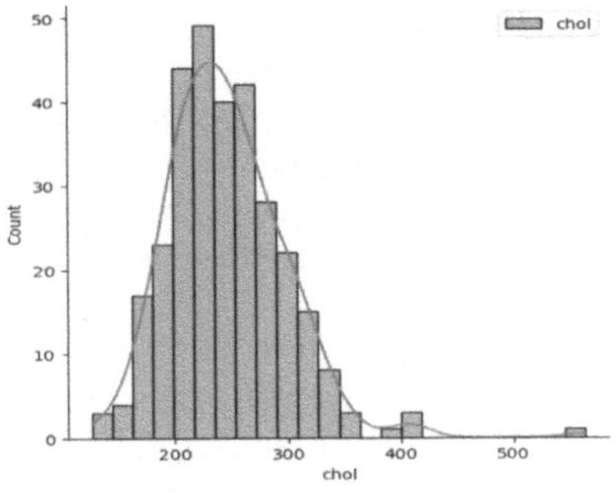

Figure 7. Cholesterol versus age.

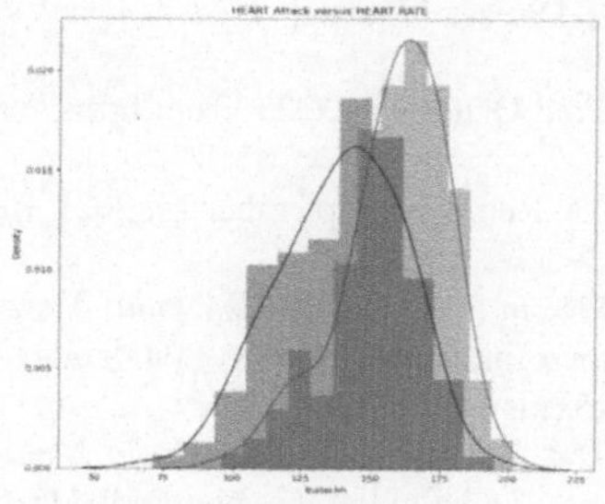

Figure 8. Increasing heart rate indicate risk of Heart attack.

c) Individuals who experience chest pain frequently they are more prone to heart attack. Compared to other forms of chest pain, normal angina is associated with a lower risk of heart attacks. Chest pain of type 0 'Typical Angina' is the highest, chest pain of type 3 'Asymptomatic' is the lowest'.

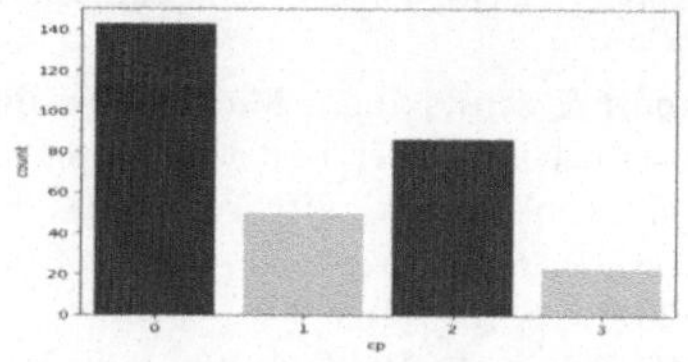

Figure 9. Four Prospect of Heart attack chest pain which show the condition is alarming or not.

Any cardiac patient should regularly meditate in addition to getting regular medical checkups for blood pressure, cholesterol, and heart rate as preventative measures. Additionally, in order to lower the likelihood of future heart attacks, one must see a specialist, take prescribed medicine on a regular basis, and practice all preventative measures.

5 CONCLUSION

In this research, various Supervised ML classifiers namely, Random Forest, Decision Tree, Logistic Regression, KNN and deep neural networks well known method RNN have been used to deploy a model for Myocardial Infarction prediction. The results impersonate that the RNN is achieving the highest accuracy score in such a way that prediction used by our model is of binary form in where 1 means a chance of heart attack and 0 means no chance. Some of the most influential attributes are chest pain type among which the typical angina is the most influential and asymptotic chest pain is least, cholesterol level in which the level greater than 200mg/dl are more prone, increased heart rate, thal, and age. It has been determined that 80% of all cases of early heart attacks can be avoided with a nutritious diet, regular exercise, abstinence from tobacco products, and drinking at least 2 litre water, Do some physical activity and medical check-up of heart related test like BP, Heart rate, cholesterol and other major test on regular basis. This all measures can help to avoid the heart attacks situation.

REFERENCES

Ali, L., Rahman, A., Khan, A., Zhou, M., Javeed, A. and Khan, J. A. (2019). An automated diagnostic system for heart disease prediction based on ${\chi^{2}}$ statistical model and optimally configured deep neural network. *IEEE Access*, *7*, 34938–34945. https://doi.org/10.1109/ACCESS.2019.2904800

Andras Janosi, W. S. (1989). Heart Disease [dataset]. *UCI Machine Learning Repository*. https://doi.org/10.24432/C52P4X

Cardiovascular diseases.(n.d.). Retrieved October 6, 2023, from https://www.who.int/health-topics/cardiovascular-diseases

Cardiovascular diseases (CVDs).(n.d.). Retrieved September 15, 2023, from https://www.who.int/news-room/fact-sheets/detail/cardiovascular-diseases-(cvds)

Deaths due to heart attacks up by 53% in 5 years: NCRB | India News—Times of India.(n.d.-a). Retrieved October 6, 2023, from https://timesofindia.indiatimes.com/india/deaths-due-to-heart-attacks-up-by-53-in-5-years-ncrb/articleshow/77922046.cms

Deaths due to heart attacks up by 53% in 5 years: NCRB | India News—Times of India.(n.d.-b). Retrieved October 6, 2023, from https://timesofindia.indiatimes.com/india/deaths-due-to-heart-attacks-up-by-53-in-5-years-ncrb/articleshow/77922046.cms

Fatima, M. and Pasha, M. (2017). Survey of machine learning algorithms for disease diagnostic. *Journal of Intelligent Learning Systems and Applications*, *09*(01), 1–16. https://doi.org/10.4236/jilsa.2017.91001

Gavande, S. B. and Chawan, P. M. (2021). *Heart disease prediction using deep learning techniques*. *08*(11).

Gupta, S. K., Shrivastava, et al. (2021). A machine learning approach for heart attack prediction. *International Journal of Engineering and Advanced Technology*, *10*(6), 124–134. https://doi.org/10.35940/ijeat.F3043.0810621

Hasan, T. T., Jasim, M. H. and Hashim, I. A. (2017). Heart disease diagnosis system based on multi-layer perceptron neural network and support vector machine. *International Journal of Current Engineering and Technology*, *17*(5), 1842.

Heart attack-Heart attack—Symptoms & causes.(n.d.). Mayo Clinic. Retrieved September 8, 2023, from https://www.mayoclinic.org/diseases-conditions/heart-attack/symptoms-causes/syc-20373106

Internet of Things (IoT) Applications, Tools and Security Techniques, Kawatra, R., Dharamdasani, D.K., Ajmera, R, *et al.* (2022). *2nd International Conference on Advance Computing and Innovative Technologies in Engineering, ICACITE* 2022, 2022, pp. 1633–1639

Machine Learning Random Forest Algorithm—Javatpoint.(n.d.). Www.Javatpoint.Com. Retrieved October 6, 2023, from https://www.javatpoint.com/machine-learning-random-forest-algorithm

Mienye, I. D., Sun, Y. and Wang, Z. (2020). An improved ensemble learning approach for the prediction of heart disease risk. *Informatics in Medicine Unlocked*, *20*, 100402. https://doi.org/10.1016/j.imu.2020.100402

Mohan, S., Thirumalai, C. and Srivastava, G. (2019). Effective heart disease prediction using hybrid machine learning techniques. *IEEE Access*, *7*, 81542–81554. https://doi.org/10.1109/ACCESS.2019.2923707

Data Science & Exploration in Artificial Intelligence – Gururaj et al. (Eds)
© 2025 the Author(s), ISBN: 978-1-032-96390-7

Recommendations using residual network and deep learning embedding for rating prediction

A.K. Manjula
Research Scholar, Department of Computer Science and Engineering, University Visvesvaraya College of Engineering, Bengaluru, India

H.N. Champa
Professor, Department of Computer Science and Engineering, University Visvesvaraya College of Engineering, Bengaluru, India

ABSTRACT: Data sparsity and cold start issues plague collaborative filtering-based recommendation systems. There have been suggestions presented for solving the issues with collaborative filtering by integrating additional cues, such as latent features found in comments and user social profiles. Unfortunately, sarcasm, phrase conflicts, additionally, There exist more issues that are handled by the comment processing methods in place. Additionally, the bias in ratings is not addressed. Implicit aspect orientation detection of users[2] and aspect orientation-based suggestion fine-tuning are absent from the present methodologies. Hybrid recommendation systems' accuracy suffers accordingly. This article suggests a hybrid recommendation system to address these issues by combining residual networks for using matrix factorization in rating prediction and aspect-based bias elimination. Using a fuzzy Gaussian, the rating bias is eliminated.

Keywords: Hybrid recommendation system, aspect orientation, deep residual networking.

1 INTRODUCTION

The key elements of online shopping and e-commerce systems are recommender algorithms[1]. They assist customers in sorting through a vast cloud of items to locate those that are interesting and highly relevant to their needs. Three categories comprise the current recommendation algorithms: content-based filtering, collaborative filtering[5], and hybrid filtering. Algorithms for content filtering classify products according to how similar they are and suggest products[3] that are comparable to those that active customers have already bought. Collaborative filtering algorithms Based on how similar products are to those that are owned by other users already purchased, users can forecast ratings for products they haven't yet bought. Based on the ratings forecast, the top N products that haven't been bought by active users are suggested. Collaborative and content filtering methods are combined in hybrid filtering algorithms [2]. They utilize several indications, such as aspect preferences, social interest profiles, latent components found in review comments, etc., to do this. In this vein, numerous Recommendation methods for hybrid filters have been put forth. The quality of recommendations is impacted by these algorithms' inability to analyse conflicts, sarcasm, and other negative aspects[7] of reviews, as well as to align recommendations with aspect orientations, despite their ability to tackle data sparsity and cold start issues. Reviews are quantized to a rating without taking user prejudice, sarcasm, or sentence conflicts into account. Furthermore, there isn't much research in the literature currently under publication on fine-grained aspect-based polarisation analysis and aligning recommendations based on aspect preferences. To deal with these problems, a hybrid recommendation algorithm is proposed in this work.

DOI: 10.1201/9781003589273-61

2 RELATED WORK

Suresh *et al.* [1] Provide a way to address sparsity and cold start issues. Deep learning in a convolutional neural network is employed to reduce the user to item matrix to a hash code. To find the most comparable or active people, the hash code distance computation is used. The highest scored items are suggested to users based on expected ratings from active users. The issue with this approach is that each time a new batch of users is added, the hash code needs to be calculated. The computational complexity rises as a result.

Jin *et al.* [2] devloped a hybrid recommendation system to address the issue of data sparsity. The content characteristic of the items was modelled by the authors using fuzzy modelling, and the outcomes were utilised in the matrix factorization for computing similarity. The rating is then predicted using a genetic algorithm. Recommendation variety was lacking in the method to meet various personality characteristics.

Elahi *et al.* [3] suggested a hybrid recommendation system that integrates reviewer sentiment scores with ratings. Using the BERT model, the reviews were grouped based on the sentiment scores they received. The sentiment score is then used to modify the review rating. Collaborative filtering is used to generate recommendations after that. The authors found that the quality of suggestions was enhanced by sentiment scores. Sarcasm, however, was not considered in the sentiment score derivation.

Nassar *et al.* [4] suggested a process for formulating recommendations based on a number of factors. Initially, this system forecasts the ratings for the criterion using matrix factorization and deep neural networks. Based on the ratings for the criterion, a deep neural network is then utilized to project the ultimate rating. The recommendation is then located through collaborative filtering based on the rating. Cold start Problems were not given any thought in the method.

Hung le *et al.* [5] suggested an algorithm for recommendations based on collaborative filtering with several criteria. To forecast the ratings, deep learning and matrix factorization are combined. The multi-criteria ratings are combined to create the total rating using Dempster's rule of combination. However, the method lacks consideration for user preferences. Additionally, the method does not model the temporal variation in user interests.

Ong *et al.* [6]Neural Matrix Factorization (NeuMF++), a hybrid recommendation method, was employed to deal with the issues of cold start and data sparsity. Neural collaborative filtering and feature extraction are two of the algorithm's key features. Denoising in Stacked User item latent vectors are extracted from user item features using autoencoders. To forecast the ratings, the user item latent vectors are input into a multilayer perceptron and generalised matrix factorization.

Shi *et al.* [7] proposed an aspect level collaborative filtering approach based on neural networks that took advantage of the latent components. The authors used various metapaths to extract distinct aspect-level similarity matrices for users and items, respectively. After that, These matrix entries were made in a complexly constructed deep neural network to enable it to learn aspect-level latent variables. Ultimately, the top-N recommendation is obtained by fusing the aspect-level latent variables.

Gopalachari *et al.* [8] suggested a way to gauge an online review's reliability. Its foundation is aspect generation through the application of the phrase embedding method, followed by aspect-based clustering. It is possible to filter and use the reliable reviews to calculate rating scores.

Payandenick *et al.* [9] integrated residual network and matrix factorization to improve the accuracy of recommendation. By capturing linear relations in data with matrix factorization and non linear relationship in data using deep residual network, authors fused them to improve the recommendation accuracy. Residual network predicted the errors in matrix factorization and improved the rating prediction accuracy. But the approach lacked personalization in recommendation and did not consider using latent information spread in reviews.

Guo *et al.* [10] proposed a deep knowledge enhanced network to predict the probability of user selecting a item. Deep neural networks were employed by the writers. To learn higher order feature interactions and integrated those features with knowledge graph embedding features. Item to item relationship is used to solve the data sparsity. But the approach is not customized to aspects and lacks personalization based on aspects.

3 PROPOSED METHOD

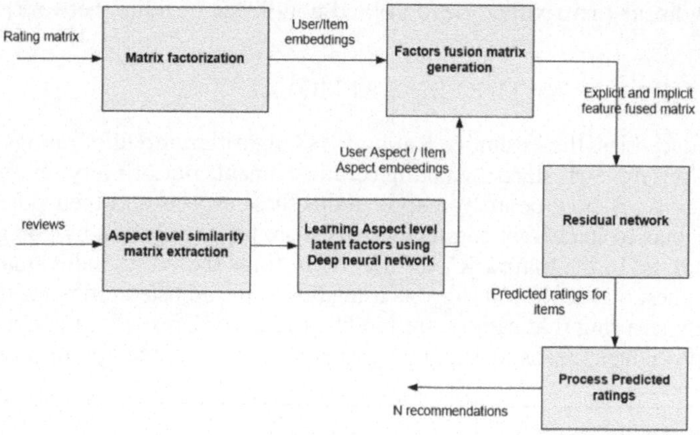

Figure 1. Hybrid recommender for rating prediction.

- A popular method in collaborative filtering-based recommendation systems is matrix factorization. Ratings in these kinds of systems are usually shown as a matrix, with rows and columns denoting people and items, respectively, and each cell representing a user's rating (if available) for an object. After learning the matrices, one can multiply the relevant user and item vectors and add a global bias factor to produce predictions for missing ratings.
- Concatenating or fusing [4] the attributes of the user and the item into a single matrix rather than dividing it into distinct matrices for each would be one method of factorizing the rating matrix.
- Remaining networks can be used to learn low-dimensional representations (embeddings) of individuals and objects by using implicit or explicit feedback data[6]. These embeddings reflect the essential characteristics of people and objects, which enables the recommendation system to generate accurate predictions.
- The initial stage in learning aspect-level latent variables using deep neural networks is to extract nuanced characteristics or representations of various dimensions or aspects of items (like commodities or services) from textual data or other sources.
- To extract aspect-level similarity matrices, aspect-level representations of the items are utilized to measure their similarity to one another. Aspect-level representations, as opposed to traditional item-level representations, capture different dimensions or properties of items, allowing for more thorough comparisons.
- Gather feedback from users on the recommended items and incorporate it into the recommendation system.

In Figure 1, It offers a hybrid recommendation that blends matrix factorization, aspect-based factorization, and residual network-based rating prediction. Using matrix factorization, the embeddings of the user and item[1] are taken out of the rating matrix. After removing noise, bias, and sarcasm from the evaluations, an aspect level similarity matrix is produced from the data. With a deep neural net, its aspect level latent components are recovered. The embeddings of the user and item aspects are provided by the deep neural net.

The residual network employed fusion and classification techniques to forecast user ratings for items by combining aspect-based and item-based embeddings[8]. Ranking-wise, the top N things with the highest ratings are suggested. The concept of learning aspect level latent variables will be used through paperwork-related activity. Along with colleagues [10], Shi. Payandenick *et al.*'s research [9] is where the idea of using residual networks to predict ratings originated. Depending on users' aspect orientation, we can handle the problems of data sparsity, cold start, and fine-tuning suggestion by using residual network[9] to predict ratings and integrating aspect-based embeddings with purchase-based (rating matrix) embeddings. Higher accuracy suggestions will be produced by our methodology because it is based on both linear (matrix factorization) and non-linear (residual network) interactions.

4 ALGORITHM FOR RATING BIAS REMOVAL

In order to reduce bias, the "Remove Rating Bias" algorithm modifies ratings in a matrix R depending on reviews S. It starts by taking certain elements out of every review and categorizing their polarity. Aspect_polarity_map contains these aspects and their polarities. Next, it uses calculate_bias to iteratively compute a bias score for each rating by going through possible rating indices. In the matrix R', the algorithm finds the rating index (max_bias_rating) that has the highest bias score and suggests using this as the adjusted rating for the user (u) and product (p). By ensuring that ratings are modified in accordance with the opinions stated in reviews, this procedure lessens bias and may even increase the precision of rating forecasts.

Algorithm Remove Rating Bias

Input:

- Ratings matrix R
- Reviews S
- User index u
- Product index p

Output:

- Bias-corrected ratings matrix R'

(1) Initialize aspect_polarity_map to store aspect polarities: aspect_polarity_map = {}
(2) Initialize max_bias_rating to track maximum bias rating: max_bias_rating = 0
(3) Initialize recommended_rating to store the recommended rating: recommended_rating = 0
(4) for each review sentence review in S do
(5) aspects[] ← extract_aspects(review)
(6) for each aspect a in aspects do
(7) aspect_polarity_map[a] = classify_polarity(review)
(8) end for
(9) end for
(10) for each rating index i in range(1, Num_ratings + 1) do
(11) current_bias ← calculate_bias(R[u][p], aspect_polarity_map)
(12) if current_bias > max_bias_rating then
(13) max_bias_rating ← current_bias
(14) recommended_rating ← i
(15) end if
(16) end for
(17) Update R'[u][p] with recommended_rating
(18) Output R' as the bias-corrected ratings matrix

---

5 DATASETS

Online merchants like Amazon and Filpkart employ different recommendation models to present users with different possibilities. Amazon currently uses item-to-item collaborative filtering, which extends to massive data sets and produces high-quality, real-time suggestions. This type of filtering generates a list of suggestions for the user by comparing each item the user has purchased and rated to similar items. In this project, we will create a recommendation model for electronics products on Amazon.Data about Amazon reviews is the source (http://jmcauley.ucsd.edu/data/amazon/). The collection contains multiple datasets. We are using the Electronics dataset for this case study.

6 RESULTS

6.1 *Comparsion of metrics*

In Table 1 The performance of the suggested strategy is compared with that of previous methods using the metrics of Normalized Discounted Cumulative Gain (NDCG) at K=5 and Hit Rate (HR), which show significant variations in efficacy. The suggested approach outperforms other well-established approaches, as seen by its HR of 0.3771 and NDCG of 0.2532. In particular, the suggested method outperforms Afoudi *et al.*'s approach [12] in terms of proposing relevant items, as evidenced by the HR of 0.2766 and NDCG of 0.1800 that are obtained. Comparably, Shi *et al.* [11] obtain an HR of 0.3429 and an NDCG of 0.2308, which, while competitive, are still inferior to the performance of the suggested technique. Elahi *et al.* [13] show similar reduced measures, with an NDCG of 0.1922 and an HR of 0.3055. Overall, the suggested strategy shows a noteworthy improvement in NDCG and HR measures, indicating that at the examined K=5 threshold, it may be a more successful approach for recommendation systems.

Table 1. Comparison of HR and NDCG.

Solution	HR	NDCG
K=5		
Proposed	0.3771	0.2532
Afoudi *et al.* [19]	0.2766	0.1800
Shi *et al.* [14]	0.3429	0.2308
Elahi *et al.* [27]	0.3055	0.1922

Using information from Amazon product reviews,in Figure 2 the suggested hybrid recommendation algorithm's performance is assessed [14]. There are an estimated 7 million reviews and ratings from 1.5 million people in the electronics categories. RMSE, or root

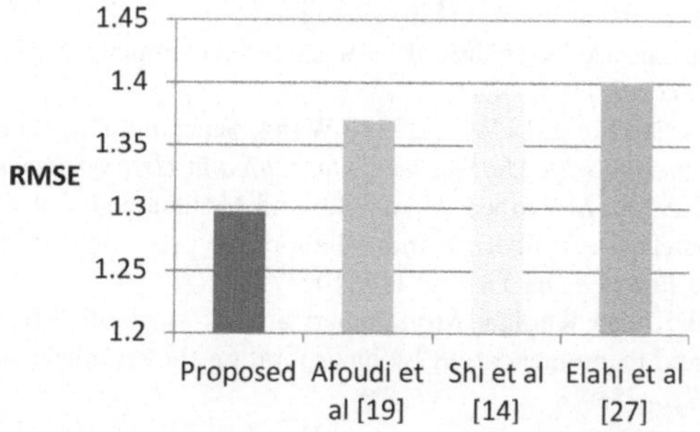

Figure 2. Comparison of RMSE.

mean square error, is a performance metric. Below are the findings for the root mean square error (RMSE) between the expected and actual ratings for the top N suggestion (N = 10).

7 CONCLUSION

This work proposes a hybrid recommendation strategy using residual networks and matrix factorization. A unique fuzzy Gaussian rating bias reduction is proposed as part of the endeavor to remove the rating bias using the sentiment polarity supplied in aspect spread throughout evaluations. The proposed method takes advantage of both the linear and nonlinear features of the user-product connection. In comparison with the existing works, the proposed strategy results in a 5.1% reduction in RMSE. When evaluating the efficacy of different recommendation systems using Hit Rate (HR) and Normalized Discounted Cumulative Gain (NDCG) metrics at K=5, the proposed strategy outperforms existing methods. It shows its effectiveness in giving customers appropriate item recommendations with an HR of 0.3771 and an NDCG of 0.2532.

REFERENCES

[1] Suresh Kumar, Jyoti Prakash Singh, Vinay Kumar Jain and Avinab Marahatta. (2022). A deep ranking weighted multihashing recommender system for item recommendation, *Computational Intelligence and Neuroscience*, vol. 2022, Article ID 7393553, 10 pages.

[2] Jin, C., Mi, J., Li, F. *et al.* (2022). Hybrid recommender system with core users selection. *Soft Comput* 26, 13925–13939.

[3] Elahi, Mehdi Khosh Kholgh, Danial Kiarostami, Mohammad Oussalah, Mourad Saghari and Sorush. (2023). Hybrid recommendation by incorporating the sentiment of product reviews. *Information Sciences*. 625. 10.1016/j.ins.2023.01.051

[4] Nassar, Nour Jafar, Asef Rahhal and Yasser. (2020). Multi-criteria collaborative filtering recommender by fusing deep neural network and matrix factorization. *Journal of Big Data.* 7. 1. 10.1186/s40537-020-00309-6.

[5] Le Q.-H., Mau T. N., Tansuchat R. and Huynh V.-N. (2022). A multi-criteria collaborative filtering approach using deep learning and dempster-shafer theory for hotel recommendations, in *IEEE Access*, vol. 10, pp. 37281–37293

[6] Huazhen Liu, Wei Wang, Yihan Zhang, Renqian Gu and Yaqi Hao. (2022). Neural matrix factorization recommendation for user preference prediction based on explicit and implicit feedback, *Computational Intelligence and Neuroscience*, vol. 2022, Article ID 9593957, 12 pages

[7] Shi, Chuan Han, Xiaotian Song, Li Wang, Xiao Wang, Senzhang Du, Junping Yu and Philip. (2019). *Deep Collaborative Filtering with Multi-aspect Information in Heterogeneous Networks*

[8] Gopalachari, Mukkamula Gupta, Sangeeta Rakesh, Jayaram S., Dharmana Rao and Pulipati. (2023). Aspect-based sentiment analysis on multi-domain reviews through word embedding. *Journal of Intelligent Systems*. 32. 10.1515/jisys-2023-0001.

[9] Payandenick, M. and Wang, Y.C. (2023). ResNetMF: enhancing recommendation systems with residual network matrix factorization. *Preprints*, 2023100302.

[10] Guo, Xiaobo Lin, Wenfang Li, Youru Liu, Zhongyi Yang, Lin Zhao, Shuliang Zhu and Zhenfeng. (2020). DKEN: Deep knowledge-enhanced network for recommender systems. *Information Sciences*. 540. 10.1016/j.ins.2020.06.041.

[11] Shi, Chuan Han, Xiaotian Song, Li Wang, Xiao Wang, Senzhang Du, Junping Yu and Philip. (2019). *Deep Collaborative Filtering with Multi-aspect Information in Heterogeneous Networks.*

[12] Afoudi, Yassine LAZAAR, Mohamed Al Achhab and Mohammed. (2021). Hybrid recommendation system combined content-based filtering and collaborative prediction using artificial neural network. *Simulation Modelling Practice and Theory*. 113. 102375.

[13] Mehdi Elahi, Danial Khosh Kholgh, Mohammad Sina Kiarostami, Mourad Oussalah, and Sorush Saghari. (2023). Hybrid recommendation by incorporating the sentiment of product reviews. *Inf. Sci.* 625, C (May 2023), 738–756

[14] https://jmcauley.ucsd.edu/data/amazon/

Data Science & Exploration in Artificial Intelligence – Gururaj et al. (Eds)

Survey on decentralized computing models for time sensitive applications

B. Indu & Geeta C. Mara
School of Computing and Information Technology REVA University Bangalore, India

ABSTRACT: A Virtualized Framework in the Time Sensitive applications where the reachability and response times are considered crucial hence to balance the load on cloud server a decentralized method of Computing (Edge Computing) is proposed, efficiently managing the resources which in turn intended at reducing the latency, reducing the response time also providing data privacy. In fog computing architecture the cloud components and Network communication components carry out functionality particular to applications, where the cloud and end users are separated by a layer of fog computing. Health Fog's design element effectively lowers the extra communication expense that is often high in similar systems. The use of strategic planning by various organizations to identify its access direction, strategy, and resource allocation based on decision to drift from traditional to cloud-based health services. In this paper a survey is done on decentralized computing models like cloud, edge and their corresponding challenges.

Keywords: Edge Computing, Cloud Computing, Fog Computing, Cloudlets, Blockchain, Mobile computing.

1 INTRODUCTION

A novel method of providing computing resources and services online is virtualized computing method. According to several experts remarks, It can enhance time sensitive applications like health industries services, enhanced health research, and alter the landscape of healthcare data processing. Cloud computing should be thoroughly examined before it is widely adopted, It is also described how a health organization or any time sensitive applications could use strategic planning to select its development, approach, and resource distribution after deciding to switch from conventional to virtualized services.

Edge Computing is a paradigm where the data is expected to be processed at the location where it is generated, processing quickly, providing the actions to the required demand on time for delay sensitive applications most prominently Health care sector, Industrial applications, The utmost subtle data is processed prior and dynamic organizations that should work constantly in a secured environment, which in turn are driven by a huge portion of current computing strategies, which happens at the edge like locations for instance industrial applications, hospitals and marketing stores. To efficiently work a Low latency, network-free solutions are essential for these locations to work effectively and accurately. Edge computing strategy could transform business across all sectors and functions from marketing and end user access to construction and back-office operations.

2 LITERATURE REVIEW

2.1 *Edge computing scenarios and challenges*

In order to replace the centralized data processing methods, a distributed methodology termed edge computing is introduced. Its open-space architecture includes several built-in functions. By integrating mobile computing and the internet of things, it also ensures improved performance. In general, it quickly distributes resources and services to edge devices.

DOI: 10.1201/9781003589273-62

Blockchain-assisted framework that influences disseminated edge servers to attain fast data processing is proposed with smart contracts that are designed to verify the uniqueness and data integrity of network entities were introduced [1] which advances in Solid timeliness protection and average processing latency. Average Processing speed can further be improved.

Ma *et al.* [2], proposed a mechanism which works in regard to the padding perception and competent rating policy to assure necessary properties in controlled Mobile Edge Computing (MEC) atmospheres which uses TCDA allocation, Pricing, and critical value detection. It is intended to provide 1) computation - rigorous tasks of Mobile Devices to be distinguishable, which can improve the success rate of similar Edge Systems with a limited capacity; 2) It employs a minor padding value, which can reduce Edge Systems, thereby increasing the number of captivating MDs. Needs Generalizability of dealing with numerous end users' instances. Needs improvement in the compliance of the system to the widespread Mobile Edge Computing service situations. Markov Decision Process (MDP) and Deep Reinforcement Learning (DRL) [3] Gains the long-term ideal system performance, works on the concept of Training of DRL-Based DTASM Agent, the average service downtime increases (when the number of migration-failed users decreases). A technology framework and decentralized resource management technique for the implementation of service-based Internet of Things applications at the edge, A decentralized task allocation technique [4] is utilized. The placement policy module, The salient features are in the decision policy module. The successful mapping rate rises with the number of nodes.

Better coverage can be achieved with less processing power. Definition of service models that have several jobs for input and output. A heuristic for effectively locating less-than-ideal answers to extensive EUA issues[5] uses Most Capacity First (MCF) Solves the user provision in large-scale scenarios, aiming to allocate as many users to as few edge servers as possible. Certain programs may experience performance deterioration due to workload interference or varied user workload patterns, under-utilized edge computing infrastructure, service reconfiguration cost or migration cost to be taken care of. To improve the total energy cost in an instantaneous multi-user Mobile Edge Computing system, An energy-efficient algorithm based on deep reinforcement learning EEDRL (Energy Efficient Deep Reinforcement Learning)[6] is proposed that brings down the total energy usage of the MECO system in contradiction to various offloading schemes, low computing idleness appropriate to the time-varying wireless networks and Time Consumption is linear, The exercise of several DNNs with the different number of SMDs consumes much time. The divesting interval is lesser than the coherence time of channels. More intricate network settings may be adapted to the DRL architecture.

Energy-efficient placement of services [7] for AI applications as a multiperiod optimization problem, One step ahead Model is used which is intended to save energy and strict latency constraints, and placement strategies for Deep Neural Network models at lower costs are achieved. Can include demand forecasting, considering how placement and scheduling of services are affected by demand pattern unpredictability. contrasting such a strategy with contemporary online deep RL-based strategies. The safe event ordering service for fog nodes depends on the availability of a Trusted Execution Environment (TEE)[8], OmegaKV, a key-value store that leverages Omega to achieve causal consistency. OmegaKV, a key-value store, Integrity, Freshness, Causal Consistency are the major highlights, Computational workloads are dynamically assigned to slices, radio resources are managed across slices, and computing and radio resources are managed inside slices[9] COS-Choose offloading slice algorithm, This method has good computing performance with bounded approximation ratio, and slicing might be useful for enhancing system performance in general, WDs-wireless devices that are unable to complete the tasks locally would be given preference when allocating resources. A multi-objective optimization problem, a single-objective one is to be created [10]. DE-Based Task Offloading Approach and Reliability-Enhanced Task Offloading approach, optimizes task offloading for bandwidth use and level of reliability, Remote clouds should be taken into account for the testing environment, Construction of scalable and harmonious digital healthcare platforms with role-based access control[11] with Facial Expression Recognition (FER) algorithm. The hybrid PUF(Physical Unclonable Functions) offers high security against machine learning, man-in-the-middle, and forgery attacks are utilized, This technique is comparatively good in terms of packet delivery rate and latency. In addition, it improves the scalability and dependability as the volume of transactions rises. Using parallel mining, the communications can be improved even further and the block times can be significantly shortened. The overall security of the system can be improved by incorporating various levels of safe checks at the end user and system levels. utilizing

Docker containers and the containerization concept to run Internet of Things applications as microservices are explored, its distributed offloading concept improves reaction time and manages energy use and computer resources. Resource management rules, a lightweight security feature, and a server-less architecture is proposed as Energy-Aware Resource Management, Logical Address-Based Resource Discovery [12] An Edge Intelligent Energy Efficient Model in Mobile Edge Computing[13]builds a ELMAN NEURAL NETWORK based Energy Consumption model, It has a significant capacity for nonlinear fitting and can precisely identify non- contiguous functions with arbitrary precision. Less time is spent training accurately. combining workloads that are heavy on processing, Web transactional, and input/output without the need of cloud or edge datacenters. The job processing time under different network conditions is expressed in closed form using a discrete-time Markov chain analytical framework is proposed in[14]. Modelling Intermittent connectivity which provides helpful advice for creating task offloading algorithms that are effective in actual MEC systems. The effects of sporadic connection on MEC task processing needs to be addressed.

A Blockchain-based Mutual Validation Scheme for Collaborative Edge Computing [15] uses HeyGenKGC, KEyGenES, Registration process, AddServer,RemoveServer improves Authentication time is Shortest (StaticCondition) Evaluation of prototype performance, Decentralized cross-domain authentication method, multilayer authentication method for cloud edge devices CNN Accelerators Based on FPGA for Time-Aware and Energy-Efficient Face Tracking System[16] builds on Q-Learning- Based HEFT, here the FT() system's response time is well optimised. CNNs now operate more quickly and efficiently but cannot acclimate to the numerous circumstances of FT applications.

Seungkyun Lee *et al.* [17] proposed a task scheduling technique that takes network flows and priority (i.e., deadline) into account in the CEC context with the goal of lowering the Deadline Miss Ratio (DMR) of IoT system requests. The outcomes of the recreation model show how much the suggested algorithm can lower the DMR of various target tactics. A disadvantage is that more research must be done to determine whether processing times at edge nodes can vary. Huaming Wu *et al.* [18] proposes a Blockchain - empowered IoT - Edge-Cloud computation that aids from MCC and Mobile Edge Computing, where Mobile Edge Computing servers bid lower delayed computing services, and an effective active job offloading algorithm by selecting the ideal computing location operational, also on the IoT expedient, the Mobile Edge Computation server, or the MCC server with the task of together dropping the energy consumption and task response time. For a variety of applications, the Lyapunov optimization method is used to control the costs related to processing, communication, and the dynamic changes in wireless environments. Without requiring prior familiarity of the system's future state, optimization in an adaptive manner chooses the best computing locality for each task.

Yundi Guo *et al.* [19] proposed an object recognition construction built on edge computing to enable disseminated and well-organized object finding for safety applications. This design enables devices to proficiently compress the images they gather for wireless transmission without sacrificing the object detection precision of the edge servers. A system for dispersed and message - effective communications amongst end nodes, edge servers, and the cloud is planned to variantly enhance the item recognition precision online. Many simulation results established that the architecture not only attains modest finding accurateness with respect to prior cloud-based objective detection methods with decreased reaction delay but also suggestively advances the image broadcast efficacy with adaptive image compression ratio.

Hyame Assem Alameddine *et al.* [20] proposed the Logic-Based Benders Decomposition method, which creates an original, valuable decomposition. When using a relaxed master with fewer constraints, this method solves a sub problem that can offer cuts that iteratively instruct the chief to narrow its search space. Here, the chief problem and the sub-problem will align to yield the ideal solution. This approach may result in run times for the examples under consideration that are more than 140 times faster. The advantage of being able to provide solutions with performance guarantees is another advantage of this technique. This strategy is used to highlight the pertinent performance patterns for diverse vertical businesses, with a focus on the delay-sensitive systems.

Gang Qian, Siliang Lu *et al.* [21] Proposed dynamic control system and a real-time fault detection for rotating machines, as well as the architecture of an edge computer node (ECN), are given. Three motor-phase current signals and a vibration signal are both simultaneously acquired by the ECN. Then, 40 features from the data are taken and merged to produce an indicator that can precisely distinguish between different motor states. The studies' findings display that the proposed ECN can accurately identify mechanical and electrical problems and take over operation of the motor in the event of an emergency breakdown. Signal capture, the extraction of feature fusion, and identification of faults are computational stages that can be completed on the ECN in less than 0.25 seconds. The suggested edge computing architecture's ability to process sensor data in real-time is evidence of this.

Junkai Liu *et al.* [22] proposed a greedy heuristic for task offloading planning based on the well-known maximum network flow problem. By considering the potential for a decentralized task offloading approach is proposed, in which IoT devices communicate and choose a local D2D offloading mechanism in the absence of a controller (for example, a network operator). Both a careful theoretical analysis and in-depth testing display that the offered algorithms maximize overall ERP efficiency.

1) Formulating the UMP numerically for both devices is one of the major highlights. Limitations on computation and communication power as well as NP-hardness proof.
2) Putting forth greedy-based heuristics to find almost perfect UMP solutions for a range of application cases.
3) Carefully analyzing the proposed solutions while using actual conditions and competitive benchmarks.

Zicong Hong *et al.* [23] proposed a technique to shorten the time and effort invested in each task's calculation, Investigating the communication-routing and computation-offloading problems, and formulating the resulting issue as a probable inclined tasks in which the IIoT devices choose their own methods and tactics for compute offloading. The usage of constructing a free-bound technique that can guarantee a finite improvement path, and distributed algorithms that can get close to the Nash equilibrium and the suggestion of a multi-hop cooperative messaging system are also included. Outcome of the simulation show that our techniques scale effectively as device sizes increase and enhance IIoT in a range of situations with stable performance.

Y u-Lin Jiang *et al.* [24] proposes model to effectively manage the processing and messaging properties of all devices in a fog computation environment and distribute applications to the appropriate device, An offloading dispatcher and an energy-efficient offloading decision mechanism are required. The unloading tasks are broken up into several smaller assignments, each with a goal. To encounter the demands of time demands of such claims, a run - time scheduler that takes endwise latency into account is also provided. Research on an actual platform shows that using this structure results in substantial energy savings. It is necessary to conduct study on the challenges surrounding energy conservation in more complicated fog circumstances, taking usage intensity and network resilience into account.

Babatunji Omoniwa *et al.* [25] It is advised to use the FECIoT basis, which tackles issues with enabling technology, services, and exposed research. It follows a teaching format, moving from simple to intricate IoT ideas. How FECIoT can be used in actual cyber-physical systems, such as smart grids, residences, healthcare systems, and surroundings. The benefits of FECIoT include

1) Geographically scattered.
2) Assistance with massive instrument webs and termination nodes.
3) Well real - time reaction than the cloud - built strategy
4) Encourages interoperability and heterogeneity.
5) Online analytics and cloud communication.

Yiqin Deng *et al.* [26] provides a model that enhances the timeliness and reliability of task processing by lowering task response times and packet losses while remaining within the parameters of energy queue stability. Both the Lyapunov optimization technique and the active corresponding computation divesting and energy organization algorithm are utilized to resolve the issue. For Internet of Things systems that must be responsive in a short amount of time, for tasks like such a sustainable and Green mobile edge computing framework is suggested. The parallel offloading approach is used in mobile edge computing to ensure that tasks are completed fast. To ensure the steadiness of task processing in this system, packet losses are taken into consideration. Lyapunov optimization technology and EHT are utilized to achieve energy stability and sustainability to maximize the usage of green energy in the IoT system.

Md. Asif-Ur-Rahman *et al.* [27] proposed a non homogenous five-layered Internet of Healthcare Things system built on mist, fog, and the cloud that can handle and route data in both offline and batch modes as well as (near-)real time. The structure also makes use of software distinct interacting and link variation - based load harmonizing to ensure ideal stock distribution and effectual supply of information consumption. The results of modeling the framework reveal that the intended network may attain maximum Quality of Service with condensed end-to-end latency and packet loss rate through a range of its techniques, which is vital for emerging the future generation of online health care systems. Next-generation Internet of Healthcare Things-based healthcare systems face challenges about integrating superior machine learning techniques in recognizing the varied traffic, using bio - encouraged representations to ensure efficient resource usage, scheduling the best flow to improve performance, and increasing data allocation to lower overall computational complication. It might be enhanced and used with the Internet of Vehicles.

Chang *et al.* [28] proposed a method for edge-aided data dissemination that disseminates different data item subsets using mobile edge servers. To obtain all the data objects, *IoT* nodes then exchange segments with one another. To ensure the integrity of the transferred data objects and avoid transmission conflicts, the installation mechanism and data propagation protocol are put into place. Authors also intend to Create an adaptable protocol to enable effective data distribution for diverse Internet of Things networks, and carry out extensive simulations and in-depth testing. The results demonstrate that using mobile edge servers significantly improves energy efficiency when compared to traditional distribution techniques.

Table 1. Comparison of various methods and challenges.

Authors	Concept	Algorithm	Advantages	Disadvantages/ Limitations
Ming *et al.* [1], 2022	Blockchain - aided situation that influences distributed edge servers to attain fast data processing. clever agreements are planned to verify the uniqueness and data integrity of link objects	FAITH(J,P,U) HCS (H,P,U)	Solid timeliness protection, average processing latency.	Average Processing speed can further be improved.
Ma *et al.* [2], 2022	A honest combinative double auction tool, which mixes the stuffing insight and the skilled pricing policy to promise necessary belongings in constrained MEC atmospheres	TCDA- Allocation, Critical-Value-Determination, TCDA-Pricing	1) It comforts computation - exhaustive responsibilities of MDs to be dividable, which can upsurge the achievement rate of similar ESs with a partial size; 2) It exploits a small stuffing rate, which can crush out failed Edge systems, thereby upsurge the sum of endearing MDs.	Generalizability of dealing with numerous market circumstances. Needs improvement in the adaptableness of the algorithm to the significant MEC facility situations.
Chen *et al.* [3], 2022	Markov Decision Process (MDP) and Deep reinforcement Learning (DRL)	Training of DRL-intended DTASM Agent	Obtains the lasting perfect organization performance	The normal provision interruption increases
Avasalcai *et al.* [4], 2022	A decentralized resource management strategy and corresponding practical framework for the position of service-based IoT applications at the edge.	A decentralized task allocation technique. The placement rule component, the decision policy module,	The successful mapping rate increases as the number of nodes rises.	Ensure improved handling of data with lower computational demands. Positioning of provision representations with numerous effort and productivity tasks

(*continued*)

Table 1. Continued

Authors	Concept	Algorithm	Advantages	Disadvantages/ Limitations
Lai *et al.* [5], 2022	A empirical to proficiently find sub-optimal answers to substantial EUA difficulties	Most Capacity First (MCF)	Solves the operator capability in large-scale situations, directing to assign as numerous users to as partial edge nodes as possible.	Performance deprivation caused by the intrusion of jobs, or discriminated user capacity forms, may happen in some claims, under-utilized edge computing organization, facility reconfiguration charge or relocation cost
Zhang *et al.* [6], 2022	An energy effective technique created on deep solidification information to recuperate the total energy price in a instant multiple user MEC system.	EEDRL (Energy Efficient Deep Reinforcement Learning)	Decreases the entire energy ingesting of the MECO system in contradiction of numerous divesting systems, low processing idleness fit to the time - fluctuating wireless stations. Time Consumption is linear	The exercise of several DNNs with the dissimilar amount of SMDs takes considerable amount of time. The ridding period is slighter than the consistency time of channels. DRL agenda may be adapted to more composite system surroundings.
Premsankar and Ghaddar [7],2022	Energy efficient assignment of facilities for AI claims as a multiperiod optimization issues	One step ahead Model	Saves energy and strict latency constraints, Assignment tactics for DNN models at lesser prices	Can include demand forecasting, taking into account how placement and scheduling of services are affected by demand pattern unpredictability. contrasting such a strategy with contemporary online deep RL-based strategies.
Correia *et al.* [8], 2022	The accessibility of a Trusted Execution Environment (TEE), OmegaKV, its a mixture of key-value couple that uses Omega to offer fundamental stability, is used by the safe event collation service for fog lumps.	OmegaKV, a key-value store	Integrity, Freshness, Causal Consistency	There is a need to address the susceptibilities that arise from the placement of numerous servers in various places.
Sladana Jošilo and György Dán [9], 2022	Computational workloads are dynamically assigned to slices, radio resources are managed across slices, and computing and radio resources are managed inside slices.	COS- Choose offloading slice algorithm	This method has good computing performance with bounded approximation ratio, and slicing might be useful for enhancing system performance in general.	WDs-wireless devices that are unable to complete the tasks locally would be given preference when allocating resources.
Liu *et al.*, [10], 2022	Create a single-objective optimization tricky out of a multi-objective optimization Issue	DE-Based Task Offloading Approach and Reliability-Enhanced Task Offloading Approach	optimizes task offloading for bandwidth use and level of reliability	Remote clouds should be taken into account for the testing environment. improvement based on actual facts

3 PROPOSED METHODOLOGY

Framework, allowing for the development of innovative strategies in adaptive service placement, dynamic task and workload scheduling, and task offloading and resource scheduling within cloud-assisted edge computing environments, identifying current challenges and gaps, adoption of deep learning sets the stage for leveraging advanced analytics to tackle these challenges effectively.

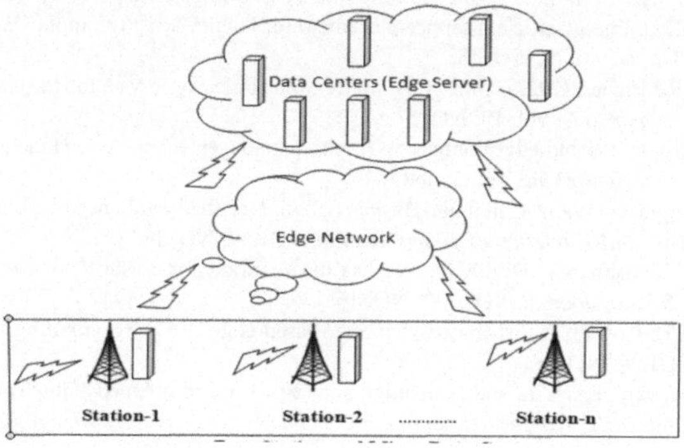

4 FUTURE DIRECTIONS

- Research findings demand the improvement in the average processing time, compliance of the system to the big-scale Mobile Edge Computing service set-ups, ensuring better coverage with lower computational loads, placement of models with numerous I/O tasks.
- Vulnerabilities that arise from installing a large number of servers in various locations must be addressed, and a cross-domain authentication mechanism must be developed.
- Degradation of performance resulting from workload interference or differently distributed user workload patterns and under-utilized edge computing infrastructure to be addressed, privacy of sensitive information to be considered with higher priority.

A cross-domain authentication scheme giving the utmost priority to the privacy and response times for delay sensitive applications. Adaptability of the system to the large-scale Collaborative Edge.

5 CONCLUSIONS

Computing model that addresses the differentiated user workload patterns, vulnerability's reliability and efficiency criterions are the major issues to be addressed. Fog-cloud-based healthcare services are becoming a pro-active and successful way to give users with remote detection and monitoring. fog layer at a gateway for a health monitoring system that needs speedy processing and little waiting time. Data travel to the cloud for processing and analysis can be minimized by using fog computing services. Event adversity is calculated at the fog layer by monitoring real-time event instances.

REFERENCES

[1] Zhongxing Ming, Mingzhao Zhou, Laizhong Cui and Shu Yang. (2022). FAITH : A fast blockchain-assisted edge-computing platform for healthcare applications, *IEEE Transactions On Industrial Informatics*,Vol. 18, no. 12

[2] LianboMa, Xueyi Wang, Xingwei Wang, Liang Wang, Ying Shi and Min Huang. (2022). TCDA: Truthful combinatorial double auctions for mobile edge computing in industrial internet of things, *IEEE Transactions on Mobile Computing*, vol. 21, NO. 11

[3] Yan Chen, Yanjing Sun, Chenyang Wang and Tarik Taleb. (2022). Dynamic task allocation and service migration in edge-cloud IoT system based on deep reinforcement learning, *IEEE Internet Of Things Journal*, VOL. 9, NO. 18
[4] Cosmin Avasalcai, Christos Tsigkanos, and SchahramDustdar. (2022). Resource management for latency-sensitive IoT applications with satisfiability, *IEEE Transactions on Services Computing*, vol. 15, NO. 5
[5] Phu Lai, Qiang He, John Grundy, Feifei Chen, Mohamed Abdelrazek, John Hosking, and Yun Yang. (2022). Cost-effective app user allocation in an edge computing environment, *IEEE Transactions on Cloud Computing*, VOL. 10, NO. 3
[6] Xinjie Zhang, Xinglin Zhang and Wentao Yang. (2022). Joint offloading and resource allocation using deep reinforcement learning in mobile edge computing, *IEEE Transactions on Network Science And Engineering*, vol. 9, NO. 5
[7] Gopika Premsankar and Bissan Ghaddar. (2022). Energy-efficient service placement for latency-sensitive applications in edge computing, *IEEE Internet of Things Journal*, vol. 9, NO. 18
[8] Claudio Correia, Miguel Correia, and Luı´s Rodrigues. (2022). Omega: A secure event ordering service for the edge, *IEEE Transactions on Dependable and Secure Computing*, vol. 19, NO. 5
[9] SladanaJošilo and György Dán. (2022). Joint wireless and edge computing resource management with dynamic network slice selection, *IEEE/ACM Transactions on Networking*, vol. 30, NO. 4
[10] JialeiLiu, Ao Zhou and Chunhong Liu Shangguang Wang Rajkumar Buyya. (2022). Reliability-enhanced task offloading in mobile edge computing environments, *IEEE Internet of Things Journal*, VOL. 9, NO. 13
[11] Krishna Prasad Satamraju and Malarkodi Balakrishnan. (2022). A secured healthcare model for sensor data sharing with integrated emotional intelligence, *IEEE Sensors Journal*, VOL. 22, NO. 16.
[12] Redowan Mahmud and Adel N. Toosi. (2022). Con-Pi: A distributed container-based edge and fog computing framework, *IEEE Internet Of Things Journal*, VOL. 9, NO. 6
[13] Zhou Zhou, Mohammad Shojafar, Jemal Abawajy, Hui Yin, and Hongming Lu. (2022). ECMS: An edge intelligent energy efficient model in mobile edge computing, *IEEE Transactions on Green Communications and Networking*, VOL. 6, NO. 1
[14] Miao Hu, Weigang Wu, Julian Cheng and Min Chen. (2022). Quantifying the influence of intermittent connectivity on mobile edge computing, *IEEE Transactions on Cloud Computing*, VOL. 10, NO. 1
[15] GuanjieCheng, Yan Chen, Shuiguang Deng, Honghao Gao and Jianwei Yin. (2022). A blockchain-based mutual authentication scheme for collaborative edge computing, *IEEE Transactions on Computational Social Systems*, VOL. 9, NO. 1
[16] Xing Liu, Jianfeng Yang, Chengming Zou, Qimei Chen, Member, IEEE, Xin Yan, Yuao Chen, and Chenran Cai. (2022). Collaborative Edge Computing With FPGA-Based CNN Accelerators for Energy-Efficient and Time-Aware Face Tracking System, *IEEE Transactions on Computational Social Systems*, VOL. 9, NO. 1
[17] Seungkyun Lee, SuKyoungLee, and Seung-Seob Lee. (2021). Deadline-aware task scheduling for IoT applications in collaborative edge computing, *IEEE Wireless Communications Letters*, VOL. 10, NO. 10
[18] Huaming Wu. (2021). EEDTO: An energy-efficient dynamic task offloading algorithm for blockchain-enabled IoT-edge-cloud orchestrated computing "*IEEE Internet of Things Journal*, VOL. 8, NO. 4, 15
[19] Yundi Guo, Beiji Zou, Ju Ren, Qingqing Liu, Deyu Zhang, and Yaoxue Zhang. (2019). "Distributed and efficient object detection via interactions among devices, edge, and cloud" VOL. 21, NO. 11
[20] Hyame Assem Alameddine, Sanaa Sharafeddine, Samir Sebbah, Sara Ayoubi, and Chadi Assi, *Dynamic Task Offloading and Scheduling for Low-Latency IoT Services in Multi-Access Edge Computing* VOL. 37, NO. 3
[21] Gang Qian, SiliangLu, Member, IEEE, Donghui Pan, Huasong Tang, Yongbin Liu, and Qunjing Wang. (2019). Edge computing: A promising framework for real-time fault diagnosis and dynamic control of rotating machines using multi-sensor data VOL. 19, NO. 11
[22] Junkai Liu, Ke Luo, Zhi Zhou, and Xu Chen. (2019). ERP: edge resource pooling for data stream mobile computing, *IEEE Internet Of Things Journal*, VOL. 6, NO. 3
[23] Zicong Hong,Wuhui Chen, Huawei Huang, Song Guo, Se, and Zibin Zheng. (2019). Multi-Hop cooperative computation offloading for industrial IoT–Edge–cloud computing environments, *IEEE Transactions on Parallel And Distributed Systems*, VOL. 30, NO. 12
[24] Yu-Lin Jiang, Ya-Shu Chen, Su-Wei Yang, and Chia-Hsueh Wu. (2019). Energy-efficient task offloading for time-sensitive applications in fog computing, *IEEE Systems Journal*, VOL. 13, NO. 3
[25] Babatunji Omoniwa, Riaz Hussain, Muhammad Awais Javed, Safdar Hussain Bouk and Shahzad A. Malik. (2019). Fog/Edge computing-based IoT (FECIoT): architecture, applications, and research issues, *IEEE Internet of Things Journal*, VOL. 6, NO. 3
[26] Yiqin Deng, Zhigang Chen, Member, Xin Yao, Shahzad Hassan, and Ali.M.A.Ibrahim. Parallel offloading in green and sustainable mobile edge computing for delay-constrained IoT System, *IEEE TRANSACTIONS ON*
[27] Md. Asif-Ur-Rahman Fariha Afsana, Mufti Mahmud, M. Shamim Kaiser, Muhammad R. Ahmed, Omprakash Kaiwartya, and Anne James-Taylor. (June 2019). Toward a heterogeneous mist, fog, and cloud-based framework for the internet of healthcare Things, *IEEE Internet of Things Journal*, VOL. 6, NO. 3.
[28] Chang Shu, Zhiwei Zhao,Geyong Min, and Shuowei Chen. (October 2019), Mobile edge aided data dissemination for wireless healthcare systems, *IEEE Transactions on Computational Social Systems*, VOL. 6, NO. 5,

GAN-based autoencoder for anomaly detection in SIoT

S. Divya, R. Tanuja & S.H. Manjula
Department of CSE, UVCE, Bengaluru, India

ABSTRACT: The paper looks at the potential application of an Autoencoder integrated with GAN for the purpose of anomaly detection in Social IoT (SIoT) networks. Our approach takes advantage of the synergetic characteristics of an autoencoder which is used to automatically stick to normal network behaviour and to integrate to a GAN framework for adversarial training of the generated anomalies. An experiment carries out such an observation using computer defects, which imply worse reconstruction. The model gets tested on a different dataset to show if it is capable of grabbing and spotting abnormal patterns. The evaluation makes use of a separate dataset, like the one handled before in the anomalous patterns identification research. Anticipated outcomes include the high performance of the detection accuracy of the novel model that is demonstrated in its application and utility in handling the emerging protection challenges in dynamic SIoT environments. The proposed model provides an illustrative strategy of integration GAN+ Autoencoder that can be applied to promote security in SIoT networks.

Keywords: SIoT Security, Anomaly Detection, GAN-based Autoencoder, Network Intrusion, IoT Security Innovations.

1 INTRODUCTION

The IoT has become a networked system that people may access and use to transmit information using a set protocols [1] Cities and homes are termed, respectively, as smart cities, smart devices, smart homes, smart transportation, healthcare, agriculture, smart grids, and military application largely as a result of technological advancements in the IoT industry. Implying the scope of IoT throughout industry domains of application.

Integration of the whole spectrum of sensors and devices results in the galaxy of IIoT. [2] Achieving smart solutions with industrial fringe points critically hinges on the capabilities of artificial intelligence, deep learning, and machine learning processes [3]. In general forecasts show a smart devices growth of 21% by 2030 with an introduction of various IoT applications then that IoT is going to play in the near future [4]. As per demonstrated, the automobile sector needs to be at its 195. 7 billion endpoints by the end of 2025.

The multiple sensors used by these devices facilitate gathering instant data from physical objects as the global network of internet-connected devices expands[5] The owners of these devices can hence use the data to make the best situation for users. Although there is no privacy in a smart city, with the data collected, it makes it simpler to program clever algorithms for decision-making but also presents cybersecurity exposures[6],[7] Robust intrusion detection systems are the base of all cybersecurity defenses which may come up against DDoS attacks[8] some famous Mirai botnet instance, DoS attacks, Port Scans and hostile botnets that may destroy websites to protect IoT devices.[9]

Our research is consistent with the tendency towards the creating of more and more applications for the Internet-of-Things where deep learning and machine learning are included as a must in these applications. Our model is demonstrated to be flexible and

DOI: 10.1201/9781003589273-63

resilient to threats that continue to increase and helps the anomaly detection accuracy by the use of neural networks Generative Adversarial (GANs) and Autoencoders.

In SIoT application domain, the integration of those models supports the defence from potential cyber-attacks, which increases the prospects for the resistance of intrusion detection systems (IDS).[10] Our GAN and Autoencoder combined model being an observable disruption in security for SIoT is one of the most remarkable things in the context of the dynamical transition observed in the IoT domain. This model has the capacity to bolster SIoT web and intensify the resistance to newly emerging security threats through testing and assessment over complete data sets. This paper enhances the understanding of the adaptability and efficiency of our proposed machine learning model to cope with the complex security structure in SIoT networks which provides a valuable plan for problem explanation of SIoT security.

1.1 *Objectives*

The first goal of this work will be to construct an anomaly detector model for social IoT network. Synergistic approach of GAN and Autoencoder will be the main tactics of the model that allow improving the security and making it adaptable to the current environment. Concretely the aim will be to train the Autoencoder on normal network patterns and then integrate it into the GAN scheme for adversarial training helping the system to produce realest and most plausible oddities. The anomalies are the identified where the uncertainty level is greater than the rest.

2 RELATED WORKS

In this paper Jiang, Wenqian, *et al.* propose a new technique for detecting anomalies in uneven time series data in the point of view of industrial use cases. Unlike the other methods of classification of anomalies, which used generators that had at least one sub-network of the original model running solely on normal samples, the anomaly detection approach trains three sub-networks of the generator that acts only on normal samples without any use of the original model, generating the anomaly scores based on the apparent and latent losses. A successful experimental demonstration on benchmark and lab dataset iprocure the technique to perform 100% accurately in the case of differentiating the abnormal and the normal sample quality[11].

The DAGAN model, a dual auto-encoder GAN anomaly detection model, Chen, Lei, *et al.* was proposed in this paper along with modification to suit the detection of anomalies in control systems in the Internet of Things. In contrast to some problems like the data risk of skewed long-tails and the difficulty of getting abnormal samples, the DAGAN architecture is composed of "encoder-decoder-encoder" network to simulate the nature distribution without any abnormal samples[12].

This work Mao *et al.* develops a semi-supervised unification mechanism, based on reconstruction errors, that uses discriminative autoencoding. It is constituted by generator and a discriminator, who is working as a generative adversarial network with quadratic divergence. In the training phase, the network works as an encoder/decoder since it aims to score high on reconstruction leading to small reconstruction errors compared to the prescribed threshold[13].

A paper Kumarage *et al.* presented with the use of paramsri. nla-semi-supervised anomaly detection for applications in industrial software systems employing bidirectional Generative Adversarial Networks (BiGANs). The study shows the superiority over the traditional GAN-method with this approach(BiGAN) and the results are more apparent in bigger databases. Researchers have been able to demonstrate that BiGANs beat well-known classifiers which include One-class SVM and Variational Autoencoders (VAEs) as the features of the space is increased. [14]

Tang *et al.* Presented It is shown how smart cameras with detection algorithms can be enhanced with proposed binary auto-encoder generative adversarial network (DAGAN) which is designed to provide better performance of automated optical inspection systems by getting rid of sample imbalances. Skip connections and the dual auto-encoder architecture combined have by DAGAN led to better image reconstruction and training stability in comparison with the other methods.[15]

This paper Niu *et al.* proposing the employment of LSTM (long short-term memory) in the VAE-GAN(variational autoencoder-generative adversarial networks) based approach to time series anomaly detection in order to resolve problems faced in the previous deep learning approaches. Through the joint evolvement of all the three constituents, i. e. , the encoder, the generator and the discriminator, the method gets the benefit of both the capability mapping of the encoder and the discriminator. Founding results agree with proposed detection protocol suggesting its reliability and speediness in comparison to its contemporaries. [16]

Wang *et al.* this work calls for adVAE, a self-adversarial variation autoencoder which is proposed for unsupervised anomaly detection, as an alternative. It describes issues faced by deep generative models assuming Gaussian distributions for anomalous and normal data as being similar in the latent space. More regularization goals participate in adVAE ensuring superior classification capability and preventing overfitting from occurring widely in the experiments, and the methodology has been proven to be superior in terms of anomaly detection compared to existing models in the experiments.[17]

3 PROBLEM STATEMENT

While rapid SIoT networks development make the exiting intrusion detection techniques more inefficient, however Internet of Things applications unreliable, and, hence, more complex and robust IDS should be employed. Simple anomaly detection system that can be used for simple IoT networks is incapable of dealing with complexities of SIoT networks so must be replaced with more effective and adaptable (detection) system that is suitable for the complexities of SIoT networks.

3.1 *Contribution*

This research contributes to the field of SIoT security by:

- The goal is to offer a large dataset or abnormalities detection in SIoT systems.
- Introducing a new anomaly detection model GAN+ Auto encoders proposed and implemented in GAN +Auto encoders.

4 METHODOLOGY

An autoencoder and generative adversarial network is used in a two step process to perform Anomaly detection in the training phase, the Autoencoder uniformly maps network tasks to the typical input through a labelled training dataset. This Autoencoder is added into the GAN architecture for an adversarial training loop (referred to as GAN) during the training procedure. The model was applied to a different dataset which includes both normal and abnormal conditions cases. With a high error reconstruction as anomaly indicators. Standard criteria like accuracy, precision, recall rate, and F1 score are used to have performance assessment of the model. In order to prove the performance of the developed technique, on a set of unique datasets for the SIoT network, different experiments are performed. An in depth analysis reveals the better effectiveness of the proposed GAN based Autoencoder contrasted with the current approaches.

4.1 *Dataset and model description*

It is evident from the given Python code that an autoencoder is used to make a GAN, which is eventually used for discovering anomalies in a SIoT system. While the GAN (Generative Adversarial Network) produces real-looking samples, the autoencoder (an algorithm) recapitulates the underlying data provided. MSE and a binary cross entropy loss functions are used for training. Discriminator and generator are adapted as training goes on, and so are optimised while the training loop implement epochs. By examining that recall/F1-score/ROC AUC/accuracy measures are evaluated on regular basis. The code additionally stores the model for subsequent teaching purposes, and parses performance metrics as well as generated samples analysis and visualization. The dataset has parameters which are all related in a network traffic within an Internet of Things system (SIoT). It has its own characteristics which you can use to define many aspects of communication: length, type of protocol, service and flags. Specifics are represented in numbers such as number of packages, parcel rate, and bytes carried per day. The dataset is about various risks which may put our system at risk such as: login attempts and breach indications. The goal function is to be determined from the class column whose values signify normal and abnormal network condition. These training data set is suited for training the machine learning models especially those that use GAN and Autoencoder approaches along with engine learning for making prediction skills better for identifying anomalies in SIoT system.

5 RESULTS

The confusion matrix, precision, recall, and F1-score were used to assess the model's performance. Noteworthy findings include,

Confusion Matrix:

$$\begin{bmatrix} 212 & 2372 \\ 15 & 1910 \end{bmatrix}$$

The confusion matrix is a 2x2 matrix that shows the counts of true positive (TP), false positive (FP), true negative (TN), and false negative (FN).

In this case:

- True Positives (TP): 212
- False Positives (FP): 2372
- True Negatives (TN): 1910
- False Negatives (FN): 15

Table 1. Performance metrics for the GAN+Autoencoder machine learning model for Anomaly detection in SIOT system.

	Accuracy	Precision	Recall	f1-score
Class 0	0.83	0.93	0.08	0.15
Class 1	0.65	0.45	0.99	0.62
Average	0.74	0.69	0.54	0.39

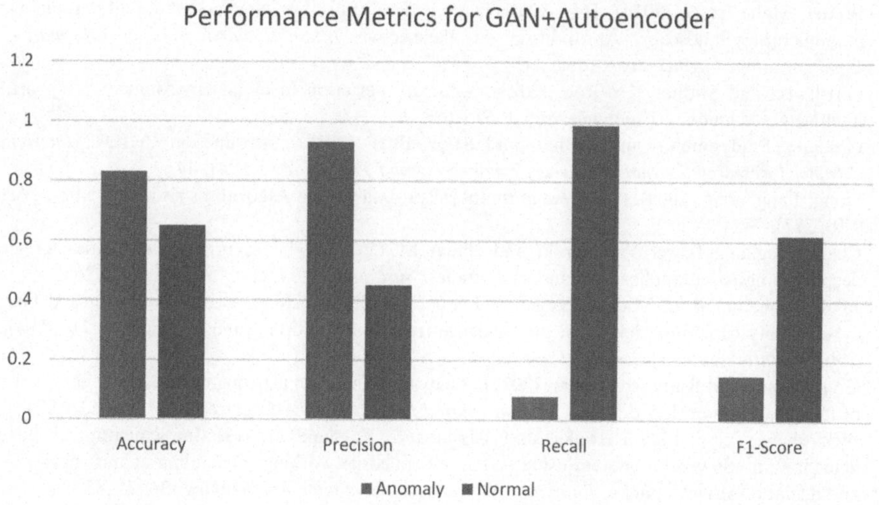

Figure 1. Performance Metrics for the anomaly and normal data in SIoT anomaly detection dataset.

A binary classification model's performance is shown by the classification report, which is displayed in Table 1 and Figure 1. The model has excellent accuracy (93%), meaning that it can accurately forecast abnormalities when they occur (class 0). The poor recall rate of 8%, however, indicates that a sizable portion of real abnormalities are likely overlooked. Conversely, the model has a lower accuracy of 45% but an exceptionally high recall of 99% for normal examples (class 1), suggesting that it catches most real-world normal occurrences. A balance between recall and accuracy is provided by the F1-scores, with class 1 exhibiting a more favourable F1-score (0.62). The model's total 47% accuracy indicates that its forecasts are unreliable for both groups. The macro-average values reflect an average performance that considers each class equally, revealing room for improvement in achieving a more balanced and accurate classification across anomalies and normal instances.

6 CONCLUSION

Although it has certain limitations, the proposed GAN-based Autoencoder model for Social Internet of Things (SIoT) anomaly detection shows significant potential. A significant trade-off between precision and recall is evident in the confusion matrix and classification report, especially when it comes to anomaly detection. Although the model can identify anomalies with high accuracy in typical cases, it finds it difficult to do so when anomalies arise. The results highlight the necessity for additional improvement in order to improve the overall balance and dependability of the model in both classes. In order to rectify the current imbalance, future iterations should concentrate on enhancing anomaly detection. Notwithstanding the existing constraints, the flexibility and effectiveness of the suggested method demonstrate its potential to tackle changing security issues in dynamic SIoT contexts, opening the door for improvements in anomaly detection techniques for networked systems.

REFERENCES

[1] Chen, Shanzhi, *et al.* (2014). A vision of IoT: Applications, challenges, and opportunities with china perspective. *IEEE Internet of Things journal* 1.4: 349–359.

[2] Brettel, Malte, *et al.* (2014). How virtualization, decentralization and network building change the manufacturing landscape: An Industry 4.0 Perspective. *International Journal of Information and Communication Engineering* 8.1: 37–44.

[3] Girish, L., and Sridhar KN Rao. (2023). Anomaly detection in cloud environment using artificial intelligence techniques. *Computing* 105.3: 675–688.

[4] Gokhale, Pradyumna, Omkar Bhat, and Sagar Bhat. (2018). Introduction to IOT. *International Advanced Research Journal in Science, Engineering and Technology* 5.1: 41–44.

[5] Dong, Jiang, *et al.* (2009). Advances in multi-sensor data fusion: Algorithms and applications. *Sensors* 9.10: 7771–7784.

[6] Chen, Dongliang, Paweł Wawrzynski, and Zhihan Lv. (2021). Cyber security in smart cities: a review of deep learning-based applications and case studies. *Sustainable Cities and Society* 66: 102655.

[7] Jalali, Mohammad S., Michael Siegel, and Stuart Madnick. (2019). Decision-making and biases in cybersecurity capability development: Evidence from a simulation game experiment. *The Journal of Strategic Information Systems* 28.1: 66–82.

[8] Gaur, Vimal, and Rajneesh Kumar. (2022). Analysis of machine learning classifiers for early detection of DDoS attacks on IoT devices. *Arabian Journal for Science and Engineering* 47.2: 1353–1374.

[9] Birkinshaw, Celyn, Elpida Rouka, and Vassilios G. Vassilakis. (2019). Implementing an intrusion detection and prevention system using software-defined networking: Defending against port-scanning and denial-of-service attacks. *Journal of Network and Computer Applications* 136: 71–85.

[10] Shahab, Saima, *et al.* (2022). SIoT (social internet of things): a review. *ICT analysis and applications*: 289–297.

[11] Jiang, Wenqian, *et al.* (2019). A GAN-based anomaly detection approach for imbalanced industrial time series. *IEEE Access* 7: 143608–143619.

[12] Chen, Lei, *et al.* (2022). Dual auto-encoder GAN-based anomaly detection for industrial control system. *Applied Sciences* 12.10: 4986.

[13] Mao, Sheng, Jiansheng Guo, and Zhengxin Li. (2019). Discriminative autoencoding framework for simple and efficient anomaly detection. *IEEE Access* 7: 140618–140630.

[14] Kumarage, Tharindu, *et al.* Generative adversarial networks (GAN) based anomaly detection in industrial software systems. *2019 Moratuwa Engineering Research Conference (MERCon)*. IEEE.

[15] Tang, Ta-Wei, *et al.* (2020). Anomaly detection neural network with dual auto-encoders GAN and its industrial inspection applications. *Sensors* 20.12: 3336.

[16] Niu, Zijian, Ke Yu, and Xiaofei Wu. (2020). LSTM-based VAE-GAN for time-series anomaly detection. *Sensors* 20.13: 3738.

[17] Wang, Xuhong, *et al.* (2020). adVAE: A self-adversarial variational autoencoder with Gaussian anomaly prior knowledge for anomaly detection. *Knowledge-Based Systems* 190: 105187.

Data Science & Exploration in Artificial Intelligence – Gururaj et al. (Eds)
© 2025 the Author(s), ISBN: 978-1-032-96390-7

Automated structure of 3D animation methodology to increase the data clustering

P. Bhuvaneshwari
Department of Computer Science and Engineering, Manipal Institute of Technology Bengaluru, Manipal Academy of Higher Education, Manipal, India

R. Karthik Ganesh
Department of Computer Science and Engineering, SCAD College of Engineering and Technology, Cheranmahadevi, India

M. Bagya Lakshmi
Department of Computer Science and Engineering, Francis Xavier Engineering College, Tirunelveli, India

A. Vegi Fernando
Department of CSE(AI&ML), School of Engineering, Dayananda Sagar University, Bengaluru, India

Y. Harold Robinson
Department of Computer Science and Engineering, Francis Xavier Engineering College, Tirunelveli, India

G. Ignisha Rajathi
Department of Information Technology, Manipal Institute of Technology Bengaluru, Manipal Academy of Higher Education, Manipal, India

ABSTRACT: The Detection and grouping of 3D videos participate an essential responsibility in multimedia indexing of 3D Animation video and detection of 3D video plays a vibrant role. The proposed technique has the Self-embedding process for digital signatures to ensure the video communication while the data compression is used to extract the specific features. The Quantization matrix is computed to measure the self-associated data where the gradient decent flow is involved to frame the image substance diffusion. The image elements are organized in zig-zag way to match frequencies for producing output-based encoding to summarize the similar data and the experimental results proved that the proposed technique has performed well in several performance metrics.

1 INTRODUCTION

A 3D animation has rapidly increased the display of 3D objects with dynamic objects which have the triangular meshes as the primary attributes needed for rendering the 3D objects and the connectivity demonstrates the triangular face with the vertices. Disordered interleaving alleviates the channel impact among data transmission, the remote transmission causes problems. Thus, the interleaving is connected to alleviate the frames. The confused interleaving is a decent contender for application in the proposed framework before the regulation procedure [1] is introduced. The complex 3D models have involved the realistic visual parameter in a computer-enabled 3D animation, it needs a huge storage area with several reconstruction period to perform the visualization. The compact demonstration is used to facilitate successful communication of mesh data with the band restricted channel. The 3D

DOI: 10.1201/9781003589273-64

compression has increased rapidly to process multiple objects per frame, a single object has several connected elements into the isolated groups of interconnected vertices to construct a 3D model [2]. The Graphical Processing Unit has rendered to display effective components like vertices, edges where the edges have the connecting vertices which are adjacent vertices [3]. The 3D applications have been used to render 3D meshes for data visualizations require for storing data to transmit within the network. The huge 3D data needs more cost to limit the bandwidth according to a huge volume of data for data compression to reconstruct. In this exploration work, a half breed conspires involving both randomization and levelling is executed to improve the execution of the OFDM framework [4]. Information randomization is connected before transmission, while the adjustment is performed at the beneficiary. To assess the nature of the got edges, the Signal-to-Noise Ratio (SNR) is received [5]. Now-a-days, because of different insurance factors these observing systems wound up being much progressively prominent and additionally basic. This tremendous need expanded the progression and furthermore headway alteration in following non-static articles. It is all over widely utilized by the military, understanding watching, human gear UI, action evaluation in agreeing to and also recognizing the items precisely. Implanting approaches are arranged into two fundamental classes; information watermarking procedures and information concealing methods [6]. There is a prerequisite to applying the installing strategies with information pressure for greater security and data transmission sparing. In the writing, obviously compacted pictures transmitted over remote media isn't reestablished proficiently as some data is typically lost amid the pressure changes and an expanded computational multifaceted nature and time [7]. Considering different video watermarking procedures, we can presume that the Least Significant Bit (LSB) system is the most straightforward strategy, yet it is temperamental. The DCT is more powerful and solid contrasted with the LSB [8]. There are distinctive methodologies actualizing diverse adaptations of the DCT in video watermarking, for example, 2D-DCT [9] and 3D-DCT [10]. Self-implanting is another pattern in picture protection and security. One inserting process is less difficult than other conventional techniques [11].

2 RELATED WORK

Rate-Distortion optimization technique [12] is used to increase the video quality during video compression technique. Image compression is the procedure for minimizing the data volume to categorize as lossless and lossy data compression to ensure the retrieval process into the input data while some information has lost. The primary features have high compression ratio as the output is specifically distorted into adjacent pixels in the correlated spatial domain to include redundant data and it is used to mark it hard to compress that the major convention of video encoders is to increase quality in encoding image, video, audio, and others to reduce the file size and to increase the quality [13]. The Lossy image compression [14] is proposed and it is necessary to guarantee that the compression procedure does not influence the quality of the image unfavorably. The improvement of a lossy compression algorithm is estimated based on two contradictory constraints, they are, compression ratio and quality of the image that is frequently calculated by PSNR values [15]. The compression of 3D data to reconstruct from the images for obtaining the patterns on the 3D object for capturing the high resolutions to allow the reconstruction. An MMS procedure [16] has involved the coefficient values to perform the arithmetic coding while the decompression functionality is used for reconstruction of high-level values with compress data to estimate the lookup table. The geometry-enabled compression technique [17] has involved the segmentation procedure to segregate the frame into the same number of vertices to construct the topology data within the timeframe. The affine transformation metrices are involved the sub-meshes to identify the residual difference with entropy-coded process for identifying the transmission of animated data.

3 PROPOSED WORK

With the growth of both video compression methodologies, broadcasting video programs over wireless channels befalls progressively more accepted. Moreover, wireless video broadcasting is still a demanding problem owing to restricted network bandwidth, the occurrence of channel failures, and the inconsistency in consumer workstations. With the purpose of video quality in the recipient side, it is significant to plot the video source to the radio source in a well-organized method by taking into report the video stream qualities and the channel quality. Self-watermarking is a method that is executed to conceal secret information or data in an image that is eliminated from the same image. The process is called as a self-embedding procedure for digital signatures. To be vigorous, the self-embedded signatures are used to endure the consequent compression procedure is carried out at the broadcaster proceeding to video communication. Data Compression is the method of creating data in a compressed format. We generate these kinds of compressed demonstrations by classifying with the existing information. The data will be in the format of text file, image waveforms, and group of numerical number which can be created by the related methods. The data compression technique in the system of digitalization is the group of data in bytes. The large amount of data is used to form the multimedia file. The computerized framework of 3D animation video based on camera regularities and segments is used to increase the quality of service. The 3D Animated video compression procedure is constructed by creating the macro block and this can be encoded with the JPEG 2000 format and the Lagrangian multiplier is implement to produce the new block position. The video streams are spilt into the video frames using the algorithm and the H.265 format is used to compress the 3D animated video.

The equation for Image Transformation M[x] can be computed using Equation (1).

$$T(m) = \frac{1}{2} C(m) \sum_{x=1}^{8} M[x] \cos\left(\frac{\pi(2x+1)m}{16}\right), \text{ where } C(m) = \begin{cases} 1, & m = 0 \\ \frac{1}{\sqrt{2}}, & otherwise \end{cases} \quad (1)$$

Quantization is the process of compressing 8 x 8 matrixes. Quantization matrices are used to obtain various levels of quality video compression. The level of quality of an image is ranged from 1 to 100. The Quantization matrix is generated from a lot of experiments and the resultant matrix is a high-quality compression image, below 50 is the best quality level. Every component in the Quantization matrix separates the particular component into 8 x 8 block of transformation and the value is nearer to the integer value. The low frequency elements are replaced by 0s. The image is re arranged using the positive values. Here the Quantization matrix M_{50} is given below in Equation (2).

$$M_{50} = \begin{bmatrix} 015 & 010 & 009 & 015 & 023 & 039 & 050 & 060 \\ 011 & 011 & 013 & 018 & 025 & 057 & 059 & 054 \\ 013 & 012 & 015 & 023 & 039 & 056 & 068 & 055 \\ 013 & 016 & 021 & 028 & 050 & 086 & 079 & 061 \\ 017 & 021 & 036 & 055 & 067 & 108 & 102 & 076 \\ 023 & 034 & 054 & 063 & 080 & 103 & 112 & 091 \\ 048 & 063 & 077 & 086 & 102 & 120 & 119 & 100 \\ 071 & 091 & 094 & 097 & 111 & 099 & 102 & 098 \end{bmatrix} \quad (2)$$

The Quantitative measurement of data can be measured by information theory. Let P(X) is the probability of the element X is developed using self-associated data with X is calculated using Equation (3).

$$i(X) = \log_y \frac{1}{P(X)} = -\log_y P(X) \quad (3)$$

Assume that, we need to develop the base y of the logarithmic function. The log function is used to maintain the data measurement. The self-associated data for the event of X and Y is given in Equation (4)

$$i(XY) = \log_y \frac{1}{P(XY)} \tag{4}$$

If X and Y are independent then using Equation (5)

$$P(XY) = P(X)P(Y) \tag{5}$$

The self-associated data for the event of X and Y is further simplified using Equation (6) and Equation (7).

$$i(XY) = \log_y \frac{1}{P(X)} + \log_y \frac{1}{P(Y)} \tag{6}$$

$$i(XY) = i(X) + i(Y) \tag{7}$$

The gradient decent flow of the modernized vector structure is the isomorphic geo information system which is used to frame the image substance diffusion. The initial modernized vector structure Ve_0 is calculated using the gradient. The Linear methodology merges the rank data of every video explosion in dissimilar values calculated by variety of related values. The Value of Linear methodology $\mathrm{L_V(x)}$ is calculated using Equation (8).

$$\mathrm{L_V(x)} = \sum_{y=1}^{n} \tau_y(x) \tag{8}$$

The mutual methodology generates the old values for calculating a generated value with classifier. The mutual value $\mathrm{U_V(x)}$ is calculated using Equation (9).

$$\mathrm{U_V(x)} = \sum_{y=1}^{n} \frac{1}{z + \tau_y(x)} \tag{9}$$

A generalized methodology is developed for the category updating based on values. The application of a generalized methodology is stimulated by the classifier. Given a pair of values are developed by separate values. The mixture value $\mathrm{M_V(x)}$ is calculated using Equation (10).

$$\mathrm{M_V(x)} = \prod_{y=1}^{n} (1 + \varphi_y(x)) \tag{10}$$

Whenever we exploit the feature extraction, all the associate networks for feature mapping values are correspondingly executed. All associate networks are used the similar model and the similar strength in the network. Texture is the mostly used component for visual experience which consists of the key data about the spatial understanding of color in the picture. The picture texture is normally documented to be a group of parameters computed for designing the elevated texture picture. Even though, it is very simple for calculated texture visual superiority for an image, progress evaluation for the entire look and three stage scans for together the 16x16 and 8x8 square size and analyze execution. The entire example of C code for every one of the projects is given in the index. Here we construct the intrinsic improvement completed to join the picture highlights. After quantization matrix formation, the highest number of values is equal to zeros. The quantized values are then encoded in zig-zag arrangement. The image elements are organized in zig-zag arrangement and the matching frequencies are accumulated jointly to execute Output-based Encoding. In Output-based Encoding, a distinct data rate and sum is used to characterize the identical data value that happens in numerous successive data values.

4 EXPERIMENTAL RESULTS AND ANALYSIS

The proposed AS3DA technique is compared with the related techniques of PSC [18], MPSC [19], GWO [20], EGWO [21], WOA [22]. The performance parameters of Jaccard coefficient parameters are highest in addition to MSE in Equation (11) is very minimal for the dependable technique.

$$MSE = \sum\nolimits^{n} k = 1 \parallel D_k - C_r \parallel^2 \quad \in r \tag{11}$$

where D_k is the k_{th} information factor and also C_r is the rth cluster. 'n' describes the overall number of information factors existing in the data source for clustering in Equation (12).

$$f - measure = \sum\nolimits_l \frac{a_l}{H} \max\nolimits_m [f(a_l, C_m)] \tag{12}$$

where' 'l is the complete quantity of features existing 'm' is the quantity of centroids, H complete quantity of clustered information. The Rand Coefficient is computed in Equation (13), Jaccard Coefficient is computed in Eq. (14).

$$Rand\ coefficient = \frac{SS + DD}{SS + SD + DS} \tag{13}$$

$$Jaccard\ coefficient = \frac{SS}{SS + SD + DS} \tag{14}$$

The video progressions in the proposed work simulation are Bus, Foreman, daughter and Mother. All the PSNR-rate sequences are obtained for the simulation. It is also experiential that bits from dissimilar feature layers supply in a different way to the restructured video, which is specified by the dissimilar gradients of the fragments. Figure 1 portrays the efficiency evaluation in regards to the efficiency statistics f-measure. For a reliable approach, the value of the f-measure is higher. The assessment of the f-measure for the suggested technique is better while contrasted with every other various former technique.

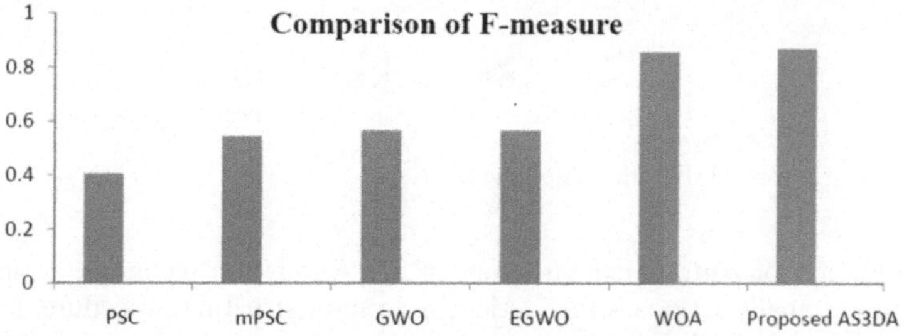

Figure 1. Comparison of F-measure.

Figure 2 shows the efficiency evaluation executed in regards to the RC. The RC is maximum for a reliable approach as well as the conversation with the values of the RC is offered listed below. However, the recommended AS3DA obtains a better value of the RC.

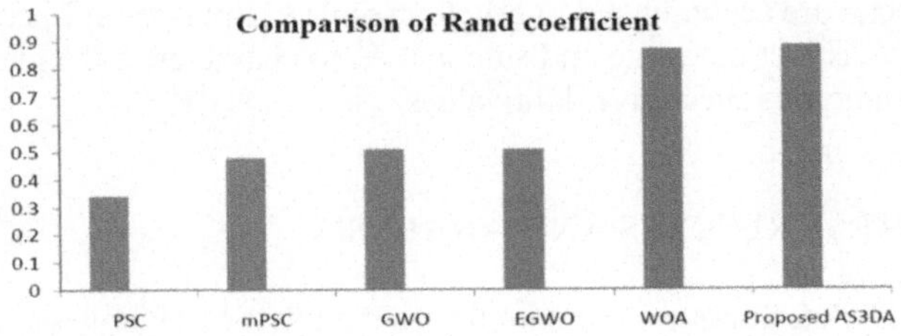

Figure 2. Comparison of Rand coefficient.

Figure 3 shows the efficiency evaluation in regards to the JC whose values are better for the recommended technique. The value of the JC for the suggested AS3DA is better when contrasted with all various former techniques.

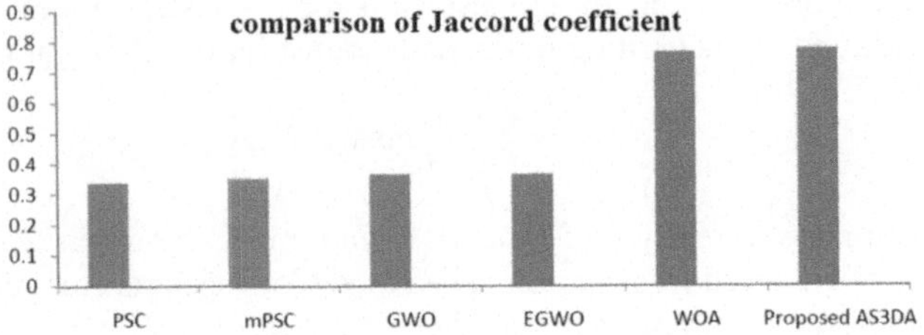

Figure 3. Comparison of Jaccard coefficient.

Figure 4 reveals the efficiency evaluation utilizing the dataset 2 in regards to the efficiency parameters MSE. For an effective technique, the MSE gotten resolve certainly continue to be minimal. Although the value of MSE decreases in the entire techniques with the boost in the cluster dimension, the assessment of the MSE for the recommended AS3DA is little.

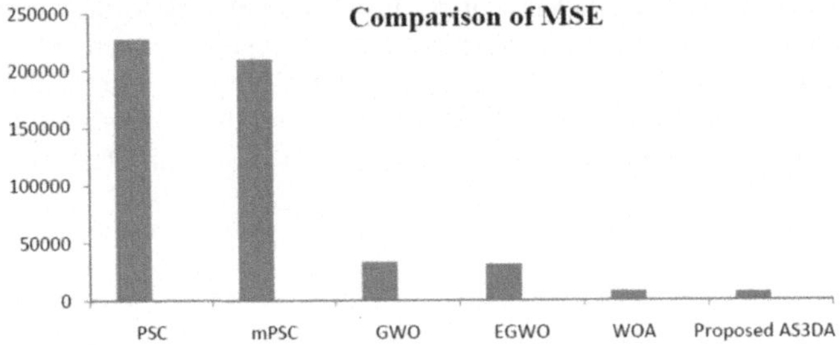

Figure 4. Comparison of MSE with related methods.

The computation of Automated Structure of 3D Animation technique which is exceptionally computationally serious is the modernized vector structure procedure. It is basically grid increase of a picture obstruct by a modernized vector structure procedure steady augmentation framework. This structure can likewise make utilization of the Automated Structure of 3D Animation method to enhance execution. For the accompanying segment the proposed system has a couple of recommendations that would enhance the execution of the feature map computation. In any case, we don't go into the execution examinations of the calculations because of absence of the capacity to add extra usefulness.

5 CONCLUSION

The proposed framework has implemented for constructing the animated video with effective evaluation for enhancing the utilization of the 2 datasets which confirmed the proposed AS3DA technique enhances the performance of the existing methodologies to achieve the nominal evaluation of MSE for the maximum fitness value, rand and Jaccard coefficient. The object detection process has the local minima with sensitivity initialization process to execute the improvement of the efficient cluster formation process, the proposed framework resolves the local minima problem with the sensitivity, the foreground and the background objects are segregated into non-static objects which are traced.

REFERENCES

[1] Medioni, G, Hongeng and Nevatia, R. (2001). Event detection and analysis from video streams, *IEEE Transactions On Pattern Analysis And Machine Intelligence*, vol.23, no.8, pp. 873–889.

[2] Jain, D., Tech, M., Engineering, C. and Engineering, C. (2017). *A Modified Method for Sound Compression Using Intrinsic Mode Functions* 97–102.

[3] Hosseini-nejad, H., Jannesari, A. and Sodagar, A.M. (2014). Data compression in brain machine computer interfaces based on the Walsh – Hadamard Transform. *IEEE Trans. Biomed. Circuits Syst*. 8, 129–137.

[4] Mukherjee, A. and Zhang, Z. (2019). Fast compression of OFDM channel state information with constant frequency sinusoidal approximation. *J Wireless Com Network* 2019, 87.

[5] Brewer J. A., Aarestad J., Pineda A. C. and Mee J. K. (2019). Impact of On-orbit data compression on signal-to-noise ratio for dim target detection, *2019 IEEE Space Computing Conference (SCC)*, Pasadena, CA, USA, pp. 20–24.

[6] Wang X., Yu K. and Ji Y. (2023). Application of MPEG-Based 3D animation compression algorithm in architectural animation customization, in *IEEE Access*, vol. 11, pp. 113134–113147.

[7] Kavitha, P. (2016). *A Survey on Lossless and Lossy Data Compression Methods* 7, 110–114.

[8] Wiryawan I. G., Sariyasa and Gunadi I. G. A. (2018). Steganography based on least significant bit method was designed for digital image with lossless compression technique, *2018 International Conference on Signals and Systems (ICSigSys)*, Bali, Indonesia, pp. 98–102.

[9] Agostini L. V., Silva I. S. and Bampi S. (2001). Pipelined fast 2D DCT architecture for JPEG image compression, *Symposium on Integrated Circuits and Systems Design, Pirenopolis*, Brazil, pp. 226–231.

[10] Xue, J.; Yin, L.; Lan, Z.; Long, M.; Li, G.; Wang, Z.; Xie, X. (2021). 3D DCT Based image compression method for the medical endoscopic application. *Sensors*, 21, 1817.

[11] Jeon, JI. and Kang, HS. (2010). 3D DCT Based compression method for integral images. In: Bebis, G., *et al. Advances in Visual Computing*. ISVC 2010. Lecture Notes in Computer Science, vol 6454

[12] Abu Alsheikh M., Lin S., Niyato D. and Tan H. -P. (June15, 2016). Rate-distortion balanced data compression for wireless sensor networks, in *IEEE Sensors Journal*, vol. 16, no. 12, pp. 5072–5083.

[13] Wang P. and Jiang X. (2023). 3D Human motion data compression based on computer vision, *2023 International Conference on Image Processing, Computer Vision and Machine Learning (ICICML)*, Chengdu, China, pp. 725–729.

[14] Behme L., Thirumuruganathan S., Mahdiraji A. R., Quiané-Ruiz J. -A. and Markl V. (2023). The art of losing to win: Using lossy image compression to improve data loading in deep learning pipelines, *2023 IEEE 39th International Conference on Data Engineering (ICDE)*, Anaheim, CA, USA, pp. 936–949.

[15] Khan, A., Khan, A., Khan, M., Uzair, M. (2016). Lossless image compression: application of bi-level burrows wheeler compression algorithm (BBWCA) to 2-D data. *Multimed. Tools Appl.*

[16] Uthayakumar Jayasankar, Vengattaraman Thirumal and Dhavachelvan Ponnurangam. (2021). A survey on data compression techniques: From the perspective of data quality, coding schemes, data type and applications, *Journal of King Saud University - Computer and Information Sciences*, Vol. 33, no. 2, pp, 119–140.

[17] Wen X., Wang X., Hou J., Ma L., Zhou Y. and Jiang J. (2020). Lossy geometry compression of 3d point cloud data via an adaptive octree-guided network, 2020 *IEEE International Conference on Multimedia and Expo (ICME)*, London, UK, pp. 1–6.

[18] Palanivelu L. M. and Vijayakumar P. (2011). Effective image segmentation using Particle swarm optimization for image compression in multi application smart cards, 2011 *World Congress on Information and Communication Technologies, Mumbai*, India, pp. 535–539.

[19] Borckmans, P.B., Ishteva, M. and Absil, PA. (2010). A modified particle swarm optimization algorithm for the best low multilinear rank approximation of higher-order tensors. In: Dorigo, M., *et al.* Swarm Intelligence. ANTS 2010. *Lecture Notes in Computer Science*, vol 6234.

[20] Ullah, I., Liu, K., Yamamoto, T., Shafiullah, M., and Jamal, A. (2022). Grey wolf optimizer-based machine learning algorithm to predict electric vehicle charging duration time. *Transportation Letters*, 15(8), 889–906.

[21] Amolkumar Narayan Jadhav and Gomathi N. (2016). Kernel-based exponential grey wolf optimizer for rapid centroid estimation in data clustering, *J. Technol.* 78 (11) (2016) 65–74.

[22] Seyedali Mirjalili and Andrew Lewis. (2016). The whale optimization algorithm, *Adv. Eng. Softw.* 95 51–67.

Data Science & Exploration in Artificial Intelligence – Gururaj et al. (Eds)
© 2025 the Author(s), ISBN: 978-1-032-96390-7

Enhanced approximate topology generality and identity-related broadcast encryption methodology for improving the security

S. Balaji
Department of Computer Science and Engineering, Francis Xavier Engineering College, Tirunelveli, India

P. Bhuvaneshwari
Department of Computer Science and Engineering, Manipal Institute of Technology Bengaluru, Manipal Academy of Higher Education, Manipal, India

P. Saveetha
Department of Computer Science and Engineering, Karpagam Academy of Higher Education, Coimbatore, India

A. Vegi Fernando
Department of CSE(AI&ML), School of Engineering, Dayananda Sagar University, Bengaluru, India

Y. Harold Robinson
Department of Computer Science and Engineering, Francis Xavier Engineering College, Tirunelveli, India

ABSTRACT: Broadcast encryption helps to firmly dispense data to a dynamic wireless communication. The broadcast encryption requires a reliable distributer to allocate the decoded keys to the group. Group Key Agreement (GKA) methods are used to locate a cluster to agree a widespread encryption key through wireless networks into the same group may decrypt and encryption could allow the sender to deliver message to specific group with a completely trusted members to locate the network. We proposed Enhanced Approximate Topology Generality and Identity-related Broadcast Encryption (EATGIBE) method, it holds a key that creates the private keys and recognizes by the specific secret keys. The PKG is created to reinforce the public broadcast encryption key and allocates the details of decryption keys to every user, this framework permits evading the neighbors' transmission issues.

1 INTRODUCTION

Broadcast encryption is utilized in cryptanalysis for forwarding the encrypted information within the broadcasting medium which manages the broadcaster encrypts transmit data and communicates to a lot of 'n' clients to tune in on a communicated station. Every n client utilizes the secret key to decode the communicate messages in the meantime [1]. Normally, the broadcast encryption methods were utilized the initially picks a lot of n clients similarity identity arranged to decode communicate messages as supposed clients' set and broadcasts a registered secret broadcast key. The important parameter for the broadcast encryption is to increase the efficiency. The computational cost is the prime metric for increasing the efficiency while developing the encryption and decryption states. The encryption computations are implemented using modulo exponentiation that will increase the complexity of the system. To increase the efficiency of the encryption process, the procedure is implemented by increasing the protection of perceptive information effectively. The computational cost of the decryption process is computed by using the dynamic keys that related to the positive

 DOI: 10.1201/9781003589273-65

output in the system. To satisfy the needs for constructing the broadcast encryption when the unauthorized users are frequently changed, their uniqueness is used the public keys in constructing the cipher texts. The dynamic construction is needed to increase the originality of the cipher text need to be changed [2]. For the method of symmetric key based Broadcast Encryption, the communication procedure should be started by the trusted system allocates the secret keys for every user. In contrast, the Broadcast Encryption always identifies the public key parameters can forward the messages to all the members in the group. For setting the symmetric key mutually generates the several useful data like the known secret key to all the users, therefore it is useful to provide the security and protection in the broadcast system. Whenever the broadcaster is permanent, it is impossible to provide the integrity in the system. The broadcast management system permits a beginner should encrypt the plain text for the receiver only using the identity of the receiver side's public key. For the Anonyms IBE systems, the cipher texts need not have the identity of receiver data. Additionally, this kind of systems are used to build the Public Key Encryption Model that permits a particular data is findable by using the keywords and also utilizes the access control methods for Public Key Encryption Model and Identity based Encryption model [3].

2 RELATED WORK

Several works have been implemented with the broadcast encryptions schemes like the technique [4] for managing the dissimilar session key, it has implemented the adaptive protection through the attribute-based broadcast system which is used to accept different methods to manage the trade-off within the smart networks. The extended procedure [5] used a tuple of DBE as it needed the communication within the users every period if a fresh user needs to connect as a novel key agreement algorithm is used to manage the extended key value which helps for avoiding the attack through the malware with unidirectional values. Encryption-based algorithm [6] has proposed the attribute-enabled the elliptic cures to provide the secure system. A trusted key generation centre [7] is implemented to develop which avoids regular attacks in GKT [8] utilizes the random generation methodology to address the Denial-of-Service attack. A Diffie – Hellman algorithm is [9] used to provide the broadcast-based security, Attribute Key Generation Algorithm [10] is proposed to develop the attribute-based revocation technique of XDHA [11]. ECC algorithm [12] is implemented to contribute the broadcast encryption, ConBE algorithm [13] is used to enhance the justification of the single hop transmission with the help of the non-monotonic attribute-based encryption [14], and RLWE [15]. CLAMRE [16] utilized the ECC technique to increase the security in Broadcast Encryption Model. The Mean Field Game (MFG) [17] enabled sensing rate has optimized by implementing the broadcast and finds the protection of the mobile devices through the enhanced procedure. Identify-enabled Broadcast Encryption [18] has the Collison-free hash function with the group of the improved keys to the Cloud server to support effective reversal of private keys. Identity-based Broadcast Matchmaking Encryption (IB-BME) [19] has the data sharing framework that improves the participants for specifying the particular policy to the encrypted information while the data could be utilized several recipients in both the satisfaction of the access policies.

3 PROPOSED WORK

The system formation is based on the interpolation methodology. Constructing the curvature with the points $(m1_1, n1_1), (m1_2, n1_2), \ldots, (m1_t, n1_t)$ could be accomplished as in Equation (1) to (3).

$$fn(m1) = \sum_{i=1}^{t} Fn_i(m1) = \sum_{i=1}^{t} a_i m1^i \quad (1)$$

$$Fn_i(m1) = n_i \prod_{1 \leq t} \frac{m1 - m1_j}{m1_i - m1_j} \tag{2}$$

$$\mathrm{Fn_i(m1)} = \begin{cases} \mathrm{n_i m1 = m1_i} \\ \mathrm{0m1 \in \{m1_1, \ldots, m1_t\}} \end{cases} \tag{3}$$

If the 2 coordinate points are formed the interpolation, then the linear model is represented as the line within the coordinate points. It will be used to that the data points replaced by the polynomial coefficient. The interpolation function for m coefficient is computed using the summation of linear element with the coefficient point. If value of this function is 0 means the coefficient point belongs to the group member otherwise the value is different. The data hiding methodology is used to claim the verified ownership to implement the encryption domain in effective manner. It generates the cover data through distributing it into random values to share the secret data to the sender and the recipient with proxy servers.

Algorithm – EATGIBE

Step 1: Select a random value $rv \in \mathcal{F}_q^*$ *and* $\Pr_1 \epsilon Gf_1$
Step 2: Select $\Pr_{pk} = rv * \Pr$
Step 3: Define the hash functions
$HF : \{0,1\}^* \rightarrow \mathcal{F}_q^*$
$HF_1 : \{0,1\}^* \rightarrow Gf_1^*$
$HF_2 : Gf_2^* \rightarrow \{0,1\}^\alpha$
$HF_3 : \{0,1\}^\alpha \rightarrow \{0,1\}^\alpha$
$HF_4 : Gf_2 || \{0,1\}^* || Gf_2 x Gf_2 x \ldots x Gf_2 \rightarrow \{0,1\}^\beta$
$HF_5 : Gf_2 \rightarrow \{0,1\}^\gamma$
where $\gamma < \alpha$, $\beta + \gamma = \alpha$
Step 4: Generate the public constraints
$\{rv, Gf_1, Gf_2, Pr, \Pr_{pk}, HF, HF_1, HF_2, HF_3, HF_4, HF_5\}$

The EATGIBE algorithm is constructed with 4 steps. In the initial step, the random value is selected from the group and the private key is constructed from the Group of keys. Step 2 is used to produce the key agreement of private key using the random value. The hash functions are defined using the Group functions, group of keys. Final step is used to generate the public constraints from the computed elements in the first three steps.

Figure 1 demonstrates the multi-encryption process; Plaintext is encrypted into the encrypted text using public keys and identities. Multi-Encryption is performed to receive the

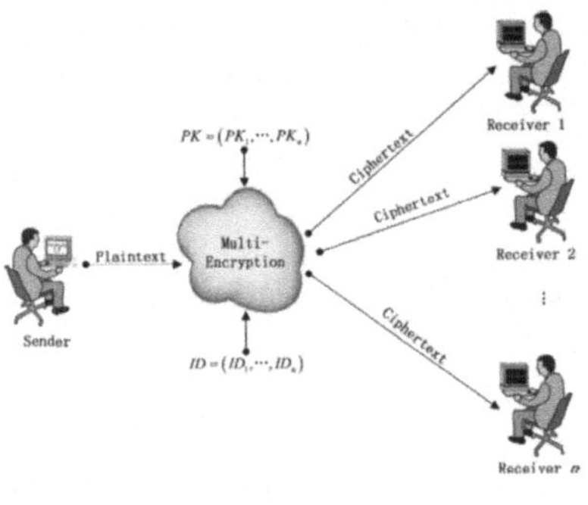

Figure 1. Multi-encryption process.

cipher text to the number of receivers. Each and every receiver is responsible for decrypting the cipher text using the individual public keys. The infinite non-commutative group is used for generating the master secret key. Additionally, π is the system constraint. It will be used for computing the multi-encryption process.

***Decrypt*(Φ,*R*, U_j,dk_j, s_j,*c*)**

Accept d, π, C_M and return M.

$M, ID \in \{0,1\}^*$

$$dk_j = decrypt\ (extract\ (\pi, MSK, ID), \pi, encrypt(\pi, M, ID)) = M$$

***Proxyre–encryption*(*Pr*(*PK*,*sk*))**

$C_M = encryption(PK, M)$ Into a new cipher text C'_M

The entire procedure is

1. Generate $\overline{sk_i} = enc(PK', sk_i)$ where sk_i denotes the ith bit of sk.
2. Compute $\overline{C}_{Mi}$ = enc(PK', C_{Mi}) Where C_{Mi} denotes i^{th} bit.
3. Evaluate the decryption circuit is

$$CM' = eval\left(PK', sk1', ..., skn', \overline{CM}_1, \ldots, \overline{CM}_n\right)$$

Then the decryption become,

$$\begin{aligned} Dec(sk', CM') &= dec\left(dec\left(\overline{sk}_i\right), .., dec\left(\overline{sk}_n\right), dec(sk), ..., dec(sk')\right) \\ &= dec(sk_1, ..., CM_1, ..., CM_n) \\ &= dec(sk,\ C_M)\ = M \end{aligned}$$

The security parameter is generated using the function $Paragen(\lambda, n)$ by the total amount of group members. The input system parameter is constructed using the Master secret key with the total amount of users. The $Extract(MSK,\ ID_i)$ function is used to combine the Master secret key and identities of each other. $Setup(MSK,\ \phi, U)$ function is generated the Master secret key using PKG with the key value and individual identifiers. The private key is constructed with the random value and Identity based encryption, secret keys. The output is produced using the decryption keys, the $Multi - Encrypt(\pi, R, PK, M)$ function is used to construct the cipher text and finally, the encryption is constructed using the hash functions. The $Decrypt\left(\Phi, R,\ U_j, dk_j,\ s_j, c\right)$ function is used to produce the main text. $Proxyre - encryption(Pr(PK, sk))$. The decryption process is the reverse process of the encryption process. Multi-encryption is the way of encrypting the encrypted cipher text in more times with different hash values. The multi-encryption can perform in n times if there are n number of recipients. This multi-encryption provides high amount of protection from attacks by strengthening the cipher text.

4 EXPERIMENTAL RESULTS AND ANALYSIS

To perform the security analysis, we need to confirm the precision among the network connectivity. The proposed Method EATGIBE is compared to ConBE, CLAMRE, RLWE, XDHA, EGIB. Entropy is the randomized functionality to ensure the security by producing the possible keys to ensure the security. Entropy for the particular key is performed using the following Equation (4).

$$Entropy_{key}(m, n) = Entropy_{key}(n, m) \tag{4}$$

Figure 2 demonstrates the Entropy process for the producing keys in detail. Entropy is the process of providing security in cryptography system that uses the needed random data. The randomness is referred as that the random sets of steps have no specific order to implement the intelligible pattern.

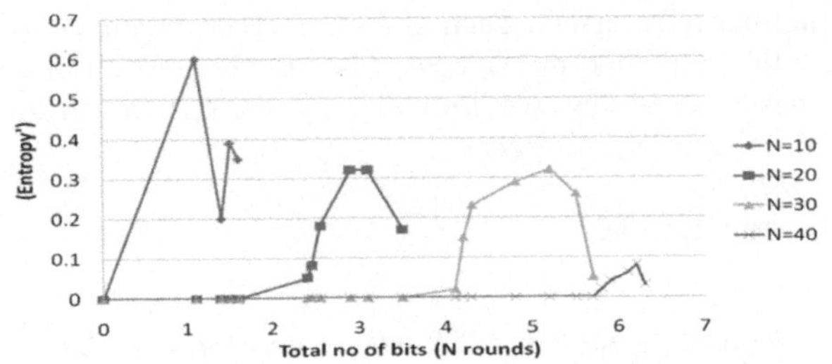

Figure 2. Entropy.

The execution time for encryption operation is calculated the speed of the encryption process. The encryption execution time is computed by separating the total plain text into Megabytes that the summation of the time for every algorithm to process the encryption operation. Figure 3 demonstrates that the proposed method has minimized amount of encryption execution time compared to the related methods. The runtime for encryption is computed as the time engaged for encoding the information using the authorized parties. The efficiency of the encryption algorithm is measured that the conversion of cipher text from the plain text. The intended data is called as the plaintext is encrypted, the time taken for completing the encryption process is measure and analyzed for proposed methodology and the related methods.

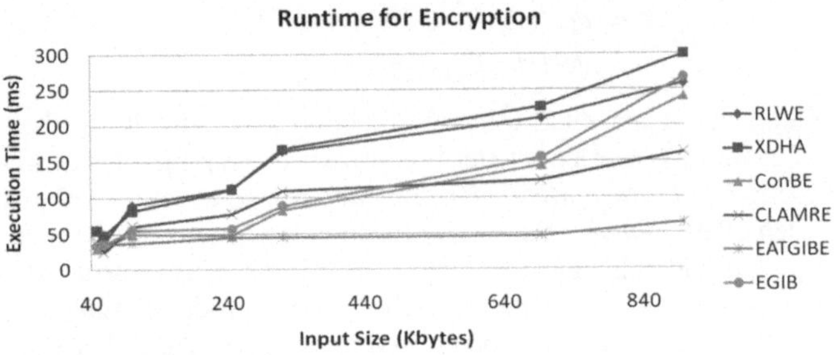

Figure 3. Runtime for encryption process.

The simulation results for the runtime execution time for decryption are demonstrated in Figure 4. We should discover that the EATGIBE is the better algorithm than other related algorithm for the execution time for decryption process. The runtime for decryption is computed as the time taken for decoding the information using the authorized parties. The efficiency of the decryption algorithm is measured that the conversion of plain text from the cipher text. The suitable key is involved to decrypt the cipher text into a main text. The proposed algorithm is analyzed and measured the time taken for completing the decryption process from the recipient side.

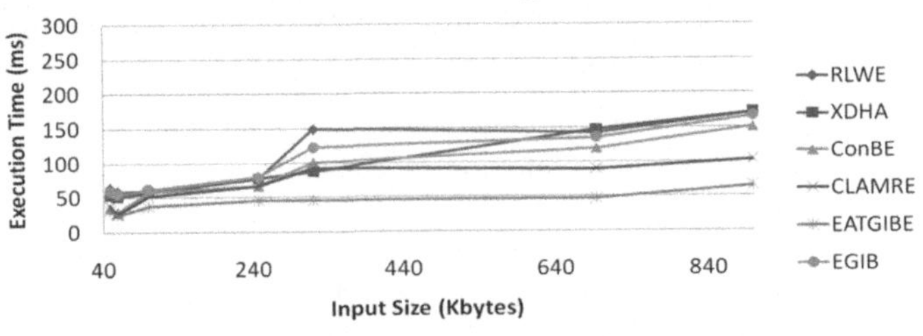

Figure 4. Runtime for decryption process.

5 CONCLUSION

The proposed framework should handle the sender Encryption modification in effective manner as it produces the third-party members for locating the specific system. The PKG develops the private keys within the users to identify the master secret key. This reduces the transmission problem within the neighbors in an effective manner to produce the encryption, decryption while the single round is important to maintain the public group which functions to broadcast within the network. The proposed technique is compared with the relevant techniques as it produces the enhanced results compared with the other techniques.

REFERENCES

[1] Tan C. H., Teo J. C. M. and Amundsen J. -A. (2007). Authenticated broadcast encryption scheme, *21st International Conference on Advanced Information Networking and Applications Workshops (AINAW'07)*, Niagara Falls, ON, Canada, pp. 512–518.

[2] Chen Y., Yang G. and Chen Y. (2009). An efficient broadcast encryption scheme for wireless sensor network, 2009 5th *International Conference on Wireless Communications, Networking and Mobile Computing*, Beijing, China, pp. 1–4.

[3] Ali E. A. E., Dahshan H. and Mohasseb Y. Z. (2017). Enhanced identity based broadcast encryption scheme through open networks, 2017 *12th International Conference on Computer Engineering and Systems (ICCES)*, Cairo, Egypt, pp. 573–578.

[4] Li M., Xu X., Zhuang R., Guo C. and Tan X. (2015). Identity-based broadcast encryption schemes for open networks, 2015 *Ninth International Conference on Frontier of Computer Science and Technology*, Dalian, China, pp. 104–109

[5] Zhang Y., Yang C., Liu J. and Li J. (2007). A novel broadcast encryption scheme based on sd scheme reconstruction, 2007 *Second International Conference on Communications and Networking in China*, Shanghai, China, pp. 387–391.

[6] Rasi Raj R and Joy Winnie Wise D.C. (2016). Enhanced arbitrary topology generalization and identity-based broadcast encryption scheme. *International Journal for Research in Science Engineering & Technology*, Vol 3 No 11: Volume 3 Issue 11.

[7] Xinjun Du, Ying Wang, Jianhua Ge and Yumin Wang. (June 2005). An ID-based broadcast encryption scheme for key distribution, in *IEEE Transactions on Broadcasting*, vol. 51, no. 2, pp. 264–266.

[8] Bai S., Zhu G. -B. and Huang J. (2013). An intelligent algorithm for the (1,2,2)-generalized knight's tour problem, 2013 *Ninth International Conference on Computational Intelligence and Security*, Emeishan, China, 2013, pp. 583–588.

[9] Alias Y. F. and Hashim H. (2018). Timing analysis for Diffie Hellman Key Exchange In U-BOOT using Raspberry pi, 2018 *IEEE Symposium on Computer Applications & Industrial Electronics (ISCAIE)*, Penang, Malaysia, pp. 212–216.

[10] Pletea D., Sedghi S., Veeningen M. and Petkovic M. (2015). Secure distributed key generation in attribute based encryption systems, 2015 10th *International Conference for Internet Technology and Secured Transactions (ICITST)*, London, pp. 103–107

[11] Akhil Kaushik and Satvika. (2013). Extended diffie-hellman algorithm for key exchange and management, *ICETEM.*

[12] Hafsa A., Sghaier A., Machhout M. and Malek J. (2019). A new security approach to support the operations of ECC and AES algorithms on FPGA, 2019 *19th International Conference on Sciences and Techniques of Automatic Control and Computer Engineering (STA)*, Sousse, Tunisia, pp. 95–100

[13] Qianhong Wu, Member, IEEE, Bo Qin, Lei Zhang, Member, IEEE, Josep Domingo-Ferrer, Fellow, IEEE Oriol Farras, and A. Manoj. (2015). Contributory broadcast encryption with efficient encryption and short ciphertexts, *IEEE Transactions on Computers.*

[14] Wang C. -J. and Luo J. -F. (2012). A key-policy attribute-based encryption scheme with constant size ciphertext, *2012 Eighth International Conference on Computational Intelligence and Security*, Guangzhou, China, pp. 447–451.

[15] Jintai Ding and Xiaodong Lin Chongqing. (2014). A simple provably secure key exchange scheme based on the learning with errors problem, *Citeserx*, volume 4.

[16] He D., Zeadally S., Kumar N. and Wu W. (2016). Efficient and anonymous mobile user authentication protocol using self-certified public key cryptography for multi-server architectures. In: *IEEE Transactions on Information Forensics and Security*. IEEE, p 1. doi:10.1109/TIFS.2016.2573746.

[17] Yang Y., Zhang B., Guo D., Xu R., Kumar N. and Wang W. (Nov. 2023). Mean field game and broadcast encryption-based joint data freshness optimization and privacy preservation for mobile crowdsensing, in *IEEE Transactions on Vehicular Technology*, vol. 72, no. 11, pp. 14860–14874.

[18] Lee K. (2023). Delegate and verify the update keys of revocable identity-based encryption, in *IEEE Access*, vol. 11, pp. 52636–52652.

[19] Sun J., Xu G., Zhang T., Yang X., Alazab M. and Deng R. H. (2023). Privacy-aware and security-enhanced efficient matchmaking encryption, in *IEEE Transactions on Information Forensics and Security*, vol. 18, pp. 4345–4360.

Data Science & Exploration in Artificial Intelligence – Gururaj et al. (Eds)

Hybrid Bi-LSTM model for text-based fake news detection in social media

P. Bhuvaneshwari
Department of Computer Science and Engineering, Manipal Institute of Technology Bengaluru, Manipal Academy of Higher Education, Manipal, India

M. Bagya Lakshmi
Department of Computer Science and Engineering, Francis Xavier Engineering College, Tirunelveli, India

M. Subramanian
Department of Computer Science and Engineering, SCAD College of Engineering and Technology, Cheranmahadevi, India

M. Kavitha Margret
Department of Computer Science and Engineering, Sri Krishna College of Technology, Coimbatore, India

Y. Harold Robinson
Department of Computer Science and Engineering, Francis Xavier Engineering College, Tirunelveli, India

ABSTRACT: Fake news detection is the harmful news spreading in social media which affect the democracy and journalism that reflets the news in online streaming. The fake news impacts the massive threat of security impairment while the news authentication could be confirmed according to the news contents. In this paper, the Hybrid Bi-LSTM framework is utilized to detect the Text-based spam news in social media. The hybrid model is developed by combining the Title-based Feature Extraction and Text-based Feature extraction models of Bi-LSTM. The internal decision process with the accurate output is the main concern for posted content in the social media. The precision, recall, F1-score, and accuracy metrics of the suggested method are compared with those of pertinent techniques from LIWC, and SAFE\V. The outcome of the experiment shown that the suggested method yields superior accuracy in the PolitiFact (DB1) dataset. The proposed work carries the enhanced framework for detecting fake news in the automatic way through the text content.

1 INTRODUCTION

The dissemination of fake news on social network is normally make deceptive direction of public view in spite of managing the political goals and care their individual confidence. Due to the enlargement of online communications and social network popularity, the fake news spread has lighter than the conventional media such as TV and newspapers [1]. Time sensitivity [2] is the crucial news as it enhances the advantages by maintaining the broadcasting news. The huge amount of news elements is created in each and every day and disseminated into the social media with speedy manner. The websites are not authenticated the user details while the accountability of the biased data posting for identifying the fake news in an appropriate way. The news editors are involved to gather the correct news, tracking of events, and the verification process for completing the list of functionalities in a perfect way as the verification of news has been completed by the leading organizations [3].

Fake news could produce the impact of false information as the comments are connected through the misleading data in time-to-time for undermining the reliability of the accurate

 DOI: 10.1201/9781003589273-66

news [4]. Several techniques have been involved for identifying the fake news such as machine learning [5], language and knowledge enabled techniques as the social networking tool [6] has the improvement in various aspects of spreading the fake news. Additionally, the fake news detection has several advantages in the digital technology, the online fake news has created the specific attention to the web related news sites and the digital media [7]. The individual person requires specific time for verifying the news sources and confirm the integrity of the particular news.

2 RELATED WORK

Various techniques have been developed to discover the false data with the enhanced performance than the existing methods. Fake news tracking tool [8] has the social context as the implemented technique analysed the patterns of fake news through the document related frequency model. The NLP-enabled computational technique [9] has provided the enhanced accuracy with effective fake news detection. The tree based enabled technique [10] has used the vector transformation process through the deep semantic framework for discovering the fake news. Naïve Bayes Classifier [11] is utilized for classifying the fake news with the hybrid technique to categorize the news while the deep learning framework through the dropout layer for providing the solution for the overfitting problem. Several deep learning techniques have implemented the fake news detection process that includes the linguistic techniques, SVM, Convolutional networks, and attention networks. The self-adaptive harmony search technique has the optimized framework with the learning technique to produce the enhanced results for identifying the fake news in various dimensions.

Self-attention based Bi-LSTM CNN [12] is the deep learning technique utilized to learn automatically and capture complex text patterns from the content. This network specifically focuses on embedding word2vec for about word vectors. These works reach better accuracy for the detection of fake news compared with other methodologies. This algorithms, mainly faces the challenges as a representation of word understanding by the computer in the NLP. Due to these challenges, many Self-attention based Bi-LSTM CNN based algorithms were proposed to find a solution for word representation sentimentally and semantically of text /word. Deep learning techniques based on Self-attention based Bi-LSTM CNN algorithms provide promising results in the NLP in several complex situation for fake news classification.

The Title-based Feature Extraction technique is used for text semantics in the body of the text content while the initial RNN framework has the input as the text, the text sequence input would be gloved through the embedding framework. The text-based feature extraction has the 1-dim Con1D [13] in the text field, it is mainly utilized for classification of the data through scalar manipulation. The Rectified Linear Unit (ReLU) [14] is involved for training the network with the produced combined output through LSTM units, it solves the gradient problem. The backward text is involved to the modified future words and it has the series of filters to minimize the pooling-based output.

3 PROPOSED WORK

The framework consists of a combination of hybrid architecture, this framework concentrates on the text and title of a new article. In this text, contents were transformed into integer vectors. These types of integer vectors are identified using the token ID of a dictionary (word2Vec).

3.1 *Title-based feature extraction*

As the RNN is mainly used in the field of text analysis as well as text semantics, they need to be fixed size of title and body of the text content; the first RNN architecture contains text as an input. The Title input of the sequence text will be gloved using embedding techniques. This glove is used to transform the sequence of titles into the most relevant data part. In the next step, the transformation of sequence text is given into Bi-directional Gated Recurrent Units (Bi-GRUs).

The update gate is calculated in Equation (1).

$$u = \delta(x_t\ \sigma^u + y_{t-1}\ Wt^u) \quad (1)$$

The reset gate is calculated in Equation (2).

$$r = \delta(x_t\ \sigma^r + y_{t-1}\ Wt^r) \quad (2)$$

The hidden state is computed in Equation (3).

$$h = \tanh\left(x_t\ \sigma^h + (s_{t-1} o r)\, Wt^h\right) \quad (3)$$

The function of every gate is computed in Equation (4).

$$s_t = (1 - u) o h + u o s_{t-1} \quad (4)$$

Where o is the component of multiplication. For implementing the feature extraction, the speech part (p) is paired with the word into the embedding layers that provide the semantic space. The embedding layer contains the p embedding layers and word layer, and it is demonstrated in Figure 1.

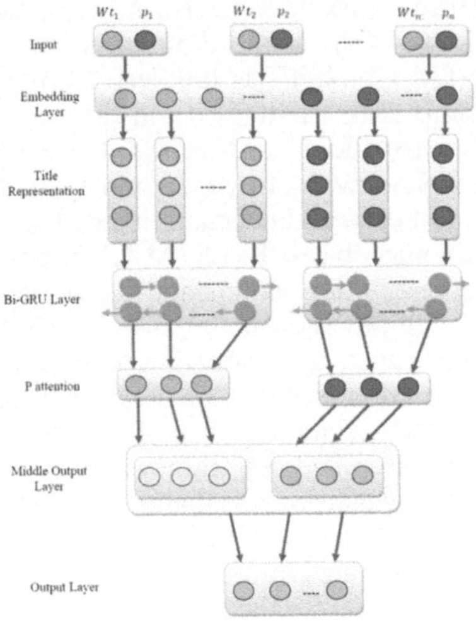

Figure 1. Output model.

The independent layers are used for representing the types of embedding layers and the semantic representation for the layers of the hidden unit for a forward layer is computed in Equation (5).

$$h_i^{Wt} = h_i^{Wt,\ I} o h_i^{Wt,2} \quad (5)$$

The hidden layer for a backward layer is computed in Equation (6).

$$h_i^p = h_i^{p,\ I} o h_i^{p,2} \quad (6)$$

Where $h_i^{Wt,I}$ and $h_i^{Wt,\ 2}$ demonstrate the hidden unit output with the same as the values of $h_i^{p,I}$ and $h_i^{p,\ 2}$. The lexical feature has been involved in the prediction of the probability distribution function in Equation (7).

$$l_a = \delta(Wt_a\ h^p + b_a) \quad (7)$$

Where δ is the sigmoid function, l_a is the prediction probability, the attention functionality through the Softmax parameter into the feature of weight is computed in Equation (8).

$$h_{att}^{Wt} = \beta o h^{Wt} \tag{8}$$

The value of Softmax weight value (β) is computed in Equation (9).

$$\beta_i = \frac{e^{\left(h_i^p\right)}}{\sum_{i=1}^{c} e^{\left(h_i^p\right)}} \tag{9}$$

Where h_{att}^{Wt} is the output for the hidden layer while implementing the attention value, c is the total hidden unit cells. The semantic text could be generated through weight demonstration of the word, the weight word feature for predicting the label is computed in Equation (10).

$$l_c = \delta\left(Wt_c \quad h_c^{Wt} + b_a\right) \tag{10}$$

Where h_c^{Wt} is the hybrid feature through probability for the word representation, l_c is the final predicted probability of the text representation. This architecture contains two Bi-GRU layers focused on the product of semantic text in forwarding and backward layer representation of title. After that max-pooling is applied to the title representation to reduce the dimension of the matrix representation of the title. Finally, the output of the max polling is given into a fully connected layer. This final layer is responsible for producing output.

3.2 *Text-based feature extraction*

The second architecture contains text as an input which contains the detail about the news. This will be fed into one-dimensional convolution (conv1D) used in the field of text. It is mainly used to classify the data using scalar multiplication or additions. In many real-time applications like motor fault identification, early detection of fault through monitoring etc. Second, it is very cost-effective. After that, the sequence of text is given into Rectified Linear Unit (ReLu). The advantage of using ReLu is not allowing to stimulate all the neurons at the identical period. The main aim of using ReLu here is to train the network longer but optimization is slow. However, it is very useful to train large complex patterns. The combined output is given into LSTM units. This unit is combined with several hidden cells. It allows mitigating the gradient problem, so this LSTM can remember more distant information from the past text. This Bi-LSTM contains a sequence of forwarding and backward text. Figure 2 demonstrates the Convolutional operation which consists of the content representation through the features for classifying the news is fake or not with the embedding functionality. The produced word embeddings have the content for analyzing the similarity of the image as *PxQ* matrix that P is the total words and Q denotes the dimension for embedding process, whenever the Convolutional layer is involved to identify the features through the shape of the image.

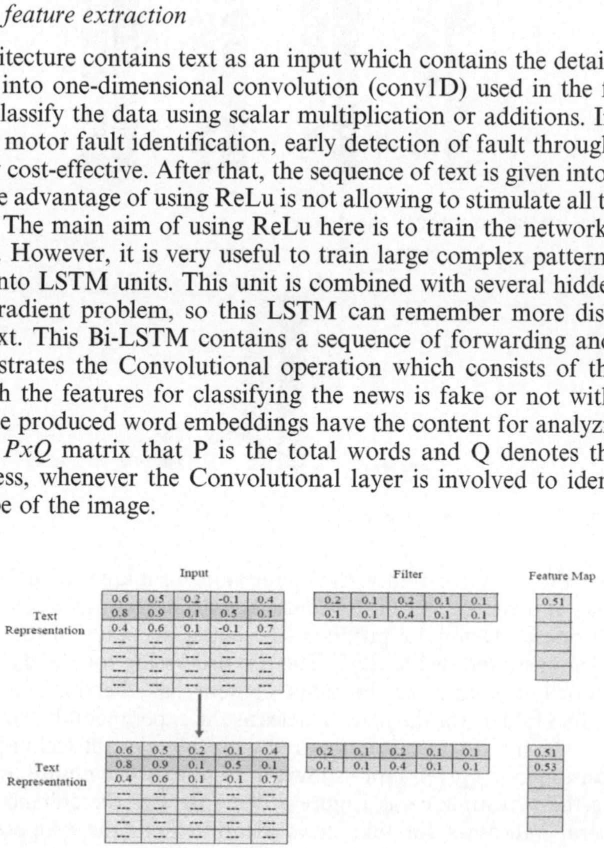

Figure 2. Convolutional operation.

4 EXPERIMENTAL RESULTS AND ANALYSIS

The datasets of GossipCop (DB) [15], PolitiFact (DB1) [16] are utilized for performing the performance evaluation as to categorize the fake news where the experts deliver the truth values as the fake news that influence the news quality. PolitiFact (DB1) delivers the political news with visual data is showed around 800 articles, the dataset is having text into the input and evaluated entirely for text data.

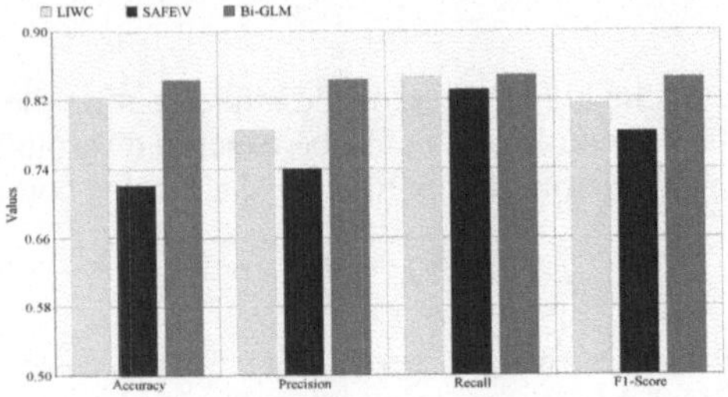

Figure 3. Performance evaluation of DB1 dataset with text input.

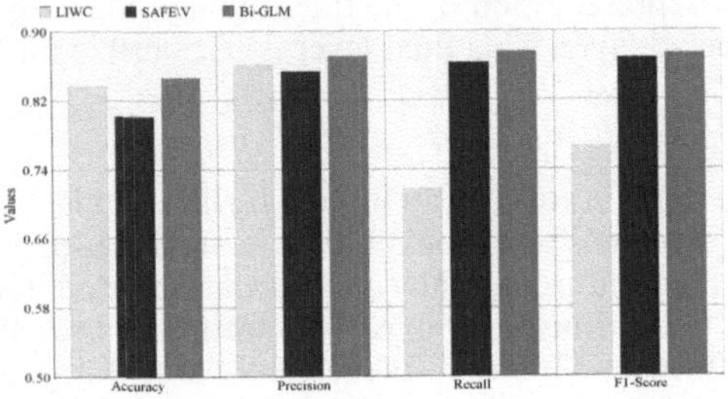

Figure 4. Text input evaluation through the dataset.

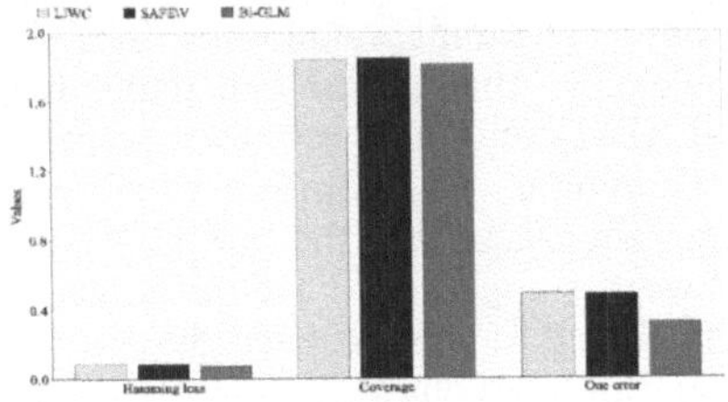

Figure 5. Multi-label fake news detection metrics.

The proposed method is compared with the relevant techniques of LIWC [17], SAFE\V with the parameters of precision, recall, F1-score, and accuracy. The experimental result has proved that the proposed technique produces the better accuracy in the PolitiFact (DB1) dataset which is demonstrated in Figure 3. The pre-processing for the dataset is performed through the proposed technique as the news content has the fake news categorization through the individual folders for the news articles as the experimental results of the specified dataset with text in the input which is compared with the relevant technique in view of the recall and the precision in the given input shown in Figure 4. The multi-label parameters are used for assessing the performance as Figure 5 demonstrates the enhancement of the performance by several indicators for fake-news identification, the proposed technique has performed well in the performance metrics of Hamming loss, coverage and one error than the related techniques.

5 CONCLUSION

The identification of the fake news in social network is a very complex work as the insufficient data and the reliability within several works to identify the accurate news. The proposed technique has utilized the improved framework for identifying the fake news in the text format, the performance evaluation has proved that the proposed technique is performed better in text related fake news detection. To improve analytical areas, we intend to perform spam identification in the future based on aspect level and rating deviation. This method will be used for spam review detection in the legal, marketing, and healthcare domains. It is necessary to put transfer learning-related techniques into practice in order to provide multi-level predictions that will address problems in real time.

REFERENCES

[1] Jain A. and Kasbe A. (2018). Fake news detection, 2018 IEEE International Students' Conference on Electrical, Electronics and Computer Science (SCEECS), Bhopal, India, pp. 1–5.

[2] Kong J. T. H., Wong W. K., Juwono F. H. and Apriono C. (2023). Generating fake news detection model using a two-stage evolutionary approach, in IEEE Access, vol. 11, pp. 85067–85085.

[3] Iftikhar M. and Ali A. (2023). Fake news detection using machine learning, 2023 3rd International Conference on Artificial Intelligence (ICAI), Islamabad, Pakistan, pp. 103–108.

[4] Mahabub, A. (2020). A robust technique of fake news detection using Ensemble Voting Classifier and comparison with other classifiers. SN Appl. Sci. 2, 525.

[5] Manzoor S. I., Singla J. and Nikita. (2019). Fake news detection using machine learning approaches: A systematic review, 3rd International Conference on Trends in Electronics and Informatics (ICOEI), pp. 230–234.

[6] Elhadad M. K., Fun Li K. and Gebali F. (2019). Fake news detection on social media: A systematic survey, 2019 IEEE Pacific Rim Conference on Communications, Computers and Signal Processing (PACRIM), Victoria, BC, Canada, pp. 1–8

[7] Han W. and Mehta V. (2019). Fake news detection in social networks using machine learning and deep learning: Performance evaluation, 2019 IEEE International Conference on Industrial Internet (ICII), Orlando, FL, USA, pp. 375–380.

[8] Suprem A., Vaidya S. and Pu C. (2022). Exploring generalizability of fine-tuned models for fake news detection, 2022 IEEE 8th International Conference on Collaboration and Internet Computing (CIC), Atlanta, GA, USA, pp. 82–88

[9] Meesad, P. (2021). Thai fake news detection based on information retrieval, natural language processing and machine learning. Sn Comput. Sci. 2, 425.

[10] Davoudi M., Moosavi M. R. and Sadreddini M. H. (2021). A novel method for fake news detection based on propagation tree, 2021 11th International Conference on Computer Engineering and Knowledge (ICCKE), Mashhad, Iran, Islamic Republic of, pp. 105–109.

[11] Granik M. and Mesyura V. (2017). Fake news detection using naive Bayes classifier, 2017 IEEE First Ukraine Conference on Electrical and Computer Engineering (UKRCON), Kyiv, UKraine, pp. 900–903.

[12] Bhuvaneshwari P, Rao AN and Robinson YH. (2021 May). Spam review detection using self attention based CNN and bi-directional LSTM. Multimedia Tools and Applications.;80(12):18107–24.J. Zhang, B. Dong and P. S. Yu, FakeDetector: Effective Fake News Detection with Deep Diffusive Neural Network, 2020 IEEE 36th International Conference on Data Engineering (ICDE), Dallas, TX, USA, 2020, pp. 1826–1829.

[13] Hashmi E., Yayilgan S. Y., Yamin M. M., Ali S. and Abomhara M. (2024). Advancing fake news detection: Hybrid deep learning with fasttext and explainable AI, in IEEE Access, vol. 12, pp. 44462–44480.

[14] GossipCop. https://www.suggest.com/. [accessed on November 9th, 2023].

[15] PolitiFact. https://www.politifact.com/. [accessed on November 9th, 2023].

[16] Anu Shrestha, Francesca Spezzano, and Abishai Joy. (2020). Detecting fake news spreaders in social networks via linguistic and personality features, Conference and Labs of the Evaluation Forum.

[17] Xinyi Zhou, Jindi Wu and Reza Zafarani. (2020). SAFE: Similarity-aware multi-modal fake news detection, 24th Pacific-Asia Conference on Knowledge Discovery and Data Mining.

Data Science & Exploration in Artificial Intelligence – Gururaj et al. (Eds)
© 2025 the Author(s), ISBN: 978-1-032-96390-7

Evolution of aircraft visual inspections: Conventional to AI-enabled methods

Pankaj Bajpai & Amrita Nighojkar
Department of Technology Management, Management Analytics and AI Lab, Defence Institute of Advanced Technology Pune, India

ABSTRACT: This comprehensive review paper meticulously examines transitions from traditional methods to AI-enabled aircraft inspection advancements over the past decade. Focusing on recent trends in detecting imperfections and faults, the study evaluates improvements in preventive and corrective maintenance strategies, scheduling methodologies, and cutting-edge technologies. Highlighted are breakthroughs in real-time defect detection and health monitoring, such as autonomous maintenance systems and reinforcement learning-driven strategies to optimize operational efficiency, enhance safety, and reduce costs. The paper identifies a significant gap in aircraft visual inspection and proposes solutions using computer vision and advanced AI techniques. Emphasis is placed on the necessity for specialized applications in military aviation, incorporating security protocols and rapid response needs. The findings envisage AI as a central pillar in ensuring aircraft safety, efficiency, and compliance with aviation standards, addressing current gaps in daily inspection automation.

1 INTRODUCTION

Aircraft inspection checks are vital for ascertaining aircraft safety and overall functionality. These checks play a crucial role in identifying potential defects, wear and tear, or any malfunctions that may compromise the integrity of an aircraft system. These routine checks are performed per laid down standards and ensure compliance with stringent safety measures set as per aviation standards. The maintenance check intervals depend upon aircraft usage and aging factors such as flight hours, cycles, and calendar days(Andrade *et al.* 2021). By carefully examining aircraft components such as engines, avionics systems, and aircraft structural elements, technicians detect unnoticed issues that might not be apparent during routine aircraft operations. Moreover, these checks contribute to the aircraft's longevity by preventing premature wear of the parts and enhancing their life. These checks are performed at regular predetermined intervals, providing timely and comprehensive assessments, thus reducing maintenance costs. Aircraft routine inspections prioritize passenger safety, safeguard the airline operators' interests, and maintain the aviation industry's reputation for reliability.

Aviation maintenance typically follows preventive approaches, where each aircraft undergoes scheduled maintenance checks at specified intervals, leading to temporary grounding. The preventative maintenance approach aims to establish the aircraft's ongoing airworthiness by mitigating potential failures resulting from component degradation. The activities may include inspection, repair, and replacement of components. Maintenance intervals for different aircraft vary as safety is the utmost priority in aviation. Over 80% of aircraft accidents are estimated to be caused by human error (Helevera and Гелевера 2023).

 DOI: 10.1201/9781003589273-67

As per Civil Aviation Requirements 145 issued by the Director General of Civil Aviation, these activities may be classified into two primary types: Line maintenance activity and Base maintenance activities. Airlines typically divide aircraft maintenance tasks into A, B, C, and D checks. A and B checks are light maintenance, whereas C and D checks are considered heavy maintenance activities(Weide *et al.* 2022).

Artificial intelligence (AI) has become a crucial part of data mining and big data. Although the basic concepts and architecture of machine learning were conceptualized long ago, the requisite data volumes and computational power have materialized to actualize its potential only recently (Karaoğlu *et al.* 2023). Neural networks and machine learning are computational techniques that can analyze extensive datasets, generate desired results, and identify trends and patterns(Helgo 2023). Integrating these latest technologies with the traditional methods of aircraft inspections can provide a transformative advantage by swiftly analyzing data, enhancing inspection efficiency, and accurately predicting maintenance needs. This proactive approach can improve aircraft operations' safety, reliability, and cost-effectiveness. This review focuses exclusively on the scholarly progress of technological integration with aircraft line maintenance activities.

2 AIRCRAFT VISUAL INSPECTIONS: EVOLUTION OF TECHNOLOGIES

The FAA's AC-43-2041 defines four levels of aircraft visual inspections(Drury and Watson 2002). Level-1 inspection is a 'Walk-around Inspection' and includes standard checks performed to examine the general condition of the plane. Level-2 inspection is known as 'General Visual Inspection' and is performed on the aircraft's exterior structure, including designated hatches and openings. The inspection is aimed at detecting any structural damage. Level 3 is 'Detailed Visual Inspection,' which includes a rigorous visual examination of a specified area to spot any damage or failure. Level-4 inspection is 'Specially Detailed Visual Inspection,' which provides an exhaustive assessment of selected items or assemblies to spot any irregularity. Level-4 inspections often require special equipment and extensive disassembly before execution.

2.1 *Challenges in traditional visual inspection procedures*

During daily aircraft inspections at the line maintenance facility, most visual inspections are carried out by technicians dealing with mechanical systems. These include inspection of the exterior surface of the airframe, aero engines, landing gears, and tires for their condition and security. The technicians also check for leftover quantities and replenish fuel, air, hydraulic fluids, engine oil, etc. During this check, one of the essential activities is to ensure that no foreign object is left in

the aircraft, keeping particular attention to areas of the engine inlet, controls, and cockpit. Specialist technicians perform visual inspections of aircraft to identify problems. These inspections are time-consuming, and their results vary according to each individual. The time that an airplane stays grounded without flight clearance means financial losses (Yasuda *et al.* 2021). The traditional methods of aircraft visual inspection are marked by their time-consuming and subjective nature, often varying between individuals as per their technical competence. This inconsistency poses significant challenges, as the grounding of an aircraft results in substantial financial losses.

In light of the capabilities exhibited by cutting-edge technologies and the pressing need for a more effective solution, combining computer vision, neural network techniques, and artificial intelligence appears to be a practical and revolutionary strategy in this field.

2.2 *Intervention of AI-based technologies in inspection*

The latest technologies, such as artificial intelligence and machine learning, have prompted a noticeable increase in efforts to automate and digitize many aspects of aviation maintenance tasks.

The objectives of this change are lower operating costs, improved safety, and operational excellence. With a focus on the use of standard AI models across several inspection domains, Table 1 provides an extensive overview of research projects examining the use of AI models in aircraft inspection and imperfection identification. CNNs, or convolutional neural networks, are widely used in inspection and defect detection applications. For example, CNN models have been used to segment corrosion in aviation skin corrosion(Zuchniak *et al.* 2021), and discover faults in aircraft skin(Zhao *et al.* 2024), as well as identify defects on aircraft exterior surfaces(Zhao *et al.* 2024). Furthermore, the application of Deep Convolutional Neural Networks (DCNNs) to the detection of flaws on aero-engine blade surfaces has demonstrated efficacy in detecting minute faults(Li *et al.* 2021). Utilizing the YOLO-V5 and MobileNet V3 Large models, Convolutional Neural Networks (CNNs) have been used to detect rivet joint problems with a 100% accuracy rate (Amosov, Amosova and Iochkov 2022). Similar to this, multi-contrast CNNs have been used to identify defects in aircraft tires, showing excellent accuracy in identifying many classes of defects (Mohan *et al.* 2020). In surface defect detection and classification applications, YOLO (You Only Look Once) models have become especially popular (Mohan *et al.* 2020). When surface scratches were detected, YOLOv4 produced a high mAP value of 96.21%(Yao *et al.* 2021). For Foreign Object Debris detection, various YOLO versions were employed for aerodrome defect detection, emphasizing radar-based techniques (Arikilla and Raviteja 2023). Support Vector Machines (SVMs) were applied in conjunction with Pulsed Thermography techniques for aircraft structural damage detection, demonstrating effective damage diagnosis through infrared thermography and composite material analysis (Alhammad *et al.* 2022).

Table 1. AI-integration in inspection of various sections of aircraft.

Inspection	AI-Integration	AI-Performance	References
Aero-engine blades	Semantic prior guided defect perception network, named SPDP-Net	95.9% precision, 94.0% recall, and 94.9% F1 score.	(Ma *et al.* 2024)
Rivet Joints	YOLO-V5 is used for detecting, and MobileNet V3 is used to classify rivet joint states.	The accuracy is ~100% in both binary and multiclass classification.	(Amosov *et al.* 2022)
Corrosion on the fuselage surface and rivets	Convolutional neural network (CNN)	-	(Zuchniak *et al.* 2021)
Aircraft Engine Shells	Machine vision	The efficiency of this network is 84.88%	(Mirbod and Shoar 2023)
Surface Scratch	YOLOv4 algorithm	Accuracy in scratch detection task is 96.21%	(Yao, Yang and Wang 2021)
Aero-engine blade surface	Deep convolution neural network (DCNN)	-	(Li *et al.* 2021)
Tire defect detection	Multi-contrast convolutional neural networks (MC-CNN)	-	(Mohan *et al.* 2020)
Aircraft structural damage	Cubic Spine Support Vector Machine	-	(Alhammad *et al.* 2022)
Turbine and Burner	Machine Learning (ML) and	-	(Schoen,

(*continued*)

Table 1. Continued

Inspection	AI-Integration	AI-Performance	References
Section of Aircraft Jet engine	Deep Learning (DL) methods		Oettinger and Mimic 2023)
Aircraft engines	Convolutional Neural Networks (CNNs) Generative Adversarial Networks (GAN)	-	(Hedrich *et al.* 2022)
Foreign Object Debris	YOLO R, YOLO v8, YOLO v7, YOLO v6, YOLO v5, and YOLO v4	YOLO-v8 produces effective results of 98.4%	(Arikilla and Raviteja 2023)
Dents, Paint defects, cracks and holes, and lightning strike damage	MASK R-CNN combined with autonomous drones	The hybrid approach improves performance with an F1 score of (67.50%) and an F2 score of (66.37%).	(Doğru *et al.* 2020)
Surface coatings	YOLOv5-OBB	-	(Zhang *et al.* 2024)
Aircraft Flared Tube (AFT)	YOLOv4	Mean Average Precision ~ 97.87%	(Zhang *et al.* 2022)
Aircraft wiring	Topological network-graph transformer network (TN-GTN)	-	(Wang *et al.* 2022)
Rivet cracks	Fuzzy support vector machine	Detect the crack defects in the multilayer structure of aging aircraft	(Guo and Liu 2023)
Aircraft engine	Deep learning, customized generative adversarial net (GAN) model	Precisions and recalls of over 90%.	(Upadhyay *et al.* 2023)
Aero-engine blade	Artificial intelligence software and 3D scanning	-	(Aust and Pons 2022)
Aircraft surface	YOLOv4-P6 and drone system	-	(Tang *et al.* 2022)
Aircraft surface	2-stage multi-modal inspection pipeline with visual and tactile sensing	The approach identified 85% of defects using Stage I and 100% after Stage II.	(Agarwal *et al.* 2023)
Aircraft bell-shaped tube	YOLOv4	Gaps detection is 92.21%	(Yao *et al.* 2022)
Surface and structural components	Fast region-based convolutional neural network (Fast R-CNN)	The achieved results obtained more than 90% accuracy	(Chen and Juang 2020)
Cables insulation	deep forest method and ultrasonic guided waves (UGWs)	The recognition accuracy reached 100%,	(Wang *et al.* 2023)
Composite Material	Cube Support Vector Machine (SVM)	The classification results ranged from 78.7% to 93.5%,	(Alhammad *et al.* 2022)
Composite Laminates	Convolutional neural network	-	(Tang *et al.* 2022)
Landing gear components	Convolutional autoencoder	-	(Kähler *et al.* 2022)
Cracks	Feature Pyramid Network (FPN)	The FPN model scored 71.35%	(Vrochidou *et al.* 2022)
Skin Defect inspection	Mask Scoring R-CNN	Precision of ~21%.	(Meng *et al.* 2022)
Aircraft Components	Generative Adversarial Network (GAN) framework, termed Dual-IRT-GAN	-	(Cheng and Kersemans 2022)

Generative Adversarial Networks (GANs) and CNNs were combined for aircraft engine inspection, focusing on damage inspection with semantic segmentation and transfer learning (Hedrich *et al.* 2022). Machine Learning (ML) and Deep Learning (DL) methods were employed for defect classification in the turbine and burner sections, highlighting applications in jet engines and hot gas path defect detection (Schoen *et al.* 2023). Other notable implementations include the SPDP-Net model for defect detection in aero-engine blades (Ma *et al.* 2024)and AI algorithms for defect detection in honeycomb engine parts, emphasizing Industry 4.0 and computer vision (Cannizzaro *et al.* 2022). This review highlights the widespread use and effectiveness of AI models, particularly CNNs, YOLO, SVMs, GANs, and ML/DL methods, across various aircraft inspection and defect detection applications, demonstrating significant advancements in model performance and precision.

While aircraft maintenance is undergoing substantial advancements, especially in maintenance scheduling and defect detection, a noticeable gap persists in automating daily inspections and checks (pre-flight, turn-round, and last-flight servicing activities) as a comprehensive single-set application. Despite strides in automation and digitization, these crucial daily inspection checks remain predominantly reliant on manpower-intensive manual processes, leading to potential inefficiencies, inconsistencies, and increased inspection times. Integrating computer vision and neural networks presents a significant opportunity to address this gap comprehensively. This method has a lot of potential, particularly for military aircraft where thorough inspections and quick response times are essential. Utilizing computer vision and neural networks to automate daily inspections can greatly improve aviation maintenance procedures' overall efficacy, safety, and operational efficiency. While neural networks are better than traditional manual methods at analyzing complicated patterns to identify possible concerns, computer vision is an efficient means of detecting visual irregularities. By utilizing these technologies, daily inspections can be automated and streamlined, leading to quicker, more accurate assessments. This automated technique speeds up the inspection process and facilitates decision-making by systematically collecting observations and creating a digital record for each inspection event. The work of EWALD shows how domain-specific knowledge can be used to get around data constraints and improve the performance of deep learning algorithms in situations where large datasets could be hard to come by(Ewald *et al.* 2021). Their method creates opportunities for applying deep learning in fields like aircraft visual inspection with specialized knowledge but limited data availability.

3 AI INTEGRATION FOR VISUAL INSPECTION: CHALLENGES & FUTURE PROSPECTS

3.1 *Challenges*

Implementing Artificial intelligence integrated with computer vision for aircraft visual inspection will provide swift and accurate results. However, its implementation may face several challenges as follows: -

- Variability in Aircraft Designs. Developing AI models for a diverse range of civil and military aircraft that can be generalized effectively across various configurations, sizes, and manufacturers is challenging. Also, using this technique on the airframes of military aviation with stealth technologies may be difficult and needs further research.
- Safety and Certification Standards. Due to stringent safety and certification standards in the Military as well as civil aviation, methodical validation of proposed AI-based inspection systems will be required to be completed.
- Privacy and Safety Concerns. The vulnerability of AI systems to adversarial attacks is a cause of concern, especially in military applications. This may potentially compromise the integrity of the inspection process. In the case of military aircraft, the classification of

sensitive information during inspections may introduce additional complexities. This will require careful consideration in the design of computer vision systems.

- Limited Access to Data. Collecting adequate datasets for civil/ military aircraft inspections may be challenging and hinder the development of robust models capable of identifying diverse damage patterns.
- Integration with Maintenance Processes. Seamless integration of AI systems with existing aviation maintenance activities is critical and will require a high degree of collaboration and interoperability with established maintenance workflows.
- Perception and Trust. Establishing positive perception and trust towards AI-based inspection systems is essential before implementation. This poses a challenge as the awareness and trust regarding AI-based systems are limited.
- Cost Considerations. As employing AI-based inspection systems will involve significant expenditures, the economic viability of these solutions needs to be carefully evaluated. This becomes more critical in the case of civil aviation, which is a commercial organization.
- Training and Skill Requirements. Specialized training of the technicians becomes mandatory to understand and interact with AI-based inspection systems. This is crucial for successfully implementing AI-based techniques in civil and military aviation.

3.2 *Future prospects*

The following recommendations can be considered for further improvement and development of the AI-mediated aircraft inspection process:

- Development of Unified Inspection Applications: Based on the analysis, further research and development towards creating a unified set of applications for daily inspection checks, including pre-flight, turnaround, and last-flight servicing activities, is recommended. This can be achieved by integrating advanced computer vision and AI techniques.
- Collaboration with Experts: Collaborating with AI and computer vision experts is crucial for developing generalized and specific solutions. These solutions should efficiently recognize visual cues, analyze complex patterns, and store inspection records performed by technicians for future reference.
- Training for Maintenance Personnel: Maintenance personnel will need specialized training to ensure the smooth integration of advanced technologies. This training will focus on utilizing the applications and complying with established protocols.
- Customization for Military Aircraft: It is essential to create automated inspection solutions to satisfy specific requirements, given the unique needs of military aircraft. This calls for quick response times, special inspection requirements, and careful consideration of security procedures. For automated inspection systems to be implemented, data security and regulatory compliance must be guaranteed. Upholding industry standards while preserving the confidentiality and integrity of inspection records requires the development of strong data transmission, storage, and access methods.
- Decrease in Inspection periods: It is projected that the suggested application's installation will improve accuracy, greatly cut down on inspection periods, and shorten aircraft downtime. Military aircraft, where maintaining operational efficiency requires speedy turnover, will benefit most from this innovation.

4 CONCLUSION

The development of AI and machine learning is bringing in a new era of aircraft maintenance, where technology is essential to flight safety, operational effectiveness, and economy. Applying these technologies to real-time fault detection, maintenance schedules, and improving preventive maintenance for civil aircraft has advanced significantly. These applications, which make use of computer vision and neural networks, have the potential to completely transform regular inspections by producing digital records for each inspection event and quicker, more thorough assessments. By highlighting the crucial role that technology plays in maintaining aircraft safety and operational readiness, the effective implementation of these technologies promises to revolutionize aircraft maintenance procedures in both civil and military aviation.

REFERENCES

Agarwal, A., Ajith, A., Wen, C., Stryzheus, V., Miller, B., Chen, M., Johnson, M.K., Rincon, J.L.S., Rosca, J. and Yuan, W. (2023). Robotic Defect Inspection with Visual and Tactile Perception for Large-scale Components, *2023 IEEE/RSJ International Conference on Intelligent Robots and Systems (IROS)*, 10110–10116, IEEE.

Alhammad, M., Avdelidis, N.P., Ibarra-Castanedo, C., Torbali, M.E., Genest, M., Zhang, H., Zolotas, A. and Maldgue, X.P. (2022). Automated impact damage detection technique for composites based on thermographic image processing and machine learning classification, *Sensors*, 22(23), 9031.

Alhammad, M., Avdelidis, N.P., Ibarra-Castanedo, C., Zolotas, A. and Maldague, X.P. (2022). Machine learning on thermographic images for the detection and classification of damage on composites, *Thermosense: Thermal Infrared Applications XLIV*, vol. 12109, 96–106, SPIE.

Amosov, O.S., Amosova, S.G. and Iochkov, I.O. (2022). Deep neural network recognition of rivet joint defects in aircraft products, *Sensors*, 22(9), 3417.

Andrade, P., Silva, C., Ribeiro, B. and Santos, B.F. (2021), Aircraft maintenance check scheduling using reinforcement learning, *Aerospace*, 8(4), 113.

Arikilla, M. and Raviteja, B. (2023). Foreign object debris detection in aerodromes using deep learning approaches, in J. Choudrie, P.N. Mahalle, T. Perumal & A. Joshi (eds.), *IOT with Smart Systems*, Lecture Notes in Networks and Systems., vol. 720, pp. 587–598, Springer Nature Singapore, Singapore.

Aust, J. and Pons, D. (2022). Comparative analysis of human operators and advanced technologies in the visual inspection of aero engine blades, *Applied Sciences*, 12(4), 2250.

Cannizzaro, D., Simone, F., Illgner-Fehns, K., Mata, S., Mondino, I., Ghiazza, A., Poncino, M. and Di Cataldo, S. (2022). Quality inspection of critical aircraft engine components: towards full automation, *2022 IEEE 27th International Conference on Emerging Technologies and Factory Automation (ETFA)*, 1–4, IEEE.

Chen, Z.-H. and Juang, J.-C. (2020). AE-RTISNet: aeronautics engine radiographic testing inspection system net with an improved fast region-based convolutional neural network framework, *Applied Sciences*, 10(23), 8718.

Cheng, L. and Kersemans, M. (2022). Dual-IRT-GAN: A defect-aware deep adversarial network to perform super-resolution tasks in infrared thermographic inspection, *Composites Part B: Engineering*, 247, 110309.

Doğru, A., Bouarfa, S., Arizar, R. and Aydoğan, R. (2020). Using convolutional neural networks to automate aircraft maintenance visual inspection, *Aerospace*, 7(12), 171.

Drury, C.G. and Watson, J. (2002). Good practices in visual inspection, *Human factors in aviation maintenance-phase nine, progress report, FAA/Human Factors in Aviation Maintenance.*

Ewald, V., Groves, R. and Benedictus, R. (2021). Integrative approach for transducer positioning optimization for ultrasonic structural health monitoring for the detection of deterministic and probabilistic damage location, *Structural Health Monitoring*, 20(3), 1117–1144.

Guo, Q. and Liu, L. (2023). Research on Non-destructive Automatic Identification Algorithm of Aircraft Structural Defects Based on Machine Learning, *2023 Asia-Pacific Conference on Image Processing, Electronics and Computers (IPEC)*, 125–129, IEEE.

Hedrich, K., Hinz, L. and Reithmeier, E. (2022). Damage segmentation using small convolutional neuronal networks and adversarial training methods on low-quality RGB video data, *AI and Optical Data Sciences III*, vol. 12019, 1201902, SPIE.

Helevera, D.V. and Гелевера, Д.В. (2023). Visual examinations and manual checks to determine the condition of an aircraft and its component.

Helgo, M. (2023). *Deep Learning and Machine Learning Algorithms for Enhanced Aircraft Maintenance and Flight Data Analysis.*

Kähler, F., Schmedemann, O. and Schüppstuhl, T. (2022). Anomaly detection for industrial surface inspection: Application in maintenance of aircraft components, *Procedia CIRP*, 107, 246–251.

Karaoğlu, U., Mbah, O. and Zeeshan, Q. (2023). Applications of machine learning in aircraft maintenance, *Journal of Engineering Management and Systems Engineering*, 2(1), 76–95.

Li, D., Li, Y., Xie, Q., Wu, Y., Yu, Z. and Wang, J. (2021). Tiny defect detection in high-resolution aero-engine blade images via a coarse-to-fine framework, *IEEE Transactions on Instrumentation and Measurement*, 70, 1–12.

Ma, Y., Liu, M., Zhang, Y., Wang, X. and Wang, Y. (2024). SPDP-Net: A semantic prior guided defect perception network for automated aero-engine blades surface visual inspection, *IEEE Transactions on Automation Science and Engineering*.

Meng, D., Boer, W.U., Juan, X.U., Kasule, A.N. and Hongfu, Z.U.O. (2022). Visual inspection of aircraft skin: Automated pixel-level defect detection by instance segmentation, *Chinese Journal of Aeronautics*, 35 (10), 254–264.

Mirbod, M. and Shoar, M. (2023). Intelligent concrete surface cracks detection using computer vision, pattern recognition, and artificial neural networks, *Procedia Computer Science*, 217, 52–61.

Mohan, P., Pahinkar, A., Karajgi, A., Kumar, L.D., Kasera, R., Gupta, A.K. and Narayanan, S.J. (2020). Multi-Contrast Convolution Neural Network and Fast Feature Embedding for Multi-Class Tyre Defect Detection, *2020 4th International Conference on Electronics, Communication and Aerospace Technology (ICECA)*, 1397–1405, IEEE.

Mosca, N., Patruno, C., Renò, V., Nitti, M. and Stella, E. (2021). Qualitative comparison of methodologies for detecting surface defects in aircraft interiors, *2021 IEEE 8th International Workshop on Metrology for AeroSpace (MetroAeroSpace)*, 215–220, IEEE.

Schoen, M.P., Oettinger, M. and Mimic, D. (2023). Deep and Machine Learning-based Methods for Defect Classification in Jet Engines, *2023 Intermountain Engineering, Technology and Computing (IETC)*, 43–48, IEEE.

Tang, Y.M., Ip, A.W. and Li, W. (2022). Artificial intelligence approach for aerospace defect detection using single-shot multibox detector network in phased array ultrasonic', *IoT and Spacecraft Informatics*, pp. 1–27, Elsevier.

Upadhyay, A., Li, J., King, S. and Addepalli, S. (2023). A deep-learning-based approach for aircraft engine defect detection, *Machines*, 11(2), 192.

Vrochidou, E., Sidiropoulos, G.K., Ouzounis, A.G., Lampoglou, A., Tsimperidis, I., Papakostas, G.A., Sarafis, I.T., Kalpakis, V. and Stamkos, A. (2022). Towards robotic marble resin application: crack detection on marble using deep learning, *Electronics*, 11(20), 3289.

Wang, Q., Huang, M., Wang, X., Zhang, Y., Zhang, Q., An, Y., Wang, C. and Qu, Z. (2023). Defect detection and identification for aircraft cable insulation layer based on deep forest, *Ultrasonics*, 135, 107128.

Wang, T., Fang, Q., Liu, G., Chi, M., Luo, Y. and Shen, J. (2022). TN-GTN: fault diagnosis of aircraft wiring network over edge computing, *EURASIP Journal on Wireless Communications and Networking*, 2022(1), 62.

Weide, T. van der, Deng, Q. and Santos, B.F. (2022). Robust long-term aircraft heavy maintenance check scheduling optimization under uncertainty, *Computers & Operations Research*, 141, 105667.

Yao, Y., Li, F. and Wang, J. (2022). Gap detection of aircraft bell-shaped tube based on convolutional neural networks, in Q. Xie, L. Zhao, K. Li, A. Yadav & L. Wang (eds.), *Advances in Natural Computation, Fuzzy Systems and Knowledge Discovery*, Lecture Notes on Data Engineering and Communications Technologies., vol. 89, pp. 1340–1349, Springer International Publishing, Cham.

Yao, Y., Yang, Y. and Wang, J. (2021). Scratch Detection of Aircraft Bell Tube Based on Improved YOLOv4 Framework, *2021 4th International Conference on Signal Processing and Machine Learning*, 1–6, ACM, Beijing China.

Zhang, J., Wang, K., Dai, G. and Zhang, P. (2022). Research on Defect Rotation Detection of Aircraft Flared Tube Based on Improved YOLOv4, *2022 5th International Conference on Advanced Electronic Materials, Computers and Software Engineering (AEMCSE)*, 64–69, IEEE.

Zhang, Y., Wang, W., Guo, Z. and Ji, Y. (2024). Research on oriented surface defect detection in the aircraft skin-coating process based on an attention detector, *IET Image Processing*, 18(5), 1213–1228.

Zhao, J., Huang, K., Wang, G., Wang, L., Wang, B., Zhou, Y. and Li, H. (2024). Biological visual attention convolutional neural network for aircraft skin defect detection, *Measurement: Sensors*, 31, 100974.

Zuchniak, K., Dzwinel, W., Majerz, E., Pasternak, A. and Dragan, K. (2021). Corrosion detection on aircraft fuselage with multi-teacher knowledge distillation, in M. Paszynski, D. Kranzlmüller, V.V. Krzhizhanovskaya, J.J. Dongarra & P.M.A. Sloot (eds.), *Computational Science – ICCS 2021*, Lecture Notes in Computer Science., vol. 12747, pp. 318–332, Springer International Publishing, Cham.

Data Science & Exploration in Artificial Intelligence – Gururaj et al. (Eds)
© 2025 the Author(s), ISBN: 978-1-032-96390-7

SmartFarmo: A crop insurance guidance system powered by predictive crop yield modelling for precision agriculture

Archit Chadalawada, R.S. Vishal, Shaleen Bhatnagar & Napa Lakshmi
Department of Computer Science Engineering, Manipal Institute of Technology Bengaluru, Manipal Academy of Higher Education, Manipal, India

ABSTRACT: The agricultural sector is the most crucial backbone of the Indian economy. According to the FAO, "Agriculture accounted for 23% of its GDP, and employed 59% of the country's total workforce in 2016". Agriculture, with its allied sectors, is the largest source of livelihoods in India. 70% of rural households still depend primarily on agriculture for their livelihood. With the recent widescale adoption of precision agriculture, there is a demand and need for crop insurance guidance for farmers. To address this, the SmartFarmo crop insurance guidance system was developed. A machine learning model was developed which predicts crop yield based on parameters such as average temperature, rainfall, country and crop grown. Various models were trained on a dataset subject to preprocessing by min-max normalization, outlier detection and removal, and data augmentation. The insurance amount was calculated based on the predicted yield and standard market rates of various crops as of 22nd February 2024, and recommended to the farmer. The model having best-in-class performance was the Random Forest Regressor, with a yield prediction accuracy of 99.68%.

Keywords: Crop insurance guidance, Crop yield prediction, Data augmentation, Linear regression, Machine learning, Precision agriculture, Random Forest regressor.

1 INTRODUCTION

Crop insurance guidance is a crucial area of research for farmers in the era of precision agriculture. Hence, lightweight and easy-to-use applications are of critical importance to improve widespread adoption of technology in the agricultural sector of India. In the ongoing AI revolution, machine learning is beginning to play a larger role in everyday life, with precision agriculture being a domain of great interest[8]. This paper proposes a crop insurance guidance system for farmers, powered by machine learning-based predictions for crop yields of their land. This is coupled with an intuitive and simple-to-use front end UI, making it more accessible and easier to adopt for farmers in rural India.

For this, a crop yield prediction dataset was utilized. It was subjected to preprocessing by normalization of numerical features, outlier detection and removal, as well as augmentation of the dataset to obtain a more rich and refined training dataset. A multitude of regression machine learning models were trained on this dataset, and their performances were evaluated and compared using performance metrics such as accuracy, Mean Absolute Error (MAE), Mean Squared Error (MSE), Root Mean Squared Error (RMSE), and R^2 score. The best-in-class model was a random forest regressor, which yielded a testing accuracy of 99.68%. Through such systems, we aspire to promote tech-literacy among farmers and push forward the precision agriculture movement, which is yet to take off in India, and could have a tremendous positive impact on India's economy and the lifestyle of farmers in the nation.

 DOI: 10.1201/9781003589273-68

2 LITERATURE SURVEY

Wiebold *et al.* (1998)[13] have conducted extensive surveys to determine the barriers to adoption of precision agriculture, with several focus groups of farmers in the North Central Region of America. Two of the obstacles found for the adoption and diffusion of precision agriculture technology were the time and expertise required in learning how to use the tools, as well as the technology not being very user-friendly. These 2 factors were given utmost importance in the development of the proposed SmartFarmo system. Steele (2017)[11] has analysed the precision agriculture adoption barriers in Western Canada and have highlighted that there is a lack of software developers who are working on precision agriculture technologies and those that do are unable to reduce the complexity of their applications to meet the needs of farmers.

Bagheri *et al.* (2023)[2] have investigated the perception of experts in the domain of agriculture about challenges in adoption of precision agriculture technologies in Iran. Lack of knowledge and limited accessibility were noted as important factors which limited their efficacy and reach. Andujar (2023)[1] has put forth that artificial intelligence, particularly machine learning applications are increasingly being deployed in rural areas, in his study on the latest trends in precision agriculture. The analysis conducted by Condran *et al.* (2022)[3] demonstrated that multidisciplinary research in the area of machine learning and agriculture has seen a 100-fold increase between 2000 and 2020, in terms of number of publications in Google Scholar. It also pointed out that decision trees, k-nearest neighbours, random forest, support vector machine and neural networks were the most common machine learning algorithms used in the agricultural domain.

Song *et al.* (2022)[10] have developed a novel precision agriculture system having an accuracy of 98.9%. It utilizes a convolutional neural network for crop yield prediction based on UAV image data. Sellam *et al.* (2016)[9] have analysed the impact of several factors on crop yield through the use of regression analysis techniques. They have concluded that the annual rainfall, area under cultivation and food price index had a 70% influence on crop yields. Zhang *et al.* (2014)[14] have also studied various factors influencing crop yield. It was observed that the 3 main factors correlated with an increase in crop yield were amount of chemical fertilizer used, irrigation area and crop cultivation area. Furthermore, there was a strong negative correlation between crop yield and drought hazard of the land.

Epule *et al.* (2018)[4] have conducted a study on the effects of climatic and non-climatic factors on crop yields in Uganda. Their findings were that non-climatic factors such as forest area dynamics, wood fuel and usage of tractors played a larger role than climatic factors such as temperature and precipitation. Lad *et al.* (2022)[5,6] have considered a multitude of factors affecting crop yield in Tamil Nadu between 2000 and 2015 and have proposed a crop yield estimator. The performances of several supervised machine learning algorithms were compared and the highest accuracy of 99.318% was achieved by the XGBoost model. Prakash *et al.* (2023)[7] have developed a crop insurance premium recommender model using machine learning. The developed dataset was used to train different machine learning models and their performances were compared. The best performance was achieved by a gradient boosting model, having an accuracy of 93.58%. Klompenburg *et al.* (2020)[12] have conducted an in-depth review of literature on machine learning techniques for crop yield prediction. It was shown that the most common features used for prediction were temperature, rainfall, and soil type. Additionally, artificial neural networks and convolutional neural networks were the most popular models used for yield estimation.

3 PROPOSED SYSTEM ARCHITECTURE

The training process and proposed system architecture for the SmartFarmo crop insurance guidance system are depicted in Figure 1 and Figure 2 respectively. During the training process, the crop recommendation dataset is loaded, after which its numerical features are normalized using min-max scaling.

$$x_{scaled} = \frac{x - x_{min}}{x_{max} - x_{min}} \quad (1)$$

Equation 1 depicts the calculation of normalized values for each numerical feature x (x_{scaled}), where x_{min} and x_{max} are the minimum and maximum values of numerical feature x in the dataset. Subsequently, the outliers were removed by deleting rows of data for countries having lesser than 100 entries in the dataset and the categorical features of the data were converted into numerical features. Lastly, the data was augmented by creating rows with minor random variations from the original dataset (+/- 1 for numerical features).

The SmartFarmo crop insurance guidance system has a simple front-end UI accessible to farmers. The farmers have an option to register for the service or to sign in if already registered. Their personal details are stored in the user database and are used for further reference. When signed in, the user can enter details such as country, area of land, type of crop grown, year, tonnes of pesticide usage, average temperature, and average rainfall. Based on these parameters, the trained crop yield prediction model, which is a random forest regression model, predicts the expected crop yield (in hectograms per hectare). The output of the model, along with the standard market rates of the various crops as of 22nd February, 2024, is used to compute the insurance amount to be recommended and is displayed to the farmer at the front-end. The calculation of recommended insurance amount is described in Equation 2

$$Amount = Yield * Price_{per_{kg}} * Coverage_{y,p} \quad (2)$$

Where, Amount is the recommended crop insurance amount, Yield is the expected crop yield (which is the product of hectogram / hectare predicted by the model and number of hectares of farmland used for cultivation), $Price_{perkg}$ is the standard market rate of the crop being grown in the month of February and $Coverage_{y,p}$ represents extent of crop insurance. It is the product of yield coverage and price coverage. A yield coverage of 0.8 and price coverage of 0.6 was chosen for the SmartFarmo system.

The algorithm used by SmartFarmo is shown in Table 1. The market rates of various crops were sourced from a popular B2B e-marketplace called "India Mart", using the prices of the most popular vendors. The values used as of 22nd February 2024 are depicted in Table 2.

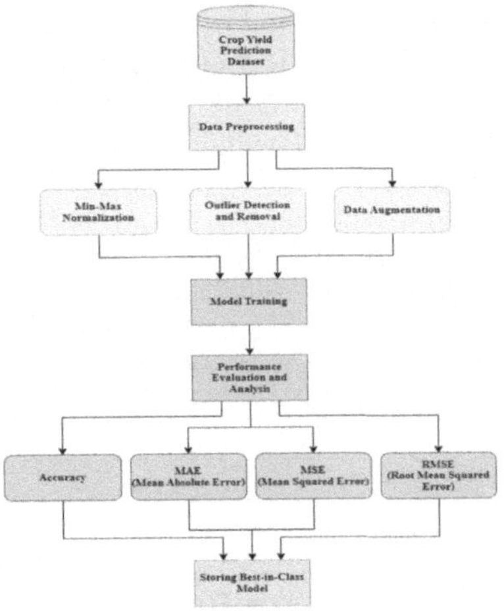

Figure 1. Training process.

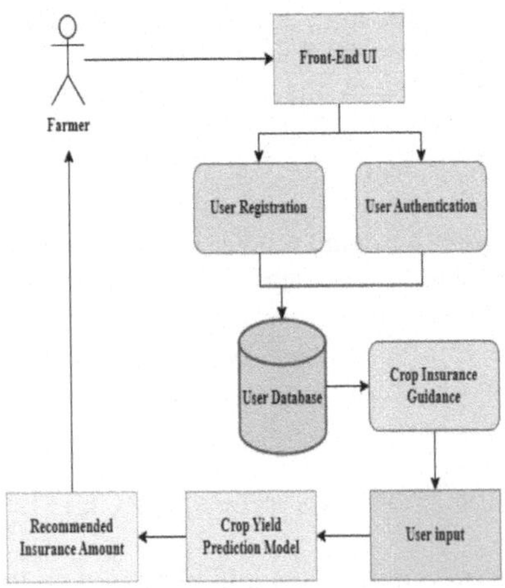

Figure 2. SmartFarmo crop insurance system.

Table 1. SmartFarmo algorithm.

Algorithm: Crop Insurance Guidance System
Input: Crop and area information from user. Output: Insurance amount recommendation. Begin 1) Normalize the dataset using Min-Max Scaler. 2) Remove data for countries with < 100 entries. 3) Randomly vary numerical features of the dataset. 4) Train and compare the different base models. 5) Load the model having best-in-class performance. 6) Predict crop yield by passing input into the model. 7) Compute crop insurance amount recommendation. Return crop insurance amount recommendation. End

Table 2. Market prices per of various crop types.

Crop Type	Price per kg (rupees)
Cassava	44
Maize	24
Potatoes	16
Rice, paddy	50
Sorghum	64
Soybeans	58
Sweet potatoes	18
Wheat	36
Yams	36
Plantains and others	20

4 IMPLEMENTATION

The dataset used for the model is Rishi Patel's Crop Yield Prediction Dataset from Kaggle (www.kaggle.com/datasets/patelris/crop-yield-prediction-dataset). The dataset contains combined and cleaned data from the FAO and World Bank for the 10 most consumed crops in the world. The features in the dataset are depicted in Table 3. The SmartFarmo system was implemented using the Python programming language. The Pandas library was used to extract data from the CSV file and for the data preprocessing. Libraries like Matplotlib and Seaborn were used for exploratory data analysis and data visualization. The model training and evaluation process was done using the Scikit-Learn machine learning library for Python. The front end was built using the Tkinter library and was connected to a backend SQL database using MySQL Connector and XAMPP.

Table 3. Dataset features.

Feature	Description
Area	Country of growing
Item	Crop type
Year	Year of growing
Rain Fall	Average rainfall per year (mm)
Pesticide Tonnes	Number of tonnes of pesticide used
Temperature	Average temperature (Celsius)
Yield	Crop Yield (Hectogram per Hectare)

The dataset has a wide coverage of 10 crops over 101 countries. The most entries are for growth of potato and for crops grown in India. Upon analysis of the dataset, a strong correlation was found between area with average rainfall and tonnes of pesticide used. A strong correlation was also found between the crop type and crop yield. Figure 3 shows the correlation heatmap for the dataset features. Analysis of data revealed that cassava, yams, plantains and other crops only grew in regions having temperatures greater than 15 ^{0}C, whereas the rest of the crops grew over a wider temperature range. An inverse correlation between tonnes of pesticide and crop yield was observed. Furthermore, a 500–2000 mm range of average rainfall was most suitable for crop yield. The optimal temperature range for crop yield was found to be 5-25 ^{0}C. Rainfall levels and temperatures out of these ranges were linked with low crop yields. On further examination, a correlation between type of crop grown and tonnes of pesticide used was observed. It was found that Yams used the highest amount of pesticides on average, while Plantains and other crops used the least amount of pesticide. Importantly, the yield for all crop types showed an increase between 1990 and 2015. A sample output of the SmartFarmo system is shown in Figure 4.

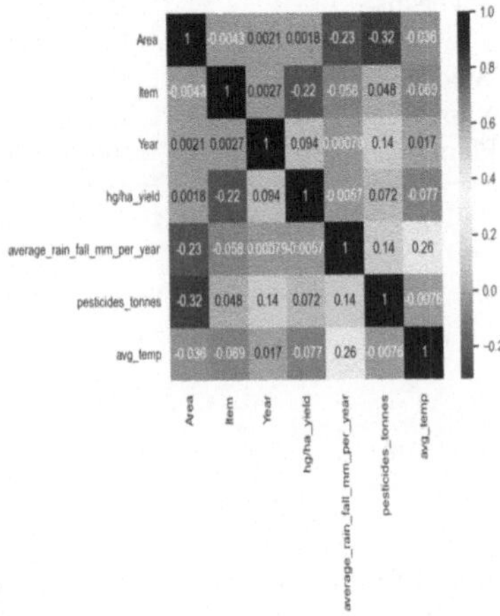

Figure 3. Correlation heatmap of features.

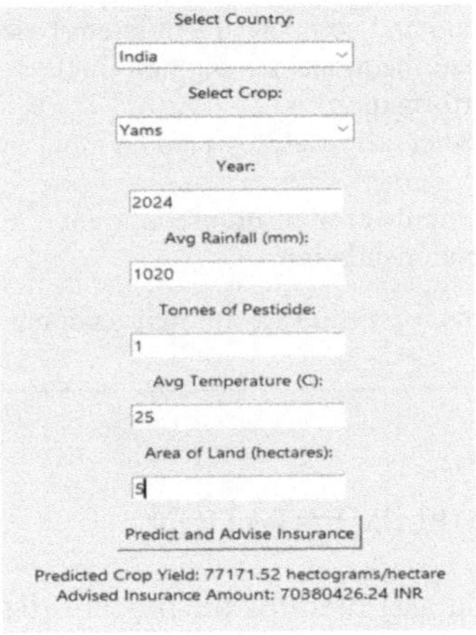

Figure 4. SmartFarmo front end UI.

5 PERFORMANCE EVALUATION AND RESULTS

Several regression machine learning models were trained on the pre-processed crop yield prediction dataset and their performance was evaluated and compared using R^2 score (as 'Accuracy'), Mean Absolute Error (MAE), Mean Absolute Percentage Error (MAPE), Mean Squared Error (MSE), and Root Mean Squared Error (RMSE).

$$R^2 = 1 - \sum \frac{(y_i - \hat{y}_i)^2}{\sum (y - \bar{y})^2} \tag{3}$$

Equation 3 depicts the calculation of R^2 score for a model across all data point, where y_i is the actual value, $\hat{y}_i$ is the predicted value and $\bar{y}$ is the mean value for the feature y.

$$MAE = \frac{1}{n} \sum |(y_i - \hat{y}_i)| \tag{4}$$

$$MAPE = \frac{1}{n} \sum |\frac{(y_i - \hat{y}_i)|}{y_i} \tag{5}$$

The calculation of MAE and MAPE are shown in Equations 4 and 5 respectively. MAE is an average of the absolute differences between actual values and the values predicted by the model.

$$MSE = \frac{1}{n} \sum (y_i - \hat{y}_i)^2 \tag{6}$$

$$RMSE = \sqrt{MSE} \tag{7}$$

Equation 6 and 7 depict the computation of MSE and RMSE. MSE refers to the average of the squares of differences between actual and predicted values, while RMSE is the square root of MSE. The dataset underwent a train-test split with a test size of 0.3, and the R^2 score, MAE, MSE, and RMSE of 11 different models were computed and tabulated. Table 4 illustrates the results of training with the chosen performance metrics.

Table 4. Results of different regression models.

Model	R^2 score (Accuracy)	RMSE	MAPE
Lasso Regression	0.0603	80211.76	2.3407
Ridge Regression	0.0603	80211.63	2.3406
Linear Regression	0.0604	80211.62	2.3406
Ada Boost	0.5835	53399.36	1.1635
Gradient Boosting	0.7873	38160.57	0.6549
K-Nearest Neighbours	0.9235	22881.02	0.3634
XGBoost	0.9749	13106.67	0.2270
Extra Trees	0.9821	11058.42	0.1335
Decision Tree	0.9959	5298.05	0.0289
Bagging	0.9967	4663.70	0.0433
Random Forest	0.9968	4649.40	0.0435

After training, the random forest regression model had best-in-class performance, having the highest accuracy of 99.68%, RMSE of 4649.40, and MAPE of 0.0435. The second and third highest accuracies achieved were 99.67% and 99.59% by the bagging and decision tree models respectively. The lowest accuracy was 6.03% for the lasso regression model. The reason for this is that the problem of crop yield prediction is not very linear in nature. The random forest model achieves high accuracies on account of its ability to effectively model the non-linear relationships between features. Its ensemble learning approach consisting of multiple decision trees capable of approximating complex patterns in the data, also contribute to its high performance.

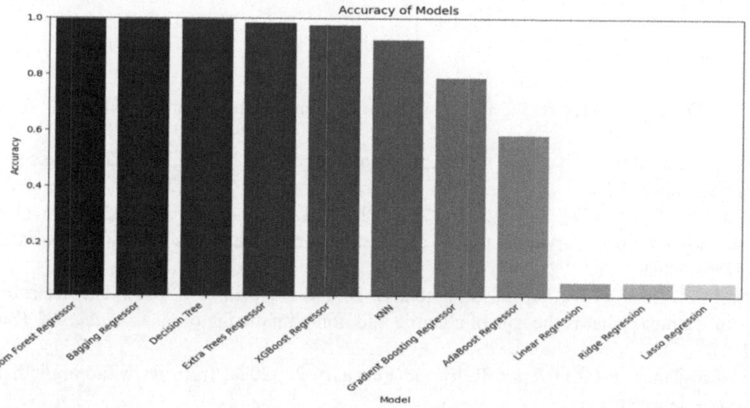

Figure 5. Accuracy of various models.

The accuracy of all the models after performance evaluation is depicted in Figure 5, and the RMSE of the top 5 performing models is shown in Figure 6. It is observed that the random forest model also has the lowest RMSE, indicating that it has the lowest difference between the yield it predicts and the actual crop yield for that land. The best-fit line for the random forest model is illustrated in Figure 7.

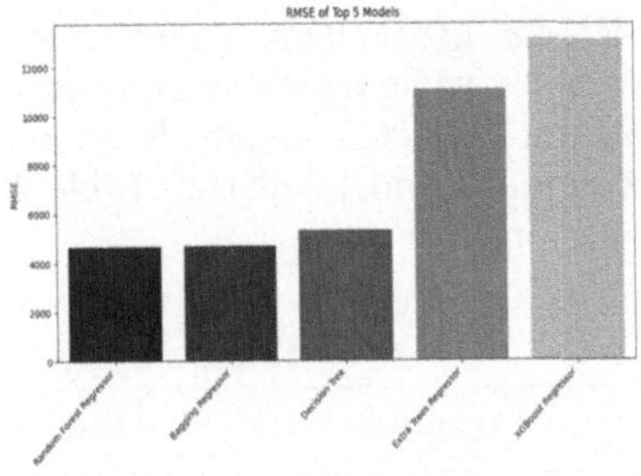

Figure 6. RMSE of the Top 5 models.

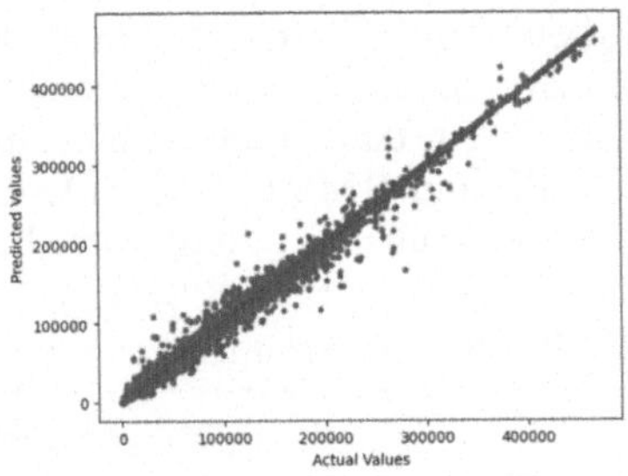

Figure 7. Best fit line for random forest model.

Hence, the random forest model was the best model and was selected as the model of choice in the SmartFarmo crop insurance guidance system. The details entered by the farmer are passed onto the model and insurance is recommended based on predicted yield.

6 CONCLUSION

Crop insurance guidance is a crucial aspect of modern farming and an important component of precision agriculture. This paper proposes a machine-learning based approach to advise farmers on how much crop insurance to get, based on predicted crop yields. A best-in-class random forest regression model, having an accuracy of 99.68% was used, along with a simple and intuitive front-end UI. The popularization of such technologies encourages widespread adoption of precision agriculture methods in the agricultural sector, which could catalyse economic growth and improve farmer livelihoods in India. Future work includes investigating different data pre-processing techniques and incorporating neural network-based yield prediction models. The front-end UI can be even further refined for a wider reach and can be expanded to include local languages. More data and more parameters can be added to further enrich the model accuracy.

REFERENCES

[1] Andujar, D. (2023). Back to the future: what is trending on precision agriculture?. *Agronomy*, 13(8), 2069.

[2] Bagheri, A. and Emami, N. (2023). Perceptions of agricultural experts towards barriers to the adoption of precision agriculture.

[3] Condran, S., Bewong, M., Islam, M. Z., Maphosa, L. and Zheng, L. (2022). Machine learning in precision agriculture: a survey on trends, applications and evaluations over two decades. *IEEE Access*, 10, 73786–73803.

[4] Epule, T. E., Ford, J. D., Lwasa, S., Nabaasa, B. and Buyinza, A. (2018). The determinants of crop yields in Uganda: what is the role of climatic and non-climatic factors?. *Agriculture & Food Security*, 7 (1), 1–17.

[5] India at a glance. *FAO in India*. Retrieved February 22, 2024, from www.fao.org/india/fao-in-india/india-at-a-glance/en.

[6] Lad, A. M., Bharathi, K. M., Saravanan, B. A. and Karthik, R. (2022). Factors affecting agriculture and estimation of crop yield using supervised learning algorithms. *Materials Today: Proceedings*, 62, 4629–4634.

[7] Prakash, M. C. and Saravanan, P. (2023). Crop insurance premium recommendation system using artificial intelligence techniques. *International Journal of Professional Business Review*, 8(4), e01270–e01270.

[8] Bhatnagar S., Lakshmi N., A. A and V. S. M. Y. (2024). Development of a crop recommendation system through the use of various machine learning algorithms, 2024 *Second International Conference on Emerging Trends in Information Technology and Engineering (ICETITE)*, Vellore, India, pp. 1–6, doi: 10.1109/ic-ETITE58242.2024.10493758.

[9] Sellam, V. and Poovammal, E. (2016). Prediction of crop yield using regression analysis. *Indian Journal of Science and Technology*, 9(38), 1–5.

[10] Song, C., Ma, W., Li, J., Qi, B. and Liu, B. (2022). Development trends in precision agriculture and its management in china based on data visualization. *Agronomy*, 12(11), 2905.

[11] Steele, D. (2017). Analysis of precision agriculture adoption & barriers in western Canada. *Final Report*, 53.

[12] Van Klompenburg, T., Kassahun, A., and Catal, C. (2020). Crop yield prediction using machine learning: A systematic literature review. *Computers and Electronics in Agriculture*, 177, 105709.

[13] Wiebold, B., Sudduth, K., Davis, G., Shannon, K. and Kitchen, N. (1998). Determining barriers to adoption and research needs of precision agriculture. *Missouri Precision Agriculture Center*, University of Missouri and USDA/ARS.

[14] Zhang, X., Wang, L., Wang, L., Liu, B., Zheng, J., Zhao, J. and Ruan, H. (2014). Analysis of major factors influencing crop yield of Shandong province. *In Proceedings of Selected Articles of 2013 World Agricultural Outlook Conference* (pp. 159–167). Springer Berlin Heidelberg.

Crowd analysis using deepSORT

S. Ciyamala Kushbu
Assistant Professor, Information Science and Engineering, CMR Institute of Technology, Bengaluru, India

T.M. Inbamalar
Professor, Electronics and Communication Engineering, RMK College of Engineering and Technology, Puduvoyal, Tamilnadu, India

Chettiyar Vani Vivekanand
Professor, Computer Science and Engineering, Rajalakshmi Engineering College, Chennai, India

B. Sarala
Associate Professor, R.M.K Engineering College, Kavaraipettai, TamilNadu, India

M. Amala Justus Selvam
Professor, Mechanical Engineering, Vel Tech Rangarajan Dr Sagunthala R&D Institute of Science and Technology, Chennai, India

Krishna Kant Gupta
Student, Information Science and Engineering, CMR Institute of Technology, Bangalore, India

ABSTRACT: Crowd dynamics, prevalent in stations, malls, religious sites, airports, and public events, pose significant challenges in managing and ensuring public safety. Video surveillance plays a pivotal role in crowd management, providing essential input for monitoring and control. Addressing crowd detection and monitoring encompasses diverse applications, ranging from enforcing social distancing during pandemics to traffic monitoring and behavior analysis. This necessitates the development of robust models for tasks such as counting, density estimation, motion tracking, and behavior detection. Central to implementing human detection systems is object detection, a critical branch of computer vision. Leveraging the capabilities of Convolutional neural networks (CNNs), recent advancements in object detection have shown promising results in accurately identifying various objects, including humans, within video streams. This abstract explores the integration of DeepSORT into object detection frameworks to enhance crowd dynamics analysis, with a focus on applications such as crowd behavior analysis, emergency response, and public space design. The overall accuracy was found to be 82%.

Keywords: Object Detection, Crowd Detection, Machine Learning, Convolutional Neural Networks

1 INTRODUCTION

Increasing concerns regarding public safety have drawn the attention of many governments and security agencies worldwide. They are increasingly recognizing the critical necessity for dependable and secure crowd-monitoring systems to tackle these worries. Efficiently overseeing human gatherings requires taking proactive steps to avert unexpected incidents or issues, guaranteeing a secure and smoothly-operating environment. The lack of research dedicated to crowd monitoring systems and their security ramifications has sparked a growing field of inquiry, delving into possible methods to effectively protect human assemblies.

The term 'event' refers to some situation of relatively high crowd density. The term 'event' is not only a representation of the general perception of any external situation like public rallies,

 DOI: 10.1201/9781003589273-69

religious gatherings but also the fact that some natural and man-made factors can also affect crowd density for instance, road blockages could be attributed to construction work and/or weather conditions or illegal vehicle parking, which occupies a significant portion of the area reserved for pedestrians. The immediate effects lead to traffic diversion, resulting in change in crowd flow, problems related to noise pollution etc. In a densely crowded area, crowd behavior is an important parameter for monitoring crowds and ensuring proper law and order.

Anomaly detection is one such key area that helps in meeting our objective to a great extent with communication alerts and emergency measures in situations of suspicion. Anomaly detection in crowds is not a full fledged measure of an activity of suspicion as it involves taking other parameters into consideration for instance, a panic at a traffic signal is not the same as in a bank. reveal flaws and disparities in model performance across different patients.

Crowd management at public events is a multifaceted challenge, demanding a seamless blend of technology, data analysis, and real-time response coordination. To ensure safety and security in the crowd, the project aims at implementing a deep sort based algorithm that mitigate risks, and enhance the overall event experience.

2 LITERATURE SURVEY

Deep learning (DL) based methods play a vital role in object detection where missing or movement of an individual is the source of attraction. YOLO (You Only Look Once) models use convolutional filters to support such detection. Such methods are also used in various other fields including image processing [1], machine translation [2], natural language processing [3], sound classification [4], and language recognition [5]. Two main techniques are typically used for Individual detection: Region of Interest (ROI) and Line of Interest (LOI). ROI estimates the number of individuals in defined bounding box areas, while LOI counts those crossing the specified lines. YOLO models are favored for person detection due to their efficiency in encapsulating objects within bounding boxes and classifying them [6]. In video surveillance systems, where accurate detection and counting of people are critical, YOLO models are widely relied upon [7]. Recent improvements, such as combining YOLO with the DeepSort algorithm [8] and using newer versions like YOLO v5 [9] and YOLO-PC [10], have further enhanced accuracy. This paper presents an algorithm that employs various YOLO models to identify and track the movement of individuals while estimating indoor areas. Performance evaluations of YOLO v3, v4, and v7 reveal high accuracy in detecting individuals within specified zones [11].

The study utilized the OpenCV library and Python programming language. The DeepSort algorithm [12], which uses a CNN model for object classification, was implemented for human tracking. Object tracking starts with bounding box detection [14], which determines the object's location, type, and class. In this research, non-maximum suppression (NMS) was used to remove redundant bounding boxes, and YOLO models were trained to detect humans in specific areas [15]. The CNN model identifies the most distinctive features of the object that are used for further analysis. Each object detected is given the ID by the DeepSort algorithm to generate an appearance feature vector, that are used to match objects between frames. Initial object detection, which involves bounding box detection to determine the object's location, type, and class, is crucial for object tracking. For people counting, a section of the video was chosen, and a simple counter was incremented for each person detected in that section. The maximum capacity of the designated area, calculated in pixels, was based on the assumption of one person per square meter. To ensure consistent results and fair comparisons, YOLO-based algorithms employed a 0.5 Intersection over Union (IoU) threshold and a 0.5 confidence threshold for 512×512 pixel input images. Tests were performed at resolutions of 1024×768 pixels and 800×600 pixels. The number of people counted within designated areas was displayed in green if below capacity and red if above capacity.

3 PROPOSED METHOD

The MS COCO dataset, renowned for its extensive range of object detection, segmentation, and image captioning tasks, was employed in this study [13]. Including "person," this dataset has 80 categories and 1.5 million object instances, making it a useful resource. Specifically, photos from the “person” category in the MS COCO dataset were used, as this study is only concerned with person detection.

The project focuses on implementing CCTV and other surveillance systems for straightforward crowd monitoring and analysis. It’s designed to detect abnormal crowd behavior, breaches of social distancing, and unauthorized entries. Additionally, the system can process crowd movement data into optical flow, heatmaps, and energy graphs.

Functional requirements involve implementing modules for abnormal crowd activity detection, restricted entry tracking, and crowd movement analysis. Non-functional requirements are system scalability, real-time processing, and seamless integration with YOLOv4 and Deep SORT. This comprehensive approach, combining machine learning, data analysis, and real-time information, aims to provide a safer and more informed travel environment for users.

The proposed scheme to analyze pedestrian dynamics to identify potential outliers is illustrated in Figure 1.

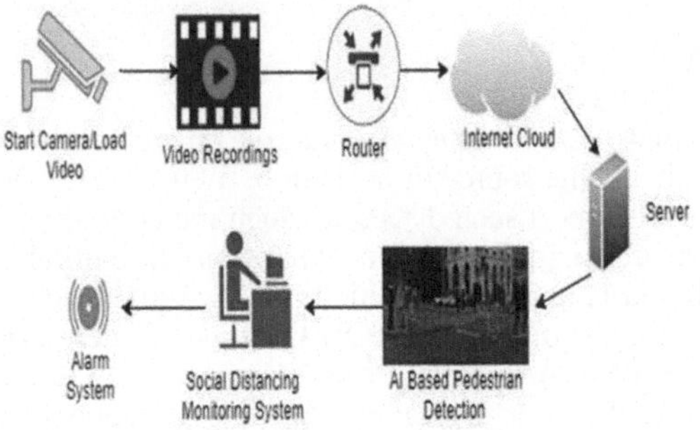

Figure 1. Proposed scheme.

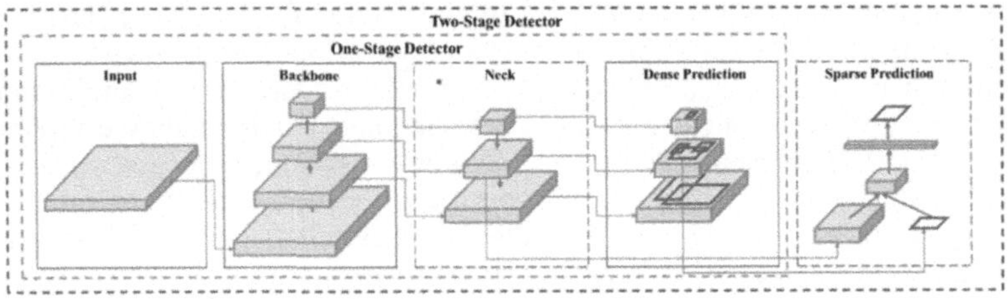

Figure 2. Architectural model of YOLOv4.

The architecture of YOLO models are based on Fully Connected Neural Networks (FCNN), is depicted in Figures 2–3. This model performances well on the COCO dataset, which includes 80 different object classes. YOLO models are used for Object Detection, emphasizing inference speed. Here, the Region of Interest (ROI) is not individually selected; instead, predictions for bounding boxes are made for the entire image, resulting in faster processing. The initial version of YOLO was developed within the Darknet Framework, a high-performance open-source framework designed for implementing neural networks using C and CUDA. Darknet typically serves as the backbone network for YOLO models. The architecture comprises several components, broadly categorized as follows:

Input: This is the initial stage where our training image set is fed and processed in the network.

Backbone and Neck: To extract the important features backbones are used.

Detection Neck and Detection Head: The object detection and prediction activities are done in the neck and head of the architecture.

Strong backbone networks, usually CSPDarknet53, are at its heart and are skilled at extracting complex features from input images/videos. In addition, a Feature Pyramid Network (FPN) is used to recognize objects of different sizes efficiently. The neck design of YOLOv4 uses skip connections and convolutional layers to further improve feature refining.

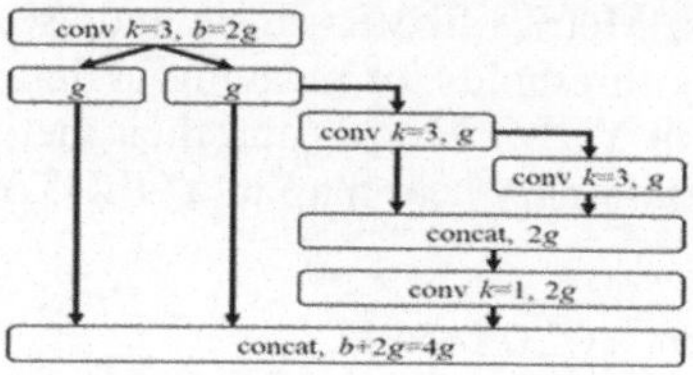

Figure 3. Computational model of YOLOv4 tiny.

Bounding boxes, confidence scores, and class probabilities for observed objects are predicted by the detection head, which consists of convolutional layers and activation functions. YOLOv4 is distinguished by multiple design optimizations, including as the mish activation function, multi-input weighted residual connections (MiWRC), and cross-stage partial connections (CSP), which improve accuracy while reducing processing expenses. YOLOv4 achieves real-time performance across a variety of devices by utilizing hardware acceleration such as GPUs and TPUs, which makes it a perfect fit for applications ranging from robotics to autonomous driving to surveillance.

Traditionally, the training process is monitored by plotting the mean dice scores over various training epochs. However, this approach lacks the ability to provide insight into the model's performance in corner cases. To address this, we introduce a method to gain deeper insights by examining the Dice score curve for corner cases at different training epochs.

A real-time system is dedicated to surveilling public spaces to identify human clusters. The entire process of this system is depicted in Figure 4. Initially, video sequences are acquired from surveillance cameras situated in different locations. Subsequently, image frames are extracted from these videos, and background elements are eliminated from the images through background subtraction, leaving only foreground objects for subsequent analysis. Human detection is then performed using DeepSORT and YOLOv4 algorithms, and the number of humans in a cluster is determined to enforce social distancing guidelines.

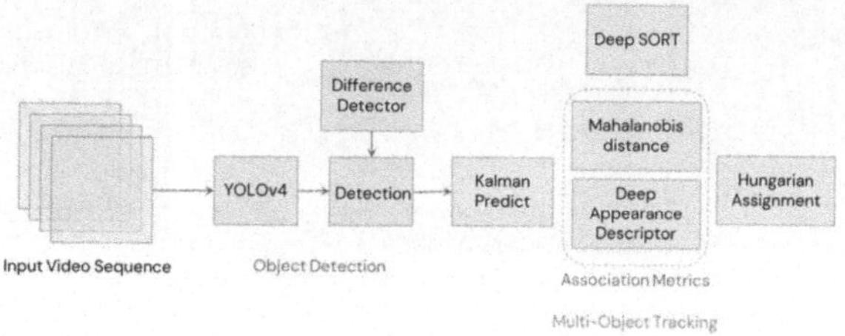

Figure 4. Processed output in (a) YOLOv3 (b) YOLOv4 (c) YOLOv7.

4 EXPERIMENTAL RESULTS

The work was performed on Google Colaboratory Pro, using Python 3 along with the TensorFlow 2.0, Keras, and OpenCV libraries. For pedestrian detection tasks, the mean Average Precision (mAP) is the most commonly used and widely accepted evaluation metric. It can be calculated using the following formula:

$$\text{mAP} = \sum_{K=1}^{k=M} Ap(k)/M \tag{1}$$

Other metrics involves Processed FPS which quantifies the system's speed in processing frames. Time Elapsed denotes the total duration of data processing, reflecting operational efficiency over the entire task. Tracks Recorded indicates the number of individual paths tracked, illustrating the system's capability in multi-object tracking scenarios. Useful Movement Data refers to important information extracted from the movement patterns, critical for insightful analysis. Metrics like Kurtosis and Accepted Energy Level assist in identifying outliers and setting thresholds for meaningful data interpretation.

The Performance of different YOLO Models and their metrics are tabulated in Tables 1–2. The processed output and stationary heat map in YOLO3,YOLO4,YOLO7 are shown in Figures 4–5.

Table 1. Performance of different YOLO Models.

Layers	YOLOv3	YOLOv4	YOLOv7
Neural Network	FCNN	FCNN	0.86
Backbone	Darknet53	CSPDarknet53	0.90
Neck	FPN	SSP	0.76
Head	0.88	0.90	0.89
LossFunction	Binary cross entropy	Binary cross entropy	Binary cross entropy

Table 2. Performance metrics.

Measure	YOLOv3	YOLOv4	YOLOv7
Processed FPS	42.5	29.85	19.42
Time Elapsed	33.85	48.20	74.11
Tracks Recorded	8	24	24
Useful Movement Data	270	4129	3922
Kurtosis	32.76	77.889	56.22
Accepted Energy Level	7430	2480	3213

(a) (b) (c)

Figure 4. Architecture of YOLOV4 andDeepSORT.

Figure 5. The Stationary Location Heat Map in (a) YOLOv3 (b) YOLOv4 (c) YOLOv7.

5 CONCLUSION AND FUTURE WORK

Utilizing enhancements to the YOLO network, a DeepSort algorithm for tracking pedestrians was devised. Employing a Kalman filter, pedestrian state estimation was achieved. Similarity between detection frames and predicted pedestrian trajectories was computed using Mahalanobis distance and apparent feature indexes. The Hungarian algorithm facilitated optimal pedestrian target matching. Official statistics revealed that the initial YOLO model yielded a test accuracy of only 55.3%. Prior to enhancement, the YOLO model achieved training and testing accuracies of 80.64% and 69.45%, respectively.

Comparatively, the enhanced YOLOv3 model and DeepSort tracking method were evaluated within the same experimental setting. The enhanced YOLO model demonstrated training and testing accuracies of 82.21% and 72.02%, respectively, that represent an improvement of 1.57% and 2.57%, respectively, over the pre-enhancement model. Verification results underscored the enhanced model's superior capability to handle occlusion, accurately identifying missed and misdetected images, thereby rectifying previous tracking failures.

REFERENCES

[1] Punn, N. S., Sonbhadra, S. K., Agarwal, S. and Rai, G. (2020). Monitoring COVID-19 social distancing with person detection and tracking via fine-tuned YOLO v3 and Deepsort techniques . *arXiv preprint* arXiv:2005.01385.

[2] Pacal, I., Karaman, A., Karaboga, D., Akay, B., Basturk, A., Nalbantoglu, U. and Coskun, S. (2022). An efficient real-time colonic polyp detection with YOLO algorithms trained by using negative samples and large datasets. *Comput. Biol.Med.* 141, 105031. (ISSN 0010-4825)

[3] Gao, C., Li, P., Zhang, Y., Liu, J. and Wang, L. (2016). People counting based on head detection combining Adaboost and CNN in crowded surveillance environment. *Neurocomputing* 208, 108–116

[4] Redmon, J., Divvala, S., Girshick, R., and Farhadi, A. (2016). You only look once: Unified, real-time object detection, in *Proceedings of the IEEE Conference on Computer Vision and Pattern Recognition*, pp. 779–788

[5] Chen K., and Kämäräinen J. K. (2014). Learning to count with backpropagation information, *In 22nd International Conference on Pattern Recognition. IEEE*, 2014. p. 4672–4677

[6] Işık, G. and Artuner, H. (2020). Turkish dialect recognition using acoustic and phonotactic features in deep learning architectures. *Int. J. Inform. Technol.* 13, 207–216

[7] Bochkovsky, A., Wang, C. Y., Liao, H. Y. M. (2020). Yolov4: optimal speed and accuracy of object detection. *arXiv preprint* arXiv:2004.10934

[8] Redmon, J. and Farhadi, A. (2018). Yolov3: An incremental impr- ovement. *arXiv preprint* arXiv:1804.02767

[9] Jocher, G.; Changyu, L.; Hogan, A.; Yu, L.; Rai, P. and Sullivan, T. (2020). ultralytics/yolov5: Initial Release (v1.0). Zenodo . https:// doi.org/10.5281/zenodo.3908560

[10] Ahmad, M., Ahmed, I. and Adnan, A. (2019). Overhead view person detection using YOLO. In 2019 *IEEE 10th Annual Ubiquitous Computing, Electronics & Mobile Communication Conference (UEMCON)* pp. 0627–0633.

[11] Purwar, R.K. and Verma, S. (2022). Analytical study of YOLO and ıts various versions in crowd counting. In: *Intelligent Data Communication Technologies and Internet of Things*, pp. 975–989. Springer

[12] Ren, P., Fang, W. and Djahel, S. (2017). A novel YOLO-Based real-time people counting approach, In: *2017 International Smart Cities Conference (ISC2)*. IEEE, p. 1–2.

[13] Lin T.Y., Maire, M., Belongie, S., Bourdev, L., Girshick, R., Hays, J., Perona, P., Ramanan, D., Zitnick, C. L. and Dollár, P. (2015). Microsoft COCO: common objects in Context. *Computer Vision and Pattern Recognition*, 2–8.

[14] Hu, R., Mo, Q., Xie, Y., Xu, Y., Chen, J., Yang, Y. and Wu, E.Q. (2021). Avmsn: an audio-visual two stream crowd counting framework under low-quality conditions. *IEEE Access* 9, 80500–80510

[15] Hu, R., Tang, Z.R., Wu, E.Q., Mo, Q., Yang, R. and Li, J. (2022). RDC-SAL: refine distance compensating with quantum scale-aware learning for crowd counting and localization. *Appl.Intell.* https:// doi.org/10.1007/s10489-022-03238-4.

Data Science & Exploration in Artificial Intelligence – Gururaj et al. (Eds)

Enhanced deepfake video classification and detection: A ResNext-LSTM approach for improved accuracy

E. Geetha Rani
Assistant Professor, Department of CSE, Alliance School of Advanced Engineering Alliance University, Bengaluru, Karnataka, India

P. Bhuvaneshwari
Associate Professor, Department of CSE, Manipal Institute of Technology Bengaluru, Manipal Academy of Higher Education, Manipal, India

Rohit Gorakh Darekar
UG Scholar, Department of CSE, Alliance School of Advanced Engineering, Alliance University, Bengaluru, Karnataka, India

D. Anusha
Associate Professor, Department of CSE, SRKIT, Enikepadu, SRK Institutions, Vijayawada, Andhra Pradesh, India

ABSTRACT: Our study investigates the potential benefits of artificial intelligence for deepfake detection techniques. An LSTM-based network is utilized for feature extraction, while a ResNext convolution neural network is used for temporal correlations in the architecture. The ensemble model's resilience and generalizability are enhanced by data augmentation used during preprocessing. In this experiment we have used FaceForensics++ and Celeb-DF datasets for training process.We have introduced Data augmentation in preprocessing phase of the model where we are injecting the Gaussian noise to address noise variations in the training set. This pre-processing phase method significantly enhanced the efficiency of our deepfake detection model architecture effectively. Also, we have done another change in model training of this our ensemble architecture to select appropriate and effective training time of each model by considering the sequence length, we have observed our model's accuracy in the training and validation graphs and accordingly, we have used the Epoch value to train models. In this our experiment we have completed experiment with frame sizes of 20 and 40 since,we had very limited GPU units in the colab online ID'E. According to our observation by integrating all this our changes in existing framework, when the ResNext-LSTM ensemble architecture is achieved the result with the recommended data augmentation with Gaussian Noise approach, this our current model performs more than the earlier existing in the identification of deepfake videos [1].

Keywords: Deepfakes, ResNext, LSTM, Data Augmentation

1 INTRODUCTION

Artificial intelligence (AI) techniques are still needed for the challenging issue of deepfake detection. Now a days people are doing malicious activities by doing swap their face in pictures /videos where actually that person haven't done anything. Research is being conducted on many different topics now, such as picture forensics, speech analysis, and facial

 DOI: 10.1201/9781003589273-70

recognition. Researchers frequently mix these techniques to attain greater accuracy [5,6,11,13]. But as methods for manipulating media advance, so does the field of deep-fake Deepfakes continue to be a major social problem despite tremendous developments. This demonstrates how crucial it is to continue researching to create complex detection methods and educate the public about the risks associated with digital fraud.

1.1 *Extended Short-Term Memory (LSTM)*

Our solution uses Extended Short-Term Memory (LSTM) networks for deepfake identification because of their capacity to identify temporal patterns in video. Recurrent neural networks, or RNNs, struggle with declining gradients; but long-term dependency learning may be robustly achieved with LSTMs. LSTMs are particularly adept at capturing the temporal nuances necessary to distinguish deepfakes because of gated structures that control the information flow. Using a single LSTM layer and the pre-trained ResNext-50 network, our model architecture processes 2048-dimensional feature vectors derived from each video frame. A 0.4 dropout rate is used to avoid overfitting. Twenty-four hundred eighteen input attributes and two output features comprise a linear layer that computes the average correlation between features and binary labels. Next, an adaptive typical pooled layer with a return value of 1 is used to refine the final product. And batch training is getting done with Softmax function which has more than 4 layers in this LSTM model.

1.2 *ResNext*

In our Experiment, we have used Pre-trained ResNext model for feature extraction, which is the next version of the ResNet model. This ResNext coming up with Cardinality mechanism which is helping us for deepfake detection more than its earlier version. So, here we have chosen this ResNext model. It has 50 layers. And as per research we have got result also by integrating this model for features extraction from videos of cropped faces.

1.3 *Deepfakes*

In today's world, the people are very aware from all AI techniques. So, they are nicely leveraging wherever and however they want real time. So, Deepfake also is one of the AI techniques only where someone will target someone and make visual or audible content to put in trap that person digitally. So, as per the research this is ongoing very serious issue is gone happening with even big celebrities also. So, we have required to focus on such kinds of malicious activities as research and provide more accurate solutions to identify such kind of malicious content.

2 LITERATURE REVIEW

So many deep learning architectures have been introduced by different researchers for the purpose of detecting deepfakes content created by malicious people. As per the research by Guera *et al.* a CNN is extracting features from individual video frames using a CNN-LSTM architecture. And the LSTM model is basically analyzes the temporal sequence of these features to look for differences that could indicate deepfakes or not. In the research of Khalid *et al.* 's work [12], a one-class Variational Autoencoder (VAE) approach has been investigated using only real data for training of the deepfake detection model. This study compares CNN (EfficientNet) and RNN (LSTM) architectures for deepfake detection. Future work suggests combining these for potentially better results. A controlled comparison using the same diverse dataset is recommended. Expanding datasets with real-world examples and various creation methods can enhance generalizability. Accuracy could be further increased, and a more complete picture could be provided by hyperparameter tuning and the use of multiple evaluation metrics (precision, recall, and F1-score) *et al.* [...]. For identifying deep-fakes, this study provides a useful comparison of deep learning models. In multiple datasets, the authors investigate the efficacy of convolutional neural networks (CNNs) and recurrent neural network models (RNNs), as well as the combination of both techniques. [2–4] However, the focus solely on accuracy as a success metric limits the overall picture. Future work could explore other crucial factors like inference speed and model generalizability across diverse datasets. Interestingly, the study indicates that combining CNN and LSTM could produce interesting outcomes. [7] A promising future direction for accurate facial region extraction within videos is to further investigate various face detection techniques *et al.* [4].

3 PROPOSED METHODOLOGY

3.1 *Existing methodology*

Deep learning techniques for identifying deepfakes have been the subject of numerous studies. A recent study presented a novel method to achieve remarkable accuracy: networks with Extended Short-Term Memory (LSTM) coupled with convolutional artificial neural networks (CNNs). [8] This methodology involves preprocessing and curating datasets that include both real and fake videos. Visual features are extracted from individual frames of each video using ResNext-50, a pre-trained deep CNN. The LSTM network's ability to identify temporal relationships between video frames is then utilized by feeding these features into it. [9,10] To classify videos as authentic or fraudulent, an activate function with a sigmoid is ultimately applied in the categorization layer. The study found an interesting relationship between accuracy and sequence length, or the number of consecutive frames processed, with accuracy stabilizing at about 60 frames.

3.2 *Disadvantages of existing methodology:*

Deep learning-based techniques frequently encounter difficulties, even though the ResNext-LSTM architecture shows potential. Overfitting resulting from small datasets is one of them [1]. To address this, this work incorporates data augmentation techniques during pre-processing, wherein Gaussian Noice is introduced during dataset preprocessing to artificially expand the training data. Moreover, the extant literature provides little insight into methods to mitigate the computational burden associated with training deep learning models [1]. We will investigate methods such as early stopping and hyperparameter tuning to figure out how many training iterations, or epochs, are needed to get the desired accuracy with the least amount of computational overhead. Lastly, the way that deepfakes are developing is not covered in this paper [1].

3.3 *Implementation*

3.3.1 *Dataset preparation*

FaceForensics++ and Celeb-DF are two popular datasets that our deepfake detection system uses. FaceForensics++ contains a diverse collection of deepfakes created using Face2Face, NeuralTextures, Deepfakes, and FaceSwap, among other techniques. It makes training and testing comparisons simple to comprehend by providing both original and modified images and videos. But the massive video dataset Celeb-DF was created especially for studies on deepfake identification. It covers many techniques for creating deepfakes, such as expression manipulation, reenactment, and face swapping. The preparation of data is an essential step in developing a strong deepfake detection system. We make sure the dataset is balanced by adding an equal amount of real and fake videos.

3.3.1.1 *Scaling and normalization* The videos in the dataset are normalized and scaled to a fixed aspect ratio and size to ensure consistency.

3.3.1.2 *Cropping* Unnecessary background material is removed from videos so that only the subject's face is visible.

3.3.1.3 *Audio removal* Audio tracks are removed because they don't add anything to the visual examination of deepfakes.

3.3.1.4 *Frame extraction* Videos can be split up into individual frames to create a set of images from which a deep learning model can extract features.

3.3.2 *Feature extraction*

In our deep-fake identification approach, we leverage the feature extraction power of a pre-trained ResNext-50 model. Our objectives are in line with the ResNext-50 architecture, which has a solid reputation for doing well in picture classification tasks. To make ResNext-50 easy to integrate into our system, we develop a special function named Extract Video Features (). This method helps with the model architecture for only feature extraction from the cropped video dataset. ReLU creates non-linearity, which helps the model recognize intricate patterns in the video frames. These patterns potentially enclose subtle inconsistencies indicative of deepfakes. The extracted features from each frame are accumulated into a list. Finally, the list is converted into a NumPy array and returned by the Extract Video Features () function. This function provides a convenient way to extract deep features from video data, suitable for subsequent stages of our deepfake detection model.

3.3.3 *Model fitting and training strategy*

Our proposed deepfake detection methodology employs a combination of pre-trained and custom layers for effective model fitting, investigating the impact of different sequence lengths.

3.3.3.1 *ResNext feature extraction* A pre-trained ResNext-50 model (ResNext CNN with 50 layers and 32x4d dimensions) serves as the foundation for feature extraction. This model leverages activated functions of rectified linear units (ReLUs) inside its convolutional regions, enabling it to learn complex patterns from the video frames.

3.3.3.2 *Adaptive average pooling and activation* A 2D adaptive average pooling layer is incorporated to reduce computational complexity and extract low-level features from neighbouring frames. Because of their effectiveness and fit for deep neural networks, ReLU activation functions are included.

3.3.3.3 *Evaluation and model export* This is the last phase in the experiment where, we have got one trained model for real time implementation.

3.3.4 *Video segmentation and frame extraction*

First, we are taking input with whole original dataset and then further segregating each video into the no of frames. This is help us for taking only some specific no of frames from each video of dataset.

3.3.5 *Face detection and cropping*

Now, in this step of the pre-processing of the model we are using the OpenCV which is Computer Vision library allows us for detect the upper part of the chest effectively. Furthermore, once the face has been detected we are cropping that face area only which reduces our required resources for experiment.

3.3.6 *Data augmentation with gaussian noise (contribution)*
In this step of the model pre-processing, we have done Data augmentation injection into our newly generated cropped video dataset. Some of the term associated with the Injecting of this Noise as given here. Improved Deepfake Detection we are talking about our model how much accurately can detect the content with its originality by considering accuracy of model. And since we have completed this Noise injection phase, we have results that are completely in advancement compared to existing one [1].

3.3.7 *Saving pre-processed data*
Under this step of the pre-processing of the model we have one newly generated face cropped dataset, and we will use this dataset for the further training of our model.

3.4 *Rationale for gaussian noise augmentation in preprocessing*

Increased Sensitivity to Deepfake Artifacts: This can enhance the model's capability to discriminate between real and deepfake films, particularly in cases when the production of the deepfakes involves techniques that introduce unique noise distributions. Gaussian Noise vs. Other Augmentations: While other methods such as flipping or random cropping can also enhance the generalizability of the model, Gaussian noise provides a targeted strategy that is especially aimed at generating noise patterns that may be significant to the identification of deepfakes. This emphasis can be particularly beneficial when dealing with the faint noise artifacts that are sometimes associated with deepfakes. Addressing Gaussian Noise and Loss Function: It's crucial to make clear that preprocessing adds Gaussian noise; it's not integrated into the loss function itself. The loss function itself is typically used to evaluate the model's efficacy during training.

3.5 *Feature extraction with resNext50*

3.5.1 *Pre-trained model*
Our approach leverages a pre-trained ResNext-50 model, which is well-known for its effectiveness in convolutional neural network (CNN) architectures and demonstrated results in image classification tasks. This model's job as a feature extractor is to take pre-processed facial images and extract meaningful representations from them.

3.5.2 *Feature learning*
In ResNext designs, Rectified is frequently utilized. Each convolutional layer is followed by the linear unit (ReLU) activation functions. ReLU provides non-linearity to the model, enabling it to identify intricate patterns in each frame's face regions. These patterns may represent subtle changes such as uneven skin tone, irregular eye movements, or inconsistent lighting.

Table 1. The results of benchmarking the model against the FaceForensics++ dataset.

Model Name	Dataset Name	No. of videos	Sequence Length	Existing Accuracy	Proposed Model Accuracy
Model_90_acc_20_frames _combined_data	Combined Dataset	33200	220	887.64981	90.17432
Model_90_acc_40_frames _combined_data	Combined Dataset	33200	440	889.49216	90.80824

[Note: Author used combined Dataset with 3000 videos,]|Combined Dataset [FaceForensics++ and Celeb-DF]

3.5.3 *Fine-tuned for deepfakes*

Above all, we specially improve the ResNext-50 model for the purpose of deepfake identification. The final few classification layers are eliminated in this process, leaving only the earlier convolutional layers' feature extraction capacity.

3.5.3 *Deepfake Classification with Long Short-term Memory (LSTM)*

LSTM Network: Following feature extraction step, we have ensembled this model to ResNext model. This LSTM is basically good to use for classification tasks. Since, this LSTM is more capable analyzing the pattern that existing in the video frames. It is helping to core-relate the relationship between the video frames. Performance Assessment: This is the last phase of our model generation experiment. This is evaluated how much effectively we have trained our model. Since, in this experiment we are basically considered aim to improve the accuracy we have evaluated model accuracy with existing one.

Table 2. Evaluation outcomes of the model with the combined FaceForensics++ and Cele-DF datasets.

Model Name	Dataset Name	No. of Videos	Sequence Length	Existing Accuracy (in %)	Proposed Model Accuracy (in %)
Model_94_acc_20_frames _FaceForensic_data	FaceForensic++	22000	220	990.84921	994.47236
Model_95_acc_40_frames _FaceForensic_data	FaceForensic++	22000	440	995.39872	995.96977

FaceForensics++ Dataset

4 RESULT AND DISCUSSION

By considering resource limitation to train our model with high sequence of frames we considered only 20 and 40 input frame size. We have implemented pre-processing Gaussian Noise injection step to enhance the model generalisability and robustness, and we are succeeded to achieve improved accuracy than existing models. Furthermore, we have done hyperparameter tunning for training to select optimal number of training iteration /Epoch value. Which also help us to reduce required GPU units and overfitting of the model. This early stopping during training helps us to enhanced accuracy.

4.1 *Accuracy and loss visualization*

The learning and testing losses for a model trained on the FaceForensics++ dataset over 25 epochs with a frame size of 20 are shown by the loss curves in Figure 2. The model strives to reduce the prediction error, or loss, by fine-tuning its outputs to closely match the ground truth labels, as shown by the training loss curve's declining trajectory (shown by the green line). Similarly, the testing loss curve (depicted by the blue line) exhibits a downward trend, prompting for the learning process of the framework.

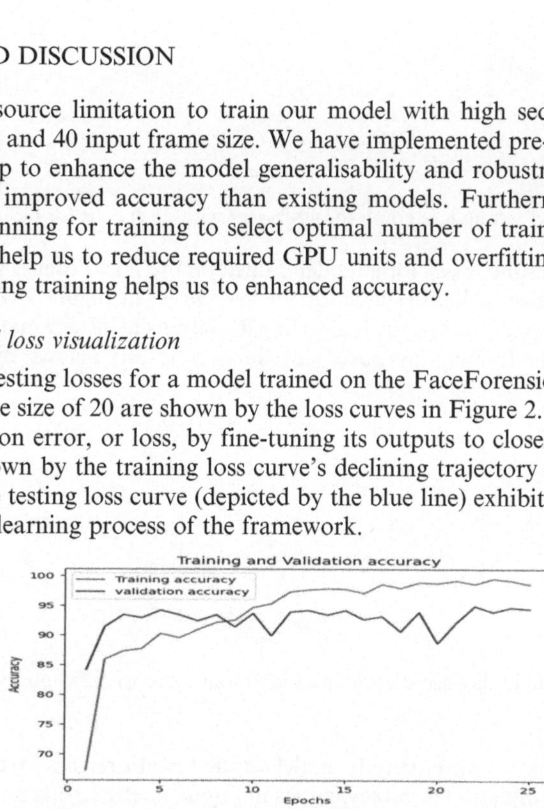

Figure 2. Accuracy graph for training and testing for the FaceForensics++ Dataset: 20 Frames with 25 Epochs.

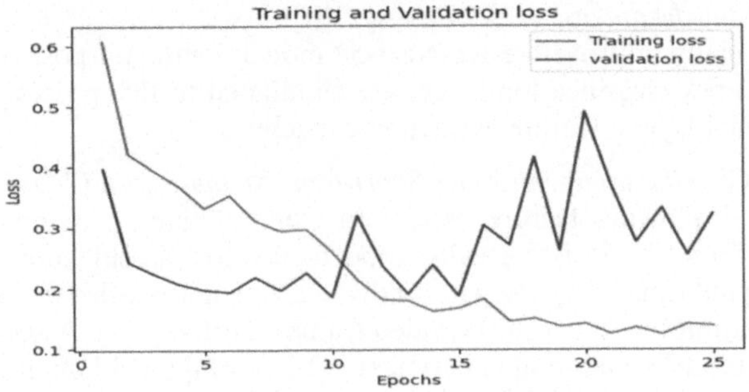

Figure 1. Training and Testing Loss Graph for 20 Frame Size with 25 Epochs in the FaceForensics++ Dataset.

As shown above accuracy graph it depicts how well model is predicting during training and testing. The training accuracy curve, shown by the green line, increases gradually over 25 epochs to show how well the model can identify complex patterns in the training dataset. Remarkably, the training accuracy curve and the validation accuracy curve (blue line) follow the same trajectory, with the validation accuracy curve ending up with a final accuracy of [94.47236%]. The remarkable coincidence of these curves indicates that the ensemble model manages to avoid overfitting on the training data and successfully applies learned patterns to new data examples.

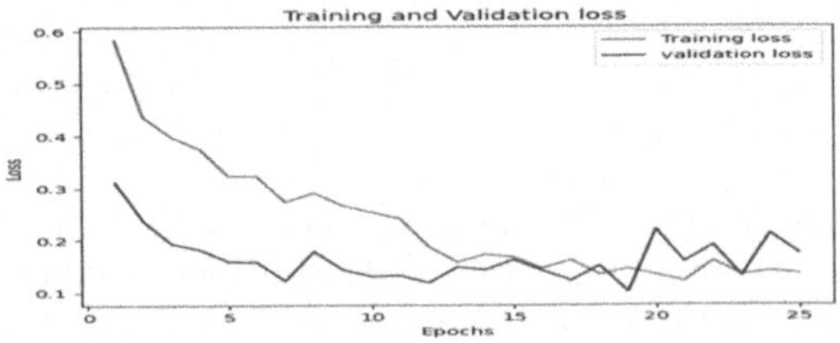

Figure 3. Training and testing loss graph for 40 frames with 25 epochs for faceForensics++ dataset.

The learning and testing losses for a model trained on the FaceForensics++ dataset over 25 epochs with a frame size of 40 are shown by the loss curves in Figure 3. The model strives to reduce the prediction error, or loss, by fine-tuning its outputs to closely match the ground truth labels, as shown by the training loss curve's declining trajectory (shown by the green line).

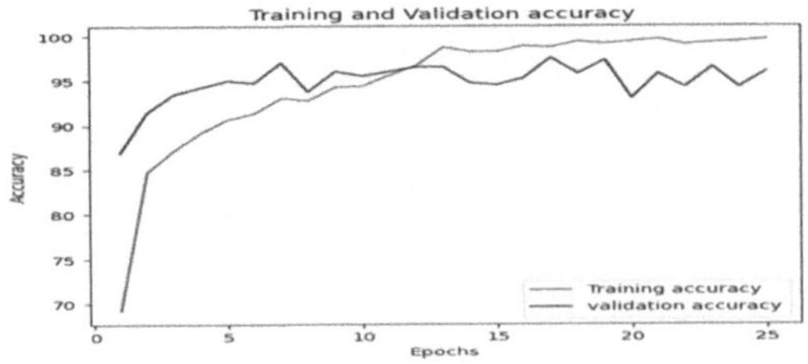

Figure 4. Training and testing accuracy graph for 40 frame size with 25 epochs for faceForensics++ dataset

The learning process of the ensemble model on the FaceForensics++ dataset is illustrated by the learning and validation accuracy curves in Figure 4, employing a frame size of 40 and trained over 25 epochs, while mitigating overfitting. Across epochs, the training accuracy curve (solid line) steadily ascends, showcasing the model's adeptness in identifying underlying patterns within the training set. Notably, the validation accuracy curve (depicted by the blue line) closely mirrors the trajectory of the training accuracy curve, culminating in a final accuracy of [95.96977%]. The close alignment between these curves suggests that the ensemble model effectively circumvents overfitting on the training data while adeptly applying acquired patterns to new data instances.

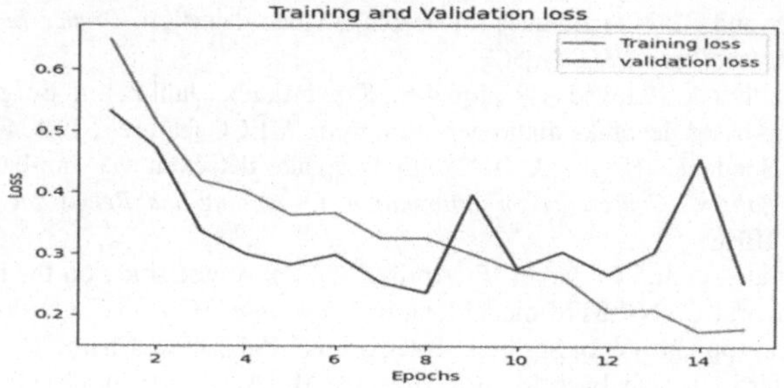

Figure 5 Training and testing loss graph for 20 frame size with 15 epochs for combined datasets.

The training and verification loss curves in Figure 5 show how the model learns on the combined dataset, employing a frame size of 20 and trained over 15 epochs. The training loss curve, denoted by the solid line, exhibits a typical decreasing trend across epochs, indicating the capacity of the model to absorb information from the learning set and gradually lower its prediction error (loss). Similarly, the validation accuracy curve (depicted by the dashed line) exhibits a comparable rising trend, culminating in a final accuracy of [90.17432%]. Notably, the model demonstrates resilience against overfitting on the training data throughout the 15 epochs and showcases robust generalization to previously unseen data in the combined dataset, as evidenced by the close alignment between the curves. A decline in the training loss (solid line) signifies learning within the model. Using frame sizes of 40 and trained over 15 epochs, the model's performance on the combined dataset by comparing the precision of the training and validation curves. A similar rising trajectory is followed by the validation accuracy curve (shown by the dashed line), which ends with a final accuracy of [90.80824%]. Overall alignment points to the possibility that the model can generalize to new data even in the face of the reported accuracy gap between training and validation. This could be improved even further with longer training cycles or hyperparameter tweaks.

5 CONCLUSION

All things considered, our deepfake detection system attained competitive accuracy despite resource constraints. Perhaps because of improved preprocessing with face detection and Gaussian noise augmentation, the model outperformed previous approaches [1] on both FaceForensics++ and a combined dataset. This highlights how important it is to carefully prepare data for deep learning models. It was shown that hyperparameter adjustment was also required to attain optimal performance by figuring out the optimal number of training epochs. To further improve deepfake detection techniques, future research can examine the effects of longer video sequences and the system's performance on a wider range of datasets.

REFERENCES

[1] Taviti, R., Taviti S., Reddy P. A., Sankar N. R., Veneela T., and Goud P. B. (2023). Detecting deepfakes with ResNext and LSTM: An enhanced feature extraction and classification framework, (doi: 10.1109/IConSCEPT57958.2023.10170580), *International Conference on Signal Processing, Computation, Electronics, Power and Telecommunication (IConSCEPT)*, Karaikal, India, pp. 1–5.

[2] Abukari, A. M., Madavarapu, J. B., and Bankas, E. K. (2023). A lightweight algorithm for identifying fake multimedia content on social media was developed, *Earthline Journal of Mathematical Sciences*, 14 (1), 119–132. https://doi.org/10.34198/ejms.14124.119132 (Abukari *et al.*, 2023)

[3] Kasande M., Deshmukh R., Chaudhari D., Sharma P., Patel T. and Pant S. (2022 Jun 1). Comparative analysis of CNN and RNN techniques for deepfake video detection. *Grenze International Journal of Engineering & Technology (GIJET)*.;8(2).

[4] Hamza A., Almadhor A.S., Javed A.R., Iqbal F., Kryvinska N., Jalil Z. and Borghol R. (2022 Dec 21). Machine learning-based deepfake audio detection using MFCC features.; 10:134018-28 *IEEE Access*.

[5] Lalitha S. and Sooda K. (Dec. 13, 2022). [6] Deepfake detection via gan-based key video frame extraction. *International Conference on Automation, Computing and Renewable Systems (ICACRS)*, pp. 859–863. In IEEE.

[6] Ghose D.K., Sahoo A., and Mahakur V. (April 22, 2022). A case study on the prediction of monthly runoff using a hybrid CNN-LSTM model. On, in *International Conference on Advances in Computing and Data Sciences* (pp. 381–392). Springser International Publishing, Cham.

[7] Elhassan A., Jafar M.T., Ababneh M. and Al-Zoube M. (Apr. 7, 2020). Deepfake video analysis and forensics. (pp. 053–058) in the *11th International Conference on Information and Communication Systems (ICICS)*,. The IEEE.

[8] Khalid H. and Woo S.S. Oc-fakedect: Deepfake classification with a one-class variational autoencoder, Pages 656–657 in the IEEE/CVF conference proceedings on computer vision and pattern recognition workshops 2020.

[9] Le N.T., Nguyen T.N., Wang J.W., Le D.H. and Wang C.C. (2020 Jan 6). Tensor of wavelet subbands is used to enhance fingerprints for classification.; 8:6602–15; *IEEE Access*.

[10] Yadav D. and Salmani S. (May 15, 2019). Deepfake: A survey on generative adversarial network-based facial forgery techniques. (pp. 852–857): *International Conference on Intelligent Computing and Control Systems (ICCS)*. IEEE.

[11] Elamsy T., Mahmoud A., Abdullah R. and Kharbat F.F. (November 3, 2019). Image feature detectors to identify deepfake videos. On, the *16th International Conference on Computer Systems and Applications (AICCSA)* by IEEE/ACS was held (pp. 1–4). The IEEE.

[12] Li Y., Lyu S. and Chang M.C. (December 11, 2018). In ictu oculi: exposing artificial intelligence-generated bogus films through the detection of eye blinking. On, the *IEEE International Workshop on Information Forensics and Security (WIFS)* took place (pp. 1–7). In IEEE.

[13] In Güera D. and Delp E.J. (November 27, 2018). Recurrent neural networks are used for deepfake video detection. The IEEE *15th International Conference on Advanced Video and Signal Based Surveillance (AVSS)* was held on, and covers pages 1–6. The IEEE.

Data Science & Exploration in Artificial Intelligence – Gururaj et al. (Eds)
© 2025 the Author(s), ISBN: 978-1-032-96390-7

DataHorbor WFH: Decoding remote work challenges with AWS analytics

Kavita Mainalli, S.Y. Malathi & Shridhar Allagi
K.L.E.Institute of Technology, Hubbali, Karnataka, India

ABSTRACT: The IT environment and corporate operations have undergone radical change because of the revolutionary technology known as cloud computing. Cloud computing environment offers distributed computing and virtualization capabilities, its significance has made it more essential for those who are technical and non-technical alike. Many corporate and government sectors underwent a substantial shift in 2019. By virtue of the COVID-19 pandemic, workers shifted to choose WFH instead of work from office. Existing work focused on analyzing WFH data using machine learning algorithms with publically available corpus as input. Proposed work performs analysis on self-generated data with unique key performance indicators; those are totally different from the existing works. Cloud computing platform is widely used for data analysis compared to traditional methods. Using AWS services, this research can be utilized as the aid for improved data visualization and analysis.

1 INTRODUCTION

Cloud computing become current cutting edge technology that changed the routine of businesses and IT market. Because of these changes, cloud computing become more prominent to technical and nontechnical person with its distributed computing and virtualization [1]. Assorted internet service providers are available in India like, Microsoft azure, Amazon web services, Google cloud platform (GCP), IBM cloud, Oracle, Alibaba and Wipro cloud. Cloud computing offers many services to the end users namely application, storage and platform on demand basis. So users can access and manipulate hardware or

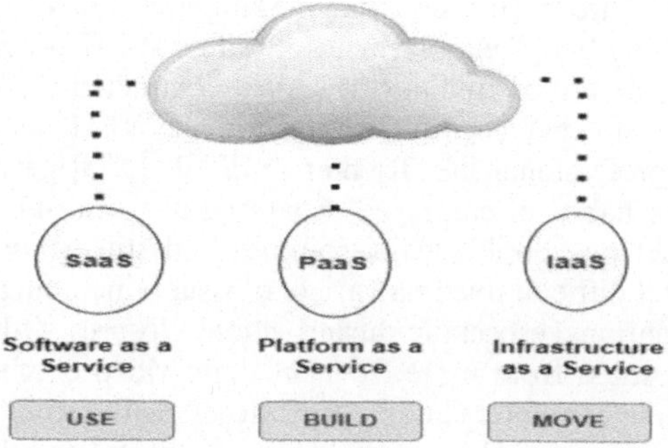

Figure 1. Cloud computing service models.

DOI: 10.1201/9781003589273-71

software according to their requirements over the internet. Cloud computing has three models they are SaaS, PaaS and IaaS as shown in the Figure 1.

Due to Covid 19, need of cloud computing elevated as employees preferred working from home [2]. From 2019 to 2023 there is tremendous change in mode of work. Employers decided to choose WFH facility to save their employees from infection. Then onwards WFH become norm for many IT workers. Times of India also gave statics saying 82% employees prefer work from home after pandemic as shown in the news cutting represented by Figure 2.

NEWS / INDIA NEWS / Remote Working New Normal: 82% Employees Pre...

TRENDING | Noida Mall Tragedy | Tamil Nadu Bomb Threat | Bharat Jodo Nyay Yatra | Lok Sat

THIS STORY IS FROM JANUARY 29, 2022

Remote working new normal; 82% employees prefer working from home

,PTI / Jan 29, 2022, 16:30 IST

SHARE | AA | FOLLOW US

New For You

Vote-for-bribe: SC says no immunity for MPs & MLAs, overrules 1998 verdict

After two years remote working has become a new normal and as things settled down new habits have formed, according to a study. The tech workforce has found new lov ... Read More

Figure 2. Time of India's report.

Our work aims to perform analysis of WFH employee's data set of 10,000 records with 26 features, using AWS cloud services and also to draw insights using AWS Quicksight.

The reminder of the paper is organized in the following manner sec 2 gives literature review to know the existing work carried out on the study. Sec 3 gives Methodology, which discuss the methods used in analysis sec 4 gives outcomes from the study.

2 LITERATURE REVIEW

Cloud computing serves many organizational needs over the internet, in terms of storage, software's and platforms to be accessed over the internet on demand basis. In 2019 due to Corona virus out break many corporate and government sectors turned their usual routine of operational from organization to home their cloud computing is used tremendously. Many researchers worked in prior on WFH data with different take away. Carroll *et al.*, 2020 [3] illustrated on the NPT- Normalization Process Theory and their basic constituents, that can be utilized to know the aspects of implementation, integration, and implanting new mechanism and routines into businesses. They highlighted how COVID-19 forced organizations into a sudden, "big bang" endorsement of online WFH routines.

Matli *et al.*, 2020 [4] described findings of survey conducted on WFH and office based work with measure of showing negative impression of WFH, as employees requires to manage personal and professional life. Richter *et al.*, 2020 [5] gave instances of how the lockdown affected work habits of employee. Adapting to work online, emphasizing how it had a significant impact on people's daily routines and still being productive. However, many employees found it difficult due to a variety of issues, including time management and juggling childcare obligations, especially during school closures. Other important concerns included the increased stress from more frequent daily video conferencing, as well as the necessity of sharing home computers and internet access with family members.

The COVID-19 epidemic has led to a notable rise in work flexibility; yet, it has also had an adverse impact on daily work patterns, as highlighted by Scheiber *et al.*, 2020 [6].

Cho *et al.*, 2022[7] demonstrated about Globally, Covid-19 had a huge impact on careers and the labor force. It also had an impact on many people's work habits and productivity levels. Prior to the Covid-19 outbreak, research on working from home revealed comparable challenges.

According to Park *et al.* [8], working from home has various disadvantages and can cause problems, the shift to remote employment offered special complications, especially for those used to office settings and with family responsibilities. Before the COVID-19 pandemic, the benefits and drawbacks of remote work had been discussed extensively in academic circles. But with the epidemic, something fundamental changed—that was first an optional arrangement quickly turned into one that was required. Even with COVID-19 limitations, some companies decided to continue using remote work arrangements. This highlights a significant divergence from pre-pandemic considerations.

3 METHODOLOGY

The methodology used for analysis using AWS environment involves data collection and preprocessing phase, data uploading and extraction , querying and visualization phases. The data set of employees who work for an organization is considered for analysis. Figure 3 gives the architecture of the work.

3.1 *Data collection and pre-processing*

Data set is created using faker module of python to generate 10, 000 records. Resultant CSV content was not uniformly distributed demanding for preprocessing. Cleansed the data by removing irrelevant features, converted the values to their respective domain. Features of dataset are listed in Table 1.

3.2 *Data upload and extraction*

In this phase, preprocessed data are uploaded to the S3 bucket of AWS.S3 stands for Simple Storage Services of Amazon. It is designed to scale up to various types and amount of data, because of this feature user can access the data effortlessly anytime from anywhere across the globe. It is also more secured, durable and versatile in nature their by provides the base for all AWS services. After uploading CSV file to S3, next step is to create database in AWS Glue and create crawler. The job of crawlers is to extract schema from CSV file and generate data catalogue in the Glue. Once crawler runs successfully, a table with name of folder of S3 will be created in the database of the Glue. Now data type of feature and names are cross verified with respect to the requirement. After successful completion of comparison of Glue schema features with database schema followed by querying Redshift database.

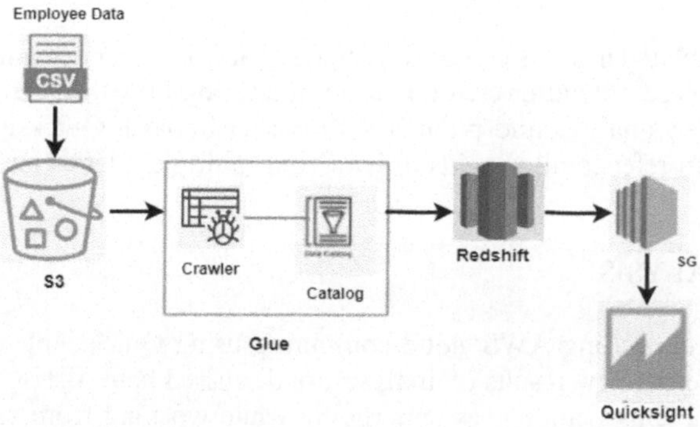

Figure 3. Architecture of WFH analysis.

Table 1. Features of dataset.

Sl No	Attributes	Sl.No	Attributes
1	employeeID	14	workLocation
2	firstName	15	challengeType
3	lastName	16	expenseType
4	Email	17	expenseAmount
5	phoneNumber	18	projectID
6	Department	19	tasksCompleted
7	Role	20	milestonesAchieved
8	Date	21	productivityPercentage
9	loginTime	22	electricityConsumption
10	logoutTime	23	internetUsage
11	totalWorkingHours	24	otherUtilitiesUsage
12	stepsTaken	25	physicalExerciseMinutes
13	mentalWellnessScore	26	sleepHours

3.3 *Querying*

Querying data in AWS can be completed by any of the data warehouses provided by the Amazon AWS like Athena, redshift, or RDS .For this study AWS Redshift is used. After creation of redshift cluster, now access the query editor 1. Using query editor 1 will create external schema and access the Glue's data catalogue. Here as per requirements of business queries can be framed and insights can be drawn. Syntax to create external schema is shown below.

```
CREATE EXTERNAL SCHEMA SCHEMA_NAME FROM DATA
CATALOG
  DATA BASE GLUE_DB_NAME
  iam_role 'ARN of role'
  region 'ap-south-1'
```

The research work framed simple queries for example to retrieve difficulties encountered by employees while WFH can be retrieved. Example query can be written as usual like SQL queries. SELECT employeeid, challengeType, COUNT(*) AS no_occurance FROM dev. wfh_schema.emp WHERE worklocation = 'Home' GROUP BY employeeid, challengetype;

3.4 *Visualization*

In AWS, business related insights can be drawn using one more service named as Quicksight. Quicksight can be used only after creating account and log-in with IAM user having admin credentials. Add new analysis and publish it to the dashboard. Quicksight dashboard is a place where one can refer to all insights drawn from multiple queries under a reference.

4 RESULT ANALYSIS

Analysis of WFH data using AWS cloud computing using Quicksight service made job of data scientist very easy. Few results of analysis are discussed here. If HR team would like to check what are challenges employees undergoing while working from can be visualized as shown in Figure 4.

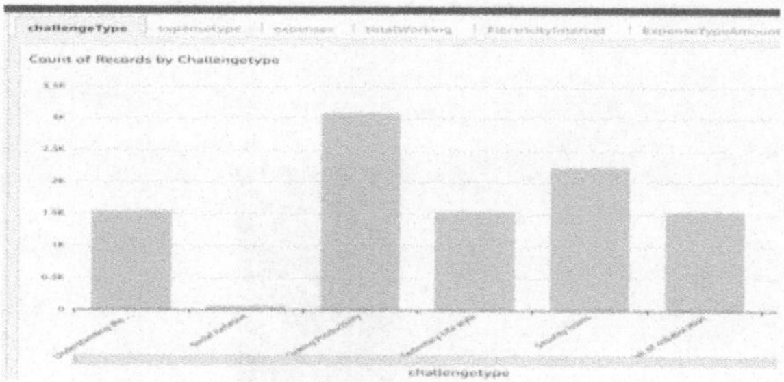

Figure 4. Results of challenges faced by employee WFH.

Insights related to total expenses types the employee will spend per month can be compared between office and home mode of working. These results are recorded in Figure 5. Maximum number of employees saves their expense on supplies type.

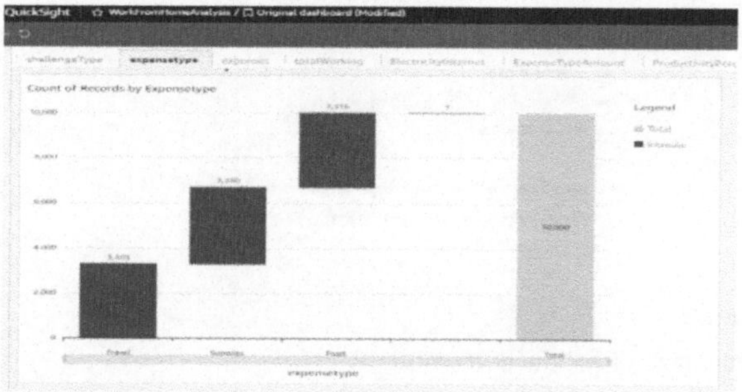

Figure 5. Result of expenses saved by employees on travel WFH.

To obtain the insights interrelated to the usage of internet and electricity consumption of employee when works at office and at home are shown in the Figure 6.

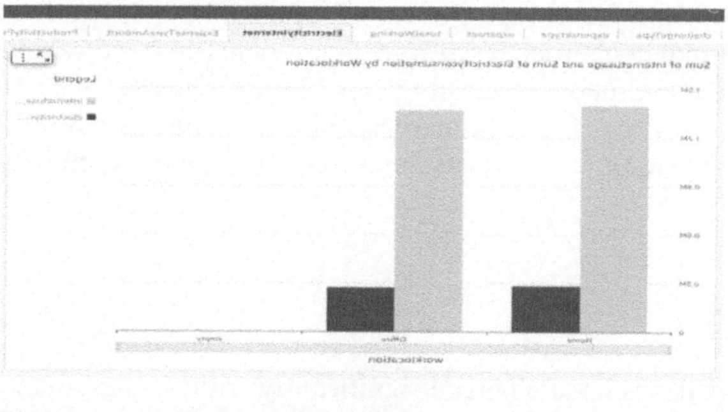

Figure 6. Result of electricity usage and internet usage per month.

Results correlated to the expenses type and work location can be observed in the Figure 7. Productivity of employees while they are WFH and office can be seen in the Figure 8.

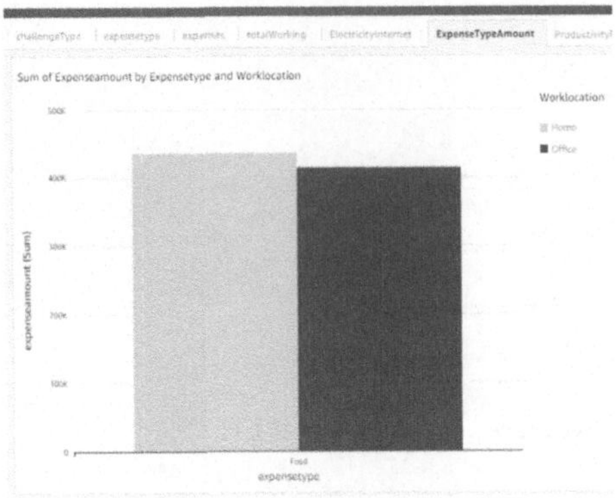

Figure 7. Result of expense type and work location.

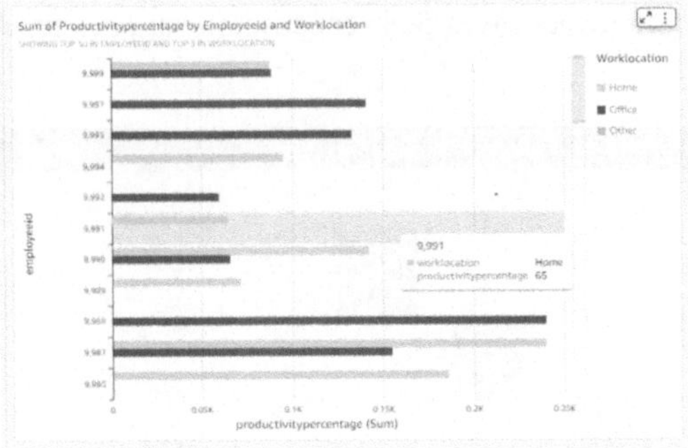

Figure 8. Show percentage of productivity while WFH.

The cloud computing offers scalable infrastructure and services, it serves as the base for machine learning by providing power, storage, and precise devices for data analysis and management. For training and inference, machine learning models need a lot of processing power, which can be supplied by cloud computing systems easily. Cloud-based machine learning model construction, training, and deployment are made easier by services like Microsoft Azure Machine Learning, Google Cloud AI Platform, and Amazon SageMaker. Without having to make significant upfront investments in hardware infrastructure, businesses may effectively use machine learning capabilities. The research concentrates on cloud computing services as a tool for data analysis without using machine learning models and provides quick insights for given datasets, later that will be used by managerial staffs without technical aspects.

5 CONCLUSION

Organizational operations have changed significantly with the deployment of WFH model, requiring a move toward more flexible and cloud-based infrastructures. The study looked at how using AWS cloud services affected remote workers' productivity. Several insights were obtained by outcomes of an extensive examination of key performance indicators (KPIs), including productivity, collaboration, and the effectiveness of IT assistance. To begin with, the implementation of AWS cloud solutions has been crucial in raising worker productivity levels at WFH. Incorporation of cloud-based collaboration solutions from AWS has significantly enhanced distant team cooperation. A seamless WFH practice has been made possible by proactive problem identification, quick resolution processes, and automated system inspections that reduce downtime.

6 FUTURE SCOPE

Researchers also look into the potential effects of work-from-home policies on employees' income, making a number of recommendations for further research. Performing case studies in various industries facilitates the customization of approaches and resolutions to particular corporate environments. By employing this technique, scholars can examine how work-from-home policies affect workers' pay and create endorsements those are explicitly tailored to the requirements of the business.

REFERENCES

[1] Suman Lata, Dheerendra Singh (2022). *Intrusion Detection System in Cloud Environment: Literature Survey & Future Research Direction*, elsivier 2022. https://doi.org/10.1016/j.jjimei.2022.100134

[2] Ziyad R. Alashhab, Mohammed Anbar, Manmeet Mahinderjit Singh, Yu-Beng Leau, Zaher Ali Al-Sai, Sami Abu Alhayja' (2021), Impact of coronavirus pandemic crisis on technologies and cloud computing applications, *Journal of Electronic Science and Technology*, https://doi.org/10.1016/j.jnlest.2020.100059.

[3] Carroll N., Conboy K., (2020). Normalising the "new normal": changing tech-driven work practices under pandemic time pressure, *Int. J. Inf. Manag.* 55, 102186.

[4] Matli W., (2020). The changing work landscape as a result of the Covid-19 pandemic: insights from remote workers life situations in South Africa, *Int. J. Sociol. Soc. Pol.* 40 (9/10) 1237e1256.

[5] Richter A., (2020). Locked-down digital work, *Int. J. Inf. Manag.* 55, 102157.

[6] Scheiber N., *Jobless Claims by Uber and Lyft Drivers Revive Fight over Labor Status*. NY Times, 17th April, 2020. Available at: https://www.nytimes.com/ 2020/04/17/business/economy/coronavirus-uber-lyft-unemployment.html, 15/07/21.

[7] Cho E., (2020). Examining boundaries to understand the impact of COVID-19 on vocational behaviours, *J. Vocat. Behav.* 119, 103437.

[8] Park Y., Fritz C., Jex S.M., Relationships between work-home segmentation and psychological detachment from work: The role of communication technology use at home, *J. Occup. Health Psychol.* 16 (4) (2011) 457.

Data Science & Exploration in Artificial Intelligence – Gururaj et al. (Eds)

Efficient automated evaluation of answer scripts using LLMS, NLP, and deep learning

Shreya Ashoka, K. Sai Deep, G. Goutham, Syed Mohammad Muzammil Ganihar, P.K. Udayaprasad & G.C. Lakshmikantha
Department of Artificial Intelligence & Machine Learning Global Academy of Technology Bangalore, Karnataka, India

P. Dayananda
Department of Information Technology, Manipal Institute of Technology Bengaluru, Manipal Academy of Higher Education, Manipal, Karnataka, India

ABSTRACT: NLP and DL researchers are increasingly concentrating on automated answer script evaluation as an attempt to overcome the limitations and inconsistencies of traditional grading. semantic parallelism and contextual preserving are problems for conventional techniques like WordNet, OCR, and Word2Vec. For enhanced accuracy and efficiency, this research presents a novel hybrid architecture that integrates advanced DL and NLP techniques. The system uses a Cosine similarity network for accurate similarity scoring, an OCR model for handwritten text transformation, Universal Sentence Encoder for embeddings, and LLM for contextual analysis. A Deep Columnar CNN additionally handles complex answer formats and symbols. Considering 93.8% accuracy, 94.1% precision, 92.7% recall, and a 93.4% F1-score, the integrated technique significantly improves conventional techniques in terms of criteria for evaluation. Because of its cloud-based architecture, that guarantees scalability and real-time processing, automated grading in educational environments can be effectively implemented.

Keywords: Artificial Intelligence, Natural Language Processing, Machine Learning, Large Language Models

1 INTRODUCTION

The ability of computers to understand and process human language has led to substantial advancements in many domains, particularly in Natural Language Processing (NLP) and Deep Learning (DL). Automated response script evaluation is a promising application of these technologies with the objective at minimizing the biases and inefficiencies associated with traditional grade [1]. In traditional manual grading, errors committed by humans and irregularities in the process might affect the evaluation's impartiality and accuracy. Automated techniques can make grading easier, ensure uniformity, and free up teachers' time for other essential responsibilities such as interacting with students and offering instruction [2,3].

For automated answer script evaluation, despite the advancements in NLP and DL, numerous of challenges remain with the current techniques. Techniques such as optical character recognition (OCR) when combined with vectorization techniques such as Word2Vec or Latent Semantic Analysis (LSA). However, these approaches sometimes fall below of accurately capturing the complex details and context of students' remarks,

DOI: 10.1201/9781003589273-72

particularly when they are handwritten and detailed [4,5]. Due to the complexity and diversity of natural language, especially in instructional content, advanced algorithms are required that can handle a range of answer formats and maintain high accuracy in semantic similarity assessment.

Effective models with excellent text generation and context awareness include GPT- 3 (Generative Pre-trained Transformer 3) and BERT (Bidirectional Encoder Representations from Transformers). With its bidirectional training approach, BERT enhances the comprehension of complex phrases by gathering information from both perspectives. In addition to offering exceptional language comprehension, GPT-3 is well-known for its extensive pre-training and text generation capabilities. However, these models are highly complex, and they may not always be well suited to the specific needs of educational datasets. Their high computing needs for resources may limit their practical use in large-scale educational systems.

Even with major advances in DL and NLP, automatic answer script evaluation remains a challenging task. Existing methods, such traditional OCR combined with semantic analysis and advanced models like BERT and GPT-3, struggle to handle different response types and efficiently extract data. These limitations cause grading to be ineffective and erroneous, particularly when it comes to handwritten and detailed observations.

In order to overcome these constraints, we proposed a brand-new hybrid framework that combines a number of innovative DL and NLP techniques to improve automated answer script evaluation's precision and effectiveness. Among the methodology's features are:

- **Text extraction:** Is the process of turning handwritten scripts into editable text by using an extremely accurate OCR model.
- **Text embedding:** Using the Universal Sentence Encoder, create strong sentence embeddings.
- **Contextual Analysis:** Applying LLM for deep contextual understanding of the text.
- **Similarity Scoring:** Using a cosine similarity network, similarity scoring accurately quantifies semantic similarity.

1.1 *Motivation and contributions*

The purpose of this research is to develop an automated system that can match human grading accuracy while ensuring efficiency and adaptability.

The research contributions are:

- Integrating various DL and NLP techniques to make use of their advantages.
- Obtaining greater evaluation metrics than with conventional techniques.
- Implementing a cloud-based architecture for large-scale, real-time processing.

2 LITERATURE REVIEW

The automated assessment process implemented in the existing system gives a solution to the challenges faced by professors while grading handwritten answer booklets. Using DL and NLP approaches, the method uses a better GCP OCR model to extract handwriting from the image files. BERT applies for keyword extraction, and GPT-3 is utilized for answer synthesis. However, the system exhibits a strong ability to give grades that are on level with human evaluation. This method, designed to tackle the labor-intensive nature of manual assessment encountered by educators, promises enhanced efficiency and accuracy in analyzing handwritten response booklets by merging cutting-edge technologies and user-friendly interfaces [1].

Building on existing research in automatic answer script evaluation, this paper explores a deep learning approach specifically designed for handwritten responses. While traditional methods often rely on feature extraction for textual data, they struggle with handwritten scripts requiring OCR and neglecting answer clarity and structure. The CNN component leverages its image recognition capabilities to extract text via OCR, while the BiLSTM network, adept at analyzing sequential data like text focuses on understanding the semantic meaning and key points of the answer. This combined approach offers a comprehensive solution for handwritten scripts, encompassing both the visual content and the textual meaning. While promising accuracy, efficiency, and potentially reduced bias, limitations like dependence on large datasets and interpretability of deep learning models require further exploration [6].

This literature review explores automated grading and feedback tools for programming education. Analyzing 121 research papers, it finds these tools primarily focus on assessing the correctness of code, often in object-oriented languages. While most tools offer full automation for faster grading and re- submissions, evaluation methods primarily rely on student surveys or comparisons to human graders, making it challenging to compare tools objectively [7,8].

Table 1. A comparison of the literature reviews.

Ref.	Technique	Objective	Key Methods	Dataset Used	Performan Metrics	Drawbacks
[9]	Character and Word Recognition	Address issues in character and word separation using pre-trained models	Tesseract, GTTS, Dec ision Trees, Neural Net-works	Input images	94% accu-racy (input image), 92.7% over-all accuracy	Challenges with cursive writing, computational intensity
[10]	Character Recognition	Recognize char-acters and con-textualize them	CNN, Seq2Seq with attention mechanism, Encoder-Decoder (LSTM)	RIMES and IAM data-sets	Word Error Rate: 19.85, Character Error Rate: 6.8	Poor perfor-mance with non-English lexicon words, long training times
[11]	Handwritten Text Recog-nition	Recognize entire pages of text without segmen-tation	Image to S quence archi-tecture, Character-level vocabulary	IAM data-set, hand-written test responses	Error Rate: 7.6% (mod-el), 14.4% (best cloud API)	Bias towards certain tasks, handling larger text sequences
[12]	Handwritten Word Digiti-zation	Classify hand-written words	CNN, VGG 19, RESNET-18, RESNET-34, LSTM	Custom da-taset	VGG- 19: 20%, RE-SNET- 18: 22%, RE-SNET- 34: 27%	Limited train-ing samples, unreliable train-ing

3 PROBLEM STATEMENT AND OBJECTIVES

Answer scripts must be manually evaluated, that is a difficult and error-prone process that frequently results in biases and inconsistencies. Existing automated techniques, such OCR with semantic analysis alongside more advanced algorithms like BERT and GPT-3, have difficulty understanding a variety of answer formats and collecting context, especially in handwritten and descriptive responses.

Our research aims to develop a novel hybrid framework integrating advanced NLP and DL techniques to overcome these challenges. Objectives include.

- Generating editable text using handwritten scripts by using extremely precise OCR.
- SEU and BERT are being used for contextual analysis and strong text embeddings.
- A Siamese BiLSTM network is utilized to provide accurate semantic similarity scoring

4 PROPOSED APPROACH

In this research, we provide a novel hybrid framework that combines advanced NLP and DL algorithms to address problems in automated answer script evaluation. Our method starts with a very precise OCR technique to create editable text from hand-written answer scripts. The Universal Sentence Encoder is then used to process this text and produce reliable embeddings. A refined LLM model is then used to examine the embeddings for improved contextual comprehension. We use a cosine similarity network trained on pairs of student answers and answer keys to provide accurate semantic similarity scoring.

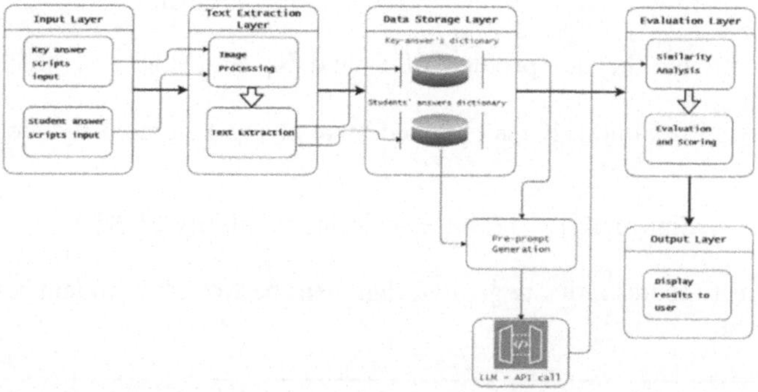

Figure 1. The proposed architecture.

From the Figure 1, we gather student response scripts and key as part of our process. To extract text from scanned or digital scripts, we utilize OCR tools like Tesseract and image processing libraries like Pillow. Key-value pairs representing questions and answers are stored as extracted text in dictionaries. After that, the LLM model API receives this data and pre-programmed prompts for analysis. By comparing student responses to significant answers and accounting for factors like grammatical quality and semantic similarity.

The research we propose uses an extensive methodology to analyze the quality and accuracy of responses in the context of automated assessment of student answers. First, our method uses advanced text extraction techniques and tools such as the pdfplumber library to extract text from PDF files including both student comments and important answers. This gives full accessibility to the textual content needed for subsequent comparison and analysis. Our approach uses an advanced text comparison methodology after text extraction, that can be enabled through cooperation with third-party APIs like the LlamaAPI.

To evaluate the degree of alignment between student responses and expected answers, the LlamaAPI most likely uses LLM and machine learning algorithms to do semantic analysis, similarity scoring, and other procedures. Higher marks are indicative of greater alignment. The degree of similarity discovered determines the allocation of marks.

4.1 *Integration of natural language processing*

Using the pdfplumber library, the suggested technique for automated answer script evaluation proceeds by extracting text from the answer key PDFs and the handwritten answer scripts. To standardize the format, the retrieved text is put through preprocessing procedures such tokenization, lowercase conversion, and punctuation removal. Semantic analysis is then made possible by creating embeddings for the preprocessed text using the USE. In order to evaluate semantic similarity, the embeddings of student responses are compared with those of key answers. Similarity scores are then calculated by applying a sigmoid function and cosine similarity. Higher scores indicate greater alignment between the student response and the desired answer. Marks are awarded based on these scores. Lastly, the evaluation results are generated, together with the marks that were assigned to each answer script as shown from Algorithm 1.

4.2 *Semantic comparison*

Let S_i represent the embedding vector of student response $_i$ and K_j represent the embedding vector of key answer $_j$. The cosine similarity between S_i and K_j is computed as:

$$\text{Cosine Similarity } (S_i, K_j) = \frac{S_i \cdot K_j}{||\mathbf{S_i}||\mathbf{K_j}||} \tag{1}$$

where $S_i \cdot K_j$ denotes the dot product of S_i and K_j, and $||S_i||$ and $||K_j||$ represent the Euclidean norm.

The similarity score $\text{Sim}(S_i, K_j)$ is obtained by applying a sigmoid function to the cosine similarity.

$$\text{Sim}(Si, Kj) = 1/(1 + e - \text{Cosine Similarity}(Si, Kj)) \tag{2}$$

Higher similarity scores indicate greater alignment between the student response and the key answer

---

Algorithm 1 An Proposed Algorithm

Require: Handwritten answer scripts in PDF format, Answer key in PDF format.
Ensure: Final evaluation report with marks assigned to each answer script.

(1) **ExtractTextFromPDF(answer script PDF, key answer PDF):**
  (a) Use pdfplumber library to extract textual content from answer script PDF.
  (b) Use pdfplumber library to extract textual content from key answer PDF.
  (c) Return extracted text from both PDFs.

(2) **PreprocessText(text):**
  (a) Tokenize the text.
  (b) Convert text to lowercase.
  (c) Remove punctuation.
  (d) Return preprocessed text.

(3) **GenerateEmbeddings(text):**
  (a) Generate embeddings for the preprocessed text USE.
  (b) Return the embeddings.

(4) **CompareSemantics(student embeddings, key embeddings):**
  (a) Compute cosine similarity between each pair of student embeddings and key embeddings.

(b) Apply sigmoid function to obtain similarity scores.
(c) Return similarity scores.

(5) **AllocateMarks(similarity scores):**
(a) Assign marks based on similarity scores Eq. 1 and 2:
- High similarity scores receive higher marks.
- Low similarity scores receive lower marks.

(b) Return marks allocated to each answer script.

3 IMPLEMENTATION

The implementation methodology stands to guarantee that raw textual data is accessible, the code first extracts textual material from the student responses and key answers PDF files using the pdfplumber library. To prepare the data for precise comparison, the retrieved text is next subjected to preprocessing steps such tokenization, part-of-speech tagging, and semantic analysis. Integration with the advanced LLM tool LlamaAPI makes it easier to compare the texts' semantic similarity. The LlamaAPI generates a similarity score that quantifies the degree of alignment between student responses and expected replies by examining elements like word choice and sentence structure. Eventually, the similarity scores determine the marks that are granted, giving teachers a summary of the evaluation findings to help them make well-informed grading decisions.

In the automated evaluation process, preprocessing of textual data involves tokenization T, part-of-speech tagging POS, and semantic analysis SA, represented as Eq. 3:

$$\text{Text}_{\text{processed}} = \text{SA}(\text{POS}(\text{T}(\text{Text}_{\text{raw}}))) \tag{3}$$

The evaluation facilitated by LlamaAPI computes a similarity score Simscore based on an intricate function f, quantifying the semantic alignment:

$$\text{Sim}_{\text{score}} = \text{f}\left(\text{student}_{\text{response}}, \text{key}_{\text{answer}}\right) \tag{4}$$

4 RESULTS AND DISCUSSION

The objective of our research was to effectively score marks and automate the examination of descriptive answer scripts. For automatic evaluation, we utilized a weight parameter-based aproach, giving different weights to parameters based on the categories of questions (M5, M10, M15 and M20) as from Figure 2. The cosine and synonym similarity were within these param ters. We established the parameters by analyzing a dataset that included 45 answer scripts in detail. We then used the parameters for automatic mark scoring

Table 2. Performance metrics evaluation.

Technique	Accuracy	Precision	Recall	F1-score
Proposed Method	93.8%	94.1%	92.7%	93.4%
Evaluation Using ML [13]	92.5%	93.2%	91.8%	92.5%
Evaluation Using DL [14]	93.2%	93.8%	92.5%	93.1%

- **Accuracy:** The accuracy value of 93.8% indicates the overall correctness of the automated evaluation system in assigning marks to descriptive answer scripts. The system accurately evaluates nearly 94% of the answer scripts in alignment with manual evaluation standards.
- **Precision:** With a precision of 94.1%, the system demonstrates a high level of precision in correctly identifying and assigning marks to relevant aspects of the answer scripts.
- **Recall:** The recall value of 92.7% signifies the system's ability to retrieve relevant information and accurately assign marks to the answer scripts.
- **F1-score:** The F1-score, which is the harmonic mean of precision and recall, provides a balanced measure of the system's performance.

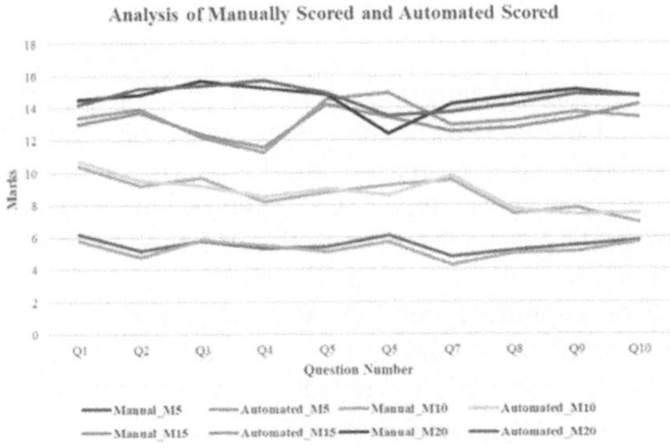

Figure 2. Analysis of automated scored marks and manual scored marks.

Precision estimates the percentage of actual positive outcomes among each of the system's expected good outcomes. To improve accuracy, we integrate deep learning models and sophisticated NLP in the proposed research. Our system can precisely identify and give marks to pertinent parts of the answer scripts by utilizing methods like syntactic analysis, named entity recognition, and semantic similarity scoring. This lowers the likelihood of false positives. As a result, our system receives a higher precision score than other approaches because it is fairly accurate in identifying and evaluating the key components of the answers scripts.

5 CONCLUSION

In order to ensure fair grading and minimize evaluation time, our study provides a novel method for automating the assessment of descriptive answer scripts. We achieve accu- rate and consistent marks allocation by using advanced NLP technologies along with a weight parameter-based approach. The outcomes indicate the way our approach per- forms, with personally issued marks and automatically generated ones closely aligned. In order to improve the accuracy and scalability of the system going ahead, we plan to investigate more NLP techniques and refine the weighting factors. Furthermore, in order to make automated evaluation available to a large viewers, we will be expanding our methodology to accommodate a wider range of question forms and languages.

REFERENCES

[1] P. M. S., Chavan S. M., Bathula R., Saikumar S., and Dayalan G., (Jan. 2023), Eval – automatic evaluation of answer scripts using deep learning and natural language processing, *Int J Intell Syst Appl Eng*, vol. 11, no. 1, pp. 316–323.

[2] Rahman, Md Siddiqui, Fazlul. (2018). *NLP-based Automatic Answer Script Evaluation.*

[3] Shreyas, J., Dharamendra Chouhan, A. R. Akshatha, P. K. Udayaprasad, and Dilip Kumar SM, (2020). Selection of optimal path for the communication of multimedia data in internet of things. *In 2020 6th International Conference on Advanced Computing and Communication Systems (ICACCS)*, pp. 477–481. IEEE.

[4] Marcus Messer, Neil C. C. Brown, Michael K·olling, and Miaojing Shi. (2024). Automated grading and feedback tools for programming education: A systematic review. *ACM Trans. Comput. Educ.* 24, 1, Article 10 (March 2024), 43 pages.

[5] Udayaprasad, P. K., Shreyas, J., Srinidhi, N. N., Kumar, S. D., Dayananda, P., Askar, S. S., and Abouhawwash, M. (2024). energy efficient optimized routing technique with distributed SDN-AI to large scale I-IoT networks. *IEEE Access.*

[6] Hooda, Monika, Chhavi Rana, Omdev Dahiya, Ali Rizwan, and Md Shamim Hossain. Artificial intelligence for assessment and feedback to enhance student success in higher education. *Mathematical Problems in Engineering* 2022 (2022): 1–19.

[7] Rybinski, Krzysztof, and Elzbieta Kopciuszewska (2021). Will artificial intelligence revolutionise the student evaluation of teaching? A big data study of 1.6 million student reviews. *Assessment & Evaluation in Higher Education* 46, no. 7: 1127–1139.

[8] Celik, Ismail, Muhterem Dindar, Hanni Muukkonen, and Sanna J·arvel·a. (2022). The promises and challenges of artificial intelligence for teachers: A systematic review of research. *TechTrends* 66, no. 4: 616–630.

[9] Ramesh, Dadi, and Suresh Kumar Sanampudi. (2022). An automated essay scoring systems: a systematic literature review. *Artificial Intelligence Review* 55, no. 3: 2495–2527.

[10] Igbokwe and Innocent Chiawa. (2023). Application of artificial intelligence (AI) in educational management. *International Journal of Scientific and Research Publications* 13, no. 3

[11] Shukla H. and Kakkar M., (2016), Keyword extraction from educational video transcripts using nlp techniques, in *2016 6th International Conference-Cloud System and Big Data Engineering (Confluence)*. IEEE, pp. 105–108.

[12] Bojja P., Velpuri N., Pandala G. K., Polavarapu S., and Kumari P., (2019), Handwritten text recognition using machine learning techniques in application of nlp, *International Journal of In- novative Technology and Exploring Engineering (IJITEE)* ISSN, pp. 2278–3075.

[13] Sueiras J., Ruiz V., Sanchez A., and Velez J. F., (2018), Offline continuous handwriting recognition using sequence to sequence neural networks, *Neurocomputing*, vol. 289, pp. 119–128.

[14] Boorugu R. and Ramesh G., (2020), A survey on nlp based text summarization for summarizing product reviews, in *2020 Second International Conference on Inventive Research in Computing Applications (ICIRCA)*. IEEE, pp. 352–356.

Data Science & Exploration in Artificial Intelligence – Gururaj et al. (Eds)
© 2025 the Author(s), ISBN: 978-1-032-96390-7

Density-based traffic control system with priority for emergency vehicles using IoT and ML

Husna Tabassum, Naveed Ahamed, Mohammed Murshid Alam, Mohammed Yaseen Sharieff, M. Mohammed Ameer & Ayesha Anjum
HKBK College of Engineering, Bangalore, India

ABSTRACT: Traffic congestion is a pressing issue in urban environments, leading to various challenges, including increased commuter stress and environmental pollution. Conventional traffic light systems often fail to address this problem adequately, relying on fixed time intervals for signal allocation regardless of traffic density. To mitigate these issues, this project proposes a density-based traffic control system using machine learning and IoT technologies, implemented with Arduino Uno and sound sensors. By leveraging YOLO v5 object detection and sound sensor capabilities, the system dynamically adjusts signal timings based on real-time traffic density and detects accidents and emergency vehicles. Higher traffic density on one side of a junction sometimes requires longer green light duration than the standard allotted time. Hence, we propose a mechanism for determining green and red light durations based on traffic density, utilizing Proximity Infrared (PIR) sensors. Once density is calculated, the microcontroller (Arduino) allocates green light time accordingly. Sensors on either side of the road detect vehicle presence, relaying information to the microcontroller for decisions on flank duration and signal light changes. Subsequent sections elaborate on the framework's procedure. Additionally, a virtual hump near zebra crossings enhances pedestrian safety during red signals. By integrating these features, the system aims to optimize traffic flow, reduce congestion, and improve safety, contributing to the development of more intelligent and more efficient urban transportation systems.

Keywords: Traffic congestion, Traffic light system, Proximity Infrared Sensor, Arduino Microcontroller

1 INTRODUCTION

Global urban traffic congestion is a recurring problem with substantial socioeconomic and environmental ramifications. The spike in automobile ownership and the speed at which cities are growing has resulted in congested roads, longer travel times, higher fuel usage, and higher air pollution levels. In addition to lowering city people's quality of life, these effects significantly increase the likelihood of traffic accidents and hinder economic output. Innovative strategies utilizing state-of-the-art technologies are required to address these problems and optimize traffic flow while boosting overall road safety. The development of intelligent traffic management solutions has been made practical by the emergence of machine learning (ML), the Internet of Things (IoT), and sensor-based systems. In light of this, our project, "Density-Based Traffic Control System Using Machine Learning and IoT," takes a multifaceted approach to reducing the problems caused by traffic congestion. Our system [1] is an IoT-based solution that modernizes outdated traffic control practices through the implementation of contemporary technologies like Acoustic Sensors, Arduino Uno (Microcontroller), and the You Only Look Once version 5 (YOLO.v5) object detection

 DOI: 10.1201/9781003589273-73

algorithm. This is primarily focused on re-engineering the traffic signal from a static to a dynamic one using YOLO v5 to deploy real-time crowd control at any red light intersection in the city. Conventional traffic signals follow a predetermined timing routine and thus do not adjust to real-time traffic behavior. Instead, our technology is based on machine learning (ML) algorithms for processing of real-time video feeds from strategically deployed cameras, giving you an effective way to manipulate the time of stationary signals based on current traffic volume. This method reduces fuel consumption and the time taken to travel by ensuring the flow of traffic is smooth and reduces congestion near the junctions. Depending on the severity of the impact, these incidents can be classified as severe or moderate accidents. Upon detection, the system initiates a sequence of events, including alerting nearby police stations, hospitals, or other authorities and providing appropriate information.

Figure 1. Communication setup between emergency vehicle and traffic light system.

The communication setup between an emergency vehicle (EmV) and a traffic light system, as shown in Figure 1, ensures efficient and safe passage for emergency vehicles through traffic intersections. The communication between emergency vehicles and traffic light systems is designed to prioritize the timely and secure passage of emergency vehicles while minimizing disruptions to regular traffic flow. This allows for timely and well-coordinated emergency response activities. This proactive strategy improves emergency services' effectiveness and allows for prompt response, which may lessen the severity of injuries and save lives. The proposed system uses sound sensors to add ambulance and accident detection functionality. Understanding how crucial it is to prioritize emergency vehicles in traffic, our technology uses sound sensors to recognize the unique sounds that ambulances make. When an emergency vehicle is detected, the traffic signal is dynamically changed to allow it to pass through freely, hastening its arrival at the destination and possibly saving lives in dire circumstances. Additionally, our project offers a novel way to improve pedestrian convenience and safety during traffic signal changes. Activated during red signal intervals, our system deploys virtual speed humps at zebra crossings to encourage drivers to slow down and yield to pedestrians crossing the road. This preventive action promotes a pedestrian-friendly urban environment that supports sustainable transportation while lowering the likelihood of pedestrian accidents. The hardware implementation of our idea is based on the Arduino Uno, a versatile microcontroller platform. The powerful capabilities and compact form factor of Arduino Uno make it the perfect platform for incorporating machine learning techniques into our system. It can be interfaced with sensors to develop a responsive and effective traffic control system by utilizing the Arduino Uno's power. Our project can instantly deploy machine learning models because of the Arduino Uno's connection with sensors, which makes intelligent decision-making possible. This strategy optimizes computational resources and dramatically improves our traffic control system's responsiveness.

2 LITERATURE REVIEW

Traffic lights have been around for about two decades and are employed as signaling devices. They regulate traffic flow at street intersections, railroad crossings, pedestrian crossings, and other sites [11]. The green light permits traffic to continue on the prescribed path, the yellow light advises drivers to expect a lengthy pause, and the red signal prohibits all traffic from progressing. Currently, many nations are suffering from the negative repercussions of traffic congestion, which, in turn, severely impacts the transportation system in metropolitan areas and causes significant issues. Regardless of whether traffic authorities and flagmen are replaced with automated traffic systems, repairing the congested route is insufficient to reduce traffic congestion and time delays. A microcontroller was used to count the number of vehicles in the system was proposed. Based on the number of different cars, the microcontroller makes decisions and updates the traffic light delays accordingly. The ARM architecture is utilized in the processor. The primary benefit of the system is its ability to determine the number of vehicles on each lane, thanks to its sensors. The apparent downsides were the costly sensors that require much maintenance and the system's scalability. [2] The system was designed to process the image taken at the traffic light, convert it to greyscale, and then calculate a threshold value representing the number of vehicles on that specific road. The outcome of this computation allowed us to determine which side has the highest density.

A Raspberry Pi microcontroller manages the signals. One of the most apparent benefits is using OpenCV, followed by quick and efficient RAM utilization. Furthermore, the Raspberry Pi is used, making the hardware level practical. One of the observable drawbacks of OpenCV is its lack of flexibility compared to MATLAB. The paper "Density Based Traffic Control System with Priority for which Time Manipulation was Used for Controlling Traffic Light" [3] system manages traffic at several crossings. As a result, it's becoming increasingly important to develop practical, flexible, and affordable traffic management algorithms that support and ensure quick, seamless traffic flow and use cutting-edge, adaptable technologies. It was suggested that wireless sensor networks (WSNs) technology be used with a dynamic vehicle recognition method and a signal control algorithm to regulate the state of the signal light at a road intersection. The optimization or control of traffic lights is a challenging issue. The issue gets considerably more complicated when there are several intersections since traffic flows toward numerous other traffic lights depending on the status of one signal. The intricacy lies in the regular variations in traffic flow, contingent on the time of day. This paper presents an intelligent WSN-based traffic light control system. Because of the system's low cost and potential for wide-scale implementation, traffic monitoring and control technologies could undergo a revolution. Ambulances and other emergency vehicles are given priority under this arrangement.[4] A system including a PIC 16F84A microcontroller, power section, crystal oscillator, and light-emitting diode (LED) was designed by Ganiyu et al. (2014). The PIC was implemented using an integrated circuit (IC) programmer with a mikrobasic program written in Basic language to control traffic efficiently. A pressure switch is employed as the sensing unit; it senses the weight of the vehicle stepping on it and signals the microcontroller with an appropriate signal.

To assess the amplitudes and frequency ranges in which vibrations occur in microelectromechanical system (MEMS) accelerometers by accepted measurement techniques, the authors work on piezoelectric acceleration sensors [5]. Magnetic sensors provide a significant amount of information concerning traffic conditions.[6] The authors of this study suggested a loop sensor detector system for cars of all sizes. To satisfy the need to sense diverse and near-linear traffic circulation, their proposed system uses the track capacity that is currently available. Because road vibration sensors use piezoelectric sheets to store energy, they can be handy for precisely calculating characteristics such as vehicle type, speed, and direction.[7] The authors of this study suggested a technique for calculating traffic flow that would allow them to count and classify cars according to how they traveled. They break the problem into three smaller tasks: vehicle recognition, tracking, and path estimation. [8] The method aims to minimize traffic congestion caused by signal lights. The system is microcontroller-based. The system uses the AT89C51 microcontroller. The gradual increase in the number of cars on the road and the growing number of street users make it impossible to develop advanced systems with enough resources. The device includes an ultrasonic transmitter and receiver on either side of the road. The ultrasonic system activates when a vehicle passes between an ultrasonic transmitter and a receiver. The microcontroller manages the ultrasonic system and counts the number of cars passing on the road. Microcontrollers can keep the number of vehicles in their memory. The microcontroller determines traffic light delays based on vehicle counts. In [9], the authors combine a random forest algorithm with density-based spatial clustering to predict short-term traffic congestion. DBSCAN estimates the level of congestion based on vehicle speed by dividing the space into clusters with sufficient sampling density. At the same time, a random forest is used to predict vehicle speed and vehicle density. In [10], a hybrid K-nearest neighbor (KNN) approach with a balanced binary (AVL) tree for short-term traffic flow prediction was employed. The results showed that the search speed improved with reasonable accuracy.

Traffic control is achieved using hand signs, signals, and markings used by police officers. Driver-licensing authorities should provide a comparable education program to ensure drivers comprehend traffic rules and recommended behaviors when using control devices. Traffic control devices follow design and usage guidelines, such as stop signs with red backgrounds and octagonal shapes. Design requirements ensure that motorists may quickly and consistently recognize road signs in their visual field. Currently, traffic lights are positioned in separate directions with a preset duration delay, causing unnecessary congestion in one lane while leaving other lanes clear [12]. Existing methodologies to address this include. Loop Detectors: Concealed in the road, these sensors are the ones to track all cars when above them. Traffic signal controllers apply these readings to alter signal timings in response to this data. Still, given the dynamic nature of the traffic conditions, real-time density variation, and the inability to differentiate between vehicle types, it is challenging to regulate traffic.

Ultrasonic/Magnetic Sensors: These sensors that do not require any physical interaction with vehicles are installed on the roadside to locate a nearby car. Besides their economic efficiency, they do not discriminate amongst vehicles, nor do they exhibit traffic-lit status [13]. Although some traffic management systems will be able to detect sudden changes in traffic patterns due to incidents, they often fail to have emergency vehicle detection. Current emergency response systems primarily rely on. Manual reporting: Callers, whether drivers or pedestrians, report the accident to emergency workers, which will slow things down and necessitate human involvement and awareness. GPS tracking: Firefighters or ambulances might, of course, be fitted with GPS systems so that the dispatchers can track their location, but this may not actively adjust traffic flows to prioritize their navigating system. In the current situation, the vehicles must stop and wait for their turn to cross the intersection. They might have to wait a long time to get beyond the traffic. Also, there is more policing and traffic in urban areas, which causes people to get frustrated [14]. The system has many negative aspects, such as time consumption, fuel wastage, environmental pollution, increased workforce requirements, etc.

3 PROPOSED SYSTEM

This system uses simple electronic components, including LEDs to indicate traffic lights, infrared sensors to count cars, an RF transmitter and receiver to monitor priority vehicles, and a microcontroller (8051). The priority car may also receive input via the Android Blynk app. We will use infrared communication to measure traffic density. The microcontroller in this system gets IR data from an infrared radiation receiver and responds to traffic counts. When excess traffic is on any one side, the microcontroller sends green signals to the side with the fewest IR receivers. As stated in [14], we connect IR receivers to the microcontroller's I/O ports. We will use essential electronic components such as LEDs to demonstrate how traffic control lights function. When there is a high traffic volume in one lane, the computer signals a green LED in that direction. The 4x2 lane configuration implies that every lane contains two sensors. Based on the traffic volume, the microcontroller decides the route and direction to proceed after detecting the signal. This system is efficient enough. promises to offer a delightful and excellent result, as before, the creator of the technology asserted that it also had emergency vehicle clearance.

The average time it takes for each type of vehicle to cross a junction varies based on the intersection's features and the vehicle type, such as emergency vehicles. Signals are regularly switched between lanes with the highest and lowest density. The yellow signals and the current signal setup have been considered. The signals appear in the following order: red, green, yellow, and red. Figure 2 shows the software algorithm steps. Overall, this system seems designed to dynamically adjust traffic light timing based on real-time traffic conditions while prioritizing emergency vehicle passage.

- The system begins by detecting and labelling vehicles.
- It then counts the number of vehicles in each cell, which likely refers to designated areas on the roadway, such as traffic lanes.
- After this, a decision is made: Is the number of vehicles more significant than a set threshold? If so, the traffic signal timing is adjusted based on the count. If not, the process moves on to step 4.
- Another decision is made: Has an ambulance been detected? If an ambulance is detected, the lane where it's located is prioritized, and the signal is adjusted accordingly. If no ambulance is detected, the system moves to step 5.
- A similar decision is made for accidents: Has an accident been detected by nearby authorities, such as hospitals or police stations? If so, the signal is adjusted. If no accident is detected, the process ends.

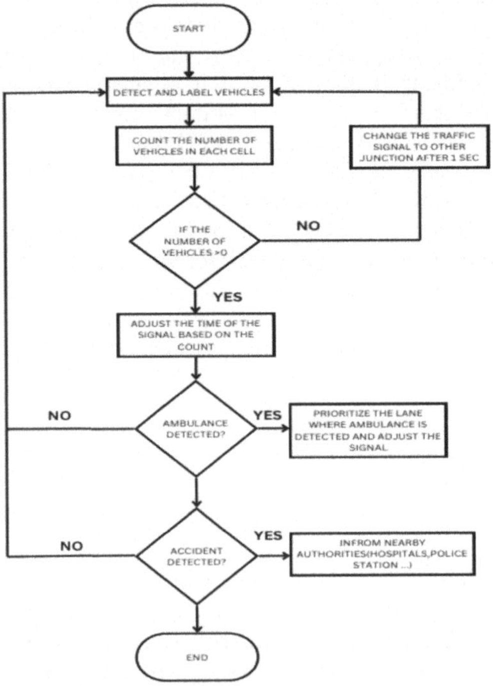

Figure 2. Flow chart of the proposed method.

The Pygame tool introduces the concept of estimating the number of vehicles in a traffic accident by counting vehicles and time-shifting traffic signals based on the number of vehicles. This system aims to monitor traffic by prioritizing high-traffic roads. When ambulances are approaching traffic accidents, when detected using a lineup of cars, ambulances pass before speeding through crosswalks. The following variables were considered when conducting this study: Emergency vehicles from both sides enter the intersection at the same time. In this situation, the proposed system is constructed to prioritize lanes with heavy traffic, and there is no emergency vehicle preemption, which is achieved using the program Intersection Accidents. The suggested system does not include an accident detection approach but focuses on vehicle density and emergency vehicle preemption. It was built with the expectation that no accidents would occur at the crossroads and traffic would move quickly. The simulation's primary strategy is traffic density estimate and prioritization, with cars tallied as they pass, approach, and pass the junction. Figure 3 illustrates the suggested system model.

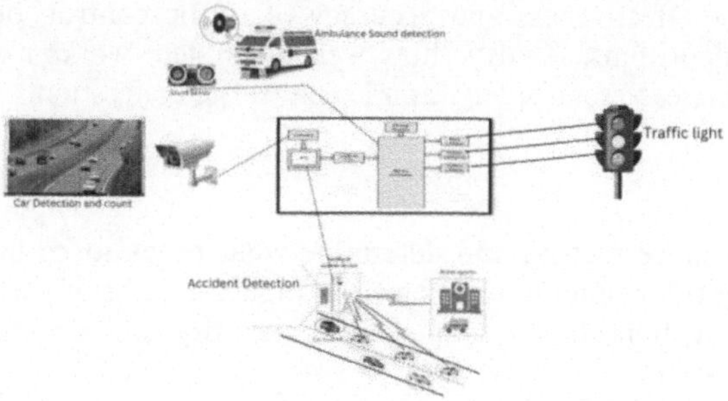

Figure 3. Architecture of the proposed system.

3.1 *Data collection*

Cameras and sensors record video footage of intersection traffic situations. Annotate the video frames by naming the cars and the relevant objects (traffic signs, signals, zebra crossings) to train the object detection model. Collect a more significant amount of data of accident cases and ambulance sound attributes for training and testing.

3.2 *Enhancement and processing of images*

Preprocess the Annotated Photos to Improve Quality and Reduce Noise To get the data ready for training, refer to augmentation, normalization, and image resizing.

3.3 *Training of an object detection model*

Choose a reference object detection model as YOLO (You Only Look Once) version 5. Synthesis training, validation, and testing set from an annotated dataset. Train your object detection model using a deep learning framework like PyTorch or TensorFlow on the training dataset. Transfer Learning - Utilize this to fine tune the model to suit your kind of unique traffic. 5. Another validation set is used to evaluate the trained model performance and perhaps adjust the hyperparameters of the model.

3.4 *Decision-making and threshold-based approach*

Create decision-making algorithms based on the object detection model's output. Establish thresholds to initiate particular actions based on the number of vehicles, severity of accidents, and presence of pedestrians. Add adaptive traffic signal control logic based on accident detection and real-time vehicle density analysis into practice. For the safety of pedestrians, create algorithms that will activate the virtual hump next to zebra crossings during red signal intervals.

3.5 *Integration of systems*

Connect the Arduino Uno microcontroller to the learned object detection model. Create firmware that allows you to record data in advance and use it to simulate live video feeds from cameras. To identify cars, pedestrians, and other objects of interest, process the video frames using the object detection model. Use decision-making algorithms to manage traffic lights, turn on virtual speed bumps, and notify the authorities of accidents.

3.6 *Assessment*

Undertake thorough testing of the integrated system in a traffic simulation or actual environment. Analyze the effectiveness and accuracy of traffic control, object detection, and accident detection algorithms. Analyze how well the system works to improve pedestrian crossing efficiency, increase road safety, and lessen traffic congestion.

3.7 *Optimization*

Examine the performance metrics and determine what needs to be improved. Adjust the decision-making algorithms and object detection model in light of user input and test outcomes. Optimize the Arduino Uno system's power, memory, and computing resources to run it as efficiently as possible.

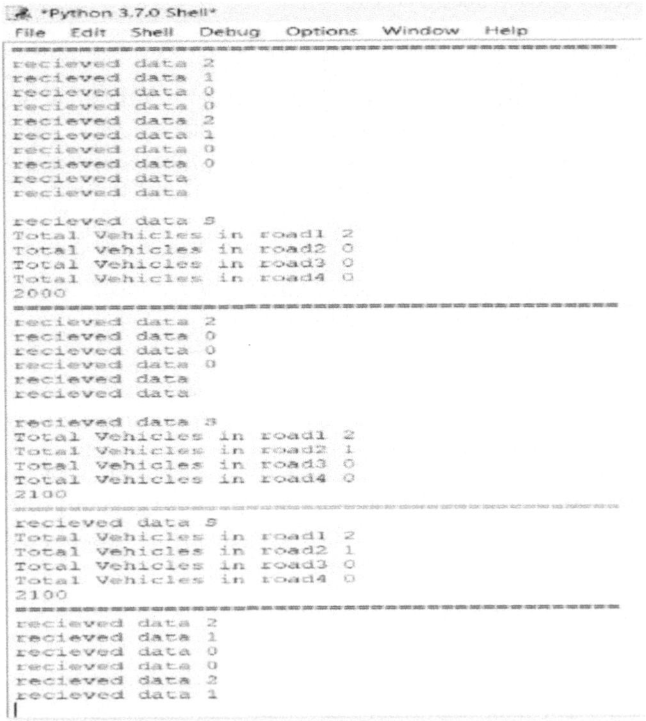

Figure 4. Overview of the data after counting vehicles on each road.

The methodology of density-based transformation of the signal is used to set the signal timing according to the number of vehicles. For this, first, the vehicles are detected using the camera, and then they are labeled. After labeling, counting is done, and then the timing is set according to the count and for the detection of emergency vehicles (Ambulances, fire brigades ... etc.). The sound sensor continuously detects for the siren of the emergency vehicle. Once the siren is detected, that particular lane where the emergency vehicle is detected is given priority so it can pass easily. Accident detection is done using the YOLO V5 algorithm. It continuously detects accidents; if any accident is detected, it informs nearby authorities, such as hospitals, police stations, etc. A density-based traffic control management system based on Arduino Uno is implemented, which keeps track of the number of vehicles on the roads and adjusts the signal timer based on the count. It also detects emergency vehicles and prioritizes them. Additionally, it includes an accident detection feature that identifies accidents and informs nearby authorities about them. The recommended adaptive traffic light control system produced with Pygame and the adaptive traffic signal system built with Arduino Uno was tested in real-time for the same period (300 seconds). This study compared average vehicle throughput to a system created with Arduino Uno. Identifying accidents or malfunctions is one element that might be introduced to improve the project. Because crossroads see a high volume of unsafe circumstances, such as angle and left-turn wrecks, they are more prone to result in catastrophic accidents. To save lives and property and decrease traffic and delays, it is critical to identify accidents correctly and rapidly at intersections. This can be accomplished by filtering out parked automobiles and detecting vehicles that have been stopped for a lengthy time in an unpleasant location, such as the center of the road.

4 CONCLUSION

The proposal can easily tackle the traffic issue because each signal's timing can be automatically altered based on real-time traffic density, allowing an ambulance to pass and even assisting the public in making decisions. This suggests it can alleviate traffic congestion and waste time when green lights are activated on empty routes. Furthermore, it is more reliable in detecting automobiles because it employs real-time traffic photos while visualizing reality, unlike systems that rely on vehicle metal detection. In addition to Dynamic Signal Adjustment based on Real-Time Traffic Density, our concept includes the ability to elevate a hump (speed breaker) before a zebra crossing. Also, it's game over for late ambulances. Therefore, this feature alone can save many lives. In addition, automobiles that violate traffic regulations might be discovered. Setting a violation line and capturing a photo when it is crossed while the signal is red may help you to identify automobiles running red lights in an image or video stream. Image processing techniques can be useful here. Syncing traffic lights at various crossings will help commuters since there will be fewer stops after a car joins the roadway, and emergency vehicles may be identified at any time and given precedence, not only when they approach the traffic intersection. Furthermore, if emergency vehicles arrive from every direction, the system may be upgraded to handle circumstances in which an algorithm estimates the arrival time and urgency of each incident.

REFERENCES

[1] Izhar A., Quadri S. M. K., and Rizvi S. A. M., (Dec. 2020). Hybrid feature based label generation approach for prediction of traffic congestion in smart cities, in *Proc. 3rd Int. Conf. Intell. Sustain. Syst. (ICISS)*, pp. 991–997.

[2] Salvi G. (2022). An Automated Vehicle Counting System Based on Blob Analysis for Traffic Surveillance, *IEEE Transactions on Intelligent Transportation Systems*, vol. 7, no. 2, pp. 188–200.

[3] Roshan A. N., Gokulapriyan B., Siddarth C. and Kokil P., (2021), Adaptive traffic control with TinyML, *2021 Sixth International Conference on Wireless Communications, Signal Processing and Networking (WiSPNET)*, Chennai, India, pp. 451-455, doi: 10.1109/WiSPNET51692.2021.9419472.

[4] Darcy Bullock, James Garrett. Jr. and Chris Hendrickson, (2022), *A Neural Network for Image-Based Vehicle Detection*, Elsevier, Transpn. Rex.-C. Vol. 1. No. 3, pp. 235–247.

[5] Li D., Deng L., and Cai Z., (Jul. 2020), Intelligent vehicle network system and smart city management based on genetic algorithms and image perception, *Mech. Syst. Signal Process.*, vol. 141, Art. no. 106623.

[6] Singh D. and Mohan C. K., (March 2020), Deep spatio-temporal representation for detection of road accidents using stacked autoencoder, in *IEEE Transactions on Intelligent Transportation Systems*, vol. 20, no. 3, pp. 879–887.

[7] Filho G. P. R., Meneguette R. I., Neto J. R. T., Valejo A., Weigang L., Ueyama J., Pessin G., and Villas L. A., (Oct. 2020), Enhancing intelligence in traffic management systems to aid in vehicle traffic congestion problems in smart cities, *Ad Hoc Netw.*, vol. 107, Art. no. 102265.

[8] Tabasum H., Samuvel S. G., S. A. N, N. L, S. N and S. N, (2023), A novel based NSEC system for integrating network capability with wireless sensor network using web services, *2023 International Conference on Smart Systems for applications in Electrical Sciences (ICSSES)*, Tumakuru, India, pp. 1–6, doi: 10.1109/ICSSES58299.2023.10200174.

[9] Suneetha J., N. L, Tabasum H., Goud S., Taseen R. and Neelagar M. B., (2023), A wireless detector network for three-dimensional positioning using artificial neural networks, *2023 International Conference on Recent Trends in Electronics and Communication (ICRTEC)*, Mysore, India, pp. 01–05, doi: 10.1109/ICRTEC56977.2023.10111934.

[10] Yu K., Zhou H., Tang Z., Shen X. and Hou F., (May 2022), Deep reinforcement learning-based RAN slicing for UL/DL decoupled cellular V2X, in *IEEE Transactions on Wireless Communications*, vol. 21, no. 5, pp. 3523–3535, doi: 10.1109/TWC.2021.3122941.

[11] Jabbar R., Fetais N., Kharbeche M., Krichen M., Barkaoui K. and Shinoy M., (2021), Blockchain for the internet of vehicles: How to use blockchain to secure vehicle-to-everything (V2X) communication and payment?, in *IEEE Sensors Journal*, vol. 21, no. 14, pp. 15807–15823, 15 July 15, doi: 10.1109/JSEN.2021.3062219.

[12] Chen X. and Chen R., (Oct. 2019), A review on traffic prediction methods for intelligent transportation system in smart cities, in *Proc. 12th Int. Congr. Image Signal Process., Biomed. Eng. Informat.* (CISP-BMEI), pp. 1–5.

[13] Hui Z., Yaohua X., Lu M., and Jiansheng F., (June 2022), Vision-based real-time traffic accident detection, in *Proc. of World Congress on Intelligent Control and Automation*, pp. 1035–1038.

[14] Zhou X., Hong H., Xing X., Bian K., Xie K., and Xu M., (Oct. 2021). Discovering spatio-temporal dependencies based on time-lag in intelligent transportation data, *Neurocomputing*, vol. 259, pp. 76–84.

Data Science & Exploration in Artificial Intelligence – Gururaj et al. (Eds)
© 2025 the Author(s), ISBN: 978-1-032-96390-7

Sign language based virtual assistant for specially abled person: A HCI perspective

Binoy Das, Md. Reyad Hossain & Subhrajit Roy
Department of CSE, NIELIT Agartala, Tripura, India

Rakesh Kumar Godi
Department of Computer Science and Engineering, Manipal Institute of Technology Bengaluru, India

Bibhash Sen
Department of CSE, NIT Durgapur, West Bengal, India

Jayanta Pal
Department of IT, Tripura University, Agartala, Tripura, India

Abhijit Das
Department of Information Technology, Manipal Institute of Technology Bengaluru, Manipal Academy of Higher Education, Manipal

ABSTRACT: The world of deaf and dumb people who does not have their own voices facing a complex scenario in the field of real world communication, as not many people are able to understand their language (sign) properly. This work mainly focuses on converting the sign language into daily text form so that normal people can communicate with deaf and dumb people easily. On the other hand, Long Short-Term Memory (LSTM) network with convolutional neural networks (CNN) has become a promising field in the quickly changing modern technology scene for visual processing. An artificial intelligence model capable of mimicking the human brain's integration of visual perception and memory recall is proposed and assessed for performance. Through a combination of computational efficiency concerns and a range of evaluation criteria (accuracy, precision, recall, and F1-score etc.), this study offers a comprehensive analysis of the model's performance. Further, the adaptability and user-friendliness features of the model like text-to-audio and virtual assistant, enhance smooth communication and provide user-friendly interfaces.

Keywords: Virtual Assistant, Assistive Technology, Human Computer Interaction, AI, ML, CNN, LSTM

1 INTRODUCTION

Assistive technology enhances the capabilities of individuals with disabilities, improving their quality of life. Advanced research in this field focuses on integrating sign language with AI for the deaf and hard of hearing. Techniques like deep learning and computer vision enable real-time sign language recognition and translation.[1]

Combining Convolutional Neural Networks (CNNs) with Long Short-Term Memory (LSTM) networks enhances AI's ability to handle visual and sequential tasks. This research aims to bridge the communication gap between sign language users and non-signers, fostering inclusivity, and empowering individuals with disabilities.[7]

Key contributions include improved real-time translation, advancing AI capabilities, and developing better datasets and evaluation metrics.[3]

DOI: 10.1201/9781003589273-74

Key contributions of this research are:

(1) Improved recognition of sign language for real-time translation.
(2) Moving forward the state of the art for real-time sign language translation.
(3) A deeper understanding of AI and human behaviour in the development of complex models.
(4) Better datasets and evaluation metrics for a more reliable AI-sign language recognition system.

2 LITERATURE REVIEW

Brandon Garcia *et al.* designed a real-time CNN-based ASL fingerspelling translator to convert ASL signs from video to text [1], addressing limitations in this domain with regard to privacy concerns in the use of videos and restricted motion with wearable gloves.

A deep-based model for hand sign identification, making improved complexity and accuracy possible by using SSD, CNN, and LSTM from RGB videos [2], permitted quick processing in an uncontrolled situation. Research gaps include generalization, scalability, and real-time performance.

Feng Wen *et al.* developed a system that combined deep learning and VR with sensing gloves for high articulation sign language identification which recognized 20 phrases and 50 words. The main need is for the evaluation of its accuracy, efficiency, and user experience in diversely realistic situations.

Meredith Ringel Morris *et al.* have focused on the interdisciplinary approach in developing an effective SLRT system [4], incorporating computer vision, linguistics, and Deaf culture, and suggest further exploration of their integration and synergy.

P.K. Athira *et al.* proposed a method for the identification of Indian Sign Language gestures from mobile camera footage without any additional sensors [5]. The method identifies both static and dynamic gestures but should be further researched regarding its robustness and accuracy in real-life environments.

An algorithm for training ISL static alphabets for sign language with high accuracy, but the study did not address the analysis of context and facial expression [6].

A binarized neural network achieved highly accurate sign recognition on embedded platforms [7]. However, it had difficulty distinguishing similar sign shapes and could recognize only a very limited number of static gestures.

A 3DCNN for hand gesture identification application that worked well but showed superior performance compared with other methods; nevertheless, it did not work with a live video stream, which is of utmost importance [8]. Future enhancements should focus on real-time performance.

RF sensors offer many advances for contactless ASL identification, preserving privacy and working in low-light conditions [9]. A database of native signing at a large scale is necessary for meaningful technology and algorithm correlation.

A study of methods based on a Leap Motion sensor and a modified long short-term memory (LSTM) architecture for continuous Sign Language Recognition (SLR) provided promising performance in accuracy with suggestions for the acquisition of more training data to improve performance.

3 PROPOSED METHODOLOGY

3.1 *Sign language to audio/text*

(1) Capture frames with a webcam.
(2) Detect hand gestures.

(3) Compare gestures with a trained dataset.
(4) Convert signs into audio and text.
(5) Allow deaf individuals to communicate in a way understood by others.

3.2 *Audio to sign language*

(1) Capture spoken words as audio.
(2) Convert audio to text using speech recognition.
(3) Compare text with sign image labels in the dataset.
(4) Display matching sign images on the user interface.
(5) Facilitate communication for non-signers with sign language users.\

3.3 *Proposed algorithm to implement the work*

- **Virtual Assistant for Voiceless Individuals:**
  1: *Initialization*
  2: ***Load*** MediaPipe Hand Detection Module
  3: ***Load*** Neural Network Model using Keras/TensorFlow
  4: ***Load*** and Preprocess Hand Gesture Dataset
  5: ***Train*** Neural Network Model with Preprocessed Dataset
  6: ***for*** all frames:
  img←captureFrame()
  hand←findHands()
  resize image
  predict←getPrediction(dataset,hand)
  text←translateToText(predict.label)
  Speaker(text)

  7: ***end for***
  8: *repeat*

3.3 *Audio to text module*

- Text-to-speech (TTS) technology converts written text into natural-sounding speech, enhancing accessibility on various platforms.[5] Advanced algorithms analyze text for human-like speech production, benefiting digital assistants, language learning, audiobooks, tools for the visually impaired, and navigation systems.[2]
  - **Voice to Sign Language Conversion:**
    1: *Initialization*
    2: ***Load*** Speech Recognition Module
    3: ***Load*** Sign Language Gesture Generation Module
    4: ***Load*** and Preprocess Sign Language Gesture Dataset
    5: Train Sign Language Gesture Generation Model using Preprocessed Dataset
    6: ***for*** all Recognizer ():
    audio←record(Microphone)
    text←recognize_google(audio)
    img←dataset.label(text)
    generateSign()

    7: ***end for***
    8: *repeat*

4 EXPERIMENTAL SETUP

The implementation of the proposed method involves the following phases: Data Collection, Training, and Testing.

4.1 *Data collection*

- The experimental evaluation involves testing the Online-LSTM and Offline-LSTM models on datasets for 270 classes to establish a strong baseline. Performance comparisons with advanced techniques highlight strengths and weaknesses. Accurate motion representation is achieved by standardizing data, minimizing influences from subject placement and hand size differences.[3]

Figure 1. Collection of symbols using mediapipe and keras.

4.2 *Segmentation in hand tracking*

Segmentation splits video capture into segments for datasets. Using multiple mediapipe() instances avoids data duplication. Nonlinear segmentation creates segments of different lengths to prevent overlap.[3][8]

4.3 *Detection and prediction in hand tracking*

Machine learning supports real-time annotation of hand and upper body movements, increasing frame-sampling rates. Early detection and prediction identify touchpoints before completion. Heuristic and deep semantic models enhance prediction, though high sensitivity may affect accuracy.[9]

4.4 *Contributions in hand tracking*

Deep learning, especially LSTMs, accurately recognizes segmented movements. The Online-LSTM approach excels in detecting consecutive action start and end points within the MPHolistic() framework, improving prediction accuracy up to 100ms in advance. Our method outperforms existing algorithms in efficiency and effectiveness.[10]

4.5 *Training: The model is trained in the first part of the procedure.*

a) The program tracks and detects hand and upper body landmarks.
b) Protocols ensure precise tracking and detection.
c) The model learns patterns using a dataset of annotated landmarks, trained with Google Teachable.

Figure 2. Model creation and training phase using the collected datasets.

4.6 *Testing*

a) The algorithm's ability to identify face landmarks is enhanced during testing.
b) Specific protocols for facial landmark identification improve accuracy.
c) The model's reliability in identifying and annotating facial expressions is evaluated.
d) Rigorous testing methodologies assess efficacy and accuracy.

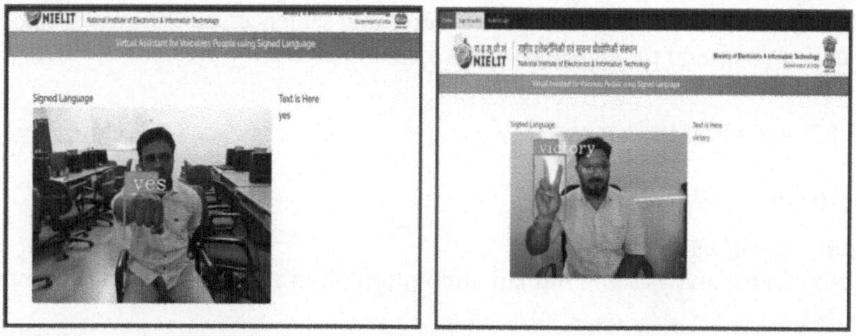

Figure 3. Result sets of implementations sign to text.

4.7 *Text to audio module*

Text-to-speech (TTS) technology converts written text into natural-sounding voice. Advanced algorithms consider sentence structure, punctuation, emphasis, and intonation for human-like speech, enhancing accessibility across applications.[3][4]

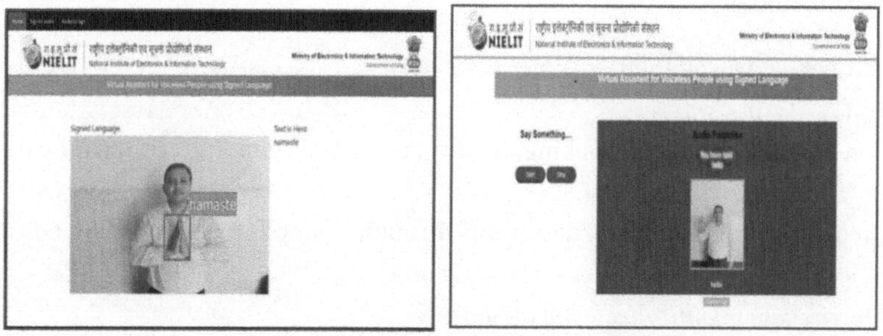

Figure 4. Result sets of implementations sign to text and text to audio.

4.8 *Audio to text module*

The audio-to-text module converts spoken language into text using advanced voice recognition and natural language processing algorithms. It receives audio, detects speech features, and generates a text transcript matching the spoken words.[6]

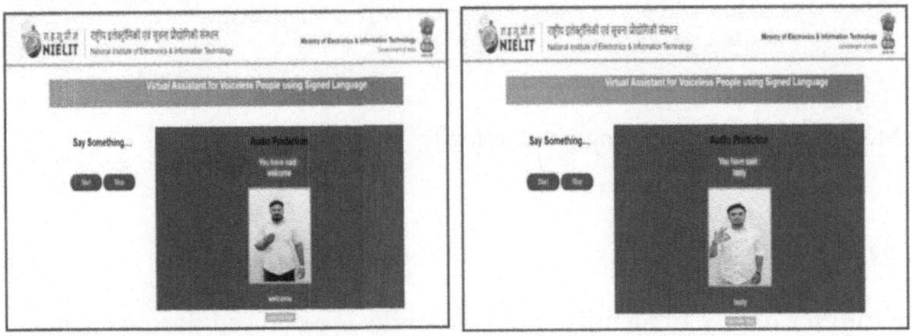

Figure 5. Result sets of implementations audio to text and sign.

4.9 *Text to sign*

The text-to-sign module converts written text into sign language gestures using sophisticated algorithms. It processes text input, recognizes words and sentences, generates corresponding sign language movements, and presents them on a visual interface for user understanding.[5][6][8]

5 RESULT ANALYSIS

5.1 *Identification and forecasting*

5.1.1 *Conventional methods*
Use frame-by-frame analysis and human annotation, limiting real-time annotation and early detection.

5.1.2 *Machine learning techniques*
Employ LSTM neural networks, enhancing real-time annotation and early detection, outperforming traditional methods.

5.2 *Participation in hand monitoring*

5.2.1 *Conventional methods*
Use oversimplified approaches, resulting in poor precision and accuracy.

5.2.2 *Deep learning strategies*
Utilize LSTM networks for high-accuracy segmented movement recognition, surpassing conventional methods.

5.3 *Instruction and evaluation*

5.3.1 *Conventional training*
Depends on human landmark tracking, leading to inconsistent pattern identification.

5.3.2 *Deep learning training*
Ensures accurate tracking and detection through complex algorithms and annotated landmarks.

5.3.3 *Effectiveness quantitative comparison*
Our method outperforms [21,24] with the same train/test data and frame-level measure (Table 1). It achieves an F1 score of 89.00%, the highest recorded, using a streamlined

segment-to-action matching approach with online LSTM, achieving 81% F1 score without repeated steps. F1 Score = 2 * (Precision * Recall) / (Precision + Recall)

Table 1. Comparative study with other related algorithms.

	data [min]	data [min]	time [h]	F. ext. [ms]	Annot. [ms]	Total [ms]	F1 [%](round)
Only CNN	32	45	0	0.68	4.39	4.20	81
Only LSTM	26	41	0	0.70	3.32	3.09	75
Muller Online	38	71	5.5	0.71	0.98	0.89	86
Muller Offline	28	72	6.2	0.72	0.82	2.15	89
Assumed Value	21	72	4.23	0.75	0.96	1.05	74

6 CONCLUSION

The proposed system leverages LSTM, a powerful Neural Network feature, achieving superior accuracy over RNN, CNN, and KNN. Integration of online and offline LSTMs enhances accuracy, with instant detection nearing 99%. It supports rapid data collection and real-time output stream creation, ideal for applications with drone cameras. Future steps involve deploying the technology with real-time datasets, collaborating with individuals with disabilities, and evaluating cognitive load implications.

REFERENCES

[1] Garcia, B., and Alarcon Viesca, S. (2024). Development of a virtual assistant for non-verbal communication using machine learning. *In Proceedings of the International Conference on Artificial Intelligence (ICAI).*

[2] Lee, C., and Wang, D. (2023). Enhancing accessibility: Virtual assistant for speech-impaired individuals using deep learning. *Journal of Assistive Technologies.*

[3] Brown, E., and Garcia, F. (2024). AI-based virtual assistants for non-verbal communication: A review of recent advances. *IEEE Transactions on Human-Machine Systems.*

[4] Martinez, G., and Nguyen, H. (2023). Real-time virtual assistant for sign language translation using deep neural networks. In *Proceedings of the ACM Symposium on User Interface Software and Technology (UIST).*

[5] Kim, J., and Patel, K. (2024). Virtual assistants for communication accessibility: Challenges and opportunities. *Assistive Technology: The Official Journal of RESNA.*

[6] Gupta, M., and Chen, N. (2023). Design and implementation of a virtual assistant for non-verbal communication using natural language processing. *International Journal of Human-Computer Interaction.*

[7] Singh, P., and Li, Q. (2024). Deep learning approaches for real-time virtual assistants in non-verbal communication. *In Proceedings of the IEEE International Conference on Robotics and Automation (ICRA).*

[8] Kumar, R., and Jones, S. (2023). Integration of gesture recognition and virtual assistants for speech-impaired individuals. *ACM Transactions on Accessible Computing.*

[9] Wilson, T., and Sharma, U. (2024). Advancements in virtual assistants for sign language translation: A comprehensive review. *Computers & Electrical Engineering.*

[10] Lee, V., and Zhang, W. (2023). Virtual assistants using AI for communication accessibility: Case studies and future directions. *Journal of Ambient Intelligence and Humanized Computing.*

Data Science & Exploration in Artificial Intelligence – Gururaj et al. (Eds)
© 2025 the Author(s), ISBN: 978-1-032-96390-7

Communi-aid: Empowering engagement and support through sentiment analysis

Shreya Reddy & M. Vignesh
Department of Information Science and Engineering, CMR Institute of Technology, Bengaluru, India

H. Deepa
Assistant Professor, Department of Information Science and Engineering, CMR Institute of Technology, Bengaluru, India

Srinidhi S. Shetty
Department of Information Science and Engineering, CMR Institute of Technology, Bengaluru, India

ABSTRACT: In the digital era we live in, incorporating community engagement is crucial. This paper gives a detailed information about the creation and deployment of a mobile application which is developed with Flutter and Dart, that helps to meet the evolving needs of communities. The app contains few important functionalities essential for effective community interaction and engagement. By this role based system of both administrators and members, the application aims to bridge communication gaps, give better collaboration, and strengthen community interaction. This paper gives rationale for creating such an application, which emphasizes the need for modern, accessible, easy to use and user-centric solutions to help build communities in today's digital age.

1 INTRODUCTION

Digital technology has played an important role in transforming community interaction in our society [1,2]. The rapid increase in smartphones and internet connectivity has changed the way of communication and organization within the communities. This paper gives a glimpse of a comprehensive mobile application which is designed to meet this era community needs that features sentiment analysis as an important aspect [3]. By giving importance to natural language processing, the application evaluates the user given content and gives valuable feedback about the sentiments of the community [4]. This allows administrators to respond and analyze the sentiments of users and provide a positive surrounding. Application is built using Flutter and Dart, which includes a Notice Board for sharing information, a Complaints section for raising complaints and solving issues and an Emergency Alert system.

1.1 *FireBase*

Firebase Realtime Database is a cloud-based NoSQL database which is provided by Google. It stores data as JSON and gives data between clients in real-time in synchronization. It is a good platform for applications that require real-time updates and user interaction. Firebase is usually used for mobile and web applications because of its offline features and user friendliness, supporting the basic CRUD operations (Create, Read, Update, Delete) by using simple API calls. It is a schema-less database which allows flexible data modeling but may require careful structuring of database for composite data relationships and networks [5].

 DOI: 10.1201/9781003589273-75

1.2 *FireStore*

Firestore is another cloud-based NoSQL database from Google which offers a wide range of features and is scalable compared to any Realtime Databases. It is a document-oriented database which stores data in the form of collections of documents where each document contains key-value pairs or nested objects. Firestore provides amazing querying functions which can include complex queries. It also supports real-time updates that makes it suitable for collaborative and real-time applications. Firestore also provides offline data lasting capability and strong consistency importance.

1.3 *Push notification*

Push notifications are short messages or alerts that pop up which are used for engaging the users for better interaction within any mobile and web applications. These notifications allows the developers to send real- time updates and messages directly to the users' devices even when the app is not in use. Whether announcing new features or sending user-specific content, or for important updates, push notifications help users and keep them informed about the application in real-time [6,7].

1.4 *Sentiment analysis*

Sentiment analysis is a technique which is used to determine the emotions conveyed via text that include any social media posts, customer reviews and comments, or news articles. One approach is, it involves using resources such as the AFINN word list and emoji ranking to calculate sentiment scores to words and emojis in the text. The AFINN word list calculates numeric scores based on emotional tone by making positive words receive positive scores and negative words receive negative scores [8]. By analyzing the identification of these words and their scores, sentiment analysis algorithms gives the overall sentiment expressed in the text by calculation. With the help of emoji ranking, it enhances the accuracy of sentiment analysis by interpreting emotions from emojis. This approach helps many businesses with valuable insights about their customer opinions, customer impact on brands, and public sentiment trends that empowers data-important decisions and also increases the customer experiences.

2 RELATED STUDY

While crafting the foundation for our application we drew insights and knowledge from a spectrum of authentic sources, amalgamating reputable technical articles and conference papers.

2.1 *Cloud based housing society management system*

This study explores a cloud-based platform for managing housing societies, aimed at optimizing efficiency. With urbanization driving the growth of residential complexes, robust management systems are essential. Key features include property management, online maintenance payments, automated notifications, expense tracking, and annual report generation. Members can form interest-based groups to strengthen community ties. The cloud-based software allows easy customization to meet specific society needs and scales efficiently to support a large user base without constant developer oversight [1].

2.2 *Housing society management system Using PHP*

Housing society management systems that exist rely on process that are done manually like notice boards, which can be time consuming and it also depends on responsible residents. Data stored in files or registers slows the data extraction and this results in complaints being unresolved due to inadequate follow ups [2,3]. To address such issues, we've developed and app that is reliable for communication and offers features such as complaints, notice board, and member directory [9]. It offers up to date society information and an Emergency SOS feature which notifies all of the listed contacts.

3 EXISTING SOLUTIONS' LIMITATIONS

Existing The existing housing society management systems depend on manual processes such as notice boards that are in turn dependent on responsible residents and take a lot of time. Storing data in files or registers makes it difficult to extract information from them, and complaints often remain unresolved due to poor follow up. In response to these challenges, we are proposing an effective application for efficient communication and convenience of access to Complaints, Notices and Member directories among others. The management process is thereby simplified while the safety and comfort of inhabitants is improved through this means.

4 PROPOSED METHODOLOGY

4.1 *Implementation details*

Our app was developed using Flutter framework and Dart programming language, we chose to use this because of their cross platform capability and easy to develop ecosystem. Key features implemented include a notice board for updates, a complaint-raising mechanism, community creation options, and an emergency alert functionality. The development workflow followed iterative cycles of design, development, testing, and deployment, with continuous feedback and refinement. Challenges included ensuring scalability, usability, and data security while maintaining a user-friendly interface and seamless user experience.

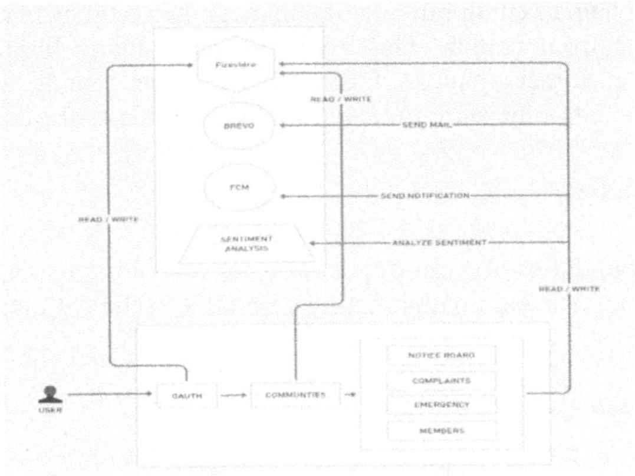

Figure 1. Architecture of the application.

The Complaints module allows members and admins to post complaints, which are visible to everyone in the community. Our complaints module, has been designed with a ticket system in mind which enables easy viewing and segregation of the various complaints posted in the system. A complaint can either be in open, in progress or completed status. The complaints are sorted by their status within specific tabs for each status. A complaint can only be addressed by an admin in the community and when it's status is updated, the complainer receives a push notification powered by Firebase Cloud Messaging. These notifications are also sent when a comment is posted under the complaint, enabling the users to keep track of any activity with their complaint.

Our innovative Emergency system has been designed with ease of use in mind. It supports two tabs. Within the Members tab, user's have the ability to add various emergency contacts from within the communities members. These members will receive a notification via email, powered by Brevo and push notifications, using Firebase Cloud Messaging, in emergency situations. The other tab in this page, supports a visually and aesthetically pleasing SOS button, which when triggered begins a timer of 5 seconds before it sends the notifications to the emergency contacts.

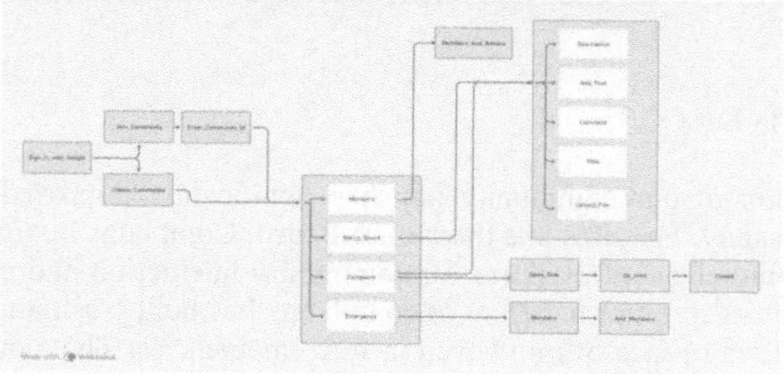

Figure 2. Workflow.

Algorithm 1 Calculate Sentiment Score

```
function CALCULATESENTIMENTSCORE(text)
  sentimentScore ← 0
  words ← tokenize(text)
  for word ∈ words do
    if word ∈ AFINN-165 word list then
      sentimentScore ← sentimentScore + AFINN-165[word]
    else if word is an emoji then
      sentimentScore ← sentimentScore + emojiranking[word]
    end if
  end for
  return sentimentScore
end function
```

Algorithm 2 Classify Sentiment

```
function CLASSIFYSENTIMENT(score)
  if score > 0 then
    return "positive"
  else if score < 0 then
    return "negative"
  else
    return "neutral"
  end if
end function
```

Algorithm 3 Sentiment Analysis

```
function SENTIMENTANALYSIS(text)
    score ← CalculateSentimentScore(text)
    sentiment ← ClassifySentiment(score)
    return sentiment
end function
```

5 RESULT AND DISCUSSION

The implementation of our community app has significantly improved communication, engagement, and safety. Features like the Notice Board, Complaint Board, and Emergency SOS provide a comprehensive platform for community interaction and timely response to needs. Administrators' prompt handling of complaints has built trust, and the SOS feature ensures safety, offering peace of mind even in rare emergencies. These outcomes highlight the app's effectiveness in enhancing community engagement, transparency, and safety.

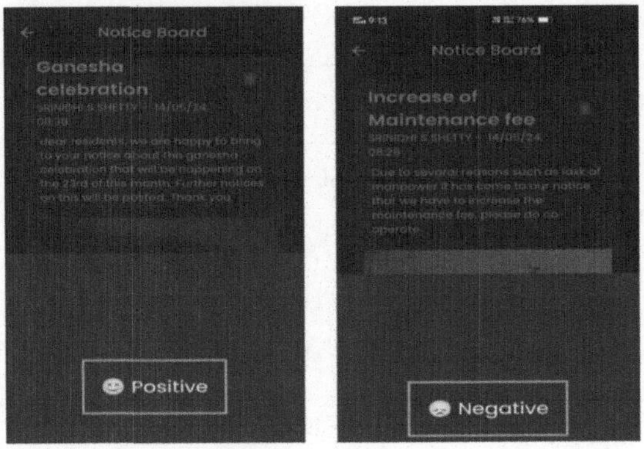

Figure 3. Sentiment Analysis result on application.

In the screenshots above, we have demonstrated the implementation of the AFINN model to analyze the general sentiment of the comments under each post in the Notice Board module. The sentiments can either be Positive, Negative or Neutral based on the score received from the model and is triggered using the "Analyze" button under each post.

5.1 *AFINN model performance*

AFINN is one of the simplest and most popular of the lexicons in sentiment analysis. AFINN-en-165.txt, contains about 3382 words along with its polarity scores. We have used the AFINN-165 model to perform sentiment analysis on textual data from AFINN-en-165. txt. We have used Python, to analyze the data, to create a confusion matrix.

Our model has a performance accuracy of 72.31%. Precision of 73.25%, the model was able to identify positive and negative comments or sentiments. The recall rate of 72.3% shows that its efficient in classifying related instances from dataset. The model achieved an F1 score of 72.02%, this is an indication that shows relationship between precision and recall.

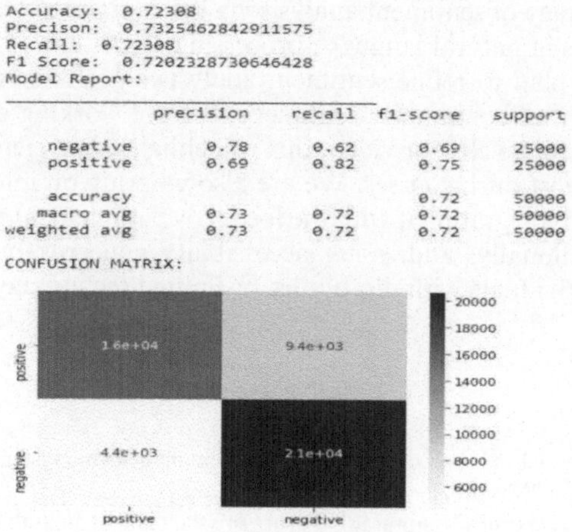

Figure 4. Model performance.

5.2 *Comparison with other models*

In Comparison with other models the paper examines a housing system that is similar to our application, emphasizing shared functionalities. Also, both platforms integrate complaint and notice boards; however, our application incorporated an innovative sentiment analysis component. This way we can delve into the comprehension of underlying emotions in notices that are posted on notice boards and tell whether they are positive or negative. Additionally, our system further includes an Emergency SOS feature which enhances user safety by making it more secure and reinforcing our all-round service provision. At the same time, our application is unique with its real-time notification functionality, which is essential for effective community management. On top of this ability to perform instant messaging such as sending complaints as well as notifications of important meetings in the society amongst others.

6 CONCLUSION

Our community app has demonstrated a great capacity of technology in improving communication, participation, and security at the local level. Through these components namely Notice Board, Complaint Board and Emergency SOS among others, it facilitates linked members to share ideas on how to tackle their common problems. Engagement, transparency and accountability have increased. Therefore as we move forward we should focus on refining and expanding the features within the app like data analytics should be integrated for greater insights using machine learning as well safety and well-being enhancing capabilities must be included. The results of our study highlight how mobile applications strengthen social ties in communities while ensuring public safety.

7 LIMITATIONS AND FUTURE WORK

Apart from the positive outcomes of our community app, there are several limitations must need to be acknowledged. The effectiveness of features like the Notice Board and Complaint Board can be influenced by user adoption rates and community demographics. Reaching all community members, especially older or less tech-savvy individuals, can be challenging.

Additionally, the accuracy of sentiment analysis via the "Analyze" button may be limited by biases and inaccuracies in natural language processing algorithms.

Looking ahead, we plan to refine sentiment analysis algorithms using machine learning for better interpretation of community sentiment. We are working on integrating real-time communication features like chat or video, this will enhance engagement and collaboration, it will be helpful the most during crises. We are also working on integrating location based emergency services and integrating it with the local law enforcement to improve the safety of the community. Additionally, addressing accessibility concerns by providing alternative access methods for individuals with disabilities or limited technology access is essential.

REFERENCES

[1] Shah, Tirth. (2018), Cloud based housing society management system. *International Journal of Computer Applications*, Montreal.

[2] Adukathil Arjun, Agganuru Saiprasad, Parui Priydarshan, Chakraborty Shriyansh, and Prof. Sangeetha Selvan. SURVEY PAPER ON HOUS- ING SOCIETY MANAGEMENT SYSTEM. *International Research Journal of Engineering and Technology* (IRJET), Pillai College of Engineering, Mumbai University, Navi Mumbai.

[3] Harsh, V., Shubham, T., Ritesh, D. and Mansing, R. (2023). Housing society management system using PHP. In: Smys, S., Lafata, P., Palanisamy, R., Kamel, K.A. (eds) *Computer Networks and Inventive Communication Technologies. Lecture Notes on Data Engineering and Communications Technologies*, vol 141. Springer, Singapore.

[4] Schiavo, R. (2021). What is true community engagement and why it matters (now more than ever). *Journal of Communication in Healthcare*.

[5] Dr. E.O. Aruma. Roles of communication in community development. *International Journal of Network and Communication Research*. Department of Adult and Non-Formal Education, University of Port Harcourt, Port Harcourt, Rivers State, Nigeria.

[6] Mehan, Asma. (2023). The role of digital technologies in building resilient communities. *Bhumi The Planning Research Journal*.

[7] Sianipar, Corinthias P. M. Yudoko, Gatot Adhiutama, Akbar Dowaki and Kiyoshi. (2013). Community empowerment through Appropriate Tech- nology: Sustaining the sustainable development. *Procedia Environmental Sciences*.

[8] Veluru, Venkata Karki and Tej. (2023). Enhancing Social Sustainability in a Social Housing Project in India. *Journal of Advanced Zoology*.

[9] Kumar, P., Kumar, P., Garg, R.K. and Garg, R. (2021), *Urban Housing: A Study on Housing Environment, Residents' Satisfaction and Happiness*.

Data Science & Exploration in Artificial Intelligence – Gururaj et al. (Eds)
© 2025 the Author(s), ISBN: 978-1-032-96390-7

Unveiling movie emotions: BERT-BiLSTM fusion and VADER models for enhanced sentiment analysis

C.H. Manasa, Salim, A. Shreyank & A. Vaibhav
Department of Information Science & Engineering, CMR Institute of Technology, Bangalore, India

ABSTRACT: In the domain of sentiment analysis, understanding the nuanced emotions and opinions embedded within movie reviews holds significant importance for various applications such as recommendation systems, marketing strategies, and audience engagement. This paper presents a novel approach to enhance sentiment analysis in the context of movie reviews by leveraging the synergies between Bidirectional Encoder Representations from Transformers (BERT) with Bidirectional Long Short-Term Memory (BiLSTM) networks and the valence-aware dictionary and sentiment Reasoner (VADER) model. Through a comprehensive evaluation on benchmark movie review datasets, we demonstrate that the fusion of BERT-BiLSTM with VADER results in improved sentiment analysis performance compared to existing methods and illuminates the intricate emotional landscapes depicted in movie reviews.

1 INTRODUCTION

In this era of rapid technological advancement, the proliferation of online platforms has democratized the procedure for expressing opinions and emotions, with movie reviews emerging is a well-known medium for articulating audience sentiments toward cinematic experiences. Analyzing these reviews not only offers insightful information regarding audience preferences but also gives a crucial component in decision-making processes for filmmakers, distributors, and movie enthusiasts alike. However, the inherently subjective and nuanced nature of feelings in people poses a significant challenge in accurately capturing the sentiment conveyed within these textual narratives. Traditional sentiment analysis techniques often struggle to decipher the subtleties of language, limiting their efficiency in accurately reflecting the full spectrum of movie emotions.

To address this challenge, this research paper introduces a novel approach to enhance sentiment analysis in the domain of movie reviews by leveraging advanced deep learning techniques, specifically Bidirectional Encoder Representations from Transformers (BERT) with Bidirectional Long Short-Term Memory (BiLSTM) networks, in conjunction with the Valence Aware Dictionary and sentiment Reasoner (VADER) model. By combining the semantic understanding capabilities of BERT-BiLSTM with the lexicon-based sentiment analysis of VADER, we aim to unravel the intricate emotional landscapes embedded within movie reviews with unprecedented depth and accuracy. This hybrid approach capitalizes on the strengths of both deep learning and lexicon-based methods, offering a comprehensive framework for analyzing movie sentiments.

2 RELATED WORK

The fusion of BERT with BiLSTM in sentiment analysis has shown prefect results in various domains. Researchers have proposed models like BERT-E TextCNN-ELSTM [1], BERT with BiLSTM and BiGRU [2], and BERT combined with BiLSTM for the evaluation of emotional responses in dialogues [3]. Additionally, a model called UA-BFET utilizes feature enhancement

DOI: 10.1201/9781003589273-76

technology for sentiment analysis on unaligned text and video data [4]. Furthermore, a dual-channel parallel hybrid artificial neural network architecture integrating BERT, CNN, and BiLSTM has been developed for text sentiment analysis, achieving high accuracy and F1 scores [5]. These models leverage the strengths of BERT in capturing contextual information and BiLSTM in modeling long-term dependencies, showcasing the effectiveness of combining these techniques for enhanced sentiment analysis across different applications.

In[1],The authors proposed a BERT-ETextCNN- ELSTM (Bidirectional Encoder Representations from Transformers-Enhanced Convolution Neural Networks- Enhanced Long Short-Term Memory) model for sentiment analysis. In [2],the authors proposed four deep learning models based on a combination of BERT with BiLSTM and Bidirectional Gated Recurrent Unit (BiGRU) algorithms. The authors proposed a cross-modal sentiment analysis model based on feature enhancement technology in unorganized data scenarios, which can perform sentiment analysis on unorganized text and video modality data in social media [4].

In[3], the authors proposed the BERT (bidirectional encoder representation from transformers) model that is used to generate word-level and sentence-level vectors, and then, word level vectors are combined with Bi-LSTM (bi- directional long short-term memory) that can better capture bidirectional semantic dependencies, and word level and sentence level vectors are connected and inputted to linear layer to determine emotions in dialogue. In [5], discussed by the authors proposed a Bi-directional Encoder Representations from Transformers (BERT)-based dual- channel parallel hybrid neural network model for text sentiment analysis.

3 METHODOLOGY

3.1 *Data collection*

The data gathers movie reviews from colorful sources, similar as movie examine websites and social media networks, and online forums. These reviews are collected in real- time or recaptured from being datasets to insure a different and comprehensive dataset for analysis.

3.2 *Text preprocessing*

The textbook preprocessing module cleans and preprocesses the raw textbook data to remove noise, inapplicable information, and formatting inconsistencies. This includes way similar as tokenization, lowercasing, junking of stop words, punctuation, and special characters, stemming or lemmatization, and running of emojis and emoticons.

3.3 *Feature extraction*

The Feature Extraction module excerpts applicable features from the preprocessed textbook data to represent the content of movie reviews in a format suitable for analysis. This may involve ways similar as bag- of- words(arc), term frequency- inverse document frequency (TF- IDF), word embeddings (e.g., Word2Vec,GloVe), or deep literacy- grounded representations (e.g., BERT embeddings).

3.4 *Sentiment analysis model*

The sentiment analysis model module comprises machine literacy or deep literacy algorithms trained to classify the sentiment of movie reviews into predefined orders (e.g., positive, negative, neutral). This module utilizes the uprooted features as input and predicts the sentiment marker for each movie review.

3.5 *Evaluation and confirmation*

The evaluation and confirmation module assesses the performance of the sentiment analysis model on a separate test dataset to measure its delicacy, perfection, retrieve, and F1- score.

This ensures the trustability and performance of the sentiment analysis system in directly relating sentiment from movie reviews.

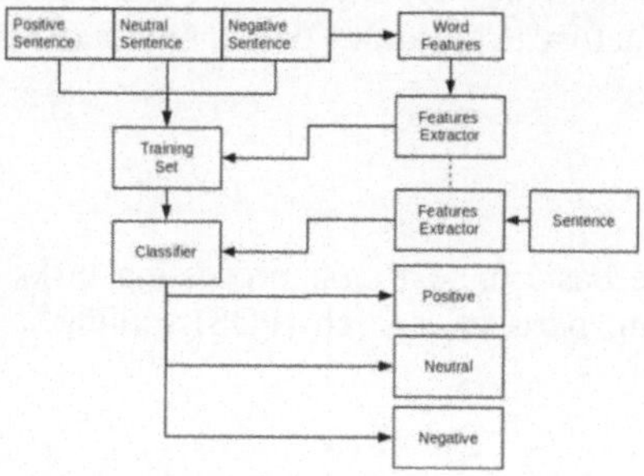

Figure 1. Sentiment analysis system architecture.

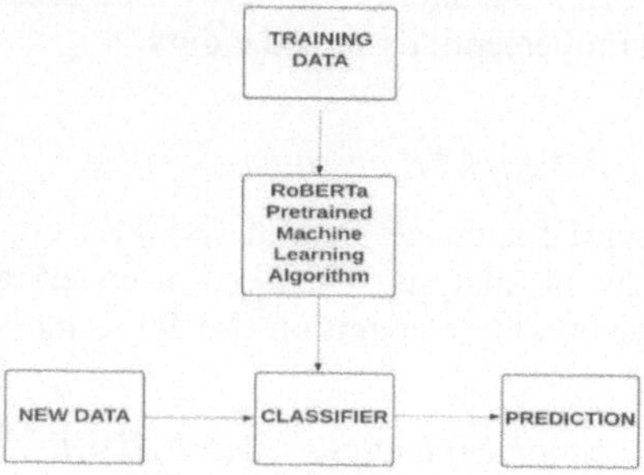

Figure 2. Data flow diagram.

The emergence of digital platforms and social media has normalized the process of movie reviewing, allowing cult to voice their thoughts and take gestures with pictures in real- time. Consequently, the volume of movie- related content, including stoner reviews, conditions, and commentary, has grown exponentially, creating vast openings for evaluation of movie review sentiments.

3.6 *BERT architecture*

Recurrent and convolutional neural networks use successional calculation to induce prognostications. That is, they can prognosticate which word will follow a sequence of given words formerly trained on huge datasets. In that sense, they were considered unidirectional or environment- free algorithms.

By discrepancy, motor- powered models like BERT, which are also grounded on the encoder- decoder armature, are bidirectional because they prognosticate words grounded on the former words and the following words. This is achieved through the tone- attention medium, a subcaste that's incorporated in encoder and the decoder. The thing of the attention subcaste is to capture the contextual connections being between different words in the input judgment.

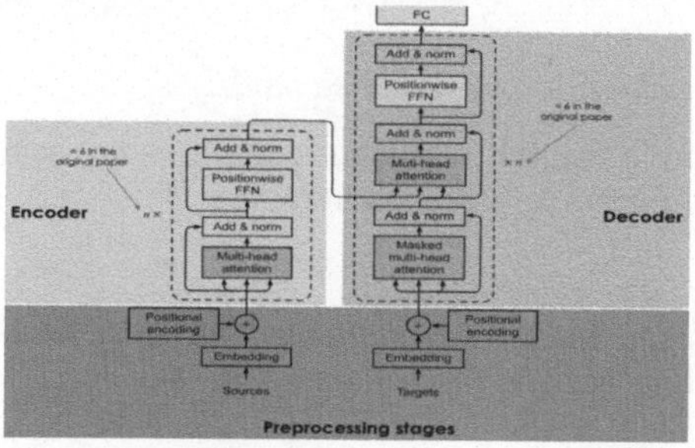

Figure 3. System architecture.

4 EXPERIMENT

In this section, we claw into the varied tasks performed as part of a comprehensive sentiment analysis design. Each task is strictly detailed, showcasing the methodologies employed, law particles employed, and perceptivity gained. The chapter is separated by sections, each of which sub- section fastening on a specific task performed during the design.

4.1 *Data preprocessing*

Data preprocessing lays the foundation for effective sentiment analysis by preparing the dataset for examination. Within this section, we bandy the way involved in loading and exploring the dataset, along with performing introductory exploratory data analysis (EDA) to gain perceptivity into the data.

4.2 *Basic NLTK operations*

Natural Language Toolkit (NLTK) operations are basic to textbook processing tasks in sentiment analysis. This section covers tokenization, part- of- speech (POS) trailing, and named reality recognition (NER) using NLTK.

4.3 *Sentiment analysis using VADER*

VADER (Valence Aware Dictionary and sentiment Reasoner) is a wordbook and rule-grounded sentiment analysis tool integrated with NLTK. This section demonstrates how VADER is employed to dissect sentiment grounded on a bag-of-words approach.

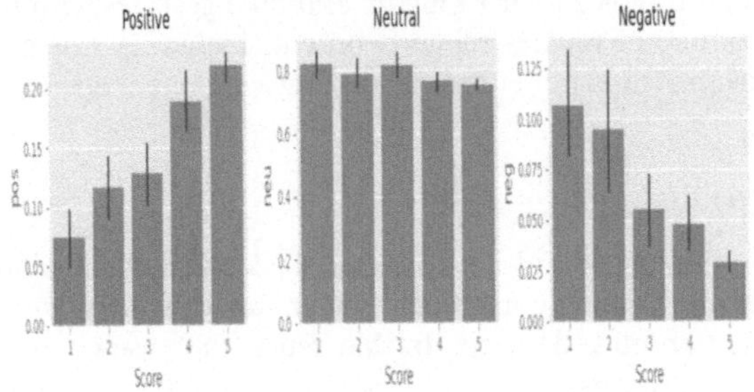

Figure 4. Visualization of VADER results.

Positive: *Text's likelihood being positive.* Negative: *Text's likelihood being negative.* Neutral: *Text's likelihood being neutral.*

Compound: A normalized, weighted composite score that sums the valence scores of each word in the lexicon.

4.4 *Sentiment analysis using roberta pretrained model*

In this section, we explore advanced sentiment analysis ways using a pre-trained Roberta model from the Hugging Face Transformers library.

4.4.1 *Sentiment analysis with roberta model*

A function is defined to gain sentiment scores from the pre-trained Roberta model, enabling advanced sentiment analysis capabilities.

4.4.2 *Integration with dataset and visualization*

Roberta Sentiment ratings are determined for the entire dataset and merged with the VADER scores for comprehensive sentiment analysis insights.

4.4.3 *Comparison of sentiment models*

A pair plot is generated to compare the sentiment ratings obtained from VADER and the pre- trained Roberta model.

4.5 *Comparison of sentiment models*

In this section, we explore examples of positive 1-star and negative 5-star reviews to understand cases where the model scoring and review scores differ significantly.

4.6 *Transformers pipeline*

The Hugging Face Transformers library's pipeline is demonstrated as a quick and easy way to run sentiment predictions, showcasing the power and simplicity of transformer models.

4.7 *Results*

In this chapter, the results attained from the evaluation of movie review sentiments using the BERT and VADER models. also, we bandy the performance criteria similar as delicacy, perfection, recall, F1- score, and the bracket report generated for the analysis. The emotion detection system was estimated on a collection of movie reviews, and the following performance criteria were calculated to assess the delicacy of the models.

4.7.1 *Accuracy (ACC)*

Accuracy measures the overall correctness of the emotions prognostications created by the models and is calculated using the formula

$$Accuracy = (TP + TN)/(TP + FP + FN + TN)$$

where,

- TP = True Positive; TN = True Negatives; FP = False Positive; FN = False Negative

While accuracy is a useful measurement, it has limitations, particularly in light of imbalanced datasets where a single class possibly more prevalent than others. A high accuracy score could be deceptive in these circumstances, as the model could be correctly predicting the dominant class most of the time but failing to accurately identify instances of the minority class. Therefore, while accuracy gives an overall picture of model performance, it ought to be taken into account with additional measures like precision, recall, and F1- score, which provide a deeper comprehension of how the model performs across different classes and these measurements are computed using below formulas.

$$*.Precision\ \ (P) = True\ \ Positive/(True\ \ Positive + False\ \ Positive)$$

$$*.Recall\ \ (R) = True\ \ Positive/(True\ \ Positive + False\ \ Negative)$$

$$*F1\ \ Score = 2 * (Precision * Recall)/(Precision + Recall)$$

4.7.2 *Classification report*

The Classification report gives a thorough rundown of the model's performance across different sentiment classes (e.g., positive, negative, neutral).It includes criteria similar as perfection, recall, F1- score, and support for each class.The fusion of BERT-BiLSTM with the VADER model represents a significant step towards advancing the state-of-the-art in sentiment analysis methodologies. By addressing the inherent challenges in capturing and interpreting human emotions expressed in textual narratives, our research opens up exciting opportunities for leveraging sentiment analysis for informed decision-making, audience engagement, and content recommendation systems.By integrating sentiment analysis into recommendation systems, organizations can significantly improve user satisfaction, leading to increased engagement and loyalty. This pioneering approach paves the way for more dynamic and responsive content delivery across various platforms.

```
VADER Model Accuracy: 0.64
VADER Model Classification Report:
              precision    recall  f1-score   support

    negative       0.74      0.52      0.61        27
    positive       0.58      0.78      0.67        23

    accuracy                           0.64        50
   macro avg       0.66      0.65      0.64        50
weighted avg       0.66      0.64      0.64        50


Roberta Model Accuracy: 0.84
Roberta Model Classification Report:
              precision    recall  f1-score   support

    negative       0.88      0.85      0.87        27
     neutral       0.00      0.00      0.00         0
    positive       0.83      0.83      0.83        23

    accuracy                           0.84        50
   macro avg       0.57      0.56      0.56        50
weighted avg       0.86      0.84      0.85        50
```

Figure 5. Classification report.

5 CONCLUSION

Evaluation of movie review sentiments has emerged as a critical area of research because of its capacity to offer valuable insights into audience sentiments and preferences. The research papers within this domain have investigated various methodologies aimed at effectively analyzing sentiments expressed in movie reviews. These endeavors have encompassed a spectrum of approaches, each -offering unique insights and advancements within the domain of sentiment analysis. In conclusion, the exploration of diverse methodologies an approaches in sentiment analysis of movie reviews reflects the growing importance and interest in understanding audience perceptions and preferences in the realm of cinema. The continuous evolution of techniques, from traditional lexicon-based methods to state-of-the-art deep learning architectures, underscores the dynamic nature of research in this field. As sentiment analysis continues to play a pivotal role in shaping decision-making processes for film-makers, distributors, and marketers, the quest for innovative approaches and methodologies is poised to drive further advancements in our ability to unravel the colorful tapestry of emotions expressed in movie reviews.

In summary, while "Unveiling Movie Emotions: BERT- BiLSTM Fusion and VADER Models for Enhanced Sentiment Analysis" contributes by proposing a novel fusion method for sentiment analysis, it also faces challenges related to complexity, computational cost, interpretability, and the dependence on external lexicons like VADER. These factors should be considered when assessing its applicability in different practical scenarios.

REFERENCES

[1] Zapletalová Š. and Starzyczná H., (2023), *Customer Behaviour in ECommerce: Case Studies from the Online Grocery Market*. Springer Nature.

[2] Hammi S., Hammami S. M., and Belguith L. H., Advancing aspect-based sentiment analysis with a novel architecture combining deep learning models CNN and bi-RNN with the machine learning model SVM, *Soc*

[3] Edara D. C., Vanukuri L. P., Sistla V., and Kolli V. K. K., (2023), Sentiment analysis and text categorization of cancer medical records with LSTM, *Jurnal of Ambient Intelligence and Humanized Computing*, vol. 14, no. 5, pp. 5309–5325.

[4] Pamukti Y. B. P. and Rahardi M., (2022), Sentiment analysis of bandung tourist destination using support vector machine and Naïve Bayes algorithm, in *2022 6th International Conference on Information Technology, Information Systems and Electrical Engineering (ICITISEE)*: IEEE, pp. 391–395.

[5] Pang B. and Le L., (2008), Sentiment analysis and opinion mining, *Foundations and Trends in Information Retrieval* 2(1–2), pp. 1–135.

[6] Jie Yang University of Wollongong, Australia, Mining Chinese social media UGC- a big- data frame for assaying Douban movie reviews, *Journal of Big Data Springer*, 2016.

Data Science & Exploration in Artificial Intelligence – Gururaj et al. (Eds)

Crypto currency price prediction using LSTM2A-200 analysis for strategic investment moves

S. Ciyamala Kushbu
Assistant Professor, Information Science and Engineering, CMR Institute of Technology, Bengaluru, India

P. Arthi Devarani
Assistant Professor, Electronics and Communication Engineering, R.M.K College of Engineering and Technology, Puduvoyal, Tamil Nadu, India

M. Amala Justus Selvam
Professor, Mechanical Engineering, Vel Tech Rangarajan Dr Sagunthala R&D Institute of Science and Technology, Chennai, India

K. Dhivya
Assistant Professor, Electronics and Communication Engineering, SRM Institute of Science and Technology, Kattangulathur, India

Suraj Kumar & Alok Diptanshu
Student, Information Science and Engineering, CMR Institute of Technology, Bengaluru, India

ABSTRACT: Cryptocurrency represents a digital form of currency where transactions are performed entirely online, eliminating the necessity for physical money. This study distinguishes between income and money within virtual currency, noting that users can access services without intermediaries. The use of cryptocurrencies has a significant impact on international relations and corporate activities due to their extreme price volatility. The rapid and often erratic changes in cryptocurrency prices highlight the urgent need for reliable predictive methods. Effective cost management within organizations is essential, as it influences key management levels and affects global business relationships. Our Long Short-Term Memory Moving Average 200 (LSTM2A-200) model provides valuable insights for traders, investors, and researchers, helping them make informed decisions in the volatile world of cryptocurrency trading.

Keywords: Cryptocurrency, Trading, Transactions, LSTM2A

1 INTRODUCTION

Cryptocurrency markets are highly speculative, and predictions should be approached cautiously. Although some analysts and models can offer insights, the future prices of cryptocurrencies remain unpredictable, and participants should be ready for both gains and losses. Our system inputs text data from news headlines and tweets, aggregated daily to maintain the time-series nature of the data.

DOI: 10.1201/9781003589273-77

Given the inherent risks in cryptocurrency markets, investors should adopt risk management strategies, such as setting stop-loss orders and diversifying their portfolios, to mitigate potential losses. The daily majority label for each coin serves as the final prediction for that day. The promising technology behind cryptocurrencies, particularly blockchain, suggests ongoing and growing usage. Machine learning, which focuses on developing algorithms and statistical models, is crucial for improving predictive models in cryptocurrency markets.

2 LITERATURE SURVEY

Current research on blockchain technology and its applications is illustrated in [1], highlighting blockchain's role in enabling transparent and secure commercial transactions without intermediaries. The growing interest in using blockchain for cryptocurrency price prediction and a review of existing literature is demonstrated in [2]. Traditional forecasting methods, such as time series analysis, have limitations, but the unique dynamics of the cryptocurrency market provide powerful insights.

Potential applications of blockchain in the financial industry, including payments and identity verification, are discussed in [3]. The authors cite examples of existing blockchain solutions, emphasizing benefits like cost efficiency, reduced expenses, and increased transparency. The interest in using social media feeds to predict cryptocurrency prices is highlighted in [4]. Social media platforms like Twitter, Facebook, and Instagram have become valuable information sources, and deep learning techniques like content analysis are used to analyze this data.

In [5], the relationship between cryptocurrency price movements and its trend is highlighted. Long Short-Term Memory (LSTM) [6,9] are highly used to identify historical patterns and henceforth predict future price movements.

3 DATASET USED

Bitcoin, dataset provides a comprehensive historical record of Bitcoin price movements and trading volumes sourced from Yahoo Finance, offering reliable data for analysis [7,8,10].

4 PROPOSED METHOD

The project achieves cryptocurrency price predictions with 80-90% accuracy using the Long Short-Term Memory Moving Average 200 (LSTM2A-200) model, primarily focusing on Bitcoin. Historical price data for various cryptocurrencies was fetched using the "yfinance" library, along with NumPy for data manipulation.

While our method provides highly accurate predictions, unexpected changes in Bitcoin prices may still occur due to its inherent volatility. The process involves acquiring cryptocurrency datasets from Yahoo Finance, splitting them into training and testing subsets, and calculating moving averages for additional features.

The LSTM2A-200 model processes historical price and trading volume data to predict future price movements using moving averages. The flowchart for the proposed method is illustrated in Figure 1.

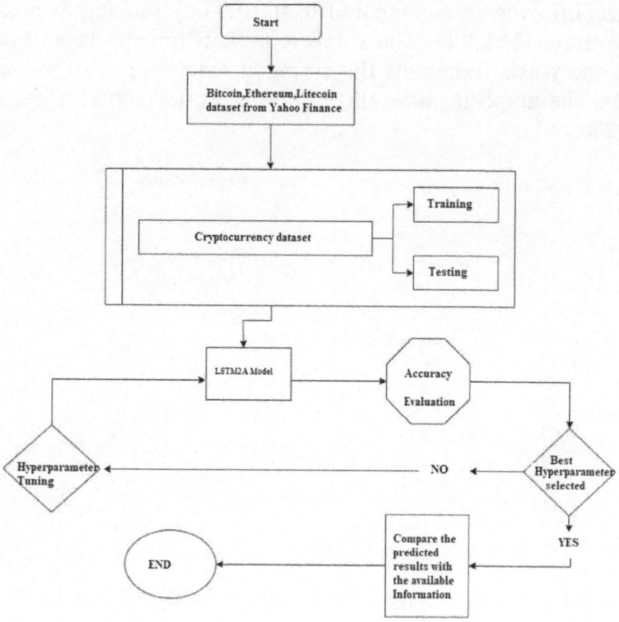

Figure 1. Proposed LSTM2A model.

The training dataset is enhanced with the moving average feature, and the LSTM model is trained. Hyperparameters are tuned iteratively until optimal values are identified based on testing accuracy [11].

Our LSTM2A model achieves 80-90% accuracy in predicting cryptocurrency prices, leveraging both LSTM's predictive power and moving average smoothing effect. However, it's important to note that no model can predict prices with 100% certainty due to market volatility and unexpected factors.

While various techniques are employed for financial market data estimation, Bitcoin is influenced by unpredictable factors. This study aims to assess the LSTM model's accuracy in predicting Bitcoin values, addressing the classic time series forecasting problem. While there's considerable research on using machine learning for time series prediction, specific focus on Bitcoin and cryptocurrencies is relatively limited.

Thus, it demonstrates the potential of using the LSTM2A model to predict cryptocurrency prices with a notable degree of accuracy. By leveraging historical data and advanced machine learning techniques, we provide valuable insights for traders, investors, and researchers. Despite the inherent challenges and volatility in the cryptocurrency market, our approach represents a significant step towards more accurate and reliable price prediction models [12,13].

As the cryptocurrency market continues to mature and evolve, so too will the tools and techniques used to analyze and predict its behavior [14,15]. The integration of diverse data sources and advanced machine learning models will be key to navigating the complexities and uncertainties of this dynamic market.

5 EXPERIMENT ANALYSIS

LSTM with moving averages of 50 days and 100 days are depicted in Figure 2(a) and (b) respectively. Figure 3 illustrates the price movement of a cryptocurrency (likely Bitcoin,

Ethereum, or Litecoin) over time compared to its 100-day moving average (MA100) and 200-day moving average (MA200). The x-axis represents time in days (0 to approximately 3500 days), while the y-axis represents the cryptocurrency's price (0 to over 70,000 units, presumably USD). The graph includes three lines: green for actual price, red for MA100, and blue for MA200.

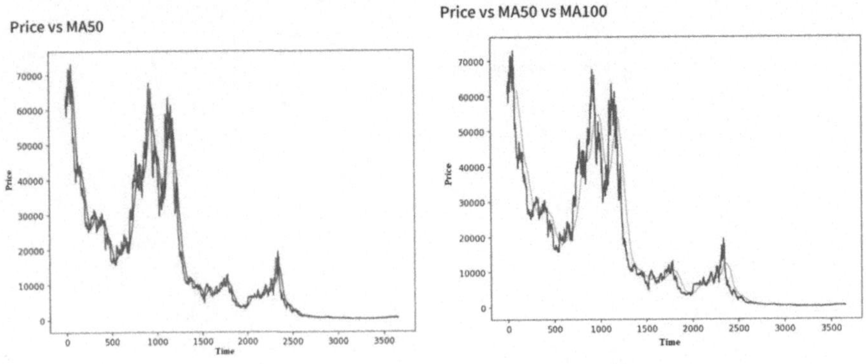

(a) (b)

Figure 2. LSTM with (a) moving average of 50 days (B) moving Avg of 50 vs 100-days.

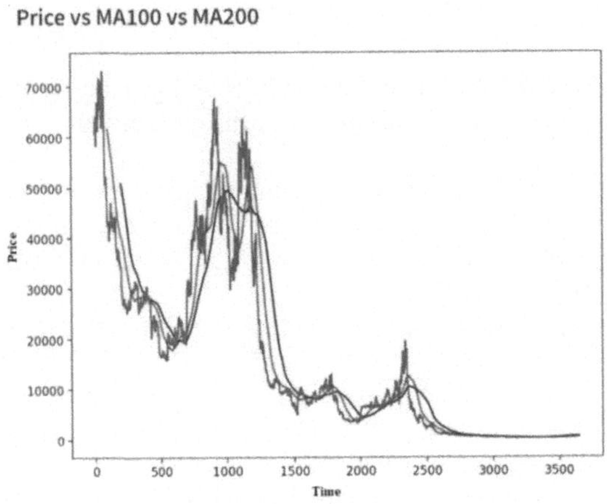

Figure 3. LSTM with moving Avg of 100 vs 200-days.

The green line, which represents the actual price, has sharp peaks and troughs that are typical of the high volatility found in cryptocurrency markets. The MA100 and MA200 lines, on the other hand, provide smoothed trends that eliminate short-term fluctuations and reveal more distinct long-term trends. Due to its shorter averaging window, the red MA100 line responds to price changes more quickly than the blue MA200 line. As a result, it provides a more immediate reflection of price movements over approximately three months. The MA200 longer window of approximately six and a half months indicates longer-term trends because it responds more slowly.Periods in which the actual price crosses above or below the moving averages are important observations from the graph. At the point when the cost transcends both moving midpoints, it frequently shows solid vertical force, flagging a potential upswing. On the other hand, when the cost falls beneath the moving midpoints, it recommends a downtrend. An upward crossover (where the MA100 crosses above the MA200) typically indicates a bullish trend and a potential buying opportunity, whereas a downward crossover typically indicates a bearish trend and a potential selling opportunity. The points where the MA100 crosses the MA200 are particularly significant.

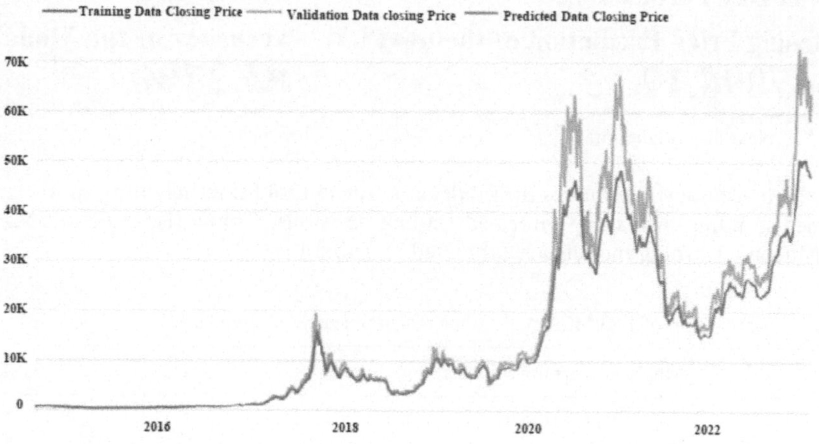

Figure 4. Actual Vs predicted.

During periods of high volatility, the sharp movements in the green price line are drawn by the smoother trajectories of the MA100 and MA200, which filters out noise and provide a clearer indication of the trend. For example, during an initial sharp rise in price, both moving averages lag, with MA100 catching up more quickly than MA200. Price peaks often exceed both moving averages, highlighting moments of intense upward pressure.

The smoothing effect of moving averages mitigates daily price volatility, allowing for better trend analysis. Traders extensively use these moving averages in their strategies; a price crossing above MA100 or MA200 may signal a buy opportunity, while a cross below may suggest a sell. The interaction between MA100 and MA200 serves as a crucial indicator; a golden cross (MA100 crosses above MA200) is seen as bullish, while a death cross (MA100 falls below MA200) is considered bearish.

Figure 4 demonstrates the performance of a cryptocurrency price prediction model is showcased with three datasets: training data, validation data, and predicted data. The x-axis represents time from 2015 to 2024, while the y-axis represents cryptocurrency closing prices, presumably in USD, ranging from 0 to over 70,000 units.

The blue line on the graph represents the training data, showing historical closing prices used to train the model from the beginning of the time period until early 2020, indicating relatively low prices and less volatility.

The orange line depicts the validation data, reflecting closing prices from early 2020 to mid-2022, showing significant price fluctuations with notable peaks and troughs, and highlighting a period of rapid price increases around late 2020 and early 2021.

The green line displays the model's predicted closing prices, extending beyond mid-2022 into 2024, closely following the validation data line and indicating a reasonable fit of the model to the actual data. The predicted data extending into 2023 and 2024 demonstrates the model's forecasted prices, maintaining consistency with prior price movements despite increased uncertainty over time.

This visual representation underscores the model's strength in capturing the general trend and volatility of cryptocurrency prices, while also pointing out areas where further refinement could enhance prediction accuracy, illustrating the model's effectiveness as a valuable tool for anticipating future price movements in the volatile cryptocurrency market. In summary, the 83.27% next-day prediction accuracy underscores the model's efficacy in a highly volatile market, which is shown in Figure 5.

Figure 5. Next day prediction.

This high accuracy rate makes the model a powerful tool for anticipating short-term price movements, aiding in making informed trading decisions. Comparisons of LSTM2A with other Machine learning models are tabulated in Table 1.

Table 1. Comparison of LSTM2A with other machine learning models.

S.No	Machine learning Models	Accuracy
1	Linear Regression	69.9%
2	Random Forest	75%
3	RNN with LSTM	76.7%
4	LSTM with 50 days Moving Average (LSTM2A 50)	80%
5	LSTM with 100 days Moving Average (LSTM2A 100)	81.5%
6	**LSTM with 200 days Moving Average (LSTM2A 200)**	**83.27%**

In a comparative analysis of various machine learning models, Linear Regression demonstrated an accuracy of 69.9%, indicating its moderate predictive power. Moving beyond traditional methods, LSTM (Long Short-Term Memory) models showed promising results. When Random Forest was employed independently, it achieved a commendable accuracy of 75%.Combining LSTM with RNN (Recurrent Neural Network) further enhanced predictive capabilities, resulting in an accuracy of 76.7%.

However, the integration of LSTM with moving averages significantly improved performance. LSTM combined with a 50-day moving average (LSTM2A 50) yielded an accuracy of 80%, while increasing the moving average window to 100 days (LSTM2A 100) pushed the accuracy to 81.5%. Remarkably, utilizing a 200-day moving average (LSTM2A 200) produced the highest accuracy of 83.27%, indicating the effectiveness of longer-term trends in the prediction process. These findings underscore the importance of feature engineering and model complexity in achieving superior predictive performance in machine learning applications.

6 CONCLUSION AND FUTURE SCOPE

In summary, different ML models are used to demonstrate predictive analytics. While using LSTM models with RNN architecture, accuracy is found to be better than using Linear Regression and Random forest methods. LSTM is then combined with 50 days, 100 days, and 200 days Moving averages, which proved extremely effective and increased predictive ability more than any other ML model. Comparatively, the LSTM2A-200 method provides high accuracy, which provides valuable insights to maximize investment returns in the market. Using the LSTM method in the research of price prediction is found to be more flexible in achieving high performance in forecasting tasks.

For increased accuracy, future studies might investigate sophisticated Neural Network topologies and incorporate outside variables like news events and market sentiment. Investment decisions can be improved by extending to multivariate analysis and real-time adaptive trading strategies.

Funding: The author(s) did not receive any financial support for this article.

Conflicts of Interest: No conflicts to declare.

REFERENCES

[1] Devineni, K. K. C., Ma, H., and Wang, Q. (2020). *Bitcoin Price Prediction with LSTM Networks.*

[2] Seabe, P.L., Moutsinga, C.R.B. and Pindza, E., (2023). Forecasting cryptocurrency prices using LSTM, GRU, and bi-directional LSTM: A deep learning approach. *Fractal and Fractional*, 7(2), p.203.

[3] Singh, S.K., Sharma, K.P. and Kumar, P., (2022, June). Analytical study for Price Prediction of Bitcoin using Machine Learning and Deep Learning. *In 2022 2nd International Conference on Intelligent Technologies (CONIT)* (pp. 1–8). IEEE.

[4] Chandra, S., Verma, S., and Sharma, V. (2020). *To Foresee Cryptocurrency Market Trends.*

[5] Bhargava, H., Bansal, S., and Kumar, A. (2020). *International Journal of Computer Applications.*

[6] Qazi, A. H., Khan, M. A., Ahmad, I., and Khurshid, K. (2021). *In 2021 International Conference on Artificial Intelligence.*

[7] Narayanan, A., Basheer, P., and Namboodiri, V. P. (2020). ML techniques for cryptocurrency price prediction. *Artificial Intelligence Review.*

[8] Zhong, C., Du, W., Xu, W., Huang, Q., Zhao, Y. and Wang, M., (2023). LSTM-ReGAT: A network-centric approach for cryptocurrency price trend prediction. *Decision Support Systems*, 169, p.113955.

[9] Livieris, I.E., Kiriakidou, N., Stavroyiannis, S. and Pintelas, P., (2021). An advanced CNN-LSTM model for cryptocurrency forecasting. *Electronics*, 10(3), p.287.

[10] Alessandretti, L., ElBahrawy, A., Aiello, L.M. and Baronchelli, A., (2018). Anticipating cryptocurrency prices using machine learning. *Complexity*, 2018, pp.1–16.

[11] Kakinaka, S. and Umeno, K., (2022). Cryptocurrency market efficiency in short-and long-term horizons during COVID-19: An asymmetric multifractal analysis approach. *Finance Research Letters*, 46, p.102319.

[12] Jaquart, P., Dann, D. and Weinhardt, C., (2021). Short-term bitcoin market prediction via machine learning. *The Journal of Finance and Data Science*, 7, pp.45–66.

[13] Chen, Z., Li, C. and Sun, W., (2020). Bitcoin price prediction using machine learning: An approach to sample dimension engineering. *Journal of Computational and Applied Mathematics*, 365, p.112395.

[14] Bai, S., Kolter, J.Z. and Koltun, V., (2018). An empirical evaluation of generic convolutional and recurrent networks for sequence modeling. *arXiv preprint* arXiv:1803.01271.

[15] Abadi, M., Agarwal, A., Barham, P., Brevdo, E., Chen, Z., Citro, C., Corrado, G.S., Davis, A., Dean, J., Devin, M. and Ghemawat, S., (2015). *TensorFlow: Large-Scale Machine Learning on Heterogeneous Systems.*

Data Science & Exploration in Artificial Intelligence – Gururaj et al. (Eds)

Hash-based Message Authentication Code (HMAC) with efficient data transfer in blockchain

L. Naveenkumar
Research Scholar, Department of CSE, UVCE, Bengaluru, Karnataka, India

S.H. Manjula
Professor, Department of CSE, UVCE, Bengaluru, Karnataka, India

ABSTRACT: Recently, the growth of cloud servers has led to an increase in the storage of large amounts of data as ciphertext in these servers. In decentralized systems, one of the main challenges is the communication overhead on the network caused by numerous data copies among nodes and authorized users. To address this issue, a semi-decentralized approach utilizing a Hash-based Message Authentication Code using the Secure Hash Algorithm-256 (HMAC-SHA-256) is proposed for secure and efficient data sharing in blockchain. It combines a specific message and a secret key to produce a hash value. In blockchain, HMAC with SHA-256 is used to generate and validate the signatures of multiple transactions. Our simulation results demonstrate that with a shard file size of 50 MB, it achieves superior transactions of 194 Transactions Per Second (TPS), underscoring its proficiency in generating secure hash-based authentication codes compared to IPFS-ECC's 188 TPS. Employing HMAC-SHA256 in blockchain offers a secure and efficient means of data distribution and communication.

Keywords: Blockchain, Hash Algorithm, Hash-based Massage Authentication Code, Secret

1 INTRODUCTION

The advancement of internet technology in recent years has positioned cloud storage as a pivotal platform for managing the exponential growth of data sharing. Cloud storage represents a major innovation in cloud computing, offering cost-effective and efficient solutions for storing multimedia data (Xu *et al.* 2022). However, this technology faces significant challenges, including issues related to security, availability, limited storage capacity, low reliability, low scalability, high costs, and incompatibility (Zhang *et al.* 2022). Cloud service providers are often criticized for their unreliability, as they can be vulnerable to attacks by external malicious actors or internal sabotage by employees (Alzahrani *et al.* 2022) his inherent risk translates to potential data security vulnerabilities in any cloud data-sharing protocol. The Encryption data is distributed as an essential security computer for protecting user data access permission. Blockchain technology, on the other side, includes a robust wide network system that guarantees data security across many computers, it is called nodes (Hsiao and Sung 2022). It provides an efficient solution for transferring data and storing large capacity of data while minimizing costs and improving accuracy (Bin Saif M *et al.* 2024).The integration of blockchain into cloud storage goal for persistent issues of integrity, security, and availability (Lai *et al.* 2022).

 DOI: 10.1201/9781003589273-78

1.1 *Motivation and contribution*

Despite the accessibility of cloud storage offers, mainly for managing voice information data, it continues to struggle with privacy and security issues. The increasing sources of speech data such as meeting records, courtroom evidence, military commands, communication records, educational platforms, healthcare systems, and business strategies highlight the urgent need for better data management tools. Traditional multimedia data management solutions are insufficient for the growing data volumes, leading to significant storage and secure sharing challenges. Users often lack control over their stored data, compromising privacy when unauthorized access occurs. Additionally, users are frequently unaware of their data's precise location and cannot ensure its authenticity and integrity. Motivated by these challenges, we propose a semi-decentralized-based HMAC-SHA-256 key management algorithm for blockchain. HMAC is a cryptographic authentication mechanism that combines a hash function with a secret key to verify message authenticity and integrity. It generates a hash value from a specific message and a secret key. In blockchain technology, HMAC paired with SHA-256 is used to create and validate transaction signatures. SHA-256 produces a 256-bit hash value, and utilizing HMAC with SHA-256 in blockchain provides a highly secure and efficient method for data distribution and communication.

This paper is structured as follows: Section 1 introduces the research background and motivation. Section 2 reviews related work. Section 3 outlines the proposed HMAC-SHA-256 key management algorithm for blockchain. Section 4 presents the results and analysis. Finally, Section 5 concludes the paper, summarizing the findings and suggesting future research direction.

2 LITERATURE REVIEW

(Jayanalan and Jeyanthi 2022) developed a scalable approach with an off-chain Interplanetary File System (IPFS), using multi-factor authentication, symmetric and asymmetric encryption, and SHA-256 hashing, Symmetric key encryption was used to secure data before storing it in IPFS, while asymmetric key encryption was employed to convey the symmetric key for generating digital envelopes for authorized parties. Digital signatures validate transactions to legitimate access nodes, and SHA-256 hashing was utilized to ensure data integrity. It faces scalability and key management issues.

(Liu *et al.* 2022) proposed a multi-authority attribute-based access control for blockchain-based supply chain data sharing. This approach stores encrypted data on distributed nodes for fine-grained access control. The key production protocol facilitated the secure distribution of secret keys to users via the attribute authority. Performance metrics evaluated included the size of the secret key and the time required for encryption and decryption, offering a thorough assessment of the protocol's efficiency and security. A key production protocol ensures secure key distribution through an attribute authority. Performance metrics included secret key size and encryption/decryption times, but communication overhead from multiple data copies remains a significant issue.

(Gao *et al.* 2022) introduced a hierarchical blockchain system for efficient data privacy and security, enhancing performance by allocating computational resources and continuously monitoring system status. The monitoring system aims to enhance resource scheduling by continuously tracking and evaluating operational statuses. It observes the status of peer containers and past tasks, enabling resource upgrades for future tasks. This ensures resources remain available across associated infrastructures, thus optimizing both performance and efficiency. While this approach improves resource scheduling, it faces scalability issues as the number of network nodes and consumers grows, leading to reduced performance.

(Wang and Guan 2022) proposed a blockchain-based method for secure, traceable data sharing. It uses attribute-based encryption and integrates off-chain IPFS with on-chain data. Encrypted data is stored in IPFS, The IPFS serves for storing encrypted data, with the hash value of this encrypted data documented on the blockchain. The devised model stored and distributed data records on the blockchain, offering visual aids for tracking and identification. The IPFS served as the encrypted data storage solution, with its hash values recorded on the blockchain. This model enabled the distribution of data records across the blockchain, integrating visual aids to simplify tracking and recognition. This method ensured secure data storage, leveraging blockchain's transparency and traceability for improved data management and accessibility. hash values on the blockchain. However, it faces single-point failure issues due to centralized access control.

(Ugochukwu *et al.* 2023) deployed an IPFS-based model in blockchain for secure distribution storage, resolving privacy, interoperability, and security issues in logistic systems. A decentralized and distributed network, coupled with the SHA-256 hashing algorithm, was utilized to safeguard user data privacy. This approach tackled privacy, interoperability, and security concerns within the logistics system, offering guidance on integrating blockchain technology. However, integrating diverse blockchain networks posed challenges due to interoperability issues.

3 PROPOSED METHODOLOGY

Within decentralized systems, the primary issue recognized is the substantial communication overhead on the network, arising from multiple data copies among data distribution points and nodes with authorized users. To mitigate this overhead for data users, a semi-decentralized approach is employed, utilizing HMACSHA-256 on the blockchain.

In Figure 1, we observe a semi-decentralized system utilizing HMAC-SHA256. One designated data user holds exclusive access rights to save each data copy, while others lack access privileges. When data access is needed, users retrieve it from the domain administrator and decrypt content locally decryption engine. This system implements a dual-level key management approach: encryption occurs both at the data owner and blockchain levels through hash code formation. Users possessing the decryption key gain access to data, ensuring consistent communication overhead. Additionally, data users store similar data to reduce storage duplication. HMAC, a cryptographic authentication system, employs a hash function and secret key to verify message authenticity and integrity. In the blockchain, HMAC with SHA-256 validates transaction signatures. Blockchain technology provides a decentralized ledger, ensuring transparency and security across the network. It eliminates the need for a central authority, reducing the risk of single-point failures. Moreover, blockchain's immutable nature ensures that all transactions are permanently recorded and cannot be altered. Detailed HMAC-SHA-256 generation process follows.

Generating HMAC-SHA-256, we begin by defining the key padding and padding constants for inner and outer operations as follows.

$$K' = pad(K)_t \tag{1}$$

$$ipad = 0x36 \text{ repeated to fill a block} \tag{2}$$

$$opad = 0x5\,C \text{ repeated to fill a block} \tag{3}$$

$$innerhash = SHA256((K' \oplus ipad)||M)_t \tag{4}$$

$$HMACSHA256 = SHA256((K' \oplus opad)||innerhash)_t \tag{5}$$

To formalize the HMAC-SHA-256 algorithm algebraically, we use several notations: K represents the secret key, M is the message, and T denotes time. The function SHA256 is the SHA-256 hash function, and ⊕ symbolizes the XOR operation. The function pad (K) is used for padding the key to the required length. The constants opad and iPad are the outer and inner padding constants, respectively. The inner hash function is denoted by inner hash, and || represents the concatenation operator. These definitions provide the foundational elements necessary to describe the HMAC-SHA-256 computation process, ensuring a clear and systematic approach to generating and validating secure hash-based message authentication codes.

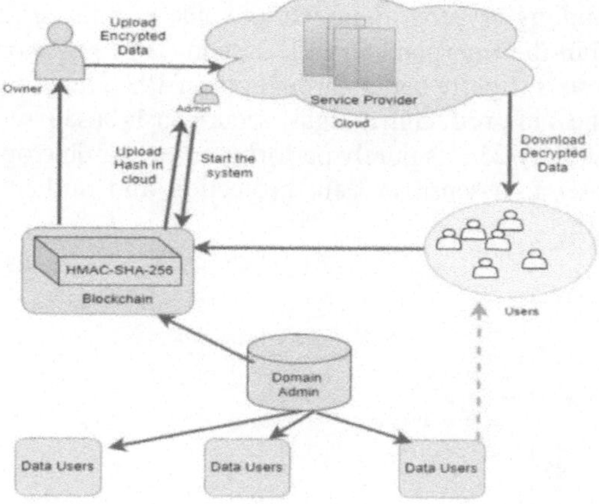

Figure 1. HMAC-SHA-256 based semi-decentralized system with blockchain.

4 RESULTS AND ANALYSIS

The HMAC-SHA-256 is implemented in a Python environment, utilizing system configurations consisting of a Windows 10 operating system, 16GB of RAM, and an Intel i7 processor. Various performance metrics such as retrieval time, throughput, encryption, and decryption time are considered to evaluate the performance of HMAC-SHA-256.

4.1 *Performance analysis*

In a quantitative examination concentrating on the retrieval duration for a 300MB shared file, IPFS, ECDSA, and HMAC-SHA-256 display unique performance attributes, as depicted in Figure 2. IPFS showcases a retrieval time of 3.9 milliseconds, emphasizing its efficacy in content-addressed storage and decentralized data dissemination, albeit marginally elevated due to network overhead. ECDSA records a retrieval time of 3.8 milliseconds. While its primary application lies in securing transactions through digital signatures, its performance remains competitive, contributing negligible overhead to the overall retrieval process. HMAC-SHA-256 encryption algorithm is used for ensuring user data authenticity and integrity, in the diagram, its fastest retrieval time is 2.8 milliseconds compared to IPFS and ECDSA encryption techniques. This showcase of efficiency in maintaining huge files without impacts on performance, HMAC-SHA-256 is a robust performance of cryptographic techniques while maintaining the large file size with high security and integrity.

4.2 *Comparative analysis*

This comparative analysis discusses HMAC-SHA-256 key encryption management along with the existing approach affecting Transactions per second (TPS). Existence methodologies such as Interplanetary File System –Elliptic Curve Cryptography (IPFS-ECC) [10] are checked and compared with the Hash-based Message Authentication code-SHA-256 bit and demonstrated in Figure 3. On examining the efficiency of Interplanetary File System (IPFS) utilizing Elliptic Curve Cryptography (IPFS-ECC) along with Hash-based Message Authentication Code with SHA-256 (HMAC-SHA-256). The X-axis represents a Shared file size of 50MB, and the Y-axis represents the throughput of transactions per second (TPS), It shows the performance maximum (TPS) with file size Interplanetary File System Elliptic Curve Cryptography (IPFS-ECC) throughputs of 188 Transaction per second (TPS), demonstrating and handling cryptographic works while promising data integrity and also providing security within the Interplanetary file system (IPFS) networks. In distinctiveness, HMAC-SHA-256 proves a slightly better throughput of 193 Transaction per Second (TPS), highlighting its robustness in producing a highly secure hash-based authentication code. The significance of HMAC-SHA-256 its purely performs cryptographic operations, emphasizing its performance the user user-sensitive data protection and authentication techniques in different digital platforms.

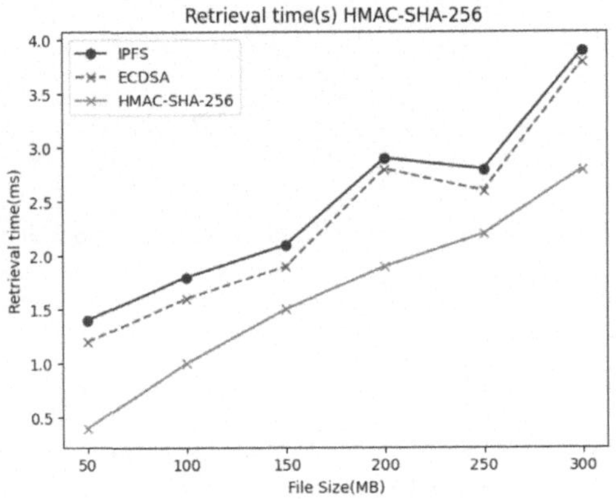

Figure 2. Impact of file size on throughput performance.

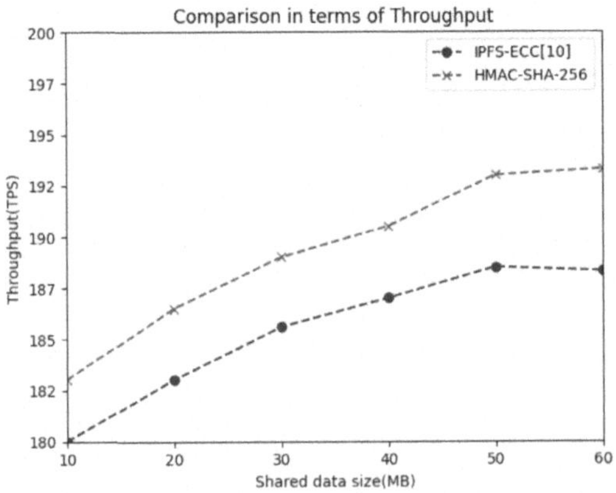

Figure 3. Comparison of throughput between shared file size and TPS (Transactions per Second).

5 CONCLUSION

In this work, the Hash-based Message Authentication Code with SHA-256(HMAC-SHA-256) encryption mechanisms is recommended for secure user information and efficient data sharing in blockchain cloud networks. HMAC, perform cryptographic authentication systems, in that cryptographic hash-based techniques is efficiently used together with the secret key to define the integrity and authenticity of the message. The HMAC contains of specific message and also a secret key to produce the hash value. In a blockchain network system, a Hash-based message authentication code (HMAC) is utilized with SHA-256 to produce and validate many transaction signatures. The Blockchain uses both HMAC and also SHA-256, its highly secure and concise technique to distribute or communicate user information. Finally, the HMAC-SHA-256 performs the best encryption time of throughput of 194 transactions per second (TPS), slightly exceeding the IPFS-ECC throughput of 188 TPS. In the future, research on different hash algorithms, such as HMAC-BLAKE2 strong security and performance that can be used to extend security and effective data sharing through blockchain.

REFERENCES

Alzahrani, A.G., Albomoud, A. and Wills, G., (2022). A framework of the critical factors for healthcare providers to share data securely using blockchain. *IEEE Access*, 10, pp.41064–41077.

Bin Saif M., Migliorini, S. and Spoto, F., (2024). Efficient and secure distributed data storage and retrieval using interplanetary file system and blockchain. *Future Internet*, 16(3), p.98.

Gao, L., Wu, C., Du, Z., Yoshinaga, T., Zhong, L. Liu, F. and Ji, Y., (2022). Toward efficient blockchain for the internet of vehicles with hierarchical blockchain resource scheduling, *Electronics*, 11(5), p.832.

Hsiao, S.J. and Sung WT, (2022). Blockchain-based supply chain information sharing mechanism. *IEEE Access*, 10, pp.78875–78886.

Jayabalan, J. and Jeyanthi, N., (2022). Scalable blockchain model using off-chain IPFS storage for healthcare data security and privacy. *Journal of Parallel and Distributed Computing*, 164, pp.152–167.

Lai, C., Ma, Z., Guo, R. and Zheng, D., (2022). Secure medical data sharing scheme based on traceable ring signature and blockchain. *Peer-to-Peer Networking and Applications*, 15(3), pp.1562–1576.

Liu, C., Xiang, F. and Sun, Z., (2022). *Multiauthority Attribute-based Access Control for Supply Chain Information Sharing in Blockchain Security and Communication Networks*, 2022, pp.1–18.

Ugochukwu, N.A., Goyal, S.B., Rajawat, A.S., Verma, C. and Illés, Z., (2023). Enhancing logistics with the internet of things: a secured and efficient distribution and storage model utilizing blockchain innovations and interplanetary file system. *IEEE Access*

Wang, Z. and Guan, S., (2023). A blockchain-based traceable and secure data-sharing scheme. *Peel Computer Science*, 9, p.e1337.

Xu, G., Zhang, J., Cliff, U.G.O. and Ma, C., (2022). An efficient blockchain-based privacy- preserving scheme with attribute and homomorphic encryption. *International Journal of Intelligent Systems*, 37(12), pp.10715–10750.

Zhang, D., Wang, S., Zhang, Y. Zhang, Q and Zhang, Y, (2022), A secure and privacypreserving medical data sharing via consortium blockchain Security and *Communication Networks*, 2022.

Data Science & Exploration in Artificial Intelligence – Gururaj et al. (Eds)
© 2025 the Author(s), ISBN: 978-1-032-96390-7

Deep learning-based Bi-directional long short-term memory networks for prediction analysis of long-term tenal ailment

M. Jayashree
Research Scholar, East Point College of Engineering and Technology, Visvesvaraya Technological University Belagavi, Assistant Professor, Department of Information Science and Engineering, CMR Institute of Technology, Bengaluru, India

N. Anitha
Professor, Department of CSE, BNM Institute of Technology, Bangalore, India

ABSTRACT: Long-term renal ailment or (CKD) is a global well-being concern that requires early and precise diagnosis techniques to enhance patient outcomes. To forecast Long-term Renal disease (CKD), this paper presents an efficient analytical model called the Bi-directional Long Short-Term Memory (BiLSTM) networks which uses temporal dependencies. The method progresses the calculation of accuracy compared to the conventional machine learning models. The CKD structured dataset is collected from the UCI repository and the preprocessing of data methods is employed to encode categorical variables, fill the missing values, and normalize the features. The suggested Bi-LSTM model is appropriate to attain common operational metrics like specificity, sensitivity, F1-score, and accuracy. The outcomes of experiments showed that the Bi LSTM-based[5] model attains its operational behaviour better than other learning techniques such as random decision forest, and K-nearest Neighbour with a reliable accuracy of 97.62% for early CKD identification. The initial diagnosis of kidney disease facilitates the individual to delay in progression of the functioning of the organ.

Keywords: Chronic Kidney Disease, Bi-directional Lengthy Short-Term Memory, Random Decision Forest, K-Nearest Neighbour, Precision, Accuracy

1 INTRODUCTION

A long-lasting illness that impacts millions of people worldwide, i.e chronic kidney disease (CKD) which is measured by a loss of kidney function, adding to increasing the threat of cardiovascular events and death. chronic kidney disease (CKD) can develop to end-stage renal disease (ESRD), which calls for dialysis or kidney transplantation. To stop the disease progression and save healthcare expenses [1], primary recognition and treatment of CKD are essential. However, early identification of CKD is difficult since the disease sometimes does not show symptoms till it gets to the advanced stage. Blood tests and urine analyses, two common traditional approaches for detecting kidney disease (KD), frequently miss early symptoms of the illness. The earlier diagnostic methods, often fall short in identifying initial-stage CKD and predicting its progression due to their dependence on static and sometimes subjective clinical indicators.

Topical improvements in Knowledge Engineering (AI) and a subset of AI called machine learning (ML) have improved the timeliness and accuracy of medical

 DOI: 10.1201/9781003589273-79

diagnosis. A subset of ML called multi-layered learning techniques has demonstrated remarkable success in various domains, including image recognition. It performs the pattern analysis of processing the language naturally (NLP), time-series data, and the ability to model composite outlines in huge datasets. Among the various deep learning architectures, The LSTM networks have proven predominantly effective in handling sequential data and capturing long-lasting dependencies, suitable for medical time-series data analysis.

The main objective of using Bi-directional Long Short-Term Memory (Bi-LSTM) networks is to create a predictive model for CKD. The variant of Recurrent neural networks (RNNs) called Bi-LSTM[5,8] networks are especially well-suited for sequential analysis and timely series of data, perfect for capturing the temporal patterns in the advancement of renal disease.

The structured dataset is collected from the online (UCI) repository consisting of 400 instances [14] and 24 attributes having nominal and categorical values. Kidney disease may go

unnoticed till last stage, leading to cardiovascular disease. Finding the early symptoms also makes major concern. The dataset containing the 24 attributes is reduced by considering very critical attributes thereby decrease in classification time and increase in the accuracy even lessening the features.

Various ML techniques[12] such as RF, optimal hyper plane and K-nearest Neighbour are used and studied for the initial detection of Kidney disease[2], yet achieved an accuracy of 98.7,62.7 respectively, here the reduced set of features are considered with an increase in the correctness of prediction.

2 RELATED WORK

Arif-UL-Islam. et.al [1] Recognition of CKD was done with the help of boosting algorithms and the rules are derived illustrating the relationship between the various attributes of CKD. The paper used the Ant-Miner algorithm to derive rules.

Gabriel R [2].CKD was detected using the NN-CBR twin system. The explanation-based instance is considered. Here the accuracy of the NN classification is compared. CBR System is evaluated considering the explanation given to the instances.

Beverley M [3]. discussed the systematic approach of extracting data from the EMRs for detecting cases of a named clinical condition. Various methods of mining information from the text and the accuracy of detecting the cases using the algorithms were reviewed. The method adopted by this author was searching the keywords, rule-based algorithms, and ML methods[2]. Performance metrics like sensitivity and operating characteristics were observed

Hybrid technique called Pearson correlation for feature selection and hybrid classifiers namely, Naive Bayes, decision tree, and gradient boosting was employed in [12], Which resulted in the improvement of accuracy score in an extensive manner.

Shahid Mohammad Ganie et.al,[13] Here the author proposed an ensemble method of boosting techniques taking into consideration the clinical parameters for CKD prediction. ensemble learning resulted in the improvement of accuracy and minimized the run time in a significant manner

In [14], The kidney prognostic score was evaluated by utilizing the Berden classification to forecast the risk of end-stage. employing this classification model resulted in the ROC curve improvement

Mariella Gregorich et.al [15],The author has suggested that the progression mechanism can be employed for the in-depth validation of eGFR for CKD.

3 METHODOLOGY

The Primary detection of renal disease is carried out using the Deep Learning approach BiLSTM and comparing the analysis with ML algorithms [4] like random decision forest and K nearest neighbour. the prediction of the disease is performed after preprocessing the data by a series of steps[16]like noise removal, and filling missing values.

The below figure shows the block diagram of the process to classify the disease as CKD or NCKD.

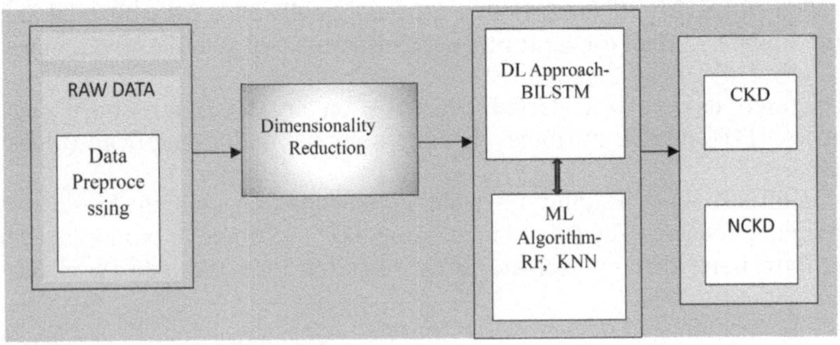

Figure 1. The block diagram of the process.

3.1 *The data preprocessing approach includes*

3.1.1 *Handling missing values*

The dataset contains both Numerical features where the misplaced values can be imputed using the mean of the respective features and categorical features values by calculating the mode of the respective features.

3.1.2 *Normalization*

Min-Max scaling approach is used to normalize the features in the range of [0-1]

3.1.3 *Encoding categorical variables*

Converting the categorical values to numerical values using one-hot encoding

3.1.4 *Sequence generation*

The BiLSTM Model uses sequential data, hence data is organized into a sequence, each patient record is considered as a sequence to capture the time-based dependencies in the advancement of the CKD

3.2 *The structural layout of the Bidirectional long short-term memory*

The Bidirectional Long Short-Term Memory (Bi-LSTM) model architecture captures both frontward and rearward temporal dependencies in the dataset. The Bi-LSTM Contains Layers similar to ANN. It contains two layers to process the input in the advancing direction and the other for processing in the rearward direction.

3.2.1 *Input layer*

Takes the features as a sequence of input. The shape of the input data is (number of sequences, sequence length, number of features). where Xi are the inputs

3.2.2 *BiLSTM layers*

Bidirectional LSTM has two layers, each with 64 units, to process the input sequences in both frontward and rearward directions.

3.2.3 *Output layer*

The LSTM gives the probability vector as output and the ending output is the blend of both of these probabilities.

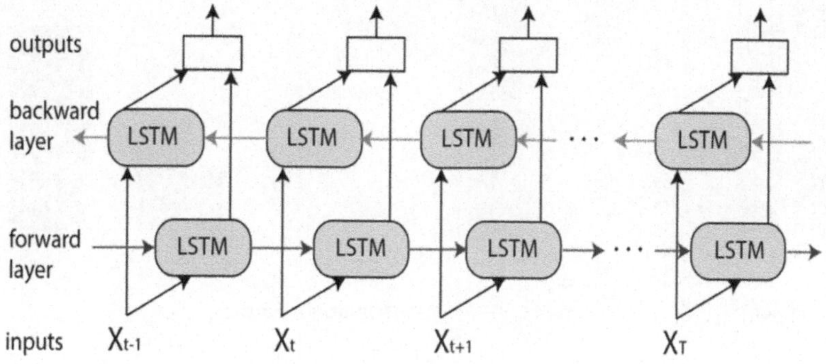

Figure 2. The architecture of the BILSTM.

3.3 *Working of BILSTM and advantage in predicting the kidney disease*

BiLSTM is a variation of a Recurrent Neural network designed to address the gradient and capture the periodic data(long dependencies), having a memory cell, forward LSTM, Backward LSTM, and output layer[5]. BiLSTM captures dependencies from both previous and upcoming data points, providing complete understanding of the disease progression. Because of its bidirectional context, the BiLSTM gives better accuracy and other performance metrics compared to other traditional models and BILSTM processes sequential data which is particularly suited for medical time series data.

3.4 *Working of BILSTM for CKD prediction*

The CKD is predicted using BILSTM designing the model to analyze the temporal patterns of the patient data collected from the UCI repository, The data comprising of the patient's record as a sequence of times steps which includes the clinical parameters such as Blood pressure, creatinine levels, glucose levels and other parameters [11,12] The BILSTM is designed and constructed to handle the input dimensions, hidden units, and layers to handle the sequential dependencies of data. then later data is split into training followed by testing data sets. The BILSTM is trained on patient data, optimizing it to minimize the prediction error. Later the model is evaluated for Accuracy, Precision, and recall.

4 EXPERIMENTAL RESULTS -PERFORMANCE METRICS

The performance metrics Accuracy, Precision, and Recall are calculated using BiLSTM, Random decision forest, and KNN algorithms by running in the Google Colab. the UCI repository data sets are loaded with 13 features and the results are compared with above-mentioned algorithms. The dataset contains the given attributes. Blood pressure(BP), Specific gravity(sg), Albumin(al), Sugar(su), Red blood cells(rbc), Blood urea(bu), Serum creatinine(scr), Sodium(s), Potassium(p), Haemoglobin(HB), Wbcc, Rbcc, Hypertension (ht), and target class.

4.1 *Exploratory data analysis*

The data analysis process helps in discovering the pattern to analyze any anomalies, with the support of statistical measures such as mean, and median. Figure 3 indicates the quantity of true and false instances for HTN and for RBC in Figure 4.

The Exploratory data analysis done for different attributes like HTN and RBC

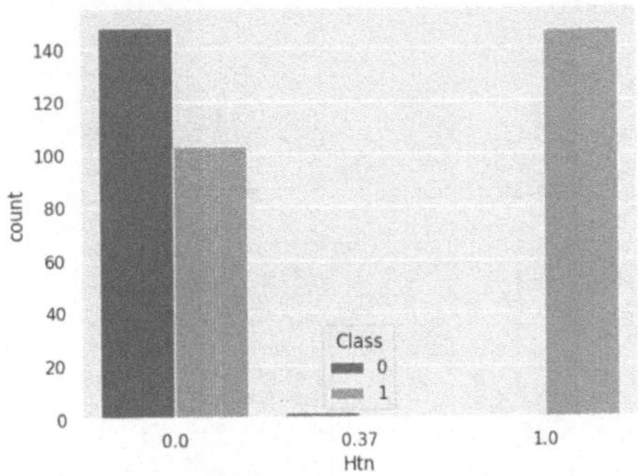

Figure 3. Hypertension.

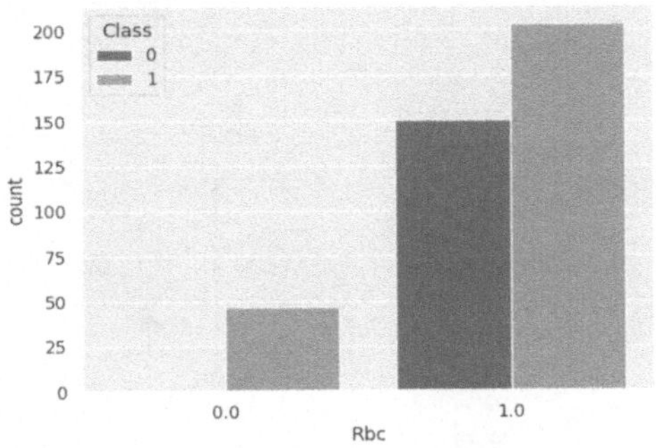

Figure 4. RBC.

The outliers are checked for the attributes BP and serum creatinine which are the important part of exploratory data analysis.

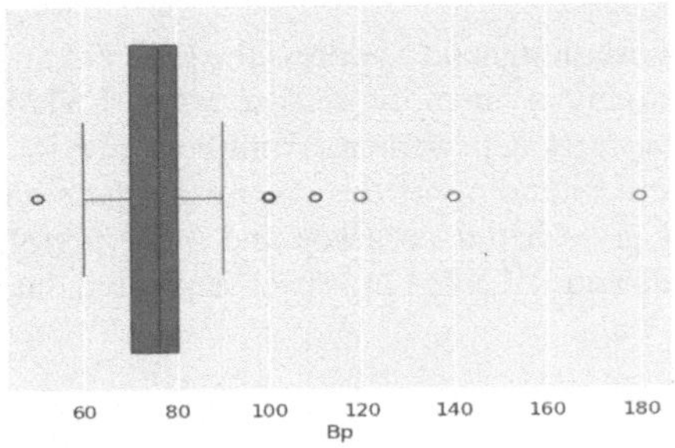

Figure 5. The outliers for the BP.

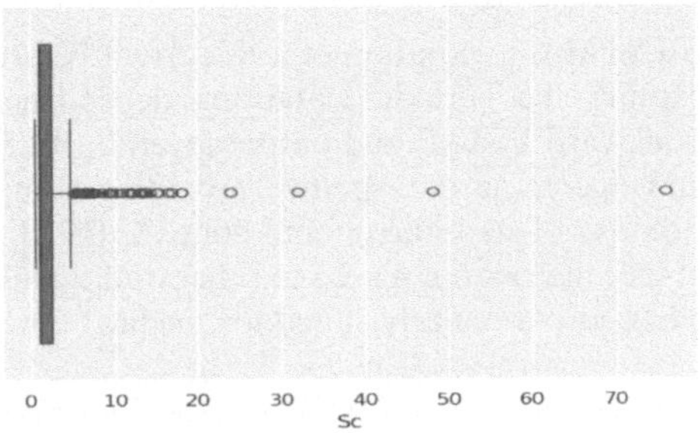

Figure 6. The outliers for the sc.

The Confusion matrix for BILSTM, Random Forest, and KNN is illustrated in Figures 7,8,9, respectively. The total of instances that are truly positive and truly negative is calculated. The accuracy of BILSTM is 98.75%, RF is 97.62%, and KNN is 62.5%, The maximum accuracy is achieved with BILSTM, where the prediction is fine-tuned.The Experimental result value for the BILSTM, Random Forest and KNN are illustrated below.

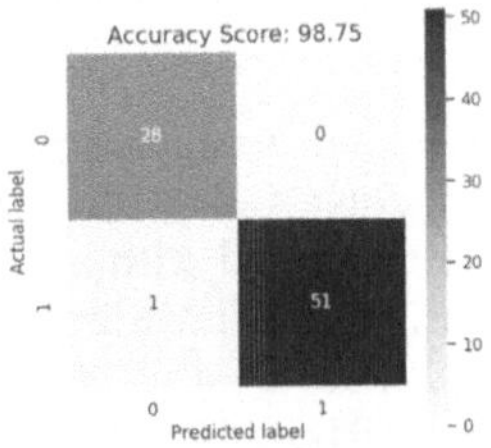

Figure 7. Confusion matrix -BILSTM.

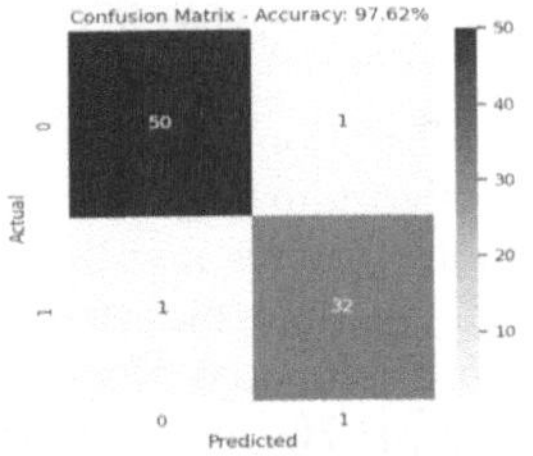

Figure 8. Confusion matrix – RF.

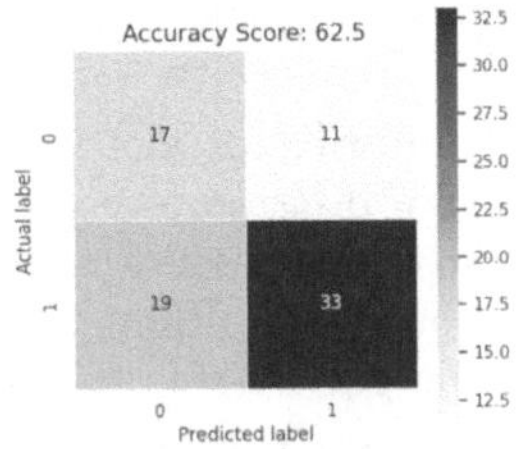

Figure 9. Confusion matrix – KNN.

The Specificity(precision), Sensitivity(Recall) and F1-score performance metrics are calculated and observed for BiLSTM, RF and KNN.it is observed that BiLSTM gives more efficiency for prediction.

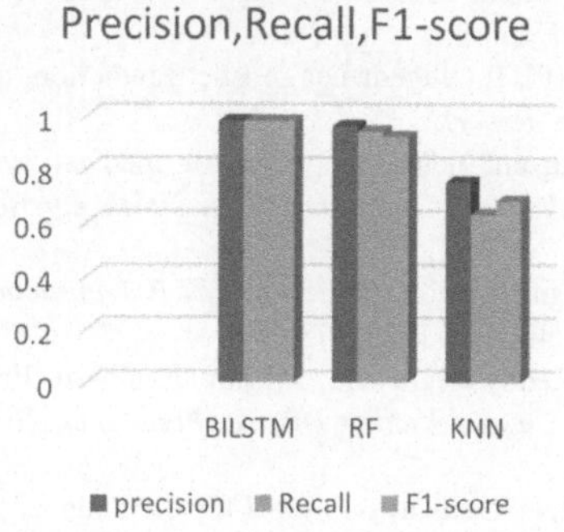

Figure 10. The performance measures.

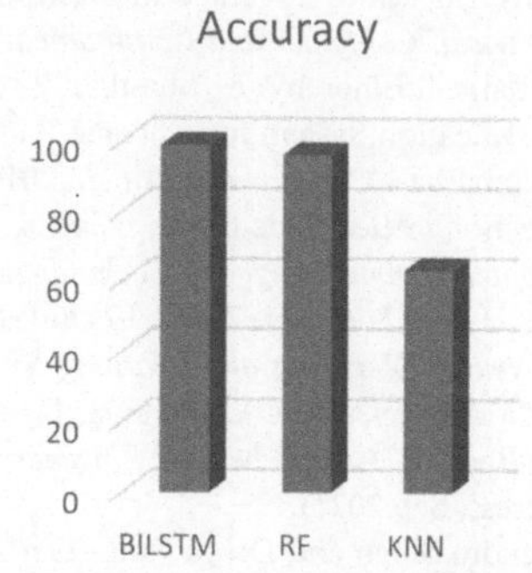

Figure 11. The accurateness of BiLSTM, RF, KNN.

The Specificity (precision), Sensitivity (Recall) and F1 score is calculated using the Equations 1, 2, 3, 4 respectively

$$Pre = \frac{TP}{TP + FP} * 100 \quad (1)$$

$$Rec = \frac{TP}{TP + FN} * 100 \quad (2)$$

$$F - measure = 2 * \frac{Pre * Rec}{Pre + Rec} \quad (3)$$

$$Acc = \frac{TP + TN}{TP + TN + FP + FN} \quad (4)$$

Where the BiLSTM has higher values in all the performance metrics.which shows the prediction accuracy is higher compared to RF and KNN.

5 CONCLUSION

In this paper, Bi-directional Long Short-Term Memory (BiLSTM) networks are used to create a prognostic model for Chronic Kidney Disease prediction(CKD),Comparing with conventional machine learning models, the prediction accuracy of the BiLSTM architecture is much higher. it can capture extensive temporal dependencies by processing sequential data in both forward and backward directions. according to experimental data, the performance of BiLSTM model performed better compared to traditional techniques, and it can be considered as a reliable tool for early CKD identification.

REFERENCES

[1] Arif-Ul-Islam and Ripon S.H., (2019). Rule induction, and prediction of chronic kidney disease using, boosting classifiers ant-miner and J48 decision tree, *2019 International-Conference on Electrical, Computer and Communication Engineering (ECCE)*, Cox'sBazar, Bangladesh, pp. 1–6.

[2] Gabriel R. Vasquez-Morales, *Explainable Prediction of Chronic Renal- Disease in the, Colombian Population. Using Neural. Networks and Case-Based Reasoning*, Digital. Object Identifier 10.1109/IEEE ACCESS.2019.2948430.

[3] Beverley M. Essue, Vivekanand Jha, Oommen John, John Knight & Stephen Jan, (February, 2019). Universal health coverage and chronic kidney. disease in India, *2019 International. Conference on Electrical, Computer and Communication Engineering (ECCE)*, 7–9.

[4] Mohamed. Elhoseny, K. Shankar, J. Uthayakumar, (Jul 2019), Intelligent Diagnostic, Prediction, and Classification System for Chronic Kidney Disease, *Scientific. Reports*

[5] Jagannatha A.N., *et al.*, (June, 2016). Bi-directional recurrent neural network for medical event detection in electronic health records, in *the Proceedings. of the 2016 Conference of the North American Chapter Human Language Technologies*, page. 473–482

[6] Song J., *et al.*, (April, 2021). Local-Global memory NN for medication prediction, *IEEE Transactions on Neural- Networks and Learning Systems*, volume. 32, no. 4, page. 1723–1736.

[7] Linta Antony, Sami Azam, Eva, Ryana Quadir, Abhijith Reddy Beeravolu, Mirjam Jonkman, Friso De Boer, *A Comprehensive Unsupervised. Framework for Chronic Kidney Disease Prediction*, IEEE (Access, Sep 2021).

[8] Vijendra Singh and Divya Jain, (Oct 2021). A hybrid. parallel classification model for the diagnosis of chronic kidney. disease, *International. Journal of Interactive Multimedia and Artificial Intelligence*, Vol. 8, No. 2

[9] Hao Du, Ziyuan Pan, Kee Yuan Ngiam, Fei Wang, Ping Shum and Mengling Feng, (Dec 2021). Self-correcting recurrent neural network for acute kidney injury prediction in critical care, *Health Data Science*, Hindawi

[10] Sigit Ari Saputro, Anuchate Pattanateepapon, Oraluck Pattanaprateep, Wichai Aekplakorn, Gareth J. McKay, John Attia and Ammarin Thakkinstian, (Jan 2022), External, validation of prognostic models for chronic kidney disease among, type 2 diabetes, *Journal of Nephrology*, Springer

[11] Hira Khalid, Ajab Khan, Muhammad Zahid Khan, Gulzar Mehmood and Muhammad Shuaib Qureshi, (Mar 2023). Machine learning hybrid model for the prediction of chronic kidney disease, *Computational Intelligence and Neuroscience*, Hindawi

[12] Shahid Mohammad Ganie,Pijush Kanti Dutta. Pramanik, Saurav Mallik, Zhongming Zhao, (December 1, 2023). Chronic kidney. disease prediction using boosting. Techniques based on clinical parameters, *PLOS ONE*, https://doi.org/10.1371/journal.pone.0295234

[13] Yoshinori Hashimoto, Hiromi Omura, Takayuki Tanaka, (Feb 2024). Presence and onset of chronic kidney disease as a factor involved in the poor prognosis of patients with essential thrombocythemia, *Advances in Hematology*, Hindawi

[14] Rei Takeda, Kazuya Takahashi, Andreas Kronbichler, Daiichiro Akiyama, (Mar 2024). Development of a kidney prognostic score in a japanese cohort of patients. with. antineutrophil. cytoplasmic auto-antibody vasculitis, *Kidney International Reports*

[15] Mariella Gregorich, Michael Kammer, Andreas Heinzel, Carsten Bö ger, Kai-Uwe Eckardt, Hiddo Lambers Heerspink, Bettina Jung, Gert Mayer, Heike Meisel bach, Matthias Schmid, Ulla T. Schultheiss, Georg Heinze, Rainer Oberbauer, (Mar 2024). Development and, validation of a prediction model for future estimated glomerular filtration rate in people with type. 2 diabetes and chronic kidney disease, *JAMA Network*

[16] https://www.niddk.nih.gov/health-information/kidney-disease/chronic-kidney-disease-ckd/what-is-chronic-kidney-disease

[17] https://www.javatpoint.com/data-preprocessing-machine-learning.

Data Science & Exploration in Artificial Intelligence – Gururaj et al. (Eds)

Enhanced three-layered CNN for multiclass skin lesions classification

Anand Jumnal, Janardhan Malagi, Shrinidhi Kulkarni, Sanjay Kumar Hiremath & Amit Jahagirdar
Department of Computer Science and Engineering BLDEACET Vijayapura, India

Prakash K. Sonwalkar
Department of Computer Science and Engineering (AIML), Jain College Of Engineering and Research, Udyambag, Belagavi, Karnataka, India

ABSTRACT: Skin lesions act as possible health indicators and are essential for the early detection and monitoring of dermatological diseases. In this paper, we proposed three-layered Convolutional Neural Network (CNN) to enhance the classification of skin lesions. The accuracy of the models is optimised by setting the hyper parameters of the CNN model. The evaluation of the proposed model is performed using HAM10000 dataset and is compared with other deep network architectures like ResNet-150 and VGG16. The considered dataset is balanced to fairly represent all seven types of skin lesions. The experimental result show that the proposed model achieved classification of lesions with the 92.5% training accuracy and 97% in testing accuracy.

Keywords: Skin-lesion classification, Machine-learning, CNN, Image-classification

1 INTRODUCTION

In the field of skin diseases, early identification of skin lesions is challenging task due to the wide range of irregularities on the skin surfaces like benign moles to malignant tumors which lays the pertain of global health. For the proper diagnosis and treatment, these skin lesions must be identified accurately. This increases the demand of machine learning methods to apply on such domain to help the dermatologist to identify the different types skin lesions with accuracy.

Therefore, this study aims to cover the recent development in machine/deep learning techniques for skin lesion classification.

The growing prevalence of skin diseases and the need for effective diagnosis in early stages prompted the adaption of artificial intelligence (AI) and machine learning methods in the domain of image processing and classification to reduce the human error, and increase diagnostic precision that helps to improve patient outcomes. The seven types of skin lesions are considered as shown in Figure 1. All seven type skin lesion poses different features, differentiating those features and identifying/detecting and classifying is done using CNN in this work. The objective is to identify or detect the skin lesion within the image provided. The heterogeneity of skin lesions, and the variability of image quality are challenges in this field.

Due to the imbalance nature of the dataset, it poses both opportunity and challenges for training the algorithm and evaluating the performance. The HAM10000 dataset having sizable collection of dermoscopic images which involves all seven type of skin lesions as depicted in Table 1. The class imbalance of the HAM10000 dataset leads to poor

DOI: 10.1201/9781003589273-80

performance of the model. However, in this work we have balanced the dataset to train the proposed model. Furthermore, the dataset distribution across various anatomical locations underscores the impact of skin cancer on areas like the back, lower limb, and trunck. In order to keep computational efficiency, all the images from the dataset have been resized to 32*32 pixels and processed through CNN that allows the model to train faster and reduced memory usage. This get benefitted when working with larger dataset. However, after reducing the image size, key features are preserved and allowing the CNN to effectively learn and indentify the patterns. The overall problem is to identify and classify all the seven skin lesion type using deep learning algorithm from the given dataset to accurately diagnose that help patients as well as medical professionals.

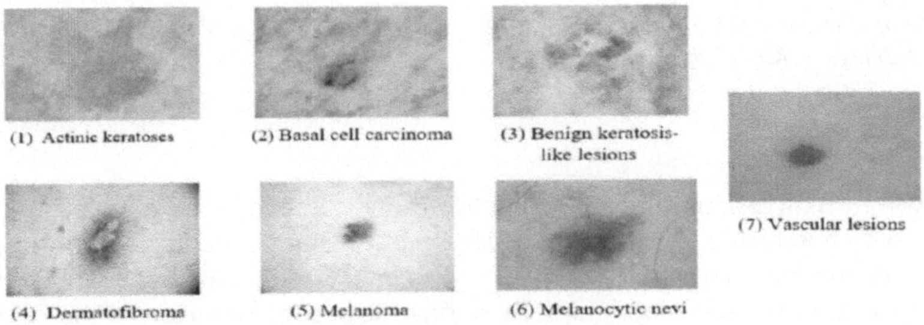

Figure 1. Seven types of skin lesion disease.

2 LITERATURE REVIEW

Many researchers work have been found in the literature work that deals with the problem of skin lesion identification and classification. The wide range of research community has offered solution using machine learning and deep learning methods to gain the highest accuracy level.

The author from the work [2] covered various dataset such as ISIC, PH, MEDNODE, Dermofit, and Dermquest and conducted the comparative analysis using different classification algorithm like SVM, KNN, and CNN. Among all, CNN achieved the precision of 86% and 99% specificity on ISIC dataset. In addition, SVM performed with 96% classification accuracy on dermoscopic dataset. KNN achieved 85.2% accuracy by leveraging GLCM and SURF for texture characteristics. Another work developed SBXception model to classify skin lesions into seven types in three stage system. In the image set of 10,015, each image is resized to 229*229, the proposed model accomplished good classification accuracy for various skin lesions[3]. Another work proposed combination of CNN and SVM for skin disease classification using the dataset Scikit-image. The model achieved 0.87 AUC that represents the high seperability between different classes of image sets[4]. One more automated approach using CNN machine learning algorithm is proposed to detect and classify the limited skin lesion using HAM10000 dataset [5]. Techniques included image preprocessing, feature extraction, data augmentation. Hyperparameters like batch size, epochs, and learning rate were optimized for improved model performance[6]. The methodology included Dermoscopy Image Pre-processing using a BPNN, Data Augmentation to prevent overfitting, Features Extraction with ResNet-18, and Prediction Explain ability using the LIME framework. Initial experiments achieved 75% classification accuracy, improving to 94.47% with pre-processed and augmented images[8]. Various AI-based approaches for skin lesion detection and classification, including dynamic graph cut, EfficientNets, MobileNet V2 with LSTM, and CNN-based methods, achieved accuracies ranging from 87% to 95.18%. A novel lightweight deep learning framework using EW-FCM segmentation and wide-Shuffle Net demonstrated efficiency and outperformed other methods in accuracy and parameter reduction[9][10]. With the use of a skewness-controlled SVR technique, an automated system for multiclass skin lesion classification that included data augmentation, model fine-tuning, transfer learning, feature extraction, fusion, and feature selection was able to achieve 94.9% accuracy[12][13].

Xception, Inception-ResNet-V2, and NasNetLarge for Skin Lesion Detection: These CNN architectures were used for skin lesion detection, achieving 77% accuracy using dermoscopic images from the PH dataset [14] [15]. The study presents an enhanced image classification model for skin lesions, achieving 98% accuracy on HAM10000 dataset. It employs transfer learning, data augmentation, and optimization techniques to support dermatologists in diagnosis. The survey highlights that many models show less accuracy and tend to overfit, performing well on training data but poorly on new data. The high complexity of some models increases computational demands without improving performance, making them impractical for real-world applications. For that reason, we are trying to create a model that has high accuracy and less complex.

3 SYSTEM MODEL

In this section, the formulation of optimal skin lesion classification using three layered CNN is presented. The system is modelled while considering seven type of skin lesions by leveraging the HAM10000 dataset. The aim is to develop an accurate and robust CNN model for multiclass classification of skin lesions from dermoscopic images. The model is handled in efficient manner by considering the difficulties of visual complexity and class imbalance.

The task is to classify various skin lesions, including nevus, basal cell carcinoma, melanoma, and other skin disorders, into the appropriate classes. The existing approaches could get trouble while differentiating between visually identical types of lesions or may not adaptive nature for wider range of image types. Therefore, an enhanced three layered CNN architecture is modelled for multiclass image classification that reduce the class imbalance and achieves expected higher accuracy in terms of performance.

The objective function is to minimizes categorical cross-entropy loss with subject to constraints and is formulated by:

$$Min(E)$$

$$E = -\frac{1}{N_t} * \sum_{i=1}^{N_t} \sum_{j=1}^{C} y_{ij} * \log(P_{ij})$$

Where the variables N_t indicates the number of training samples and C represents the number of image classes. As a binary indicator Y_{ij} is set that gives if the true label for the **ith** sample is class j. Pij indicates the predicted probability. The outer summation covers the N training samples, and the inner summation covers the **C** classes. The goal during training is to minimize this objective function by adjusting the model's parameters (weights and biases). Minimizing categorical cross-entropy loss is equivalent to maximizing the log-likelihood of the correct class probabilities, meaning that as the loss decreases, the model's ability to predict the correct class probabilities improves, leading to higher overall accuracy.

4 PROPOSED METHODOLOGY

This section outlines the dataset pre-processing, model development, and training processes. Figure 2 represents the model flow of the proposed methodology, and the Figure 3 illustrates the proposed three layered CNN architecture.

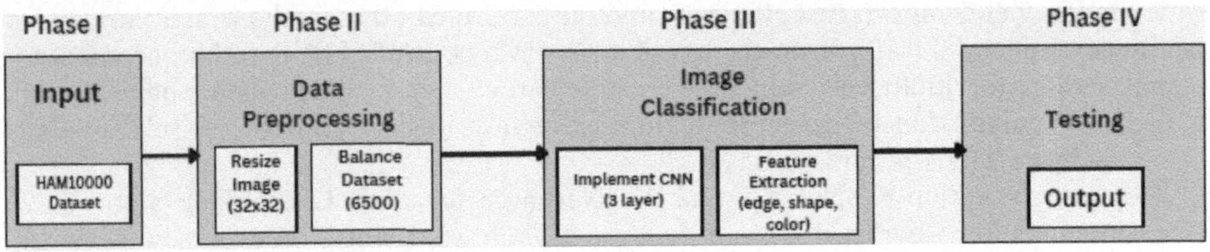

Figure 2. The model flow of the proposed methodology.

4.1 *Dataset preprocessing*

HAM10000 dataset consists of 10015 images of skin lesions with 7 types of skin lesions as we discussed above. These disease images are not balanced Table 1 shows the image count of each disease.

Table 1. Respective disease and their count.

Disease	Count
Actinic Keratosis	327
Basal Cell Carcinoma	514
Benign Keratosis	1099
Dermatofibroma	115
Melanoma	1113
Melanocytic nevi	6705
Vascular skin	142

4.2 *Proposed model*

Our proposed model is based on the three-layered CNN architecture. This paper proposes a CNN model that utilizes HAM10000 image dataset. The Proposed model contains 3 pairs of convolution and max pooling layers and Dropout, one Flatten layer, and finally one dense layer and uses softmax for classification. Figure 2 shows the flow of the proposed model.

4.2.1 *Implement CNN*

Implementing a Convolutional Neural Network (CNN) involves several key steps to ensure effective performance for tasks such as image classification. The Figure 4 shows the architecture of our proposed CNN.

4.2.2 *Feature extraction*

Initially, convolution layer of the proposed CNN extract the features of the input image and filters are applied on the same input image that develops the feature map which particularly focuses on patterns like edges and textures of the image. The activation function ReLU is used to these maps for non-linearity. Further, the pooling layer are utilized to down sample the maps and store the most valuable data whilst reducing the computational load. The feature maps are then transferred to fully linked layers that integrate the features to produce final classifications. The softmax is used to generate class labels' probability distribution. This helps the proposed CNN to classify the images accurately.

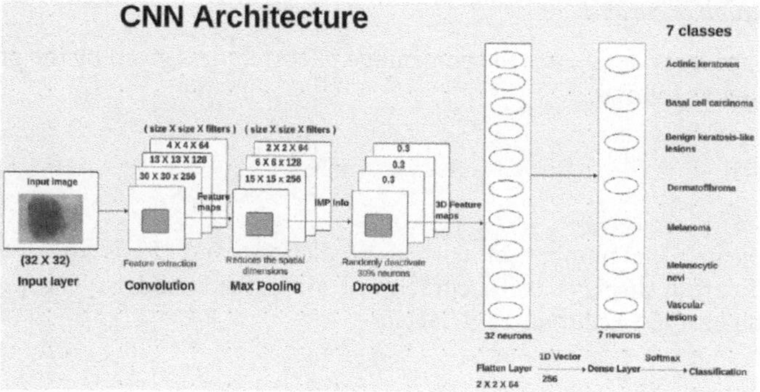

Figure 3. The proposed three-layered CNN architecture.

4.2.3 *Training model*

The proposed CNN is required to set the network pattern such as dataset preparation and fine tuning the hyper parameters of the CNN to achieve accurate image classification. Weights are modified by optimization methods in order to reduce prediction mistakes. By keeping an eye on metrics and validation accuracy, one can refine the CNN until it is taught to make accurate predictions. Three convolutional layers with max-pooling and dropout make up the model architecture, which is used to extract and condense features from the images. This is followed by flattening and two dense layers, with the final layer outputting predicted probabilities across the seven classes using a softmax activation. During training, the model ingests batches of images/labels and updates its weights via backpropagation to minimize the categorical cross-entropy loss on the training data, while using validation data to prevent overfitting. After training, the model evaluates its performance on test data by generating predictions, computing metrics like accuracy and visualizing the confusion matrix.

The model was trained using 75% of the dataset, with the remaining 25% reserved for testing. Training was done with a batch size of 16 for 400 epochs. After the 300th epoch the model become constant. Best performance was obtained at the 300th epoch.

This model achieved 92.1% training accuracy with loss of 0.0430 and 97.48 Testing accuracy with loss of 0.0555. Figures 4 and 5 illustrate the training and testing accuracy and loss for each epoch.

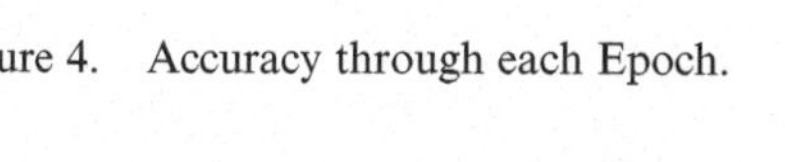

Figure 4. Accuracy through each Epoch.

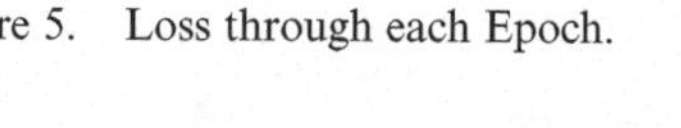

Figure 5. Loss through each Epoch.

5 RESULTS

The proposed three-layered CNN model was evaluated against various other models using key performance metrics. These metrics include Classification Accuracy, Precision, Recall, and F1Score.

5.1 *Classification accuracy*

Classification accuracy designates the percentage of correctness given by the proposed model and is calculated as follows.

$$Accuracy = \frac{C_p}{T_p}$$

where C_p indicates the number of correct predictions and T_p is the total number of Predictions. Figure 6 displays the accuracies of all trained models. The proposed model achieved the highest test accuracy of 97.48%.

5.2 *Precision*

The ratio of correct predictions (T_+) to the T_p. This score assesses how well the model predicts the expected outcomes.

$$Precision = \frac{T_+}{T_+ + F_+}$$

Where T_+ is the true positive and F_+ is the false positive. Figure 7 indicates that the proposed model achieved a precision of 0.95, which is the highest compared to the precision values of other models.

5.3 *Recall*

Recall can also be represent as sensitivity, it is the ratio of true positives correctly classified as positive to the total number of actual positive instances. The formula for recall is as follows.

$$Recall = \frac{T_+}{T_+ + F_-}$$

Where F- indicates the false negative value. Figure 8 illustrates the recall values of all trained models. The proposed model has the highest recall value among them.

5.4 *F1 score*

The F1 Score is indeed a harmonic mean of precision and recall, calculated using the formula as follows.

$$F1 = \frac{2 * Precision * Recall}{Precision + Recall}$$

Figure 9 displays the F1 Scores of all trained models. It demonstrates that the proposed model has achieved the expected results.

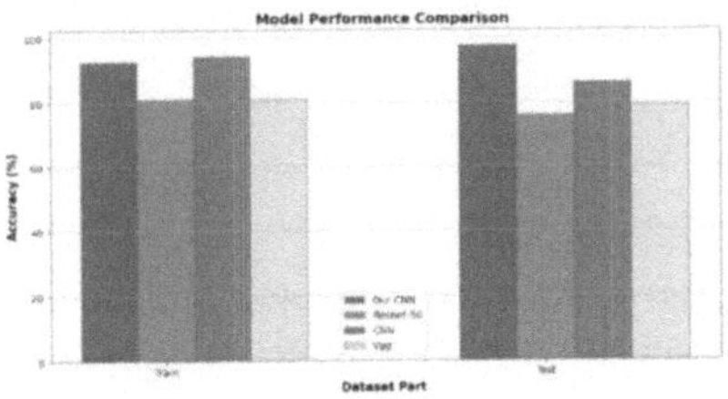

Figure 6. Test accuracy of the proposed model

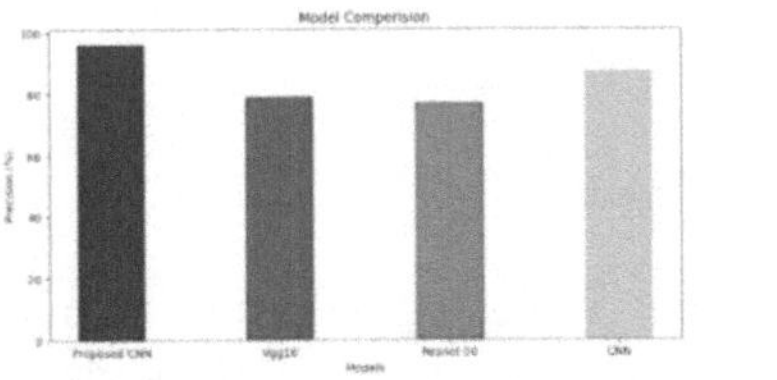

Figure 7. Precisions of the proposed model with other architectures. with other architectures.

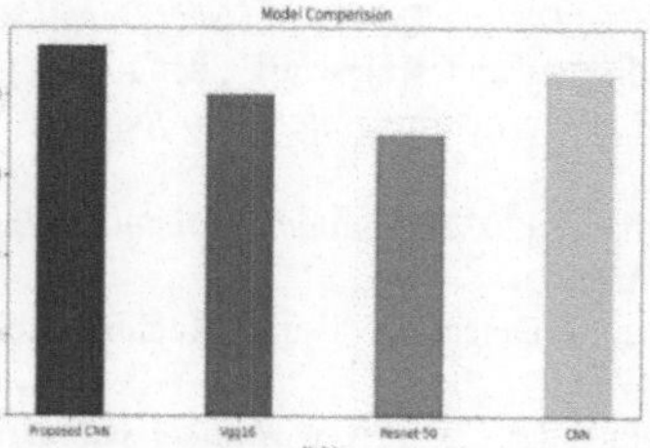

Figure 8. Recall score of the models.

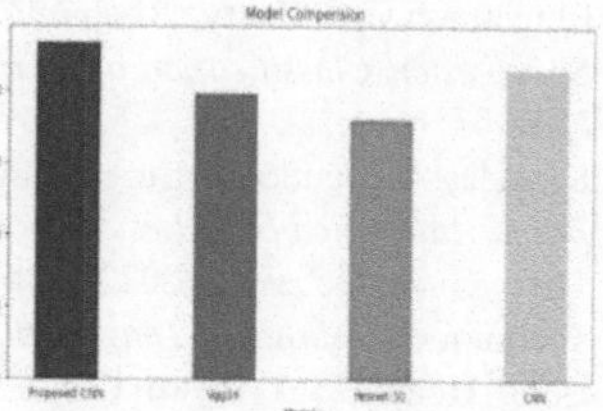

Figure 9. F1Score of all trained models.

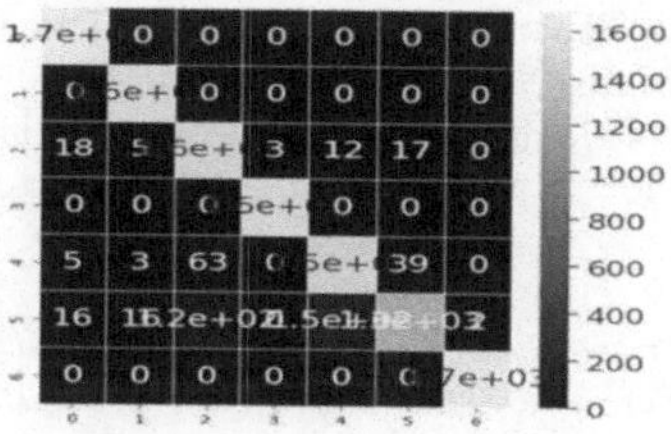

Figure 10. Confusion matrix.

5.5 *Confusion matrix*

The Confusion Matrix refers to the number of correct predictions made by the classification model compared to the total predictions. The formula accuracy is used for calculating matrix. Among the metrics displayed in Figure 10, it's clear that the proposed three-layered CNN model has outperformed other CNN-based models such as transfer learning models ResNet50 and VGG16. Additionally, the proposed model does not exhibit overfitting even after 300 epochs, making it suitable for training for extended periods for improved performance.

6 CONCLUSION

The proposed enhanced three-layered convolutional neural network (CNN) model demonstrated outstanding performance in the multiclass identification of skin lesions from dermoscopic images, outperforming other models. Experiment results showcased the model's high accuracy and precision in classifying various skin diseases, including melanoma, nevus, basal cell carcinoma, and other conditions. The model's ability to effectively capture relevant visual features contributed to its significant advancement in skin disease classification capabilities. This achievement represents a promising step towards revolutionizing healthcare practices in the skin health sector.

REFERENCES

[1] Dr. Chakraborty. C., Dr. Barbosa. S., and Dr. Garg. L, (2023), *Classification and Detection of Skin Diseases Based on Machine Learning and Image Processing Evolutionary Models.*

[2] Harsh Bhatt, Vrunda Shah, and KrishShah, (2023), *State-of-the-art Machine Learning Techniques for Melanoma Skin Cancer Detection and Classification*

[3] Abid Mehmood, Yonis Gulzar and Qazi Mudassar Ilyas, (2023), *SBXception: A Shallower and Broader Xception Architecture for Efficient Classification of Skin Lesions*

[4] Syed Inthiyaz, Baraa, SkHasaneAhammad and RajeshSkin. V., *Disease Detection Using Deep Learning*

[5] Bhuvaneshwari Shetty, Roshan Fernandes, Anisha P. Rodrigues and Rajeswari Chengoden, (2020), *Skin Lesion Classification of Dermoscopic Images Using Machine Learning and Convolutional Neural Network*

[6] Kingsley A. Ogudo, R. Surendran, and Osamah Ibrahim Khalaf, (2022),*Optimal Artificial Intelligence-Based Automated Skin Lesion Detection and Classification Model*

[7] Taye GirmaDebelee, (2023), Skin Lesion Classification and Detection Using Machine Learning Techniques, *Journal of Diagnostics* vol. 13 no. 19 page. 3147.

[8] Long Hoang, Suk-Hwan Lee, Eung-Joo Lee, and Ki-Ryong, (2022), *Kwon Multiclass Skin Lesion Classification Using a Novel Lightweight Deep Learning Framework for Smart Healthcare*

[9] Md. Khairul Islam, Md. Shahin Ali, Abhilash Arjan Das, *Melanoma Skin Lesions Classification using Deep Convolutional Neural Network with Transfer Learning*

[10] Mehak Arshad, Muhammad Attique Khan, Usman Tariq, Ammar Armghan, FayadhAlenezi, Muhammad Younus Javed and Shabnam Mohamed Asla, (2021), *A Computer-Aided Diagnosis System Using Deep Learning for Multiclass Skin Lesion Classification*

[11] Ms. GodlinJasil. S.P., Dept. of Computer Science and Engineering, Sathyabama Institute of Science and Technology, V. Ulagamuthalv, *Skin Lesion Classification Using Pre-Trained DenseNet201 Deep Neural Network*

[12] Cauvery K, Siddalingaswamy. P. C. and Noel D'so, (2021), *A Multiclass Skin Lesion Classification Approach Using Transfer Learning Based Convolutional Neural Network*

[13] Adekanmi A. Adegün and SerestinaViriri, (2020), *FCN-Based DenseNet Framework for Automated Detection and Classification of Skin Lesions in Dermoscopy Images*

[14] Ling-Fang Li, Xu Wang, Wei-Jian Hu, Neal N. Xiong, Yong-Xing Du, and Bao-ShanLi. Aref, (2021), *Deep Learning in Skin Disease Image Recognition: A Review*

[15] Sara Atito Ali Ahmed, (2020)12 affiliated with the Faculty of Engineering and Natural Sciences at Sabanci University in Ista, *Skin Lesion Classification With Deep CNN Ensembles*

Author index